# ADVANCES IN PLANT PHYSIOLOGY,

## *VOLUME – 9*

**ADVANCES IN PLANT PHYSIOLOGY**, VOLUME 9, 2006
An International Treatise Series

MOLECULAR PHYSIOLOGY AND BIOLOGY OF PLANTS

# ADVANCES IN PLANT PHYSIOLOGY

[ AN INTERNATIONAL TREATISE SERIES ]

VOLUME - 9

2006

*Editor*
**A. Hemantaranjan**
Department of Plant Physiology
Institute of Agricultural Sciences
Banaras Hindu University,
Varanasi – 221 005 INDIA

SCIENTIFIC PUBLISHERS (INDIA)
P.O. BOX 91 JODHPUR

*Published by:*
PAWAN KUMAR
Scientific Publishers (India)
5-A, New Pali Road, P.O. Box 91
JODHPUR - 342 001
E-mail: info@scientificpub.com
www.scientificpub.com



ISSN: 0972-9917
ISBN: 81-7233-459-1

Lasertype set: Rajesh Ojha
Printed in India

# PREFACE

The plant physiology and plant molecular biology research community has evidently endorsed the new directions taken by the treatise, accounting for the steady ability of ***Advances in Plant Physiology*** to attract the pre-eminent scientists in plant biology. Certainly, the configuration of *Volume 9* of the ***International Treatise Series*** has been done absolutely due to commendable contributions from World Scientists of eminence in unambiguous fields. I am delighted to declare that within the time span of nine years, now this treatise has been duly recognized through ISI Web of Knowledge – *Current Contents* in the hearts of distinguished readers and has beyond doubt achieved the international status. I once again reiterate that this programme has been undertaken with a view to reinforce the identical efforts to recognize the outcome of meticulous research in some of the very sensible and stirring areas of ***Plant Physiology-Biochemistry-Plant Molecular Physiology/Biology.*** In order to sustain and further advance *Plant Physiology*, I am dedicated to continue the originality and the introduction of spanking new ideas, ensure that the treatise welcomes the best science done across the full extent of modern plant biology, in general, and plant physiology, in particular, persevere on advancing the quality of what is published, place high value on the quality of production, and be highly attentive and responsive to the rapidly changing face of academic publishing. The new initiatives that I have in mind to further advance *Plant Physiology Treatise Series* come out from my faith that the treatise would substantially benefit by becoming more positive about illustrating the significance of the science that we bring out.

Indeed, our objective is to publish innovative science of the highest quality across the broad disciplinary scope of the treatise. In our Instructions for Authors, we state that work reported in international journals/ treatises including the ***Advances in Plant Physiology Series*** should be strongly executed, provide new information, and move the field to the next level.

The high visibility and influence of review articles published in *Advances in Plant Physiology* offer the opportunity to further broaden the disciplinary arena in which the treatise receives high-calibre submissions. For example, *Advances in Plant Physiology* offers the opportunity for high-calibre reviews in environmental, evolutionary, and agricultural plant biology to be published in a treatise series with higher impact and greater penetration than would otherwise be possible in more specialized disciplinary publications. In conclusion, I hearten you to share your suggestions and criticisms about what we are doing and how we are doing it with me; they may not always result in a change, but all vital suggestions will be given serious thought.

In *Volume 9,* with inventive applied research, attempts have been made to bring together much needed **twenty review articles** by Forty-six contributors from Australia, Belgium, France, Germany, India, Italy and Spain dispersed in *Nine Sections* as described below:

Since plants frequently encounter stresses when exposed to different environments and adversely affect growth, development, or productivity of plants in manifold ways, utmost research emphases in this important area have been put down at the global scale and, therefore, The first and foremost *Section I* necessarily allocates ten vital chapters related to *Molecular Physiology of Plants under Environmental Stresses: Tolerance Mechanism and Responses.* In ***Chapter 1,*** veteran Spanish scientists summarize the role of hydrogen peroxide ($H_2O_2$) in the regulation of plant responses to the interaction between drought and oxidative stress. Drought stress is a complex syndrome involving not only water deprivation, but also nutrient limitation and salinity and may upset the balance between antioxidant defences and the amount of reactive oxygen species (ROS), resulting in oxidative stress. The review in ***Chapter 2*** aims to discuss various changes occurring in plants at physiological and molecular levels under conditions of drought besides covering various responses of vital living processes to drought and development of tolerance. An eminent Australian Scientist in ***Chapter 3*** has given an account of the increasing understanding of the $Ca^{2+}$ signatures which are proposed to encode the specificity of the ozone-induced response, and in so doing, lead to the appropriate downstream response. Besides this, the chemistry behind the "natural" production of reactive oxygen species during "normal" photosynthetic and respiratory processes has been briefly summarised and the various mitigating defence mechanisms of the plant outlined. ***Chapter 4*** has been adequately dealt with physiological, biochemical and growth responses of plants to tropospheric ozone by an excellent team from the Banaras Hindu University, India.

In spite of the rapid pace made in the past years into the understanding of reactive oxygen species and the mechanisms of their detoxification, a lot of questions and gaps remain in our understanding of how these affect stress response in plants. Superoxide dismutase (SOD) catalyzes the destruction of the $O_2^-$ free radical and thus protects oxygen-metabolizing cells against the harmful effects of superoxide free radicals. An interesting review with similar aim has been illustrated in ***Chapter 5.*** Subsequently in ***Chapter 6,*** exclusive details of salt stress adaptations of mangroves, their physiological and molecular biological aspects including genetics of mangroves have been dealt comprehensively by a team of illustrious workers of high stature who have no doubt largely contributed their original laboratory work taken up to study the general physiology of stress and possible adaptive features of tree mangroves. Indeed, this extensive and amazingly productive review very reasonably enriches the *Volume 9* for that the three Indian authors are especially congratulated. Similarly, coconut palm responds to the stress conditions at morphological, anatomical, physiological and biochemical levels. The atmospheric stress influences the photosynthetic rates and if prolonged with soil moisture stress, dry matter production and its partitioning are influenced. Consequently, a very demanding review on adaptive strategies of coconut palm under stressful conditions has been extensively written in ***Chapter 7*** where experienced authors from the Central Plantation Crops Research Institute, Kasaragod have suggested that since coconut responds to the stressful environments at morphological, anatomical, physiological and biochemical levels, attention should be given to develop the molecular markers linked to desirable traits and to understand the inheritance patterns of these traits. The genotypes with desirable traits for tolerance to stress conditions can be used in breeding strategies for abiotic stress tolerance in future crop improvement strategies.

It is important to develop transgenic plants that can accumulate metals for phytoremedation of metal polluted environment. With this view, the molecular physiology of heavy metal stress in plants has been interestingly well illustrated in ***Chapter 8*** by renowned Indian and Japanese scientists. The review presented in ***Chapter 9*** summarises the progress of heavy metal research using duckweeds during the past few years primarily since 1997. Remarkably duckweeds take up mineral elements as well as heavy metals and other (partially toxic) compounds – as long as the extent of the toxic effects does not prevent further vegetative growth of the plants. In ***Chapter 10,*** a very young scientist from India has exceptionally endeavoured to overview jasmonic acid as a stress relieving plant cell signaling molecule. After that, in the *Section II,* two renowned scientists from Italy concentrate primarily on recent advances in the biosynthesis and functions of cuticular waxes, with emphasis on *Arabidopsis* and maize in ***Chapter 11***. An update on recently identified endoproteolytic processes in chloroplasts and plant mitochondria has been wonderfully consolidated by three eminent German workers in ***Chapter 12*** of the *Section III*. Next to these, in ***Chapter 13*** of the *Section IV,* two dedicated Japanese workers have very scientifically discussed predominantly their own experiences with regard to identification of rice root proteins regulated by gibberellin using proteome analysis. The *Section V* on *Crop Modeling* consists of highly needed review on crop simulation modelling and its implications and potentials for rubber, which has been included for the first time in this *International Treatise Series* from India as ***Chapter 14.*** The important review in the *Section VI* as ***Chapter 15*** superbly written by nine distinguished and authoritative French scientists discusses the early steps necessary to establish symbiosis, from the recognition mechanisms between rhizobacteria and their host plants, to the cellular and morphological events in different root tissues associated with nodule primordium formation. The *Section VII* on *Molecular Basis of Metabolism* consists of two outstanding reviews as ***Chapter 16*** and ***17,*** which focus on the enzymes catalyzing fructan breakdown in plants: fructan exohydrolases on the one hand and strong auxin like activity of 4-chloroindole-3-acetic acid (4-Cl-IAA) in different bioassays besides its well established role in plant growth on the other hand respectively. Subsequently in *Section VIII* related to *Crop Physiology and Biochemistry,* attempt has been made in ***Chapter 18*** to review the researches pertaining to role played by diverse physiological parameters in grain growth and development and the relationship that exists between the yield of grains by a very dedicated and veteran Indian Plant Physiologist. Out of two essential chapters in *Section IX* especially dealing with *Post-Harvest Physiology* and new to this *Treatise Series,* ***Chapter 19*** summarizes that recalcitrant seeds should be kept at moisture contents above their lowest-safe values and ventilation is needed to remove toxic gases and to prevent anoxia. However further research is needed for prolonging the life of these seeds. Eventually, in ***Chapter 20,*** a very thoughtful and distinguished scientist from Division of Biotechnology, Institute of Himalayan Bioresource Technology, aimed to review the changes in plant growth substances especially auxins, gibberellins, cytokinins, abscisic acid and ethylene during ripening and post harvest periods in fruits. Altogether, ***Volume 9*** is a vast wealth of applicable knowledge in the field of biological sciences.

In this dedicated endeavour, I am overjoyed to state my authentic appreciation to all distinguished and talented ***Members of the Advisory Committee***, for bringing up this unsurpassed, realistic, thoughtful and far-reaching treatise up to the international standard. In addition, I am extremely grateful to the ***Fellow Members of the Indian Society for Plant Physiology, New Delhi*** for their trustworthy moral support and valued suggestions from tin.e to time. My esteem is very much due to the Hon'ble Vice-Chancellor Prof. Panjab Singh, Banaras Hindu University, the Director, Institute of Agricultural Sciences Prof.Shiv Raj Singh, the Dean, Faculty of Agriculture Prof. Janardan Singh, and my admired colleagues of the Banaras Hindu University and other universities and institutes of the world for their consistent moral support in this gigantic task. Besides these, I have my sincere thanks to Mr. A. Ishan Rahul, N.T.S.E. *Fellow* (Govt. of India) and student of National Institute of Technology, for his incredible perfectionism in the conception of the very sensible and exceptional cover page of this *Volume.*

Last but not the least, I am indebted to my affectionate mother and my all family members for their blessings and good wishes in this wide mission. I have my profound esteem to all of them. Besides these, I am extremely thankful to the outstanding staff members of the Scientific Publishers, Jodhpur, India, for their genuine competence in the perfect printing and worldwide circulation of this treatise. Above all, I supremely recall that this undertaking is in advance dedicated to the adoring memory of my revered father Late Dr. A. Chittaranjan Sahay.

**Dr. A. Hemantaranjan**
Fellow, I.S.P.P.
Editor-in-Chief
Advances in Plant Physiology *Series*

Department of Plant Physiology
Institute of Agricultural Sciences
Banaras Hindu University
Varanasi 221 005
INDIA
E-mails: hemantaranjan@gmail.com; hemantaranja@satyam.net.in

# CONTRIBUTORS

## AUSTRALIA

**T. Ann Cuin**, School of Agricultural Science, University of Tasmania, Private Bag 54, Hobart, Tasmania, 7001, Australia. E-mail: Tracey.Cuin@utas.edu.au

## BELGIUM

**A. Van Laere,** Laboratory for Molecular Plant Physiology, Kasteelpark Arenberg 31, B-3001 Leuven (Heverlee), Belgium

**W. Van den Ende,** Laboratory for Molecular Plant Physiology, Kasteelpark Arenberg 31, B-3001 Leuven (Heverlee), Belgium

## FRANCE

**B. Péret,** Rhizogenesis lab. IRD (Institut de Recherche pour le Développement). 911 Avenue Agropolis, BP 64501, 34394 Montpellier Cedex 5, FRANCE.

**C. Franche,** Rhizogenesis lab. IRD (Institut de Recherche pour le Développement). 911 Avenue Agropolis, BP 64501, 34394 Montpellier Cedex 5, FRANCE.

**D. Autran,** Rhizogenesis lab. IRD (Institut de Recherche pour le Développement). 911 Avenue Agropolis, BP 64501, 34394 Montpellier Cedex 5, FRANCE.

**D. Bogusz,** Rhizogenesis lab. IRD (Institut de Recherche pour le Développement). 911 Avenue Agropolis, BP 64501, 34394 Montpellier Cedex 5, FRANCE. E-mail: bogusz@mpl.ird.fr

**F. Auguy,** Rhizogenesis lab. IRD (Institut de Recherche pour le Développement). 911 Avenue Agropolis, BP 64501, 34394 Montpellier Cedex 5, FRANCE.

**L. Laplaze,** Rhizogenesis lab. IRD (Institut de Recherche pour le Développement). 911 Avenue Agropolis, BP 64501, 34394 Montpellier Cedex 5, FRANCE.

**M. Oureye-Sy,** Rhizogenesis lab. IRD (Institut de Recherche pour le Développement). 911 Avenue Agropolis, BP 64501, 34394 Montpellier Cedex 5, FRANCE.

**M. Obertello,** Rhizogenesis lab. IRD (Institut de Recherche pour le Développement). 911 Avenue Agropolis, BP 64501, 34394 Montpellier Cedex 5, FRANCE.

**V. Hocher,** Rhizogenesis lab. IRD (Institut de Recherche pour le Développement). 911 Avenue Agropolis, BP 64501, 34394 Montpellier Cedex 5, FRANCE.

## GERMANY

**K.-J. Appenroth,** [1]University of Jena, Institute of General Botany and Plant Physiology, Dornburger Str. 159, D-07743 Jena, Germany

## INDIA

**A. Ahmad,** Deptt. of Botany, Aligarh Muslim University, ALIGARH-202002, India. E.mail: aqil_ahmad@lycos.com

**A. B. Das,** Regional Plant Resource Center, Nayapalli, Bhubaneswar, Orissa-751015, India, E-mail: a_b_das@hotmail.com

**A. Hemantaranjan,** Department of Plant Physiology, Institute of Agricultural Sciences, Banaras Hindu University, Varanasi 221 005, India

**B. Ali,** Deptt. of Botany, Aligarh Muslim University, ALIGARH-202002, India

**B. K. Sarma,** Department of Mycology and Plant Pathology, Institute of Agricultural Sciences, Banaras Hindu University, Varanasi 221005, India

**Davood Eradatmand Asli,** Department of Botany, Panjab University, Chandigarh, 160 014, India

**I. S. Dua,** Department of Botany, Panjab University, Chandigarh, 160 014, India

**J.P. Singh,** Department of Plant Physiology, [2]Department of Agronomy, Institute of Agricultural Sciences, Banaras Hindu University, Varanasi 221 005, India

**K.V. Kasturi Bai,** Central Plantation Crops Research Institute, Kudlu. P. O. Kasaragod, 671 124, Kerala, India

**K. Shah,** North Eastern Hill University, Department of Biochemistry, School of Life Sciences, Shillong, India

**M. Agrawal,** Department of Botany, Banaras Hindu University, Varanasi 221 005, India

**P. Mohanty,** Regional Plant Resource Center, Nayapalli, Bhubaneswar, Orissa-751015, India. E-mail: prasanna37@hotmail.com

**P.K. Nagar,** Division of Biotechnology, Institute of Himalayan Bioresource Technology, Palampur, 176061, India

**Q. Fariduddin,** Deptt. of Botany, Aligarh Muslim University, ALIGARH-202002, India

**S. Hayat,** Deptt. of Botany, Aligarh Muslim University, ALIGARH-202 002, India

**S. K. Dey,** Rubber Research Institute of India, Regional Research Station, Agartala-799 006 , India

**S. Mishra,** Regional Plant Resource Center, Nayapalli, Bhubaneswar, Orissa-751 015, India

**S. Naresh Kumar,** Central Plantation Crops Research Institute, Kasaragod, 671 124, India Email: nareshkumar.soora@gmail.com

**S. Choudhury,** School of Life Sciences, Assam (Central) University, Silchar 788011, India

**S. R. Ambika,** Dept. of Botany, Bangalore University, Bangalore-560 056, India

**S. Azeez,** Central Plantation Crops Research Institute, Kudlu. P.O. Kasaragod, 671 124, Kerala, India

**S. Tiwari,** Department of Botany, Banaras Hindu University, Varanasi 221 005, India. E-mail: supriyabhu@yahoo.co.in

**V. Rajagopal,** Central Plantation Crops Research Institute, Kasaragod, 671 124, India

## ITALY

**M. Motto,** CRA-Istituto Sperimentale per la Cerealicoltura, Sezione di Bergamo Via Stezzano 24, 24126 Bergamo, ITALY.

**M. Sturaro,** CRA-Istituto Sperimentale per la Cerealicoltura, Sezione di Bergamo Via Stezzano 24, 24126 Bergamo, ITALY. E.mail: sturaro@iscbg.it

## JAPAN

**H. Matsumoto,** Research Institute for Bioresources, Okayama University, Kurashiki 710-0046, Japan

**H. Konishi,** National Institute of Agrobiological Sciences, Tsukuba 305-8602, Japan

**S. K. Panda[1,2],** [1]Research Institute for Bioresources, Okayama University, Kurashiki 7100046, Japan & [2]School of Life Sciences, Assam (Central) University, Silchar 788011, India

**S. Komatsu**, National Institute of Agrobiological Sciences, Tsukuba 305-8602, Japan

## SPAIN

**L. Alegre,** Departament de Biología vegetal, Facultat de Biologia, Universitat de Barcelona, Avinguda Diagonal 645, 08028-Barcelona, Spain. E-mail: alegre@ub.edu

**S. Munné-Bosch,** Departament de Biología vegetal, Facultat de Biologia, Universitat de Barcelona, Avinguda Diagonal 645, 08028-Barcelona, Spain.

**T. Jubany-Marí,** Departament de Biología vegetal, Facultat de Biologia, Universitat de Barcelona, Avinguda Diagonal 645, 08028-Barcelona, Spain.

# VOLUME - 9

# CONTENTS


*Preface* *v*

*Contributors* *ix*


## *Section I:*
## Molecular Physiology of Plants under Environmental Stresses: Tolerance Mechanism and Responses


1 Drought and Oxidative Stress: The Role of Hydrogen Peroxide
— *Tana Jubany-Marí, Sergi Munné-Bosch and Leonor Alegre* 1

2 Molecular Physiology of Drought Tolerance Mechanism in Plants
— *A. Hemantaranjan and J. P. Singh* 13

3 Calcium and Oxidative Stress
— *Tracey Ann Cuin* 41

4 Physiological, Biochemical and Growth Responses of Plants to Tropospheric Ozone
— *Supriya Tiwari and Madhoolika Agrawal* 67

5 Superoxide dismutase - Scavengers of Reactive Oxygen Species
— *K.V. Kasturi Bai and Shamina Azeez* 87

6 Physiology of Mangroves: Strategies on Stress Adaptation with Special Emphasis on Tolerance to High Salinity
— *A. B. Das, S. Mishra and P. Mohanty* 101

7 Adaptive Strategies of Coconut Palm Under Stressful Conditions
— *S. Naresh Kumar, V. Rajagopal and K.V. Kasturi Bai* 155

8 Molecular Physiology of Heavy Metal stress in Plants
— *S.K. Panda, S.Choudhury and H. Matsumoto* 169

9 Plants For Heavy Metal Toxicity Assessment : Duckweeds (*Lemnaceae*)
— *Klaus-J. Appenroth and Kavita Shah* 193

10 Jasmonic Acid - A Stress Relieving Plant Cell Signaling Molecule
— *B.K. Sarma* 205

## Section II:
## Recent Advances in the Biosynthesis and Functions of Cuticular Waxes

11 Plant Cuticular Waxes: Biosynthesis and Functions
— *M. Sturaro and M. Motto* 229

## Section III:
## Plant Biogenesis

12 Proteolysis in Plant Mitochondria and Chloroplasts
— *P. F. Huesgen, H. Schuhmann and I. Adamska* 255

## Section IV:
## Root Growth Regulation

13 Identification of rice root proteins regulated by gibberellin using proteome analysis
— *Setsuko Komatsu and Hirosato Konishi* 297

## Section V:
## Crop Modeling

14 Crop Simulation Modeling: Implications and Potentials for Rubber
— *S. K. Dey* 315

## Section VI:
## Actinorhizal and Legumes Nitrogen Fixing Symbioses

15 Make your way to Nodules. Early Events in Actinorhizal and Legumes Nitrogen Fixing Symbioses
— *Daphné Autran, Laurent Laplaze, Valérie Hocher, Florence Auguy, Mame Oureye-Sy, Mariana Obertello, Benjamin Péret, Claudine Franche and Didier Bogusz* 327

## Section VII:
## Molecular Basis of Metabolism

16 Fructan Exohydrolases (FEHS) in Fructan and Non-Fructan Plants
— *Wim Van den Ende and André Van Laere* 347

17 4-Chloroindole-3-Acetic Acid : Metabolism and Bioactivity in Plants
— *B. Ali, S. Hayat, Q. Fariduddin and A. Ahmad* 369

## Section VIII:
## Crop Physiology & Biochemistry

18 A Study into Grain Growth of *Triticum aestivum* L. Wheats as Steered by Anatomical, Biochemical and Physiological Parameters
— *I. S. Dua and Davood Eradatmand Asli* 379

## *Section IX:*
## Post-Harvest Physiology

19 Recalcitrant Seed Storage Behaviour
— *S.R. Ambika* 449

20 Changes in Plant Growth Substances During Fruit Ripening and Post Harvest Periods
— *P.K. Nagar* 475

**INDEX** 493

**Contents of Previous Volumes (1 to 8)** 505

*SECTION — I*

# *MOLECULAR PHYSIOLOGY OF PLANTS UNDER ENVIRONMENTAL STRESSES: TOLERANCE MECHANISM AND RESPONSES*

*Advances in Plant Physiology*, Vol. 9
Ed. A. Hemantaranjan
Scientific Publishers (India), Jodhpur, 2006 pp. 1-12
E-mail: **info@scientificpub.com** www.scientificpub.com

# 1

# DROUGHT AND OXIDATIVE STRESS: THE ROLE OF HYDROGEN PEROXIDE

Tana Jubany-Marí, Sergi Munné-Bosch and Leonor Alegre*

Departament de Biología vegetal, Facultat de Biologia, Universitat de Barcelona, Avinguda Diagonal 645, 08028-Barcelona, Spain
* Corresponding autor, e-mail: lalegre@ub.edu, fax +34-934112842

## INTRODUCTION

In this chapter we will summarize the role of hydrogen peroxide ($H_2O_2$) in the regulation of plant responses to the interaction between drought and oxidative stress. Several reviews (Smirnoff, 1993; Mano, 2002; Alegre and Munné-Bosch, 2003; Apel and Hirt, 2004; Desikan *et al.*, 2004) and other reports (Neill *et al.*, 2002; Noctor *et al.*, 2002; Pastori and Foyer, 2002) document the relationship between drought-oxidative stress and the role of antioxidants and reactive oxygen species (ROS) in plant responses to stress. Here, we build on those studies, incorporating our own personal experience with Mediterranean plants, to update current understanding of the double role $H_2O_2$ plays in (i) causing oxidative damage and (ii) as a signaling molecule in the sensing and control of plant responses to drought-stress.

Plants frequently encounter stresses; i.e., external conditions that adversely affect growth, development, or productivity. Drought stress is a complex syndrome involving not only water deprivation, but also nutrient limitation and salinity. Furthermore, drought-stress may upset the balance between antioxidant defences and the amount of reactive oxygen species (ROS), resulting in oxidative stress (Munné-Bosch *et al.*, 2001; Desikan *et al.*, 2003). Moreover, levels of light that are optimal for photosynthesis in well-watered plants can become excessive in plants subjected to drought. Research on ROS has emphasized the oxidative damage that results from exposure to environmental stresses, as well as on the role ROS plays in defending against pathogens. More recently, it has become apparent that ROS have important roles as

$$^{3}O_2 + e^- \rightarrow O_2 \ + e^- \rightarrow H_2O_2 + e^- \rightarrow OH \ + e^- \rightarrow H_2O$$

Oxygen Superoxide hydrogen peroxide hydroxyl radical water

↓

$^{1}O_2$

Singlet Oxygen

**Figure 1.** Generation of ROS by energy transfer or sequential univalent reduction of ground state triplet oxygen.

signaling molecules both in the control of plant development and in the sensing of the external environment. A complex network of enzymatic and small molecule antioxidants controls the concentration of ROS and repairs oxidative damage.

### 1.1 Generation of ROS

Ground state oxygen may be converted to the much more reactive ROS forms, either by energy transfer or by electron transfer reactions. The former leads to the formation of singlet oxygen, whereas the latter results in the sequential reduction to superoxide, hydrogen peroxide, and hydroxyl radical (Fig. 1).

In plants, ROS are continuously produced as byproducts of various metabolic pathways localized predominantly in chloroplasts, but also in peroxisomes and mitochondria (Fig. 2).

In mammalian cells, mitochondria are the major source of ROS. However, the relative contribution of mitochondria to ROS production in green tissues is very low. One reason for the lower production of $H_2O_2$ could be the presence in plant mitochondria of alternative oxidase (AOX), which catalyzes the reduction of $O_2$ by ubiquinone. AOX competes with the cytochrome $bc_1$ complex for electrons and thus may reduce ROS production in mitochondria.

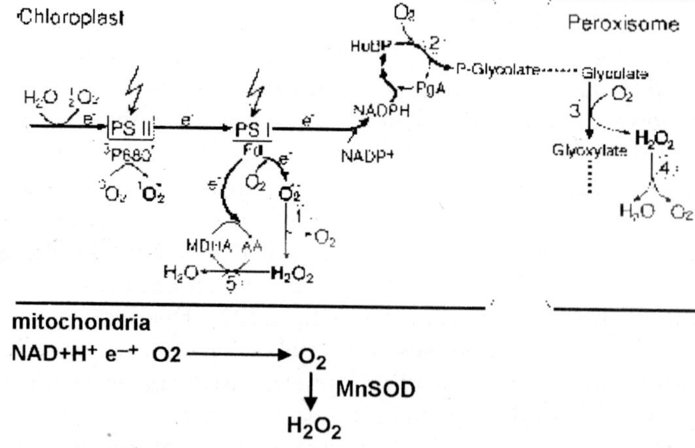


**Figure 2. Main pathways of ROS generation in chloroplast, peroxisome and mitochondria.** (1) superoxide dismutase, (2) Rubisco, (3) glycolate oxidase, (4) catalase, (5) ascorbate peroxidase. MnSOD Mn Superoxide dismutase.

Plants also generate ROS from dedicated enzymes such as plasma-membrane NADPH oxidases or cell wall peroxidases at high pH values (Bolwell *et al* 1995). These enzymes produce $O^{\cdot -}_2$ that is dismutated to $H_2O_2$ by superoxide dismutase.

In this chapter we will focus on the hydrogen peroxide molecule.

## 1.2 Hydrogen Peroxide Scavenging

Under physiological steady state conditions, $H_2O_2$ is scavenged by a range of antioxidative defense components that are often confined to particular compartments.

Scavenging of $H_2O_2$ in plants is essential for cell protection and cell signaling (Dat *et al.* 2000; Mittler 2002; Neill *et al.*, 2002; Apel and Hirt, 2004).

Plants contain at least five different enzymes capable of rapid and efficient $H_2O_2$ removal. They are encoded by five distinct gene families: ascorbate peroxidases (APXs), catalases (CATs), glutathione peroxidases (GPXs), peroxiredoxins (PrxRs), and type III peroxidases (Prxs). Together with the antioxidants ascorbic acid (AA) and glutathione (GSH), these enzymes provide cells with efficient machinery to detoxify $H_2O_2$. The function of the five $H_2O_2$-removal enzymes, together with the activity of superoxide dismutase (SOD), is particularly important in preventing the formation of the highly reactive hydroxyl radical ($OH^{\cdot}$) via the Haber-Weiss/Fenton reaction. Ascorbate peroxidase is found in almost every compartment of the plant cell and it participates in the removal of $H_2O_2$ as part of the ascorbate-glutathione or Asada-Halliwell pathway (Fig. 3).

Moreover, hydrogen peroxide is eliminated by catalases (CAT). This enzyme rapidly destroys most of the $H_2O_2$ produced by metabolism, although it allows low steady-state levels to persist, presumably to maintain redox signaling pathways (Noctor and Foyer, 1998). Catalase activity is essential for the removal of $H_2O_2$ produced in the peroxisomes by photorespiration (Noctor *et al.*, 2000).

The enzyme catalase converts hydrogen peroxide to oxygen and water:

$$2H_2O_2 \xrightarrow{\textbf{CAT}} 2H_2O + O_2$$

The different affinities of APX (µM range) and CAT (mM range) for $H_2O_2$ suggest that they belong to two different classes of $H_2O_2$-scavenging enzymes: APX might be responsible for the fine modulation of $H_2O_2$ for signaling, whereas CAT might be responsible for the removal of excess $H_2O_2$ during stress.

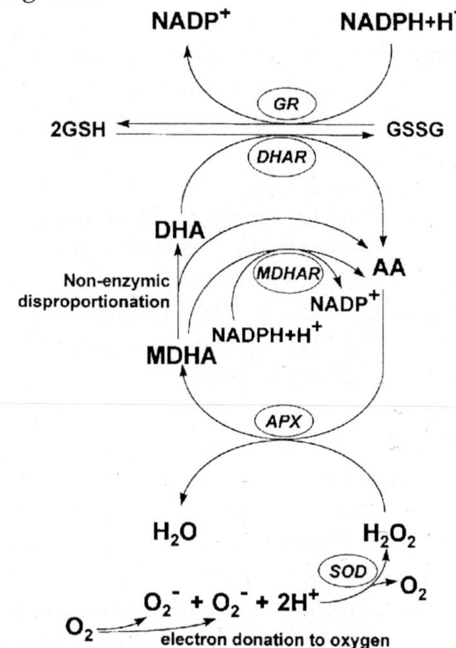


**Figure 3. The ascorbate-glutathione cycle.** Ascorbate is used by ascorbate peroxidase (APX) to detoxify $H_2O_2$, which itself is generated from $O_2^{\cdot}$ by the action of SOD. Ascorbate is oxidized to monodehydroascorbate (MDHA), which can be reduced to AA through the action of monodehydroascorbate reductase (MDHAR), but as a radical, MDHA also disproportionates to ascorabte and dehydroascorbate (DHA). DHA is then converted to ascorbate through the action of dehydroascorbate reductase (DHAR), which requires glutathione. (Asada, 1999).

The coordinated function of all reactive scavenging enzymes is crucial for controlling signal transduction pathways that utilize $H_2O_2$ in plants. These pathways are involved in the control and regulation of processes such as stomatal signaling, stress responses, and hormonal signaling.

## 2. DROUGHT STRESS ENHANCES THE ACCUMULATION OF $H_2O_2$

The equilibrium between the production and scavenging of $H_2O_2$ can be disturbed by drought-stress. As a result of such disurbances, intracellular levels of $H_2O_2$ may rapidly rise. While under normal growth conditions the production of $H_2O_2$ in cells is low (a steady-state level of 0.5 μM $H_2O_2$ in chloroplasts), drought disrupts cellular homeostasis and enhances the steady-state level of $H_2O_2$ to 15 μM (Mittler, 2002). The production of $H_2O_2$ during a drought results from (i) a dysfunction in the water-water cycle of chloroplasts (Asada, 1999; Alegre and Munné-Bosch, 2003), (ii) photorespiration (Noctor *et al.*, 2002), and/or (iii) respiration in mitochondria (Mittler, 2002).

In previous studies (Munné-Bosch *et al.*, 2001, 2003), we have shown an accumulation of $H_2O_2$ in the cell walls and plasma membranes of drought stressed plants (Figure 4). In this case, however, the source of $H_2O_2$ has not yet been determined.

The sub-cellular locations in which $H_2O_2$ is generated and perceived are important in determining the spectrum of activated cellular responses. Furthermore, $H_2O_2$ may diffuse through biological membranes; thus, $H_2O_2$ produced at a specific cellular site (e.g., chloroplast during stress) can affect other cellular compartments.

One of the first responses to drought is stomata closure, which allows plants to conserve water despite its declining availability in soil. Consequently, photosyn-thesis is particularly sensitive to water deficit. Stomatal closure reduces $CO_2$ availability for photosynthesis and decreases carbon assimilation; under these conditions, a dysfunction in the water-water cycle of chloroplasts can lead to $O^{-}_2$ and $H_2O_2$ accumulation. Furthermore, enhancement of the photorespiratory pathway leads to an increase in $H_2O_2$ production. Hydrogen peroxide is also generated as a secondary messenger in abscisic acid (ABA)-mediated stomatal closure (Pei *et al.*, 2000).

The chloroplast has been considered to be the main source of $H_2O_2$, and consequently, one of the main targets for $H_2O_2$ damage during stress. However, it has recently been suggested that $H_2O_2$ produced at the chloroplast during drought stress may be just a re-directioning of electrons from an overreduced photosynthetic electron transfer chain. Mitochondria are another cellular site of $H_2O_2$ production and enhanced $H_2O_2$ can, in fact, trigger PCD. Both mitochondria and chloroplasts contain $H_2O_2$–scavenging mechanisms. By contrast, little is known about the $H_2O_2$–scavenging properties of the nucleus, where $H_2O_2$ could accumulate under drought-stress conditions (personal observation).

Because $H_2O_2$ is toxic but also participates in signalling events, plants require several mechanisms to regulate $H_2O_2$ levels. These mechanisms include the avoidance and/or scavenging of $H_2O_2$ generation. Mechanisms that might prevent $H_2O_2$ generation during drought include anatomical adaptations (such as leaf movement and curling), development of a refracting epidermis, chloroplasts movement and molecular mechanisms that rearrange the photosynthetic apparatus and its antennae in accordance with light quality and intensity. Different scavenging mechanisms enable (i) a fine-tuned modulation of low levels of $H_2O_2$ for signalling purposes, and (ii) detoxification of excess $H_2O_2$ during stress.

Understanding the exact role $H_2O_2$ plays in the responses of plants to drought is complex, since it depends on a multitude of factors: the interaction between $H_2O_2$ and

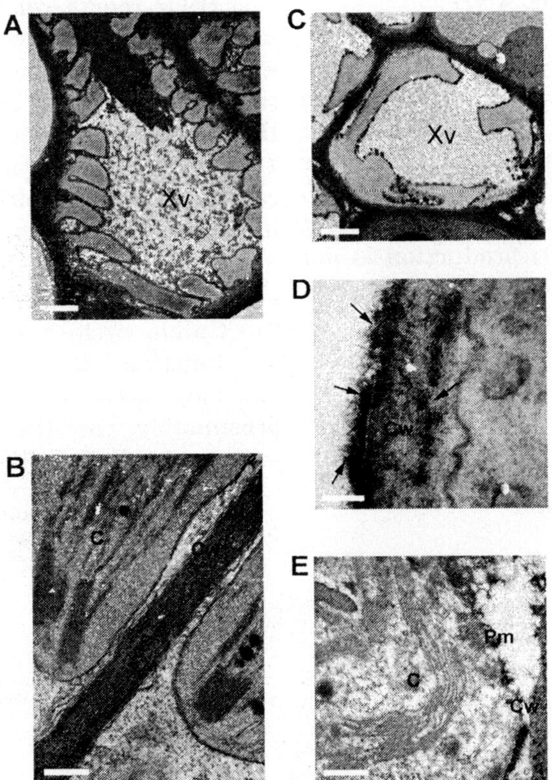


**Figure 4. Subcellular localization of $H_2O_2$ in drought-stressed leaves of *Salvia officinalis* plants**. Ultrastructure of xylem vessels and mesophyll cells of non-stressed leaves (March) (A,B) and drought-stressed leaves (July) (C,D,E). (A) Arrangement of spiral thickening in xylem vessels (bar = 0.8 μm). (B) Detail of mesophyll cells (bar = 0.6 μm). (C) Staining with $CeCl_3$ to localize $H_2O_2$ (arrows) from electron-dense deposits of cerium perhydroxides in the spiral thickening of the xylem vessels (Xv) and parts of the associated cell wall (bar = 0.8 μm). (D) Localization of $H_2O_2$ in mesophyll cell walls (cw) (bar = 80 nm). (E) Distortion of thylakoids and accumulation of $H_2O_2$ in mesophyll cell walls and plasma membranes (pm) (bar = 0.4 μm).

other ROS molecules, the complex and subtle interplay between ROS and antioxidants, and plant ontogeny. Moreover, the intracellular site of its production affects the specificity of its biological activity. Furthermore, there are problems with the methodology for detecting $H_2O_2$; e.g., intracellular real-time imaging using fluorescent dyes such as dichlorofluorescein diacetate ($H_2$-DCFDA) is typically used to assay $H_2O_2$. However, as this method is not specific, we cannot assess whether the effects of a physiological process is due to $H_2O_2$ or to another ROS.

CAT is one of the most rapidly turned over proteins in leaf cells, increasing its activity under severe drought conditions. Indeed, it may be indispensable for $H_2O_2$ detoxification during stress (Willekens *et al.*, 1997).

In this chapter, we will focus on the role of $H_2O_2$, whose clear participation has been reliably demonstrated. Other physiological processes, such as senescence and programmed cell death, which are difficult to discern between the effects of $H_2O_2$ and those of other ROS, have been omitted.

## 3. HYDROGEN PEROXIDE AND OXIDATIVE DAMAGE

$H_2O_2$ is a less reactive oxidant when compared to other ROS. Nevertheless, $H_2O_2$ may be particularly harmful, because it is relatively stable and may therefore spread within or among cells by diffusion and affect compartments where $H_2O_2$ production is not under tight control. $H_2O_2$ can give rise to more ROS, which greatly increases its toxicity. In the presence of transition metals, such as $Fe^{2+}$ or $Cu^{1+}$, the extremely reactive hydroxyl radical can be formed in the Fenton reaction.

$H_2O_2$ reacts relatively weakly with the majority of biological molecules. Nevertheless, photosynthesis has been shown to be rapidly inactivated by low concentrations of $H_2O_2$ since it affects $CO_2$ fixation (Asada, 1992, 1994). A concentration of 10 μM $H_2O_2$ caused a 50% inhibition of $CO_2$ assimilation in isolated chloroplasts (Keiser, 1976).

The ferredoxin/thiorredoxin (Fd/Trx) system plays a significant role in regulating the Calvin cycle and related processes, and its function also extends to modulating enzyme activity in mitochondria, and presumably, in the nucleus as well. In summary, during light-dependent electron

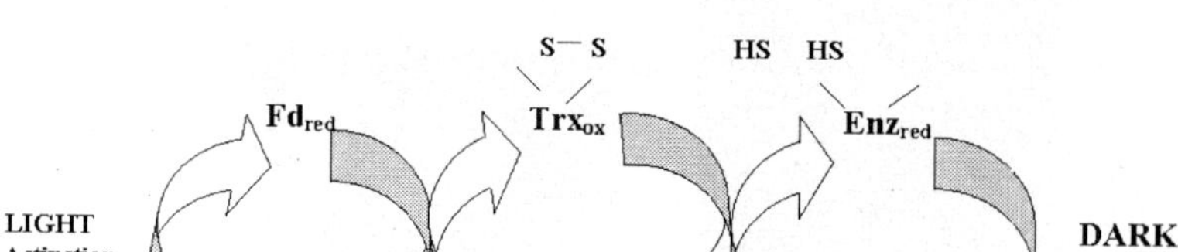


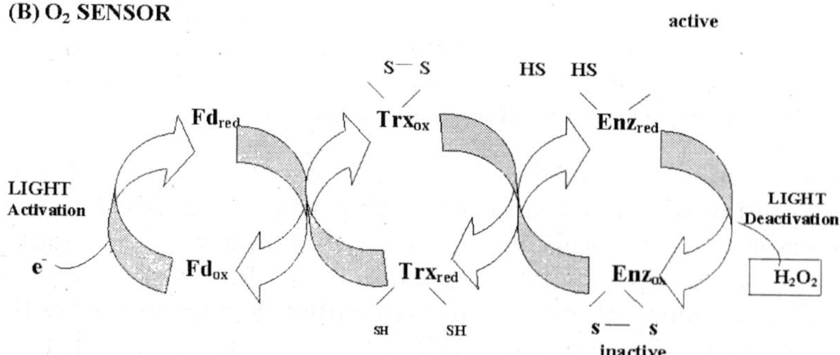


**Figure 5. Role of the ferredoxin/thioredoxin system in chloroplasts.** (A) Function of the Fd/Trx system in nonstressed plants. Photosynthetic electron transfer by PSI leads to the reduction of ferredoxin. Reduced ferredoxin in turn reduces thioredoxin, a regulatory disulfide protein. Reduced thioredoxin can reduce the disulfide bonds of numerous target proteins, modulating its enzymatic activity. (B) Under drought-stress, $H_2O_2$ could oxidize the target enzymes deactivating enzyme activity. (Buchanan and Balmer, 2005).

transport in chloroplasts, reduced ferredoxin ($Fd_{red}$) can reduce disulphide groups of oxidized thioredoxin ($Trx_{ox}$). The reduced form of Trx ($Trx_{red}$) can in turn reduce the oxidized forms of the enzyme (inactive), forming a reduced enzyme, which contains a thiol group, thereby activating the enzyme.

The main targets for $H_2O_2$ are the thiol protein groups, which could be oxidized, thereby inactivating the enzyme (Fig. 5); e.g., oxidizing cysteine and methionine protein residues. Oxidation of cysteine leads to the formation of disulfide bonds, which can consequently block an enzyme's active sites.

Even thiols can be oxidized to sulfinic ($SO_2H$) and sulfonic acids ($SO_3H$) as a result of oxidative stresses, a reaction that appear to be irreversible (Fig. 6).

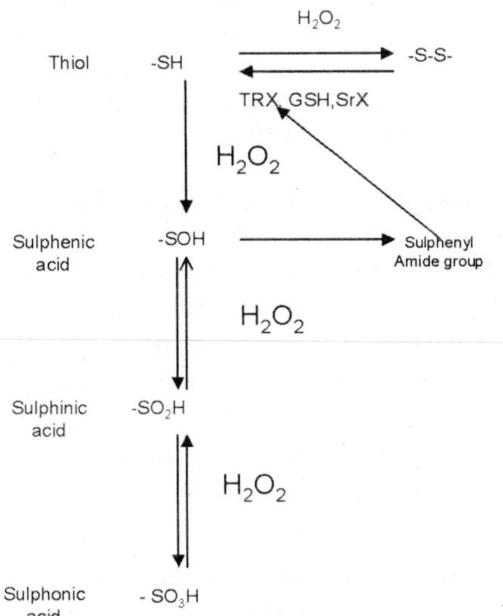


**Figure 6. Protein modification by $H_2O_2$.** Thiol groups can be oxidized, leading to conformational changes. Potential modifications can lead to the formation of cystine bridges, or sulfenic acid, sulfinic acid or sulfonic acid groups. Reversal of such oxidations might involve reduced glutathione (GSH), thioredoxin (TRX) or sulfiredoxin (Srx). (adapted from Desikan *et al.*, 2005)

In chloroplasts, the sensitive targets of $H_2O_2$ inhibition are those enzymes of the Calvin cycle regulated by the ferredoxin/thioredoxin system, such as NADP-sedoheptulose-1,7-biphosphate, phosphoribulokinase or transketolase. Furthermore, thiorredoxin had been linked to ATP synthesis, as well as the oxidative pentose phosphate cycle.

By also participating in the oxidative pentose phosphate cycle, transketolase thus contributes to the synthesis of erythrose-4 phosphate, which in turn serves as the starting point for several other pathways, including that leading to the synthesis of shikimate. Therefore, its activity affects not only the Calvin cycle, but also carbohydrate, amino acid, and natural product metabolism.

Moreover, many other enzymes are regulated by the ferredoxin/thioredoxin system, including: certain enzymes of the $C_4$ cycle; nitrogen metabolism, sulphur metabolism, fatty acid biosynthesis, assembly/folding; protein degradation; starch degradation; glycolysis, plastid division, DNA replication. Then, the processes range from DNA replication, transcription, and plastid division to metabolism and biosynthesis, protein assembly/folding, and degradation, to $HCO_3^-/CO_2$ equilibration.

$H_2O_2$ can affect enzyme activity, either directly via the oxidant-induced formation of disulfide bonds, or through the change in the ratio of reduced and oxidized thioredoxin (or other reducing agents). A small variation in the redox status of the thioredoxin pool can impede its ability to efficiently catalyze the cleavage of target disulfide.

Due to its direct and indirect cytotoxicity, plant survival depends on the efficient scavenging of $H_2O_2$.

## 4. HYDROGEN PEROXIDE AS A SIGNALING MOLECULE

Earlier concepts that $H_2O_2$ exert its effects through physicochemical damage have been replaced by one that recognizes the active role of $H_2O_2$ in cell signaling.

Hydrogen peroxide has been implicated in (i) ABA stomatal closure and in (ii) gene expression.

### 4.1 $H_2O_2$- and ABA-Mediated Stomatal Closure

Recent work has shown that ROS are essential signals mediating ABA-induced stomatal closure. Stomatal closure is a rapid plant response to drought stress that attempts to prevent the excess transpiration that would lead to water deficit. It has been shown that under drought-stress conditions, a rapid enhancement of ABA occurs that activates the synthesis of $H_2O_2$, which in turn, induces the closure of stomata. In summary, ABA activates the synthesis of $H_2O_2$ in guard cells, apparently via a plasma membrane NADPH oxidase. This increase in $H_2O_2$ activates the plasma membrane $Ca^{2+}$ inward channels, inactivates the $K^+$ inward channels, and causes cytosolic alkalinization (Figure 7). Moreover, the rise in intracellular calcium causes membrane depolarization, thereby inhibiting the plasma membrane proton pump. Membrane depolarization activates the $K^+$ outward channels. The efflux of $K^+$ causes a decrease in guard cell turgor and the closure of stomata.

The requirement of $H_2O_2$ in ABA-induced stomatal closure has been previously demonstrated (Pei *et al.*, 2000). Tobacco plants overexpressing DHAR, which have higher ascorbic acid levels, smaller proportions of DHA, and low $H_2O_2$ levels, have impaired drought ABA-induced stomatal closure and wilt more readily. Antisense reduction of DHAR activity had the opposite effect (Chen and Gallie, 2004). Furthermore, the level of $H_2O_2$ and the AA redox state in guard cells and whole leaves are diurnally regulated such that the former increases during the afternoon, whereas the later decreases. DHAR serves to maintain a basal level of AA recycling in guard cells that is insufficient to scavenge the high rate of $H_2O_2$ produced in the afternoon, thus resulting in stomatal closure.

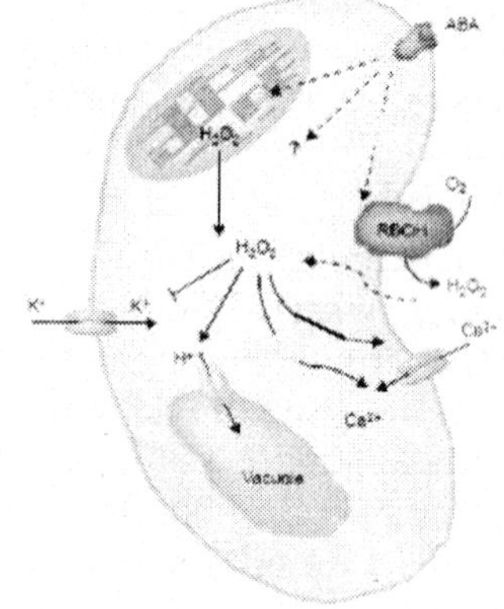


**Figure 7. $H_2O_2$ signaling in guard cells.** $H_2O_2$-mediated ABA signaling in guard cells. ABA induces $H_2O_2$ synthesis via plasma membrane NADPH oxidase (RBOH) or in the chloroplast. $H_2O_2$ inactivates the inward $K^+$ channels, causes cytosolic alkalinization, and activates plasma membrane $Ca^{2+}$. (Adapted from Neill *et al.*, 2002).

### 4.2 $H_2O_2$ Regulated Gene Expression

$H_2O_2$ production induces a rapid increase in the expression of genes involved in stress signal transduction, antioxidant biosynthesis, and cell protection mechanisms concomitant with repression of genes encoding photosynthesis components. However, little information is available of the nature of the redox sensors involved.

The principal modes of $H_2O_2$ action affecting gene expression are summarized in Figure 8.

First of all, $H_2O_2$ interacts selectively with a putative target molecule ($H_2O_2$ sensor) that perceives the increase in $H_2O_2$ concentration, and then translates this information into a change of gene expression. Plants contain two-component histidine kinases. Whether these proteins can function as $H_2O_2$ sensors in plants is currently under investigation.

Once $H_2O_2$ has interacted with a sensor, the information is translated through a signal transduction pathway until the gene expression is modified. Such a change in transcriptional activity may be achieved through: (a) the activation of mitogen protein kinase (MPK) cascade; (b) by the oxidation of signaling pathway components that subsequently activate transcription factors; or (c) by targeting and modifying the activity of transcription factors. However, neither the mechanism of activation nor the downstream targets of the MAPK pathways are yet known. ROS-induced activation of MAPK appears to be central in mediating cellular responses to multiple stresses.

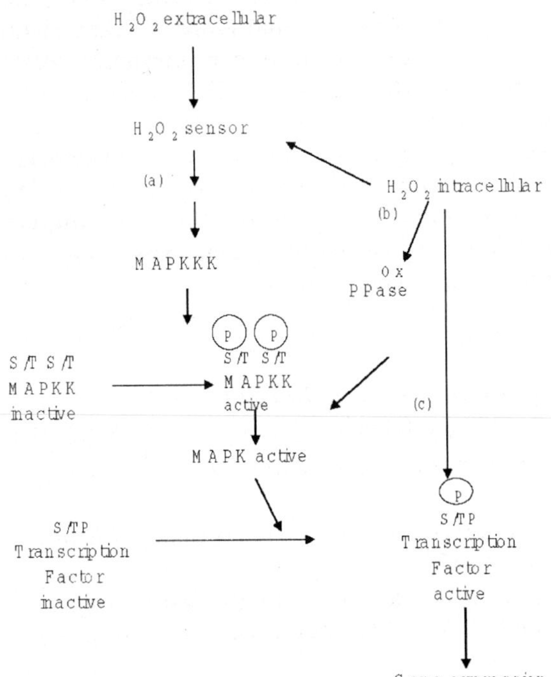


**Figure 8. Schematic representation of cellular $H_2O_2$-sensing and -signaling mechanisms**. $H_2O_2$ sensors such as membrane-localized histidine kinases can sense extracellular and intracellular $H_2O_2$ (a). Intracellular $H_2O_2$ can also influence the $H_2O_2$-induced mitogen-activated protein kinase (MAPK) signaling pathway through the inhibition of MAPK phosphatases (PPases) (b) or downstream transcription factors (c). S, serine; T, treonine. (Adapted from Apel and Hirt, 2004)

## 5. HYDROGEN PEROXIDE AND CELL WALL LIGNIFICATION

In plants, $H_2O_2$ plays a role in cell wall lignification that is particularly important in xylem vessels, which confers resistance against the tensile force of water columns. Lignins are a three-dimensional, amorphous hetero-polymer that result from the oxidative coupling of three hydroxycinnamyl alcohols: coumaryl, coniferyl, and sinapyl alcohols. The classical view of lignin polymerization is that cell wall oxidases convert these hydroxycinnamyl alcohols into free radicals, which then spontaneously polymerize to produce lignin. Both laccase ($H_2O_2$-independent) and peroxidase ($H_2O_2$-dependent) activities are responsible for free radical formation (Pomar *et al*., 2002).

The presence of $H_2O_2$ in the cell wall of plants subjected to drought-stress conditions has been demonstrated by histochemical staining (Fig. 4, and Munné-Bosch *et al*., 2001, 2003), but the mechanism underlying its generation remains unclear.

One question that rises from our results is from where does all this hydrogen peroxide originate? Electron microscope images (Fig. 4) showed that in *Cistus clusii* plants hydrogen peroxide accumulation was mainly located in the cell wall closest to the intercellular spaces, especially during midday when irradiation is high and there exists a greater possibility of chloroplast ROS generation. Interestingly, under drought stress conditions, chloroplasts are known to move within the cytosol to gain closer proximity to intercellular spaces. In other words, hydrogen peroxide accumulation in cell walls usually occurs close to chloroplasts. Are chloroplasts the source of hydrogen peroxide in *Cistus clusii* cell walls when these plants are exposed to drought?

Much accumulated evidence indicates the involvement of a membrane-bound NADPH oxidase producing superoxide ($O_2^{\cdot -}$), which in turn spontaneously forms $H_2O_2$.

## 6. CONCLUSIONS AND OPEN QUESTIONS

Hydrogen peroxide ($H_2O_2$) is continuously produced in plants as byproducts of aerobic metabolism. $H_2O_2$ is a less reactive oxidant than other ROS, even though is able to diffuse through membranes and affect other compartments and neighbouring cells. Furthermore, it inactivates enzymes regulated by the ferredoxin/ thioredoxin system, and can give rise to additional ROS, which greatly increases its toxicity. $H_2O_2$ can be scavenged by various cellular enzymatic and nonenzymatic mechanisms. The equilibrium between the production and scavenging of $H_2O_2$ may be disturbed by drought-stress conditions, thereby increasing $H_2O_2$ levels. In only a few years, the understanding of $H_2O_2$ has changed from one that saw it as toxic metabolic side-product requiring antioxidant mechanisms to protect cells from its effects, to one that places it at the centre of plant biology as a key signal regulating various processes including redox regulation of electron transfer chains (photosynthesis and respiration), enzyme regulation, stomatal behaviour, gene expression, and cell wall lignification, all of which induce resistance to drought-stress conditions.

Although certain functions of $H_2O_2$ have been elucidated, many questions have yet to be answered. What are the concentrations of $H_2O_2$ in various subcellular compartments? What contributions to the cellular $H_2O_2$ pool are made by the various sources? Can $H_2O_2$ produced in one cellular compartment exert an effect in others? How is $H_2O_2$ perceived by the cell? Is the current methodology employed sufficiently precise to fully characterize $H_2O_2$? Further work more is necessary to improve our understanding of the complex functions of $H_2O_2$ in plant responses to drought-oxidative stress.

## REFERENCES

Alegre, L. and Munné-Bosch, S. 2003. Regulation of plant responses to drought: function of plant hormones and antioxidants. In: *Advances in Plant Physiology*. (Hemantaranjan, A., ed). **5**: 267-285

Apel, K. and Hirt. H. 2004. Reactive oxygen species: metabolism, oxidative stress, and signal transduction. *Annual Review of Plant Biology*, **55**: 373-399.

Asada, K. 1992. Production and scavenging of active oxygen in chloroplasts. In: *Molecular Biology of free radical scavenging systems*. (Scandalios, J.G., ed), pp. 173-192.

Asada, K. 1994. Production and action of active oxygen species in photosynthetic tissues. In: *Causes of photooxidative stress and amelioration of defense systems in plants*. (Foyer, C.H. and Mullineaux, P.M. eds), pp: 77-104.

Asada, K. 1999. The water-water cycle in chloroplasts: scavenging of active oxygens and dissipation of excess photons. *Annual Review of Plant physiology and Plant Molecular Biology*, **50**: 601-639.

Bray, E.A., Bailey-Serres, J. and Weretilnyk, E. 2000. Responses to abiotic stresses. In: *Biochemistry & Molecular Biology of Plants*. (Buchanan, B.B., Gruisem, W., R. Jones, eds). American Society of Plant Physiologists, pp. 1158-1203.

Buchanan, B.B. and Balmer, Y. 2005. Redox regulation: A broadening horizon. *Annual Review of Plant Biology*, **56**:187-220

Bolwell, G.P., Butt, V.S., Davies, D.R. and Zimmerlin, A. 1995. The origin of the oxidative burst in plants. *Free Rad. Res*, **23**: 517-532

Chen, Z. and Gallie, D.R. 2004. The ascorbic acid redox state controls guard cell signaling and stomatal movement. *The Plant Cell*, **16:** 1143-1162.

Dat. J., Vandenabeele, S., Vramová, E., van Montagu, D., Inzé, D., and van Breusegem, F. 2000. Dual action of the active oxygen species during plant stress responses, *Cellular and Molecular Life Science*, **57:** 779-795.

Desikan R., Hancock J.Y. and Neill, S. (2004) Oxidative stress signalling. In: *Plant responses to abiotic stress*. (Hirt, H., Shinozaki K., eds). Springer, pp: 121-148

Desikan, R., Hancock, J., Neill, S. 2005. Reactive oxygen species, a signalling molecule. In: *Antioxidants and reactive oxygen species in plants.* (Smirnoff, N., ed), pp: 169-196.

Finkel, T. 2000. Redox-dependent signal transduction. *FEBS Letters*, **476**: 52-54.

Keiser, W.M. 1976. The effect of hydrogen peroxide on $CO_2$- fixation of isolated chloroplasts, *Biochimicha et Biophysica Acta* , **440**: 476-482.

Larcher, W. (2003) *Physiological Plant Ecology.* 4th Edition. Springer, pp: 401-416.

Luna, C.M., Pastori, G.M., Driscoll, S., Groten, K., Bernard, S. and Foyer, C.H 2005. Drought controls on $H_2O_2$ accumulation, catalase (CAT) activity and CAT gene expression in wheat. *Journal of Experimental Botany*, **56**: 417-423.

Mano, J. 2002. Early events in environmental stresses in plants: induction mechanisms of oxidative stress, In: *Oxidative stress in plants.* (Inzé, D., M. Van Montagu, M., eds), pp: 217-245.

Meyer, Y., Verdoucq, L. and Vignols, F. 1999. Plant thioredoxins and glutaredoxins: identity and putative roles. *Trends in Plant Science*, **4**: 388-394.

Mitler, R. 2002. Oxidative stress, antioxidants and stress tolerance, *Trends in Plant Sciences*, **7**: 405-410.

Munné-Bosch, S., Jubany-Marí, T. and Alegre, L. 2001. Drought-induced senescence is characterized by a loss of antioxidant defences in chloroplasts. *Plant, Cell and Environment*, **24**: 1319-1327.

Munné-Bosch, S., Jubany-Marí, T. and Alegre L. 2003. Enhanced photo-and antioxidative protection, and hydrogen peroxide accumulation in drought-stressed *Cistus clusii* and *Cistus albidus* plants, *Tree Physiology*, **23:** 1-12.

Neill, S., Desikan, R. and Hancock, J. 2002. Hydrogen peroxide signalling. *Current opinion in Plant Biology*, **5**: 388-395.

Neill, S.J., Desikan, R., Clarke, A, Hurst, R.D and Hancock, J.T. 2002. Hydrogen peroxide and nitric oxide as signalling molecules in plants. *Journal of Experimental Botany*, **53**: 1237-1247.

Noctor, G. and Foyer, C. 1998. Ascorbate and glutathione: keeping active oxygen under control. *Annual Review of Plant Physiology Molecular Biology*, **49**: 249-279.

Noctor, G., Veljovic-Jovanovic, F. and Foyer, C.H. 2000. Peroxide processing in photosynthesis: antioxidant coupling and redox signalling. *Procedings of the Royal Society of London, B*, **355**: 1465-1475.

Noctor, G., Veljovic-Jovanovic, S.D., Driscoll, S., Novitskaya, L. and Foyer, C.H. 2002. Drought and oxidative load in the leaves of $C_3$ plants: a predominant role for photorespiration?. *Annals of Botany*, **89**: 841-850.

Pastori, G.M. and Foyer, C.H. 2002. Common components, networks and pathways of cross-tolerance to stress. The central role of "redox" and abscidsic acid mediated controls. *Plant Physiology*, **129**: 460-468.

Pei, Z.M., Murata, Y., Benning, G., Thomine, S., Klüsener, B., Alle, G.J., Grill, E. and Schroeder, J.I. 2000. Calcium channels activated by hydrogen peroxide mediate abscisic acid signalling in guard cells. *Nature*, **406**: 731-734.

Pomar, F., Caballero, N., Pedreño, M.A. and Ros-Barceló, A. 2002. $H_2O_2$ generation during the auto-oxidation of coniferyl alcohol drives the oxidase activity of a highly conserved class III peroxidase envolved in lignin biosynthesis. *FEBS Letters*, **529:** 198-202.

Smirnoff, N. 1993. The role of active oxygen in the response of plants to water deficit and dessication. *New Phytologist* , **125:** 27-58.

Willekens, H., Chamnongpol, S., Davey, M., Schraunder, M., Langebartels, C., von Montagu, M., Inzé, D. and van Camp, W. 1997. Catalase is a sink for $H_2O_2$ and is indispensable for stress defence in $C_3$ plants. *Embo Journal*, **16:** 4806-4816.

*Advances in Plant Physiology*, Vol. **9**
Ed. A. Hemantaranjan
Scientific Publishers (India), Jodhpur, 2006 pp. 13-39
E-mail: **info@scientificpub.com** www.scientificpub.com

# 2

# MOLECULAR PHYSIOLOGY OF DROUGHT TOLERANCE MECHANISM IN PLANTS

A. Hemantaranjan[1] and J. P. Singh[2]

[1]Department of Plant Physiology and
[2]Department of Agronomy Institute of Agricultural Sciences, Banaras Hindu University, Varanasi 221 005, India
[1]E-mail: hemantaranja@satyam.net.in

## INTRODUCTION

An understanding of the responses of crops to their environment is fundamental to minimize the deleterious impact of unfavourable climatic conditions and to manage them for maximum productivity. In the long run, the advances of *all* applied science are preceded by the knowledge base generated by pure-basic science. Hence, applied research will eventually weaken and stagnate without advances in basic sciences. The complex regulation of stomatal conductance is related to important differences among species and genotypes in the response of stomatal to leaf water potential and relative water content. This fact is difficult to compare among different studies when the severity of drought is quantified on a water potential or water content basis.

One way of increasing productivity in stressful environments is to breed crops that are more tolerant to stress. However, success in breeding for tolerance has been limited because (a) tolerance to stress is controlled by many genes, and their simultaneous selection is difficult (Richards, 1996; Yeo, 1998; Flowers *et al.*, 2000); (b) tremendous effort is required to eliminate undesirable genes that are also incorporated during breeding (Richards, 1996); and (c) there is a lack of efficient selection procedures particularly under field conditions (Ribaut *et al.*, 1997). Genetic engineering offers an alternative approach for developing tolerant crops. Unlike classical breeding, genetic engineering is a faster and more precise means of achieving improved tolerance (Cushman and

Bohnert, 2000) because it avoids the transfer of unwanted chromosomal regions. Moreover, through genetic engineering, multiple genes can be assembled and simultaneously introduced to the crop of interest. There are many functional targets for engineering tolerance to water stress and salinity, one of them being accumulation of osmoprotectants (Rathinasabapathi, 2000).

Many authors use the terms water stress and drought as synonyms. Properly, water stress occurs in plant tissues as a consequence of a water deficit. This can occur very frequently even in well-irrigated plants, and in the short term (minutes-hours), whenever the evaporative demand is higher than the xylem capacity for refilling leaves. For instance, the incidence of a dry wind can cause a temporary water deficit in the leaf. The so-called 'midday depression' of photosynthesis is another example of temporary, short-term water deficit, that seems involving a complex interaction of water deficit, photosynthetic down-regulation and circadian rhythms (Lange *et al.*, 1982; Wise *et al.*, 1991; Muraoka *et al.*, 2000). Drought, by contrast, refers to a water deficit that is sustained in the long term (days-months). It occurs when soil is partially or totally water-depleted (for instance, due to low precipitation), while the evaporative demand is high, so a long-term imbalance is produced between water demand and supply. Drought is usually accompanied by high irradiance and temperature. Among the physiological functions that are affected by drought, there is a reduction of the germination capacity of seeds (Chen *et al.*, 1968), limited plant growth (Hsiao, 1973), decreased grain and fruit yields (Gallagher *et al.*, 1976), limited plant nutrient uptake (Heckathorn *et al.*, 1997) and increased plant predisposition to suffer diseases (Boyer, 1995a). Moreover, drought interacts with other stresses like high temperature, provided that the drought-induced decrease in transpiration did modify the leaf energy balance, so the leaf temperature usually increases during drought (Schulze *et al.*, 1987; Tyree, 1999).

In the present chapter we will focus mainly in the effects of drought, rather than in short-term water deficit. This review aims to discuss various changes occurring in plants at physiological and molecular levels under conditions of drought besides covering various responses of vital living processes to drought and development of tolerance. With these views in mind, the whole text has been divided into five major sections describing photosynthetic responses to drought, $CO_2$ assimilation, photorespiration, and mitochondrial respiration under drought stress; desiccation tolerance in the resurrection plants; and last but not the least, the physiology of drought tolerance mechanism in plants up to molecular level, which justifies considerably the main heading of this review.

## 2. PHOTOSYNTHETIC RESPONSES TO DROUGHT

### 2.1 Parameters Indicative of Drought

Leaf water potential ($\Psi$) and leaf relative water content (RWC) are largely the most used parameters to assess the degree of water stress (Boyer, 1995b). However, as stated by Kramer (1988), no single measure of plant water status can be expected to correlate with the numerous effects of water stress. In the particular case of photosynthesis and stomatal conductance, their response to leaf water potential is extremely variable among different species, the specific response being considered as reflective of particular adaptations to drought (Schulze and Hall, 1982; Tyree, 1999). Under field conditions, water stress effects on stomatal conductance and photosynthesis are detectable before any substantial change in leaf relative water content and/or water potential occurs. Water deficit limits plant growth and productivity because it decreases net $CO_2$ assimilation due to reduced stomatal conductance for $CO_2$ and/or because of non-stomatal effects like inhibition of enzymatic processes by changes in ionic or osmotic conditions (Lawlor, 1995). At high light intensities the lowered consumption of redox equivalents in the Calvin cycle makes it necessary to degrade photosynthetic electrons in processes other than $CO_2$ fixation to avoid photo-inhibition. Sharkey *et al.* (1988) showed that the activity of photosystem II can be regulated in a way that the rate of electron transport matches the capacity of the electron consuming reactions and that linear electron transport depends not only on light intensity and $CO_2$ concentration but also on the $O_2$ concentration. Oxygen can function as alternative electron acceptor directly in the Mehler reaction or indirectly in photorespiration (Badger, 1985).

## 2.2 Drought Stimulate Stomatal Closure

Stomata close progressively as drought progresses, followed by parallel decreases of net photosynthesis. As $CO_2$ availability in the chloroplasts depends on stomatal conductance, it has been many times assumed that drought decreases photosynthesis simply by closing stomata and limiting $CO_2$ availability. It is important to emphasise that stomatal conductance always correlate to photosynthesis (Wong *et al.*, 1979, 1985), but stomatal conductance is controlled not only by soil water availability, but by a complex interaction of factors internal and external to the leaf. Indeed, under water stress, a high correlation is often observed between leaf water potential and stomatal conductance, but the precise relationship is strongly dependent on the species or cultivar (Kriedemann and Smart, 1969; Schulze and Hall, 1982; Tardieu and Simonneau, 1998; Tyree, 1999), drought history of the individuals studied (Larcher *et al.*, 1981), the size of pots in which the plants are rooted (Day and Sinclair, 1998), the height of the plant at which a leaf is placed (Schäfer *et al.*, 2000), the environmental conditions during drought (Schulze and Hall, 1982), the velocity of drought imposition (Flexas *et al.*, 1999a), the development of osmotic adjustment (Morgan, 1984; Ludlow, 1989), etc. Many of these interactions are related to acclimation and adaptation processes, but others did affect the immediate response to water stress.

Stomata close as leaf-to-air vapour pressure deficit (VPD) increases (Raschke, 1979; Dai *et al.*, 1992), although stomatal sensitivity to VPD is strongly species-dependent (Oren *et al.*, 1999). Other factors influencing stomatal conductance are $CO_2$ concentration in air (Raschke, 1979), leaf photosynthetic activity (Wong *et al.*, 1979; Farquhar and Sharkey, 1982), farnesyltransf-erase activity (Pei *et al.*, 1998), etc. An interesting hypothesis stressed recently that a direct negative feedback link exists between shoot cavitation and stomatal conductance (Salleo *et al.*, 2000).

Most of the direct evidences in favour of stomatal regulation of photosynthesis under drought come from experiments in which stomatal function has been modified and/or overcome. Troughton and Slatyer (1969) found no response of photosynthesis to desiccation once the $O_2$ and $CO_2$ concentrations in air were lowered to reduce respiration and to ensure $CO_2$-limited photosynthesis. Also, by removing the lower leaf epidermis, the photosynthetic rates were almost completely recovered under mild drought (Dietz and Heber, 1983; Schwab *et al.*, 1989), but not in severely stressed plants (Graziani and Livne, 1971). Increasing $CO_2$ concentration sufficiently to overcome stomatal limitations recovered almost completely the maximum photosynthetic rates in some studies (Robinson *et al.*, 1988; Chaves, 1991; Quick *et al.*, 1992), but not in others (Graan and Boyer, 1990; Quick *et al.*, 1992; Flexas *et al.*, 1999b, 2001b). Kaiser (1987) demonstrated that simulating the osmotic conditions that prevail in chloroplasts under mild drought did not affect the photosynthetic capacity, thus supposing that reduced photosynthesis under mild drought was due to stomatal closure. Foliar application of glycinebetaine, that increases stomatal conductance and reduces photorespiration, has also been shown to recover maximum photosynthetic rates in water stressed plants (Mäkelä *et al.*, 1999).

## 3. $CO_2$ ASSIMILATION, PHOTORESPIRATION, AND MITOCHONDRIAL RESPIRATION UNDER DROUGHT STRESS

Under drought conditions, increased capacity for photorespiration and non-photochemical quenching is essential to avoid photoinhibitory damage under excess light. That makes chlorophyll fluorescence from light-adapted samples a valuable tool in screening plants for drought stress tolerance. The most interesting hypothesis is that photorespiration may serve as a way to dissipate excess electron transport, especially under reduced $CO_2$ availability, as occur under drought (Osmond *et al.*, 1980; Osmond, 1981). The rate of photorespiration increases in response to either decreased $CO_2$ or increased temperature (Leegood, 1995). Both decreased $CO_2$ availability and increased leaf temperature occur under drought due to stomatal closure. Early studies showed that photorespiration increased in response to mild water stress, at least as a proportion of total photosynthesis (Lawlor and Fock, 1975, 1977a,b; Lawlor, 1976a,b; Lawlor and Pearlman, 1981; Thomas and André, 1982; Stuhlfauth *et al.*, 1988, 1990; Scheuermann *et al.*, 1991). There are well-recognised difficulties to accurately estimate the rate of photorespiration, which impose uncertainties about these estimates. Moreover, Gerbaud and André (1987) pointed out that the measurements of photorespiratory $CO_2$ release underestimate the actual rates of photorespiration by 20-100%, because these measurements did not take into account the recycling of $O_2$ produced in photosynthesis and of $CO_2$ released in photorespiration. It is clear now that between 80 and 100% of the $CO_2$ released in photorespiration is recycled by photosynthesis in $C_3$ plants (Takeba and Kozaki, 1998; Loreto *et al.*, 1999). When recycling is taken into account, or when photorespiration is estimated by combined measurements of gas-exchange and chlorophyll fluorescence, it is concluded that photorespiration increases under mild water stress, not only in relative to photosynthesis, but also in absolute values (Krampitz *et al.*, 1984 as recalculated in Gerbaud and André, 1987; Renou *et al.*, 1990; Di Marco and Tricoli, 1993; Brestic *et al.*, 1995; Tourneux and Peltier, 1995; Flexas *et al.*, 1999b; Flexas *et al.*, 2001a). A recent study using mutants of barley defective in some photorespiratory enzymes has also demonstrated that photorespiration rates are enhanced by mild drought (Wingler *et al.*, 1999, 2000). Under moderate to severe water stress, the rate of photorespiration progressively decreases, but to a lower extent than photosynthesis, thus still increasing the ratio of photorespiration to $CO_2$ assimilation (Björkman and Schäfer, 1989; Björkman and Demmig-Adams, 1994; Wingler *et al.*, 2000).

Studies by Wu *et al.* (1991) and Heber *et al.* (1996) demonstrated that photoinhibition did occur in leaves deprived of $O_2$ in air at high irradiance and atmospheric $CO_2$ concentration, but did not occur if $O_2$ was present in a 21% concentration. Moreover, they demonstrated that photorespiration was more effective than the Mehler reaction (see section 2.4.) as a photoprotective mechanism (Wu *et al.*, 1991). An important contribution by Kozaki and Takeba (1996) showed that mutants of tobacco with increased glutamine synthetase and photorespiration capacity were less susceptible to photoinhibition than the wild-type, and mutants with reduced glutamine synthetase were more susceptible to photoinhibition than the wild-type. However, Brestic *et al.* (1995) showed that,

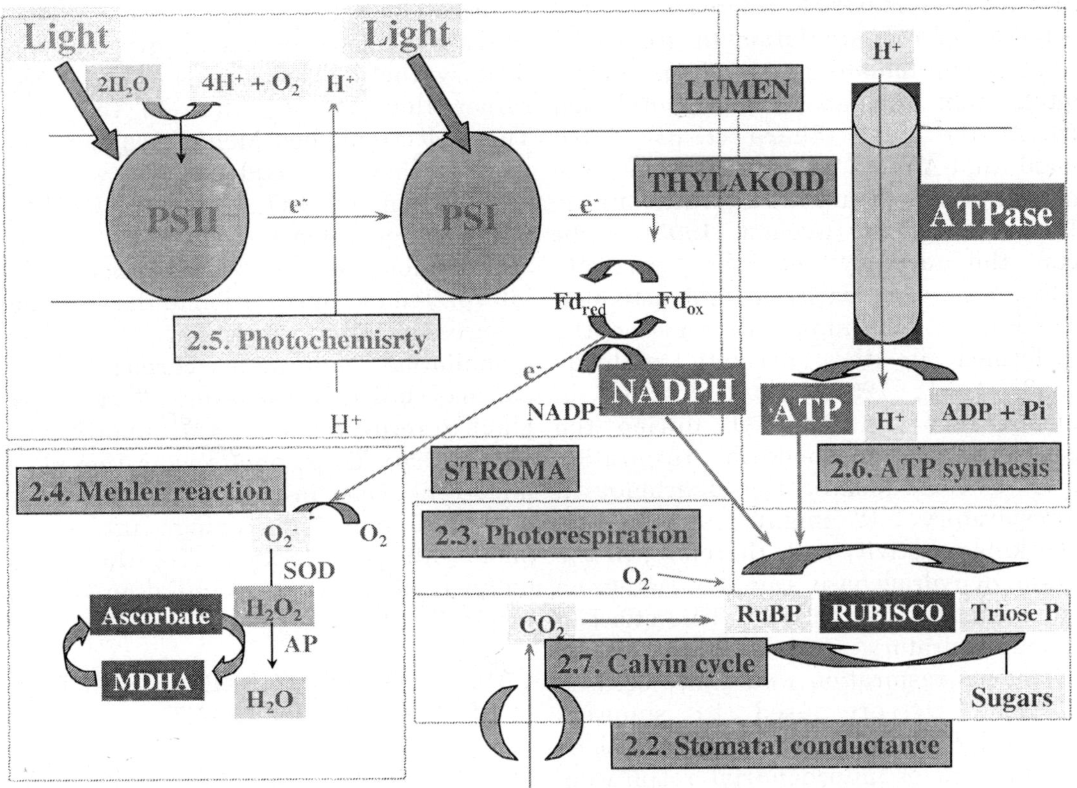


**Figure 1:** Depicts different processes in relation to photosynthesis that are affected by drought. Lumen, thylakoid membranes and stroma, as well as PSII, PSI and chloroplastic ATPase are indicated in colour.

although depriving bean plants of $O_2$ did reduce electron transport rate during drought development, it did not change the extent of PSII photoinhibition.

Studies with altered concentrations of $CO_2$ and $O_2$ around the leaves have yielded most of the current estimates of the relative rates of the Mehler reaction-driven rate of photosynthetic electron transport. These studies have yielded contradictory results, showing Mehler reaction-driven electron transport rates either lower than 10% of total ETR (Egneus *et al.*, 1975; Gerbaud and André, 1980; Genty *et al.*, 1989; Krall and Edwards, 1992; Ghashghaie and Cornic, 1994; Valentini *et al.*, 1995; Lovelock and Winter, 1996; Miyake and Yokota, 2000) or up to 30-40% of total ETR (Osmond and Grace, 1995; Park *et al.*, 1996). Direct measurements of $^{18}O_2$ uptake at saturating light and $CO_2$ in leaf discs, using mass spectrometry; also suggest Mehler reaction-driven electron transport rates lower than 10% of total ETR (Tourneux and Peltier, 1995; Biehler and Fock, 1996; Biehler *et al.*, 1997; Flexas *et al.*, 1999b; Badger *et al.*, 2000; Ruuska *et al.*, 2000).

Since drought induces stomatal closure, it has been assumed many times that most of the photosynthetic limitation under drought was due to stomatal closure. The simplest way to ascertain whether stomatal closure is the main factor controlling photosynthesis under drought is to follow the changes of intrinsic water-use-efficiency. Drought stress results in a remarkable increase of the contribution of re-assimilation to gross $CO_2$ assimilation in tomato plants. It is widely accepted that oxidative phosphorylation occurs in the light (Sharp *et al.*, 1984; Gerbaud and André, 1987). However, the magnitude of mitochondrial respiration in the light is still unclear (Krömer, 1995). Light affects the activity of the pyruvate dehydrogenase complex by a light activated protein kinase. This kinase depends on the $NH_3$ formed in the glycolate pathway (Randall *et al.*, 1996). The inhibition of photorespiration by high $CO_2$ during the measurement of mitochondrial respiration and, as a consequence, the deficiency in photorespiratory $NH_3$ might result in less protein kinase activity and, therefore, in the pyruvate dehydrogenase complex being no longer inactivated in the light. This effect of non-photorespiratory conditions on mitochondrial respiration would also occur if 20 mL $L^{-1}$ $O_2$ are used to suppress photorespiration but, under these conditions light inhibition of mitochondrial respiration has often been observed (Krömer, 1995).

At high $CO_2$ concentrations, Laisk and Sumberg (1994) detected carboxylation of a substrate other than RuBP, in addition to RuBP carboxylation, that may be caused by phospho*enol*pyruvate carboxylase activity. In our respiration measurements this non-RuBP carboxylation would be included in TPS and the rate of respiration in the light calculated from this TPS value is independent from the type of $CO_2$ assimilation. But mitochondrial respiration in the light determined by the $^{12}CO_2/^{13}CO_2$ technique could be accompanied by $CO_2$ evolution from the decarboxylation of malate or pyruvate (Laisk and Sumberg, 1994).

Studies on tomato plants showed that in addition to A and TPS the activity of photosystem II as well as gross $O_2$ uptake decrease in relation to water deficit (Haupt-Herting and Fock, 2000) and that the reduction of A cannot solely be caused by reduced internal $CO_2$ concentration. In stressed tomato plants a greater part of photosynthetic electrons flows to oxygen rather than to $CO_2$ than in the controls. These electrons feed Mehler reaction and the photosynthetic oxidation cycle. Re-assimilation of $CO_2$ consumes ATP and reducing equivalents, and higher rates of re-assimilation under water deficit were interpreted as contribution to the degradation of excess electrons (Fock *et al.*, 1992). Also, re-assimilation maintains carbon flux and enzyme substrate turnover, which helps the plant to recover after rewatering (Stuhlfauth *et al.*, 1990). Thus, photorespiration plays an important role in protecting plants from photoinhibition by using up excessive photosynthetic electrons in the glycolate pathway and by re-assimilation of (photo) respiratory $CO_2$.

## 4. DESICCATION TOLERANCE IN THE RESURRECTION PLANT

Unfavorable environmental conditions restrict the productivity and the range of habitats available to plants. This represents a severe economic constraint on agricultural production. Plants as sessile organisms have evolved a wide spectrum of adaptations to cope with the challenges of environmental stress. Quite often, however, adaptation mechanisms themselves adversely affect yield parameters, and a compromise between biomass production and environmental fitness has to be accepted. One major factor that limits the productive potential of higher plants is the availability of water. Water deficit can affect plants in different ways. A mild water deficit leads to small changes in the water status of plants, and plants cope with such a situation by reducing water loss and/or by increasing water uptake (Bray, 1997). The most severe form of water deficit is desiccation—when most of the protoplasmic water is lost and only a very small amount of tightly bound water remains in the cell.

Therefore, it has become imperative for plant biologists to understand the mechanisms by which plants can adapt to water deficit while retaining their capacity to serve as sources of food and other raw materials. Most flowering plants cannot survive exposure to a water deficit equivalent to less than 85% to 98% (v/v) relative humidity during their vegetative growth period; although desiccation is an integral part of the normal developmental program of higher plants in the context of seed formation. Only a few plants possess desiccation-tolerant vegetative tissues; these include a small group of angiosperms, termed resurrection plants (Gaff, 1971), some ferns, algae, lichens, and bryophytes.

Resurrection plants can be revived from an air-dried state and are often poikilohydrous, i.e. their water content varies with the relative humidity in the environment. Resurrection plants are found in ecological niches with limited seasonal water availability, preferentially on rocky outcrops at low to moderate elevations in tropical and subtropical zones (Porembski and Barthlott, 2001).

Many dehydration-specific gene products have been isolated but very few rehydration-specific proteins are known (Bernacchia *et al.*, 1996). This is in contrast to observations made on *T. ruralis*, a representative of the desiccation-tolerant bryophytes, which survives rapid desiccation. Here, the major changes in gene expression occur during the first hours of rehydration (Wood and Oliver, 1999).

Water deficit leads to closure of stomata and at the same time to a decrease in intercellular $CO_2$ concentration. As a consequence of the lower $CO_2$ availability, carbon assimilation is inhibited and ultimately photosynthetic capacity is lost. This observation is common to dehydration-sensitive and drought-tolerant species (Schwab *et al.*, 1989). These physiological changes are reflected at the molecular level: The steady-state levels of transcripts related to photosynthesis are down-regulated in response to water stress, e.g. the transcript for the small subunit of the Rubisco enzyme in *C. plantagineum* (Bernacchia *et al.*, 1996). In many organisms, bacteria and yeasts in particular, high concentrations of carbohydrates are observed in dry tissues, and a contribution of carbohydrates to desiccation tolerance has been proposed. In seeds of higher plants, a correlation has been observed between the accumulation of soluble sugars and the acquisition of desiccation tolerance (Leprince *et al.*, 1993). The accumulation of Suc in dehydrated tissues seems to be a common theme in different resurrection plants, although different metabolic routes may be used for the synthesis of Suc. The tetrasaccharide stachyose is the dominant sugar, comprising more than 50% of the total sugar content, and octulose is only found in small amounts in untreated and in dehydrated roots. Octulose is probably a product of photosynthesis that accumulates in leaves during the light period, but is partially metabolized at night. Because octulose has been found in the phloem sap, it is likely that it is transported from leaves to roots (Norwood *et al.*, 2000). To be aware of the biochemical basis for this control in sugar metabolism, the expression of genes encoding sugar-metabolizing enzymes was investigated. *C. plantagineum* possesses a set of genes encoding isoenzymes of Suc synthase and Suc phosphate synthase (Ingram *et al.*, 1997; Kleines *et al.*, 1999), which are differentially expressed. Besides these enzymes and the dehydration-induced expression of cytosolic glyceraldehyde dehydrogenase, it has been proposed that transketolase contributes to the conversion of octulose to Suc. In addition to a constitutively expressed, plastid-localized transketolase, induction of two genes for transketolases during rehydration has been reported (Bernacchia *et al.*, 1995).

The identified genes encode compounds that can be assigned to diverse metabolic pathways, although their precise function has not yet been demonstrated. The following main groups can be distinguished: (a) genes encoding proteins with protective properties, (b) genes encoding membrane proteins involved in transport processes, (c) genes encoding enzymes related to carbohydrate metabolism, (d) genes encoding regulatory molecules, such as transcription factors, kinases, or other putative signaling molecules, and (e) genes that show no homologies to known sequences. Studies on tissue-specific expression patterns and subcellular localizations have revealed specific cellular distributions of RNAs and proteins that appear to correlate with their predicted functions (Phillips *et al.*, 2001).

### 4.1 Proteins with Protective Properties

The most important representatives include late embryogenesis-abundant (LEA) proteins (Schneider *et al.*, 1993), some proteins with enzymatic function, e.g. an aldehyde dehydrogenase (Kirch *et al.*, 2001a), and also some small heat shock proteins (Alamillo *et al.*, 1995). LEA proteins are a heterogenous group of proteins found universally in plants, which were first discovered because they accumulate to high levels during late stages of embryo development. Synthesis of LEA proteins is associated with dehydration in seeds and with water deficit in vegetative tissues (Cuming, 1999). Dehydration-elicited expression of *LEA genes* is not restricted to resurrection plants but has been extensively reported also for non-tolerant plants.

LEA proteins are hydrophilic and have a biased amino acid composition, mostly lacking Cys and Trp. They are typically highly water soluble, and they often remain soluble after boiling. Despite extensive studies, our knowledge of the biochemical function of LEA proteins is still rudimentary, but molecular and biochemical features strongly suggest a protective role for them. A protective function correlates with their cellular distribution: LEA polypeptides are found in all cell types, accumulating abundantly in cytoplasm or plastids (Schneider *et al.*, 1993). This protective role is further corroborated by Wolkers *et al.* (1998), who showed that LEA proteins may be anchors in a structural network stabilizing cytoplasmic components during pollen drying. Two approaches have been used to demonstrate that LEA proteins function as cellular protectants: in vitro protection assays with purified proteins and in planta studies in which LEA proteins were overexpressed. The results generally support a protective role (Xu *et al.*, 1996), although contradictory observations have also been reported (Iturriaga *et al.*, 1992). One conclusion that can be drawn from functional analysis indicates that individual LEA proteins make only a small contribution to dehydration tolerance, whereas the coordinated synthesis of the complete set of proteins probably plays a central role. Biochemical experiments, including structural analysis of native proteins, are needed to understand the role of each specific LEA protein. It is remarkable that *LEA* genes contain sequence motifs that are conserved in all higher plants. This strict conservation during evolution indicates that these motifs define functional units within these proteins. The difference between tolerant and non-tolerant plants is likely to reside in the expression patterns and is likely to be at least in part a quantitative characteristic with respect to LEA genes.

### 4.2 Dehydration and Regulation of Gene Expression

Drought and salt stresses induce ABA biosynthesis largely through transcriptional regulation of ABA biosynthetic genes because blocking transcription by using transcription inhibitors impairs stress-induced ABA biosynthesis. Therefore, transcriptional regulation of ABA biosynthetic genes holds the key to understanding how ABA biosynthesis is regulated, although regulation of the specific activities of ABA biosynthesis enzymes also exists. The regulation of ABA biosynthetic genes may vary not only between different plant parts and developmental stages but also between different plant species. Like the ZEP genes in tobacco and tomato, the Arabidopsis *ZEP* gene also had a basal transcript level under non-stressful conditions. On the other hand, drought, salt, and polyethylene glycol clearly increased its expression level both in the shoot and in the root (Xiong *et al.*, 2002), demonstrating that the *AtZEP* gene is under stress regulation. Variation in the regulation of the ZEP genes observed in different experiments may be partly related to the relative basal transcript levels. A high basal transcript level may cover stress inducibility of the genes.

Gene regulation is particularly important in the case of a multigenic trait like desiccation tolerance because different regulatory pathways determine the expression of a whole set of genes. Most information is derived from promoter analyses and from differential screening procedures. One molecule that is central to dehydration-regulated gene expression is the plant hormone ABA. Exposure of *C. plantagineum* plants and callus to exogenous ABA induces genes that are otherwise activated by dehydration. Specifically in callus tissue, ABA is required to induce the genes needed for the expression of tolerance. The regulation of gene expression by dehydration and ABA involves several signaling pathways and different cis-acting elements in the stress-responsive genes (Bartels and Salamini, 2001). Promoters of LEA-type genes and of genes encoding dehydration-inducible enzymes have been analyzed and compared. Comparisons of the promoter sequences did not reveal obvious common motifs. Therefore, several promoters were analyzed in transgenic tobacco (*Nicotiana tabacum*) and Arabidopsis plants to define functional cis-elements. The promoters tested were found to be highly active in seeds and pollen, but two out of three promoters were not active in vegetative tissues of Arabidopsis or tobacco (Furini *et al.*, 1996; Velasco *et al.*, 1998). Only the promoter of the gene *CDeT6-19* was inducible by dehydration or ABA in vegetative tissues of a heterologous plant. However, ectopic expression of the Arabidopsis ABI-3 gene product leads to activation of the other two promoters in Arabidopsis leaves upon ABA treatment (Furini *et al.*, 1996; Velasco *et al.*, 1998). *ABI-3* encodes a transcription factor that is active during seed development in Arabidopsis. In view of the fact that *C. plantagineum* synthesizes in vegetative tissues many LEA-type transcripts that are closely related to those of *LEA* genes expressed during seed maturation, an *ABI-3* homolog was isolated from *C. plantagineum* to test the possibility that such a gene might be responsible for transcriptional activation of LEA-type genes in vegetative tissues of *C. plantagineum*. A gene closely related to the Arabidopsis *ABI-3* gene was isolated, and its product was indeed able to transactivate *LEA* genes in transient expression assays (Chandler and Bartels, 1997). In many genes regulated by ABA and osmotic stress, one or more ABA response elements (ABREs) play a key role in promoter activity. The ABREs have a core ACGT-containing G-box motif. It is assumed that transcription factors of the basic Leu zipper type bind as dimers to the ABREs. Modified core ABRE elements are present in LEA gene promoters in *C. plantagineum*, but they do not seem to be the major determinant of ABA responsiveness.

Genes for several potential transcription factors were isolated, and it was shown that members of diverse gene families participate in the response of *C. plantagineum* to dehydration. These include genes for Myb and a heat shock transcription factor, members of the homeodomain Leu zipper (HDZIP) family, a phospholipase D (PLD), a kinase, and the gene *CDT-1*. HDZIP genes are specific for plants and are thought to regulate developmental processes, and responses to environmental cues ranging from light perception to pathogen-induced and abiotic stress. HDZIP proteins are characterized by the presence of a DNA-binding homeodomain adjacent to a Leu zipper motif that mediates protein-protein interactions (Ruberti *et al.*, 1991). The activity of HDZIP proteins resides primarily in binding of the homeodomain to specific HDE recognition sequences present in promoters of target genes. The gene targeted by the T-DNA, which led to the constitutive activation of desiccation-related genes, was isolated and named *CDT-1* (Furini *et al.*, 1997). *CDT-1* is present in multiple copies in the genome of *C. plantagineum* and its expression can be induced by ABA in non-transformed callus and plants. The most surprising feature of the gene is that it has only a very short open reading frame encoding a putative polypeptide of 22 amino acids. To date, it has not been possible to demonstrate the presence of a polypeptide encoded by *CDT-1*, and further experiments support the hypothesis that CDT-1 may act as a regulatory RNA. Besides *CDT-1*, another example for a gene with such properties is the *ENOD40* gene from *Medicago sativa*, which controls the organogenesis of *Rhizobium meliloti*-induced $N_2$-fixing nodules (Crespi *et al.*, 1994).

## 5. MOLECULAR PHYSIOLOGY OF DROUGHT TOLERANCE MECHANISM

Studies on abiotic stresses in diverse group of plants including economic crops have advanced considerably in recent years. However, the majority of experiments testing the response of plants to changes in environmental conditions have focused on a single stress treatment applied to plants under controlled conditions. In contrast, in the field, a number of different stresses can occur simultaneously. These may include conditions such as high irradiance, low water availability, extreme temperature, or high salinity and may alter plant metabolism in a novel manner that may be different from that caused by each of the different stresses applied individually. Recent advances in abiotic stress responses of plants have been made under conditions of drought stress and water stress situations especially related to molecular physiology, binding proteins, changes in the redox state of α-tocopherol, ascorbate, and the diterpene carnosic acid in chloroplasts of labiatae species differing in carnosic acid contents etc. in the understanding of drought tolerance mechanism. Within their natural habitat, plants are subjected to a combination of abiotic conditions that include stresses such as drought and heat. Drought and heat stress have been extensively studied; however, little is known about how their combination impacts plants. Some of the facts emerged out with reference to the ***bases of stress tolerance*** in diverse plant group are described below in eleven specific sections covering summary of reports from 1999 to 2005.

### 5.1 Oligosaccharide Metabolism in Coleus

*Coleus blumei* Benth. plants were exposed to a restricted water supply for 21 d. The relative water content in leaf tissues of variegated coleus was reduced from 80% (control) to 60% (drought-stressed). Under drought conditions, the stomatal conductance and leaf photosynthetic rate were reduced. In green leaf tissues, drought stress also greatly decreased the diurnal light-period levels of the raffinose family oligosaccharides (RFOs) stachyose and raffinose, as well as those of other non-structural carbohydrates (galactinol, sucrose, hexoses, and starch). However, drought had little effect on soluble carbohydrate content of white, non-photosynthetic leaf tissues. In green tissues, galactinol synthase activity was depressed by drought stress. An accumulation of *O*-methyl-inositol was also observed, which is consistent with the induction of myoinositol-6-*O*-methyltransferase activity seen in the stressed green tissues. In source tissues, RFO metabolism is apparently reduced by drought stress through a combined effect of decreased photosynthesis and reduced galactinol synthase activity. Moreover, a further reduction in RFO biosynthesis may have been due to a switch in carbon partitioning to *O*-methyl-inositol biosynthesis, creating competition for myoinositol, a metabolite shared by both biochemical pathways (Pattanagul and Madore, 1999).

### 5.2 Enhanced Accumulation of BiP in Transgenic Plants

The binding protein (BiP) is an important component of endoplasmic reticulum stress response of cells. Despite extensive studies in cultured cells, a protective function of BiP against stress has not yet been demonstrated in whole multicellular organisms. Alvim *et al.* (2001) have obtained transgenic tobacco (*Nicotiana tabacum* L. cv Havana) plants constitutively expressing elevated levels of BiP or its antisense cDNA to analyze the protective role of this endoplasmic reticulum lumenal stress protein at the whole plant level. Elevated levels of BiP in transgenic sense lines conferred tolerance to the glycosylation inhibitor tunicamycin during germination and tolerance to water deficit during plant growth. Under progressive drought, the leaf BiP levels correlated with the maintenance of the shoot turgidity and water content. The protective effect of BiP overexpression against water stress was disrupted by expression of an antisense BiP cDNA construct. Although overexpression of BiP prevented cellular dehydration, the stomatal conductance and transpiration rate in droughted sense leaves were higher than in control and antisense leaves. The rate of photosynthesis under water deficit might have caused a degree of greater osmotic adjustment in sense leaves because it remained unaffected during water deprivation, which was in marked contrast with the severe drought-induced decrease in the $CO_2$ assimilation in control and antisense leaves. In antisense plants, the water stress stimulation of the antioxidative defenses was higher than in control plants, whereas in droughted sense leaves an induction of superoxide dismutase activity was not observed. These results suggest that overexpression of BiP in plants may prevent endogenous oxidative stress (Alvim *et al.*, 2001).

### 5.3 Nitric Oxide and the Adaptive Plant Responses against Drought Stress

Nitric oxide (NO) is a very active molecule involved in many and diverse biological pathways where it has proved to be protective against damages provoked by oxidative stress conditions. In this work, we studied the effect of two NO donors, sodium nitroprusside (SNP) and *S*-nitroso-*N*-acetylpenicillamine SNP-treated on the response of wheat (*Triticum aestivum*) to water stress conditions. After 2 and 3 h of drought, detached wheat leaves pretreated with 150 $\mu$M SNP retained up to 15% more water than those pretreated with water or $NO_2^-/NO_3^-$. The effect of SNP treatment on water retention was also found in wheat seedlings after 7 d of drought. These results were consistent with a 20% decrease in the transpiration rate of SNP-treated detached wheat leaves for the same analyzed time. In parallel experiments, NO was also able to induce a 35%, 30%, and 65% of stomatal closure in three different species, *Tradescantia* sp. (monocotyledonous) and two dicotyledonous, *Salpichroa organifolia* and fava bean (*Vicia faba*), respectively. In SNP-treated leaves of *Tradescantia* sp., the stomatal closure was correlated with a 10% increase on RWC. Ion leakage, a cell injury index, was 25% lower in SNP-treated wheat leaves compared with control ones after the recovery period. Carboxy-PTIO (2-(4-carboxyphenyl)-4,4,5,5-tetramethylimidazoline-1-oxyl-3-oxide), a specific NO scavenger, reverted SNP action by restoring the transpiration rate, stomatal aperture, and the ion leakage to the level found in untreated leaves. Northern-blot analysis showed that SNP-treated wheat leaves display a 2-fold accumulation of a group three late embryogenesis abundant transcript with respect to control leaves both after 2 and 4 h of drought periods. All together, these results suggest that the exogenous application of NO donors might confer an increased tolerance to severe drought stress conditions in plants (Mata and Lamattina, 2001).

### 5.4 Ureides in Nitrogen Fixation Inhibition in Soybean

The sensitivity of $N_2$ fixation to drought stress in soybean (*Glycine max* Merr.) has been shown to be associated with high ureide accumulation in the shoots, which has led to the hypothesis that $N_2$ fixation during drought is decreased by a feedback mechanism. The ureide feedback hypothesis was tested directly by measuring the effect of 10 mM ureide applied by stem infusion or in the nutrient solution of hydroponically grown plants on acetylene reduction activity (ARA). An almost complete inhibition of ARA was observed within 4 to 7 d after treatment, accompanied by an increase in ureide concentration in the shoot but not in the nodules. The inhibition of ARA resulting from ureide treatments was dependent on the concentration of applied ureide. Urea also inhibited ARA but asparagine resulted in the greatest inhibition of nodule activity. Because ureides did not accumulate in the nodule upon ureide treatment, it was concluded that they were not directly inhibitory to the nodules but that their influence mediated through a derivative compound, with asparagine being a potential candidate. Ureide treatment resulted in a continual decrease in nodule permeability to $O_2$ simultaneous with the inhibition of nitrogenase activity during a 5-d treatment period, although it was not clear whether the latter phenomenon was a consequence or a cause of the decrease in the nodule permeability to $O_2$ (Serraj *et al.*, 1999).

### 5.5 Developing Maize Kernels in Response to Water Deficit

The early post-pollination phase of maize (*Zea mays*) development is particularly sensitive to water deficit stress. Using cDNA microarray, Yu and Setter (2003) studied transcriptional profiles of endosperm and placenta/pedicel tissues in developing maize kernels under water stress. At 9 d after pollination (DAP), placenta/pedicel and endosperm differed considerably in their transcriptional responses. In placenta/pedicel, 79 genes were significantly affected by stress and of these 89% were up-regulated, whereas in endosperm, 56 genes were significantly affected and 82% of these were down-regulated. Only nine of the stress-regulated genes were in common between these tissues. Hierarchical cluster analysis indicated that different sets of genes were regulated in the two tissues. After rewatering at 9 DAP, profiles at 12 DAP suggested that two regulons exist, one for genes responding specifically to concurrent imposition of stress, and another for genes remaining affected after transient stress. In placenta, genes encoding recognized stress tolerance proteins, including heat shock proteins, chaperonins, and major intrinsic proteins, were the largest class of genes regulated, all of which were up-regulated. In contrast, in endosperm, genes in the cell division and growth category represented a large class of down-regulated genes. Several cell wall-degrading enzymes were expressed at lower levels than in controls, suggesting that stress delayed normal advance to programmed cell death in the central endosperm. It was further suggested that the responsiveness of placenta to whole-plant stress factors (water potential, abscisic acid, and sugar flux) and of endosperm to indirect factors may play key roles in determining the threshold for kernel abortion (Yu and Setter, 2003).

Further, expression of soluble (*Ivr2*) and insoluble (*Incw2*) acid invertases was analyzed by Anderson *et al.* (2002) in young ovaries of maize (*Zea mays*) from 6 d before (−6 d) to 7 d after pollination (+7 d) and in response to perturbation by drought stress treatments to distinguish their roles in early kernel development and stress. The *Ivr2* soluble invertase mRNA was more abundant than the *Incw2* mRNA throughout pre- and early post-pollination development (peaking at +3 d). In contrast, *Incw2* mRNAs increased only after pollination. Drought repression of the *Ivr2* soluble invertase also preceded changes in *Incw2*, with soluble activity responding before pollination (−4 d). Distinct profiles of *Ivr2* and *Incw2* mRNAs correlated with respective enzyme activities and indicated separate roles for these invertases during ovary development and stress. In addition, the drought-induced decrease and developmental changes of ovary hexose to sucrose ratio correlated with activity of soluble but not insoluble invertase. Ovary abscisic acid levels were increased by severe drought only at −6 d and did not appear to directly affect *Ivr2* expression. In situ analysis showed localized activity and *Ivr2* mRNA for soluble invertase at sites of phloem-unloading and expanding maternal tissues (greatest in terminal vascular zones and nearby cells of pericarp, pedicel, and basal nucellus). This early pattern of maternal invertase localization is clearly distinct from the well-characterized association of insoluble invertase with the basal endosperm later in development. This localization, the shifts in endogenous hexose to sucrose environment, and the distinct timing of soluble and insoluble invertase expression during development and stress collectively indicate a key role and critical sensitivity of the *Ivr2* soluble invertase gene during the early, abortion-susceptible phase of development.

## 5.6 Drought-Induced Changes and Acquisition of Stress Tolerance

Munné-Bosch and Alegre (2003) measured endogenous concentrations of these antioxidants, their redox states, and other indicators of oxidative stress in chloroplasts of three Labiatae species, differing in their CA contents, exposed to drought stress in the field to assess antioxidative protection by carnosic acid (CA) in combination with that of other low-molecular weight ($M_r$) antioxidants ($\alpha$-tocopherol [$\alpha$-T] and ascorbate [Asc]) in chloroplasts. Damage to the photosynthetic apparatus was observed neither in CA-containing species (rosemary [*Rosmarinus officinalis*]) and sage [*Salvia officinalis*]) nor in CA-free species (lemon balm [*Melissa officinalis*]) at relative leaf water contents between 86% and 58%, as indicated by constant maximum efficiency of photosystem II photochemistry ratios and malondialdehyde levels in chloroplasts. The three species showed significant increases in $\alpha$-T, a shift of the redox state of $\alpha$-T toward its reduced state, and increased Asc levels in chloroplasts under stress. Lemon balm showed the highest increases in $\alpha$ -T and Asc in chloroplasts under stress, which might compensate for the lack of CA. Besides, whereas in rosemary and sage, the redox state of CA was shifted toward its oxidized state and the redox state of Asc was kept constant, lemon balm displayed a shift of the redox state of Asc toward its oxidized state under stress. In vitro experiments showed that both CA and Asc protect $\alpha$-T and photosynthetic membranes against oxidative damage. These results are consistent with the contention that CA, in combination with other low-$M_r$ antioxidants, helps to prevent oxidative damage in chloroplasts of water-stressed plants, and they show functional interdependence among different low-$M_r$

antioxidants in chloroplasts (Munné-Bosch and Alegre, 2003).

Amiard *et al.* (2003) reported that in perennial ryegrass drought stress also caused fructans to accumulate in leaf tissues, mainly in leaf sheaths and elongating leaf bases. This increase was mainly due to the accumulation of long-chain fructans (degree of polymerization >8) and was not accompanied by a Suc increase. Interestingly, Suc but not fructan concentrations greatly increased in drought-stressed roots. Putative roles of fructans and sucrosyl-galactosides have been also stated in relation to the acquisition of stress tolerance (Amiard *et al.*, 2003).

Recently, plant-emitted ethylene has also received considerable attention as a stress hormone and is considered to play a major role at low concentrations in the tolerance of several species to biotic and abiotic stresses. However, airborne ethylene at high concentrations, such as those found in polluted areas (20–100 nL $L^{-1}$) for several days, has received far less attention in studies of plant stress tolerance, though it has been shown to alter photosynthesis and reproductive stages (seed germination, flowering, and fruit ripening) in some species. The extent of oxidative stress, photo- and antioxidant protection, and visual leaf area damage were evaluated by Munné-Bosch *et al.* (2004) in ethylene-treated (approximately 100 nL $L^{-1}$ in air) and control (without ethylene fumigation) holm oak (*Quercus ilex*) plants exposed to heat stress or to a combination of heat and drought stress to assess the potential effects of airborne ethylene on plant stress tolerance in polluted areas,. Control plants displayed tolerance to temperatures as high as 50°C, which might be attributed, at least in part, to enhanced xanthophyll de-epoxidation and 2-fold increases in $\alpha$-tocopherol, and they suffered oxidative stress only when water deficit was superimposed on temperatures above 45°C. By contrast, ethylene-treated plants showed symptoms of oxidative stress at lower temperatures (35°C) than the controls in drought, as indicated by enhanced malondialdehyde levels, lower $\alpha$-tocopherol and ascorbate concentrations, and a shift of the redox state of ascorbate to its oxidized form. In addition, ethylene-treated plants showed higher visual leaf area damage and greater reductions in the maximum efficiency of the PSII photochemistry than controls in response to heat stress or to a combination of heat and drought stress. These results demonstrate for the first time that airborne ethylene at concentrations similar to those found in polluted areas may reduce plant stress tolerance by altering, among other possible mechanisms, antioxidant defenses.

## 5.7 Response of Arabidopsis to a Combination of Drought and Heat Stress

The response of Arabidopsis plants to a combination of drought and heat stress was found to be distinct from that of plants subjected to drought or heat stress. Transcriptome analysis of Arabidopsis plants subjected to a combination of drought and heat stress revealed a new pattern of defense response in plants that includes a partial combination of two multigene defense pathways (i.e., drought and heat stress), as well as 454 transcripts that are specifically expressed in plants during a combination of drought and heat stress. Metabolic profiling of plants subjected to drought, heat stress, or a combination of drought and heat stress revealed that plants subject to a combination of drought and heat stress accumulated sucrose and other sugars such as maltose and gulose. In contrast, Pro that accumulated in plants subjected to drought did not accumulate in plants during a combination of drought and heat stress. Heat stress was found to ameliorate the toxicity of Pro to cells, suggesting that during a combination of drought and heat stress sucrose replaces Pro in plants as the major osmoprotectant. The results highlight the plasticity of the plant genome and demonstrate its ability to respond to complex environmental conditions that occur in the field.

Rizhsky *et al.* (2004) performed an initial analysis of the molecular and metabolic response of Arabidopsis to a combination of drought and heat stress. Our study revealed a new pattern of defense response in plants that includes a partial combination of two multigene defense pathways (drought and heat stress), as well as 454 transcripts that are specifically expressed in cells during a combination of drought and heat stress. In addition, plants subjected to a combination of drought and heat stress accumulated high levels of sucrose and other sugars, but did not accumulate Pro.

Transcripts specific for a combination of drought and heat stress (cutoff 2 $log_2$) belonged to a number of different groups including HSPs, proteases, starch degrading enzymes, and lipid biosynthesis enzymes. The steady-state level of different transcripts encoding signal transduction proteins was also elevated during a combination of drought and heat stress. These included receptor-like kinases, small GTP-binding proteins, MYB transcription factors, and protein kinases. In addition, the expression of at least five different transcripts encoding membrane channels was elevated in plants subjected to a combination of drought and heat stress (CLC-b chloride channel, aquaporin membrane intrinsic protein (MIP), potassium transporter, $Na^+/Ca^{2+}$-antiporter, and an ABC-type transporter). Defense transcripts specifically elevated in cells during a combination of drought and heat stress included thioredoxin and 2-peroxiredoxin, important for the prevention of oxidative stress, P450s, and a salt-inducible protein.

It was suggested that a moderate level of water stress is accompanied by the accumulation of compounds such as Pro and Gly-betaine, whereas a severe level of water stress is accompanied by the accumulation of sugars such as sucrose (Hoekstra *et al.*, 2001). Although the relative water content of plants subjected to drought and plants subjected to a combination of drought and heat stress was not significantly different, plants subjected to drought accumulated Pro whereas plants subjected to a combination of drought and heat stress accumulated sucrose. This difference may suggest that a combination of drought and heat stress imposes on plants a different type of internal stress (compared to drought or heat stress), that requires sucrose rather than Pro as an osmoprotectant. Alternatively, Pro may be toxic to cells during a combination of drought and heat stress. Thus, sucrose may be required to replace Pro as the major osmoprotectant of cells during the stress combination. Results (Pnueli *et al.*, 2003) also suggested that plant HSFs function as a network of transcription factors that controls the expression of HSPs during different stresses. Thus, we identified oxidative- and light-stress-specific HSFs (Pnueli *et al.*, 2003) as well as heat-stress- and drought-specific HSFs. Future analysis of this gene family, including measurements of HSF activity in cells, might provide an initial insight into how plants compensated during evolution for their sessile nature by developing complex and specialized gene families to control their response to environmental conditions. These were most likely created by gene duplication, however, acquired specific roles related to specific pathways or stresses, as well as their combination (Arabidopsis Genome Initiative, 2000). Similar results were also obtained with several members of the MYB transcription factor gene family, and MYB-At1g26580 was identified as specifically elevated during a combination of drought and heat stress.

### 5.8 Drought Tolerance at Reproductive Stage in Rice

Drought is a major constraint to rice (*Oryza sativa*) yield and its stability in rainfed and poorly irrigated environments.

Identifying genomic regions influencing the response of yield and its components to water deficits will aid in our understanding of the genetics of drought tolerance and development of more drought tolerant cultivars. Drought tolerance has been considered as a valid breeding target to partially compensate for the loss in yield. Phenotypic traits associated directly with drought tolerance are unclear; however, several investigations noted that deep rooting (Lilley and Fukai, 1994; Pantuwan *et al.*, 1996; Wade *et al.*, 1996) and osmotic adjustment (Ludlow and Muchow, 1990; Jongdee and Cooper, 1998) are associated with drought tolerance. Until recently, lack of concept, direction, and protocol has remained a significant obstacle to genetic improvement of drought tolerance (Blum *et al.*, 1996). Selecting rice for drought tolerance based on unknown genetic mechanisms will result in inefficient genetic improvement. Standard assays used to measure drought tolerance must be enhanced since they are important to success in genetic improvement. The line source sprinkler system can be used to maintain variable water application rates across plots, thus allowing the assessment of different levels of water supply at specific growth stages (Hanks *et al.*, 1976).

Water stress-induced spikelet sterility limits rice production under upland conditions. The causes of spikelet sterility under drought stress are poorly understood. Selote and Khanna-Chopra (2004) report on a test of two rice varieties under drought stress. The susceptible variety had greater panicle sterility than the resistant one under stress. The susceptible variety also had less antioxidant activity than the resistant variety.

In this study the role of antioxidant defence management in drought-induced spikelet sterility was investigated in two rice (*Oryza sativa*) genotypes differing in drought resistance (Selote and Khanna-Chopra, 2004). Drought-resistant N22 genotype showed less water stress-induced spikelet sterility when compared to the susceptible N118 genotype under upland conditions. The N22 panicles maintained higher RWC and turgor potential and lower $H_2O_2$ levels across the developmental stages under water stress than that of N118 panicles. Drought-induced enhancement in superoxide dismutase (SOD, EC 1.15.1.1) activity coupled with higher ascorbate (AsA), glutathione (GSH) content and enhanced ascorbate peroxidase (APX, EC 1.11.1.11) and glutathione reductase (GR, EC 1.6.4.2) activities resulted in lower $H_2O_2$ levels in N22 panicles. In contrast, insufficient enhancement in SOD, APX and GR activities resulted in relatively higher $H_2O_2$ levels under water stress in N118 panicles. The N22 panicles exhibited a higher number of SOD and APX isozymes in comparison with N118 panicles that might provide better reactive oxygen species scavenging. Hence it is concluded that well-equipped antioxidant defence plays an important role in minimizing water stress-induced spikelet sterility in upland rice (Selote and Khanna-Chopra, 2004). The authors therefore concluded that "inefficient antioxidant defense" was the likely cause of rice panicle sterility under drought stress. This far reaching conclusion was made while in the same paper the authors also report that the susceptible variety had relatively lower panicle relative water content (RWC) and turgor under stress.

Painstaking work in the era of technological advances indicate that molecular tools facilitate the identification and genomic locations of genes controlling traits related to drought tolerance using quantitative trait loci (QTL) analysis. Quantitative trait loci (QTL) for grain yield and its components and other agronomic traits were identified by Lanceras *et al.* (2004) using a subset of 154 doubled haploid lines derived from a cross between two rice cultivars, CT9993-510 to 1-M and IR62266-42 to 6-2. Drought stress treatments were managed by use of a line source sprinkler irrigation system, which provided a linearly decreasing level of irrigation coinciding with the sensitive reproductive growth stages. The research was conducted at the Ubon Rice Research Center, Ubon, Thailand. A total of 77 QTL were identified for grain yield and its components under varying levels of water stress. Out of the total of 77 QTL, the number of QTL per trait were: 7-grain yield (GY); 8-biological yield (BY); 6-harvest index (HI); 5-d to flowering after initiation of irrigation gradient (DFAIG); 10-total spikelet number (TSN); 7-percent spikelet sterility (PSS); 23-panicle number (PN); and 11-plant height (PH). The phenotypic variation explained by individual QTL ranged from 7.5% to 55.7%. Under well-watered conditions, it was observed a high genetic association for BY, HI, DFAIG, PSS, TSN, PH, and GY. However, only BY and HI were found to be significantly associated with GY under drought treatments. QTL flanked by markers RG104 to RM231, EMP2_2 to RM127, and G2132 to RZ598 on chromosomes 3, 4, and 8 were associated with GY, HI, DFAIG, BY, PSS, and PN under drought treatments. The aggregate effects of these QTL on chromosomes 3, 4, and 8 resulted in higher grain yield. These QTL will be useful for rainfed rice improvement, and will also contribute to our understanding of the genetic control of GY under drought conditions at the sensitive reproductive stage. Close linkage or pleiotropy may be responsible for the coincidence of QTL detected in this experiment. Digenic interactions between QTL main effects for GY, BY, HI, and PSS were observed under irrigation treatments. Most (but not all) DH lines have the same response in measure of productivity when the intensity of water deficit was increased, but no QTL by irrigation treatment interaction was detected. The identification of genomic regions associated with GY and its components under drought stress will be useful for marker-based approaches to improve GY and its stability for farmers in drought-prone rice environments.

### 5.9 Rice Genes under Stresses and Abscisic Acid Application

Rabbani *et al.* (2003) prepared a rice cDNA microarray including about 1,700 independent cDNAs derived from cDNA libraries prepared from drought-, cold-, and high-salinity-treated rice plants to identify cold-, drought-, high-salinity-, and/or abscisic acid (ABA)-inducible genes in rice (*Oryza sativa*). They confirmed stress-inducible expression of the candidate genes selected by microarray analysis using RNA gel-blot analysis and finally identified a total of 73 genes as stress inducible including 58 novel unreported genes in rice. Among them, 36, 62, 57, and 43 genes were induced by cold, drought, high salinity, and ABA, respectively. They observed a strong association in the expression of stress-responsive genes and found 15 genes that responded to all four treatments. Venn diagram analysis revealed greater cross talk between signaling pathways for drought, ABA, and high-salinity stresses than between signaling pathways for cold and ABA stresses or cold and high-salinity stresses in rice. The rice genome database search enabled to identify possible known cis-acting elements in the promoter regions of several stress-inducible genes and also to expect the existence of novel cis-acting elements involved in stress-responsive gene expression in rice stress-inducible promoters. Comparative analysis of Arabidopsis and rice showed that among the 73 stress-inducible rice genes, 51 already have been reported in Arabidopsis with similar function or gene name. Transcriptome analysis revealed novel stress-inducible genes, suggesting some differences between Arabidopsis and rice in their response to stress.

### 5.10 Drought Stress and Heat Shock on Gene Expression in Tobacco

In nature, plants encounter a combination of environmental conditions that may include stresses such as drought or heat shock. Although drought and heat shock have been extensively studied, little is known about how their combination affect plants. Rizhsky *et al.* (2003) used cDNA arrays, coupled with physiological measurements, to study the effect of drought and heat shock on tobacco (*Nicotiana tabacum*) plants. A combination of drought and heat shock resulted in the closure of stomata, suppression of photosynthesis, enhancement of respiration, and increased leaf temperature. Some transcripts induced during drought, e.g., those encoding dehydrin, catalase, and glycolate oxidase, and some transcripts induced during heat shock, e.g. thioredoxin peroxidase, and ascorbate peroxidase, were suppressed during a combination of drought and heat shock. In contrast, the expression of other transcripts, including alternative oxidase, glutathione peroxidase, phenylalanine ammonia lyase, pathogenesis-related proteins, a WRKY transcription factor, and an ethylene response transcriptional co-activator, was specifically induced during a combination of drought and heat shock. Photosynthetic genes were suppressed, whereas transcripts encoding some glycolysis and pentose phosphate pathway enzymes were induced, suggesting the utilization of sugars through these pathways during stress. Their results demonstrate that the response of plants to a combination of drought and heat shock, similar to the conditions in many natural environments, is different from the response of plants to each of these stresses applied individually, as typically tested in the laboratory. This response was also different from the response of plants to other stresses such as cold, salt, or pathogen attack. Therefore, improving stress tolerance of plants and crops may require a reevaluation, taking into account the effect of multiple stresses on plant metabolism and defense (Rizhsky *et al.*, 2003).

### 5.11 Enhanced Tolerance to Environmental Stress in Transgenic Plants

Abiotic stresses cause extensive losses to agricultural production worldwide. Acclimation of plants to abiotic conditions such as drought, salinity, or heat is mediated by a complex network of transcription factors and other regulatory genes that control multiple defense enzymes, proteins, and pathways. Associated with the activity of different transcription factors are transcriptional coactivators that enhance their binding to the basal transcription machinery. Although the importance of stress-response transcription factors was demonstrated in transgenic plants, little is known about the function of transcriptional coactivators associated with abiotic stresses. Very recently, Suzuki *et al.* (2005) report that constitutive expression of the stress-response transcriptional coactivator multiprotein bridging factor 1c (MBF1c) in Arabidopsis (*Arabidopsis thaliana*) enhances the tolerance of transgenic plants to bacterial infection, heat, and osmotic stress. Moreover, the enhanced tolerance of transgenic plants to osmotic and heat stress was maintained even when these two stresses were combined. The expression of MBF1c in transgenic plants augmented the accumulation of a number of defense transcripts in response to heat stress. Transcriptome profiling and inhibitor studies suggest that MBF1c expression enhances the tolerance of transgenic plants to heat and osmotic stress by partially activating, or perturbing, the ethylene-response signal transduction pathway. Present findings suggest that MBF1 proteins could be used to enhance the tolerance of plants to different abiotic stresses.

### 5.12 Drought Stress Tolerance by Chemical Priming

A very recent evidence reveals that drought and salt stress tolerance of Arabidopsis (*Arabidopsis thaliana*) plants increased following treatment with the nonprotein amino acid β-aminobutyric acid (BABA), known as an inducer of resistance against infection of plants by numerous pathogens (Jakab *et al.,* 2005). BABA-pretreated plants showed earlier and higher expression of the salicylic acid-dependent *PR-1* and *PR-5* and the abscisic acid (ABA)-dependent *RAB-18* and *RD-29A* genes following salt and drought stress. However, *non-expressor of pathogenesis-related genes 1* and *constitutive expressor of pathogenesis-related genes 1* mutants as well as transgenic NahG plants, all affected in the salicylic acid signal transduction pathway, still showed increased salt and drought tolerance after BABA treatment. On the contrary, the *ABA deficient 1* and *ABA insensitive 4* mutants, both impaired in the ABA-signaling pathway, could not be protected by BABA application. BABA-induced water stress tolerance is based on enhanced ABA accumulation resulting in accelerated stress gene expression and stomatal closure. There is a possibility to increase plant tolerance for these abiotic stresses through effective priming of the preexisting defense pathways without resorting to genetic alterations (Jakab *et al.,* 2005).

## CONCLUSION AND FUTURE PERSPECTIVE

Studies on abiotic stresses in diverse group of plants including economic crops have advanced considerably in recent years. However, the majority of experiments testing the response of plants to changes in environmental conditions have focused on a single stress treatment applied to plants under controlled conditions. In contrast, in the field, a number of different stresses can occur simultaneously. These may include conditions such as high irradiance, low water availability, extreme temperature, or high salinity and may alter plant metabolism in a novel manner that may be different from that caused by each of the different stresses applied individually.

Recent advances in abiotic stress responses of plants have been made under conditions of drought stress and water stress situations especially related to molecular physiology - binding proteins, changes in the redox state of α-tocopherol, ascorbate, and the diterpene carnosic acid in chloroplasts of labiatae species differing in carnosic acid contents etc. in the understanding of drought tolerance mechanism. Conclusively, the binding protein (BiP) is an important component of endoplasmic reticulum stress response of cells. Elevated levels of BiP in transgenic sense lines conferred tolerance to the glycosylation inhibitor tunicamycin during germination and tolerance to water deficit during plant growth. Under progressive drought, the leaf BiP levels correlated with the maintenance of the shoot turgidity and water content. It has been suggested that overexpression of BiP in plants may prevent endogenous oxidative stress.

Nevertheless, the sensitivity of $N_2$ fixation to drought stress in soybean (*Glycine max* Merr.) has been shown to be associated with high ureide accumulation in the shoots, which has led to the hypothesis that $N_2$ fixation during drought is decreased by a feedback mechanism. A moderate level of water stress is accompanied by the accumulation of compounds such as Pro and Gly-betaine, whereas a severe level of water stress is accompanied by the accumulation of sugars such as sucrose. Interestingly, recent reports point out that the exogenous application of NO donors might confer an increased tolerance to severe drought stress conditions in plants.

Further, studies on developing maize kernels in response to water deficit indicates that the early post-pollination phase of maize (*Zea mays*) development is particularly sensitive to water deficit stress. It was suggested that the responsiveness of placenta to whole-plant stress factors (water potential, abscisic acid, and sugar flux) and of endosperm to indirect factors may play key roles in determining the threshold for kernel abortion.

Evidently, one way of increasing productivity under drought stress is to breed crops that are more tolerant to stress. However, success in breeding for tolerance has been limited because (a) tolerance to stress is controlled by many genes, and their simultaneous selection is difficult (b) incredible effort is required to eliminate undesirable genes that are also incorporated during breeding; and (c) there is a lack of efficient selection procedures particularly under field conditions. Genetic engineering offers an alternative approach for developing tolerant crops. Unlike classical breeding, genetic engineering is a faster and more precise means of achieving improved tolerance. There is possibility of enhanced tolerance to environmental stress in transgenic plants. The identification of genomic regions associated with grain yield (GY) and its components under drought stress will be useful for marker-based approaches to improve GY and its stability for farmers in drought-prone rice environments. It is also concluded that well-equipped antioxidant defence plays an important role in minimizing water stress-induced spikelet sterility in upland rice.

Last but not the least; the response of plants to a combination of drought and heat shock, similar to the conditions in many natural environments, is different from the response of plants to each of these stresses applied individually, as typically tested in the laboratory. Hence, improving stress tolerance of plants and crops may require a reevaluation, taking into account the effect of multiple stresses on plant metabolism and defense. In fact, investigations made so far in this direction, as illustrated in part in the preceding pages, eventually makes us aware with intricate facts in drought tolerance mechanism, leaving immense scope of research for this crucial global problem. Indeed, physiologists too need extensive studies to contribute basic details in this direction.

## REFERENCES


Alamillo, J.M., C. Almoguera, D. Bartels and J. Jordano, 1995. *Plant Mol. Biol.,* **29**: 1093-1099.

Alvim F. C., M.B.S. Carolino, J.C.M. Cascardo, C.C. Nunes, C.A. Martinez, W.C. Otoni and E.P.B. Fontes, 2001. Enhanced accumulation of bip in transgenic plants confers tolerance to water stress. *Plant Physiol.,* **126**: 1042-1054.

Amiard, V., A.M. Bertrand, J.P. Billard, C. Huault, F. Keller and M.P. Prud'homme, 2003. Fructans, But not the sucrosyl-galactosides, raffinose and loliose, are affected by drought stress in perennial ryegrass. *Plant Physiol.*, **132**: 2218-2229

Arabidopsis Genome Initiative, 2000. Analysis of the genome sequence of the flowering plant Arabidopsis thaliana. *Nature*, **408:** 796–815

Badger, M.R., S. von Caemmerer, S. Ruuska and H. Nakano, 2000. Electron flow to oxygen in higher plants and algae: rates and control of direct photoreduction (Mehler reaction) and rubisco oxygenase. *Philos. T. Roy. Soc. B*, **355**: 1433-1446.

Badger, M.R. 1985. Photosynthetic oxygen exchange. *Annu. Rev. Plant Phys.,* **36:** 27-53.

Bartels, D. and F. Salamini, 2001. Desiccation tolerance in the resurrection plant *Craterostigma plantagineum*. A contribution to the study of drought tolerance at the molecular level. *Plant Physiol.,* **127**: 1346-1353.

Bernacchia, G., F. Salamini and D. Bartels, 1996. *Plant Physiol.,* **111:** 1043-1050.

Bernacchia, G., G. Schwall, F. Lottspeich, F. Salamini and D. Bartels, 1995. *EMBO J.,* 14: 610-618.

Biehler, K. and H. Fock, 1996. Evidence for the contribution of the Mehler-peroxidase reaction in dissipating excess electrons in drought-stressed wheat. *Plant Physiol.*, **112**: 265-272.

Biehler, K., S. Haupt, J. Beckmann, H. Fock, and T.W. Becker, 1997. Simultaneous $CO_2$- and $^{16}O_2/^{18}O_2$-gas exchange and fluorescence measurements indicate differences in light energy dissipation between the wild type and the phytochrome-deficient *aurea* mutant of tomato during water stress. *J. Exp. Bot.*, **48**: 1439-1449.

Björkman, O. and B. Demmig-Adams, 1994. Regulation of photosynthetic light energy capture, conversion, and dissipation in leaves of higher plants. In: *Ecophysiology of Photosynthesis* (eds. E.-D. Schulze and M.M. Caldwell), pp. 17-47. Springer-Verlag. Berlin.

Björkman, O. and C. Schäfer, 1989. A gas exchange-fluorescence analysis of photosynthetic performance of a cotton crop under high-irradiance stress [Extended abstract.]. *Philos. T. Roy. Soc. B*, **323**: 309-311.

Blum, A., R. Munns, J.B. Passioura, N.C. Turner, R.E. Sharp, J.S. Boyer, H.T. Nguyen and T.C. Hsiao, 1996. Letters to the editors: genetically engineered plants resistant to soil drying and salt stress: how to interpret osmotic relations. *Plant Physiol.* **110**: 1051–1053.

Boyer, J.S. 1995a. Biochemical and biophysical aspects of water deficits and the predisposition to disease. *Annu. Rev. Phytopathol.*, **33**: 251-274.

Boyer, J.S. 1995b. Measuring the water status of plants and soils. Academic Press, San Diego.

Bray, E.A. 1997. Plant responses to water deficit. *Trends Plant Sci.,* **2**: 48-54.

Brestic, M., G. Cornic, M.J. Fryer, and N.R. Baker, 1995. Does photorespiration protect the photosynthetic apparatus in French bean leaves from photoinhibition during drought stress? *Planta*, **196**: 450-457.

Chandler, J. and D. Bartels, 1997. *Mol. Gen. Genet.,* **256:** 539-546.

Chaves, M.M. 1991. Effects of water deficits on carbon assimilation. *J. Exp. Bot.***,** **42**: 1-16.

Chen, D., S. Sarid, and E. Katchalski, 1968. The role of water stress in the inactivation of messenger RNA of germinating wheat embryos. *P. Natl. Acad. Sci. USA,* **61**: 1378-1383.

Crespi, M.D., E. Jurkevitch, M. Poiret, Y. D'Aubenton-Carafa, G. Petrovics, E. Kondorosi and A. Kondorosi, 1994. *EMBO J.*, **13:** 5099-5112

Cuming, A. 1999. *In* PR Shewry, R. Casey, eds, Seed Proteins. Kluwer Academic Publishers, Dordrecht, The Netherlands, pp 753-780.

Cushman, J.C., H.J. Bohnert, 2000. Genomic approaches to plant stress tolerance. *Curr. Opin. Plant Biol.,* **3**: 117-124.

Dai, Z., G.E. Edwards, and M.S.B Ku, 1992. Control of photosynthesis and stomatal conductance in *Ricinus communis* L. (castor bean) by leaf to air vapor pressure deficit. *Plant Physiol.*, **99**: 1426-1434

Di Marco, G. and D. Tricoli, 1993. Effect of water deficit on photosynthesis and electron transport in wheat grown in a natural environment. J. Plant Physiol. 142: 156-160.

Dietz, K.J. and U. Heber, 1983. Carbon dioxide gas exchange and the energy status of leaves of *Primula palinuri* under water stress. *Planta,* **158**: 349-356.

Egneus, H., U. Heber, U. Matthiesen and M. Kirk, 1975. Reduction of oxygen by the electron transport chain of chloroplasts during assimilation of carbon dioxide. Biochim. Biophys. Acta 408: 252-268.

Farquhar, G.D. and T.D. Sharkey, 1982. Stomatal conductance and photosynthesis. *Annu. Rev. Plant Phys.*, **33**: 317-345.

Flexas, J., J.M. Escalona, and H. Medrano 1999a. Water stress induces different levels of photosynthesis and electron transport rate regulations in grapevines. *Plant Cell Environ.*, **22**: 39-48.

Flexas, J., M. Badger, W.S. Chow, H. Medrano and C.B. Osmond 1999b. Analysis of the relative increase in photosynthetic $O_2$ uptake when photosynthesis in grapevine leaves is inhibited following low night temperatures and/or water stress. *Plant Physiol.*, **121**: 675-684.

Flexas J., J. Bota, J.M. Escalona, B. Sampol and H. Medrano 2001a. Stomatal conductance indicates a threshold between predominant stomatal and non-stomatal limitations to photosynthesis in water stressed grapevines. *Plant Cell Environ.*, **14:** pp.

Flexas, J., J.M. Escalona, S. Evain, J. Gulías, I.Moya, C.B. Osmond and H. Medrano 2001b. Non-photochemical quenching of chlorophyll fluorescence, photosynthesis and stomatal conductance can be estimated from steady-state chlorophyll fluorescence (Fs) measurements in water-stressed $C_3$ plants. *Plant Physiol.* (c.f. *Advances in Plant Physiology,* Volume **4**, 2004, Ed. A. Hemantaranjan)

Flowers T.J., M.L. Koyama, S.A. Flowers, C. Sudhakar, K.P. Singh and A.R. Yeo, 2000. QTL: their place in engineering tolerance of rice to salinity. *J. Exp. Bot.,* **51:** 99-106.

Fock, H.P., K. Biehler and T. Stuhlfauth, 1992. Use and degradation of light energy in water stressed *Digitalis lanata*. *Photosynthetica*, **27:** 571-577.

Furini, A., C. Koncz, F. Salamini, and D. Bartels, 1997. *EMBO J.*, **16**: 3599-3608.

Furini, A., F. Parcy, Salamini and D. Bartels, 1996. *Plant Mol. Biol.,* **30**: 343-349.

Gaff, D.F. 1971. *Science*, **174**: 1033-1034.

Gallagher, J.N., P.V. Biscoe and B. Hunter, 1976. Effects of drought on grain growth. *Nature,* **264**: 541-542.

Genty, B., J.M. Briantais and N.R. Baker, 1989. The relationship between the quantum yield of photosynthetic electron transport and quenching of chlorophyll fluorescence. *Biochim. Biophys. Acta*, **990**: 87-92.

Gerbaud, A. and M. André, 1980. Effect of $CO_2$, $O_2$, and light on photosynthesis and photorespiration in wheat. *Plant Physiol.*, **66**: 1032-1036.

Gerbaud, A. and M. André, 1987. An evaluation of the recycling measurements of photorespiration. *Plant Physiol.*, **83**: 933-937.

Ghashghaie, J. and G. Cornic, 1994. Effect of temperature on partitioning of photosynthetic electron flow between $CO_2$ assimilation and $O_2$ reduction and on the $CO_2/O_2$ specificity of Rubisco. *J. Plant Physiol.*, **143**: 643-650.

Graan, T. and J.S. Boyer, 1990. Very high $CO_2$ partially restores photosynthesis in sunflower at low water potentials. *Planta,* **181**: 378-384.

Graziani, Y. and A. Livne, 1971. Dehydration, water fluxes, and permeability of tobacco leaf tissue. *Plant Physiol.*, **48**: 575-579.

Hanks, R.J., Keller, J., V.P. Rasmussen and G.D. Wilson, 1976. Line source sprinkler for continuous variable irrigation-crop production studies. *Soil Sci. Soc. Am. J.,* **40**: 426–429.

Haupt-Herting, S. and H.P. Fock, 2000. Exchange of oxygen and its role in energy dissipation during drought stress in tomato plants. *Physiol. Plant.,* **110**: 489-495.

Heber, U., R. Bligny, P. Streb, and R. Douce, 1996. Photorespiration is essential for the protection of the photosynthetic apparatus of $C_3$ plants against photoinactivation under sunlight. *Bot. Acta*, **109:** 307-315.

Heckathorn, S.A., E.H. de Lucia and R.E. Zielinski, 1997. The contribution of drought-related decreases in foliar nitrogen concentration to decreases in photosynthetic capacity during and after drought in prairie grasses. *Physiol. Plantarum,* **101**: 173-182.

Hoekstra F.A., E.A. Golovina and J. Buitink, 2001. Mechanisms of plant desiccation tolerance. *Trends Plant Sci.*, **6:** 431–438.

Hsiao, T.C. 1973. Plant responses to water stress. *Annu. Rev. Plant Physiol.*, **24**: 519-570.

Ingram, J., J. Chandler, L. Gallagher, F. Salamini and D. Bartels 1997. *Plant Physiol.*, **115**: 113-121.

Iturriaga, G., K. Schneider, F. Salamini and D. Bartels 1992. *Plant Mol. Biol.,* **20**: 555-558.

Jakab, G., J. Ton, V. Flors, L. Zimmerli, J.P. Métraux and B. Mauch-Mani 2005. Enhancing arabidopsis salt and drought stress tolerance by chemical priming for its abscisic acid responses. *Plant Physiol.*, **139**: 267-274.

Jongdee, B. and M. Cooper 1998. Genetic variation for grain yield of rice under water deficit condition. *In* Proceedings of the 9th Australian Agronomy Conference, Wagga Wagga, Australia.

Kaiser, W.M. 1987. Effects of water deficit on photosynthetic capacity. *Physiol. Plantarum,* **71**: 142-149.

Kleines, M., R.C. Elster, M.J. Rodrigo, A.S. Blervacq, F. Salamini and D. Bartels 1999. *Planta* **209**: 13-24.

Kozaki, A. and G. Takeba 1996. Photorespiration protects $C_3$ plants from photooxidation. *Nature,* **384**: 557-560.

Krall, J.P. and G.E. Edwards 1992. Relationship between photosystem II activity and $CO_2$ fixation in leaves. *Physiol. Plantarum,* **86**: 180-187.

Kramer, P.J. 1988. Changing concepts regarding plant water relations. *Plant Cell Environ.*, 11: 565-568.

Krampitz, M.J., K. Klüge, and H.P. Fock, 1984. Rates of photosynthetic $CO_2$ uptake, photorespiratpry $CO_2$ evolution and dark respiration in water-stressed sunflower and bean leaves. *Photosynthetica,* **18**: 322-328.

Kriedemann, P.E. and R.E. Smart 1969. Effects of irradiance, temperature, and leaf water potential on vine photosynthesis. *Photosynthetica,* **5**: 6-15.

Krömer, S. 1995. Respiration during photosynthesis. *Annu. Rev. Plant Physiol. Plant Mol. Biol.,* **46:** 45-70

Laisk, A. and A. Sumberg 1994. Partitioning of the leaf $CO_2$ exchange into components using $CO_2$ exchange and fluorescence measurements. *Plant Physiol.*, **106**: 689-695.

Lanceras, J. C., G. Pantuwan, B. Jongdee and T. Toojinda 2004. Quantitative trait loci associated with drought tolerance at reproductive stage in rice. *Plant Physiol.*, **135**: 384-399.

Lange, O.L., J.D. Tenhunen and M. Braun 1982. Midday stomatal closure in Mediterranean type sclerophylls under simulated habitat conditions in an environmental chamber. *Flora,* **172**: 563-579.

Larcher, W., J.A.P.V. de Moraes and H. Bauer 1981. Adaptive responses of leaf water potential, $CO_2$-gas exchange and water use efficiency of *Olea europaea* during drying and rewatering. In: *Components of Productivity of Mediterranean-Climate region – Basic and applied aspects* (eds. N.S. Margaris and H.A. Mooney), pp. 77-83. Dr W. Junk Publishers, The Hague-Boston-London.

Lawlor, D.W. and H. Fock 1975. Photosynthesis and photorespiratory $CO_2$ evolution of water-stressed sunflower leaves. *Planta,* **126**: 247-258.

Lawlor, D.W. and H. Fock 1977a. Photosynthetic assimilation of $^{14}CO_2$ by water-stressed sunflower leaves in two oxygen concentrations and the specific activity of products. *J. Exp. Bot.*, **28**: 320-328.

Lawlor, D.W. and H. Fock 1977b. Water stress induced changes in the amounts of some photosynthetic assimilation products and respiratory metabolites of sunflower leaves. *J. Exp. Bot.*, **28**: 329-337.

Lawlor, D.W. and J.G. Pearlman 1981. Compartmental modelling of photorespiration and carbon metabolism of water stressed leaves. *Plant Cell Environ.*, **4**: 37-52.

Lawlor, D.W. 1976a. Water stress induced changes in photosynthesis, photorespiration, respiration and $CO_2$ compensation concentration of wheat. Photosynthetica 10: 378-387

Lawlor, D.W. 1976b. Assimilation of carbon into photosynthetic intermediates of water-stressed wheat. *Photosynthetica*, **10**: 431-439.

Lawlor, D.W. 1995. The effects of water deficit on photosynthesis. In: *Environment and Plant Metabolism. Flexibility and Acclimation* (ed. N. Smirnoff), pp. 129-160. BIOS Scientific Publisher, Oxford.

Leegood, R.C. 1995. Effects of temperature on photosynthesis and photorespiration. In: *Environment and Plant Metabolism. Flexibility and Acclimation* (ed. N. Smirnoff), pp. 45-62. BIOS Scientific Publisher, Oxford

Leprince, O., G.A.F. Hendry and B.D. McKersie 1993. *Seed Sci. Res.,* **3**: 231-246.

Lilley, J.M. and S. Fukai, 1994. Effects of timing and severity of water deficit on four diverse rice cultivars. III. Phenological development, crop growth and grain yield. *Field Crops Res.*, **37**: 225–234.

Loreto, F., S. Delfine and G. Di Marco, 1999. Estimation of photorespiratory carbon dioxide recycling during photosynthesis. *Aust. J. Plant Physiol.*, **26**: 733-736.

Lovelock, C. and K. Winter, 1996. Oxygen-dependent electron transport and protection from photoinhibition in leaves of tropical tree species. *Planta*, **198**: 580-587.

Ludlow, M.M. and R.C. Muchow, 1990. A critical evaluation of traits for improving crop yields in water-limited environments. *Adv. Agron.,* **43**: 107–153.

Ludlow, M.M. 1989. Strategies of response to water stress. In: *Structural and Functional Responses to Environmental Stresses* (eds. K.H. Kreeb, H. Richter and T.M. Hinckley), pp. 269-282. Academic Publishing, The Hague, Netherlands.

Mäkelä, P., M. Kontturi, E. Pehu and S. Somersalo, 1999. Photosynthetic response of drought- and salt-stressed tomato and turnip rape plants to foliar-applied glycinebetaine. *Physiol. Plantarum*, **105**: 45-50.

Mata, C.G. and L. Lamattina, 2001. Nitric oxide induces stomatal closure and enhances the adaptive plant responses against drought stress. *Plant Physiol.*, **126**: 1196-1204.

Miyake, C. and A.Yokota, 2000. Determination of the rate of photoreduction of $O_2$ in the water-water cycle in watermelon leaves and enhancement of the rate by limitation of photosynthesis. *Plant Cell Physiol.*, **41**: 335-343.

Morgan, J.M. 1984. Osmoregulation and water stress in higher plants. *Annu. Rev. Plant Physiol.*, **35**: 299-319.

Munné-Bosch, S. and L. Alegre 2003. Drought-induced changes in the redox state of $\alpha$-tocopherol, ascorbate, and the diterpene carnosic acid in chloroplasts of labiatae species differing in carnosic acid contents. *Plant Physiol.*, **131**: 1816-1825.

Munné-Bosch, S., J. Peñuelas, D. Asensio and J. Llusià 2004. Airborne ethylene may alter antioxidant protection and reduce tolerance of holm oak to heat and drought stress. *Plant Physiol.*, **136**: 2937-2947.

Muraoka, H., Y. Tang, I. Terashima, H. Koizumi, and I. Washitani 2000. Contibutions of diffusional limitation, photoinhibition and photorespiration to midday depression of photosynthesis in *Arisaema heterophyllum* in natural high light. *Plant Cell Environ.*, **23**: 235-250.

Norwood, M., M.R. Truesdale, A. Richter, P. Scott, 2000. *J. Exp. Bot.,* **51:** 1-6.

Oren, R., J.S. Sperry, G.G. Katul, D.E. Pataki, B.E. Ewers, N. Phillips and K.V.R. Schäfer 1999. Survey and synthesis of intra- and inters`pecific variation in stomatal sensitivity to vapour pressure deficit. *Plant Cell Environ.*, **22:** 1515-1526.

Osmond, C.B. and S.C. Grace 1995. Perspectives on photoinhibition and photorespiration in the field: quintessential inefficiencies of the light and dark reactions of photosynthesis? *J. Exp. Bot.*, **46**, 1415-1422.

Osmond, C.B., K. Maxwell, O. Björkman, M. Badger and R. Leegood 1997. Too many photons: photorespiration, photoinhibition and photooxidation. *Trends Plant Sci.*, **4**: 119-121.

Osmond, C.B. 1981. Photorespiration and photoinhibition. Some implications for the energetics of photosynthesis. *Biochim. Biophys. Acta*, **639**: 77-98.

Pantuwan, G., K.T. Ingram and P.K. Sharma 1996. Rice root system development under rainfed conditions. Proceedings of the Thematic Conference on Stress Physiology, Rainfed Lowland Rice Research Consortium, Lucknow, India, February 28–March 5, 1994. International Rice Research Centre, Manila, Philippines, pp 198–206.

Park, Y.-I., W.S. Chow, C.B. Osmond and J.M. Anderson 1996. Electron transport to oxygen mitigates against the photoinactivation of Photosystem II *in vivo*. *Photosynth. Res.*, 50: 23-32.

Pattanagul, W. and M.A. Madore 1999. Water deficit effects on raffinose family oligosaccharide metabolism in coleus. *Plant Physiol.*, **121**: 987-993.

Pei, Z-M., M. Ghassemian, C.M. Kwak, P. McCOurt, and J.I. Schroeder 1998. Role of farnesyltransferase in ABA regulation of guard cell anion channels and plant water loss. *Science,* **282:** 287-290.

Phillips, J.R., M.J. Oliver and D. Bartels 2001. *In* M Black, H Pritchard, eds, Desiccation and Survival in Plants: Drying without Dying CABI Publishing, Wallingford, UK.

Porembski S, and W. Barthlott 2001. *Plant Ecol.,* **151:** 19-28.

Quick, W.P., M.M. Chaves, R. Wendler, M. David, M.L. Rodrigues, J.A. Passaharinho, J.S. Pereira, M.D. Adcock, R.C. Leegood and M. Stitt 1992. The effect of water stress on photosynthetic carbon metabolism in four species grown under field conditions. *Plant Cell Environ.*, **15**: 25-35.

Rabbani, M. A., K. Maruyama, H. Abe, M. A. Khan, K. Katsura, Y. Ito, K. Yoshiwara, M. Seki, K. Shinozaki and K. Y.-Shinozaki 2003. Monitoring expression profiles of rice genes under cold, drought, and high-salinity stresses and abscisic acid application using cdna microarray and rna gel-blot analyses. *Plant Physiol.*, **133**: 1755-1767.

Randall, D.D., J.A. Miernyk, N.R. David, J. Gemel and M.H. Luethy 1996. Regulation of leaf mitochondrial pyruvate dehydrogenase complex activity by reversible phosphorylation. *In* PR Shewry, NG Helford, eds, Protein Phosphorylation in Plants. Clarendon Press, Oxford, pp 87-103.

Raschke, K. 1979. Movements of stomata. In: *Encyclopedia of Plant Physiology New Series vol. 7, Physiology of Movements* (eds. W. Haupt and M.E. Feinleib), pp. 383-441. Springer-Verlag, Berlin.

Rathinasabapathi, B. 2000. Metabolic engineering for stress tolerance: installing osmoprotectant synthesis pathways. *Ann. Bot.*, **86:** 709-716.

Renou, J.L, A. Gerbaud, D. Just and M. André 1990. Differing substomatal and chloroplastic $CO_2$ concentrations in water-stressed wheat. *Planta,* **182**: 415-419.

Ribaut J-M, C. Jiang, D. Gonzalez-de-Leon, G.O. Edmeades, D.A. Hoisington 1997. Identification of quantitative trait loci under drought conditions in tropical maize: 2. Yield components and marker-assisted selection strategies. *Theor Appl Genet.,* **94**: 887-896.

Richards, R.A. 1996 Defining selection criteria to improve yield under drought. *Plant Growth Regul.*, **20**: 157-166.

Rizhsky, L., H. Liang and R. Mittler, 2003. The water-water cycle is essential for chloroplast protection in the absence of stress. *J. Biol. Chem.,* **278**: 38921–38925.

Rizhsky L., H. Liang, J. Shuman, V. Shulaev, S. Davletova and R. Mittler. 2004. When defense pathways collide. the response of arabidopsis to a combination of drought and heat stress. *Plant Physiol.*, **134**: 1683-1696.

Robinson, S.P., W.J.R. Grant and B.R. Loveys, 1988. Stomatal limitation to photosynthesis in abscisic acid-treated and in water-stressed leaves measured at elevated $CO_2$. *Aust. J. Plant Physiol.*, **15**: 495-503.

Ruberti, I., G. Seea, S. Lucchetti, G. Morelli, 1991. *EMBO J.,* **10**: 1787-1791.

Ruuska, S. A., M.R. Badger, T.J. Andrews, and S.von Caemmerer, 2000. Photosynthetic electron sinks in transgenic tobacco with reduced amounts of Rubisco: little evidence for significant Mehler reaction. *J. Exp. Bot.*, **51**: 357-368.

Salleo, S., A. Nardini, F. Pitt and M.A. Lo Gullo, 2000. Xylem cavitation and hydraulic control of stomatal conductance in Laurels (*Laurus nobilis* L.). *Plant Cell Environ.*, **23**: 71-79.

Schäfer, K.V.R., R. Oren and J.D. Tenhunen, 2000. The effect of tree height on crown level stomatal conductance. *Plant Cell Environ.*, **23**: 365-375.

Scheuermann, R., K.Biehler, T.Stuhlfaulth, and H.P. Fock, 1991. Simultaneous gas exchange and fluorescence measurements indicate differences in the response of sunflower, bean and maize to water stress. *Photosynth. Res.*, **27**: 189-197.

Schneider, K., B. Wells, E. Schmelzer, F. Salamini, D. Bartels, 1993. *Planta*, **189:** 120-131.

Schulze, E.-D. and A.E. Hall, 1982. Stomatal responses, water loss and $CO_2$ assimilation rates of plants in contrasting environments. In: *Encyclopedia of Plant Physiology Vol. 12 B. Physiological Plant Ecology II* (eds. O.L. Lange, P.S. Nobel, C.B. Osmond and H. Ziegler), pp. 263-324. Springer-Verlag, Berlin-Heidelberg-New York.

Schulze, D.-E., R.H. Robichaux, J.Grace, P.W. Rundel, and J.R. Ehleringer, 1987. Plant water balance. *Bioscience,* **37**: 30-37.

Schwab, K.B., U. Schreiber, and U. Heber, 1989. Response of photosynthesis and respiration of resurrection plants to desiccation and rehydration. *Planta,* **177**: 217-227.

Serraj, R., V. Vadez, R. Ford Denison, and T. R. Sinclair, 1999. Involvement of Ureides in Nitrogen Fixation Inhibition in Soybean. *Plant Physiol.,* **119**: 289-296.

Selote and R. Khanna-Chopra, 2004. Drought-induced spikelet sterility is associated with an inefficient antioxidant defence in rice panicles. *Physiologia Plant.,* **121**:462-471

Sharkey, T.D., J.A. Berry and R.F. Sage, 1988. Regulation of photosynthetic electron-transport in *Phaseolus vulgaris* L., as determined by room-temperature chlorophyll a fluorescence. *Planta,* **176**: 415-424.

Sharp, R.E., M.A. Matthews, J.S. Boyer, 1984. Kok effect and the quantum yield of photosynthesis. *Plant Physiol.,* **75**: 95-101.

Stuhlfaulth, T., D.F. Sültemeyer, S. Weinz, and H.P. Fock, 1988. Fluorescence quenching and gas exchange in a water stressed $C_3$ plant, *Digitalis lanata*. Plant Physiol. 86: 246-250.

Stuhlfaulth, T., R. Scheuermann and H.P. Fock, 1990. Light energy dissipation under water stress conditions. Contribution of reassimilation and evidence for additional processes. *Plant Physiol.,* **92**: 1053-1061.

Takeba, G. and A. Kozaki, 1998. Photorespiration is an essential mechanism for the protection of $C_3$ plants from photooxidation. In: *Stress Responses of Photosynthetic Organisms* (eds. K. Satoh, N. Murata), pp. 15-36. Elsevier Science B.V., The Netherlands.

Tardieu, F. and T. Simmonneau, 1998. Variability among species of stomatal control under fluctuating soil water status and evaporative demand: modelling isohydric and anisohydric behaviours. *J. Exp. Bot.*, **49**: 419-432.

Thomas, D.A. and M.André, 1982. The response of oxygen and carbon dioxide exchanges and root activity to short term water stress in soybean. *J. Exp. Bot.*, **134**: 309-405.

Tourneux, C. and G. Peltier, 1995. Effect of water deficit on photosynthetic oxygen exchange measured using $^{18}O_2$ and mass spectrometry in *Solanum tuberosum* L. leaf discs. *Planta,* **195**: 570-577.

Troughton, J.H. and R.O. Slatyer, 1969. Plant water status. leaf temperature, and the calculated mesophyll resistance to carbon dioxide of cotton leaves. *Aust. J. Biol. Sci.*, **22**: 815-827.

Tyree, M.T. 1999. Water relations and hydraulic architecture. In: *Handbook of Functional Plant Ecology* (eds. F.I. Pugnaire and F. Valladares), pp. 221-268. Marcel Dekker Inc., New York.

Valentini, R., D. Epron, P. De Angelis, G. Matteucci, and E. Dreyer, 1995. *In situ* estimation of net $CO_2$ assimilation, photosynthetic electron flow and photorespiration in Turkey oak (*Quercus cerris* L.) Leaves: diurnal cycles under different levels of water supply. *Plant Cell Environ.*, **18**: 631-640.

Velasco, R., F. Salamini, D. Bartels, 1998. *Planta,* **204**: 459-471.

Wade, L.J., C.G. McLaren, B.K. Samson, K.R. Regmi, S. Sarkarung, 1996. The importance of site characterization for understanding genotype by environment interactions. *In* M Cooper, GL Hammer, eds, Plant Adaptation and Crop Improvement. CAB International, Wallingford, UK, pp 549–562.

Wingler, A., P.J. Lea, W.P. Quick, and R.C. Leegood, 2000. Photorespiration: metabolic pathways and their role in stress protection. *Philos. T. Roy. Soc. B.* **355**: 1517-1529.

Wingler, A., W.P. Quick, R.A. Bungard, K.J. Bailey, P.J. Lea, and R.C. Leegood, 1999. The role of photorespiration during drought stress: an analysis utilizing barley mutants with reduced activities of photorespiratory enzymes. *Plant Cell Environ.*, **22**: 361-373.

Wise, R.R., D.H. Sparrow, A. Ortiz-Lopez and D.R. Ort, 1991. Biochemical regulation during the mid-day decline of photosynthesis in field-grown sunflower. *Plant Sci.*, **74**: 45-54.

Wolkers, W.F., M.G. van Kilsdonk and F.A. Hoekstra, 1998. *Biochim. Biophys. Acta,* **1425**: 127-136.

Wong, S.C., I.R. Cowan, and G.D. Farquar, 1979. Stomatal conductance correlates with photosynthetic capacity. *Nature,* **282**: 424-426.

Wong, S.C., I.R. Cowan, and G.D. Farquar, 1985. Leaf conductance in relation to rate of $CO_2$ assimilation. III. Influences of water stress and photoinhibition. *Plant Physiol.*, **78**: 830-834.

Wood, A.J. and M.J. Oliver, 1999. *Plant J.,* **18**: 359-370.

Wu, J., S. Neimanis and U. Heber, 1991. Photorespiration is more effective than the Mehler reaction in protecting the photosynthetic apparatus against photoinhibition. *Bot. Acta,* **104**: 283-291.

Xiong, L., H. Lee, M. Ishitani and J. K. Zhu, 2002. Regulation of osmotic stress-responsive gene expression by the *LOS6/ABA1* locus in Arabidopsis. *J. Biol. Chem.,* **277:** 8588-8596.

Xu, D., X. Duan, B. Wang, B. Hong, T.H.D. Ho and R. Wu, 1996. Expression of a late embryogenesis abundant protein gene, HVA1, from barley confers tolerance to water deficit and salt stress in transgenic rice. *Plant Physiol.,* **110**: 149-156.

Yeo, A. 1998. Molecular biology of salt tolerance in context of whole-plant physiology. *J. Exp. Bot.,* **49**: 915-929.

Yu, L.-Xi and T. L. Setter, 2003. Comparative transcriptional profiling of placenta and endosperm in developing maize kernels in response to water deficit. *Plant Physiol.,* **131**: 568-582.

*Advances in Plant Physiology*, Vol. 9
Ed. A. Hemantaranjan
Scientific Publishers (India), Jodhpur, 2006 pp. 41-66
E-mail: **info@scientificpub.com** www.scientificpub.com

# 3

# CALCIUM AND OXIDATIVE STRESS

Tracey Ann Cuin

School of Agricultural Science, University of Tasmania, Private Bag 54, Hobart, Tasmania, 7001, Australia. E-mail: Tracey.Cuin@utas.edu.au

## INTRODUCTION

The calcium ion is now firmly established as a ubiquitous signalling molecule in plants (Hepler and Wayne, 1985, Rudd and Franklin-Tong, 2001, Sanders *et al.*, 2002, Hetherington and Brownlee, 2004). An increase in the cytosolic concentration of $Ca^{2+}$ ($[Ca^{2+}]_{cyt}$) couples a diverse array of signals and receptors. Numerous plant signal transduction pathways have been shown to use $Ca^{2+}$ as an integral signalling component (Sanders *et al.*, 1999, 2002). At their simplest, $Ca^{2+}$ based signalling systems are composed of a receptor, a system for generating the increase in $[Ca^{2+}]_{cyt}$, downstream components that are capable of reacting to the increase $[Ca^{2+}]_{cyt}$ and other cellular systems responsible for returning $[Ca^{2+}]_{cyt}$ back to its pre-stimulus level. Downstream from the stimulus-induced $[Ca^{2+}]_{cyt}$ increase, or "$Ca^{2+}$ signal", the cell possesses an array of proteins that can respond to the message, such as calmodulin (CaM) (Snedden and Fromm, 2001), $Ca^{2+}$-dependent protein kinases (Harmon *et al.*, 2001, Cheng *et al.*, 2002) and CaM binding proteins (Reddy *et al.*, 2002, Zhang and Lu, 2003).

Accordingly, $Ca^{2+}$, acting as an intracellular second messenger, connects a wide range of external stimuli to their characteristic intracellular response (Webb *et al.*, 1996, Trewavas and Malhó, 1998). In plants, transient elevations in $[Ca^{2+}]_{cyt}$ have been reported in response to numerous stimuli including abiotic, reactive-oxygen producing stresses (Fig. 1). Such events encompass low-temperature signals (Knight *et al.*, 1991, 1992, 1996, van Der Luit *et al.*, 1999), hyperosmotic shock (Taylor *et al.*, 1996, Takahashi *et al.*, 1997, Cessna *et al.*, 1998), light (Shacklock *et al.*, 1992, Baum *et al.*, 1999), chemically-imposed oxidative stress such as herbicides (Price *et al.*, 1994, Pei *et al.*, 2000), ozone (Clayton *et al.*, 1999, Evans *et al.*, 2005), hormones (Gilroy and

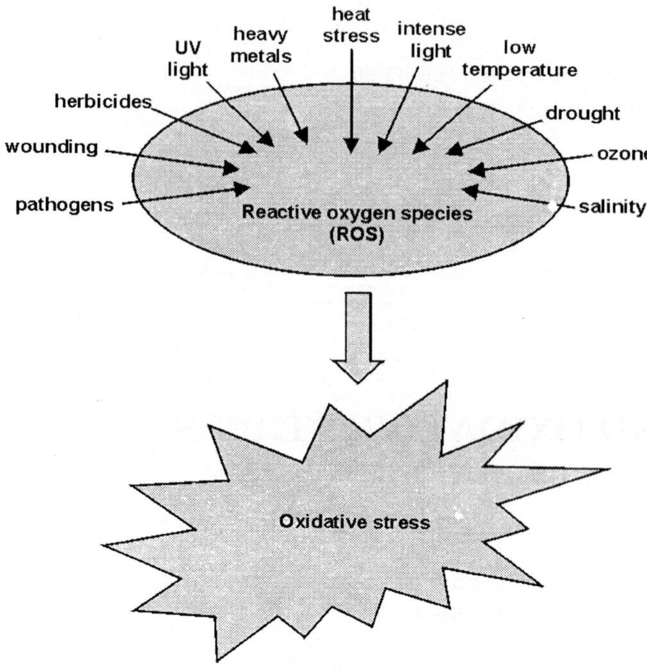


**Figure 1.** Environmental factors that increase the production of ROS in plant cells

Jones, 1992, McAinsh *et al.*, 1992), NOD factors (Ehrhardt *et al.*, 1996) and elicitors (Knight *et al.*, 1991, Mithöfer *et al.*, 1999, Blume *et al.*, 2000). Much research has been carried out over the past couple of decades into the association of $Ca^{2+}$-signalling with these abiotic-, oxidative-, reactive oxygen producing-stresses.

The wide diversity of stimuli that can elicit an increase in $[Ca^{2+}]_{cyt}$, and in so doing, trigger a signal, raises the question of how a cell is able to distinguish one $Ca^{2+}$-mobilising signal from another in order to produce the appropriate response in the plant. This question has been a highly active area of research over the past decade. The current hypothesis regarding this is that the information encrypted in the stimulus induced increase in $[Ca^{2+}]_{cyt}$, i.e., the temporal and spatial nature and the amplitude of the $[Ca^{2+}]_{cyt}$, its so-called "calcium signature", defines the outcome of the response (Trewavas, 1999). Although much effort has been directed towards answering this question, the manner in which the signal specificity is encoded and transduced is still far from clear. There are, nonetheless, many examples of diversity in the $Ca^{2+}$ signature within plant cells as a result of different oxidative stress events.

In this review, the chemistry behind the "natural" production of reactive oxygen species during "normal" photosynthetic and respiratory processes will be briefly summarised and the various mitigating defence mechanisms of the plant outlined. This will be followed by a description of the various abiotic oxidative stresses responsible for the production of reactive oxygen species (ROS). The review will then turn to the perturbations of $Ca^{2+}$ homeostasis elicited by oxidative stress. This will include examples taken from the literature of research which has shown disruption to $Ca^{2+}$ homeostasis after various oxidative stress events. Finally, there will be an account of the increasing understanding of the $Ca^{2+}$ signatures which are proposed to encode the specificity of the ozone-induced response, and in so doing, lead to the appropriate downstream response.

## 2. BIOTIC OXIDATIVE STRESS

### 2.1. Production of Endogenous Reactive Oxygen Species

Chemical activation of molecular oxygen has shaped the biochemistry of life. The formation of reactive oxygen species (ROS) accompanies every normal metabolic process in all aerobic organisms. The association of oxygen in respiratory processes and the production of oxygen during photosynthesis results in the generation of ROS. Among these are superoxides ($O_2^{\bullet-}$), hydrogen peroxide ($H_2O_2$), hydroxyl radicals ($HO^{\bullet}$) plus other free radicals that can react with, and damage DNA, proteins and lipids, potentially lead to cell death (Bowler *et al.*, 1992). A simplified pathway for the production of various ROS are outlined in figure 2.

Most parts of the plant are responsible for the production of ROS (Fig. 3), but it is within the mitochondria, and especially the chloroplast, where the production is greatest. Superoxide, an initial oxy-radical product, can further react within the cell to form more ROS such as $HO^{\bullet}$ and singlet oxygen ($^1O_2$). Superoxide itself is a charged molecule and cannot cross biological membranes, but it is readily dismutated to $H_2O_2$. By themselves, neither $O_2^{\bullet-}$ radicals nor $H_2O_2$ are very reactive. Indeed, several processes actually use such activated oxygen species in a beneficial manner (for review see Inzé and Van Montagu, 1995). For example, $H_2O_2$ can act as a signal molecule that induces the transcription of defence-related genes (Chen *et al.*, 1993, 1995), such as during the oxidative burst generated during hypersensitive plant-pathogen interactions when the $H_2O_2$ produced can function as a local trigger for programmed cell death and rapid cross-linking of cell wall proteins, both of which serve as a protective mechanism (Bradley *et al.*, 1992, Levine *et al.*, 1994).

However, $O_2^{\bullet-}$ radicals and $H_2O_2$ also react to produce additional activated oxygen species that are damaging to essential cellular components. Superoxides, for instance, can reduce quinones and transition metal complexes of $Fe^{3+}$ and $Cu^{2+}$, thus affecting the activity of metal-containing enzymes. Hydroperoxyl radicals ($HO_2^{\bullet}$), formed from $O_2^{\bullet-}$ by protonation in aqueous solutions, can cross biological membranes and subtract hydrogen atoms from polyunsaturated fatty acids and lipid hydroperoxides, thus initiating lipid auto-oxidation. This severely compromises membrane integrity (Halliwell and Gutteridge, 1999). Furthermore, $H_2O_2$ can diffuse some distance from its production site where it may inactivate enzymes by oxidising their thiol groups. For example, enzymes of the Calvin cycle, copper/zinc superoxide dismutase (SOD) and iron super oxide dismutase are all inactivated by $H_2O_2$ (Charles and Halliwell, 1980, Bowler *et al.*, 1994). By far, the most reactive of all the ROS is the hydroxyl radical ($OH^{\bullet}$), formed from $H_2O_2$ by the Haber-Weiss or Fenton reactions, using metal catalysts (Halliwell and Gutteridge, 1999). This $OH^{\bullet}$ can potentially react with all biological molecules, and because cells have no enzymatic mechanism to eliminate this highly reactive species, its excess production ultimately leads to cell death. This phenomenon of reactive oxygen species production is described as oxidative stress, and complex mechanisms have evolved in plants to mitigate and repair the damage initiated by such potentially damaging free radicals.

### 2.2. Defence Mechanisms: Limiting ROS-mediated Damage

The primary constituents of protective mechanisms against ROS-mediated damage include enzymes such as the superoxide dismutases (SOD), catalases and peroxidases and free radical scavengers, such as carotenoids, ascorbate, tocopherols, and

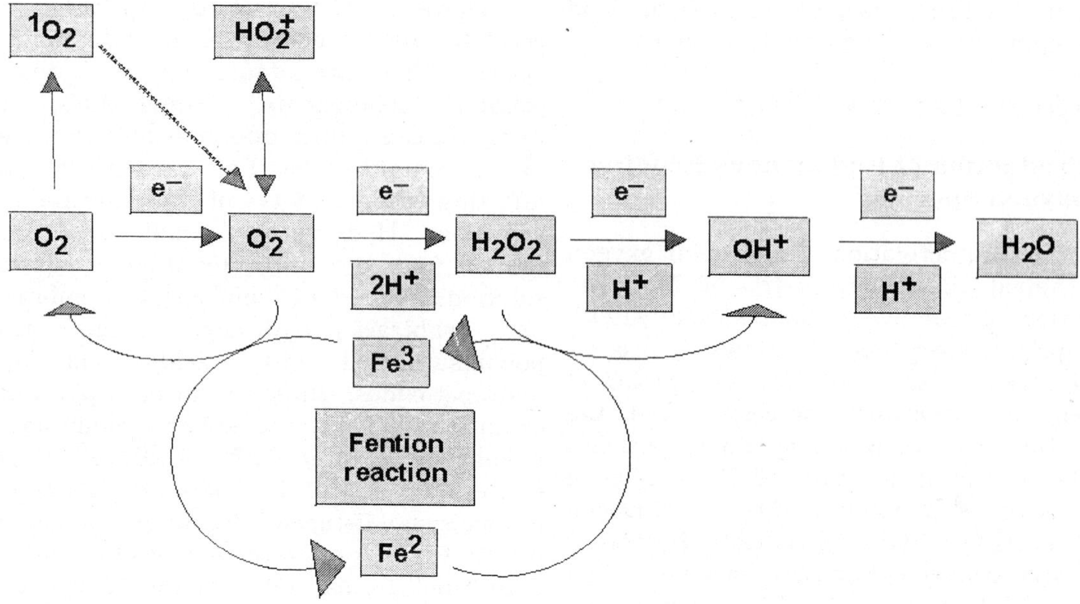


**Figure 2.** Production of reactive oxygen species (ROS) from oxygen.

oxidised and reduced glutathione (GSSG and GSH respectively). These are located at various sites throughout the cell (Fig. 3). Enzymes of scavenger metabolism such as ascorbate peroxidase (APX), glutathione reductase (GR) and dehydroascorbate reductase (DHAR), that together form the Halliwell-Asada pathway (Fig. 4), also contribute (Halliwell and Gutteridge, 1999). One important feature of these protective mechanisms is that their activity is enhanced when plant cells are exposure to conditions resulting in increased free-radical production. For instance, cellular activities of SOD and glutathione reductase increase under conditions of higher ROS production, and the levels of low molecular weight protectants may also increase (Bowler *et al.*, 1992).

Nevertheless, the mechanisms by which plant cells are able to perceive increased free-radical production and compensate by increasing the level of oxidative protection are not well understood, although the so-called prooxidant/antioxidant ratio is thought to be of special significance. The primary prooxidant constituents are dehydroascorbic acid, quinones and GSSG and the antioxidants are the equivalent reduced forms. The prooxidant/antioxidant ratio is an indictor of the redox state of the cytoplasm and considerable amounts of non-protein thiols are present in plants. These function as antioxidants by keeping other redox-active molecules in a reduced state (Noctor and Foyer, 1998). Glutathione, is the major thiol compound in the plant cell. Under most physiological conditions, glutathione is found largely in the reduced form (GSH) (Foyer *et al.*, 1997), and consequently acts as a redox buffer within the cell. The ratio of reduced to oxidised glutathione (the GSH/GSSG ratio) is accepted to be indicative of the cellular redox balance, and accordingly, is involved in ROS perception (Dröge, 2002, Foyer and Noctor, 2003). For example, the bacterial OxyR protein can be activated either directly by $H_2O_2$ or alternatively, by changes in the intracellular GSH/GSSG balance, suggesting that the increase in $H_2O_2$ is sensed in this manner (Dröge, 2002). In plants, evidence from treatments that affect the GSH/GSSG ratios in opposite ways while having the

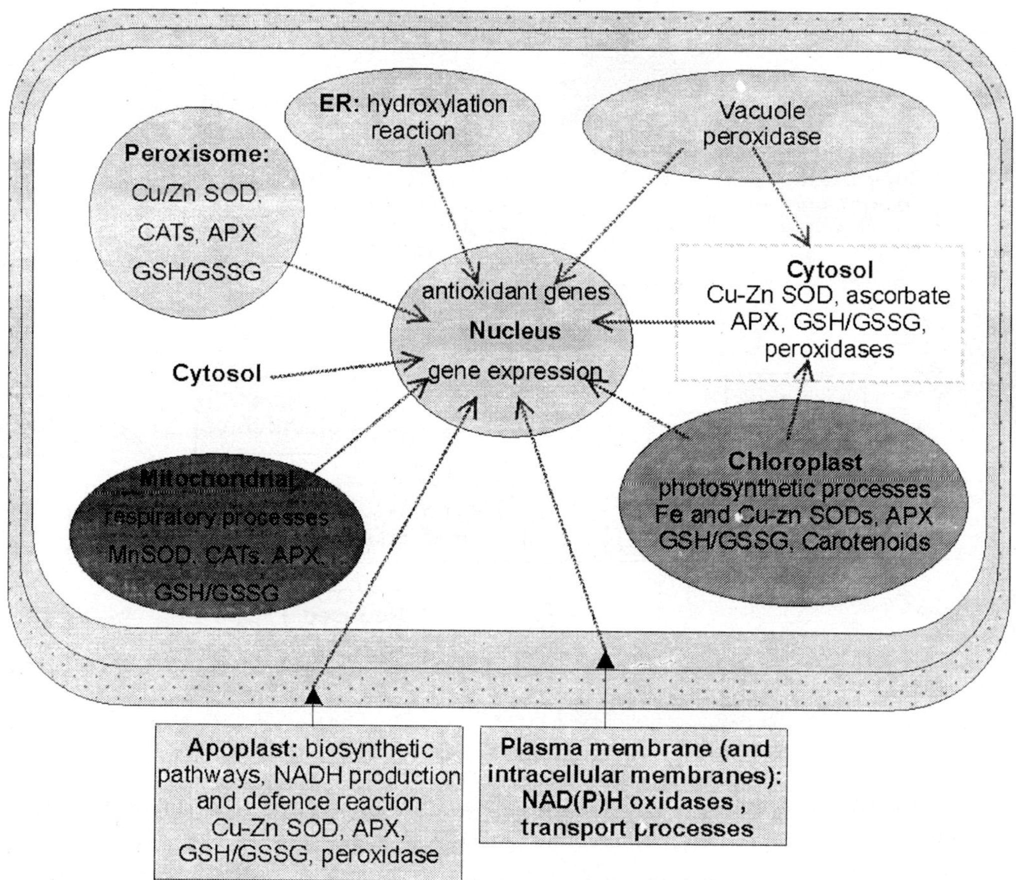


**Figure 3.** Reactive oxygen species (ROS) arise throughout the cell in response to normal metabolic processes. The various oxidative defence mechanisms are distributed in various locations throughout the cell. *Abbreviations:* APX: ascorbate peroxidase; CAT: catalyse; GSH/GSSG: reduced/oxidised glutathione; SOD: superoxide dismutase; ER: endoplasmic reticulum

converse effect on the $H_2O_2$ triggered rise in the $[Ca^{2+}]_{cyt}$, (Price *et al.*, 1994), provide support for this hypothesis.

As outlined above, ROS are continuously produced as by-products of various metabolic pathways, during certain redox reactions and during incomplete reduction of oxygen or oxidation of water by the mitochondria or chloroplast electron transfer chains (Foyer and Harbinson, 1994). The distinct subcellular localisation (Fig. 3) and the biochemical properties of antioxidant enzymes, their differential inducibility at the enzyme and gene expression level and the plethora of non-enzymatic scavengers, render the antioxidant systems a very versatile and flexible unit that can control ROS accumulation temporally and spatially (Alscher and Hess, 1993, Bowler *et al.*, 1994, Scandalios, 1994, Van Bresegem *et al.*, 1998). Consequently, under an unstressed, steady-state condition, the free-radical oxygen species are scavenged by the different antioxidative defence components (Alscher *et al.*,1997), keeping the formation and removal of ROS in balance. However, the equilibrium between the production and scavenging of ROS may be perturbed by a number of

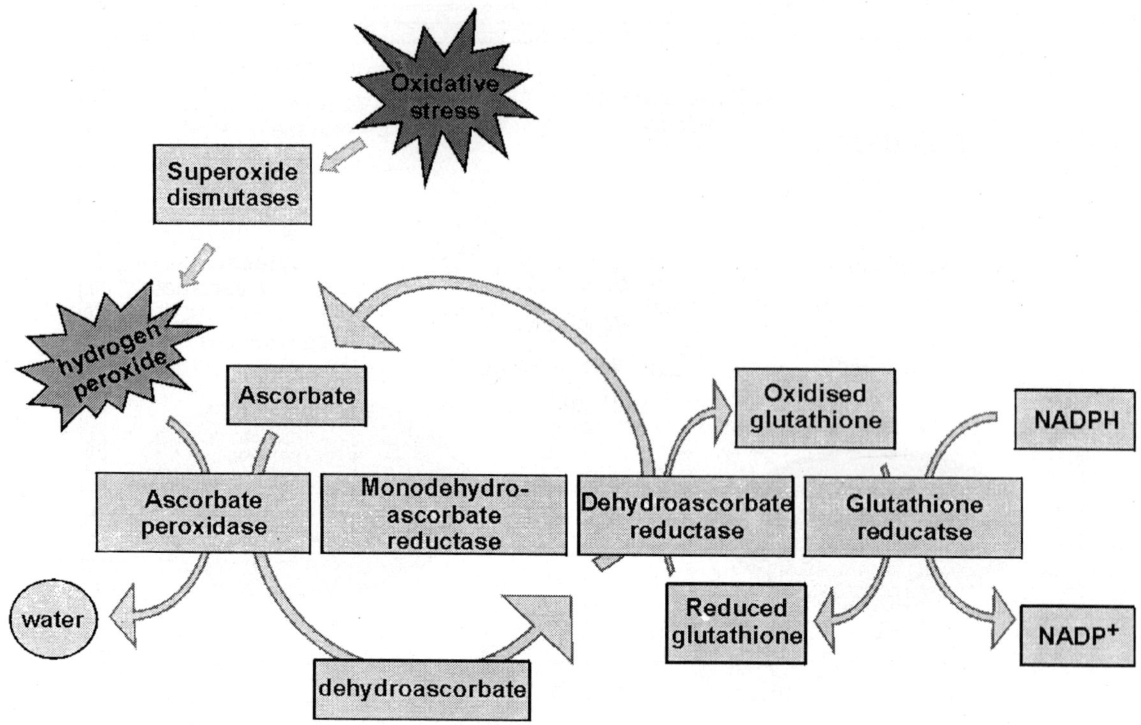


**Figure 4.** The Halliwell-Asada pathway; an important mechanism for the removal of hydrogen peroxide.

adverse ROS producing environmental factors. This leads to the defence system becoming rapidly overwhelmed, the result of which causes the intracellular levels of ROS to rise (Malan *et al.*, 1990, Elstner, 1991, Prasad *et al.*, 1994, Tsugane *et al.*, 1999).

## 3. ABIOTIC PRODUCTION OF REACTIVE OXYGEN SPECIES

As outlined above, cells are protected by the endogenous molecular systems that mitigate the effects of oxidative stress. However, certain abiotic stresses result in the enhanced production of ROS (Fig. 1) (reviewed by Bowler *et al.*, 1992, Foyer *et al.*, 1994, Inzé and Van Montagu, 1995). Such exogenous stresses can easily overwhelm the cellular defence mechanisms of plants, and due to the rapid imbalance between the production and scavenging of ROS, can severely damage target molecules, setting in motion a catastrophic cascades of events that can ultimately lead to cell death. Such potentially damaging ROS producing oxidative stresses include photoinhibition, herbicide treatment, metal toxicity, ultra-violet radiation, salt stress, drought stress, high and low temperatures, chilling and freezing and exposure to pollutant gases. The ROS producing mechanisms of these oxidative stresses are outlined below.

### 3.1. Photoinhibition

Excessive light energy causes over-reduction of the photosynthetic electron transport chain, leading to both increased production of $O_2^{\bullet -}$ on the acceptor side of Photosystem I (PSI), and increased reduction of $^1O_2$ (Asada, 1996). Oxidative stress due to high irradiance is often demonstrated in combination with other unfavourable conditions such as low temperature, low $CO_2$, drought and salt stress, which exacerbate the production of ROS.

### 3.2. Herbicides

The herbicide methyl viologen (MV) or paraquat exerts its toxicity by promoting photo-production of $O_2^{\bullet-}$ in chloroplasts and has frequently been used to experimentally induce oxidative stress in plants. When the plant tissue is exposed to light, MV becomes an electron acceptor from PSI, subsequently reducing dioxygen to $O_2^{\bullet-}$ (Van Breusegem *et al.*, 2002). It appears that the production of $O_2^{\bullet-}$ exceeds the endogenous SOD capacity (Foyer, 2002), thus leading to the rapid production of ROS which overwhelms the plant. Methyl viologen also accepts electrons from the respiratory electron transport chain in the mitochondria and forms $O_2^{\bullet-}$ in the dark (Van Breusegem *et al.*, 2002). When MV-administered chloroplasts are illuminated, APX activity is lost, leading to a loss of $H_2O_2$ scavenging capacity (Mano *et al.*, 1998), thus further increasing ROS production.

### 3.3. Metal Toxicity

Phytotoxic metals such as zinc, copper, iron and cadmium are widespread pollutants and are involved in free radical production (Foyer *et al.*, 1994). A link between metal toxicity, oxidative stress and defence responses in Arabidopsis and *Nicotiana* has been demonstrated (Kampfenkel *et al.*, 1995, Xiang and Oliver, 1998). Plants exposed to elevated levels of copper ions exhibit peroxidation and pigment bleaching (Sandmann and Gonzales, 1989). Prolonged exposure to copper results in chlorophyll degradation and a decline in the endogenous level of catalase. Copper and iron ions are redox active and catalyse the Fenton reaction producing free-radical species. Lipid peroxidases also originate from the induction of lipoxygenase in the presence of copper (Foyer *et al.*, 1994). Treating plants with cadmium decreases the chlorophyll and heme levels of germinating mung bean seedlings by inducing lipoxygenase activity with the simultaneous inhibition of the antioxidative enzymes (Van Assche and Clijsters, 1990, Somashekaraiah *et al.*, 1992, Gallego *et al.*, 1996).

### 3.4. Ultraviolet Radiation

Increasing fluxes of UV-B (290-320 nm) radiation are reaching the Earth's surface as a consequence of stratospheric ozone depletion (Kerr and McElroy, 1993). The deleterious effects of UV-B on plants have been extensively studied (Teramura and Sullivan, 1994, Bornman and Sundby-Emanuelsson, 1995), although, except for the radical production on the donor side of PSII, the sources of the ROS in UV-B treated plants have not been well identified. Damage by UV-B to Photosystem II (PSII) involves the impairment of electron transport (Hideg *et al.*, 1993) and structural damage of the reaction centre proteins (Greenberg *et al.*, 1989). In addition, UV-B irradiation can also initiate radical-yielding reactions that can be detected in leaves minutes after the cessation of treatment (Hideg and Vass, 1996).

### 3.5. Salt Stress

Salinity affects important metabolic processes located in chloroplasts and mitochondria (Cheeseman, 1988), but little is known about its effect on ROS production in these organelles. Hernandez *et al.* (1995), hypothesised that the decrease in $CO_2$ concentration in chloroplasts brought about by salinity induced stomatal closure, resulted in $NADP^+$ shortage and oxygen reduction. Experiments with leaf organelles from NaCl treated pea plants, have demonstrated a salinity-induced enhancement in $O_2^{\bullet-}$ production, as well as a strong decrease in mitochondrial MnSOD (Hernandez *et al.*, 1993). Salt-tolerant pea cultivars subjected to

NaCl stress had increased Cu/ZnSOD and APX activities, as well as an increased ascorbate pool. A similarly treated salt-sensitive pea cultivar had increased $H_2O_2$ content and lipid peroxidation products without a change in the enzymatic activities (Hernandez *et al.*, 1995). Likewise, in NaCl treated radish plants, the activity of APX increased two-fold (Lopez *et al.*, 1996) and in *Citrus*, Cu/ZnSOD and APX were induced by NaCl (Gueta-Dahan *et al.*, 1997), as did the SOD isozymes of the halophyte *Mesembryanthemum* (Miszalski *et al.*, 1998).

### 3.6. Water Stress

Water stress encompasses both direct drought stress and also water stress resulting from high soil salinity. The primary target sites of water stress have not been narrowed down to specific tissues or organelles (Alscher *et al.*, 1997). The evidence for ROS production in water stressed plants it is rather indirect. In water stressed pea plants, the amount of metal ions available for the Fenton reaction and oxidative damage of lipids, proteins and DNA increased (Moran *et al.*, 1994). During drought stress an abscisic acid signal causes stomatal closure and the light-exposed, over-reduced photosynthetic apparatus may experience oxidative stress due to reduced $CO_2$ absorption by the leaf. In addition, many of the drought- and salt-induced increases in the production of ROS appear to be due to the shortage of water for photosynthetic reactions. For example, Price and Hendry (1991), found that a water deficit caused an overall inhibition of protein synthesis, inactivation of several chloroplast enzymes, impairment of electron transport, increased membrane permeability and increased activity of enzymes of the $H_2O_2$ scavenger system. The recovery from drought appears also to be a delicate stage that may be accompanied by oxidative stress, requiring the induction of defence systems (Mittler and Zilinskas, 1994).

### 3.7. Temperature

Chilling appears to increase ROS output from the chloroplasts indirectly by inhibiting the free-radical scavenging enzymes. This is in addition to the acceleration of oxygen photo-reduction due to the over-reduction of the photosynthetic electron transport chain caused by inhibition of $CO_2$ assimilation, particularly at high irradiance. Chloroplast functions are more sensitive to chilling induced photoinhibition than those of mitochondria and disintegration of the ultra-structure of chloroplasts is caused by a combination of low temperature and high light intensity (Wise and Naylor, 1987). PSII photoinhibition is also observed (Hetherington *et al.*, 1989), and the capacity to accept electrons from PSII appears to govern the light sensitivity of PSII (Öquist and Huner, 1993). Under weak light, it is reported that PSI is inhibited in cold sensitive plants (Havauz and Davaud, 1994, Terashima *et al.*, 1994). The inhibition of PSI requires both oxygen and electrons from PSII, suggesting that the inhibitory species are the ROS photo-produced on the reducing side of PSI, namely, $O_2^{\bullet -}$ or $HO^{\bullet}$. An accumulation of $H_2O_2$ can also occur due to a decreased APX activity at low temperatures and PSI photoinhibition can contribute to this suppression of APX (Terashima *et al.*, 1998). The early stages of chilling stress may also result in a decrease of SOD activity (Gianinetti *et al.*, 1993) which will lead to an increase in $O_2^{\bullet -}$. Also, at low temperatures the water-water cycle enzymes that protect the PSI reaction centre (Asada *et al.*, 1998) will be less active so the production of ROS may surpass the scavenging capacity.

Heat stress becomes apparent in most plants over 40 •C. It causes uncoupling of electron transport with ATP synthesis, inhibition of PSII water oxidase (Katoh and San Pietro, 1967, Mohanty *et al.*, 1987) and lipid peroxidation (Mishra and Singhal, 1992). Heat treatment produces oxidative stress (Gong *et al.*, 1997a,b, Kraus and Fletcher, 1994), for example, the generation of $H_2O_2$ in tobacco (Foyer *et al.*, 1997) and mustard seedlings (Dat *et al.*, 1998). However, the source of $H_2O_2$ and the initial target of heat stress are not well elucidated. Heat stress often results in the production of specific families of proteins known as heat shock proteins (HSPs) (Howarth and Ougham, 1993). These accumulate during mild heating (Nover *et al.*, 1983) and their appearance correlates with the survival of the plant. There is considerable evidence that oxidative stress induces pathways that result in accumulation of some HSPs (Dat *et al.*, 1998, Storozhenko *et al.*, 1998, Schett *et al.*, 1999). It has also been shown that thermotolerance can be induced by compounds that induce oxidative bursts (Dat *et al.*, 1998), and that very short heat pulses can induce such bursts of $O_2^{\bullet -}$ and/or $H_2O_2$ (Vallelian-Bindschedler *et al.*, 1998).

### 3.8. Air Pollution

Atmospheric pollutants such as ozone and sulphur dioxide have been implicated in the formation of free radicals (Cross *et al.*, 1998). Of these, ozone is considered to be most damaging (Heagle, 1989). Exposure to high concentrations of ozone is one of the best characterized causes of abiotic oxidative stress. Stratospheric ozone is beneficial because it shields the earth from UV irradiation, thus preventing damage due to UV-B. However, tropospheric ozone is harmful to life because it is a highly reactive oxidant which results in the direct generation of ROS.

Tropospheric ozone is formed through photochemical reactions between nitrogen oxides, carbon monoxide and hydrocarbons released primarily by burning fossil fuels (Mauzerall and Wang, 2001). Anthropogenic hydrocarbons and oxides of nitrogen (NO, $NO_2$) react with solar UV radiation to generate ozone ($O_3$). Ozone production is particularly high in summer months due to strong sunlight, high temperature and stagnant high-pressure systems, thus concentrations tend to be at their highest during the growing season of most crop plants. Furthermore, ozone and its precursors can be transported hundreds of kilometres from cities affecting rural, crop producing areas (Cleveland *et al.*, 1976). Indeed, concentrations downwind of the sources of ozone pollution are often significantly higher than those found in urban areas (Cox *et al.*, 1975). Ozone is considered to be the most phytotoxic of all the common air pollutants and has been estimated to be responsible for up to 90% of the crop losses that result from air pollution (Heck *et al.*, 1982).

Ozone is toxic to plants because, as a powerful oxidising agent, it is able to react directly with lipids and proteins (Halliwell and Gutteridge, 1999). In aqueous environments like the apoplast, ozone can lead to the production of ROS such as $HO^{\bullet}$, $^1O_2$, $H_2O_2$ (Mehlhorn, 1990, Kanofsky and Sima, 1991, Byvoet *et al.*, 1995). Ozone has also been shown to elicit an apoplastic oxidative burst in sensitive species and cultivars (Schraudner *et al.*, 1998, Rao and Davis, 1999). It is this active generation of secondary ROS combined with the ROS produced as a result of ozonolysis and reactions with cellular components that accounts for the ability of ozone to switch on the signal transduction cascade that can lead to damage and potentially, cell death.

## 4. CALCIUM AND ABIOTIC OXIDATIVE STRESS

The mechanisms by which plants perceive oxidative stress and reactive oxygen species are poorly understood. There is evidence that ROS may be perceived indirectly by sensing changes in the cellular redox potential (Price *et al.*, 1994). Alternatively, the cell may respond to the ROS-inflicted changed in the total GSH concentration (Gomez *et al.*, 2004). A further suggestion is that the cell may detect the products of ROS-

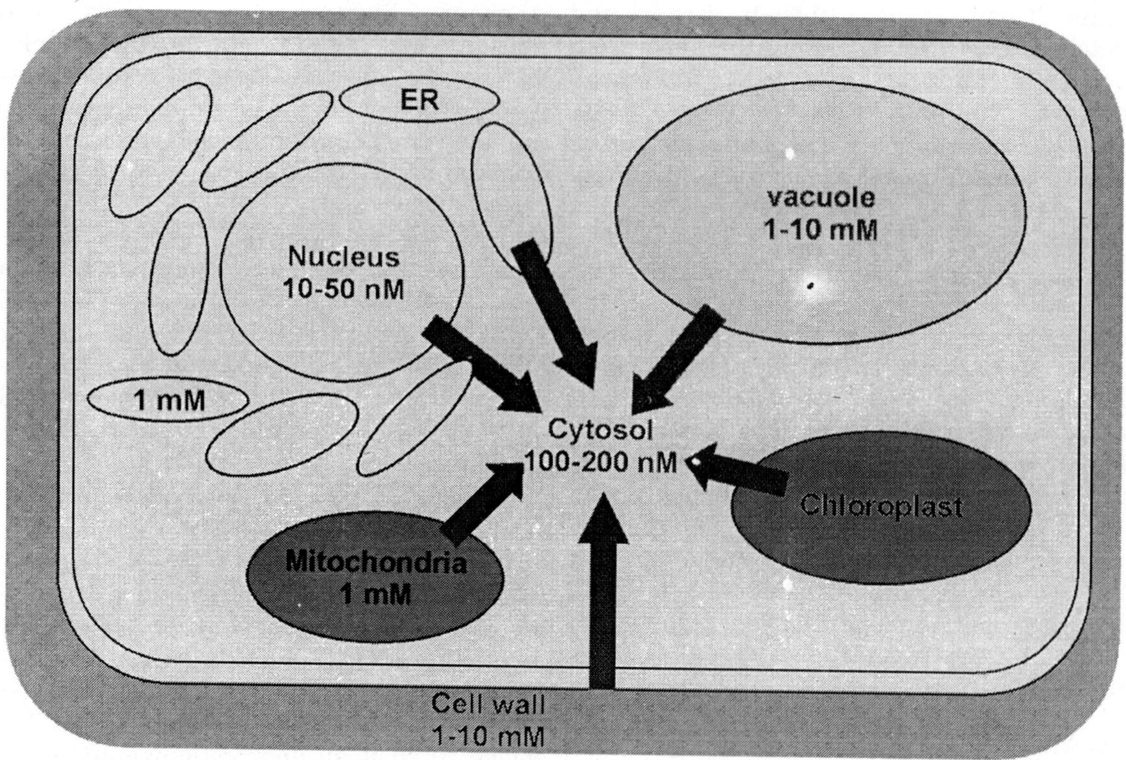


**Figure 5.** Origins and locations of calcium stores and the calcium influx into the cytosol in response to specific stimuli.

inflicted cell wall damage (Wiese and Pell, 2003). Nevertheless, the first detectible response in plants is an increase in the level of $Ca^{2+}$ within the cytosol, occurring within seconds of an oxidative stress event (Evans *et al.*, 2005), suggesting that an increase in the cytosolic calcium concentration ($[Ca^{2+}]_{cyt}$) is most likely to represent the first, upstream response.

There are now many reports in the literature of rapid increases in $[Ca^{2+}]_{cyt}$ resulting from a wide variety of causes of oxidative stress events. The following sections are an outline of the response of $Ca^{2+}$ in a number of these abiotic stresses.

## 4.1. UV-Radiation

Calcium has been implicated as a candidate as a second messenger in UV-B induced oxidative stress. Both $Ca^{2+}$ and CaM are required for the induction of chalcone synthase (CHS; the key enzyme in flavonoid biosynthesis) gene expression in Arabidopsis cell cultures (Christie and Jenkins, 1996). Similar results have been obtained in parsley protoplasts and cultured soybean cells (Frohnmeyer *et al.*, 1997, 1998), where measurements of $[Ca^{2+}]_{cyt}$ in a parsley cell culture showed that millisecond flashes of UV-B radiation were sufficient to raise the cellular $Ca^{2+}$ level and activate expression of a CHS promoter-luciferase reporter construct.

## 4.2. Low-Temperature

The role of $Ca^{2+}$ as a mediator in low temperature responses has been well documented and many studies support the hypothesis that an alteration of the $[Ca^{2+}]_{cyt}$ is a required component for the low-temperature signal transduction pathway. Low temperature initiated an increase in $[Ca^{2+}]_{cyt}$ in Arabidopsis (Knight *et al.*, 1996, Polisensky and Braam, 1996, Henriksson and Trewavas, 2003), tobacco (Knight *et al.*, 1991) and the moss *Physcomitrella* (Russell *et al.*, 1996). The increase in $[Ca^{2+}]_{cyt}$ was transient, returning to resting levels within a few minutes, even during low temperature caused by a slow cooling (Knight *et al.*, 1996, Plieth *et al.*, 1999). The expression of low temperature induced genes has been found to correlate with low temperature induced changes in $[Ca^{2+}]_{cyt}$ (Monroy *et al.*, 1993, Knight *et al.*, 1996, Sangwan *et al.*, 2001). Likewise, the role of $[Ca^{2+}]_{cyt}$ in low temperature signalling leading to the induction and expression of a low temperature regulated gene; *LTI78*, has been reported (Henriksson and Trewavas, 2003). Further evidence supporting a role for $Ca^{2+}$ signalling in low temperature responses was shown using a $Ca^{2+}$ ionophore. This resulted in the induction of low temperature responsive genes without any low temperature exposure (Monroy and Dhindsa, 1995, Sangwan *et al.*, 2001) and treatment with these substances affected the capacity of alfalfa plants to cold acclimate (Monroy *et al.*, 1993).

## 4.3. Heat Stress

Calcium transients in response to heat treatment have been detected using the $Ca^{2+}$-dependent luminescent protein aequorin in tobacco (Gong *et al.*, 1998). Calcium signalling inhibitors and CaM inhibitors limited survival and increased electrolyte leakage from membrane after heat treatment in maize (Gong *et al.*, 1997b). Similar $Ca^{2+}$/CaM inhibitors reduced the survival of Arabidopsis plants after a mild heat treatment and increased the oxidative damage caused by heat (Larkindale and Knight, 2002). Larkindale and Knight (2002) also suggested that some process other than heat shock protein (HSP) induction is required for survival of plants after heat stress and that $Ca^{2+}$ may be involved in a signalling pathway, acting between the perception of heat stress and this unknown process. Furthermore they suggest that a flux of $Ca^{2+}$ ions is required to switch on some mechanism by which plants prevent or repair oxidative damage caused by heating, and that a $Ca^{2+}$ flux is required for the plant to survive even mild increases in temperature (Larkindale and Knight, 2002).

## 4.4. Osmotic Stress

To combat the effects of osmotic stresses such as salinity and drought, plants express a number of genes which encode proteins of a protective nature including, for example, proteins which increase the accumulation of compatible solutes within the cell (Savouré *et al.*, 1995). Osmotic stress resulting from drought or salt stress mediated rapid elevations in $[Ca^{2+}]_{cyt}$ in Arabidopsis seedlings and these changes in $[Ca^{2+}]_{cyt}$ mediated increases in the expression of drought-induced genes encoding proteins which have a protective function (Knight *et al.*, 1997). A clear and essential interaction between $Ca^{2+}$ and ROS in short term signalling in response to hyperosmotic stress has also been reported for *Fucus serratus* embryos (Coelho *et al.*, 2002) and for hypoosmotic signalling in Fucus rhyzoid cells (Goddard *et al.*, 2000).

A role for $Ca^{2+}$ has also been proposed in the Salt-Overly-Sensitive (SOS) signal transduction pathway which induces expression of, and activates an $H^+/Na^+$ antiporter on the plasma membrane of plant cells (Zhu, 2002). The upstream-most event in this pathway is triggered by a rise in $[Ca^{2+}]_{cyt}$ upon salinity stress. Whether ROS is involved in this elevation of $[Ca^{2+}]_{cyt}$, or whether it is a direct effect of sodium is unknown.

## 5.0 CALCIUM AND ANTHROPOGENIC OXIDATIVE STRESS

### 5.1. Herbicide Treatment and Chemically Induced ROS

Herbicide treatments and other ROS producing chemicals such as $H_2O_2$ have been frequently used to investigate the role of elevated $[Ca^{2+}]_{cyt}$ in the response of plants to oxidative stress. Methyl viologen (MV), which generates $O_2^{\bullet -}$ radicals (Foyer *et al.*, 1994) and $H_2O_2$ both provide attractive model systems for studying the effects of oxidative stress in plants and much information on $Ca^{2+}$-signalling has been gained from the use of such systems. In addition, MV and $H_2O_2$ are important oxidative stresses. Methyl viologen has been widely used as a herbicide (i.e. Paraquat), and tropospheric $H_2O_2$, formed from $HO_2^{\bullet}$ is becoming an increasing problem (Gunz and Hoffmann, 1990).

The effects of oxidative stress on guard cells of *Commelina communis* were investigated by application of $H_2O_2$ (which generates $OH^{\bullet}$ radicals directly) and MV (the herbicide paraquat which generates $OH^{\bullet}$ as a secondary ROS after the formation of $O_2^{\bullet -}$) (McAinsh *et al.*, 1996). Using fluorescence ratio photometry of the $Ca^{2+}$-sensitive dye fura-2, the authors measured changes in guard cell $[Ca^{2+}]_{cyt}$ in response to these two chemicals. Both MV and $H_2O_2$ were found to inhibit stomatal opening and promote stomatal closure. EGTA completely abolished the effect of both MV and $H_2O_2$ on the opening and closing processes, when these chemicals were applied at low concentrations thus implicating a role for $Ca^{2+}$. At relatively higher levels, EGTA had little or no effect on the oxidative stress-mediated alterations in opening and closure. These data indicate that the effects of oxidative stress on guard cells were mediated by $Ca^{2+}$, except at high concentrations of MV or $H_2O_2$, when the severe doses caused a reduction in membrane integrity and guard cell viability. The oxidants MV and $H_2O_2$ therefore appear to affect stomatal behaviour in a concentration dependent manner, an affected that is mediated by $[Ca^{2+}]_{cyt}$.

The effects of chemically induced oxidative stress have also been studies on a whole plant level. For example, using recombinant aequorin techniques, it has been demonstrated that $H_2O_2$ stimulates a transient increase in whole-plant $[Ca^{2+}]_{cyt}$ in tobacco seedlings (Price *et al.*, 1994). This rapid transient elevation of $[Ca^{2+}]_{cyt}$ could be inhibited both by lanthanum and by ruthenium red, which act to hinder release from internal $Ca^{2+}$ stores, indicating that both intracellular and extracellular $Ca^{2+}$ stores contributed to the $Ca^{2+}$ elevation (Price *et al.*, 1994).

### 5.2. Calcium Signalling Resulting from Ozone Pollution

Ozone has a marked effect on $Ca^{2+}$ homeostasis of plants, leading to major $Ca^{2+}$ redistribution within various tissues. Ozone pollution resulted in a major redistribution of calcium oxalate crystals in Norway spruce (*Picea abies* ) needles (Fink, 1991). Crystals were restricted to the epidermal walls of needles in control trees, but in trees exposed to ozone they were "relocated" to the vacuole of the epidermal, subsidiary cells and guard cells of needles. Using single-cell X-ray microanalysis (Tomos *et al.*, 1994) higher concentrations of $Ca^{2+}$ were recorded within epidermal cell vacuoles from ozone treated barley plants, and the increase, compared to the control plants, was much higher in the cells closer to the stomata (Cuin, 1996). X-ray microanalysis studies of guard cells from Norway spruce showed that $Ca^{2+}$ concentrations increased as a result of ozone treatment and this was found to correlate with a decrease in stomatal functioning, thus implying that the role for $Ca^{2+}$ in the control of stomatal aperture is perturbed by oxidative stress (Le Thiec *et al.*, 1994). Disruption in the $Ca^{2+}$ distribution has also been reported for the calcicole *Leontodon hispidus*. Ozone treatment decreased the $Ca^{2+}$ levels in the trichomes, the epidermal and palisade cells. This was accompanied by a significant increase in the levels of $Ca^{2+}$ in the guard cells and other tissues of the leaf (De Silvia *et al.*, 2001). These data strongly indicate that ozone disrupts the $Ca^{2+}$ homeostasis and distribution within plants and suggests that it affects the ability of the plant to regulate $Ca^{2+}$. As many of the increases were seen within the cell vacuole, it implies a massive flux of $Ca^{2+}$ across the cytosol in response to ozone, which could have major implications for the $Ca^{2+}$ based signalling mechanism.

Using Arabidopsis and tobacco seedlings expressing the $Ca^{2+}$-reporting protein aequorin, ozone treatment was shown to induce an increase in whole-seedling $[Ca^{2+}]_{cyt}$ (Clayton *et al.*, 1999). The ozone-induced $Ca^{2+}$-signature revealed, was biphasic; an initial spike followed by a more prolonged elevation. The second peak was observed only a concentrations of ozone 70 ppb or above and the magnitude of the ozone induced $Ca^{2+}$-signature was highly dose-dependent. A functional dissection of the of the ozone-$Ca^{2+}$ signature of the two peaks showed that although a role could not be assigned for the first peak, the second peak of the ozone-$Ca^{2+}$ signature was required for the ozone induced expression of the antioxidant enzyme glutathione *S*-transferase (GST) (Clayton *et al.*, 1999). Similar biphasic increases in $[Ca^{2+}]_{cyt}$ in Arabidopsis have also been recently reported (Evans *et al.*, 2005) and may be important in coding the specificity of the oxidative stress (discussed below).

## 6. THE AFFECTS OF OXIDATIVE STRESS ON ION CHANNEL ACTIVITY

It has been proposed that ROS activation of $Ca^{2+}$ channels may be a central step in many ROS mediated processes which leads to the elevation of $[Ca^{2+}]_{cyt}$ (Mori and Schroeder, 2004). Early evidence for the affect of ozone on $Ca^{2+}$ channel activity came from studies of $Ca^{2+}$ transport in plasma membrane vesicles from pinto beans (*Phaseolus vulgaris*) where ozone was found to induce $^{45}Ca^{2+}$ efflux from inside-out vesicles (Castillo and Heath, 1990). This corresponded to a $Ca^{2+}$ influx into cells across the plasma membrane, indicating a direct effect of ozone on the $Ca^{2+}$ permeability of the plant cell membrane.

More recently, an effect of ozone on guard cell ion channels in broad bean (*Vicia faba*) using the patch clamp technique has been demonstrated (Torsethaugen *et al.*, 1999). Treatment of guard cell protoplasts with ozone resulted in inhibition of the plasma membrane inward-rectifying potassium channel ($K^+_{in}$) in guard cells which was associated with a disruption to stomatal functioning. Such channels mediate $K^+$ uptake required for stomatal opening and are known to be inhibited by $Ca^{2+}$ (Schroeder and Hagiwara, 1989). Thus, Torsethaugen *et al.* (1999) suggested that the action of ozone on $K^+_{in}$ may be a consequence of ozone-induced increases in guard cell $[Ca^{2+}]_{cyt}$, although they did not rule out a direct oxidation of the channel proteins.

Pei *et al.* (2000), used patch clamping to examine the effects of $H_2O_2$ on guard cell ion channels. They identified a $Ca^{2+}$ permeable channel in the plasma membrane of Arabidopsis guard cells that was activated by $H_2O_2$. Importantly, the activation of this channel was found to precede the $H_2O_2$ induced increase in guard cell $[Ca^{2+}]_{cyt}$, suggesting a role for this channel in mediating $Ca^{2+}$ influx in response to exposure to $H_2O_2$. Such a direct role of ROS on $Ca^{2+}$ channel function is supported by patch-clamp recordings of the activation of $Ca^{2+}$-permeable non-selective cation channels by $OH^\bullet$ in Arabidopsis root cells (Demidchik *et al.*, 2003), and by using non-invasive ion flux measurements (the MIFE technique, Newman 2001) of $OH^\bullet$ activated $Ca^{2+}$ influxes into Arabidopsis roots (Demidchik *et al.*, 2003).

## 7. CALCIUM SIGNATURES - ENCODING SPECIFICITY

As stated in the introduction, $Ca^{2+}$-based signalling systems are composed of a receptor, a system for generating the increase in $[Ca^{2+}]_{cyt}$, downstream components that are capable of reacting to the increase in $[Ca^{2+}]_{cyt}$, and other cellular systems responsible for returning $[Ca^{2+}]_{cyt}$ to its pre-stimulus level. The important role of $Ca^{2+}$-signalling in the transduction of environmental change into a plant response has been documented for a wide range of stimuli, including many that are additional to those resulting from oxidative stress. The fact that so many different stimuli can all generate increases in $[Ca^{2+}]_{cyt}$ raises the question as to how this common use of $Ca^{2+}$ can give such varied and specific responses.

There is now evidence that the $Ca^{2+}$ source (i.e. intra- or extra-cellular, Fig. 5), temporal and spatial nature and the amplitude of $[Ca^{2+}]_{cyt}$ changes - the so called "calcium signature" caused by a given stimuli contributes to the specificity of the response (Webb *et al.*, 1996, Trewavas, 1999, Knight, 2000). Reported different types of $[Ca^{2+}]_{cyt}$ elevations that occur in plants as a result of various stimuli can take the form of spikes, hot-spots, puffs, sparks, oscillations and waves. These are important in the stimulus-specific $[Ca^{2+}]_{cyt}$ signatures involved in encoding information necessary for eliciting the appropriate physiological response in plant cells (McAinsh and Hetherington, 1998, Ng and McAinsh, 2003). Although there are strong indications in mammals that $[Ca^{2+}]_{cyt}$ signatures are responsible for the activation of specific genes (Dolmetsch *et al.*, 1997, 1998, Li *et al.*, 1998), the involvement of $[Ca^{2+}]_{cyt}$ signatures in plants is not yet fully conclusive (see Scrase-Field and Knight, 2003). Thus, a major outstanding question regarding $Ca^{2+}$-signalling is not what stimuli trigger $Ca^{2+}$ signal transduction, but how the necessary specificity for the adaptive response is encoded. This final section discusses some of the recent work concerning how the calcium signal may code for the particular oxidative stress and send the "correct" signal for the "appropriate" response for the currently imposed stress.

An insight into the $[Ca^{2+}]_{cyt}$ signatures required for specific cellular responses has been obtained by correlating the perturbations of $[Ca^{2+}]_{cyt}$ with applied oxidative stresses. This has enabled investigations into the effect of oxidative stress on the $[Ca^{2+}]_{cyt}$ response and attempts to address the characteristics of the resulting $Ca^{2+}$ signatures often in relation to the final downstream response of the plant. From these studies, evidence of specific patterns of alterations in $[Ca^{2+}]_{cyt}$ in response to different stimuli are beginning to emerge, although a final conclusion is still to be drawn.

The response of intact Arabidopsis plants to ozone stimulates a characteristic biphasic $Ca^{2+}$ signature that is dose dependent (Clayton *et al.*, 1999). The first phase was a pronounced $Ca^{2+}$ spike that peaked within minutes of exposure to ozone, returning to basal levels within 7 - 10 minutes. This was followed by a second gradual increase in $[Ca^{2+}]_{cyt}$ that did not reach the absolute concentration of the initial spike, but was perpetuated for up to 60 minutes. Since it was shown that only the second increase in $[Ca^{2+}]_{cyt}$ was $La^{3+}$ sensitive (a $Ca^{2+}$ channel blocker), it suggests that the latter increase in $[Ca^{2+}]_{cyt}$, but not the initial spike resulted from extracellular sources. Downstream of the $Ca^{2+}$ signals, expression of *GST* was triggered, and it was only the second, sustained increase in $[Ca^{2+}]_{cyt}$ that was required for *GST*-expression.

Studies on the response of aequorin-transformed tobacco cell cultures to hypo-osmotic shock (Cessna *et al.*, 1998) have demonstrated that the source of mobilised $Ca^{2+}$ can vary according to the stimulus applied. Interestingly, in this study the source of the $Ca^{2+}$ was the complete reverse of that seen in ozone treated Arabidopsis (Clayton *et al.*, 1999). Again, a biphasic increase in $[Ca^{2+}]_{cyt}$ was observed. A $[Ca^{2+}]_{cyt}$ spike lasting about 30 seconds resulted from an almost instantaneous influx of extracellular $Ca^{2+}$. This was followed by a second spike of $[Ca^{2+}]_{cyt}$ that lasted a few minutes, reaching approximately the same concentration as the first increase. Since this second phase of $[Ca^{2+}]_{cyt}$ increase was sensitive to heparin, it suggested a role for intracellular $Ca^{2+}$ pools that may be initially triggered by calcium-induced-calcium-release (CICR), or similar mechanisms. The implication of this is that there was some sort of communication between the two $Ca^{2+}$ release pathways induced by osmotic shock, since perturbations that altered the first $Ca^{2+}$ transient had knock-on effect on the second phase of changes in $[Ca^{2+}]_{cyt}$. This suggested that there was a sequential opening of plasma membrane located $Ca^{2+}$ channels followed by release from intracellular stores and that these $Ca^{2+}$ release pathways can somehow compensate for each other.

Kiegle *et al* (2000), utilised aequorin targeted to different tissue types of Arabidopsis roots and revealed that $[Ca^{2+}]_{cyt}$ increases induced by drought, salt and cold vary with cell type. This targeted expression of aequorin revealed a new level of complexity in root $Ca^{2+}$ signalling. Although the responses to all the imposed stresses involved a biphasic response of some sort, the response of different tissues to the same stimulus, and also the same tissue to different stimuli were different. In particular, the endodermis and pericycle responded to osmotic and salt stress by displaying prolonged oscillations in $[Ca^{2+}]_{cyt}$ that were distinct from the responses of other cell types tested. This implies that there could be specific roles for certain cell types in the detection and/or response to these stimuli and suggests that different cell types generate particular $Ca^{2+}$ signatures that are dependent on their ability to respond to particular signals.

Independent sensing and/or signal transduction mechanisms were found to exist for $H_2O_2$ in the roots and aerial tissues of Arabidopsis seedlings (Rentel and Knight, 2004). Again, at the whole plant level, a characteristic biphasic $Ca^{2+}$ response to $H_2O_2$ was generally seen. The first peak was confined to the green tissue. In contrast, there was no second peak in the cotyledons and leaves of Arabidopsis seedlings in response to $H_2O_2$, the second peak was confined to the roots. The reverse pattern was found after ozone was applied. A second peak of $[Ca^{2+}]_{cyt}$ was found in the leaves, but no second peak within the roots (Evans *et al.*, 2005). Thus, while these two ROS producing oxidative stresses both elicit biphasic $Ca^{2+}$ elevations in Arabidopsis on a whole plant basis, there are significant tissue specific differences in the responses depending on the oxidative stress imposed. This leads one to speculate that the seedlings are able to distinguish between these two abiotic stress and that multiple perception and/or signal transduction mechanisms exist in Arabidopsis for responding to ROS.

The magnitude of the ozone-induced $Ca^{2+}$ response is dependent on the ozone concentration (Clayton *et al.*, 1999). When the rate of increase of ozone is slow, the first peak in the biphasic $Ca^{2+}$ response is diminished or abolished, while the second peak is unaffected (Evans *et al.*, 2005). A similar response as a result of cold stress where the induced increase in $[Ca^{2+}]_{cyt}$ was a function of the cooling rate, with a greater response during faster cooling regimes has also been reported (Plieth *et al.*, 1999). This flexibility in the $Ca^{2+}$ response would potentially enable plants to switch on different defence responses to ozone, depending on the speed at which the stress was imposed.

In addition to different $Ca^{2+}$ signatures resulting from the imposition of different oxidative stresses, there is also considerable evidence that these responses are modified by previous $Ca^{2+}$-signalling experiences. A diminished $[Ca^{2+}]_{cyt}$ elevation upon repetitive stimulation and/or a refractory period during which an increase in $[Ca^{2+}]_{cyt}$ cannot be elicited by the same abiotic stress is commonly observed. For example, plant cells challenged with $H_2O_2$ failed to respond again to $H_2O_2$ for several hours (Price *et al.*, 1994). An anoxic treatment reduced the magnitude of the initial $[Ca^{2+}]_{cyt}$ peak and delayed the sustained elevation of $[Ca^{2+}]_{cyt}$ in response to a second anoxic treatment (Sedbrook *et al.*, 1996), and the magnitude of $[Ca^{2+}]_{cyt}$ perturbations were diminished by continued chill- and re-warming (Plieth *et al.*, 1999). Such observations may indicate a desensitisation of signalling cascades. By contrast, the magnitude of the second $[Ca^{2+}]_{cyt}$ elevation, thought to be from vacuolar $Ca^{2+}$ observed during slow cooling, was increased following a period of cold acclimation (Knight and Knight, 2000), and a pre-treatment with mannitol, increased the magnitude of the $[Ca^{2+}]_{cyt}$ perturbations subsequently induced by hyperosmotic shock (Knight *et al.*, 1998).

As well as a modification of the $[Ca^{2+}]_{cyt}$ signature by a previous exposure to a particular oxidative episode, there is also experimental evidence that $[Ca^{2+}]_{cyt}$ signatures elicited by one environmental challenged can be modified by prior exposure to a contrasting one. For example, the magnitude of the $[Ca^{2+}]_{cyt}$ perturbations in response to osmotic stress was reduced by prior exposure to hypoosmotic stress (Knight *et al.*, 1997) and the magnitude of the $[Ca^{2+}]_{cyt}$ perturbations in response to hyperosmotic stress was reduced by a prior exposure to $H_2O_2$ (Knight *et al.*, 1998). Such observations may imply cross-talk between the signalling mechanisms. However, prior $H_2O_2$ treatment did not affect the $[Ca^{2+}]_{cyt}$ transient observed upon cold shock or touch (Knight *et al.*, 1998, Price *et al.*, 1994), suggesting that alternative signalling cascades and/or $Ca^{2+}$ stores and channels are recruited by these challenges. Similarly during the refractory period following heat shock, during which additional heat shocks fail to elevate $[Ca^{2+}]_{cyt}$, $[Ca^{2+}]_{cyt}$ can be increased by cold shock or mechanical perturbation (Gong *et al.*, 1998), and during the refractory period following wind-induced motion, cells can still raise $[Ca^{2+}]_{cyt}$ in response to cold-shock (Knight *et al.*, 1992), suggesting different $Ca^{2+}$-signalling mechanisms are involved and these signalling mechanisms vary with different stimuli.

Thus, pre-treatment with osmotic stress or oxidative stress causes a modification of the $Ca^{2+}$ response elicited by a subsequent challenge with a stress event. This suggests that information relating to environmental signals may be transduced differently depending on the stress history of the plant. A reduction in the magnitude of the $Ca^{2+}$ response to a stress event was also found to correlate with a reduction in stress-induced gene expression. Changes in the signal-transduction machinery which result in a $Ca^{2+}$ response may therefore be stored as a type of chemical "memory" of a previous stress event. This information is retrieved upon subsequent stimulation, resulting in an altered $Ca^{2+}$ signal and ensuring that the end responses are adjusted accordingly (Knight *et al.*, 1998). Such subtle alterations in $Ca^{2+}$ responses have been shown to have profound effects on the downstream events mediated by $Ca^{2+}$ (Dolmetsch *et al.*, 1997). The concept of signal information storage in plants has been proposed previously (Verdus *et al.*, 1996) and $Ca^{2+}$ has been implicated (Verdus *et al.*, 1997). Indeed the $[Ca^{2+}]_{cyt}$ signalling network has been likened to a basic cellular "memory" (Trewavas, 1999).

The above examples illustrate how different abiotic stresses results in often subtle, but distinct $Ca^{2+}$ signatures within a plant and such studies highlight the importance of $Ca^{2+}$ signalling in a wide range of plant responses to changes in environmental conditions. Nevertheless, it remains to be elucidated how these different chemical messages are interpreted downstream of the $Ca^{2+}$ signal, the receptors and signal transduction cascades involved and also, importantly, how they affect the final response of the plant.

## 8. FUTURE PERSPECTIVES

It is now well established knowledge that the reactive oxygen species produced as a result of oxidative stress events result in the elevation of cytosolic calcium concentrations within a plant cell. This increase has been shown to trigger downstream responses that alter and/or induce gene expression. This results in changes in cellular metabolism designed to prevent or minimise the deleterious effects of the abiotic stress. However, how the information regarding the identification and the appropriate responses to a particular oxidative stress is encoded into the calcium signal has not yet been fully clarified.

This review has presented the growing recognition of the importance of the spatial-temporal-concentration variations in $Ca^{2+}$ signal, the so-called "calcium signature" that results from the response to various abiotic stresses. It is currently thought that each $Ca^{2+}$ signature generated by a particular stimulus may be unique to a specific response, and although this has not been covered here, there are a number of examples where it has been shown that different phases of a calcium signature are required for triggering particular responses (see Rudd and Franklin-Tong, 2001, White and Broadley, 2003). This clearly suggests that certain portions of the $Ca^{2+}$ message contribute to specifying in some way the signal-transduction cascade triggered. From the above examples of different $[Ca^{2+}]_{cyt}$ signatures to different stress events, it can be seen that to investigate all the possible signatures would be a massive undertaking. Moreover, emerging data from some systems indicate that there is a further layer of complexity and that different cell types within a single plant can exhibit different $[Ca^{2+}]_{cyt}$ responses to the same signal, indeed even individual cells of the same cell type can respond differently. This would provide a means of "fine-tuning" which may be dependent on the state of the responsiveness of individual cells. Dissection and "interpretation" of such mechanisms of signalling have not even commenced, thus opening up the possibility for almost a whole new area of research. In addition, we are also beginning to uncover a complex network, with evidence of "cross-talk" and plant "memory" of previous stress events, a fascinating area, about which we know and understand so little. Despite much progress in the last few years, there are still many gaps in our knowledge.

How these calcium signatures can be so precisely controlled at an intracellular level is not currently known, although many of the cellular transport systems for $Ca^{2+}$ at the cellular level has now been identified, at least for Arabidopsis (see White and Broadly 2003 for review). This will undoubtedly be an area for future research, both for oxidative stress and other stimuli leading to elevations in $[Ca^{2+}]_{cyt}$. Nevertheless, the generation of the appropriate $Ca^{2+}$ signature from a particular stimuli and how the signalling pathways produce the suitable response to mitigate or alleviate the oxidative stress situation will certainly be an area for future research.

**Acknowledgements**

The author wishes to thank Sergey Shabala for the excellent help and advice in preparing the manuscript and Michael Hansen and David Cuin for critical reading of the text.

## REFERENCES


Alscher R.G., Hess J.L. (1993) *Antioxidants in higher plants.* Boca Raton: CRC Press.

Alscher R.G., Donahue J.L., Cramer C.L. (1997) Reactive oxygen species and antioxidants: Relationships in green cells. Physiologia Plantarum. **100:** 224-233).

Asada K. (1996) Radical production and Scavenging in chloroplasts. **In:** N.R. Baker (eds). *Photosynthesis and the Environments,* Kluwer Academic Publishers, Dordrecht, pp. 123-150.

Asada K., Endo T., Mano J., Miyake C. (1998) Molecular for relaxation and protection from light stress. **In:** N. Murata, K. Satoh. (eds). *Stress Responses of Photosynthetic Organisms.* Elsevier, Amsterdam. pp. 37-52.

Baum G., Long J.C., Jenkins G.I., Trewavas A.J. (1999) Stimulation of the blue light receptor NPH causes a transient increase in cytosolic $Ca^{2+}$. *Proceedings of the National Academy of Sciences,* USA, **96:** 13554-13559.

Blume B., Nurnberger T., Nass N., Scheel D. (2000) Receptor-mediated increase in cytoplasmic free calcium required for activation of pathogen defense in parsley. *The Plant Cell,* **12:** 1425-1440.

Bornman J.F., Sundby-Emanuelsson C. (1995) Response of plants to UV-B radiation: some biochemical and physiological effects. **In:** N. Smirnoff (ed). *Environment and Plant Metabolism: Flexibility and Acclimation.* Bios Scientific, Oxford. pp. 245-262.

Bowler C., Van Camp W., Van Montagu M., Inzé D. (1994) Superoxide dismutase in plants. *Critical Review of Plant Science,* **13:** 199-218.

Bowler C., Van Montagu M., Inzé D. (1992) Superoxide dismutase and stress tolerance. *Annual Review in Plant Physiology and Plant Molecular Biology,* **43:** 83-116.

Bradley D.J., Kjellbom P., Lamb C.J. (1992) Elicitor- and wound-induced oxidative cross-linking of a proline-rich plant cell-wall protein: A novel, rapid defense response. *Cell,* **70:** 21-30.

Byvoet P., Balis J.U., Shelley S.A., Montgomery M.R., Barber M.J. (1995) Detection of hydroxyl radicals upon interaction of ozone with aqueous media of extracellular surfactant: The role of trace iron. *Archives of Biochemistry and Biophysics,* **319:** 464-469.

Castillo F.J., Heath R.L. (1990) $Ca^{2+}$ transport in membrane vesicles from Pinto bean leaves and its alteration after ozone exposure. *Plant Physiology,* **56:** 723-727.

Cessna S.G., Chandra S., Low P.S. (1998) Hypo-osmotic shock of tobacco cells stimulates $Ca^{2+}$ fluxes driving first from external sources and then internal $Ca^{2+}$ stores. *Journal of Biological Chemistry,* **273:** 27286-27291.

Charles S.A., Halliwell B. (1980) Effect of hydrogen peroxide on spinach (*Spinacea oleracea*) chloroplast fructose bisphosphate. *Biochemical Journal,* **189:** 373-376.

Cheeseman J.M. (1988) Mechanisms of salinity tolerance in plants. *Plant Physiology,* **87:** 547-550.

Chen Z., Silva H., Klessig D.F. (1993) Active oxygen species in the induction of plant systemic acquired resistance by salicylic acid. *Science,* **262:** 1883-1886.

Chen X., Wang B., Wu R. (1995) A gibberellin-stimulated ubiquitin-conjugated enzyme gene is involved in alpha-amylase gene expression in rice aleurone. *Plant Molecular Biology,* **29:** 787-795.

Cheng S.H., Willmann M.R., Chen H.C., Sheen J. (2002) Calcium signaling through protein kinases. The Arabidopsis calcium-dependent protein kinase gene family. *Plant Physiology,* **129:** 469-485.

Christie J.M., Jenkins G.I. (1996) Distinct UV-B and UV-A/blue light signal transduction pathways induce chalcone synthase gene expression in *Arabidopsis* cells. *The Plant Cell,* **8:** 1555-1567.

Clayton H., Knight M.R., Knight H., McAinsh M.R., Hetherington A.M. (1999) Dissection of the ozone-induced calcium signature. *The Plant Journal,* **17:** 575-579.

Cleveland W.S., Kleiner B., McRae Y.E., Warner J.L. (1976) Photochemical air pollution: Transport from New York City area into Connecticut and Massachusetts. *Science*, **191:** 179-181.

Coelho S.M., Taylor A.R., Ryan K.P., Sousa-Pinto I., Brown M.T., Brownlee C. (2002) Spatiotemporal patterning of reactive oxygen production and $Ca^{2+}$ wave propagation in *Fucus* rhizoid cells. *The Plant Cell*, **14:** 2369-2381.

Cox P.A., Eggleton A.E.J., Derwent R.G., Lovelock J.E., Pack D.H. (1975) Long range transport of photochemical ozone in North Western Europe. *Nature*, **255:** 118-121.

Cross C.E., van der Vliet A., Louie S., Thiele J.J., Halliwell B. (1998) Oxidative stress and antioxidants at biosurfaces: plants, skin and respiratory tract surfaces. *Environmental Health Perspectives*, **106:** 1241-1251.

Cuin T.A. (1996) The effect of ozone-fumigation on single-cell water and solutes relations of barley leaves. PhD thesis, University of Wales, Bangor, UK.

Dat J.F., Lopez-Delgado H., Foyer C.H., Scott I.M. (1998) Parallel changes in $H_2O_2$ and catalase during thermotolerance induced by salicylic acid or heat acclimation in mustard seedling. *Plant Physiology*, **116:** 1351-1357.

Demidchik V., Shabala S.N., Coutts K.B., Tester M.A., Davis J.M. (2003) Free oxygen radicals regulation plasma membrane $Ca^{2+}$ and $K^+$-permeable channels in plant root cells. *Journal of Cell Science*, **116:** 81-88.

De Silva D.L.R., Mansfield T.A., McAinsh M.R. (2001) Changes in stomatal behaviour in the calcicole *Leontodon hispidus* due to the disruption by ozone of the regulation of apoplastic $Ca^{2+}$ by trichomes. *Planta*, 214: 158-162.

Dolmetsch R.E., Lewis R.S., Goodnow C.C., Healy J.I. (1997) Differential activation of transcription factors induced by Ca**2+** response amplitude and duration. *Nature,* **386:** 855-858.

Dolmetsch R.E., Xu K., Lewis R.S. (1998) Calcium oscillations increase the efficiency and specificity of gene expression. *Nature*, **392:** 933-936.

Dröge W. (2002) Free radicals in the physiological control of cell function. Physiological Reviews. **82:** 47-92.

Ehrhardt D.W., Wais R., Long S.R. (1996) Calcium spiking in plant root hairs responding to rhizobium nodulation signals. Cell. **85:** 673-81.

Elstner E.F. (1991) Mechanisms of oxygen activation in different compartments of plant cells. **In:** E.J. Pelland, K.L. Steffen (eds). *Active Oxygen/Oxidative Stress and Plant Metabolism*. American Society of Plant Physiologists., Rockville. pp. 13-25.

Evans N.H., McAinsh M.R., Hetherington A.M., Knight M.R. (2005) ROS perception in Arabidopsis thaliana: the ozone-induced calcium response. *The Plant Journal*, **41:** 615-626.

Fink S. (1991) Unusual patterns in the distribution of calcium oxalate in spruce needles and their possible relationships to the impact of pollutants. *New Phytologist*, **119:** 41-51.

Foyer C.H. (2002) The contribution of photosynthetic oxygen metabolism to oxidative stress in plants. **In:** D. Inzé, M. Van Montagu (eds). *Oxidative Stress in Plants*. Taylor and Francis, London. pp. 33-68.

Foyer C.H., Harbinson J. (1994) Oxygen metabolism and the regulation of photosynthetic electron flow. **In:** C.H. Foyer and P.M. Mullineaux (eds). *Causes of Photooxidative Stress and Amelioration of Defense Systems in Plants.* CRC Press, Boca Raton. pp. 1-42.

Foyer C.H., Lelandais M., Kunert K. J. (1994) Photooxidative stress in plants. *Physiologia Plantarum*, **92:** 696-717.

Foyer C.H., Lopez-Delgado H., Dat J.F., Scott I.M. (1997) Hydrogen peroxide- and glutathione-associated mechanisms of acclimatory stress tolerance and signalling. *Physiologia Plantarum*, **100:** 241-254.

Foyer C.H., Noctor G. (2003) REDOX sensing and signalling associated with reactive oxygen in chloroplasts, peroxisomes and mitochondria. *Physiologia Plantarum*, **119:** 355-364.

Frohnmeyer H., Bowler C., Schäfer E. (1997) Evidence for some signal transduction elements involved in UV-light-dependent responses in parsley protoplasts. *Journal of Experimental Botany*, **48:** 739-750.*et al.*, 1997,

Frohnmeyer H., Bowler C., Zhu J.K., Yamagata H., Schafer E., Chua N.H. (1998) Different roles for calcium and calmodulin in phytochrome- and UV-regulated expression of chalcone synthase. *The Plant Journal*, **13:** 763-772.*et al.*, 1998

Gallego S.M., Benvades M.P., Tomaro M.I. (1996) Oxidative damage caused by cadmium chloride in sunflower (*Helianthus annuus* L.) plants. *Journal of Experimental Botany*, **58:** 41-52.

Gianinitti A., Lorenzoni C., Marocco A. (1993) Changes in superoxide dismutatase and catalase activities in response to low temperature. *Journal of Genetics and Breeding*, **47:** 353-356.

Gilroy S., Jones R.L. (1992) Gibberellic acid and abscisic acid coordinately regulate cytoplasmic calcium and secretory activity in barley aleurone protoplasts. *Proceedings of the National Academy of Sciences*, USA, **89:** 3591-2209.

Goddard H., Manison N.F.H., Tomos D., Brownlee C. (2000) Elemental propagation of calcium signals in response-specific patterns determined by environmental stimulus strength. *Proceedings of the National Academy of Sciences,* USA, **97:** 1932-1937

Gomez L.D., Noctor G., Knight M.R., Foyer C.H. (2004) Regulation of calcium signalling and gene expression by glutathione. *Journal of Experimental Botany,* **55:** 1851-1859.

Gong M., Chen S.N., Song Y.Q., Li Z.G. (1997b) Effect of calcium and calmodulin on intrinsic heat tolerance in relation to antioxidant systems in maize seedlings. *Australian Journal of Plant Physiology*, **24:** 371-379.

Gong M., Li Y.J., Dai X., Tian M., Li Z.G. (1997a) Involvement of calcium and calmodulin in the acquisition of heat-shock induced thermotolerance in maize seedling. *Journal of Plant Physiology*, **150:** 615-621.

Gong M., van der Luit A.H., Knight M.R., Trewavas A.J. (1998) Heat-shock-induced changes in intracellular $Ca^{2+}$ level in tobacco seedlings in relation to thermotolerance. *Plant Physiology*, **116:** 429-437.

Greenberg B.M., Gaba V., Canaani O., Malkin S., Edelman M. (1989) Separate photosensitizers mediate degradation of the 32 kDa reaction centre II protein in visible and UV spectral regions. *Proceedings of the National Academy of Science,* USA, **86:** 6616-6620.

Gueta-Dahan Y., Yaniv Z., Zilinskas B.A., Ben-Hayyim G. (1997) Salt and oxidative stress: similar and specific responses and their relation to salt tolerance in *citrus*. *Planta*, **203:** 460-469.

Gunz D.W., Hoffmann M.R. (1990) Atmospheric chemistry of peroxides: a review. *Atmosphere and Environment*, **24A:** 1601-1633.

Halliwell B., Gutteridge J.M.C. (1999) *Free Radicals in Biology and Medicine*, 3rd ed., Oxford University Press, Oxford.

Harmon A.C., Gribskov M., Gubrium E., Harper J.F. (2001) The CDPK superfamily of protein kinases. *New Phytologist*, **151:** 175-183.

Havauz M., Davaud A. (1994) Photoinhibition of photosynthesis in chilled potato leaves is not correlated with a loss of photosystem II activity. Preferential activation of photosystem I. *Photosynthesis Research*, **40:** 75-92.

Heagle S.A. (1989) Ozone and crop yield. *Annual Review of Phytopathology*, **27:** 397-423.

Heck W.W., Taylor O.C., Adams R., Bingham G., Miller J., Preston E., Weinstein L. (1982) Assessment of crop loss from ozone. *Journal of Air Pollution Control Association*, **32:** 353-362.

Henriksson K.N., Trewavas A.J. (2003) The effect of short-term lo-temperature treatments on gene expression in *Arabidopsis* correlates with changes in intracellular Ca2+ levels. *Plant Cell and Environment*, **26:** 485-496.

Hepler P.K., Wayne R.O. (1985) Calcium and plant development. *Annual Review of Plant Physiology*, **36:** 397-439.

Hernandez J.A., Corpas F.J., Gomez L.A., del Rio L.A., Sevilla F. (1993) Salt induced oxidative stress mediated by activated oxygen species in pea leaf mitochondria. *Plant Physiology*, **89:** 103-110.

Hernandez J.A., Olmos E., Corpas F.J., Sevilla F., del Rio L.A. (1995) Salt induced oxidative stress in chloroplasts of pea plants. *Plant Science*, **105:** 151-167.

Hetherington A.M., Brownlee C. (2004) The generation of $Ca^{2+}$ signals in plants. *Annual Review of Plant Biology*, **55:** 401-427.

Hetherington S.E., He J., Smillie R.M. (1989) Photoinhibition at low temperature in chilling-sensitive and -resistant plants. *Plant Physiology*, **90:** 1609-1615.

Hideg E., Sass L., Barbato R., Vass I. (1993) Inactivation of photosynthetic oxygen evolution by UV-B irradiation: a thermoluminescence study. *Photosynthesis Research*, **38:** 455-462.

Hideg E., Vass I. (1996) UV-B induced free radical production in plant leaves and isolated thylakoid membranes. *Plant Science*, **115:** 251-260.

Howarth C.J., Ougham H.J. (1993) Gene expression under temperature stress. *New Phytologist*, **125:** 1-26.

Inzé D., Van Montagu M. (1995) Oxidative stress in plants. *Current Opinion in Biotechnology*, **6:** 153-158.

Kampfenkel K, Van Montagu M., Inzé D. (1995) Effects of iron excess on *Nicotiana plumbagifolia* plants. *Plant Physiology*, **107:** 725-735/

Kanofsky J.R., Sima P. (1991) Singlet oxygen production from the reactions of ozone with biological molecules. *Journal of Biological Chemistry*, **266:** 9039-9042.

Katoh S. San Pietro A. (1967) Ascorbate-supported NADP photoreduction by heated *Euglena* chloroplasts. *Archives of Biochemistry and Biophysics*, **122:** 144-152.

Kerr JB., McElroy C.T. (1993) Evidence for large upwards trends of ultraviolet-B radiation linked to ozone depletion. *Science*, **262:** 1032-1034.

Kiegle E., Moore C.A., Haseloff J., Tester M.A., Knight M.R. (2000) Cell-type-specific calcium responses to drought, salt and cold in the *Arabidopsis* root. *The Plant Journal*, **23:** 267-278.

Knight H. (2000) Calcium signaling during abiotic stress in plants. *International Review of Cytology*, **195:** 269-324.

Knight H., Brandt S., Knight M.R. (1998) A history of stress alters drought calcium signalling pathways in Arabidopsis. *The Plant Journal*, **16:** 681-687.

Knight M., Campbell A., Trewavas A. (1991) Transgenic plant aequorin reports the effect of touch and cold-shock and elicitors of cytoplasmic calcium. *Nature*, **352:** 524-526.

Knight H., Knight M.R. (2000) Imaging spatial and cellular characteristics of low temperature calcium signature after cold acclimation in *Arabidopsis*. *Journal of Experimental Botany*. **51:** 1679-1686.

Knight M.R., Smith S.M., Trewavas A.J. (1992) Wind-induced plant motion immediately increases cytosolic calcium. *Proceedings of the National Academy of Sciences*, USA, **89:** 4967-4971.

Knight H., Trewavas A.J., Knight M.R. (1996) Cold calcium signalling in Arabidopsis involves two cellular pools and a change in calcium signature after acclimation. *The Plant Cell*, **8:** 489-503.

Knight H., Trewavas A.J., Knight M.R. (1997) Calcium signalling in *Arabidopsis thaliana* in response to drought and salinity. *The Plant Journal*, **12:** 1067-1078.

Kraus T.E., Fletcher R.A. (1994) Paclobutrazol protects wheat seedlings from heat and paraquat injury. Is detoxification of active oxygen involved? *Plant Cell Physiology*, **35:** 45-52.

Larkindale J., Knight M.R. (2002) Protection against heat stress-induced oxidative damage in Arabidopsis involves calcium, abscisic acid, ethylene , and salicylic acid. *Plant Physiology*, **128:** 682-695.

Le Thiec D., Rose C., Garrec J.P., Laffray D., Louguet P., Galaup S., Loosveldt P. (1994) Alteration of element contents in guard cells of Norway spruce (*Picea abies*) subjected to ozone fumigation and (or) water stress: X-ray microanalysis study. *Canadian Journal of Botany*, **72:** 86-92.

Levine A., Tenhaken R., Dixon R., Lamb C. (1994) $H_2O_2$ from the oxidative burst orchestrates the plant hypersensitive disease resistance response. *Cell*, **79:** 583-593.

Li W.-H., Llopsis J., Whitney M., Zlokarnik G., Tsien R.Y. (1998) Cell-permeant caged $InsP_3$ ester shows that $Ca^{2+}$ spike frequency can optimize gene expression. *Nature*, **392:** 936-941.

Lopez R., Vansuyt G., Casse-Delbart F., Fourcroy P. (1996) Ascorbate peroxidase activity, not the mRNA level, is enhanced in salt-stressed *Raphamus satica* plants. *Physiologia Plantarum*, **97:** 13-20.

Malan C., Greyling M.M., Gressel J. (1990) Correlation between Cu/Zn superoxide dismutase and glutathione reductase, and environmental and xenobiotic stress tolerance in maize inbreds. *Plant Science*, **69:** 157-166.

Mano J., Ohno C., Asada K. (1998) Loss of $H_2O_2$-scavenging capacity due in inactivation of ascorbate peroxidase in methylviologen-fed leaves, an estimation by electron spin resonance spectrometry. **In:** G. Garab (ed). *Photosynthesis, Mechanisms and Effects*, **Vol. V.** Kluwer Academic Publishers, Dordrecht. pp. 3909-3912.

Mauzerall D.L., Wang X. (2001) Protecting agricultural crops from the effects of tropospheric ozone exposure: reconciling science and standard setting in the United States, Europe and Asia. *Annual Review of Energy and the Environment*, **26:** 237-268.

McAinsh M.R., Brownlee C., Hetherington A.M. (1992) Visualising changes in cytosolic free calcium during the response of stomatal guard cells to abscisic acid. *The Plant Cell*, **4:** 1113-1122.

McAinsh M.R., Clayton H., Mansfield T.A., Hetherington A.M. (1996) Changes in stomatal behaviour and guard cell cytosolic free calcium in response to oxidative stress. *Plant Physiology*, **111:** 1031-1042.

McAinsh M.R., Hetherington A.M. (1998) Encoding specificity in calcium signalling systems. *Trends in Plant Science*, **3:** 32-36.

Mehlhorn H. (1990) Ethylene-promoted ascorbate peroxidase activity protects plants against hydrogen peroxide, ozone and paraquat. *Plant Cell and Environment*, **13:** 971-976.

Mishra R.K., Singhal G.S. (1992) Function of photosynthetic apparatus of intact wheat leaves under high light and heat stress and its relationship with peroxidation of thylakoid lipids. *Plant Physiology*, **98:** 1-6.

Miszalski Z., Slesak I., Niewiadomska E., Baczekkwinta R., Lüttge U., Ratajczak R. (1998) Subcellular localization and stress response of superoxide-dismutase isoforms from leaves in the C-3-CAM intermediate halophyte *Mesembryanthemum crystallinum* L. *Plant Cell and Environment*, **21:** 169-179.

Mithöfer A., Ebel J., Bhagwat A.A., Boller T., Neuhaus-Url G. (1999) Transgenic aequorin monitors cytosolic calcium transient in soybean cells challenged with a-glucan or chitin elicitors. *Planta*, **207:** 566-574.

Mittler R., Zilinskas B.A. (1994) Regulation of pea cytosolic ascorbate peroxidase and other antioxidant enzymes during the progression of drought stress and following recovery from drought. *The Plant Journal*, **5:** 397-405.

Mohanty N., Murthy S.D.S., Mohanty P. (1987) Reversal of heat induced alterations in photochemical activities in wheat primary leaves. *Photosynthesis Research*, **14:** 259-267.

Monroy A.F., Dhindsa R.S. (1995) Low-temperature signal transduction: Induction of cold acclimation-specific genes of alfalfa by calcium at 25 degrees C. *The Plant Cell*, **7:** 321-331.

Monroy A.F., Sarhan F., Dhindsa R.S. (1993) Cold-induced changes in freezing tolerance, protein phosphorylation and gene expression: Evidence for a role of calcium. *Plant Physiology*, **102:** 1227-1235.

Moran J.F., Becana M., Iturbe- Ormatexe I., Frechilla S., Klucas R.V., Aparicio-Tejo P. (1994) Drought induces oxidative stress in pea plants. *Planta*, **194:** 346-352.

Mori I.C., Schroeder J.I. (2004) Reactive oxygen species activation of plant Ca2+ channels. A signaling mechanism in polar growth, hormone transduction, stress signaling and hypothetically mechanotransduction. *Plant Physiology*, **135:** 702-708.

Newman I.A. (2001) Ion transport in root: measurement of fluxes using ion-selective microelectrodes to characterize transporter function. Plant Cell and Environment. **24:** 1-14.

Ng A.K.-Y., McAinsh M.R. (2003) Encoding specificity in plant calcium signalling: Hot-spotting the ups and downs and waves. Annals of Botany. **92:** 477-485.

Noctor G., Foyer C.H. (1998) Ascorbate and glutathione: Keeping active oxygen under control. *Annual Review of Plant Physiology, Plant Molecular Biology*, **49:** 249-279.

Nover L., Scharf K.D., Neumann D.(1983) Formation of cytoplasmic heat shock granules in tomato cell cultures and leaves. *Molecular Cell Biology*, **3:** 1648-1655.

Öquist G., Huner N.P.A. (1993) Cold-hardening-induced resistance to photoinhibition of photosynthesis in winter rye is dependent upon an increased capacity for photosynthesis. *Planta*, **189:** 150-156.

Pei Z.M., Murata Y., Benning G., Thomine S., Klusener B., Allen G.J., Grill S., Schroeder J.I. (2000) Calcium channels activated by hydrogen peroxide mediate abscisic acid signalling in guard cells. *Nature*, **406:** 731-734.

Plieth C., Hansen U.-P., Knight H., Knight M.R. (1999) Temperature sensing by plants: the primary characteristics of signal perception and calcium response. *The Plant Journal*, **18:** 491-497.

Polisensky D.H., Braam J. (1996) Cold-shock regulation of the Arabidopsis TCH genes and the effects modulating intracellular calcium levels. *Plant Physiology*, **111:** 1271-1279.

Prasad T.K., Anderson M.D., Martin B.A., Stewart C.R. (1994) Evidence for chilling induced oxidative stress in maize seedlings and a regulatory role of hydrogen peroxide. *The Plant Cell*, **6:** 65-74.

Price A.H., Hendry G.A.F. (1991) Iron-catalysed oxygen radical formation and its possible contribution to drought damage in nine nature grasses and three cereals. *Plant Cell and Environment*, **14:** 477-484.

Price A.H., Taylor A., Ripley S.J., Griffiths A., Trewavas A.J., Knight M.R. (1994) Oxidative signals in tobacco increase cytosolic calcium. *The Plant Cell*, **6:** 1301-1310.

Rao M.V., Davis K.R. (1999) Ozone-induced cell death occurs via two distinct mechanisms in Arabidopsis: the role of salicylic acid. *The Plant Journal*, **17:** 603-614.

Reddy V.S., Ali G.S., Reddy A.S.N. (2002) Genes encoding calmodulin-binding proteins in the Arabidopsis genome. *Journal of Biological Chemistry*, **277:** 9840-9852.

Rentel M.C., Knight M.R. (2004) Oxidative stress-induced calcium signaling in *Arabidopsis thaliana*. *Plant Physiology*, **135:** 1471-1479.

Rudd J.J., Franklin-Tong V.E. (2001) Unravelling response-specificity in $Ca^{2+}$ signalling pathways in plant cells. *New Phytologist,* **151:** 7-33.

Russell A.J., Knight M.R., Cove D.J., Knight C.D., Trewavas A.J., Wang T.L. (1996) The moss *Physcomitrella patens*, transformed with apoaequorin cDNA responds to cold shock, mechanical perturbation and pH with transient increases in cytoplasmic calcium. *Transgenic Research*, **5:** 167-170.

Sanders S., Brownlee C., Harper J.F. (1999) Communicating with calcium. *The Plant Cell*, **11:** 691-706.

Sanders D., Pelloux J., Brownlee C., Harper J.F. (2002) Calcium at the crossroads of signalling. *The Plant Cell*, **14:** S401-417.

Sandmann G., Gonzales H.G. (1989) Peroxidative processes induced in bean leaves by fumigation with sulphur dioxide. *Environmental Pollution*, **56:** 145-154.

Sangwan V., Foulds I., Singh J., Dhindsa R.S. (2001) Cold-activation of *Brassica napus BN115* promoter is mediated by structural changes in membranes and cytoskeleton, and requires $Ca^{2+}$ influx. *The Plant Journal*, **27:** 1-12.

Savouré A., Jauae S., Hua X.-J., Ardiles W., Van Montague M., Verbruggen N. (1995) Isolation, characterization and chromosomal location of a gene encoding the •[1]-pyrroline-5-carboxylate synthetase in *Arabidopsis thaliana*. *FEBS Letters*, **372:** 13-19.

Scandalios J.G. (1994) Regulation and properties of plant catalases. **In:** C.H. Foyer, P.M. Mullineaux (eds). *Causes of Photooxidative Stress and Amelioration of Defence Systems in Plants.* CRC Press, Boca Raton. pp. 275-315.

Schett G., Steiner C.W., Groger M., Winkler S., Graninger W., Smolen J., Xu Q., Steiner G. (1999) Activation of Fas inhibits heat induced activation of HSF1 and upregulation of HSP70. *FASEB Journal*, **13**: 833-842.

Schraudner M., Moeder W., Wiese C., Van Camp W., Inzé D., Langebartels C., Sandermann H., Jr. (1998) Ozone-induced oxidative burst in the ozone biomonitor plant, tobacco Bel W3. *The Plant Journal*, **16:** 235-245.

Schroeder J.I., Hagiwara S. (1989) Cytosolic calcium regulates ion channels in the plasma membrane of *Vicia faba* guard cells. *Nature*, **338:** 427-430.

Scrase-Field S.A.M.G., Knight M.R. (2003) Calcium: just a chemical switch? *Current Opinion in Plant Biology*, **6:** 500-506.

Sedbrook J.C., Kronebusch P.J., Borisy G.G., Trewavas A.J., Masson P.H. (1996) Transgenic aequorin reveals organ-specific cytosolic $Ca^{2+}$ responses to anoxia in Arabidopsis thaliana seedlings. *Plant Physiology*, **111:** 243-257.

Shacklock P.S., Read N.D., Trewavas A.J. (1992) Cytosolic free calcium mediates red light-induced photomorphogenesis. *Nature,* **358:** 753-755.

Snedden W.A., Fromm H. (2001) Calmodulin as a versatile calcium signal transducer in plants. *New Phytologist*, **151:** 35-66.

Somashekaraiah B.V., Padmaja K., Prasad A.R.K. (1992) Phototoxicity of cadmium ions on germinating seedlings of mung bean (*Phaseolus vulgaris*): Involvement of lipid peroxidase in chlorophyll degradation. *Physiologia Plantarum*, **85:** 85-59.

Storozhenko S., De Pauw P., Van Montagu M., Kushnir S. (1998) The heat shock element is a functional component of the *Arabidopsis APX1* gene promoter. *Plant Physiology*. **116:** 1005-1014.

Takahashi K., Isobe M., Knight M.R. (1997) Hypo-osmotic shock induces increases in cytosolic $Ca^{2+}$ in tobacco suspension culture cells. *Plant Physiology*, **113:** 587-594.

Taylor A.R., Manison N.F.H., Fernandez C., Wood J., Brownlee C. (1996) Spatial organization of calcium signalling involved in cell volume control in the *Fucus* rhyzoid. *The Plant Cell*, **8:** 2015-2031.

Teramura A.H., Sullivan J.H. (1994) Effects of UV-B radiation on photosynthesis and growth of terrestrial plants. *Photosynthesis Research*, **39:** 463-473.

Terashima I., Funayama S., Sonoike K. (1994) The site of photoinhibition in leaves of *Cucumis sativus* L. at low temperatures is photosystem I, not photosystem II, *Planta*, **193:** 300-306.

Terashima I., Noguchi K., Itoh-Nemoto T., Park Y.M., Kubo A., Tanaka K. (1998) The cause of PSI photoinhibition at lower temperature in leaves of *Cucumis sativus*, a chilling-sensitive plant. *Physiologia Plantarum*, **103:** 295-303.

Tomos A.D., Hinde P., Richardson P., Pritchard J., Fricke W. (1994) Microsampling and measurements of solutes in single cells. **In:** *Plant Cell Biology, A Practical Approach.* N. Harris, K.J. Oparka (eds). Oxford University Press. Oxford. Chapter 14.

Torsethaugen G., Pell E.J., Assmann S.M. (1999) Ozone inhibits guard cell $K^+$ channels implicated in stomatal opening. *Proceedings of the National Academy of Sciences*, USA. **96:** 13577-13582.

Trewavas A. (1999) Le calcium, c'est la vie: calcium makes waves. *Plant Physiology*, **120:** 1-6.

Trewavas A.J., Malhó R. (1998) $Ca^{2+}$ signalling in plants cells: The big network! *Current Opinion in Plant Biology*, **1:** 428-433.

Tsugane K., Kobayashi K., Niwa Y., Ohba Y., Wada K., Kobayashi H. (1999) A recessive Arabidopsis mutant that grows enhanced active oxygen detoxification. *The Plant Cell*, **11:** 1195-1206.

Vallelian-Bindschedler L., Schweizer P., Mosinger E., Metraux J.-P. (1998) Heath induced resistance in barley to powdery mildew (*Blumeria graminis* f. sp. *hordei*) is associated with a burst of AOS. Physiology *and Molecular Plant Pathology*, **52:** 185-199.

Van Assche F., Clijsters H. (1990) Effects of metals on enzyme activities in plants. *Plant Cell and Environment*, **13:** 195-206.

Van Breusegem F., Kushnir S., Slooten L., Bauw G., Botterman J., Van Montagu M., Inzé D. (1998) Processing of a chimeric protein in chloroplasts is different in transgenic maize and tobacco plants. *Plant Molecular Biology*, **38:** 491-496.

Van Breusegem F., Van Montagu M., Inzé D. (2002) Engineering stress tolerance in maize. **In:** D. Inzé, M. Van Montagu (eds). *Oxidative Stress in Plants*. Taylor and Francis, London. pp. 191-215.

van Der Luit A.H., Olivari C., Haley A., Knight M.R., Trewavas A.J. (1999) Distinct calcium signaling pathways regulate calmodulin gene expression in tobacco. *Plant Physiology*, **121:** 705-714.

Verdus M.C., Cabin Flaman A., Ripolic C., Thellier M. (1996) Calcium-dependent storage/retrieval of environmental signals in plants development. C.R. Academic Science Series III **319:** 779-782.

Verdus M.C., Thellier M., Ripoll C. (1997) Storage of environmental signals in flax. Their morphogenetic effects as enabled by a transient depletion of calcium. *The Plant Journal*, **12:** 1399-1410.

Webb A.A.R., McAinsh M.R., Mansfield T.A., Hetherington A.M. (1996) Carbon dioxide induces increases in guard cell cytosolic free calcium. The Plant Journal. **9:** 297-304.

White P.J., Broadley M.R. (2003) Calcium in plants. *Annals of Botany*, **82:** 487-511.

Wiese C.B., Pell E.J. (2003) Oxidative modification of the cell wall in tomato plants exposed to ozone. *Plant Physiology and Biochemistry*, **41:** 375-382.

Wise R.R., Naylor A.W. (1987) Chilling enhanced peroxidation: evidence for the role of singlet oxygen and superoxide in the breakdown of pigments and endogenous antioxidants. *Plant Physiology*, **83:** 278-282.

Xiang C., Oliver D.J. (1998) Glutathione metabolic genes co-ordinately respond to heavy metals and jasmonic acid in arabidopsis. The Plant Cell, **10:** 1539-1550.

Zhang L., Lu Y.T. (2003) Calmodulin-binding protein kinases in plants. *Trends in Plant Science,* **8:** 123-127.

Zhu J.-K. (2002) Salt and drought stress signal transduction in plants. *Annual Review of Plant Biology*, **53:** 247-274.

*Advances in Plant Physiology*, Vol. 9
Ed. A. Hemantaranjan
Scientific Publishers (India), Jodhpur, 2006 pp. 67-86
E-mail: info@scientificpub.com www.scientificpub.com

# 4

# PHYSIOLOGICAL, BIOCHEMICAL AND GROWTH RESPONSES OF PLANTS TO TROPOSPHERIC OZONE

[1]Supriya Tiwari and [2]Madhoolika Agrawal

[1]Senior Research Fellow, Department of Botany, Banaras Hindu University, Varanasi; supriyabhu@yahoo.co.in
[2]Professor, Department of Botany, Banaras Hindu University, Varanasi; madhoo58@yahoo.com

## INTRODUCTION

Ozone is a widespread secondary air pollutant, which occurs naturally at ground level in low concentrations ranging from 5-10 ppb. In the last few decades, increase in concentrations of tropospheric $O_3$ due to industrial combustion and traffic emissions has drawn the attention of environmentalists worldwide. $O_3$ concentrations have gradually been increasing during the past 100 years with the greatest shift in the recent years. Background $O_3$ concentrations in the mid latitude northern hemisphere is increasing by 0.5- 2% per year. Concentrations of $O_3$ of anthropogenic origin vary significantly with time and geographic location. Elevated levels of $O_3$ are found around urban areas, but rural and remote regions also show elevated $O_3$ levels due to transport of its precursors. Tropospheric $O_3$ has long been known as phytotoxic air pollutant (Richard *et al.*, 1958) causing greatest amount of damage to vegetation including crops as compared to any other air pollutant. Its relative importance may still increase due to decline in the occurrence of other air pollutants in view of stringent air quality guidelines throughout the world (Stockwell *et al.*, 1997). $O_3$ is considered as the most widespread of all atmospheric pollutants (Chameides *et al.*, 1994).

## 2. TRENDS IN TROPOSPHERIC $O_3$ CONCENTRATION

World scenario: Surface $O_3$ concentrations for the period 1881- 1900 in Western Europe were about 10 ppb measured in Parc Montsouris in Paris and 9- 12 ppb in Moncalieri near Torino (Guiherit and Roemer, 2000). At about 50° N in Europe, $O_3$

levels in 1930- 1940 were about 20 ppb, in 1950- 1953 about 25 ppb and in early 1970's about 30 ppb (Guiherit and Roemer, 2000). Present day background concentrations are typically about 40-50 ppb (Guiherit and Roemer, 2000). In Northern hemisphere, $O_3$ levels in the troposphere have increased by 35% over the past century (Chameides *et al.*, 1994). Pronounced diurnal cycles are found at low elevation sites, showing low concentrations at night and highest concentrations for few hours after midday (Stockwell *et al.*, 1997). A particular trend apparent in global $O_3$ levels is the decrease in maximum 1 h concentration in the Northern Hemisphere (Lefohn *et al.*, 2001). For a period of 1981-2001, the downward trend (>10%) in 1h maximum $O_3$ levels occurred in every geographic area of USA. In Canada, national ambient $O_3$ levels (as fourth highest daily maximum 8 h $O_3$) showed a decrease from 70 ppb in 1991 to 62 ppb in 1993.

In Karachi, maximum $O_3$ concentrations were reported between 12.00 and 15.00 hours, irrespective of climatic conditions and site location. The concentrations varied as low as 10.2-25.5 ppb at upward sites, and as high as 40.8-55 ppb at the downward sites (Ghauri *et al.*, 1992). Air monitoring conducted by Wahid *et al.* (1995 a & b) in Pakistan Punjab showed variations in mean $O_3$ concentrations from 27 ppb (December to January) to 39.5 ppb (March to April).

Indian scenario: $O_3$ concentration is highly sensitive to climate forcing in the tropics. The production of hydroxy (OH) radicals is highest in the tropical region due to intense solar radiation and large water vapour content (80% of the global budget). In spite of the importance of tropical troposphere, there have been no systematic simultaneous measurements of surface $O_3$ and its precursor gases over the Indian region except those at Ahemedabad, an urban site (Lal *et al.*, 2000). Measurements performed at this site showed that inspite of higher levels of precursors, $O_3$ concentrations exceeding 80 ppbv are rarely observed (Lal *et al.*, 2000). In another study conducted by Central Road Research Institute at seven sites in Delhi, 8 h $O_3$ concentrations during day exceeded the WHO standard of 51- 102 ppb by 10- 40% (Singh, 1997). A detailed study conducted in Varanasi revealed that monthly 24 h average concentration varied from 6.0-35 ppb from relatively clean to polluted areas during 1989- 1991 (Pandey *et al.*, 1992). Studies conducted during 1999-2001 in Varanasi city, however, showed higher $O_3$ concentrations varying from 10- 48 ppb in summer and 7- 45 ppb during winter in suburban areas (Agrawal *et al.*, 2003).

## 3. SOURCES OF TROPOSPHERIC OZONE

Two major sources of natural ground level $O_3$ are the hydrocarbons that are released by the plants and the soil, and small amount of $O_3$, which occasionally migrates down to earth surface from the stratosphere. However, neither of these sources are enough to cause the present day background $O_3$ concentration, which is a threat to human health and environment. Increase in the concentration of tropospheric $O_3$ is mainly due to human activities. Tropospheric $O_3$ has been formed by photochemical reactions involving precursors like nitrogen oxides (NOx) and hydrocarbons (HCs) generated by anthropogenic activities. $O_3$ formation occurs due to photolysis of $NO_2$, which acts as a source of atomic oxygen at • < 420 nm. In the free troposphere, $O_3$ formation also depends on reaction of methane ($CH_4$), carbon monoxide (CO) and non-methane organic compounds with NOx (Lefohn *et al.*, 2001).

## 4. OZONE EXPOSURE INDICES FOR VEGETATION

One of the difficulties in assessing the impacts of $O_3$ on vegetation is the wide range of methods used to describe $O_3$ exposure.

Concentration, duration, shape of exposure and frequency of peaks are all important in determining plant response to $O_3$ (Musselman, 1994). Examples of exposure indices are 7 or 8 h seasonal mean, the sum of all hourly mean concentrations using a threshold (SUM0, total dose), AOT30, 40, 50 and 60 (equal to the sum of all hourly mean concentrations above a threshold of 30, 40, 50 and 60 ppb over a defined time period), and SUM06, 08 (equal to the sum of all hourly average concentrations above 60, and 80 ppb) (Stockwell *et al.*, 1997). For Europe, exposure index is AOT40 (accumulated exposure over the threshold of 40 ppb), calculated as the sum of differences between the hourly $O_3$ concentrations and 40 ppb for each hour when the concentration exceeds 40 ppb over a defined period of time (Karenlampi and Skarby, 1996). Researchers in Europe have agreed to use this concept as a way to identify areas at risk and try to relate $O_3$ profiles to the biological response. Due to continuous increasing $O_3$ concentrations in Europe, UN-ECE has set critical levels for $O_3$ with respect to three major types of vegetation: forest trees, crops and semi natural vegetation (Karenlampi and Skarby, 1996).

## 5. EFFECTS OF OZONE ON PLANTS

The potential of $O_3$ to damage vegetation has been known for more than 45 years, but its major contribution in vegetation loss including crop yield has been identified only during last two decades due to $O_3$ episodes occurring frequently during high pressures periods. $O_3$ at ambient concentration causes a range of effects including reduced photosynthesis (Fangmeier *et al.*, 1993), altered carbon and nitrogen metabolism (Soja and Soja, 1995; Agrawal and Agrawal, 1990) and yield reductions (Fangmeier *et al.*, 1994; Pleijel *et al.*, 1991). $O_3$ injury to crops and trees has been commonly reported in North America and Europe. Evidence of $O_3$ impacts on crops and forests in other regions are very limited.

## 6. EFFECTS OF OZONE ON PHYSIOLOGICAL PROCESSES

### 6.1 Stomatal Conductance

Gaseous uptake of $O_3$ occurs during uptake and release of carbon dioxide and oxygen, involved in the process of photosynthesis and respiration, and the release of water vapour. Without entry into the internal sites of reaction, $O_3$ is relatively harmless to vegetation. Uptake of $O_3$ is a key process in determining the effects on metabolism and physiology. Plants can vary in their response to a particular $O_3$ dose because of differences in $O_3$ uptake, or absorbed dose ($O_3$ flux), reflecting differences in leaf conductance (Reich, 1987). Plant injury is directly correlated with the $O_3$ flux. Stomatal uptake is estimated to be four orders of magnitude larger than the uptake through fractures in leaf cuticle (Guderian, 1985). The flux of $O_3$ into the plants depends on different resistances at various levels i.e. aerodynamic resistance depending on atmospheric turbulences, a boundary layer resistance caused by a layer of laminar air adjacent to the leaf, the stomatal resistance exerted by the stomatal pores, and an internal resistance in plant leaf (Mansfield and Pearson, 1996). $O_3$ concentrations are generally high during summer and spring when high temperature, high irradiance and low wind speed leads to high boundary layer and stomatal resistance causing limitation to $O_3$ uptake.

$O_3$ exposure generally results in decline in stomatal aperture (Grantz and Yang, 1996). Studies conducted on Pima cotton showed that $O_3$ caused 4% reduction in stomatal conductance over a range of $O_3$ concentrations upto 111 ppb. In field exposure chambers stomatal conductance declined by about 25 % over a smaller range of $O_3$ exposures (Grantz *et al.*, 1999). Similarly, Reiling and Davison (1994) showed that stomatal conductance reduced significantly when *Plantago major* was

Table 1. Changes in stomatal conductance of selected plant species upon different concentrations of $O_3$ exposure

| Plant Species | $O_3$ concentration | % Change | Source |
|---|---|---|---|
| *Fagus sylvatica* | 32 ppb | (+) 4.1 | Paludan- Miller *et al.* 1991 |
| *Nicotiana tabaccum* L. var. Bel W3 | 150 ppb | (+) 21.7 | Degl'Innocenti *et al.* (2002) |
| *Nicotiana tabaccum* L. var. Bel B | 150 ppb | (-) 57 | Degl'Innocenti *et al.* (2002) |
| *Triticum aestivum* L. cv. HD 2329 | 40.60 ppb | (-) 21.6 | Rajput and Agrawal (2005) |
| *Vigna radiata* L. cv. Malviya Jyoti | 55.70 ppb | (-) 26.5 | Agrawal *et al.* (2003) |
| *Beta vulgaris* L. cv. Allgreen | " | (-) 80.9 | " |
| *Pisum sativum* L. var. Arkel | 40 ppb | (+) 19.5 | Rajput and Agrawal (2004) |
| *Glycine max* var. PK 472 | 70 ppb | (-) 21.0 | Singh, 1998 |
| | 100 ppb | (-) 26.3 | " |
| *Glycine max* var. Bragg | 70 ppb | (-) 61.0 | " |
| | 100 ppb | (-) 66.1 | " |
| *Triticum aestivum* var. M234` | 70 ppb | (-) 41.0 | " |
| | 100 ppb | (-) 56.5 | " |
| *Triticum aestivum* var. HP 1209 | 70 ppb | (-) 30.0 | " |
| | 100 ppb | (-) 45.1 | " |
| *Lactuca sativa* L. cul. Candela RZ | 17.34 ppb | (-) 34.2 | Calatayud *et al.* (2002) |

exposed to $O_3$ concentration of 70 ppb for 7 h $d^{-1}$. This closure could be due to disrupted membrane permeability of guard cells (Dominy and Heath, 1985), increased internal $CO_2$ caused by impaired photosynthesis or direct stress induced changes within the chloroplast affecting abscisic acid levels (Mansfield and Pearson, 1996). Exposure to acute doses of $O_3$ results in rapid stomatal closure of plants, affecting other physiological and biochemical processes. Singh (1998) observed reductions in stomatal conductance of two varieties of soybean due to exposure to 70 and 100 ppb $O_3$ for 4 h daily (Table 1). Scanning electron micrograph study further revealed that the proportion of open stomata declined in $O_3$ exposed varieties of *Glycine max* as compared to unexposed ones (Singh, 1998). At 100 ppb $O_3$ concentration, few stomata showed collapse of guard cells in sensitive variety PK 472, whereas resistant variety Bragg did not show such structural damage. Hassan *et al.* (1994) reported loss of turgor and collapse of epidermal cells adjacent to stomata in $O_3$ exposed radish leaves leading to wide stomatal opening and higher stomatal conductance, but in turnip stomatal conductance declined in response to $O_3$ exposure (Table 1).

Transpiration rate was found to reduce in $O_3$ exposed plants due to partial closure of stomata (Mansfield and Pearson, 1996; Singh, 1998). Agrawal *et al.* (1983) observed a reduction of 21 % in transpiration rate of *Solanum melongena* plants when exposed to 80 ppb $O_3$ for 2 h $d^{-1}$ for 21 days. Low concentrations of $O_3$ act upon membrane permeability of subsidary cells leading to water flow in guard cells, which open wide due to turgidity, but at higher concentration $O_3$ causes changes in permeability of guard cells and stomata close. Therefore, the differential response of stomata to $O_3$ depends upon concentration, species, genotype and climatic conditions.

Stomatal conductance was negatively correlated with internal $CO_2$ concentration in

*G. max* var PK 472, but did not show any relationship in var. Bragg, which clearly suggests that changes in stomatal conductance may not always correspond to photosynthetic response (Singh, 1998). Decline in stomatal conductance provides a mechanism for minimizing $O_3$ uptake and subsequent adverse effects on plants. Hence stomatal conductance is very important factor of physiological control on $O_3$ absorption. Butler and Tibbitts (1979) compared the stomatal conductance in two *Phaseolus vulgaris* genotypes having differential $O_3$ sensitivity, and found that $O_3$ tolerant cultivars closed stomata in response to $O_3$ exposure, while $O_3$ sensitive cultivar had small increase in stomatal conductance. Stomata of $O_3$ resistant cultivars closed more rapidly and to a greater extent, when exposed to $O_3$ than the stomata of a sensitive cultivar. Stomatal closure may appear to provide a mechanism for minimizing uptake, and the subsequent adverse effects of $O_3$, Such action, however, is stressful, since it limits $CO_2$ uptake. The water stress and ambient relative humidity also controls stomatal response to $O_3$. Water stressed plants exhibited more rapid stomatal closure in response to $O_3$ than unstressed plants (Reich and Amundson, 1985). High relative humidity causes lower $O_3$ induced stomatal closure than at lower humidities.

## 6.2 Photosynthesis

$O_3$ exposure has been shown to reduce photosynthetic rate in spring barley (Rowland- Bamford *et al.*, 1989), soybean (Miller *et al.*, 1991), white bean (Sanders *et al.*, 1992), spring wheat (Meyer *et al.*, 1997), radish and turnip (Hassan *et al.*, 1994), wheat (Singh, 1998) and pinto bean (Reiling and Davison, 1994). Table 2 shows changes in photosynthetic rate in selected plants upon $O_3$ exposure. The reduction in net photosynthesis may be attributed to $O_3$ induced changes in photophosphorylation, electron transport system and carbon fixation processes. A linear negative photosynthetic response was reported at increasing $O_3$ concentrations (Singh, 1998). Diurnal variations in photosynthetic rate did not differ between control and $O_3$ exposed plants (Singh, 1998; Rowland- Bamford *et al.*, 1989).

$O_3$ induced reduction in photosynthetic capacity of plants is often associated with the decline in quantity of an important photosynthetic enzyme, ribulose bisphosphate/ oxygenase (RUBISCO) (Dann and Pell, 1989; Pell *et al.*, 1992). RUBISCO is a major storage protein and during normal leaf ontogeny, it is synthesized until full leaf expansion, when it attains maximum concentration. Under conditions of $O_3$ stress, the concentration of RUBISCO declines more rapidly than in non stressed foliage (Dann and Pell, 1989). Photosynthesis under $O_3$ stress is inhibited to a greater extent in fully expanded or mature foliage than in young expanding leaves. Fully expanded leaves showed the strongest decrease in quantity and activity of RUBISCO in leaves of $O_3$ fumigated potato (Eckardt and Pell, 1994). $O_3$ induced decline in RUBISCO quantity could result from inhibition of synthesis and/ or enhancement of degradation. Reddy *et al.* (1993) correlated the decline in RUBISCO mRNA content in potato with $O_3$ exposure, indicating the inhibition of RUBISCO synthesis. $O_3$ might also cause enhanced degradation of RUBISCO, which could occur as a result of enhanced proteolytic activity, or greater acceptability of RUBISCO to proteases and oxidation (Pell *et al.*, 1997). RUBISCO is a sulphydryl (SH) enriched enzyme and SH group is highly vulnerable to oxidation, and hence one of the targets of $O_3$ attack. Several reports have suggested that $O_3$ may induce modification of RUBISCO in vivo (Landry and Pell, 1993). $O_3$ induced reduction in the pool of RUBISCO would result in reduced net photosynthesis (Reich and Amundson, 1985).

Decrease in photosynthetic rate due to $O_3$ stress may not be visually evident, but can

Table 2. Changes in photosynthetic rate of selected plant species upon different concentrations of $O_3$ exposure

| Plant Species | $O_3$ concentration | % Change | Source |
|---|---|---|---|
| *Fagus sylvatica* | 32 ppb | (-) 2.0 | Paludan- Miller *et al.* 1991 |
| *Nicotiana tabaccum* L. var. Bel W3 | 150 ppb | (-) 57 | Degl'Innocenti *et al.* (2002) |
| *Nicotiana tabaccum* L. var. Bel B | 150 ppb | (-) 3.26 | " |
| *Triticum aestivum* L. cv. HD 2329 | 40.60 ppb | (-) 31.0 | Rajput and Agrawal (2005) |
| *Vigna radiata* L. cv. Malviya Jyoti | 55.70 ppb | (-) 16.84 | Agrawal *et al.* (2003) |
| *Beta vulgaris* L. cv. Allgreen | " | (-) 9.22 | " |
| *Pisum sativum* L. var. Arkel | 40 ppb | (-) 56.90 | Rajput and Agrawal (2004) |
| *Glycine max* var. PK 472 | 70 ppb | (-) 19.8 | Singh, 1998 |
| | 100 ppb | (-) 40.4 | " |
| *Glycine max* var. Bragg | 70 ppb | (-) 25.6 | " |
| | 100 ppb | (-) 32.3 | " |
| *Triticum aestivum* var. M234` | 70 ppb | (-) 36.0 | " |
| | 100 ppb | (-) 42.9 | " |
| *Triticum aestivum* var. HP 1209 | 70 ppb | (-) 29.0 | " |
| | 100 ppb | (-) 41.0 | " |
| *Lactuca sativa* L. cul. Candela RZ | 17.34 ppb | (-) 17.7 | Calatayud *et al.* (2002) |
| *Hordeum vulgare* | 150 ppb | (-) 6 | Rowland- Bamford *et al.* (1989) |

result in subtle changes in biomass production that may be recognized only at harvest. Depression in photosynthetic rates is, therefore, an important criterion in the evaluation of $O_3$ stress on plants. Decreased photosynthesis is also associated with reductions in chlorophyll content, soluble protein, adenylate and reduced carboxylation efficiency, resulting in reduced $CO_2$ uptake (Farage *et al.*, 1991).

Path dependent analysis of photosynthetic limitations has been reviewed by Jones (1985). $O_3$ may directly affect stomata first causing an immediate increase in gas phase resistance to $CO_2$. The net $CO_2$ fixation remains unchanged initially, but decline in internal $CO_2$ causes reduction in net photosynthesis later. $O_3$ may affect both stomatal and mesophyll functions simultaneously causing impairment of photosynthesis immediately after exposure. Effects of $O_3$ on net photosynthesis is also modified by environmental variables such as light, humidity, ambient $CO_2$ concentration and other biological factors such as leaf age, plant water status, respiration rate, etc. In summary, short term effects of $O_3$ may be reversible, but long term chronic exposure cause some degree of adaptation from the interaction between effects on net photosynthesis and stomatal function. Chronic exposure impaired photosynthetic ability is closely related to accelerated senescence and decreased overall growth.

Chlorophyll fluorescence transients (CFTs) have been commonly used as a diagnostic tool to detect early stress conditions in plants and are measured during dark light transition of photosynthetically active leaves (Schreiber *et al.*, 1978). Two phases of fluorescence induction kinetics have been used to describe the functioning of photosynthetic apparatus: (i) a rapid phase characterized by a fast fluorescence rise to a maximum fluorescence (Fm), which is completed in less than 0.5 s, and (ii) a slow phase characterized by a slow decrease in fluorescence (Fd) to a steady state level (Fs) where photosynthesis and fluorescence are in equilibrium, which takes 3- 5 minutes. Under $O_3$ stress conditions, where photosynthetic quantum is disturbed, chlorophyll fluorescence remains high and decrease from $F_m$ to $F_s$ is slow or does not occur at all. $O_3$ fumigated wheat plants showed reductions in photosynthetic efficiency due to $O_3$ impaired PS II activity (Grandjean and Fuhrer, 1992). Schreiber *et al.* (1978) observed that degree of changes in CFTs was greater for bean leaves following long term exposure to low $O_3$ concentration than during brief exposures to high ones.

Reiling and Davison (1994) working on *Plantago major* L. exposed to 70 ppb $O_3$ for 7 h $d^{-1}$ observed significant reductions in Fv/Fm ratio (variable to maximum fluorescence) indicating direct photosynthetic disruption. Changes in Fv/Fm were caused mainly by reduction in Fm indicating increased non-photochemical and/ or decreased photochemical quenching. Calatayud *et al.* (2002) showed that Fv/Fm ratio decreased when plants of *Lactuca sativa* L. were exposed to non-filtered air as compared to filtered air. Changes in chlorophyll fluorescence may well occur before any physical signs of deterioration. Lee *et al.* (1984) have used chlorophyll fluorescence induction technique for evaluating the feasibility of characterizing sensitive (cv. BBL 290) and resistant (cv. Astro) cultivars of snapbean plants to $O_3$ exposure. It was observed that resistant soybean cultivars showed minor changes in chlorophyll fluorescence, whereas value of Fv/Fm ratio showed drastic decrease in $O_3$ sensitive cultivar than in $O_3$ resistant ones. Singh (1998) has reported that changes in chlorophyll fluorescence were minor in $O_3$ resistant than sensitive cultivars of wheat and soybean. Table 3 shows changes in the Fv/Fm ratio of selected plants upon $O_3$ exposure.

## 7. EFFECTS OF OZONE ON CELLULAR LEVEL

### 7.1 Cell Permeability

$O_3$ enters the mesophyll cells via the stomata and dissolves in the water phase of the apoplast, where it is converted into superoxide anion, hydroxyl radicals and $H_2O_2$ (Mehlhorn *et al.*, 1990) which initiates lipid peroxidation, thus destroying the membranes (Heath, 1988). $O_3$ also reacts with unsaturated fatty acids in the plasmalemma with the formation of aldehyde (i.e. malondialdehyde) and $H_2O_2$ (Hippelli and Elstner, 1996). $O_3$ reacts in the substomatal cavity with hydrocarbons such as ethylene and some terpenes to produce $H_2O_2$ (Hewit *et al.*, 1990), which is an important oxidant. $H_2O_2$ dissolves very well in the water phase and can be transported through the membranes and circulated in plants. Further reactions of $H_2O_2$ lead to the formation of reactive oxygen species (ROS). The initial site of $O_3$ action is alteration of membrane permeability, which allows the cell content leakage into extracellular spaces leading to osmotic or ionic imbalance within the cell (Dominy and Heath, 1985).

The condensation of amino acid residues by the oxidation of sulphydryl moieties to the disulphide bridges or further oxidation of sulphydryl moieties to sulphenic ($-SO_3H$) and sulphonic ($-SO_4H$) groups are feasible, the latter being an irreversible oxidation (Mudd *et al.*, 1969). Also the cleavage of aromatic amino acid ring structures by $O_3$ has been demonstrated (Mudd *et al.*, 1969). It has been suggested that membrane lipids may be one of the primary sites of $O_3$ injury, since $O_3$ in-vitro rapidly reacts with double bonds leading to the production of oxidation products. The chemical and the physical changes in membrane lipid occur due to breakdown of phospholipids and loss of sterols from the membrane. Peroxidation of neutral lipids may raise the lipid phase transition temperature and promote the building of gel

Table 3. Changes in chlorophyll fluorescence (Fv/ Fm) of selected plant species upon different concentrations of $O_3$ exposure

| Plant Species | $O_3$ concentration | % change | Source |
|---|---|---|---|
| *Nicotiana tabaccum* L. var. Bel W3 | 150 ppb | (-) 4.37 | Degl'Innocenti *et al.* (2002) |
| *Nicotiana tabaccum* L. var. Bel B | 150 ppb | (-) 8.43 | Degl'Innocenti *et al.* (2002) |
| *Glycine max* var. PK 472 | 70 ppb | (-) 1.94 | Singh, 1998 |
| | 100 ppb | (-) 30.3 | " |
| *Glycine max* var. Bragg | 70 ppb | (-) 2.51 | " |
| | 100 ppb | (-) 5.87 | " |
| *Triticum aestivum* var. M234` | 70 ppb | (-) 1.72 | " |
| *Triticum aestivum* var. HP 1209 | 100 ppb | (-) 8.10 | " |
| | 70 ppb | (-) 1.09 | " |
| | 100 ppb | (-) 5.69 | " |
| *Lactuca sativa* L. cul. Candela RZ | 17.34 ppb | (-) 1.33 | Calatayud *et al.* (2002) |

phase lipid domain in the membrane matrix leading to leakiness of membranes (Pauls and Thompson, 1984). Dominy and Heath (1985) reported a substantial loss in ATPase activity of the plasma membrane, when leaves of *Phaseolus vulgaris* were exposed to $O_3$. ATPase is an energy dependant enzyme(s), which plays an important role in active solute transport. Damage to active transport enzymes or facilitated diffusion channels would result in dramatic changes in normal transport processes associated with the tissue.

Inhibition of glycolipid biosynthesis upon $O_3$ exposure is also reported (Mudd *et al.* 1971). This study confirmed that the percent inhibition of the synthesis of digalactosyl diglyceride (DGDG) was much greater than the synthesis of monogalactosyl diglyceride (MGDG). In another set of experiment, it was observed that synthesis of steryl glucoside and acetylated steryl glucoside was relatively resistant to inhibition by $O_3$ (Mudd *et al.*, 1971). Nouchi and Toyama (1988) showed that when morning glory leaves were exposed to 150 ppb $O_3$ for 8 h, fatty acid content (linoleic and linolenic acid) decreased significantly, while that of malonaldihyde (MDA) increased slightly. These results suggest that $O_3$ or oxy radicals cause oxidation of unsaturated fatty acids of the membranes (Nouchi and Toyama, 1988). Oxy radicals formed from $O_3$ decomposition in aqueous solution, enzymatically, or the oxygen reduction in chloroplasts can cause lipid peroxidation and lipid deestrification (Pauls and Thompson, 1984).

## 7.2 Antioxidants

$O_3$ induced ROS governs the phytotoxicity of $O_3$ (Runeckles and Vaartnou, 1997). Oxyradicals can cause serious damage to cells by attacking almost every type of molecule found in the living cells including DNA, proteins and membrane lipids (Halliwell and Gutteridge, 1989). Damage to membrane lipids can especially be serious because once initiated, peroxidation proceeds by spontaneous chain reaction, which quickly destroys membrane integrity (Paul and Thompson, 1984). Consequently antioxidant systems that are responsible for controlling the concentrations of ROS in plant tissue are considered to play an important role in mediating $O_3$ resistance.

### *7.2.1. Ascorbate*

In plants, the most abundant antioxidant is ascorbate, which is synthesized in the leaf cells and then transported across plasma membrane into leaf apoplast. The pool of apoplastic ascorbate (ASC) represents only a small fraction (1%) of ascorbate in the bulk leaf. However, it is found to be sufficient to afford a significant degree of protection against $O_3$ and/ or its toxic reaction products (Castillo and Greppin, 1988). Ascorbate is readily oxidized by $O_3$ and several other ROS to yield dehydroascorbic acid (DHA). This compound must be reduced to regenerate ASC, or is rapidly and irreversibly hydrolysed to yield 2,3 diketogluconic acid (DKG), and subsequently an array of degradation products, including oxalate (Smirnoff, 1996). Since DHA cannot be reduced efficiently in the apoplast, it is believed to be returned to the cytosol for regeneration (Castillo and Greppin, 1988). This view is supported by the presence of a carrier- mediated system on the plasma membrane for the transport of ASC/ DHA, which shows a higher affinity for DHA than ASC (Horemans *et al.*, 1997).

The importance of AA is demonstrated in the VTCI mutant of *Arabidopsis*, where low AA content in leaf tissue was associated with increased $O_3$ sensitivity (Conklin *et al.*, 1996). There is also evidence that $O_3$ tolerant genotype has elevated AA content (Lee *et al.*, 1984; Burkey *et al.*, 2000). The effects of $O_3$ treatment on leaf AA content are not consistent (Guzy and Heath, 1993; Burkey *et al.*, 2000) reflecting both species differences in response, as well as the use of acute vs. chronic exposure regimes. Foliar AA contents either increased (Burkey and Gwendolyn, 2002) or decrease (Singh, 1998; Burkey and Gwendolyn, 2002) upon exposure to $O_3$ (Table 4).

### *7.2.2 Peroxidase*

The apoplast contains both ascorbate specific and non-specific peroxidases, whose activities are stimulated by environmentally relevant $O_3$ concentrations (Ranieri *et al.*, 1996). The increments in peroxidase activity following $O_3$ exposure vary between different plant species, and are regarded as a function of the resistance of plants to $O_3$ (Castillo *et al.*, 1984). Since activity of apoplastic peroxidases is generally increased by $O_3$ (Lyons *et al.*, 1999), prior exposure to $O_3$ may modify the diffusion pathway for the pollutant leading to 'acclimatory' changes in leaf anatomy.

### *7.2.3 Superoxide dismutase (SOD)*

SOD is a highly efficient scavenger of $O_2$• radicals generated by the breakdown of $O_3$ in the apoplast under environmentally relevant conditions (Heath, 1988). There are however, conflicting reports of the effects of $O_3$ on apoplastic SOD activity. Castillo *et al.* (1987) found that apoplastic SOD activity was stimulated by $O_3$ in *Picea abies* and suggested a forward defensive role for the enzyme against the pollutant. On the other hand, Lyons *et al.* (1999) reported a decline in SOD activity in *Plantago major* in response to controlled exposure to 70 ppb $O_3$ for 7 h $d^{-1}$. Casano *et al.* (1994) found that the inducibility of SOD was significantly lowered in old than in young leaves rendering older foliage more sensitive to $O_3$ stress. In young leaves of $O_3$ exposed, high fertilized plants, SOD activity was diminished as compared with control, even though these levels did not show visible injury (Maurer *et al.*, 1997). In high-fertilized plants, elevated SOD activities appeared with development of foliar symptoms of injury.

### *7.2.4 Glutathione (GSH)*

GSH is involved in detoxification of xenobiotics mediated by glutathione- S-transferase, and destruction of $H_2O_2$ via the

Table 4. Changes in the ascorbic acid content of selected plant species upon different concentrations of $O_3$ exposure

| Plant species | $O_3$ concentration | % Change | Source |
|---|---|---|---|
| *Phaseolus vulgaris* var. Provider | 29 ppb | (-) 31.1 | Burkey and Eason (2002) |
| *Phaseolus vulgaris* var. Tenderette | " | (+) 20 | " |
| *Phaseolus vulgaris* var. R123 | " | (-) 10.5 | " |
| *Phaseolus vulgaris* var. R142 | " | (+) 3.33 | " |
| *Phaseolus vulgaris* var. Oregon- 91 | " | (-) 2.94 | " |
| *Phaseolus vulgaris* var. S144 | " | (-) 14.2 | " |
| *Phaseolus vulgaris* var. S156 | " | (+) 20.8 | " |
| *Triticum aestivum* var. M234 | 100 ppb | (-) 34.0 | Singh (1998) |
| *Triticum aestivum* var. HP 1209 | 100 ppb | (-) 20 | " |
| *Glycine max* var. Bragg | 100 ppb | (-) 4 | " |
| *Glycine max* var. PK472 | 70 ppb | (-) 15 | " |
| | 100 ppb | (-) 15 | " |

ascorbate- glutathione cycle (Foyer and Halliwell, 1976), as well as direct scavenging of ROS. Under $O_3$ exposure, level of reduced GSH in foliar tissue has been shown to decrease over unexposed levels (Lee *et al.*, 1997).

### *7.2.5 Polyamines*

Conjugated forms of polyamines are known to be associated with plant cell walls and may play a direct role in scavenging of ROS (Bors *et al.*, 1989), as well as acting as substrate for peroxidases. Apoplastic concentrations of caffeoly-putrescine have been shown to increase markedly in response to $O_3$ exposure in *Nicotiana tabacum* Bel- B, an $O_3$ tolerant cultivar (Langebartels *et al.*, 1991). $O_3$ exposure however inhibits the synthesis of arginine decarboxylase (Langebartels *et al.*, 1991), a key enzyme in plant polyamine biosynthesis, which increases the extent of visible $O_3$ injury (Rowland- Bamford *et al.*, 1989). Arginine decarboxylase activity was reported to increase in $O_3$ treated tobacco cultivar Bel B and this increase preceeded the accumulation of free and conjugated putrescine (Langebartels *et al.*, 1991). Van Buuren *et al.* (2002) found that Bel W3 plants when exposed to $O_3$ at 130 ppb for 7 h showed a 4.5 fold increase in free putrescine content immediately after the end of $O_3$ exposure compared with control plants.

## 8. EFFECTS OF OZONE ON PHOTOSYNTHATE ALLOCATION

There is good evidence that $O_3$ reduces resource allocation to roots or other non-photosynthetic organs of crop plants (Cooley and Manning, 1987). $O_3$ causes physiological changes in leaves that affect source strength, which is the amount of carbon available for allocation to sink tissues. Out of 20 diverse plant species surveyed by Cooley and Manning (1987), 17 established a lower root: shoot ratio. $O_3$ stress influences the photosynthetic allocation by altered stomatal conductance, decreased activity and concentration of RUBISCO, and by reducing leaf longevity (Dann and Pell, 1989; Zheng et al., 2002). Control fumigations of 190 and 250 ppb $O_3$ decreased the root weight by 32 to 46% in *Daucus carota* while leaf weight rose slightly (Bennett and Oshima, 1976). Similarly reduced biomass allocation to roots was observed in Pima cotton at 111 ppb

(Grantz and Yang (1996, 2000). As roots are dependant on photosynthate for their structural development, carbon-limiting stress such as $O_3$ can have a rapid and significant effect on root growth.

Decreased carbon allocation leads to reduced carbohydrate levels and storage pools in non photosynthetic parts of $O_3$ exposed plants (Cooley and Manning, 1987; Anderson and Seagel, 1997). Grulke *et al.* (1998) found decreased biomass of coarse, medium and fine root with increased pollutant load across a gradient of $O_3$ concentration in Southern California. Coarse and fine root starch concentration was also lowest in mature trees at the most polluted site (Grulke *et al.*, 2001). Decreased carbohydrate storage pools were associated with decreased root growth during the spring following exposure to ambient $O_3$ even in the absence of additional $O_3$ exposure (Anderson and Seagel, 1997).

Many other experiments suggest certain different results as far as photosynthetic allocation is concerned. Warwick and Taylor (1995) found that $O_3$ had no effect on the root: shoot allometric co-efficient of a legume, *Lotus corniculatus.* The greatest decrease of allocation in clover is not to the roots, but to the stem, which acts as storage organ. Bergmann *et al.* (1995) exposed 17 herbaceous species from seedling stage to flowering stage to two $O_3$ concentrations: CF+70 ppb, for 8h $d^{-1}$, and CF+60% ambient +30ppb. $O_3$ response varied with exposures regime, and the weight of species was reduced to 60% of the controls, but the striking differences were observed in resource allocation. Most of the species showed a proportionate change between shoot mass and reproductive organ, but *Chenopodium album and Matricaria discoidea* showed a greater vegetative shoot weight and reduced reproductive allocation. Conversely, *Papaver dubium* and *Trifolium arvense* has reduced shoot mass and increased seed / flowering allocation. This explains the stimulatory role in some cases.

## 9. FOLIAR INJURY

Research with plants exposed to $O_3$ under controlled experimental conditions clearly showed that $O_3$ could cause a wide range of symptoms in crops, forest trees and naturally growing plants (Chappelka and Samuelson, 1998). Some of these are quite distinct, while some resemble those to normal senescence (Krupa and Manning, 1988). Manning *et al.* (2002) categorized symptoms for broad-leaved plants as:

| | |
|---|---|
| Stipple: | Small punctate spots on upper leaf surfaces of older leaves delineated by leaf veins. Dead palisade parenchyma cells having range of colours from white to red-to red purple to brown to black. |
| Bronzing: | Diffuse, upper leaf surface reddish brown pigmentation. |
| Reddening: | Older leaves showing reddening or purpling on upper surface. |
| Chlorosis: | Older leaves yellow and senescence. |
| Premature Senescence: | Premature leaf fall, usually associated with chlorosis, stipple and bronzing |

$O_3$ induced visible injury is often used to assess forest damage although the extent of foliar injury does not necessarily correlate with physiological damage or reductions in growth (Pye, 1988). For conifers, chlorotic mottle and chlorotic banding on second year needles were used as diagnostic symptoms of $O_3$ injury. Acute $O_3$ stress will generally result in visible symptoms, but low concentrations over long periods may lead to hidden damage without the appearance of visible foliar injury (Pye, 1988). Biochemical and physiological injury may occur before the appearance of any visible symptoms (Frederickson *et al.*, 1996). Manning and Godzik (2004) exposed different plant species of Central and Western Europe to controlled $O_3$ concentrations (60 or 80 ppb $O_3$ for 7 h $d^{-1}$,

7 d week[-1]) to categorise the degree of susceptibility of plants to $O_3$. Forbs showed larger variations in foliar injury percentage, being maximum in *Centaurea nigra* (81.25%) and lowest in *Gentiana asclepiadea* (9.37%). Certain other forbes such as *Centaurea scabiosa* (62.5%) and *Impatiens parviflora* (62.5%) also showed high injury percentage. Observations showed that certain trees like *Sorbus aucuparia* (81.25%) and *Alnus incana* (62.5%) showed high percent injury in response to $O_3$ stress (Manning and Godzik, 2004).

## 10. EFFECT OF OZONE ON GROWTH AND YIELD

Plant growth is a result of integration and coordination of various physiological and biochemical processes, which are controlled by genetic composition of plants as well as various environmental factors. Therefore, growth response of plants to $O_3$ shows a great range of variations due to plant species, cultivars and stage of development. Increased $O_3$ exposures have deleterious effects on morphological, growth and yield characteristics of a number of plant species (Heggestad and Lee, 1990; Fangmeier *et al.*, 1994; Lehnherr *et al.*, 1987; Fuhrer *et al.*, 1992; Kress and Miller, 1983; Schnone *et al.*, 1992). It has been concluded that ambient $O_3$ concentrations reduce the yield of some major crops in United States (Heck *et al.*, 1983), Switzerland (Fuhrer *et al.*, 1989), Sweden (Pleijel *et al.*, 1991) and Germany (Adaros *et al.*, 1990). Wheat is among one of the most sensitive species. Studies conducted on cotton (Grantz and Yang, 1996) showed that $O_3$ reduces plant growth, biomass and leaf area and suppresses commercial yield of seeds. Similar yield reductions were observed in wheat (Rajput and Agrawal, 2005; Wahid *et al.*, 1995a; Maggs *et al.*, 1995) and *Vigna radiata* (Agrawal *et al.*, 2003).

Biomass accumulation in plants also depends upon the available leaf area, which provides greater surface for photosynthetic processes, thus influencing the production of plants. Grantz and Yang (1996) observed that leaf area was reduced by 83% and plant dry weight by 88% in Pima cotton relative to $O_3$ free air, following chronic exposure to a 12 h mean $O_3$ concentration of 111 ppb. The reductions in plant biomass were substantial among all plant components, with root system biomass reduced to very low levels (Grantz and Yang, 1996). Held *et al.* (1991) have shown an apparent compensatory response to $O_3$ by increasing the leaf production rate at the expense of root and hypocotyls growth as compared to control plants.

Various open top chamber studies have shown significant reductions in growth and biomass of wheat (Temmerman *et al.*, 1992) and soybean plants (Heggestad and Lee, 1990) exposed to $O_3$. Open top experiments have shown crop yield reductions up to 20% for potato and 30% for wheat, when grown in ambient air as compared to filtered air (Heagle, 1989). Although sufficient information exist regarding the adverse effects of $O_3$ on plant growth in temperate climate, but little information is available about responses of tropical plant species to enhanced levels of $O_3$. Field studies conducted by Agrawal (1982) and Agrawal *et al.* (1983 a and b) on tropical plant species revealed that $O_3$ caused growth reductions in plants such as *Vicia faba*, *Oryza sativa*, *Panicun miliaceum*, *Cicer arietinum* and *Solanum melanogena* in decreasing order. Similarly, wheat, mustard and mung showed yield reductions of 29.5, 20.4 and 73.4%, respectively at a site having a seasonal mean $O_3$ concentration of 58.50 ppb under ambient condition. Singh, (1998) showed reductions in yield of wheat and soybean when exposed to different concentrations (70 and 100 ppb) of $O_3$ (Table 5).

Table 5. Changes in yield of selected plants upon $O_3$ exposure

| Plant species | $O_3$ concentration | % Change | Source |
|---|---|---|---|
| *Triticum aestivum* L. var. Albis | 35 ppb | (-) 11.76 | Lehnherr *et al.* (1987) |
| | 100 ppb | (-) 56.6 | " |
| *Triticum aestivum* L. var. Albis | 35.7 ppb | (-) 1.85 | Fuhrer *et al.* (1992) |
| | 49.47 ppb | (-) 9.81 | " |
| | 62.22 ppb | (-) 21.4 | " |
| *Vicia faba* | 40 ppb | (-) 53.3 | Agrawal *et al.* (1988) |
| *Cicer arietinum* | 40 ppb | (-) 22.1 | " |
| *Triticum aestivum* L. Pak- 81 | 35.6 ppb | (-) 11.2 | Wahid *et al.* (1995a) |
| *Triticum aestivum* L. Chakwal- 86 | 35.6 ppb | (-) 5.6 | " |
| *Oryza sativa* L. IRRI-6 | 35 ppb | (-) 3.5 | " |
| *Oryza sativa* Basmati- 385 | 35 ppb | (-) 6.5 | " |
| *Triticum aestivum* L. Pak- 81 | 52 ppb | (-) 9.1 | Maggs *et al.* (1995) |
| *Triticum aestivum* L. Chakwal- 86 | 52 ppb | (-) 5.9 | " |
| *Oryza sativa* L. IRRI-6 | 61 ppb | (-) 1.3 | " |
| *Oryza sativa* Basmati- 385 | 61 ppb | (-) 11.6 | " |
| *Glycine max* L. Merr.cv. Corsoy | 42 ppb | (-) 5.0 | Kress and Miller, 1983 |
| | 64 ppb | (-) 23.0 | " |
| | 89 ppb | (-) 39 | " |
| | 115 ppb | (-) 52 | " |
| *Phaseolus vulgaris* L. cv. Taylors Horticultural | 44 ppb | (-) 24.30 | Schenone *et al.* (1992) |
| | 48 ppb | (-) 31.47 | " |
| *Phaseolus vulgaris* L. cv. Lingua di Fuoco | 44 ppb | (-) 4.65 | " |
| | 48 ppb | (-) 28.62 | " |
| *Phaseolus vulgaris* L. cv. Saluggia | 44 ppb | (-) 19.14 | " |
| | 48 ppb | (-) 30.81 | " |
| *Triticum aestivum* var. HUW 468 | 40.60 ppb | (-) 31.6 | Rajput & Agrawal (2005) |
| *Vigna radiata* L. cv. Malviya Jyoti | 55.70 ppb | (-) 34.3 | Agrawal *et al.* (2003) |
| *Glycine max* var. PK 472 | 70 ppb | (-) 13.9 | Singh, 1998 |
| | 100 ppb | (-) 35.5 | " |
| *Glycine max* var. Bragg | 70 ppb | (-) 10.10 | " |
| | 100 ppb | (-) 25.0 | " |
| *Triticum aestivum* var. M234 | 70 ppb | (-) 4.67 | " |
| | 100 ppb | (-) 15.78 | " |
| *Triticum aestivum* var. HP 1209 | 70 ppb | (-) 8.3 | " |
| | 100 ppb | (-) 17.1 | " |

## CONCLUSIONS

Ozone has long been recognized as a gaseous pollutant in the troposphere. Unlike other gaseous pollutants, $O_3$, a secondary gaseous air pollutant occuring naturally at relatively high background concentrations. Background $O_3$ concentration has been steadily rising from 10-20 ppb at the beginning of the 20th century to the values in the range of 20- 40 ppb in the recent years. Two major sources of tropospheric $O_3$ are transport from troposphere to stratosphere, and photochemical reactions in troposphere. The latter involves $O_3$ formation from the recombination of atomic and molecular oxygen via the photolysis of $NO_2$, and $O_3$ destruction by reaction with NO to form $NO_2$. $O_3$ concentration tends to be higher in suburban and rural locations as compared to urban regions as reactions with NO leads to local scale depletion of $O_3$.

Effects of elevated $O_3$ on plants can be observed in form of visible foliar injury, however, visible injury appears at a high concentration of $O_3$. Chronic doses of $O_3$ directly affect the physiological and biochemical processes in plants. $O_3$ uptake by plants via the leaf cuticle is negligible and is almost entirely through the stomata. On entry to the sub stomatal cavity, $O_3$ reacts with the constituents of the aqueous matrix associated with the cell wall to form other derivatives that result in the oxidation of sensitive components of plasmalemma and subsequently the cytosol. Significant negative effects of $O_3$ have been observed in the wide range of characteristics such as decreased photosynthetic assimilation, altered stomatal behavior, decreased growth and productivity, reduced carbon allocation, etc. $O_3$ inhibits photophosphorylation by reducing electron transport. The reduction in net photosynthetic rate and carboxylation efficiencies by $O_3$ has been related to reduced levels and activity of RUBISCO, impaired electron transport and problems in regeneration of ribulose bisphosphate. The decline in stomatal conductance induced by $O_3$ is also related to the $CO_2$ assimilation. $O_3$ stress leads to variations in the levels of antioxidants, which scavenge the reactive oxygen species produced by $O_3$, which can destroy the cell membrane by inducing lipid peroxidation. Besides, $O_3$ also disturbs photosynthetic allocation.

A better understanding of $O_3$ and its derived oxidants is necessary for a more detailed insight into the impacts of $O_3$ on plant growth and development. Concentration of $O_3$ of anthropogenic origin varies significantly with time and geographical locations. Pronounced diurnal cycles are found, showing low concentrations at night and highest concentration for a few hours after midday. Monitoring of $O_3$ concentration at different locations in rural areas supporting large agriculture fields is essential in view of significant negative impact of $O_3$ on plants. More emphasis should be given to air filtration studies in different regions. These studies help in depicting the future prospects of increasing $O_3$ concentration in the troposphere and its impacts on crop yield. Biomonitoring of $O_3$ should also be done to recognize the hot spots of $O_3$ pollution. As rise in the levels of tropospheric $O_3$ is likely to continue, problems related to crop yield loss may become important especially in tropical and subtropical countries having higher $O_3$ formation.

## ACKNOWLEDGEMENTS

The authors gratefully acknowledge that this publication is an output from a research project funded by the United Kingdom Department for International Development (DFID) for the benefit of developing countries. Financial assistance to Supriya Tiwari by CSIR, New Delhi is also acknowledged.

## REFERENCES

Adaros, G., H.J. Weigel and H.J. Jager. 1990. Effects of incremental ozone concentrations on the yield of bush beans (*Phaseolus vulgaris* var. *nanus* (L.) Ascher). *Gartenbauwissenschaft*, **4**: 162- 167.

Agrawal, M. and S.B. Agrawal. 1990. Effects of $O_3$ exposure on enzymes and metabolites of nitrogen metabolism. *Scientia Horticulturae*, **43**: 169- 177.

Agrawal, M., B. Singh, M. Rajput, F. Marshall and J. N. B. Bell. 2003. Effect of air pollution on peri urban agriculture: a case study. *Environ. Pollut,* **126**: 323-329.

Agrawal, M., P. K. Nandi and D. N. Rao. 1983(a). $O_3$ and $SO_2$ effects on *Panicum miliaaceum* plants. *Bull. Terry Bot. Club,* **110**: 435- 441.

Agrawal, M., P.K. Nandi and D.N. Rao. 1983(b). Ecophysiological responses of eggplants to $O_3$, $SO_2$ and a mixture of these two pollutants. *Ind. J. Air Poll. Cont.,* **4(1&2)**: 27- 32.

Agrawal. M. 1982. A study of phytotoxicity of ozone and sulphur dioxide pollutants. Ph.D Thesis submitted at Banaras Hindu University, Varanasi, India.

Andersen, C.P. and C.F. Seagel. 1997. Nutrient availability alters belowground respiration of ozone exposed ponderosa pine. *Tree Physiol,* **17**: 377- 387.

Bennett, J.P. and R.J. Oshima. 1976. Carrot response to ozone. *J. Amer. Soc. Hort. Sci.*, **101**: 638- 639.

Bergmann, E., J. Bender and H. Weigwl 1995. Growth responses and foliar sensitivities of native herbaceous species to ozone exposure. *Water, Air and Soil Poll,* **85**: 1437- 1442.

Bors, W., C. Langebartels, C. Michel and H. Sandermann. 1989. Polyamines as radical scavengers and protectants against ozone damage. *Phytochem.,* 28: 1589- 1595.

Burkey, K.O., C. Wei, G. Palmer, P. Ghosh, G.P. Fenner. 2000. Antioxidant metabolic levels in ozone sensitive and tolerant genotypes of snap beans. *Physiol. Plant.*, **110**: 195- 200.

Burkey, K.O., E. Gwendolyn. 2002. Ozone tolerance in snap bean is associated with elevated ascorbic acid in the leaf apoplast. *Physiol. Plant.*, **114**: 387- 394.

Butler, L.K. and T.W. Tibbitis. 1979. Stomatal mechanisms determining genetic resistence to ozone in *Phaseolus vulgaris* L. *J. Am. Soc. Hort. Sci.*, **104**: 213- 216.

Calatayud, A., J.W. Ramirez, J.I. Domingo and E. Barreno. 2002. Effects of ozone on photosynthetic $CO_2$ exchange, chlorophyll a fluorescence and antioxidant systems in lettuce leaves. *Physiol. Plant.*, **116**: 308-316.

Casano, L.M., M. Martin, and B. Sabater. 1994. Sensitivity of superoxide dismutase transcript levels and activities to oxidative stress is lower in mature senescent than in young barley leaves. *Plant Physiol.*, **106**: 1033.

Castillo, F.J. and H. Greppin. 1988. Extracellular ascorbic acid and enzyme activities related to ascorbic acid metabolism in *Sedum album* L. leaves after ozone exposure. *Environ. Exp. Bot.*, **28**: 231- 238.

Castillo, F.J., C. Penel and H. Greppin. 1984. Peroxidase release induced by ozone in *Sedum album* leaves. *Plant Physiol.*, **74**: 846.

Castillo, F.J., P.R. Miller and H. Greppin. 1987. Extracellular biochemical markers of photochemical oxidant air pollution to Norway spruce. *Experimentia*, **43**: 111- 220.

Chameides, W.L., P.S. Kasibhata, J. Yienger and H. Levy. 1994. Growth of continental scale metro-agro-plexes, regional ozone pollution and world food production. *Science*, **264**: 74- 77.

Chappelka, A.H. and L.J. Samuelson. 1998. Ambient ozone effects on forest trees of the eastern United States: a review. *New Phytol.*, **139**: 91- 108.

Conklin, P.L., E.H. Williams and R. L. Last. 1996. Environmental stress sensitivity of an ascorbic acid deficient *Arabidopsis* mutant. *Proceedings of the National Academy of Sciences, USA*, **93**: 9970-9974.

Cooley, D.R. and W.J. Manning. 1987. The impact of ozone on assimilates partitioning in plants: a review. *Environ. Pollut.*, **47**: 95- 113.

Dann, M.S. and E.J. Pell. 1989. Decline of activity and quantity of ribulose bisphosphate carboxylase/ oxygenase and net photosynthesis in ozone treated potato foliage. *Plant Physiol.*, **91**: 427- 432.

Degl'Innocenti, E., L. Guidi and G.F. Soldatini. 2002. Characterization of the photosynthetic response of tobacco leaves to ozone: $CO_2$ assimilation and chlorophyll fluorescence. *J. Plant Physiol.*, **159**: 845- 853.

Dominy, P.J. and R.L. Heath. 1985. Inhibition of $K^+$ stimulated ATPase of the plasmalemma of pinto bean leaves by ozone. *Plant Physiol.*, **77**: 43- 45.

Eckardt, N.A. and E.J. Pell. 1994. Ozone induced degradation of Rubisco protein and loss of Rubisco mRNA in relation to leaf age in *Solanum tuberosum* L. *New Phytol.*, **127**: 741.

Fangmeirer, A., F. Kanbach and H.J. Jager. 1993. Response of wheat photosynthesis to ozone: exposure- response relationships achieved in greenhouse chamber fumigations. *Angewandte Botanik,* **67**: 199- 203.

Fangmeirer, A., U. Brockerhoff, U.Gruters and H.J. Jager. 1994. Growth and yield responses in spring wheat (*Triticum aestivum* L. cv. Turbo) grown in open top chambers to ozone and water stress. *Environ. Pollut.*, **83**: 317- 325.

Farage, R.K., S.Y. Long, E. Lechner and N.R. Baker. 1991. The sequence of change within the photosynthetic apparatus of wheat following short-term exposure to ozone. *Plant Physiol.*, **95**: 529-535.

Foyer, C.H. and B. Halliwell. 1976. Presence of glutathione and glutathione reductase in chloroplasts: a proposed role in ascorbic acid metabolism. *Planta*, **133**: 21.

Fredericksen, T.S., J.M. Skelly, K.C. Steiner, T.E. Kolb and K.B. Konterick. 1996. Size mediated foliar response to ozone in black cherry trees. *Environ. Pollut.*, **91**: 53.

Fuhrer, J., A. Grandjean Grimm, W. Tschannen and H. Shariet- Madari. 1992. The responses of spring wheat (*Triticum aestivum* L.) to ozone at higher elevations. II. Changes in yield, yield components and gram quality in response to ozone flux. *New Phytol.*, **121**: 211- 219.

Fuhrer, J., E. Egger, B. Lehnherr, A. Grandjean and W. Tschannen. 1989. Effects of ozone on the yield of spring wheat *Triticum aestivum* L. cv. Albis grown in OTC. *Environ Pollut.*, **60**: 273- 289.

Ghauri, B.M.K., M. Salem and M.I. Mirza. 1992. Surface ozone in Karachi. pp. 169-177, In: M. Ilyas (ed.), *Ozone depletion implications for the tropics*, University of Science Malaysia/United Nations Environmental Programme, Nairobi.

Grandjean, G.A. and J. Fuhrer. 1992. The response of spring wheat (*Triticum aestivum* L.) to ozone at higher elevations, III. Responses of leaf and canopy gas exchange nd chlorophyll fluorescence to ozone flux. *New Phytol.*, **122**: 321.

Grantz, D.A. and S. Yang. 1996. Effects of ozone on hydraulic architecture in Pima cotton: Biomass allocation and water transport capacity of roots and shoots. *Plant Physiol.*, **112**: 1649- 1657.

Grantz, D.A. and Yang, S. 2000. Ozone impacts on allometry and root hydraulic conductance are not mediated by source limitation on developmental age. *J.Exp. Bot.*, **51** 919-927.

Grantz, D.A., X. Zhang and T.N. Carlson. 1999. Observations and model simulations link stomatal inhibition to impaired hydraulic conductance following ozone exposure in cotton. *Plant, Cell Environ.*,**22**: 1201- 1210.

Grulke, N.E., C.P. Andersen, M.E. Fenn and P.R. Miller. 1998. Ozone exposure and nitrogen deposition lowers root biomass of ponderosa pine in the San Bernardino Mountains, California. *Environ. Pollut.*, **103**: 63- 73.

Grulke, N.E., D. Olszyk, D. Grantz. 2001. Background pollution effects on $CO_2$ enhancement studies in forested ecosystems. *Bull. Ecol. Soc. Am.*, **82**: 54.

Guderian, R., D.T. Tingey and R. Rabe. 1985. Effects of photochemical oxidants on plants. pp. 129- 297, In: R.Guderian (ed.), *Air pollution by photochemical oxidants*, Springer, Verlag, Berlin.

Guiherit, R. and M. Roemer. 2000. Tropospheric ozone trends. *Chemosphere- Global Change Science*, **2**: 167- 183.

Guzy, M.R., R.L. Heath. 1993. Responses of ozone to varieties of common bean (*Phaseolus vulgaris* L.). *New Phytol.*, **124**: 617- 625.

Halliwell, B. and J.M.C Gutteridge. 1989. Free radicals in medicine and biology, 2$^{nd}$ edn Oxford: Clarendon Press, pp. 277- 289.

Hassan, I.A., M.R. Ashmore and J.N.B. Bell. 1994. Effects of ozone on the stomatal behaviour of Egyptian varieties of radish (*Raphnus sativus* L. cv. Baladey) and turnip (*Brassica rapa* L. cv. Sultani). *New Phytol.*, **128**: 243- 249.

Heagle, A.S. 1989. Ozone and crop yield. *Ann Rev. Phyto. Pathol.*, **27**: 397- 423.

Heath, R.L. 1988. Biochemical mechanism of pollution stress. pp. 259-286, In: W.W. Heck, D.T. Tingey and O.C. Taylor (eds.), *Assessment of crop loss from air pollutants*, Elsevier, London.

Heck, W.W., R.M. Adams, W.W. Cure, A.S. Heagle, H.E. Heggestad, R.J. Kohut, L.W. Kress, J.O. Rawlings and O.C. Taylor. 1983. A reassessment of crop loss from ozone. *Environ. Sci. Tech.*, **12**: 572A- 518A.

Heggestad, H.E. and E.H. Lee. 1990. Soybean root distribution, top growth and yield response to ambient ozone and soil moisture stress when grown in soil columns in greenhouses. *Environ. Pollut.*, **65**: 195- 207.

Held, A.A., H.A. Mooney and J.N. Gorham. 1991. Acclimitation to ozone stress in radish: leaf demography and photosynthesis. *New Phytol.*, **118**: 417- 423.

Hewitt, C.N., G.L. Kok, and R. Fall. 1990. Hydroperoxides in plants exposed to ozone mediate air pollution damage to alkene emitters. *Nature*, **344**: 56- 58.

Hippelli, S. and E.F. Elstner. 1996. Mechanisms of oxygen activation during plant stress: biochemical effects of air pollutants. *J. Plant Physiol.*, **148**: 249- 257.

Horemans, N., H. Asard and R.J. Caubergs. 1997. The ascorbate carrier of higher plant plasma membranes preferentially translocates the fully oxidized (dehydroascorbate) molecule. *Plant Physiol.*, **114**: 1247- 1253.

Jones, H.G. 1985. Partitioning stomatal and non stomatal limitations to photosynthesis. *Plant Cell Environ.*, **8**: 95- 104.

Karenlampi, L. and L. Skarby. (eds.), 1996. Critical levels for ozone in Europe: Testing and finalizing the concepts. UN-ECE workshop report. Department of Ecology and environmental Sciences, University of Kupio, Finland.363

Kress, L.W. and J.E. Muller. 1983. Impact of ozone on soybean yield. *J. Environ. Qual.*, **12(2)**: 276- 281.

Krupa, S.V. and Manning, W.J. 1988. Atmospheric ozone: formation and effects on vegetation. *Environ Pollut.*, **50**: 101- 137.

Lal, S., M. Naja and B.H. Subbaraya. 2000. Seasonal variations in surface ozone and its precursors over an urban site in India. *Atmos. Environ.*, **34**: 2713- 2724.

Landry, L.G. and E.J. Pell. 1993. Modification of Rubisco and altered proteolytic activity in ozone stressed hybrid poplar (*Populus maximowizii*). *Plant Physiol.*, **101**: 1355- 1362.

Langebartels, C., K. Kerner, S. Leonardi, M. Schraudner, M. Trost , W. Heller and H. Sandermann Jr. 1991. Biochemical plant responses to ozone. I. Differential induction of polyamine and ethylene biosynthesis in tobacco. *Plant Physiol.*, **95**: 882- 889.

Lee, E.H., A. Upadhyaya, M. Agrawal and R.A. Rowland. 1997. Mechanisms of ethylenediurea (EDU) induced ozone protection: reexamination of free radical scavenger systems in snap bean exposed to ozone. *Env. Exp. Bot.*, **38**: 199.

Lee, E.H., J.A. Jersey, C. Gifford and J.H. Bennett. 1984. Differential ozone tolerance in soybean and snapbean: analysis of ascorbic acid in ozone susceptible and ozone resistant by high performance liquid chromatography. *Env. Exp.Bot.*, **24**: 331- 341.

Lefohn, A.S., S.J. Oltmans, T. Dann and H.B. Singh. 2001. Present day variability of background ozone in the lower troposphere. *J. Geophysical Res.*, **106**: 9945- 9958.

Lehnherr, B., A. Grandjean, F. Machler and J. Fuhrer. 1987. The effects of ozone in ambient air on ribulose bisphosphate carboxylase/ oxygenase activity decreases photosynthesis and grain yield in wheat. *J. Plant Physiol.*, **130**: 189- 200.

Lynos, T., J. Ollerenshaw and J. Barnes. 1999. Impacts of ozone on *Plantago major*: apoplastic and symplastic antioxidant status. *New Phytol.*, **141**: 253.

Maggs, R.A., S.R.A. Shamsi and M. R. Ashmore. 1995. Effect of ambient air pollution on wheat and rice yield in Pakistan. *Water, Air and Soil Poll.*, **85**: 1311- 1316.

Manning, W.J. and B. Godzik. 2004. Bioindicator plants for ambient ozone in Central and Eastern Europe. *Environ. Pollut,* **130**: 33-39.

Manning, W.J., B. Godzik, R. Musselman. 2002. Potential bioindicator plant species for ambient ozone in forested mountain areas of Central Europe. *Environ. Pollut.*, **119**: 283- 290.

Mansfield, T.A. and M. Pearson. 1996. Disturbances in stomatal behaviour in plants exposed to air pollution. pp. 179- 194, In: M. Yunus and M. Iqbal (eds.), *Plant response to air pollution*, John Wiley and sons limited, Chichester.

Maurer, S., R. Matyssek, M.S. Gunthardt Georg, W. Landolt and W. Einig. 1997. Nutrition and the ozone sensitivity of birch (*Betula pendula*). I. Responses at leaf level. *Tree Struc. Func.*, **12**: 1- 10.

Mehlhorn, H., B. Tabner and A.R. Wellburn. 1990. Electron spin resonance evidence or the formation of free radicals in plants exposed to $O_3$. *Physiol. Plant.*, **79**: 377- 383.

Meyers, U., B. Kollner, J. Willenbrink and G.H.M. Krause. 1997. Physiological changes on agricultural crops induced by different ambient $O_3$ exposure regimes. *New Phytol.*, **136**: 645- 652.

Miller, J.E., W.A. Pursley, S.F. Vozzo and A.S. Heagle. 1991. Response of net carbon exchange rate of soybean to $O_3$ at different stages of growth and its relation to yield. *J. Env. Qual.*, **20**: 571- 575.

Mudd, J.B., R. Leavitt, A. Ongun and T.T. McManus. 1969. Reaction of ozone with amino acids and proteins. *Atmos Environ.*, **3**: 669- 682.

Mudd, J.B., T.T. McMAnus and A. Ongun. 1971. Inhibition of lipid metabolism in chloroplasts by ozone. In: *Proceedings of the Second International Clean Air Congress*, Academic press, Inc., New York.

Musselman, R.C., P.M. McCool and A.S. Lefohn. 1994. Ozone descriptors for an air quality standard to protect vegetation. *J. Air Waste Manag. Assoc.*, **44**: 1383- 1387.

Nouchi, I. and S. Toyama. 1988. Effects of ozone and PAN on polar lipids and fatty acid in leaves of Morning Glory and Kidney bean. *Plant Physiol.*, **87**: 638- 646.

Paludan- Miller, S., H. Saxe and J.W. Leverenz. 1999. Responses of ozone in 12 provenances of European beech (*Fagus sylvatica*): genotypic variations and chamber effects on photosynthesis and dry matter partitioning. *New Phytol.,* **144**: 261-273.

Pandey, J., M. Agrawal, N. Khanam, D. Narayan and D.N. Rao. 1992. Air pollution concentrations in Varanasi, India. *Atmos. Environ.*, **24B**: 91-98.

Paul, K.P. and J.E. Thompson. 1984. Evidence of the accumulation of peroxidized lipids in the membranes of senesing cotyledons. *Plant Physiol.*, **75**: 1152-1157.

Pell, E.J., C.D. Schlagnhaufer and R.N. Arteca. 1997. Ozone induced stress: mechanism of action and reaction. *Physiol. Plant,* **100**: 364-373.

Pell. E.J., N. Eckardt and A. J. Enyedi. 1992. Timmings of ozone stress and resulting status of ribulose bisphosphate carboxylase/ oxygenase and associated net photosynthesis. *New Phytol.*, **120**: 397-405.

Pleijel, H., L. Skarby, G. Wallin and G. Sellden. 1991. Yield and grain quality of spring wheat (*Triticum aestivum* L. cv. Darbant), exposed to different concentrations of ozone in open top chambers. *Environ Pollut.*, **69**: 151-168.

Pye, J.M. 1988. Impact of ozone on growth and yield of trees, a review. *J. Environ. Qual.*, **17**: 347-360.

Rajput, M. and M. Agrawal. 2004. Physiological and yield responses of pea plants to ambient air pollution. *Indian J. Plant Physiol.*, **9(1)**: 9-14.

Rajput, M. and M. Agrawal. 2005. Biomonitoring of air pollution in a seasonally dry tropical suburban area using wheat transplants. *Env. Mon, Assess.,* **101**: 39- 53.

Ranieri, A., G. D'Usro, C. Nali, G. Lorenzini and G.F. Soldatini. 1996. Ozone stimulates apoplastic antioxidant systems in pumpkin leaves. *Plant Physiol.*, **97**: 381.

Reddy, G.N., R.N. Arteca, Y-R Dai, H.E.Flores, F.B. Negm, E.J.Pell. 1993. Changes in ethylene and polyamines in relation to mRNA levels of the large and small subunits of ribulose bisphosphate carboxylase/ oxygenase in ozone stressed potato foliage. *Plant, Cell and Environment,* **16**: 819-826.

Reich, P.B. 1987. Quantification plant response to ozone: a unifying theory. *Tree Physiol.*, **3**: 63- 91.

Reich, P.B. and R.G. Amundson. 1985. Ambient levels of ozone reduce net photosynthesis in tree and crop species. *Science*, **230**: 566- 570.

Reiling, K. and A.W. Davison. 1994. Effects of exposure to ozone at different stages in development of *Plantago major* L. on chlorophyll fluorescence and gas exchange. *New Phytol.*, **128**: 509- 514.

Richards, B.L., J.T. Middleten and W.B. Hewitt. 1958. Air pollution with relation to agronomic crops: oxidant stipple of grapes. *Agron J.*, **50**: 410- 415.

Rowland- Bamforh, A.J., A.M. Borland, P.J. Lea and T.A. Mansfield. 1989. The role of arginine decarboxylase in modulating the sensitivity of barley to $O_3$. *Environ. Poll.*,**61**: 95- 106.

Runeckles, V.C., M. Vaartnou. 1997. EPR evidence for superoxide radicals formation in leaves during exposure to low levels of ozone. *Plant, Cell and Environment*, **20**: 306- 314.

Sanders, G.E., J.J. Colls and A.G. Clark. 1992. Physiological changes in *Phaseolus vulgaris* in response to long term $O_3$ exposure. *Ann. Bot.*, **69**: 123- 133.

Sanders, G.E., N.D. Turnbull, A.G. Clark and J.J. Colls. 1990. The growth and development of *Vicia faba* L. in filtered and non filtered open top chambers. *New Phytol.,* **116**: 67- 78.

Schenone G., G. Botteschi, I. Fumagelli and F. Montinaro. 1992. Effects of ambient air pollution in open top chambers on beans (*Phaseolus vulgaris* L.) I. Effects on growth and yield. *New Phytol.*, **122**: 689- 697.

Schreiber, U., W. Vidaver, V.C. Runeckles and P. Rosen. 1978. Chlorophyll fluorescence. Assay for ozone injury in intact plants. *Plant Physiol.*, **61**: 80- 84.

Singh, A., S.M. Sarin, D. Shannugham, N. Sharma, A.K.Attri and W. K. Jain. 1997. Ozone distribution in urban environment of Delhi during winter months. *Atmos. Environ.*, **31**: 3421- 3427.

Singh, E. 1998. Effect of $O_3$ pollution on selected crop plants. Ph.D Thesis submitted at Banaras Hindu University, Varanasi, India.

Smirnoff, N. 1996. The function and metabolism of ascorbic acid in plants. *Ann. Bot.*, **78**: 661.

Soja, G. and A.M. Soja. 1995. Ozone effects on dry matter partitioning and chlorophyll fluorescence during plant development of wheat. *Water, Air and Soil Pollution,* **85**: 1461- 1466.

Stockwell, W.R., G. Kramm, H.E. Schcel, V.A. Mohnen and W. Sciler. 1997. Ozone formation, destruction and exposure in Europe and United States. In: H. Sandermann, Wellburn, A.R. and Heath, R.L. (eds.), *Forest decline and ozone*, Springer-verlag, Berlin, Germany.

Temmermann, L., K. de Vandermeiren and M. Guns. 1992. Effects of air filtration on spring wheat grown in open top chambers at a rural site, I. Effect on growth yield and dry matter partitioning. *Environ Pollut.*, **77**: 1-5.

Van Buuren, M.L., L. Guidi, S. Formale, F. Ghetti, M. Franceschetti, G.F. Soldatini and N. Bangi. 2002. Ozone response mechanisms in tobacco implications of polyamine metabolism. *New Phytol*, **156**: 389-398.

Wahid, A., R. Maggs, S.R.A. Shamsi, J.N.B. Bell and M.R. Ashmore. 1995(a). Air pollution and its impacts on wheat yield in Pakistan Punjab. *Environ. Pollut.*, **88**: 147-154.

Wahid, A., R. Maggs, S.R.A. Shamsi, J.N.B. Bell and M.R. Ashmore. 1995(b). Effects of air pollution on rice yield in Pakistan Punjab. *Environ. Pollut.*, **90**: 323- 329.

Warwick, K.R. and G.Taylor. 1995. Contrasting effects of tropospheric ozone on five native herbs which co exist in calcareous grassland. *Global Change Biology*, **1**: 143- 151.

Zheng, Y., H. Shimizu and J.D. Barnes. 2002. Limitations to $CO_2$ assimilation in ozone exposed leaves of *Plantago major*. *New Phytol.*, **155**: 67- 78.

*Advances in Plant Physiology*, Vol. **9**
Ed. A. Hemantaranjan
Scientific Publishers (India), Jodhpur, 2006 pp. 87-100
E-mail: **info@scientificpub.com** www.scientificpub.com

# 5

# SUPEROXIDE DISMUTASE - SCAVENGERS OF REACTIVE OXYGEN SPECIES

K.V. Kasturi Bai and Shamina Azeez

Principal Scientist Plant Physiology and Scientist (Senior Scale) Biochemistry
Central Plantation Crops Research Institute, Kudlu. P.O. Kasaragod, 671 124, Kerala, India

## INTRODUCTION

SOD is widespread in archeae, bacteria, fungi, plants and animals and present in all oxygen-metabolizing cells (Gregory *et al.*, 1974); it has also been reported in anaerobic bacteria (Hewitt and Morris, 1975). One of the exceedingly rare exceptions is *Lactobacillus plantarum* and related lactobacilli.

SOD has been purified and characterized from a number of sources: fungi (Rapp *et al.*, 1973); green pea (Sawda *et al.*, 1972; Guo *et al.*, 1996); *Streptococcus mutans* (Vance *et al.*, 1972); wheat germ (Beauchamp and Fridovich, 1973); *E. coli* (Gregory *et al.*, 1973); *Saccharomyces cerevisiae* (Goscin and Fridovich, 1972), *Neurospora crassa* (Misra and Fridovich, 1972), *Capsicum annuum* (Kim *et al.*, 1997), oilseed plants (Corpas *et al.*, 1998), *Citrus limonum* (Almansa *et al.*, 1994), spinach leaves (Kitagawa *et al.*, 1991), brussel sprouts (Walker *et al.*, 1991) and many more.

This enzyme may be located in the cytosol, mitochondria, apoplast or peroxisomes and may be of the Cu (II)-Zn (II) or Mn (III) or Fe (III) types. A blue-green Cu(II)-Zn(II) enzyme comes from human and bovine erythrocytes, a wine-red Mn(III) protein is found in *E. coli,* and in chicken, and rat (Peeters-Joris *et al.*, 1975) and a yellow Fe(III) enzyme from *E. coli* (Villafranca *et al.*, 1974).

The cytosols of almost all eukaryotic cells contain Cu-Zn-SOD. The Cu-Zn-SOD available commercially is normally that found in the bovine erythrocyte. Chicken liver (and nearly all other) mitochondria, and many bacteria (such as *E. coli*) contain Mn-SOD. (Eg: Mn-SOD found in human mitochondria). *E. coli* and many other bacteria also contain Fe-SOD; some bacteria contain Fe-SOD, others Mn-SOD, and some contain both. The three types of this metallo enzyme are described in greater detail below.

In humans, three types of SOD are present. SOD1 is located in the cytoplasm, SOD2 in the mitochondria and SOD3 is extracellular. The former is a dimer, while the others are tetramers. SOD1 and SOD3 contain copper and zinc, while SOD2 has manganese in its reactive center. (http://www.answers.com/topic/superoxide-dismutase).

The presence of SOD has been shown to protect many types of cells from the free radical damage that is important in aging, senescence, and ischemic tissue damage. SOD also helps protect cells from DNA damage, lipid peroxidation, ionizing radiation damage, protein denaturation, and many other forms of progressive cell degradation. Mutations in the first SOD enzyme (SOD1) in humans have been linked to familial amyotrophic lateral sclerosis (ALS, a form of motor neurone disease). The other two types have not been linked to any disease.

SOD has been cloned and characterized from the callus of sweet potato (Lin *et al.*, 1997). The nucleotide sequence of the gene encoding the cytosolic Cu/Zn SOD from sweet potato (Lin *et al.*, 1995) and pea (White and Zilinskas, 1991), and the cDNA encoding mitochondrial manganese SOD (Wong *et al.*, 1991) have been reported. Bowler *et al.* (1990) have cloned the sequence encoding bacterial MnSOD into an yeast expression vector and subsequently introduced it into a Cu/ZnSOD-deficient mutant of the yeast. Functional expression of the protein was demonstrated, and complementation tests revealed that the protein was able to provide tolerance at wild-type levels to conditions that are normally restrictive for this mutant.

*Agrobacterium tumefaciens* harboring a plasmid vector containing the *H*evea *brasiliensis* superoxide dismutase gene was used to mediate genetic transformation and the regeneration of transgenic plants in *H. brasiliensis* (Jayashree *et al.*, 2003). Sunanda *et al.* (1998) have concluded that *Piper betle* exhibits both oxidant and antioxidant properties from the observations that hepatic SOD activity was not altered while renal SOD activity increased in mice treated with 0.1 g *P. betle*/kg.

SOD-null mutants of *Escherichia coli*– SOD A, SODB - exhibit enhanced mutagenesis, reflecting $O_2$-sensitive pathways and DNA damage. Yeast lacking either Cu.ZnSOD or MnSOD are oxygen intolerant and the double mutants are hypermutable, defective in sporulation and had a higher requirement for Met and Lys. Cu.ZnSOD-null *Drosophila melanogaster* had shorter lifespans (Fridovich, 1995).

## 2. REACTIVE OXYGEN SPECIES

The main source of reactive oxygen species (ROS) *in vivo* is aerobic respiration, although ROS are also produced by peroxisomal •-oxidation of fatty acids, microsomal cytochrome P450 metabolism of xenobiotic compounds, stimulation of leukocyte phagocytosis by pathogens (Allen *et al.*, 1974; DeChatelet *et al.*, 1974) or lipopolysaccharides, arginine metabolism, and tissue specific enzymes (Nicholls and Budd, 2000).

In an aerobic cell, cell respiration involves electron transport and oxidative phosphorylation, culminating in the oxidative degradation of carbohydrates, fats and amino acids. Here electron flow from organic substrates to oxygen yields energy for generation of the energy currency of the cell - Adenosine triphosphate (ATP) - from Adenosine diphosphate (ADP) and phosphate.

It is vital that the resultant oxygen molecule of aerobic respiration and oxidative phosphorylation be completely reduced and thus neutralized to two $H_2O$ by accepting four electrons. Partial reduction of $O_2$ by two electrons generates hydrogen peroxide ($H_2O_2$)

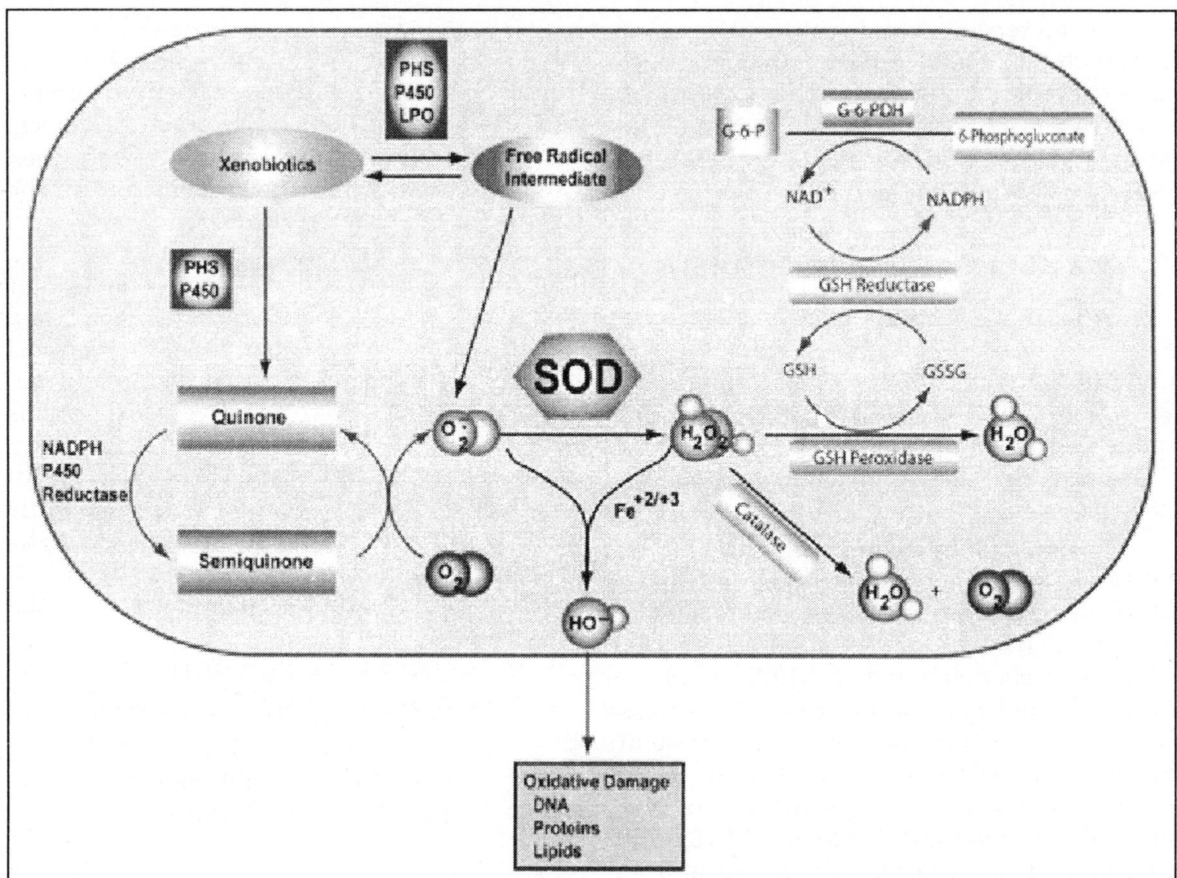


(Reproduced from: http://www.sigmaaldrich.com/Area_of_Interest/Biochemicals/ Enzyme_ Explorer/ Cell_Signaling_Enzymes/Superoxide_Dismutase.html)

and acceptance of only one electron by $O_2$ results in the superoxide radical ($O_2^-$). Both hydrogen peroxide and superoxide radical are extremely toxic to cells because they attack the unsaturated fatty acid components of membrane lipids, thus damaging their structure, as also essential proteins and DNA. Clinically, oxidative stress and ROS have been implicated in aging, lipid peroxidation and the peroxidative hemolysis of red blood cells (Fee and Teitelbaum 1972), disease states such as Alzheimer's disease, Parkinson's disease, cancer and aging. SOD deficiency might lead to Heinz body hemolytic anemia (Winterbourn *et al.*, 1975). Fridovich (1986) reports on the biological effects of the superoxide radical.

## 3. SUPEROXIDE DISMUTASE

Aerobic cells protect themselves against superoxide and peroxide attack by the protective action of the enzyme superoxide dismutase (SOD), which converts superoxide radical to hydrogen peroxide, and catalase, which converts hydrogen peroxide to water and molecular $O_2$:

$$2O_2^- + 2H^+ \xrightarrow{\text{Superoxide dismutase}} H_2O_2 + O_2$$

$$2H_2O_2 \xrightarrow{\text{Catalase}} 2H_2O + O_2$$

In the reaction catalyzed by SOD, it acts as both oxidant and reductant. The disproportionation of superoxide radical ($2O_2^-$ + $2H^+$ ↔ $H_2O_2$ + $O_2$) is accomplished by Fe-, Mn- and Cu,Zn-SODs in two steps, both first order with respect to $O_2^-$:

$$Ez_{ox} + O_2^- + H^+ \longleftrightarrow Ez_{red}(H^+) + O_2$$

$$Ez_{red}(H^+) + O_2^- + H^+ \longleftrightarrow Ez_{ox} + H_2O_2$$

Where $Ez_{ox}$ and $Ez_{red}$ denote the $Cu^{2+}Zn^{2+}$ and $Cu^+Zn^{2+}$, $Fe^{3+}$ and $Fe^{2+}$, $Mn^{3+}$ and $Mn^{2+}$, $Ni^{3+}$ and $Ni^{2+}$ states of the Cu,Zn-, Fe-, Mn- and Ni-SODs respectively. Proton uptake coupled to enzyme reduction is presumed but has not been experimentally demonstrated for Ni-SOD (Miller, 2004).

Cyanide inhibits cupro-zinc SOD but has no effect on the mangano enzyme of chicken liver mitochondria (Weisiger and Fridovich, 1973). SOD is inactivated by $H_2O_2$ (Symonyan and Nalbandyan, 1972; Fielden *et al.*, 1973) and may be protected by catalase (Bray *et al.*, 1974) with which it is usually associated. SOD is an unusually stable enzyme although its apoenzyme is very unstable (Forman and Fridovich, 1973). SOD can retain its activity for up to a year at 5 °C.

## 3.1 SOD Protein Structure

The bovine enzyme exists as a dimeric copper-zinc containing protein with a molecular weight of 2 x 16,300 (Keele *et al.*, 1971). The *E. coli* enzyme exists as a dimeric manganese or iron containing glycoprotein with a molecular weight of 2 x 22,000 (Cass, 1985; Beyer, 1991). The human enzyme exists as a tetrameric copper-zinc or manganese containing glycoprotein with a molecular weight of 2 x 23-28,300 (Keele *et al.*, 1971; Matsuda *et al.*, 1990).

Characteristic amino acid sequences were detected in the sequences of chloroplast and cytosol isozymes of Cu/Zn SOD in spinach, rice and horsetail, indicating that the rate of mutation of the cytosolic enzyme was higher than the chloroplast enzyme (Kanematsu and Asada, 1990).

### *3.1.1 Cu-Zn SOD*

'Erythrocuprein' was the first of the SODs to be identified (McCord and Fridovich, 1969). Bovine erythrocyte SOD has been extensively studied. It is identical to the enzyme from human erythrocytes and from beef heart (Bannister *et al.*, 1971; Keele *et al.*, 1971 and Nyman, 1960). It has a molecular weight of 32,500 (Keele *et al.*, 1971) and consists of two subunits of identical molecular weight joined by a disulfide bond. There are two Cu(II) and two Zn(II) atoms per molecule (Bannister *et al.*, 1971). According to Forman and Fridovich (1973) zinc has a structural, stabilizing role, while $Cu^{2+}$ is directly involved in the catalytic activity. Forman *et al.* (1973) and Rigo *et al.* (1975) indicate histidine to be involved at the active site.

$$Cu^{+2}\text{-SOD} + O_2^- \bullet\ Cu^{+1}\text{-SOD} + O_2$$

$$Cu^{+1}\text{-SOD} + O_2^- + 2H^+ \bullet\ Cu^{+2}\text{-SOD} + H_2O_2.$$

In this reaction the oxidation state of the copper changes between +1 and +2.

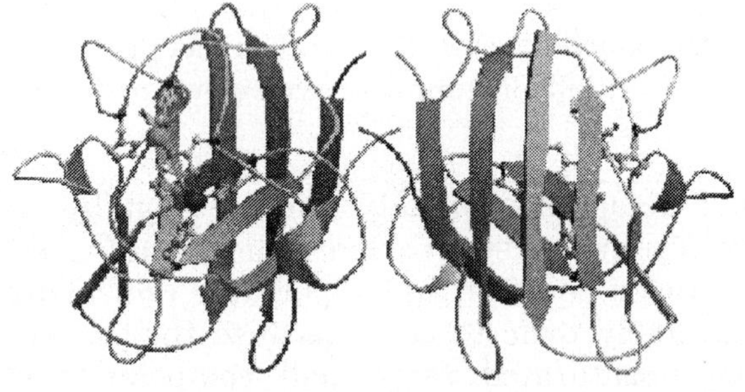

The protein is peculiar in that it does not show absorption maximum at 280 nm (McCord and Fridovich, 1969; Bannister *et al.*, 1971). The extinction coefficient of SOD is $A_{259}/A_{680}$ = 30 (Symonyan and Nalbandyan, 1972) and isoelectric point is 4.95 (Bannister *et al.* 1971).

The active site of bovine Cu/Zn-SOD lies at the base of a 15Å deep cavity formed by two loops. These loops contain charged residues, which facilitate the electrostatic guidance of superoxide to the active site. The active site of the oxidized Cu/Zn-SOD has a $Cu^{2+}$ ion coordinated in a distorted square pyramid by a solvent molecule and bridged by the imidazole ring of His61, a feature unique to enzymes of this class. The Cu is coordinated by a further three His ligands, while Zn is ligated by two additional His and an Asp. Upon reduction, Cu is presumed to release the His ligand it shares with the $Zn^{2+}$ ion, as well as departing $O_2$, to become trigonal and roughly planar (Banci *et al.*, 2002). The Cu (II) ion is directly involved in the catalytic activity, while zinc has a structural, stabilizing role (Forman and Fridovich, 1973).

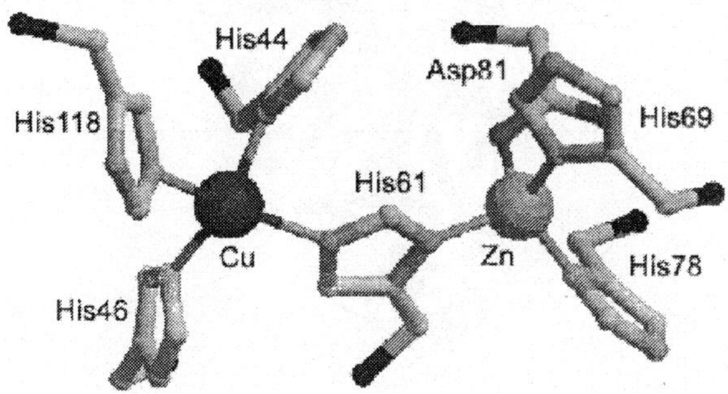


Immuno-gold electron microscopic analysis with specific antibodies has been used to localize the enzyme in the intra- and extra-cellular regions (Ogawa *et al.*, 1996). The intra-organellar distribution of SOD in two types of plant peroxisomes - glyoxysomes and leaf peroxisomes was determined by solubilization assays (Sandalio and Del-Rio, 1988).

### *3.1.2 Fe-SOD*

Fe and Mn-SODs occur as homodimers or homotetramers. Fe-SOD is found in obligate anaerobes and aerobic diazotrophs (exclusively), facultative aerobes (exclusively or together with Mn-SOD), in the cytosol of cyanobacteria, in the chloroplast stroma of higher plants and in protozoa. Fe-SOD and Mn-SOD from some organisms (e.g. *E. coli*) exhibit almost absolute metal specificity (Beyer *et al.*, 1989), while other enzymes, such as `cambialistic' SOD from *Propionibacterium shermanii*, i.e., active with either metal, Fe or Mn (Sehn, and Meier, 1994).

$$Fe^{3+} + O_2^{\cdot -} \rightarrow Fe^{2+} + O_2$$

$$Fe^{2+} + O_2^{\cdot -} + 2H^+ \rightarrow Fe^{3+} + H_2O_2$$

The active site iron is pentacoordinate, with the metal ligands ($N^{\bullet}$ of three conserved His residues, $O^{\bullet}$ of the conserved Asp residue and a water molecule) arranged in distorted trigonal bipyramidal geometry. The first His residue and a solvent molecule fill the two axial positions. In the azide-$Fe^{3+}$-SOD complex, the iron becomes hexacoordinate with distorted octahedral geometry, with azide coordinated *trans* to Asp ligand (Lah *et al.*, 1995).

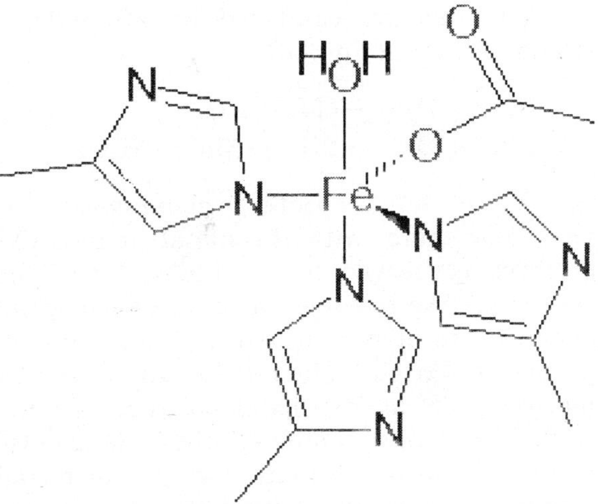


The 3D structures of several Fe-SODs have been determined (Lim *et al.*, 1997, Lah *et al.*, 1995, Cooper *et al.*, 1995, Stoddard *et al.*, 1990, Schmidt *et al.*, 1996). The monomers fold into two domains. The N-terminal domain consists of two long antiparallel α helices. The C-terminal domain contains a central ß sheet formed by three antiparallel ß strands with 4-6 surrounding α helices. The iron atom is liganded by two residues from each of N-terminal helices and two residues from the loops in the C-terminal domain.

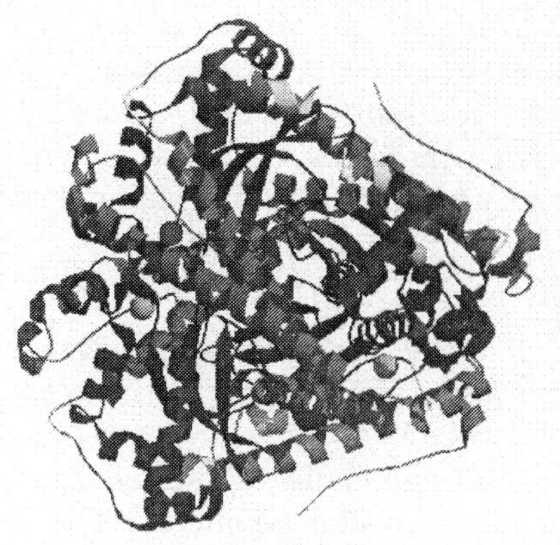

*G152a mutant of mycobacterium tuberculosis iron-superoxide dismutase at 2.90Å resolution*

### *3.1.3 Mn-SOD*

The reaction catalyzed by Mn-SOD is similar to that of Fe-SOD:

$$Mn^{3+} + O_2\cdot^{-} \rightarrow Mn^{2+} + O_2$$

$$Mn^{2+} + O_2\cdot^{-} + 2H^{+} \rightarrow Mn^{3+} + H_2O_2$$

The active site manganese is pentacoordinate, with the metal ligands ($N^{\varepsilon}$ of three conserved His residues, $O^{\delta}$ of the conserved Asp residue and a water molecule) arranged in distorted trigonal bipyramidal geometry. The first His residue and a solvent molecule fill the two axial positions. In the azide-$Mn^{3+}$-SOD complex, the manganese becomes hexacoordinate with distorted octahedral geometry, with azide coordinated *trans* to Asp ligand (Borgstahl *et al.*, 1992).



The 3D structures of several Mn-SODs have been determined (Edwards *et al.*, 1998, Borgstahl *et al.*, 1992, Schmidt *et al.*, 1996, Lah *et al.*, 1995). The monomers fold into two domains. The N-terminal domain consists of two long antiparallel α helices. The C--terminal domain contains a central ß sheet formed by three antiparallel ß strands with 4-6 surrounding α helices. The manganese atom is liganded by two residues from each of N-terminal helices and two residues from the loops in the C-terminal domain.

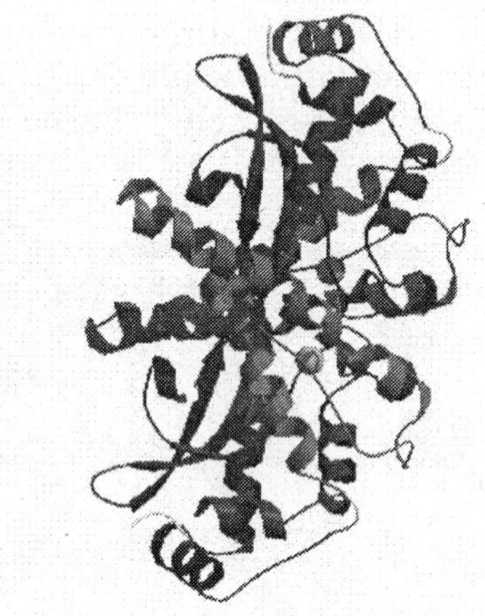

*Human manganese superoxide dismutase mutant q143n of 2.30Å resolution*

### *3.1.4 Ni-SOD*

Several *Streptomyces* spp. produce Ni-SOD which is completely unrelated to any of the above three SODs (Youn *et al.*, 1996). Ni-SOD purified from the cytosol of *Streptomyces coelicolor* is homotetramer of 13.5 kDa, stable at pH 4-8, and up to 70 °C. It is inhibited by cyanide and $H_2O_2$ but little inhibited by azide. They have no genetic or structural homology with the already characterized Mn, Fe-, or CuZn-SODs. The apoenzyme, lacking the metal, had no ability to mediate the reaction, indicating a strong requirement for $Ni^{3+}$ (Hunt *et al.*, 1998). Ni-SOD contains one Ni per monomer, but controversy exists as to whether the protein is tetrameric or hexameric. X-ray absorption spectroscopy studies reveal a penta-coordinate $Ni^{3+}$ coordinated by multiple S donors and N/O ligands (Choudhury *et al.*, 1999). There are only two Cys per monomer and the single Met is not a ligand (Bryngelson *et al.*, 2004), leading to the proposal of a dinuclear $Ni_2$ site, consistent with the $Ni^{3+}$ EPR signal intensity (Youn *et al.*, 1996; Choudhury *et al.*, 1999). Further revelations of the structure of Ni-SOD are sure to provide exciting information.

## 4. SOD ASSAY

Since plants cannot move around freely, they have developed highly sophisticated defense mechanisms toward environmental alterations. For example, the SOD activity of a plant is increased by the use of herbicides such as paraquat, increase in $SO_2$ concentration in the atmosphere, drought, or exposure to high concentrations of zinc and magnesium. These phenomena cause ROS to be generated in the plant. Findings also suggest that the reduction in the toxicity of $O_2^-$ is a very important defense system against oxidative stress. Therefore, the role of SOD assay techniques is very important in the study of plant physiology.

Different methods of assay of SOD have been standardized. The one most frequently used is based on the ability of SOD to inhibit the reduction of nitro blue tetrazolium by superoxide (Winterbourn *et al.*, 1975). A modified technique has been developed by Chen *et al.* (1996), which combines native polyacrylamide gel electrophoresis and laser densitometry.

Activity of SOD has been determined in two ways:

1. Inhibition by the enzyme of an $O_2^-$ dependent reaction.
2. Pulse radiolytic methods (Rigo *et al.*, 1975).

The method of assay of SOD (Winterbourn *et al.*, 1975) is based on the ability of superoxide dismutase to inhibit the reduction of nitro-blue tetrazolium by superoxide. One unit is defined as that amount of enzyme causing half the maximum inhibition of NBT reduction. The reaction velocity will depend largely on somewhat variable assay conditions such as light intensity and reaction temperature. Calibration of the method in individual laboratories is recommended.

The most important technical aspect of SOD assay is the applicability of the assay to different samples, with high selectivity and little interference from other components. For the production of $O_2^-$, the substrate of SOD, a xanthine-xanthine oxidase reaction is used. A probe for the detection of $O_2^-$ is included in the reaction solution. The change of the probe without any sample acts as a blank control, and the suppression ratio of the change of the probe by the sample solution is indicated as the inhibition ratio. Generally, 50% inhibition by the sample solution is used for the activity determination ($IC_{50}$). On the other hand, $O_2^-$ generated by the xanthine-xanthine oxidase reaction is spontaneously transformed to oxygen and hydrogen peroxide. This spontaneous dismutation reaction occurs rapidly in acidic conditions, and at the rate of $8.5x10^5$ - $8.5x10^4$ $M^{-1}s^{-1}$ at physiological pH (pH 7 - 8). Therefore, the second-order rate constant of the reaction between $O_2^-$ and the probe should exceed the rate constant of the dismutation reaction. In the case where the rate constants are almost the same, the concentration of the probe should be increased.

There are several different types of probes: those which change their color (colorimetric probes) when they react with $O_2^-$, those which emit light (chemiluminescence probes) when they react with $O_2^-$, and those which produce specific radicals (spin trap agents) when they react with $O_2^-$.

### 4.1 Spectrophotometric Detection

The most typical SOD detection method is the one based on spectrophotometric detection. This method uses either cytochrome C or nitroblue tetrazolium (NBT) (Fig. 1). The detection of $O_2^-$ by the cytochrome C reducing method is based on the color change to purple color from reduced cytochrome C. This is the most commonly utilized method since the discovery of SOD.

$$Cyt(Fe^3) + O_2^- \longrightarrow Cyt(Fe^2) + O_2$$

Since cytochrome C is easily reduced by reductases such as NADPH reductase and other reducing agents, it is necessary to consider contaminants in the samples. Moreover, this method requires continuous monitoring in 1.5-minute intervals, so it is not suitable for high-throughput detection.

The NBT method is based on the generation of water-insoluble blue formazan dye (• max: 560 nm) by a reaction with $O_2^-$. Because the dye is not water-soluble, a non-homogeneous suspension is created during long-term analysis that causes problems in the reproducibility of the data. In order to solubilize formazan, alternative methods such as addition of BSA have been developed. However, the addition of unnecessary proteins may make the data analyses much more complicated. Moreover, NBT is reduced by various reducing agents. This characteristic of NBT is utilized for the detection of keto-amine, which is a marker of diabetes and is the intermediate in the Maillard reaction. The most significant disadvantage of the NBT method is that 100% inhibition cannot be achieved even with the addition of excessive amount of SOD into assay solution. The direct interaction between NBT and xanthine oxidase is speculated to be the cause.

### 4.2 Chemiluminescence Method

The chemiluminescence probe used for $O_2^-$ detection can also be applied for SOD assay detection. There are two types of these probes. One is a lucigenin, and the other is a luciferin derivative (MCLA). These chemiluminescence reactions are highly pH dependent. For example, lucigenin shows extremely intense chemiluminescence at pH 9.0 and higher. Therefore, SOD detection by chemiluminescence under physiological pH conditions is not feasible. On the other hand, MCLA emits strong luminescence under physiological conditions, and is therefore used for the detection of Cu, Zn-SOD activity in the human brain. However, MCLA is not a suitable SOD assay probe since it reacts not only with $O_2^-$ but also with singlet oxygen. Moreover, MCLA reacts with dissolved oxygen to emit background luminescence, and the transitional metal ions accelerate the oxidation reaction.

### 4.3 Electron Spin Resonance Spectroscopy (ESR) Method

At room temperature, the ESR signal of $O_2^-$ in solution cannot be detected directly, but can be indirectly detected by a spin trap method. The most common trapping agent is 5,5-dimethyl-1-pyrroline N-oxide (DMPO). Since $O_2^-$ trapping DMPO indicates a particular ESR spectrum pattern, ESR detection is the most specific method for $O_2^-$ detection. However, the second-order rate constant between DMPO and $O_2^-$ is relatively lower than the reaction constant of the spontaneous reaction of $O_2^-$. Therefore, a large amount of DMPO should be added to the solution (e.g., using a final concentration of 0.45 M). Unfortunately, this large volume of DMPO increases the cost per assay. Another problem with this method is the requirement of a relatively expensive ESR instrument.

### 4.4 New SOD Assay by Use of Water-soluble Tetrazolium Salts

In order to overcome the above-mentioned problems, the following aspects need to be considered for the development of SOD assay: an economical method using a simple instrument, a less pH-sensitive method, and a highly $O_2^-$-specific method. If a general spectrophotometer is applied as a simple instrument, colorimetric probes similar to cytochrome C and NBT are preferable. The most important point for the assay is to be able to determine 100% inhibition by SOD without interference from other components. New tetrazolium salts that generate water-soluble formazan by reduction in order to establish a new SOD assay have been studied: XTT is a water-soluble tetrazolium salt first reported in 1988. Since then, it has been utilized as a substrate for the electron transfer system of bacteria cells or mammalian cells. While NBT has a *bis*-tetrazolium structure, the XTT structure is monotetrazolium and it has two sulfonic acid groups.

When XTT was used for the SOD assay, 100% inhibition was observed in proportion to the increase in the SOD concentration. This 100% inhibition could not be observed by NBT method before. At an optimized condition, 100% inhibition was still observed at various pH. The observation of 100% inhibition by XTT means that XTT is specifically reduced by $O_2^-$, and XTT overcomes the problem associated with NBT.

The reaction mixture should include 2.5 ml of 50 mM carbonate buffer (pH 9.4; and 10.2) or 50 mM phosphate buffer (pH 8.0) and 0.1 ml each of 3 mM EDTA, 3 mM xanthine, 56.1 mU/ml xanthine oxidase, 0.75 mM XTT or NBT, and sample solution containing SOD. In the NBT method, 0.1 ml of BSA is also added. It is also reported that NBT directly interacts with glucose oxidase except for xanthine oxidase. Even in the glucose-glucose oxidase reaction, which does not generate $O_2^-$, NBT was reduced to generate its formazan in a time dependent manner, but no O.D. increase was observed in the case of XTT. Initial O.D. increased with an increase in the XTT concentration. This is due to the background O.D. from XTT tetrazolium. From these results, XTT does not seem to have any direct interaction with the reduced form of some enzymes that are generated during the oxidase reaction process, and XTT was shown to be a suitable probe for SOD assay. The assay mixture (2.8 ml) should contain the following components as the final concentration: 50 mM glycine-NaOH (pH 9.5), 0.1 mM glucose, and 0.1 mM NBT, 0.1 mM XTT or 0.2 mM XTT. The reaction was initiated by the addition of glucose oxidase solution. The absorbance at 470 (XTT) or 560 nm (NBT) was monitored at 25°C. The applicability of XTT for practical samples was confirmed by the high correlation coefficient of 0.954 shown by the XTT method with the NBT method data. The value obtained by XTT method is almost double the NBT value, reflecting the difference in the sensitivity of these assays.

Food samples are very complicated multi-component mixtures. Samples such as red wine, green tea, coffee and cocoa were chosen to determine superoxide anion scavenging activity (SOSA) by XTT because these samples are already known to have SOSA. As expected, the color from food samples with no dilution interfered with the assay, and the materials in the samples reduced XTT directly. To achieve 10% or less absorbance change by these materials, tea, red wine, instant coffee or cocoa need to be diluted 100 times, 10 times, 50 times or 10 times, respectively at pH 8. Each diluted sample solution had over 50% inhibition activity at each dilution rate, thus it was shown that XTT assay was useful for the determination of SOSA in food samples. These results were consistent with the data obtained by ESR method. From these results, it was shown that XTT method is applicable for use with biological samples and food samples. XTT overcomes the problems associated with NBT and it seems that XTT is an ideal reagent for SOD assay. However, one of the problems is the pH dependent sensitivity change. The data obtained by NBT was stable in the range between pH 8 and 10.2, however the sensitivity in the case of XTT decreased as the pH was lowered. Therefore, the sensitivity of XTT at pH 8.0

was lower than that of NBT. Another problem is the water-solubility of XTT. The water-solubility of XTT is around 2 mM and a heating process was necessary to prepare an optimal XTT solution (0.75 mM). Moreover, the high background O.D is another problem. Thus, different types of tetrazolium salts were tried to further improve the assay.

From 1993 to 1998, Ishiyama and his group (1993, 1996, 1997) developed several types of water-soluble tetrazolium salts. Since the water-solubilities of these compounds are from several 10 mM to several 100 mM, these water-soluble tetrazolium salts were applied for SOD assay in place of XTT. For this assay, WST-1 and WST-8, both mono-tetrazolium salts having sulfonate group(s) were used. The reaction mixture contained 2.5 ml of a phosphate buffer (A, pH 8.0) or 50 mM carbonate buffer (B, pH 9.4; C, pH 10.2) and 0.1 ml of 3 mM EDTA, 3 mM xanthine, 58 mU/ml of xanthine oxidase, 0.75 mM WST and the sample solution containing SOD. Both WST-1 and WST-8 showed 100% inhibition as seen in XTT assay at high concentrations of SOD. The most remarkable point was to be able to obtain almost the same $IC_{50}$ at different pH solutions. This result indicated that WSTs could overcome several shortcomings associated with XTT and NBT, making them ideal reagents for SOD assay. WST-1 and WST-8 showed no formazan production by the glucose-glucose oxidase reaction. Moreover, no background increase was observed with the increase in the concentration of WST. Therefore, WSTs enable us to determine the SOD activity at lower background conditions. The sensitivity of WST was compared by comparing the $IC_{50}$ (ug/ml) data obtained using NBT assay and XTT assay with that of WST assay at pH 10.2:WST-10.22, WST-80.75, XTT-0.26, NBT-0.55. These $IC_{50}$ values indicated that WST-1 could determine SOD activity with the highest sensitivity among other tetrazolium salts. A good linear correlation between WST-1 method and XTT method was observed with the correlation coefficient of 0.968 (n=7). Winterbourn and his group (1975) have recognized the usefulness of WST-1 in SOD activity detection and developed a microplate assay for SOD activity detection. They used a microplate assay to determine the SOD activity of human erythrocytes and rat liver and heart homogenates, and reported that the result obtained by this assay was consistent with the data reported previously. Thus, it seems that WST-1 has a wide range of applicability for biological samples. Currently, its applications for other samples such as plant tissues and food samples are being investigated (http://www.dojindo.com/ newsletter/ review_vol3-3.html)

## Conclusions

Despite the rapid strides made in the past years into the understanding of reactive oxygen species and the mechanisms of their detoxification, a lot of questions and gaps remain in our understanding of how these affect stress response in plants and disease states in animals. ROS is generally understood to react with and affect numerous biomolecules, leading to irreversible tissue damage and death of the organism. Nature has evolved its own line of defense in favor of life over disease and death, by ROS scavenging mechanisms, both nonenzymatic and enzymatic. The former include cellular redox buffers, ascorbate, glutathione, tocopherols, flavonoids, alkaloids and carotenoids. The enzymatic ROS scavengers in plants are superoxide dismutase, ascorbate peroxidase, glutathione peroxidase and catalase.

SODs are the first line of defense against ROS, dismuting superoxide to $H_2O_2$, from where catalase takes over. Now four types of this metallo enzyme are known: Cu/Zn-SOD, Fe-SOD, Mn-SOD and Ni-SOD and insight into their structure and function is progressing rapidly and the answers revealed thereby has enhanced our understanding of this protein.

## REFERENCES


Allen, R., Yevich, S., Orth, R., and Steele, R. 1974. The Superoxide anion and singlet molecular oxygen: Their role in the mitochondrial activity of the polymorphonuclear Leukocyte, *Biochem Biophys Res Commun.*, **60**: 909.

Almansa, M.S., Rio, L.A., Sevilla, F. and Del-Rio, L.A. 1994. Characterization of an iron-containing superoxide dismutase from a higher plant *Citrus limonum*. *Physiologia Plantarum*, **90** (2): 339-347.

Banci, L., Bertini, I., Cramaro, F., Del Conte, R. and Viazzoli, M. S. 2002. The solution structure of reduced dimeric copper zinc superoxide dismutase. *Eur. J. Biochem.*, **269**: 1905-1915.

Bannister, J., Bannister, W., and Wood, E. 1971. Bovine Erythrocyte Cupro-zinc Protein. I. Isolation and General Characterization, *Eur J Biochem.*, **18:** 178.

Beauchamp, C., and Fridovich, I. 1973. Isozymes of superoxide dismutase from wheat germ, *Biochim Biophys Acta,* **317**: 50.

Beyer, W.; Imlay, J.; Fridovich, I. 1991. SODs: varieties and distributions. X-ray crystallography of Mn-SODs and Fe-SODs, *Prog. Nucleic Acid Res. Mol. Biol,* **40:** 221-253.

Beyer, W. F.,Jr. Reynolds, J. A. and Fridovich, I. 1989. Differences between the manganese and the ironcontaining superoxide dismutases of *Escherichia coli* detected through sedimentation equilibrium, hydrodynamic, and spectroscopic studies. *Biochemistry*, **28**: 4403-4409.

Borgstahl, G. E. O., Parge, H. E., Hickey, M. J., Beyer, W. F. Jr., Hallewell, R. A. and Tainer, J. A. 1992. The structure of human mitochondrial manganese superoxide dismutase reveals a novel tetrameric interface of two 4helix bundles. *Cell,* **71**: 107-118.

Bowler, C., Van Kaer, L., Van Camp, W., Van Montagu, M., Inze, D., and Dhaese, P. 1990. Characterization of the *Bacillus stearothermophilus* manganese superoxide dismutase gene and its ability to complement copper/zinc superoxide dismutase deficiency in *Saccharomyces cerevisiae. J Bacteriol.*, **172** (3):1539-46.

Bray, R., Cockel, S., Fielden, E., Roberts, P., Rotilio, G., and Calabrese, L. 1974. Reduction and Inactivation of Superoxide Dismutase by Hydrogen Peroxide, *Biochem J.*, **139**: 43.

Cass, A.E.G. 1985. Superoxide dismutases, *Top. Mol. Struct. Biol.,* **6**: 121-156

Chen, C.N. and Pan, S.M. 1996. Assay of superoxide dismutase activity by combining electrophoresis and densitometry. *Botanical Bulletin of Academia Sinica,* **37** (2): 107-111.

Choudhury, S. B., Lee, J-W., Davidson, G., Yim, Y-I., Bose, K., Sharma, M. L., Kang, S-O., Cabelli, D. E. and Maroney, M. J. 1999. Examination of the nickel site structure and reaction mechanism in *Streptomyces seoulensis* superoxide dismutase. *Biochemistry*, **38**: 3744 -3752.

Cooper, J. B., Mclntyre, K., Badasso, M. O., Wood, S. P., Zhang, Y., Garbe, T. R. and Young, D. 1995. X-ray structure analysis of the irondependent superoxide dismutase from *Mycobacterium tuberculosis* at 2.0 Å resolution reveals novel dimer-dimer interactions. *J. Mol. Biol.,* **246:** 531-544.

Corpas, F.J., Sandalio, L.M, Rio, L.A., Trelease, R.N. and Del-Rio L.A. 1998. Copper-zinc superoxide dismutase is a constituent enzyme of the matrix of peroxisomes in the cotyledons of oilseed plants. *New Phytologist*, **138** (2): 307-314.

DeChatelet, L., McCall, C., McPhail, L. and Johnston, R. 1974. Superoxide dismutase activity in Leukocytes, *J Clin Inves,* **53**: 1197.

Edwards, R.A., Baker, H.M., Whittaker, M.M., Whittaker, J.W., Jameson, G.B. and Baker, E.N. (1998) Crystal structure of *Escherichia coli* manganese superoxide dismutase at 2.1Å resolution. *J. boil. Inorg. Chem.*, **3**: 161-171.

Fee, J. and Teitelbaum, H. 1972. Evidence that superoxide dismutase plays a role in protecting red blood cells against peroxidative hemolysis. *Biochem. Biophys. Res.Commun.*, **49**: 150.

Forman, H. and Fridovich, I. 1973. On the stability of bovine superoxide dismutase. The effects of metals. *J Biol. Chem.*, **248**: 2645.

Forman, H., Evans, H., Hill, R., and Fridovich, I. 1973. Histidine at the active site of superoxide dismutase, *Biochem.*, **12:** 823.

Fridovich, I. 1972. Superoxide Radical and Superoxide Dismutase, *Acc Chem Res.,* **5:** 321.

Fridovich, I. 1973. Superoxide Radical and Superoxide Dismutase, *Biochem Soc Trans.*, **1**: 48.

Fridovich, I. 1986. Biological Effects of the Superoxide Radical, *Arch Biochem Biophys* **247**: 1.

Fridovich, I. 1995. Superoxide radical and superoxide dismutases. *Annu. Rev. Biochem.* **64**: 97-112

Goscin, S., and Fridovich, I. 1972.The Purification and Properties of Superoxide Dismutase from *Saccharomyces cerevisiae*, *Biochim Biophys Acta*. **289:** 276.

Gregory, E., Goscin, S. and Fridovich, I. 1974. Superoxide dismutase and oxygen toxicity in a eukaryote. *J Appl. Bacteriol.* **117**: 456.

Gregory, E., Yost, F. and Fridovich, I. 1973.Superoxide Dismutases of *Escherichia coli*: Intracellular localization and functions, *J Appl Bacteriol*. **115**: 987.

Guo, Z.F., Lu, S.Y and Li, M.Q. 1996. Purification and characterization of superoxide dismutase from pea seeds. *J South China Agrl Univ.*, **17**( 2): 83-87.

Hewitt, J. and Morris, J. 1975. Superoxide dismutase in some obligately anaerobic bacteria, *FEBS Lett.,* **50**: 315.

Hunt, C. J. E., Sim, S. J., Sullivan, T., Featherstone, W., Golden, C. V., Kapp-Herr, R. A., Hock, R. A., Gomez, A. J., Parsian and Spitz, D. R. 1998. Genomic instability and catalase gene amplification induced by chronic exposure to oxidative stress. *Cancer Res.*, **58**: 3986-3992.

Ishiyama, M., Shiga, M., Sasamoto, K., Mizoguchi, M. and He, P. 1993.*Chem. Pharm. Bull.*, **41**: 118.

Ishiyama, M., Miyazono, Y., Shiga, M., Sasamoto, K., Ohkura Y., and Ueno, K. 1996. *Anal. Sci.*, **12**, 515

Ishiyama, M., Miyazono, Y., Sasamoto, K., Ohkura Y. and Ueno, K. 1997. *Talanta*, **44**: 1299.

Jain, A.P., Mohan, A., Gupta, O.P., Jajoo, U.N., Kalantri, S.P., Srivastava L.M. 2000. Role of oxygen free radicals in causing endothelial damage in acute myocardial infarction. *J Assoc Physicians India*. **48**(5): 478-80.

Jayashree, R., Rekha, K., Venkatachalam, P., Uratsu, S.L ., Dandekar, A.M., Kumari Jayasree, P., Kala, R.G., Priya, P., Sushma Kumari, S., Sobha, S., Ashokan, M.P., Sethuraj, M.R. and Thulaseedharan, A. 2003. Genetic transformation and regeneration of rubber tree (Hevea brasiliensis Muell. Arg) transgenic plants with a constitutive version of an anti-oxidative stress superoxide dismutase gene. *Plant Cell Rep*. **22**(3): 201-9.

Kanematsu, S. and Asada, K. 1990. Characteristic amino acid sequences of chloroplast and cytosol isozymes of Cu/Zn-superoxide dismutase in spinach, rice and horsetail. *Plant Cell Physiol*. **30** (1): 99-112.

Keele, B.B., McCord, J.M. and Fridovich, I. 1971. Further characterization of bovine superoxide dismutase and its isolation from bovine heart, *J. Biol. Chem.*, **246**: 2875-2880.

Kim,Y.K., Kwon, S.I. and An,C.S. 1997. Isolation and characterization of cytosolic copper/zinc superoxide dismutase from *Capsicum annuum* L. *Molecules and Cells*, **7** (5): 668-673.

Kitagawa, Y., Tanaka, N., Kusunoki, M., Lee, G.P., Katsube, Y., Asada, K., Aibara, S. and Morita, Y. 1991. Three-dimensional structure of Cu,Zn-superoxide dismutase from spinach leaves at 2.0Å resolution. *J Biochem.*, **109**(3): 477-485.

Lah, M. S., Dixon, M. M., Pattridge, K. A., Stallings, W. C., Fee, J. A. and Ludwig, M. L. 1995. Structure-function in *Escherichia coli* iron superoxide dismutase: Comparisons with the manganese enzyme from *Thermus thermophilus*. *Biochemistry*, **34:** 1646-1660.

Lavelle, F., Michelson, A., and Dimitrijevic, L. 1973. Biological Protection by Superoxide Dismutase, *Biochem Biophys Res Commun.*, **55**: 350.

Lim, J. H., Yu, Y.G., Han, Y. S., Cho, S., Ahn, B. Y., Kim,S. H.,and Cho,y. 1997. The crystal structure of an Fe- superoxide dismutase from the hyper thermophiphi lic *Aquifex pyrolus* at 1.9 Å resolution: Structural basis of thermostability. *J. Mol. Biol.*, **270**: 259-274.

Lin, C.T., Lin, M.T., Chen, Y.T. and Shaw, J.F. 1995. The gene structure of Cu/Zn superoxide dismutase from sweet potato. *Plant Physiol.*, **108**: 827-828.

Lin, C.T., Lin, M.T. and Shaw, J.F. 1997. Cloning and characterization of a cDNA for superoxide dismutase from callus of sweet potato. *J Agril. Food Chem.*, **45**: 521-525.

Matsuda, Y.; Hagashiyama, S.; Kijima, Y.; Suzuki, K.; Kawano, K.; Akiyama, M.; Kawata, S.; Tarui, S.; Deutsch, H.F.; Taniguchi, N. 1990. Human liver manganese superoxide dismutase. Purification and crystallization, subunit association and sulfhydryl reactivity, *Eur. J. Biochem.*, **194**: 713-720.

McCord, J., and Fridovich, I.1969. Superoxide Dismutase. An Enzym:c Function for Erythrocuprein (Hemocuprein), *J Biol Chem.*, **244**: 6049.

Miller, A.F. 2004. Superoxide dismutase: active sites that save, but a protein that kills. *Current Opinion in Chemical Biology*, **8**: 162-168.

Misra, H., and Fridovich, I. 1972. The Purification and Properties of Superoxide Dismutase from *Neurospora crassa*, *J Bacteriol.*, **247**: 3410.

Nicholls, D.G., and Budd, S.L. 2000. Mitochondria and neuronal survival. *Physiol. Rev.*, **80**: 315-360.

Nyman, P. 1960. A Modified Method for the Purification of Erythrocuprein: *Biochim Biophys Acta*, **45:** 387.

Ogawa, K., Kanematsu, S. and Asada, K. 1996. Intra- and extra-cellular localization of 'cytosolic' CuZn-superoxide dismutase in spinach leaf and hypocotyls. *Plant Cell Physiol.*, **37** (6): 790-799.

Paschen, W. and Weser, U. 1973. Singlet oxygen decontaminating activity of Erythrocuprein (Superoxide Dismutase), *Biochim Biophys Acta*, **327**: 217.

Peeters-Joris, C., Vandevoorde, A., and Baudhuin, P. 1975. Subcellular Localization of Superoxide Dismutase in Rat Liver. *Biochem J.*, **150:** 31.

Petkau, A., Chelack, W., Pleskach, S., Meeker, B., and Brady, C. 1975. Radioprotection of Mice by Superoxide Dismutase. *Biochem Biophys Res Commun.*, **65:** 886.

Rapp, U., Adams, W., and Miller, R. 1973. Purification of Superoxide Dismutase from Fungi and Characterization of the Reaction of the Enzyme with Catechols by Electron Spin Resonance Spectroscopy. *Can J Biochem.*, **51**: 158.

Rigo, A., Viglino, P., and Rotilio, G. 1975. Kinetic Study of $O_2$-Dismutation of Bovine Superoxide Dismutase. Evidence for Saturation of the Catalytic Sites by $O_2$, *Biochem Biophys Res Commun.*, **63:** 1013.

Rigo, A., Viglino, P., Rotilio, G., and Tomat, R. 1975. Effect of Ionic Strength on the Activity of Bovine Superoxide Dismutase, *FEBS Lett.*, **50:** 86.

Salin, M., and McCord, J. 1975. Free radicals and inflammation. protection of phagocytosing leukocytes by superoxide dismutase. *J. Clin. Invest.*, **56:** 1319.

Sandalio, L.M. and Del-Rio, L.A. 1988. Intraorganellar distribution of superoxide dismutase in plant peroxisomes (glyoxisomes and leaf peroxisomes). *Plant Physiol.*, **88** (4): 1215-1218.

Sawda, Y., Ohyama, T., and Yamazaki, I. 1972. Preparation and Physicochemical Properties of Green Pea Superoxide Dismutase, *Biochim. Biophys. Acta*, **268**: 305.

Schmidt, M., Meier, B. and Parak, F. 1996. Xray structure of the cambialistic superoxide dismutase from *Propionibacterium shermanii* active with Fe or Mn. *J. Biol.Inorg. Chem.,* **1**: 532-541.

Sehn, A. P. and Eier, B. 1994. Regulation of an *in vivo* metalexchangeable superoxide dismutase from *Propionibacterium shermanii* exhibiting activity with different metal cofactors. *Biochem. J.*, **304**: 803-808.

Stoddard, B. L., Howell, P. L., Ringe, D. and Petsko, G. A. 1990. The 2.1Å resolution structure of iron superoxide dismutase from *Pseudomonas ovalis*. *Biochemistry*, **29**: 8885-8893.

Panda, S. and Kar, A. 1998. Betel leaf extract can be both antiperoxidative and peroxidative in nature. *Current Science*, **74** (4): 284-285.

Symonyan, M., and Nalbandyan, R. 1972. Interaction of Hydrogen Peroxide with Superoxide Dismutase from Erythrocytes. *FEBS Lett.,* **28**: 22.

Vance, P., Keele, B., and Rajagopalan, K. 1972. Superoxide Dismutase from Isolation and Characterization of Two Forms of the Enzyme. *J Biol Chem.*, **247:** 4782.

Villafranca, J., Yost, F., and Fridovich, I. 1974. Magnetic Resonance Studies of Manganese (III) and Iron (III) Superoxide Dismutases. *J Biol Chem.*, **249**: 3532.

Walker, J.L., McLellan, K.M. and Robinson, D.S. 1991. Isolation and purification of superoxide dismutase purified from Brussels sprouts (*Brassica oleracea* L.var. bullata sub var. gemmifera). *Food Chem.*, **41**(1): 1-9.

Weisiger, R. and Fridovich, I. 1973. Superoxide Dismutase. Organelle Specificity. *J Biol Chem.,* **248:** 3582.

White, D.A. and Zilinskas, B.A. 1991. Nucleotide sequence of a complementary DNA encoding pea cytosolic copper/zinc superoxide dismutase. *Plant Physiol.*, **96** (4): 1391-1392.

Winterbourn, C., Hawkins, R., Brian, M. and Carrell, R. 1975. The estimation of red cell superoxide dismutase activity. *J. Lab. Clin. Med.*, **85**: 337.

Wong, V.L., Burke, J.J. and Allen, R.D. 1991. Isolation and sequence analysis of a cDNA that encodes pea manganese superoxide dismutase. *Plant Mol. Biol.*, **17** (6): 1271-1274.

Youn, H-D., Kim, E-J., Roe, J-H., Hah, Y. C. and Kang, S.D. 1996. A novel nickel-containing superoxide dismutase from *Streptomyces* spp. *Biochem. J.*, **318**: 889-896.

*Advances in Plant Physiology*, Vol. **9**
Ed. A. Hemantaranjan
Scientific Publishers (India), Jodhpur, 2006 pp. 101-154
E-mail: **info@scientificpub.com** www.scientificpub.com

# 6

# PHYSIOLOGY OF MANGROVES: STRATEGIES ON STRESS ADAPTATION WITH SPECIAL EMPHASIS ON TOLERANCE TO HIGH SALINITY

A.B. Das*, S. Mishra and P. Mohanty

Regional Plant Resource Center, Nayapalli, Bhubaneswar, Orissa-751015
E-mail: a_b_das@hotmail.com; Fax: 00-91-674-2550274

## INTRODUCTION

Mangroves are constituent plants of tropical intertidal forest community. They include woody trees and shrubs that flourish in inhospitable zone between land and sea interface along the tropical coastline of the globe. The world is estimated to have a total of about 24 million hectare mangrove area spread over 30 countries including island nations (Fig. 1). Mangroves occur approximately between 32$^0$N and 38$^0$S and mostly on the eastern costal boarder of India. The worlds largest mangrove ecosystems exists in the Sundarbans of West Bengal followed by different pockets of Indian costs including Bhitarkanika and Mahanadi delta of Orissa (Figs. 1 & 2). True mangroves occur only in mangrove habitat and their existence is rare elsewhere because of their unsuitable ecological niches primarily for variation of salinity gradients. They also include non-exclusive mangrove species occurring in the landward margin of mangal and often in non-mangal habitats such as rainforest, salt marsh or lowland freshwater swamps. Mangroves are taxonomically diverse; true mangroves include about 54 species in 20 genera belonging to 16 families (Tomlinson, 1986). They are exclusively tropical occupying two separate hemispheric regions and are more abundant in old world tropics than the new world. Geographic distribution of mangroves varies longitudinally and latitudinally and depends on air and surface temperature, coastal aridity. While the latitudinal range of mangroves is limited by temperature, longitudinal distribution of species is affected primarily by physical barriers. As tropical coastlines of the world

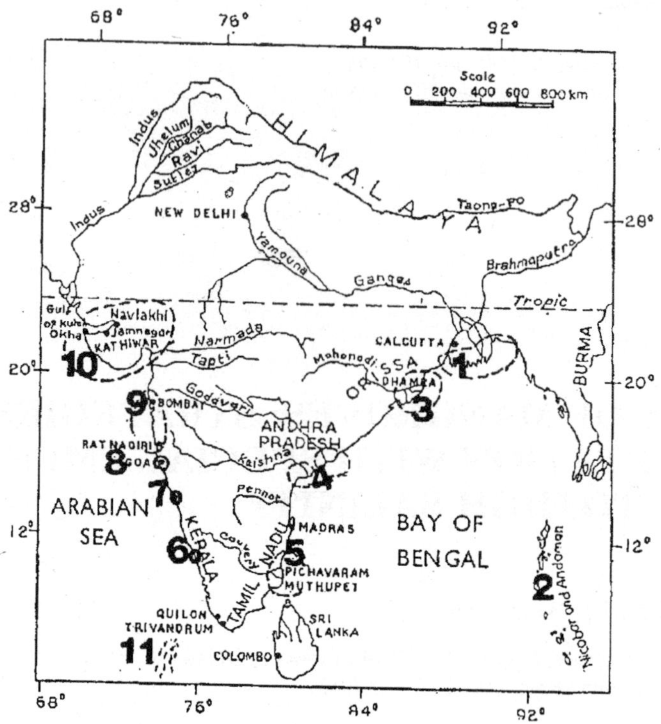


Fig. 1a. Geographic map showing distribution of mangroves in India

are separated by the great land masses and by open oceans into two major regions, accordingly, the primary separation of mangrove distribution is between Indo-West-pacific and the Atlantic-East Pacific regions.

Mangroves are characterized by their morphological adaptations, such as specialized respiratory roots, the pneumata-phores and root buttress in most of the tree mangroves and aerating devices, the lenticels and presence of salt glands on stems and leaves, specialized mechanisms for coping with excess salt concentrations that adapt them to their environment, specialized seed and propagules dispersal mechanisms i.e vivipary (Fig. 2b), restricted to the mangrove environment, taxonomic isolation (at least at the generic level) from terrestrial relatives. Though total global area occupied by mangroves is relatively small, their ecological and sociological importance is immense. They are the energy traps and nurse ecosystems and thus are primary producers and reproductive zones which shelter and feed juvenile populations of many off-shore species, including those which determine coral reef dynamics and many which are heavily exploited commercially (Redfield 1982; Tomlinson, 1986). Through-out much of the tropics, mangroves also have significant local economic importance as sources of timber, pulp and firewood, as natural manipulated fisheries, and as a potential agricultural land (Tomlinson, 1986).

Mangroves are defined and discussed more from ecological point of view as compared to that for their physiological significance and that is why mangrove physiology has received inadequate attention. Since they possess a peculiar habitat, their physiology is of considerable interest. They are essentially halophytes and therefore, their physiology should be comparable to other halophytes. Over the years, they have developed physiological mechanisms to withstand fluctuations in salinity as well as

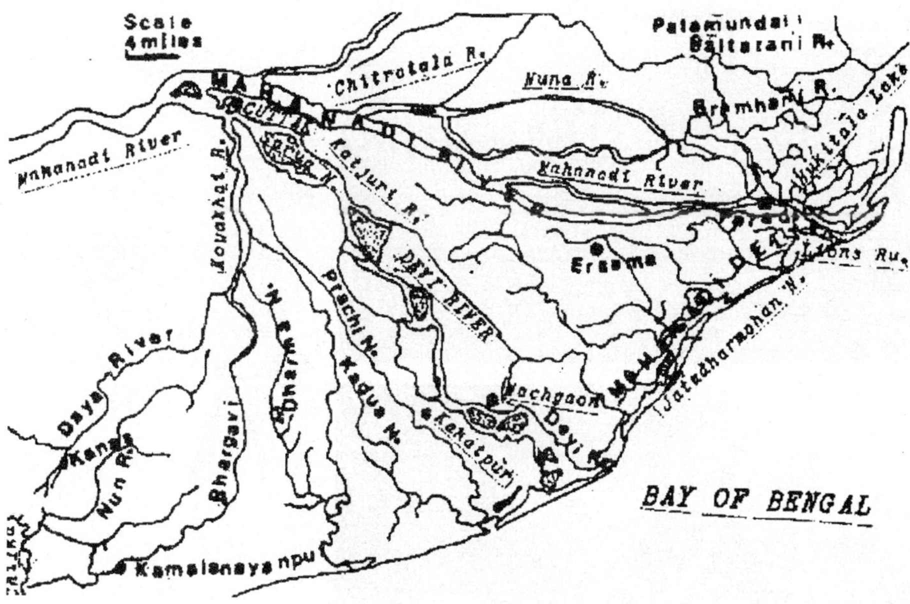


Fig. 1b. Geographic map showing distribution of mangroves in Orissa

high salinity. Though voluminous data are available on mangroves, information on most of physiological aspects particularly on salt tolerance mechanism is rather inadequate. The existence of mangroves greatly affected by the salt and that is one of the prime stress in their survivability in their habitat. Therefore, our laboratory has taken up to study the general physiology of stress and possible adaptive features of tree mangroves. In this chapter, comprehensive details of salt stress adaptations of mangroves, their physiological and molecular biological aspects including genetics of mangroves have been dealt with.

## 1.1 General Features of Mangroves

### *1.1.1. Mangrove environment*

Mangroves grow in intertidal zone between land and sea. They are periodically inundated by tide leading to permanent water logging and fluctuation in salinity. Like marine organisms, they are exposed to the air and thus to the risk of desiccation and overheating. Under high temperature condition in tropics above problems become worse. Firstly, at low tide, overheating and desiccation is greater and secondly, through evaporation, any water that remains may become even more highly saline than open sea. At high tide, warmth of water lowers oxygen in water. Thus, mangroves form an important characteristic complex in tropical and sub-tropical plant communities surviving water logging i.e. anoxic condition caused due to periodic submergence by tides, poor aeration, wide fluctuating salinity, high humidity, high solar radiation particularly of UV and strong wind, a combination of characters, not tolerated by other plants.

### *1.1.2 Water relations and gas exchange characteristics*

Carbon can not be gained without uptake of water. Mangroves grow in environment with an abundant of water supply, but they transpire slowly and maintain high water use efficiencies for $C_3$ plants (Ball and Farquahar, 1984). These water use characteristics become increasingly conservative with increase in salinities in which the plants are grown and with increase

Fig. 2a & b. Mangroves showing pneumataphore and vivipary

in the salt tolerance of the species (Ball 1988a).

Water potential and osmotic potential of plants became more negative with an increase in salinity, whereas turguor pressure increases with increasing salinity. The roots of mangroves in tropical coastal environments are bathed in salt water whose osmotic potential is –2.5 MPa (Scholander *et al.,* 1964), but at times which may be saltier. At low elevations in the intertidal zone, the soil is typically saturated with water so that the suction in the soil is close to zero and the soil water potential is therefore equal to osmotic potential. During transpiration, it requires the flow of water from roots to leaves and therefore, osmotic potential of the leaves is lower to maintain turguor. Typical values of water potential in field-grown mangroves range from –2.5 MPa to –6 MPa (Scholander *et al.,* 1964, 1968; Rada *et al.,* 1989; Smith *et al.,* 1989) and these water potential is much lower than those of well-watered plants growing in fresh water.

Low transpiration rate in mangroves is attributed to four possible reasons (Ball and Passioura, 1995). Firstly, the roots exclude most of the salts from the transpiration stream that may be so constituted that they present a very large resistance to the flow of water through them. Secondly, water potential below about 6 MPa may so tax the integrity of xylem sap that embolism occur (Sperry and Tyree, 1988). A low stomatal conductance may evolve that keeps the water potential above a value that is on the verge of inducing substantial embolism (Jones and Sutherland, 1991). In *Rhizophora mangle*, stems loss little hydraulic conductivity at xylem pressures less than –6 MPa. Thirdly, high degree of control of salt transport in relation to the expansion rate of growing tissues (Ball and Munns, 1992) and fourth reason is that concerns the increase in soil salinity that arises when the roots extract water but exclude the salt.

Conservative water use has adaptive significance for survival of mangroves in saline environments and this has far-reaching consequences for plant functioning. Maximizing carbon gain relative to water use is a whole plant phenomenon involving complex balance between several levels of plant function; stomatal behaviour in relation to photosynthesis. Variation in leaf properties in relation to light interception and evaporative demand, and the partitioning of the carbon between structures supplying and those consuming carbon-based assimilates (Cowan and Farquhar, 1977; Cowan 1986). Differences in water use characteristics find expression in all levels of plant form and function, and are major determinant of mangrove forest structure along natural salinity gradients (Ball 1988a).

### *1.1.3 Photosynthesis in mangroves*

The photosynthetic characteristics of mangroves are clearly those of plants utilizing $C_3$ photosynthetic biochemistry (Ball, 1986). Compared to non-halophytic $C_3$ plants, mangroves have low-photosynthetic rate. In *Rhizophora stylosa*, electron flow through photosystem II is more than 3 times higher than electron transport through Calvin cycle activity even with photo-respiration and temperature-dependent Rubisco specificities taken into account (Cheeseman *et al.*, 1997).

Photosynthesis in mangroves is affected by temperature, light and salinity. Quantum efficiency is much reduced by unfavourable environmental conditions such as high leaf temperatures, high salinity and water stress. Interactive effect of salinity and irradiance determines mangrove forest structure (Ball, 2002). Under optimum conditions, (leaf temperature <30$^0$C, salinity <20$^0/_{00}$, adequate nutrition) quantum efficiency is of the order of 0.05$\mu$mole $CO_2$ per $\mu$mol quanta (Ball and Critchely, 1982), but may be as low as 0.02 under less favourable conditions (high leaf temperature, high salinity and nutrient deficiency: Attiwill and Clough, 1980). Photosynthesis is strongly photoinhibited in mangrove leaves exposed to bright direct sunlight, particularly at high salinities, leading to a marked reduction in quantum efficiency. This results in a significant reduction in the rate of photosynthesis in leaves exposed to direct sunlight at the top of the canopy. Iron deficiency and other nutrient deficiency that lead to leaf chlorosis also reduce net photosynthesis. The light compensation point appears to be relatively constant at about 40-50 $\mu$mol quanta $m^{-2}$ $s^{-1}$ (Ball and Critchely, 1982, Andrews *et al.*, 1984).

The relative quantum yield measured as variable Chl a fluorescence/maximal Chl a fluorescence ($F_v/F_m$) of photosystem II of mangroves decreases at higher photon flux density and the degree of reduction varies among different species (Kao and Tsai, 1999). There are also differences among mangrove species in coefficients of non-photochemical and photochemical quenching (Kao and Tsai, 2000) of Chl a fluorescence (For a review see Papageorgeiou and Govindjee, 2004)

The temperature response of mangroves is similar to that of other $C_3$ plants and unlike that of $C_4$ plants, which generally have a higher optimum temperature for photosynthesis (Berry and Bjorkman, 1980). The rate of photosynthesis of mangroves is unaffected by leaf temperature over the range of 17$^0$C-30$^0$C, but falls sharply much above 35$^0$C and is close to 40$^0$C.

Photosyhetic rate varies in mangrove species (Das *et al.*, 2002). Net photosynthesis has a positive correlation with reaction center chlorophyll and light –harvesting complex (LHC) in mangroves (Moorthy and Kathiresan, 1999) and can be used as an index for measuring the productivity in mangrove species. Net photosynthetic rate is also affected by UV-B radiation in *Rhizophora apiculata* (Moorthy and Kathiresan, 1997).

### *1.1.4 Respiration in mangroves*

Mangroves grow in a soil that is more or less permanently waterlogged and in water whose salinity fluctuates and may be as high as that of open sea. In a waterlogged soil, the space between soil particles is filled with water. Even when water is saturated with oxygen, its oxygen concentration is far below that of air, and diffusion rate of oxygen is much less in water than through air (Ball, 1988) and therefore, waterlogged soils are low in oxygen. Mangrove trees adapt to survive such uncompromising surroundings. To adapt to submersion, they develop various forms of aerial root (Tomlinson, 1986). The roots of most trees branch off from the background and develop special aerating devices. Looping aerial roots are typical of *Rhizoophora*. In *Bruguiera* and *Xylocarpus*, a shallow horizontal root breaks the soil surface and submerges again forming a knee root. In *Avicennia* and *Sonneratia*, pnuematophores like the column roots of *Rhizophora* provide extensive gas spaces through lenticels to allow gas exchange with the underground tissues. In *Rhizophora mangle*, the aerial part of root component is divided structurally into horizontal arches and vertical columns. These have no problem in achieving adequate gas exchange at least at low tide. The columns supply oxygen to the underground roots. Air passes into column roots through numerous tiny pores or lenticels which are abundant close to the point at which the column root enters soil surface. It can then pass along roots through air spaces.

Respiration in plants includes both growth and maintenance respiration. Maintenance respiration includes the processes that maintain cellular structures, metabolites and gradients of ion. About 20% of total plant respiration is used for uptake and transport of ions. As halophytes, mangroves have high ion content and therefore, the energy devoted to maintenance of ion gradient is significant.

### *1.1.5. Nutritional status*

In addition to water, mangrove plants require adequate supply of nutrients. Possible sources of nutrients are rainfall, fresh water from rivers or as runoff from the land, tide-borne soluble or particle bound nutrients, the bacterial fixation atmospheric nitrogen and release by microbial decomposition of organic material (Hoagrath, 1999). The nutrient contribution made by rainfall is probably small. Mangroves also have a nutrient source in regular tidal inundations. Most nutrients available to mangroves are of terrestrial origin (Lugo *et al.,* 1976). Generally, the mangrove environment is poor in nutrients, to the extent that growth of the trees is restricted. Acquisition of sufficient nitrogen and phosphorus and other essential substances depends on complex interactions between the soil environment, microbial activity within it and on its aerobic state. Mangroves have anaerobic rhizosphere substrates poor in nitrogen and phosphorus (Boto, 1982). Ability to fix atmospheric nitrogen by root-associated bacteria in mangroves is not very much clear. However, it appears that some nitrogen-fixing ability is associated with the roots of *Rhizophora mangle*, *Avicennia germinans*, *Laguncularaia racemosa* (Zuberer and Silver, 1975). Some species of mangrove (*Laguncularaia*, *Lumintzera* and *Aegiceras*) form leaf nodules containing bacteria and these may capable of nitrogen fixation in other angiospermic species (Hutchings and Saenger, 1987). Although the nitrogen fixing bacteria and other organisms contribute to availability of inorganic nitrogen to the mangroves, but they often are nitrogen limited. Nitrogen and phosphorus availability limit the growth of *Kandelia candel* (Kao and Chang, 1998). Response to nutrients (N, P, K and Na) varies among mangrove species (Yates *et al.,* 2002).

Uptake of nutrients in mangroves is also affected by UV-B. Na/K ratio is lower in leaves, but higher in stems and roots of *Rhizophora apiculata* exposed to high dose of UV-B (Moorthy and Kathiresan, 1998a).This is a fertile area for future research because much literature is not available on these areas.

### *1.1.6 Nitrate and phosphate metabolism*

The availability of inorganic nitrogen uptake depends on a complex pattern of activity of bacteria within the soil. Mangrove soil is anoxic, besides very thin aerobic zone at the surface. In the anoxia zone, ammonia is produced either by nitrogen fixation or decomposition of organic matter. It diffuses upwards to the aerobic zone. In this process, some amount lost to the atmosphere and most are oxidized by aerobic bacteria, first to nitrite and then to nitrate ions by the process of nitrification. Nitrate then diffuses back through the anoxic layer. The fate of nitrate depends on the situation prevails then. Nitrate may be taken up by mangrove roots, it may be assimilated by bacteria and immobilized or it may be reduced further (anaerobia) bacterial action into either gaseous nitrogen or nitrous oxide ($N_2O$) [Boto, 1982].

### *1.1.7 Organic constituents*

Organic constituents (Photosynthetic pigments, polyphenols, proteins, tannin and TAN (Titrable acid number) varies among different mangrove species (Basak *et al.*, 1999, Das *et al.*, 2002, see Table 1). Mangroves are known to synthesize more polyphenols and tannins in response to salinity (Basak *et al.*, 1996). Phenolic compounds present in the epidermis of mangrove leaves not only varies among species, but between sites and between sun and shade leaves. More saline sites have plants with greater levels of phenolic compounds than less saline site (Lovelock *et al.*, 1992). The contents of phenolic compounds, amino acids, lipids, saccharides and protein are effected by high dose of UV-B (Moorthy and Kathiresan, 1997, 1998b). Mangroves possess different amounts and forms of carotenoids as photosynthetic pigments (Das *et al.*, 2002). They possess high UV-absorbing substances in their leaves and thylakoids (Das *et al.*, 2002, Fig. 3). The presence of UV-absorbing substances in thylakoids is an interesting observation and needs further study.

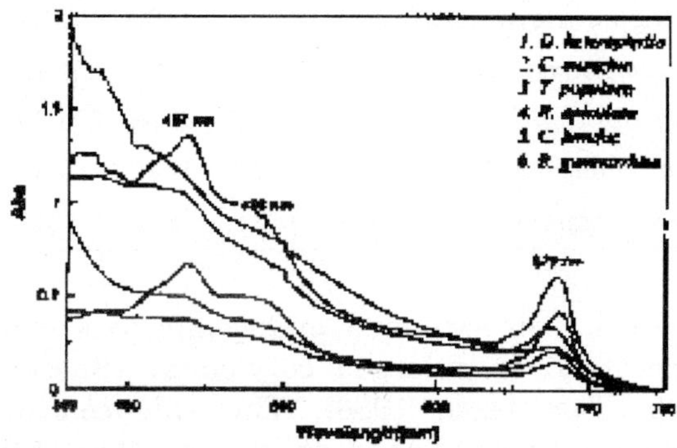

Fig. 3. Thylakoid spectra of different species of mangroves. For more details see, Das *et al.*, 2002.

### *1.1.8. Biology of reproduction*

The mode of pollination in mangroves is generally inferred from observation of flower (Fig. 4) and pollen morphology and of visits to flowers by possible pollen vectors. Accordingly, pollination is through wind (*Rhizophora*), bird (*Bruguiera*) and insects (*Ceriops and Kandelia*), bees (*Acanthus, Aegiceras, Avicennia, Exocoecaria* and *Xylocarpus*: Tomlinson, 1986) and flies (Nypa: Hamilton and Murphy, 1988).

All mangroves disperse their off springs by water. After pollination the growing embryo remains on the parent tree, and is dependent on it, for a period that continues for many months. This phenomenon is known as vivipary. Contrast to events of reproduction in orthodox species of plants, in mangrove species which show vivipary, little ABA is present, the developing propagule does not dehydrate and DNA synthesis and cell division continue

Table 1. Variation in organic constituents (% dry wt.) in leaves of nine mangrove species

| Species | Chla | Chlb | Chl a+b | Chla/b | Carotenoids | Polyphenol | Tannin | Protein | TAN |
|---|---|---|---|---|---|---|---|---|---|
| *Acanthus illicifolius* | 0.81 | 0.21 | 1.02 | 3.85 | 0.76 | 17.57 | 21.65 | 19.08 | 32.86 |
| *Acanthus Ebracteatus* | 0.55 | 0.21 | 0.86 | 2.61 | 0.38 | 16.86 | 20.45 | 17.66 | 26.25 |
| *Acanthus volubili* | 0.76 | 0.29 | 1.05 | 2.62 | 0.59 | 15.65 | 19.88 | 16.25 | 30.16 |
| *Bruguiera cylindrica* | 0.14 | 0.07 | 0.21 | 2.00 | 0.12 | 22.38 | 33.05 | 13.26 | 23.8 |
| *Bruguiera gymnorhiza* | 0.28 | 0.09 | 0.37 | 3.11 | 0.17 | 26.11 | 38.56 | 15.55 | 20.61 |
| *Brownlowia tersa* | 0.11 | 0.05 | 0.16 | 2.20 | 0.08 | 16.32 | 18.95 | 18.8 | 28.66 |
| *Ceriops tagal* | 0.16 | 0.13 | 0.29 | 1.23 | 0.09 | 38.64 | 40.11 | 18.56 | 18.05 |
| *Rhizophora apiculata* | 0.31 | 0.12 | 0.43 | 2.58 | 0.11 | 19.78 | 36.85 | 16.87 | 28.56 |
| *Xylocarpus mekongenesis* | 0.37 | 0.10 | 0.47 | 3.70 | 0.11 | 23.62 | 14.56 | 21.05 | 32.5 |

producing an extended embryonic axis (the hypocotyls) or enlarged cotyledons (Sussex, 1975, Tomlinson, 1986). The salt concentration in the hypocotyls is less than the adult plant (Joshi *et al*,. 1972, Bhosale and Shinde, 1983). These features enable them to adapt to the harsh environment (Osborne and Berjak, 1997, Farnworth and Farrant, 1998, Farnworth, 2000). The chromosome maps (mitotic and meiotic) of mangroves are available.

Fig. 4. Flowering of mangrove *Rhizophora stylosa* - an endengered species

When attached to the plant, the developing seedling develops chlorophyll and photosynthesis. The water and necessary nutrients are supplied by the parent plant. Carbohydrates are also translocated from the parent and this is more important in later stage of development, as the rate of starch accumulation increases at a time when seedling photosynthesis is actually decreasing (Pannier and Fraino de Pannier, 1975; Gunasekar, 1993).

The advanced form of vivipary is shown by *Rhizophora* and other members of the family Rhizophoraceae. *Aegiceras*, *Avicennia* and other mangrove species show a similar form of reproduction called cryptovivipary; unlike conventional seeds, there is no period of dormancy and growth is continuous.

### *1.1.9 Somatic chromosome structure in mangroves*

Somatic chromosome numbers have been reported only in 40 out of 94 species of mangroves and their associates (Appendix V). However, no meiotic reports have been found yet because of the short meiotic metaphase period due to the adverse environmental condition. Somatic chromosomes are in general too small and number varies from 2n

= 22 to 130. Some species showed the polyploidy among the ecotypes and may be that is a mechanism of adaptation in high saline condition. The existence of cytotypes is also common in these special groups of plants.

### *1.1.10 Nuclear DNA content and genome size in mangroves*

The Cytophometric estimation of nuclear DNA content has been reported in 30 out of 94 species of mangroves in India (See, Appendix V). DNA content varied from ~8 to 56 pico grams among the species. Significant interspecific and intergeneric variation of DNA content were noted. The variation of DNA content in ecotypic level might be due to the variation repetitive sequences that helps in the adaptation. The size of the genome varied from ~8000Mb in *Acanthus volubilis* to ~54200Mb in *Avicennia alba* i.e. 6.5 fold variation.

### *1.1.11 DNA markers in mangroves*

Recent studies on DNA markers in different species of mangroves have been made and many reports have appeared (See, Appendix VI). These studies establish a large number of DNA markers which are of use in a genetic transformation studies in mangroves. Prof. M.S. Swaminathan Institute in Chainai has reported to have cloned species for slat tolerance in *Avecennia*. We further note that the DNA marker vary widely on different ecotypes of the mangrove species. The initial observation suggests these are of extreme importance while making phylogenetic map of the mangrove (Jena *et al.*, 2004).

## 2. STRATEGY FOR STRESS ADAPTATION

Mangroves not only grow in a high and variable salinities, but their growth conditions are also characterized by anaerobic rhizosphere substrates poor in nitrogen and phosphorus (Boto, 1982), and prolonged periods of high irradiance and supraoptimal leaf temperatures (Bjorkman *et al.,* 1988). In these harsh environmental conditions, mangroves survive and dominate in that harsh ecosystem. Even in the areas with pronounced seasonality, most mangrove species adapt and maintain active leaves under conditions that are expected to reduce severely the photosynthetic capacity.

### 2.1 Light Adaptation

Compared to non-halophytic plants, mangroves have low light-saturated photosynthetic rate (Anderws and Muller, 1985). Rates of photosynthesis in field-grown mangrove leaves generally become light saturated at incident quantum flux densities ranging from 25 to 50% full sunlight (Ball and Critchely, 1982; Bjorkman *et al.,* 1988; Cheeseman *et al.,* 1991), consistent with the light levels received in their normal orientations under field conditions (Ball *et al.,* 1988). Because of their low-photosynthetic rate, light requirements for maximal photosynthesis are considerably less than the amounts of light available on bright, sunny days. Therefore, mangroves received an abundance of excess excitation energy which may cause light-dependent loss in photosynthetic functioning of photosystem II i.e. photoinhibition.

There are both coarse and fine adjustments operated by mangroves to irradiance regimes. The coarse adjustments occur primarily through the display and physical properties of the leaves (see 2.2) to avoid prolonged exposure to excessive irradiance, whereas fine adjustments are made at the biochemical level of chloroplast functioning.

Naturally displayed sun leaves can have lower quantum yields than shaded leaves (Bjorkman *et al.,* 1988; Cheeseman *et al.,* 1991). For example, there are large decrease in $F_v/F_m$ (measure of quantum yield) when leaves of a number of mangroves including *Rhizophora stylosa* are directly illuminated in the field (Bjorkman, Demmig and Andrews,

1988; Tuffers *et al.,* 1999). The depression in quantum yield of leaves naturally receiving high irradiance is correlated with the concentration per unit leaf area of zeaxanthin (a special type of carotinoid which is involved in photo-protection and photo inhibition) pigment (Lovelock and Clough, 1992; Lovelock and De'ath, 1993), implying that the loss of photosynthetic activity at limiting irradiance is due to protective dissipation of excessive excitation energy through the xanthophyll cycle (Demmig-Adams *et al.,* 1989).

Protection of mangrove photosynthetic machinery against the potentially damaging effects of excess photons which generates active oxygen species due to phoinhibition of PSII, is provided by enzymatic and non-enzymatic means in *Rhizophora stylosa* (Cheeseman *et al.,* 1997). The protection is effective even when excess electron transport is indicated and likely to be mediated by the processes beginning with the Mehler reaction, followed either by non-enzymatic scavenging of active oxygen species by ascorbate or by enzymatic detoxification through superoxide dismutase. $H_2O_2$ is subsequently consumed either by the antioxidative enzymes like, ascorbate peroxidase, glutathione reductase, monodehydroascorbate reductase in chloroplasts or by peroxidase-mediated oxidation elsewhere in the leaf (for a review, Foyer and Noctor, 2000, Loggini *et al.,* 1999).

Though different mechanisms are there for protecting mangroves from excessive irradiance, there are also limits to the extent to which a leaf can provide protection. When absorbed light exceeds the capacities for both useful photochemistry and protective dissipation, then the leaf becomes vulnerable to another form of photoinhibition which results in irreversible photodamage of photosystem II (Kyle, 1984).

The phenomenon of photoinhibition has been reported in different species of mangroves, but there are also no evidence of photoinhibition in exposed leaves of *Bruguiera parviflora* under natural illumination (Cheeseman *et al.,* 1991) or in *Rhizophora mangle* under greenhouse, water-stressed conditions (Cheeseman, 1994).

Tropical mangroves receive high level of ultraviolet radiation relative to plants inhabiting in temperate regions. The depletion of the atmospheric ozone layer and the associated increases in UV radiation at earth's surface may also increase the flux of UV radiation in mangrove environments. Protection to UV radiation is provided by presence of phenolic compounds (found in the epidermis of leaves), i.e., carotenoids (Violaxanthin, antheraxanthin and zeaxanthin etc.), as well as leaf succulence. Tolerance to UV radiation varies among mangroves species and it is proposed to be influenced by their ability to dissipate excess visible solar radiation (Lovelock *et al.,* 1992)

### 2.2 Temperature Adaptation

Mangrove species differ in their response to raised temperature. Photosynthesis and other variables such as leaf production usually show temperature optima. The optimal temperature varies among species and localities (Hutchings and Saenger, 1987). In some species, the optimal temperature for photosynthesis alters in individual plants with seasonal temperature changes. As mangrove species experience range of temperature daily ranging from 20$^0$C to 30$^0$C, the rise of temperature due to global worming may affect their productivity.

Optimal leaf temperatures for photosyntheis in mangroves are very close to average air temperature in the tropical and sub-tropical environments in which the plants are grown. Both assimilation rate and stomatal conductance are maximal at leaf temperatures ranging from 25$^0$C to 30$^0$C and decline with increase in leaf temperature above 35$^0$C. The transpiration rate at optimal temperature is not enough to prevent heating of leaves above ambient air temperatures.

Leaves with high transpiration rate can take advantage of high irradiance required for high photosynthetic rate and maintain minimal increase in temperature over air temperature. As mangroves are conservative water user and has low transpiration rate, it must avoid high irradiance in order to keep the leaf temperature within physiologically acceptable limits. There are three properties that contribute to maintenance of favourable temperatures with minimum evaporative cooling in leaves of mangroves, i.e., increasing leaf angle, reduction in leaf size and increasing heat capacity (Ball *et al.*, 1988).

Leaf angle affects the radiant heat loading on the leaf. The greater the leaf angle, the lower the proportion of leaf area projected on a horizontal surface. Leaf angle is regarded as a compromise between the requirements for illumination and reduction in temperature. In the Rhizphoraceae, leaf angle in foliage fully exposed to the sun was greater and hence the proportion of projected leaf area is smaller (Ball *et al.*, 1988). Although, many mangrove species possess photoprotective mechanisms, avoidance of radiation by steep angles is also more important in species like *Bruguiera gymnorhiza*.

A second leaf property influencing leaf temperature is that of leaf size. Heat convection between a leaf and its environment depends on resistance to transfer imposed by a boundary layer, the characteristics of which are a function of leaf geometry and wind speed. Decrease in leaf size enhances boundary layer conductance and results in the temperature of the leaf being closer to ambient air temperature. Leaf size in the Rhizophoraceae is smallest in the most salt-tolerant species and decreasing with increasing exposure (Ball *et al.*, 1988). Leaves of mangrove species that dominate humid low salinity wetlands, e.g., *Heritiera littoralis, Bruguiera gymnorhiza, Rhizophora apiculata*, *Rhizophora mucronata* and *Xylocarpus granatum* are much larger than those of species that dominate hypersaline environments, e.g., *Avicennia marina, Ceriops australis, Excoecaria ovalis, Lumintzera racemosa* and *Osbornia octodonta*. So, mangrove leaves are smaller when there is greatest heat load on the leaf under conditions of limitations to evaporative cooling.

Heat capacity per unit area which increases with increase in the dry weight and water content per unit area is a third leaf property influencing leaf temperature. Increase in heat capacity of leaves buffers against rapid changes in temperature. Among the Rhizophoraceae, specific leaf weight and succulence, and thus, also heat capacity, increase with salinity (Camilleri and Ribi, 1983) exposure (Ball *et al.*, 1988) and with increase in salinity tolerance of the species. The heat capacities of *Bruguiera gymnorhiza* and *Ceriops australis* are 1.1 and 2.2 x $10^3$ $Jm^{-2}$ $C^{-1}$ respectively (Camilleri and Ribi, 1983).

Reduction in leaf temperature has benefits photosynthesis. The cooler the leaf, the larger is the stomatal conductance corresponding to any given evaporation rate. Thus, conductance to $CO_2$ transfer is also greater. The effect on total conductance of $CO_2$ is more marked when reduction in temperature is associated with avoidance in radiation than when it is associated with reduction in leaf size. When there is reduction in leaf size, it is at the expense of light harvesting in that a larger leaf is required to intercept a given amount of light. Decrease in leaf size enhances the heat transfer rates, but this requires greater investment in supportive and conductive tissue per unit of exposed leaf area than in large leaves. Increase in heat capacity is at the expense of leaf carbon which might otherwise be invested in expansion in leaf area. Thus, maintenance of leaf temperature has two major consequences for the carbon and water economy of the leaves (Andrews and Muller, 1985). Firstly, the reduction in assimilative capacity and secondly maintenance of water cost of carbon gain at a minimum level.

## 2.3. Salt Adaptation

### *2.3.1. Salinity effect- A general view*

Salinity of soil and water is caused by the presence of excessive amounts of salts. Most commonly the stress caused by high $Na^+$ and $Cl^-$ in the soil solution. Salt stress has three fold effects viz. reduces water potential, causes ion imbalance or disturbances in ion homeostasis and toxicity. Altered water status brings out initial growth reduction that leading to limitation of plant productivity. Since salt stress involves both osmotic stress and ionic stress (Hagemann & Erdmann 1997, Hayashi & Murata 1998), the growth suppression is directly related to total concentration of soluble salts or osmotic potential of soil water (Flower *et al.,* 1977, Greenway & Munns, 1980). The detrimental effect is observed at whole plant level as death of plants or decreases in productivity. Suppression of growth occurs in all plants, but their tolerance level and rate of growth reduction at lethal concentration of salt vary widely among different plant species. Although the change in water status is the cause of growth suppression, the contribution of subsequent processes to inhibition of cell division and expansion and acceleration of ell death has not been well documented Hasegawa *et al.,* 2000).

Salt stress affects all the major processes such as photosynthesis, protein synthesis, energy and lipid metabolism. The immediate response is reduction in rate of leaf surface expansion leading to cessation of expansion as salt concentration increases.

#### *2.3.1.1 Effect on photosynthesis*

Plant growth is the result of integrated and regulated physiological processes. Physiological processes are affected by number of environmental factors and they determine the response of plants to stress. Limitation of plant growth by environmental factors cannot be assigned to single physiological process. The dominant physiological process is photosynthesis. Plant growth as biomass production is measure of net photosynthesis and therefore, environmental stress affecting growth also affects photosynthesis.

Salt stress causes either short or long term effect on photosynthesis. Short term effect is mainly after a few hours or within one or two days of the onset of exposure (Rawson and Munns, 1984). Short term response is important as there is complete cessation of carbon assimilation within hours (Heuer, 1991). Long term effect is after several days of exposure to salt and reduction in carbon assimilation is due to the salt accumulation in developing leaves (Munns and Termatt, 1986). Although there are reports of suppression of photosynthesis upon salt stress (Soussi *et al.,* 1998; Kao *et al.,* 2001), there is also report that photosynthesis is not slowed down by salinity and is even stimulated by low salt concentration (Kurban *et al.,* 1999).

##### *2.3.1.1.1 Mechanism of salinity effects on photosynthesis*

Photosynthesis involves long chain of mechanisms, enzymes and intermediate products and regulated by number of external and internal factors. Photosynthetic efficiency depends on sequence of metabolic events such as photochemical reactions, enzymes involved in carbon assimilation, structure of photosynthetic apparatus and transport of photosynthetic intermediates between sub-cellular compartments.

Photosynthetic rate is lower in salt-treated plants, but the photosynthetic potential is not greatly affected when rates are expressed in terms of chlorophyll or leaf area basis. Decrease in photosynthetic rate is due to several reasons viz. (i) dehydration of cell membranes which reduce their permeability to $CO_2$, (ii) salt toxicity, (iii) reduction of $CO_2$ supply because of hydroactive closure of stomata (iv), enhanced senescence induced by salinity (v) changes of enzyme activity induced by changes in cytoplasmic structure (vi) negative feedback by reduced sink activity (Iyengar and Reddy, 1996).

Photosynthetic activity decreases as water potential of leaves decreases (Iyengar and Reddy, 1996). The reduction in photosynthetic activity depends on two aspects of salinization the total concentration of salt and their ionic composition. High salt concentration in soil and water create high osmotic potential, which reduces the availability of water to plants. Decrease in water potential causes osmotic stress which reversibly inactivates photosynthetic electron transport via shrinkage of intercellular space (which is due to efflux of water through water channel in the plasma membrane, Allakhverdiev *et al.,* 2000). Increase in osmotic potential under high salt condition causes $Na^+$ ions to leak into the cytosol (Papageorgiou *et al.,* 1998) and inactivate both photosynthetic and respiratory electron transport (Allakhverdiev *et al.,* 1999).

The reduction in photosynthetic rate is also due to the reduction in stomatal conductance resulting in restricted availability of $CO_2$ for carboxylation reactions (Brugnoli and Bjorkman, 1992). Stomatal closure minimizes loss of water by transpiration and this affects chloroplast light-harvesting and energy-conversion system thus leading to alteration in chloroplast activity (Iyengar and Reddy, 1996). The extent to which stomatal closure affects photosynthetic capacity depends on the magnitude of partial pressure of $CO_2$ inside the leaf. There are also reports of non-stomatal inhibition of photosynthesis under salt stress. This non-stomatal inhibition is due to increased resistance to $CO_2$ diffusion in the liquid phase from the mesophyll wall to the site of $CO_2$ reduction in the chloroplast and reduced efficiency of RUBPcase (Iyengar and Reddy, 1996).

Reduction in photosynthetic capacity is also a consequence of inhibition of certain carbon metabolism processes by feedback from other salt-induced reactions (Greenway and Munns, 1980). Under reduced water potential, stromal levels of the substrate fructose 1,6, bisphosphate accumulate and the FBPase product fructose-6-phosphate is reduced so that FBPase becomes rate limiting to photosynthesis (Heuer, 1996). To cope with salt stress plants respond with physiological and biochemical changes that aim at retention of water in spite of high external osmoticum and the maintenance of photosynthetic activity and these make them become tolerant.

#### *2.3.1.2. Effects of salinity on antioxidative enzymes*

Salt stress is complex imposing a water deficit because of osmotic effects on a wide variety of metabolic activities (Greenway & Munns 1980, Cheeseman, 1988). This water deficit causes oxidative stress (Logini *et al.,* 1999) because of formation of reactive oxygen species (Superoxides, hydroxy and peroxy radicals). There is increase in antioxidative enzymes (Gossett *et al.,* 1994) in response to salt stress and correlation of antioxidant enzyme level and salt tolerance (Hernandez *et al.,* 1995, Sehmer *et al.,* 1995). These antioxidative enzymes detoxify reactive oxygen species (ROS).

#### *2.3.1.3. Effects on salinity on photosynthetic pigments and proteins*

In general, the chlorophyll and carotenoid content of leaves decreases under salt stress, (Agastian *et al.,* 2000), but there is also report of increase in total chlorophyll content (Wang and Nil, 2000). A significant amount of increase in specific protein (s) in response to salt stress has been reported in many species (Benhayyim *et al.,* 1993; Yen *et al.,* 1997; Elshintinawy and Elshourbagy, 2001).

*2.3.1.4. Salinity effect on nitrogen metabolism*

Nitrate reductase activity of leaves decreases in many plants under salt stress. The primary cause of reduction of a NRA in the leaves is a specific effect associated with the presence of $Cl^-$ salts in the external medium. This effect of $Cl^-$ seems to be due to a reduction in $NO_3^-$ uptake and consequently a lower $NO_3^-$ concentration in leaves, although a direct effect of $Cl^-$ on the activity of enzyme may be there (Smith 1973). There is also inhibition of nodulation and $N_2$ fixation (Bekki *et al.,* 1987). Severe salt stress reduces leghaemoglobin content and nitrogen activity (Comba *et al.,* 1998)

*2.3.1.5. Effects of salinity on lipids*

Lipids are the most effective source of storage energy, functions as insulator of delicate internal organs and hormones and play an important role as the structural constituents of most of the cellular membranes (Singh *et al.,* 2002). Lipid oxidation is problematic as enzymes do not control many oxidative chemical reactions and some of the products are highly reactive species that modify proteins and DNA (Singh *et al.,* 2002). Salt stress causes the water deficit and tolerance to water deficit relies on phospholipid bilayers, which are stabilized by sugars, especially by trehalose. Water deficit is also counteracted by unsaturation of fatty acids. Salt stress causes decrease in lipid content (Hassanein, 1999) and molar percentage of sterol and phospholipid (Wu *et al.,* 1998).

*2.3.1.6 Effect on leaf anatomy and chloroplast ultrastructure*

Salnity increases in epidermal thickness, mesophyll thickness, palisade cell length, palisade diameter and spongy cell diameter (Longstreth and Nobel, 1979). Salinity also reduces intercellular spaces in leaves (Delphine *et al.,* 1998) and causes vacuolation, development and partial swelling of endoplasmic reticulum, decrease in mitochondrial cristae and swelling of mitochondria, vesiculation and fragmentation of tonoplast and degradation of cytoplasm and vacuolar matrices (Mitsuya *et al.,* 2000). There are also rounding off of cells, smaller intercellular spaces and a reduction in chloroplast number (Bruns and Hecht-Buchholz, 1990), reduction in plant leaf area and stomatal density (Romeroaranda, 2001).

Under salt stress, thylakoid structure of the chloroplasts becomes disorganized, the number and size of plastoglobuli increases and their starch content decreases (Hernadez *et al.,* 1995, 1999). Chloroplasts aggregated, the cell membranes are distorted and wrinkled and there is no sign of grana structure (Khavarinejad and Mostofi, 1998), swelling of chloroplasts, reduction in number and depth of grana stacks (Bruns and Hecht-Buchholz, 1990)

**2.3.2. Salt tolerance**

Salt tolerance is the ability of plants to grow and completes life cycle on a saline substrate that contain high concentration of soluble salt. Plants that can survive on high concentration of salt in the rhizosphere and grow well are called halophytes. Depending on their salt tolerating capacity, halophytes are either obligate and characterized by low morphological and taxonomical diversity with relative growth rate increasing up to 50% sea water and facultative halophytes found in less saline habitats along the border between saline and non-saline upland and characterized by border physiological diversity which enable them to cope up with saline and non-saline condition.

*2.3.2.1 Mechanisms of salt tolerance*

Plants develop plethora of biochemical and molecular mechanisms to cope with salt stress. Biochemical pathways leading to products and processes that improve salt tolerance are likely to act additively and probably synergistically. Biochemical strategies include (i) selective accumulation or exclusion of ions (ii) control of ion uptake by roots and transport into leaves (iii) compartmentalization of ions at the cellular and whole plant level (iv) synthesis of compatible solutes (v) change in photosynthetic pathway (vi) alteration in membrane structure (vii) induction of plant hormones.

Salt tolerance mechanism is either a low complexity or high complexity mechanism. Low complexity mechanism appears to involve changes in single biochemical pathway. High complexity mechanisms are changes that protect major processes such as photosynthesis and respiration, e.g., water use efficiency and those that preserve such as cytoskeleton, cell wall, or plasma membrane-cell wall interactions (Botella *et al.,* 1994), chromosome and chromatin structure changes i.e DNA methylation, polyplodization, amplification of specific sequences, or DNA elimination (Wallbot and Cullis, 1985). It is believed that for the protection of higher order processes, low-complexity mechanisms are induced coordinately (Bohnert *et al.,* 1995).

*2.3.2.1.1. Ion regulation and compatmentalization*

Ion uptake and compartmentalization are not only crucial for normal growth, but also for growth under saline condition (Adams *et al.,* 1992) because the stress disturbs in ion homeostasis. Plants, whether it is a glycophyte or halophyte can not tolerate large amount of salts in the cytoplasm and therefore, they either restrict the excess salts in the vacuole or compartmentalize the ions in different tissues (Reddy *et al.,* 1992) to facilitate their metabolic functions under saline conditon. Glycophytes limit sodium uptake or partition sodium in older tissues that serve as storage compartments that are eventually sacrificed (Cheeseman, 1988). Removal of sodium from the cytoplasm or compartmentalization in the vacuoles is maintained by a salt inducible enzyme $Na^+/H^+$ antiporter (Reuveni *et al.,* 1990). Besides the sodium and potassium, calcium has also a role in salt adaptation. Calcium reduces the toxic effects of NaCl by facilitating high $K^+$ /$Na^+$ selectivity (Lauchli, 2000). High salinity causes increased cytosolic $Ca^{2+}$ and these increases potentiate stress signal transduction and lead to salt adaptation (Knight *et al.,* 1997).

Other mechanisms of salt regulation are salt secretion and selective salt accumulation or exclusion. Salt secretion occurs through development of unique cellular structures called salt glands that secrets salt (especially NaCl) from leaves and maintain internal ion concentration at lower level (Liptz and Waisel, 1974). Salt exclusion occurs through roots to regulate the salt content of their leaves in many halophytes (Levitt, 1980). Selective accumulation of ions or solutes confers the plants for osmotic adjustments which occurs through mass action and results in increased water retention and/or sodium exclusion.

*2.3.2.1.2 Induced biosynthesis of compatible solutes*

To accommodate the ionic balance in the vacuoles, cytoplasm accumulates low molecular mass compounds termed compatible solutes because they do not interfere with normal biochemical reactions (Bartels and Nelson, 1994), rather they replace water in biochemical reaction. These compatible solutes include mainly nitrogen containing compounds (Mansour, 2000) like amino acids, amides, proteins, quaternary ammonium compounds (QAC), polyamines, sugars (Kereepsi and Galiba, 2000) and polyols (Bieleski, 1982, Bohnert *et al.,* 1995). Proline (Singh *et al.,* 2000), glycine betaine (Wang and Nil, 2000)) constitute the most important nitrogen containing osmolytes. Osmotic adjustment, protection of cellular macromolecules, storage form of nitrogen maintaining cellular pH, detoxification of the cells and scavenging of free radicals are the main functions of these nitrogen containing compatible solutes. Polyols make up a considerable percentage of all assimilated $CO_2$ (Bieleski, 1982) and serve several functions: as compatible solutes, as low-molecular weight chaperones and as scavengers of stress-induced oxygen radicals (Smirnoff & Cumbes, 1989). Polyols function in two ways that are difficult to separate mechanistically: osmotic adjustment and osmoprotection. In osmotic adjustment, they act as osmolytes facilitating the retention of water in the cytoplasm and allowing sodium sequestration to the vacuole or apoplast. These osmolytes protect cellular structures by interacting with membranes, protein complexes or enzymes. These compounds have hydrogen bonding characteristics that allow them protect macromolecule from the adverse effects of increasing ionic strength in the surrounding media (Crowe *et al.*, 1992) by tight association with protein and membrane components to compensate for water loss during stress (Yancey *et al.*, 1982). Proline, quaternary ammonium and tertiary sulfonium osmolytes are zwitter ions at physiological pH. Although they are ionic, they have no net charge. Their osmotic function is due to their unique chemistry. Those polyols that are non-reducing sugars may also store excess carbon under environmental stress conditions (Vernon *et al.*, 1993).

*2.3.2.1.3 Induction of plant hormones*

High salt concentration triggers an increase in levels of plant hormones such as abscissic acid and cytokinin (Thomas *et al.*, 1992; Aldesuquy 1998; Vaidyanathan *et al.*, 1999). Rise of ABA under salt stress is associated with $Ca^{2+}$ uptake that contributes to membrane integrity maintenance which enables plant to regulate uptake and transport under high levels of external salinity in the long term (Chen *et al.*, 2001). Abscisic acid level is responsible for the alteration of salt-stress induced genes (de Bruxelles *et al.*, 1999). Abscisic acid is found to alleviate inhibitory effect of NaCl on photosynthesis, growth and translocation of assimilates (Popova *et al.*,, 1995). ABA also promotes switch from $C_3$ to CAM. In response to salt stress, ethylene also increases (GomezCadenas *et al.*, 1998).

*2.3.2.1.4 Change in photosynthetic pathway*

Salt stress inhibits photosynthesis by reducing water potential. So the main aim of salt tolerance is to increase water use efficiency under salinity. To this end, facultative halophytic plant like *Mesembryanthemum crystallinum* shifts its $C_3$ mode of photosynthesis to CAM (crassulacean acid metabolism). This change allows the plant to reduce water loss by opening stomata at night and thus decreasing transpiratory water loss under prolonged salinity condition. There is also for a shift from $C_3$ to $C_4$ pathway in response to salinity in salt-tolerant plant species (Cushman *et al.*, 1989).

*2.3.2.1.5 Molecular mechanism of salt tolerance*

Salt tolerance is a multigenic trait. Number of genes for salt tolerance in plants is categorized into different functional groups (Table 2) viz. (i) genes for photosynthetic enzymes (ii) genes for synthesis of compatible solutes (iii) genes for vacuolar sequestering enzymes (iv) genes for radical scavenging enzymes. The genes involved in salt tolerance glycophytes has been given in Table 2.

### *2.3.3. Adaptive features of Mangroves*

*2.3.3.1. Morphological features*

In response to salinity, leaves tend to be smaller and thicker in tolerant species where water economy is stringent (Ball, 1988a, Ball *et al.,* 1988). Small leaf design enhanced cooling as small leaves loss more heat by convection than large ones (see section 2.2). Leaf area of salt-stressed plants is less than half that of salt-tolerant species (Ball and Pidsley, 1995). Under the condition of high salinity, there is increase of leaf thickness from youngest to oldest leaves along shoot in *Laguncularia racemosa* (Biebl and Kintzel, 1965). Root to shoot ratio become relatively high and increases with increasing concentration of salinity in *Avicennia* and *Aegiceras* (Ball, 1988b, Saintilan, 1997).

Table 2. Functional groups of genes/proteins activated in salt stress with potential for providing salt tolerance (Adapted from Winicov, 1998)

1. Carbon metabolism and energy production/ photosynthesis
2. Cell wall/membrane structural components
3. Osmoprotectants and molecular chaperones e.g BADH, CMO (for glycine betaine)
4. Water channel proteins Aquaporins, Dehydrins, Osmotins
5. Ion transport, e.g., $H^+$ ATPase, $Na^+/H^+$ antiporter, and $Na^+/K^+$ selectivity
6. Oxidative stress defenses
7. Detoxifying enzymes
8. Proteinases
9. Proteins involved in signal perception and transduction
10. Transcription factors

#### *2.3.3.2. Anatomical features*

Many species of mangrove possess salt gland in their leaves (Fig. 5). Leaves of *Aegiceras* and *Avicennia* have deposits of salt crystals as a result of salt leak from the salt glands. The lower leaf surface of *Avicennia* is densely covered with hairs which raise the secreted droplets of salt water away from the leaf surface, preventing the osmotic withdrawal of water from the leaf tissues (Osborne & Berjak, 1997). In species of *Rhizophora, Sonneretia, Avicennia* and *Xylocarpus*, salt is also deposited in the bark of stem and roots. Several species deposit salt in senescent leaves which are then shed. This helps removing salt from metabolic tissues. Salinity causes increase in epidermal and mesophyll thickness, palisade cell length, palisade diameter and spongy cell diameter in leaves of *Atriplex*. In contrast, both epidermal and mesophyll thickness and intercellular spaces decrease significantly in leaves of salt treated *Bruguiera parviflora* (Table 3). There is also change in chloroplast ultrastructure i.e. loss of distinct granal and stromal regions and increase in number of plastglobuli (Fig. 6) There is also change in leaf anatomy i.e. development of kranz anatomy and dimorphism of chloroplast (Tomlinson, 1986), reduction in number of stomata per leaf area (Table 3, Fig. 7) and wide opening of stomata which is a typical feature of succulence (Biebl and Kinzel, 1965). High salinity causes high osmotic potential of the leaf cells of mangroves (Walter and Steiner, 1936). High osmotic potential is essential if mangroves are to draw water from the sea with its high negative water potential. As salt concentration is constant and independent of age, there is accumulation of salt by increasing the volume of leaf cells and thus inducing leaf succulence (Roth, 1992, Parida *et al.,* 2004). Increased leaf succulence enables the plants to sequester large amounts of solutes without adversely increasing cell osmotic pressure (Suarez and Sobrado, 2000) in *Avicennia germinans* (Appendix IV). Salt induced succulence can lower the resistance to $CO_2$ uptake and this increases the photosynthetic rate by increasing the internal leaf surface for gas exchange. Although the leaf succulence is often observed as a result of salt treatment (Longstreith and Nobel, 1979, Robinson, Downton and Millhouse, 1983), studies on leaf succulence of *Rhizophora mangle* from natural habitat are contracdictory (Biebl and Kinzel 1965, Camilleri and Ribi, 1983).

Table 3. Anatomical features of leaves of *B. parviflora* exposed for 45 days to zero or 400 mM NaCl; values are mean±SE (Parida *et al.*, 2004)

| Leaf tissue | [NaCl] in culture | |
|---|---|---|
| | 0 | 400 mM |
| Thickness (• m) | | |
| Upper epidermis | 40±2 | 36±3 |
| Palisade | 124±4 | 100±7 |
| Spongy | 103±5 | 92±2 |
| Lower epidermis | 32±1 | 28±3 |
| Total thickness | 299±16 | 256±12 |
| Palisade cell length (• m) | 64±3 | 52±1 |
| Palisade cell diameter (• m) | 28±2 | 27±2 |
| Spongy cell diameter (• m) | 34±3 | 30±1 |
| Inter-cellular spaces (% of total leaf area) | 21±2 | 16±2 |
| Number of stomata ($10^6$ $m^{-2}$ lower leaf area) | 137±8 | 122±12 |
| Relative stomatal pore area (%) | 3.8±0.2 | 1.4±0.2 |

There is also expansion of hypodermal cells below the upper epidermis, hypodermal storage tissues, mesophyll cells are larger and more turgescent, spongy parenchyma and palisade parenchyma are slightly loosely arranged in *Rhizophora mangle* (Werner and Stelzer, 1990). Root elongation and suberisation of cell walls are two possible pathways to avoid an invasion of these cells by salt. Root elongation maintains the capacity for accumulating ions by providing new cells (Stelzer, Kuo and Koyro, 1988); suberisation seals the ion-invaded cells and blocks the apoplasmic pathway (Lawton *et al.*, 1981, Peterson 1988, Taura *et al.*, 1988). In *Suaeda maritima*, hypodermal cells are absent and therefore, membrane surfaces of all root cortex cells are used for ion uptake (Yeo and Flower, 1986).

### *2.3.3.3 Physiological features*

Mangroves grow in an environment where salt concentration is in between that of sea water and fresh water. Sea water contains 35g/litre of salt (3.5%), 483 mM $Na^+$ and 558 mM $Cl^-$) with a osmotic potential of –2.5 MPa. So, water must be taken against this much of pressure. Besides this, mangroves experience hypersaline condition as a result of evaporation (which raises the salinity level) and fluctuation of salinity caused by tide. High salinity and salinity variation pose a problem for man-groves. To cope with the uncompromising situation, a number of physiological mechan-isms are adopted by them viz. Salt excretion (Hegemeyer, 1997), salt accumulation (Popp, 1994) and salt secretion Roth, (1992) (Table 4).

Table 4. Mechanism of salt adaptation and their known distribution in some mangrove species (Reproduced from P. J. Hograth, 1999)

| Species | Exclude | Secrete | Accumulate |
|---|---|---|---|
| *Acanthus* | + | + | |
| *Aegilatis* | + | + | |
| *Aegiceras* | + | + | |
| *Avicennia* | + | + | + |
| *Bruguiera* | + | | |
| *Ceriops* | + | | |
| *Excoecaria* | + | | |
| *Laguncularia* | | + | |
| *Osbornia* | + | | + |
| *Rhizophora* | + | | + |
| *Sonneratia* | + | + | + |
| *Xylocarpus* | | | + |

Mangroves excrete salt by ultrafiltration at the root cell membranes of cortical cells (Hegemeyer, 1997). Furthermore, hypocotyls function as an additional filter to retain salt from the shoot is considered for *B. gymnorhiza* (Lawton *et al.*, 1981).They maintain ionic balance through $K^+/Na^+$ exchange at the xylem parenchyma cells in the basal part of the plants and vacuolar $K^+/Na^+$ exchange in transpiring leaves and circulation of exchanged ions within the plants. The ultrafiltration process and $K^+/Na^+$ exchange contribute to keeping the xylem sap concentration of salt treated plants low with

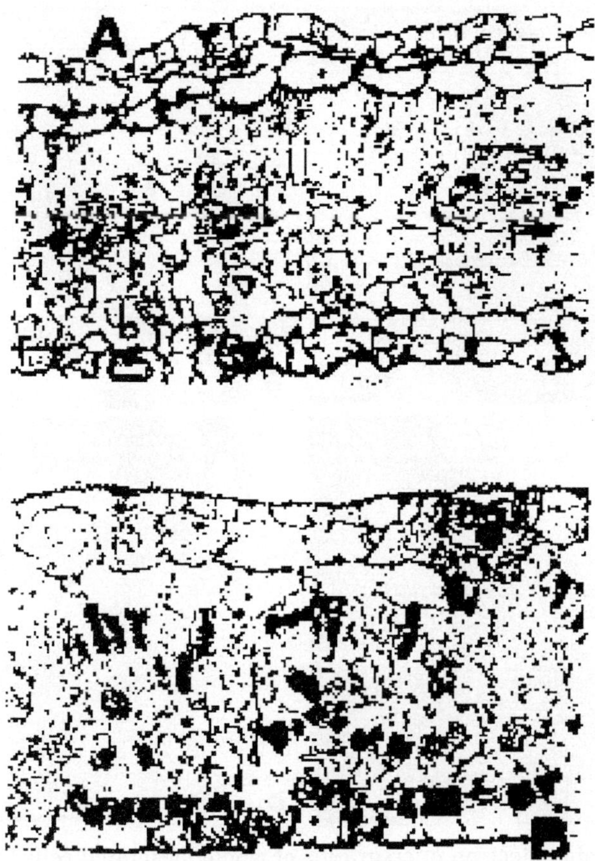


Fig. 5 & 7. T.S. of the leaf of *Aegiceras corniculatum* showing sunken stomata and salt gland respectively.

mean rates of net ion transport into the shoot being even lower due to recirculation of exchanged ions (Mallery and Teas 1984, Werne and Stelzer 1990). Salt accumulation occurs with the sequestration of $Na^+$ and $Cl^-$ into the vacuoles of the hypodermal storage tissue of the leaves (Walter and Steiner, 1936, Stewart and Ahmed 1983, Werner and Stelzer, 1990, Popp, 1994, Aziz and Khan, 2001). Salt secretion occurs through glands (Roth, 1992) and it is facilitated by efficient leaf turn over for salt shedding.

Depending on their salt eliminating mechanism mangrove plants are more often classified as excretors, accumulators and secretors. Salt non-secretors avoid salt damage by efficient sequestering of ions to the vacuoles in the leaf, (Stewart and Ahmed, 1983), translocation outside the leaf, possible cuticular transpiration (Tomlinson, 1986) and efficient leaf turnover to salt shedding. There are species which can employ more than one mechanism (Table 4) to protect against adverse effects of salinity. *Aegiceras* and *Avicennia* which are provided with salt glands also exclude 90-97% salt through a process called ultrafiltration. The concentration of xylem sap is about one tenth than that of sea water (Tomlinson, 1986). Though exact mechanism of this is not well characterized, it is understood that the process is a physical one. Negative hydrostatic pressure developed in plants by transpiration is enough to overcome negative

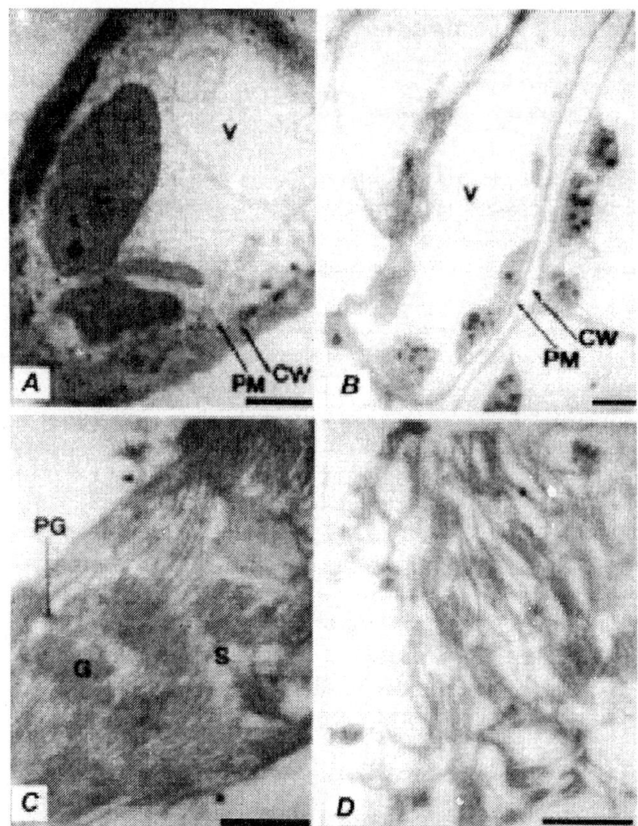


Fig. 6. Transmission electron micrographs of whole mesophyll cells (A,B) (bar = 1 μm) and chloroplasts (C,D) (bar = 0.5 μm) of control leaves (A,C) and leaves of 400 mM NaCl treated plants (B,D) C = chloroplast; CW = cell wall; EM = envelope membranes; G = granum; PG = plastoglobuli; PM = plasma membrane; S = stroma; V = vacuole.

osmotic pressure in the environment of the roots. Water is therefore drawn in, unwanted ions and other substances are excluded. *Rhizophora mangle* is not provided with salt glands, but it keeps the xylem sap essentially free of NaCl by ultrafiltration at the membranes of root cells (Walter and Steiner, 1936, Scholander *et al.,* 1966, Scholander 1968). Hydraulic properties of leaf specific conductivity of *Avicennia marina* and *Avicennia germinans* decrease at high salinity (Sobrado, 2001b). Water transport constraints imposed by whole shoot and leaf blade at high salinity are balanced by stomatal regulation of water loss which maintain stem water potentials above embolism levels (Sobrado, 2001b).

*2.3.3.4. Biochemical features*

Besides the anatomical and physiological mechanisms, mangroves also possess certain biochemical features to ameliorate the effect of high osmolarity of salt. One among them is accumulation of compatible solutes. Pinitol and mannitol are the most common compatible solutes. Proline is found mostly in *Avicennia* species and methylated quaternary ammonium compounds in *Xylocarpus* species, *Acanthus illicifolius*, *Heritiera littoralis* and *Hibiscus tiliaceus* and other species have

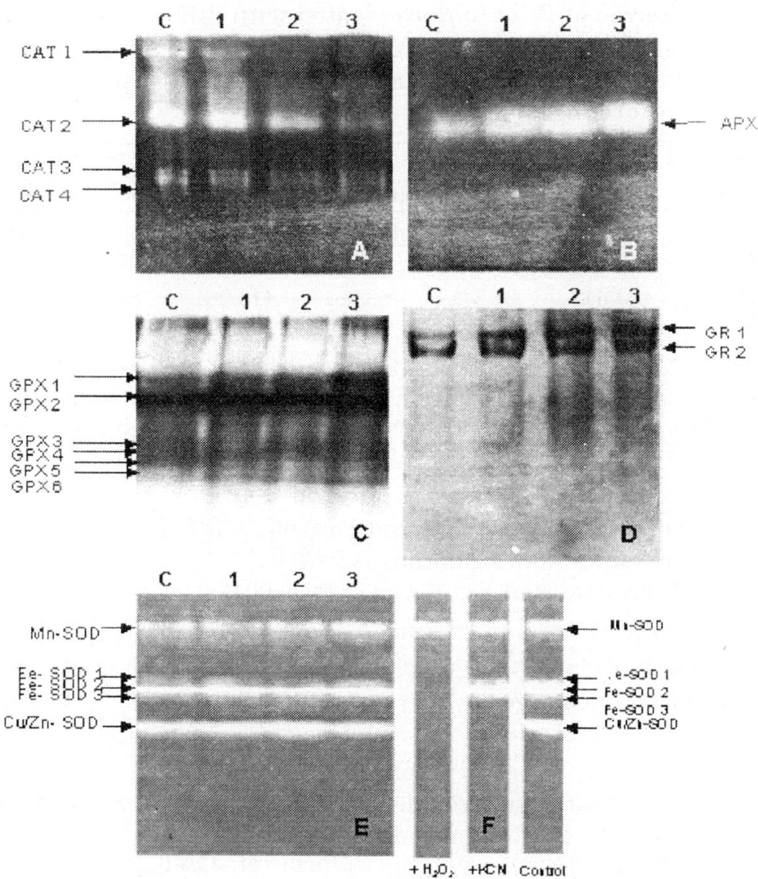


Fig. 8. Effect of NaCl on the activity staining of different scavenging enzyme in *Bruguiera parviflora*. Note the differential changes of the isoforms of scavenging enzymes due to salt stress. Details as given in Reference Parida *et al.* 2004c

small carbohydrate as the most osmo-regulating compounds (Popp *et al.,* 1985). In response to salinity, accumulation of proline and changes in polyphenol content occurs in leaves of *Bruguiera parviflora* (Parida *et al.,* 2002). Glycinebetaine, the most common compatible solute is also found in *A. marina* (Ashihara *et al.,* 1997).

Photosynthetic pigments decreases upon salt treatment for short term period (Mishra and Das, 2003) in secretor mangrove *Aegiceras corniculatum*, but decreases upon long term exposure to salt of non-secretor mangrove *Bruguiera parviflora* (Parida *et al.,* 2002). There is also decrease in amino acid and soluble protein content of the leaves (Parida *et al.,* 2002).

The seedlings of *Avicennia marina* and *Sonneretia alba* contain high concentration of $Na^+$ and $Cl^-$. NMR studies reveals that there is an accumulation of glycinebetaine, asparagines, and stachyose in *A. marina*, the highest concentration of glycinebetaine is in young leaves, while the distribution of stachyose is restricted in stems and roots. The cellular level of glycinebetaine is also increased by the salt stress (Ashihara, 2000).

Table 5. Ion levels in leaves of *B. parviflora* treated with different concentrations of NaCl for a period of 45 days. Na, K, Ca, Mg and Cl content are expressed as mmol $m^{-2}$ leaf surface Area whereas Fe, Cu and Mn content are expressed as • mol $m^{-2}$ leaf surface area (Parida *et al.* 2004)

| [NaCl] (mM) | Na mmol $m^{-2}$ | K mmol $m^{-2}$ | Ca mmol $m^{-2}$ | Mg mmol $m^{-2}$ | Fe (• mol $m^{-2}$) | Cu • mol $m^{-2}$ | Mn • mol $m^{-2}$ | Cl • mol $m^{-2}$ |
|---|---|---|---|---|---|---|---|---|
| 0 | 46.01± 2.31** | 20.52± 0.28 | 22.36± 0.63*** | 467.21± 4.21 | 424.04± 2.61 | 36.09± 0.55*** | 136.65± 1.84** | 26.92± 0.25 |
| 100 | 109.31± 1.50** | 23.41± 1.52ns | 21.24± 0.55* | 498.98± 4.70* | 413.06± 2.20ns | 33.47± 0.35* | 132.34± 1.20** | 81.11± 1.23*** |
| 200 | 134.92± 1.28** | 22.15± 1.24ns | 19.93± 1.26** | 478.68± 4.20** | 363.93± 1.80* | 25.69± 0.70** | 77.24± 0.81*** | 90.71± 0.80*** |
| 400 | 140.55± 0.75 | 21.64± 0.52ns | 13.96± 1.28 | 345.33± 4.10*** | 303.98± 2.10ns | 20.71± 0.75 | 61.43± 0.68* | 97.89± 0.87*** |

ns Not significant

*Differences from control values were significant at *P*¡Ü0.05

**Differences from control values were significant at *P*¡Ü0.01

***Differences from control values were significant at *P*¡Ü0.001

In response to salinity, $Na^+$ and $Cl^-$ concentration increases, but $Ca^{2+}$, $Mg^{2+}$, $Cu^{2+}$ and $Mn^{2+}$ decreases without change in K in leaves of *Bruguiera parviflora* (Table 5).

Upon exposure to salinity, the activity of antioxidative enzymes like superoxide dismutase and catalase is enhanced (Takemura *et al.,* 2000) in *Bruguiera gymnorhiza*. However, these enzymes differed from the seven enzymes from the leaves of secretor mangrove, *Avicennia marina*, six of which decreases in activity as salinity increases and only one of them aldolase, is stimulated to peak activity at 250 mM salt. All seven enzymes are inhibited at 500 mM salt. In the leaves of *B. parviflora*, salt treatment preferentially enhances the content of $H_2O_2$ as well as the activity of ascorbate peroxidase (APX), guaiacol peroxidase (GPX), glutathione reductase (GR), and superoxide dismutase (SOD), whereas it induces the decrease of total ascorbate and glutathione (GSH+GSSG) content as well as catalase (CAT) activity (Fig 8). Leaves of *B. parviflora* have one isoform each of Mn-SOD and Cu/Zn-SOD and three isoforms of Fe-SOD. Expression of Mn-SOD and Fe-SOD-2 was preferentially elevated by NaCl (Fig. Similarly, out of the six isoforms of GPX, the GPX-1, 2, 3 and 6 are enhanced by salt treatment, but the levels of GPX-4 and -5 changes minimally, only one prominent isoform of APX and two isoforms of GR (GR-1 and GR-2) increases, out of 4 isoform of CAT, CAT-2 isoform decreases upon salt exposure (Fig. 8). The differential changes in the levels of the isoforms due to NaCl treatment is suggested to be useful as markers for recognizing salt tolerance in mangroves. In *Bruguiera gymnorhiza*, Cu/Zn-SOD increases upon salt stress (Takemura *et al.,* 2002). However, in roots of *Aegiceras corniculatum*, activity of antioxidative enzymes CAT, APX, GPX decreases upon salt treatment for a short period for 4 days (Mishra and Das, 2003).

*2.3.3.5. Salinity effect on growth*

Growth of upland plants of *Hibiscus tiliaceus* is not affected by high salinity although the there is reduction in photosynthesis because of increased allocation to photosynthetic tissue increases LAR (leaf area ratio) to compensate for inhibition of photosynthesis by salinity (Santiago *et al.,* 2000). Plant height, leaf area, fresh weight of leaf, stem, root are decreased in response to high salinity (400 mM) in *Bruguiera parviflora* (Parida *et al.,* 2004)

*2.3.3.6 Salinity effect on water relations*

Mangroves are osmoregulators. Most of the earlier data dealing with salinity effect on physiology of mangroves related to water relations. *Salicornia europea* when exposed to salinity, develop a more negative water and osmotic potential. Small perturbations in salinity have little effect on them (Karim, 1984). To maintain osmotic balance, there is a progressive increase in tissue water and osmotic potential with increase in salinity in *Ceriops tagal* (Aziz and Khan, 2001). Salinity stress causes low stomatal conductance (Parida *et al.,* 2004) which decreases the rate of $CO_2$ accumulation and uptake, rate of transpiration and increase in xylem tension (Ball and Farquhar, 1984, Aziz and Khan, 2001). The low transpiration rates are a common feature for all mangroves (Walter and Steiner, 1936, Scholander *et al.,* 1962), but increase in transpiration in response to increase in salinity in *Bruguiera gymnorhiza* (Takemura *et al.,* 2000) is also reported and this is suggested to be aim at increasing internal salt concentration and eventually balance the increased external salt concentration. Leaf water and osmotic potentials and xylem tension increases with an increase in salinity in *Rhizophora mucronata* (Aziz and Khan, 2001). With increasing salt concentration, leaf water potential and evaporation rate decrease significantly in *Suaeda salsa* while there are no changes in relative water content (Lu *et al.,* 2002).

*2.3.3.7 Salinity effect on mangrove metabolism*

Salinity effects on photosynthesis in mangroves are studied mostly in relation to transpiration and stomatal conductance (Aziz and Khan, 2001). Ball and Farquhar (1984 a, b) have shown that unlike in *Aegiceras corniculatum*, in the most tolerant mangrove *Avicennia marina*, photosynthetic rate is barely affected by salinity. A depression of photosynthesis is also reported in *Kandelia candel*, *Bruguiera gymnorhiza* and *Rhizosphora stylosa* in response to increased salinities (Kawamitsu *et al.,* 1995). The depression of $CO_2$ assimilation is due to restriction of $CO_2$ access through stomata as there is linear proportionality relation between photosynthesis and respiration and leaf conductance (Ball, 1988, Kawamitsu *et al.,* 1995). High salinity depresses photosynthesis directly probably through partial inhibition of the activity of RUBISCO (Kotmire and Bhosale, 1985, Nazaenko, 1992). Photochemical dysfunctioning in response to salinity stress has also been observed in grey mangrove, *Avicennia marina* and the loss in photochemical activity attributed to decline in atrazine-binding polypeptide (Ball *et al.,* 1987). High salt (NaCl) uptake competes with the uptake of other nutrient ions, especially $K^+$, leading to $K^+$ deficiency. Under such condition of high salinity and $K^+$ deficiency, a reduction in quantum yield of oxygen evolution due to malfunctioning of photosystem II occurs ((Ball *et al.,* 1987). There is also a shift from $C_3$ to $C_4$ pathway in response to salinity in salt tolerant plant species like *Atriplex lentiformis* (Zhu and Meinzer, 1999). Salinity does cause down-regulation of photosynthesis in *Avicennia germinans* (Sobrado, 2000). High salinity also reduces photosynthesis in leaves of *B. parviflora*, primarily by reducing diffusion of $CO_2$ to the chloroplast, both by stomatal closure and by changes in mesophyll structure, which decreased the conductance to $CO_2$ within the leaf, as well as by affecting the photochemistry of the leaves (Parida *et al.,* 2004).

Plant growth amounts to the balance sheet of photosynthesis gains after deduction of respiratory losses and therefore, whatever effect salinity has on these processes is reflected in the growth rate of the plant integrated over time. Any stress exerted on the organism increases maintenance costs as reflected in its respiration. High salinity elicits an increase in respiration in number of mangrove plant species. Salt-induced increase in respiration is found in *Aegiceras corniculatum* and *Avicennia marina*, but with higher increase in *Aegiceras corniculatum* (Burchett *et al.*, 1989). Burchett *et al.* (1984) postulate that for *Avicennia* grown on very high salinities the maintenance processes in the leaves will compete with plant growth and maintenance processes in the roots for carbohydrate, to the extent that growth of the whole plant and growth and maintenance respiration of the roots will decline. Demand for oxygen for respiration increases under high salinity in *Avicennia marina* and there is a diversion of sucrose from macromolecule biosynthesis to respiration (Fukushima *et al.*, 1997).

Effect of salinity on nitrogen metabolism have been studied in a mangrove associate (Appendix I) *Suaeda maritima* by Stewart *et al.* (1972). They have reported that growth of *S. maritima* in the upper salt-marsh appears nitrogen-limited, going to upper to lower marsh, there is three-fold increase in total nitrogen of the plants. Nitrogen status and the activity of nitrate are related. Plants from the lower-marsh contain a fifty times higher level of enzyme than those from the upper marsh. By feeding nitrate to plants at different sites in the marsh, it is possible to induce high nitrate reductase activity in plants that previously contained low levels of the enzyme. At all sites, there is an equal potential to induce nitrate reductase. The induction experiments with plants *in situ* suggest that in this marsh nitrogen availability is the main factor limiting the growth of *S. maritima* in the upper marsh. Decrease in activity of NR, ACP, ALP as well as nitrate and phosphorus uptake, and total nitrogen levels have been reported in leaves of *Bruguiera parviflora* (Fig. 9).

#### 2.3.4 *Genetics and molecular biology of salt tolerance in mangroves*

The reason to why mangroves can grow in fluctuating and high saline conditions has led many workers to investigate at molecular level. It is found that mangrove plants have got specific proteins designated as mangrin, essential for the salt tolerance in its evolutionary process (Yamada *et al.*, 2000). The cDNAs for this novel salt-resistant factor isolated from a mangrove cDNA library using *E. coli* expression system. Mangrin cDNA driven by 35S promoter when introduced into tobacco plants, the transformants showed remarkable well growth compared to non-transformants. Our group also has found a novel protein of molecular mass 23 kDa in *B. parviflora* whose intensity gradually decreases with increasing concentration of salinity (Fig. 10). In another species of *Bruguiera* i.e *B. gymnorhiza*, several proteins are induced by NaCl treatment, but the induction of 33 kDa Mn-stabilizing protein is most important mechanism to maintain the capacity of PSII under NaCl stress (Fig. 11). This 33kDa protein is identified with oxygen evolving enhancer protein 1 (Sugihara *et al.*, 2000). However, the levels of this protein do not seem to vary in *Bruguiera parviflora*. The reason for this discrepancy is not known and would require comparative experimentation.

Among four *Avicennia* species, pinitol is the main osmoprotectant in three species, but glycinebetaine is detected only in *Avicennia marina* under high salinity. Betaine-aldehyde dehydrogenase (BADH) is detected only in three species, but choline monooxygenase (CMO) is not detected. Two types of BADH cDNAs have been isolated (Hibino *et al.*, 2001) from *A. marina* and these exhibit similar kinetic and stability properties, but the one homologue to spinach chloroplast BADH increases at high salinity.

In a high salt condition, plant cells excrete ions to environment and vacuoles to remove ions from a cytoplasm to reduce ionic stress and to maintain high ionic pressure in the vacuoles. A $Na^+$ /$H^+$ antiporter catalyzes the exchange of $Na^+$ for $H^+$ across the membrane and plays an important role in $Na^+$ removal (section 2.3.2.1.1). This antiporter is present in the mangrove, *B. sexangula* and cDNA for the vacuolar $Na^+$ /$H^+$ antiporter (BsNHx) is cloned and characterized (Tanaka *et al.,* 2000). It has been found that in *B. sexangula*, chaperonin-containing TCP1alpha (CCTalpha) has important role in stress tolerance (Yamada *et al.,* 2002). Several anti-stress genes such as chaperonin-60, c1pP protease of the lp/HSP family of chaperones, ubiquitin, eEF1A, AtDi19 and secretory peroxidase has been successfully isolated from *Avicennia marina* (Tanaka *et al.,* 2002).

In order to find out salt-responsive genes that are differentially expressed, identification and characterization of mRNA transcripts expressed in response to salinity have been done in *Bruguiera gymnorrhiza* by differential display method (Banzai *et al.,* 2002). The expression of twelve transcripts, whose corresponding cDNA fragments differentially appeared on differential display, was then analyzed by northern hybridization. Nine of them were affirmed to be up-regulated or induced by salt treatment. By screening a cDNA library, they have cloned five full-length cDNAs and identified putative protein products for three of them by homology with known protein. The RNA transcript encoding fructose-6-phosphate, 2-kinase/fructose-2,6-bisphophatase increased at 6 h following salt treatment. Transcripts encoding dihydrolipoamide dehydrogenase and lipoic acid synthase distinctly increased from 1 to 3 days after salt treatment regardless of light condition in characterization of mRNA transcripts expressed in response to salinity have been done in *Bruguiera gymnorrhiza* (Banzai *et al.,* 2002).

Salt stress leads to the generation of superoxide in the cytosol and to detoxify this, there is oxygen-scavenging system in the cytosol (section 2.3.1.2). To analyze the potential of the active oxygen-scavenging system of the cytosol in leaves of salt-stressed *Bruguiera gymnorrhiza,* full-length cDNA encoding a 153-amino acid sequence of cytosolic Cu/Zn superoxide dismutase (SOD) and a partial cDNA encoding catalase have been isolated. Northern blot analyses have showed that the transcript level of cytosolic Cu/Zn-SOD increased after 1 and 5 days NaCl treatment, but no significant change occurs in the expression of the catalase gene (Takemura *et al.,* 2002).

Salt-stress has its effect through water stress or desiccation stress. Four genes encoding BURP domain-containing proteins [two as cDNA (BgBDC2 and BgBDC3) and the other two as genomic DNA (S2 and L3)], similar to RD22 gene (desication-responsive gene in *Arabidopsis thaliana*) also have been identified, cloned and characterized from *Bruguiera gymnorhiza*. Although the four genes and RD22 are similar in respect of their repeated units of consensus sequence, the regulatory mechanisms in *B. gymnorrhiza* for the expression of RD22 homologues are different from that of RD22 (Banzai *et al.,* 2002).

These results demands that a through study on the gene expression salt-sensitive elememts be investigated in different mangrove species such that these genes paeticularly involved in salt tolerance be identified and characterized. In this context, we have tabulated some of the salt inducible genes that has been well characterized in glycophytes, but not in mangroves (Appendix-II).

## Conclusion and Future Perspective

The importance of mangroves is globally felt as this unique group of plants and their natural environment are constantly threatened. The conservation if these plants and study of their unique characteristics and their niche are currently pursued in many laboratories. In this overview, we have discussed some of the unique features of the mangroves and their ability of resisting and tolerating a variety of environmental stresses. The mangroves use different mechanisms to avoid, exclude and tolerate salt stress. We note that the mangroves not only habit saline, marshy environment, but they adapt to rapid changes in salinity level-from salinization to desalinization condition. This aspect has been studied using some typical true mangroves like *Bruguiera* and *Aegiceras* in our laboratory. We believe that such type of study will enable development of new methodologies for their preservation and conservation in saline and non-saline conditions. For example, the growth performance and vegetative propagation of ten species has been successfully demonstrated in experimental sites Mahanadi Delta (Das *et al.,* 1997, See Fig 12). Importantly, many mangrove species belonging to different mangrove families are now maintained in non-saline condition in our experimental nursery and and successfully culture hydroponically to study their physiological and reproductive behaviour (Fig. 13).

The economic and commercial values of mangroves are well established. We are their therapeutic uses. However, the studies on Their physiology, anatomy, cytology and biochemistry is in the incipient stage. We have in our study clearly seemed many important traits in mangroves that can be used as markers for salt tolerance in plants. Further studies would establish the genetic basis of these changes associated with salinization and desalinization and if these traits could be used for developing salinity tolerance in agriculturally important plants. Further, mangroves in general, tolerate low level of oxygen tension which is of common occurrence during flooding. These plants also harbour some unique fungi and such association makes them produce unusual amino acids that may have commercial value. Research into these areas is likely yield not only new products, but also would give new sights into the mechanisms of adaptive features of mangroves. India has rich biodiversity in mangroves and studies of their biogeography and biology demands immediate attention.

## Abbreviations

Chl: Chlorophyll; Rubisco: Ribulose bisphosphate carbooxylase oxygenase; PS II: photosystem II; CAM: Crassulacean Acid Metabolism; LHC: Light Harvesting Complex; ABA: Abscissic Acid; NR: Nitrate reductase; APX: Ascorbate peroxidase; CAT: Catalase; GR: Glutathione reductase; POX: non-specific peroxidase; SOD: Superoxide dismutase; ACP: Acid phosphatase; ALP: Alkaline phosphatase; LAR: Leaf area ratio; BADH: Betaine aldehyde dehydrogenase; CMO: Choline monooxygenase

## Acknowledgement

The authors wish to express their gratitude to Prof. A.K. Sharma, FNA, University of Calcutta, for inducing research on biology of mangroves in Regional Plant Resource Center, Bhubaneswar, Orissa. They also wish to thank Prof. P. Das, Former Director, Regional Plant Resource Centre for encouragemnets. Thanks are due to Prof. T. Thangaraj, Director, Regional Plant Resource Centre for providing facilities. Financial support for this works discussed in this paper at Regional Plant Resource Centre, were received from Council of Scientific and Industrial Research, New Delhi. P.M. Thanks INSA for Honorary Scientist Scheme at DAVV, Indore.

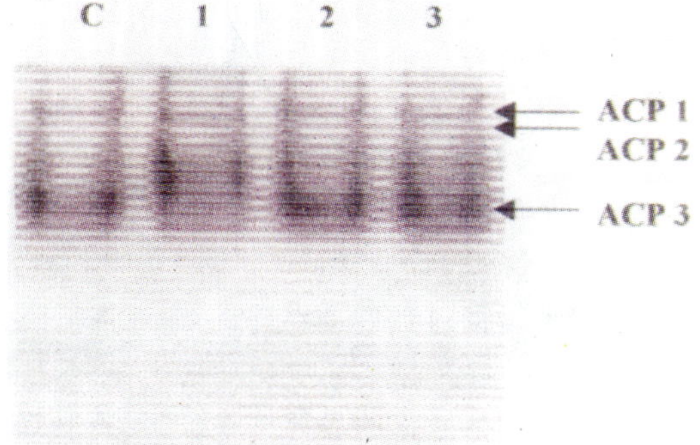


Fig. 9. Changes in isoform profile of Acid Phosphatase due to NaCl stress on *Bruguiera parviflora*. Other details has been in Parida *et al.* (2004a)

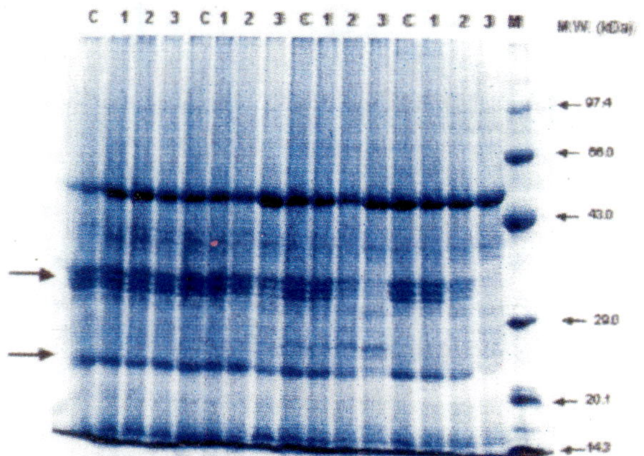


Fig. 10. Changes in protein profile during incubation with NaCl and its removal in *Bruguiera*. Changes in the protein profile has been shown as days of incubation. Other details has been in Reference Parida et al. 2004d.

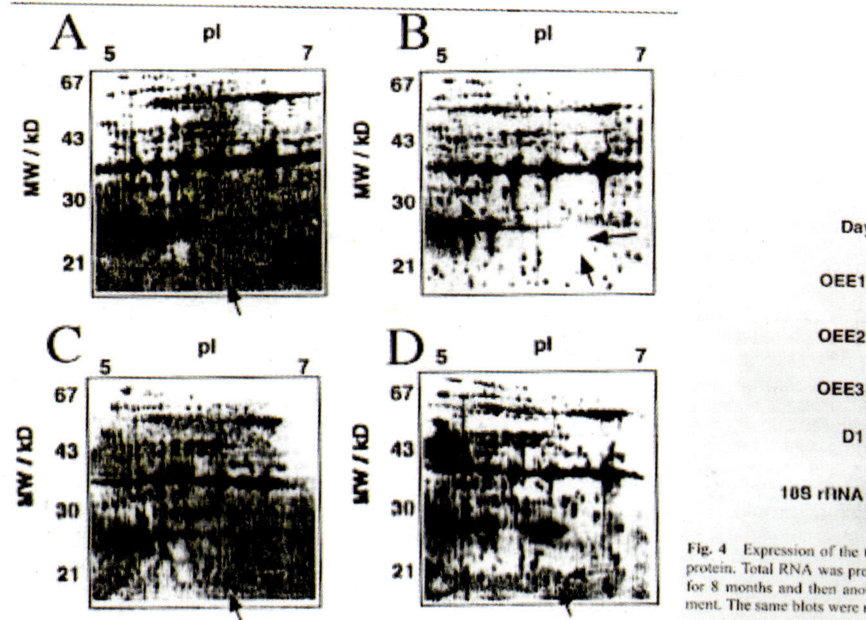


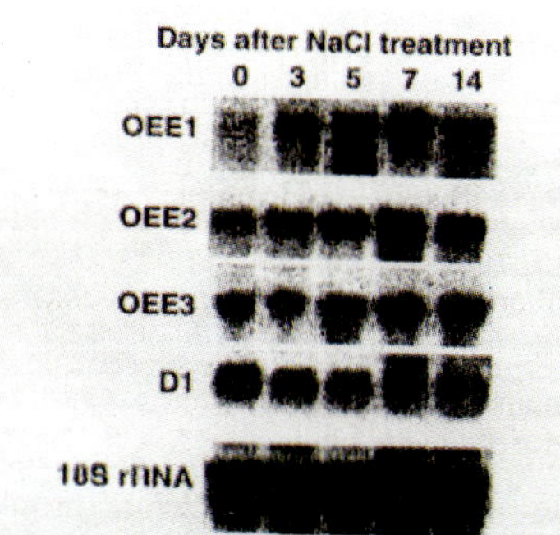


Fig. 11a. Changes in the protein profile as seen through 2D gel by NaCl stress in *Bruguiera gymnorhiza*.

Fig. 11b. Expression of the transcripts of OEE1 (33kDa Mn stabilizing protein) in *B. gymnorhiza* for details see reference Sugihara *et al.*, 2000.

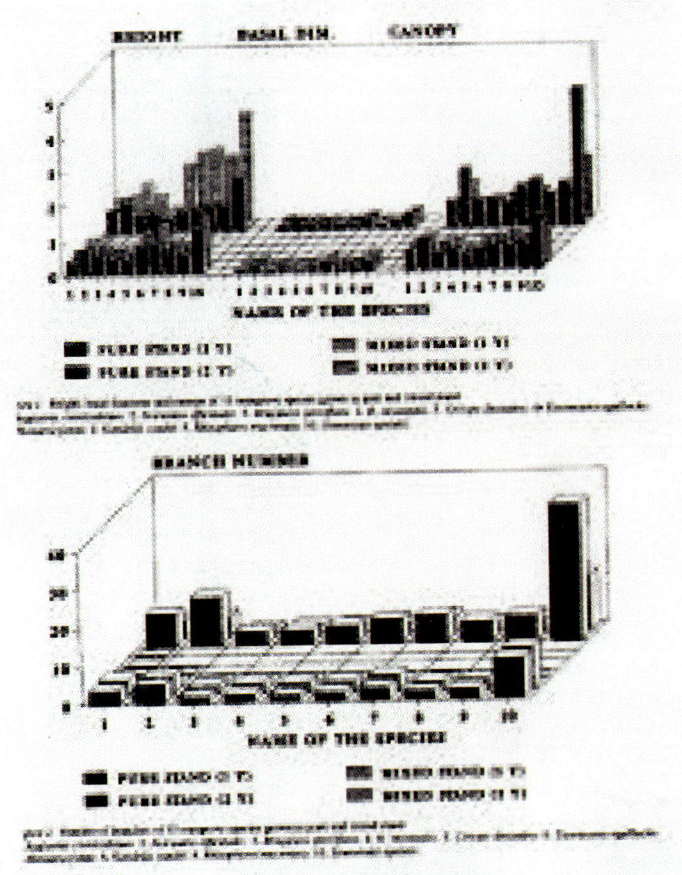

Fig. 12. Histograms showing the growth behavior of different mangrove species after restoration in the Mahanadi delta, Orissa.

Fig. 13. Hydroponic culture of *Bruguiera parviflora and Aegiceras corniculatum*

## REFERENCES


Abdulrahman FS, Williams GJ (1981) Tempeature and salinity regulation of growth and gas exchange of *Salicornia fruticosa* (L.) L. *Oecologia* (Berl) 48: 346-352

Adams P, Thomas JC, Vernon DM, Bohnert HJ, Jensen RG (1992) Distinct cellular and organismic responses to salt stress. *Plant Cell Physiol*. 33:1215-1223

Agastian P, Kingsley SJ, Vivekanandan M (2000) Effect of salinity on photosynthesis and biochemical characteristics in mulberry genotypes. *Photosynthetica* 38:287-290

Aldesuquy HS (1998) Effect of sea water salinity and gibberllic acid on abscisic acid, amino acids and water use efficiency of wheat plants. *Agrochimica* 42: 147-157

Allakhverdiev SI, Nishiyama Y, Suzuki I, Tasaka Y, Murata N (1999) Genetic engineering of the unsaturation of fatty acids in membrane lipids alters the tolerance of *Synechocystis* to salt stress. *Proc Natl Acad Sci USA* 96: 5862-5867

Allakhverdiev SI, Sakamoto A, Nishiyama Y, Murata N (2000) Inactivation of photosystems I and II in response to osmotic stress in *Synechococcus*. Contribution of water channels. *Plant Physiol* 122: 1201-1208

Anderws TJ and Muller GJ (1985) Photosynthetic gas exchange of the mangrove *Rhizophora stylosa* Griff. in its natural environment. *Oecologia* 65:449-455.

Ashihara H (2000) Characterization of the mangrove metabolism related to the salt tolerance. *Plant Cell Physiol* 41 Supplement S12 (S2-05)

Ashihara H, Adachi K, Otawa M, Yasumoto E, Fukushima Y, Kato M, Sano H, Sasamoto H and Baba S (1997) Compatible solutes and inorganic ions in the mangrove plant *Avicennia marina* and their effects on the activities of enzymes. *Z. Naturforsch*., 52c: 433-440.

Attiwill PM, Clough BF (1980) Carbon dioxide and water vapour exchange in the white mangrove. *Photosynthetica* 14:40–47

Aziz I, Khan MA (2000) Physiological adaptations to sea water concentration in *Avicennia marina* from Indus Delta, Pakistan. *Pak J Bot* 32: 171-189

Aziz I, Khan MA (2001) Effect of seawater on the growth, ion content and water potential of *Rhizophora mucronata* Lam. *J Plant Res* 114: 369-373

Ball MC (1986) Photosynthesis in mangroves. *Wetlands* (Aust.) 6:12-22

Ball MC (1988a) Ecophysiology of mangroves. *Trees* 2:129-142.

Ball MC (1988b) Salinity tolerance in the mangroves *Aegiceras corniculatum* and *Avicennia marina* I. Water use in relation to growth, carbon partitioning and salt balance. *Australian Journal of Plant Physiology* 15: 447-64.

Ball MC (1998) Mangrove species richness in relation to salinity and waterlogging: a case study along the Adelaide river floodplain, northern Australia. *Global Ecol Biogeogr Lett* 7:73-82

Ball MC (2002) Interactive effects of salinity and irradiance on growth: implications for mangrove forest structure along salinity gradients. *Trees- Structure and Function* 16: 126-139

Ball MC and Anderson JM (1986) Sensitivity of Photosystem II to NaCl in relation to salinity tolerance. Comparative studies with thylakoids of the salt-tolerant mangrove, *Avicennia marina* and the salt-sensitive pea, *Pisum sativum. Australian Journal of Plant Physiology* 13:689-698.

Ball MC and Critchely C (1982) Photosynthetic responses to irradiance by the grey mangrove, *Avicennia marina,* grown under different light regimes. *Plant Physiology* 74:7-11.

Ball MC and Farquhar GD (1984a) Photosynthetic and stomatal responses of Mangrove species, *Aegiceras corniculatum* and *Avicennia marina* to long term salinity and humidity. *Plant Physiol*., 74:1-6.

Ball MC and Munns R (1992) Plant responses to salinity under elevated atmospheric concentrations of $CO_2$. *Australian Journal of Botany* 40:515-525.

Ball MC and Passioura JB (1995) Carbon gain in relation to water use: photosynthesis in mangroves. In: Schulze E-Dad, Cladwell MM (Eds.) *Ecophysiology of Photosynthesis*, Springer Verlag, Berlin.

Ball MC, Chow WS and Anderson J.M, (1987) Salinity-induced potassium deficiency causes loss of functional photosystem II in leaves of the grey mangrove, *Avicennia marina,* through depletion of the atrazine-binding polypeptide. *Aust. J. Plant Physiol.*, 14:351-36.

Ball MC, Chow WS and Anderson JM (1987) Salinity-induced potassium deficiency causes loss of functional photosystem-II in leaves of the grey mangrove, *Avicennia marina*, through depletion of the atrazine-binding polypeptide. *Australian Journal of Plant Physiology* 14:351-361.

Ball MC, Cowan IR and Farquhar GD (1988) Maintenance of leaf temperature and the optimization of carbon gain in relation to water loss in a tropical mangrove forest. *Australian Journal of Plant Physiology* 15:263-276.

Ball MC, Farquhar GD (1984b) Photosynthetic and stomatal responses of the grey mangrove to transient salinity conditions. Plant Physiol 74:7–11

Ball MC, Pidsley SM (1995) Growth responses to salinity in relation to distribution in two mangrove species, *Sonneratia alba* and *Sonneratia lanceolata*, in northern Australia. *Funct Ecol* 9:77–85

Banzai T, Hershkovits G, Katcoff DJ, Hanagata N, Dubinsky Z, Karube I (2002) Identification and characterization of mRNA transcripts differentially expressed in response to high salinity by means of differential display in the mangrove, *Bruguiera gymnorrhiza*. *Plant Science* 62: 499-505.

Banzai T, Sumiya K, Hanagata N, Dubinsky Z, Karube I (2002) Molecular cloning and characterization of genes encoding BURP domain-containing protein in the mangrove, *Bruguiera gymnorrhiza*. *Trees - Structure and Function* 16: 87-93.

Bartels D, Nelson D (1994) approaches to improve stress tolerance using molecular genetics. *Plant Cell Environ* 17: 659-667

Basak UC, Das AB, Das P (1996) Chlorophylls, carotenoids, proteins and secondary metabolites in 14 species of mangroves *Bull J Mar Sci* 58: 654-659

Basak UC, Das AB, Das P (1999) Organic constituents in leaves of 9 mangrove species of Orissa coast, India. *Pak J Bot* 31(1: 55-62)

Basak UC, Das AB, Das P 1998. *In situ* quantitization of DNA and karyotype analysis in four threatened mangrove species found in Bhitarkanika forests of Orissa. *Cytobios* 93: 147-155.

Bekki K, Trinchant JC, Rigaut J (1987) Nitrogen fixation by *Medicago sativa* nodules and bacteroids under sodium chlorides stress. *Physiol. Plant* 71: 61-67

Bemn Hayyim G, Faltin Z, Gepstein S, Camoin L, Strosberg AD, Eshdat Y (1993) Isolation and characterization of salt-associated protein in *Citrus*. *Plant Sci*. 88: 129-140

Berry JA, Bjorkman O (1980) Photosynthetic response and adaptation to temperature in higher plants. *Ann Rev Plant Physiol* 31:491-543.

Bhosale LJ, Shinde LS (1983) Significance of cryptovivipary on *Aegiceras corniculatum* (L.) Blanco In: Tasks for vegetation science 8 (Ed. Teas HJ) pp 123-129, W. Junk, The Hague

Biebl R, Kinzel H (1965) Blattbau und Salzhaushalt von *Laguncularia racemosa* (L.) Gaertn.f. und anderer Mangrovebaume auf Puerto Rico. *Osterreichische Botanische Zeitschrift* 112: 56-93.

Bjorkman O, Demmig TJ and Andrews TJ (1988) Mangrove photosynthesis: response to high irradiance stress. *Australian Journal of Plant Physiology* 15:43-61.

Bohnert HJ, Nelson DE, Jensen RG (1995) Adaptations to Environmental stresses. *Plant Cell* 7: 1099-1111

Bostock RM, Quatrano RS (1992) Regulation of em gene expression in Rice: interaction between osmotic stress and ABA. *Plant Physiol* 98: 1356-1363

Botella MA, Quesada MA, Kononowicz AK, Bressan RA, Pliego F, Hasegawa PM, Valpuesta V (1994) Characterization and in situ localization of a salt-induced tomato peroxidase gene. *Plant Mol Biol* 25: 105-114.

Boto KG (1982) Nutrient and organic fluxes in mangroves. In: Clough BF (Ed) Mangrove ecosystems in Australia. pp 239-258, ANU Press, Canbera.

Briens M, Larher F ( 1982) Osmoregulation in halophytic higher plants: a comparative study of soluble carbohydrates, polyols, betaines and free proline. *Plant Cell Environ* 5: 287-292

Brugnoli E, Bjorkman O (1992) Growth of cotton under continuous salinity stress: influence on allocation pattern, stomatal and non-stomatal components of photosynthesis and dissipation of excess light energy. *Planta* 187:335–347

Bruns S, Hecht-Buchholz C (1990) Light and electron microscope studies on the leaves of several potato cultivars after application of salt at various developmental stages. *Potato Res* 33: 33-41

Burchett MD, Field CD and Pulkownik A (1984) Salinity, growth and root respiration in the grey mangrove, *Avicennia marina*. *Physiologia Plantarum* 60:113-118.

Burchett MD, Clarke CJ, Field CD and Pulkownik A (1989) Growth and redpiration in two mangrove species at a range of salinities. *Physiol Plant* 75:299-303

Camilleri JC, Ribi G (1983) Leaf thickness in mangroves (*Rhizophora mangle*) growing in different salinities. *Biotropica* 15:139-141.

Casas AM, Bressan RA, Hasegawa PM (1991) Cell growth and water relations of the halophyte, *Atriplex nummularia* I in response to NaCl. *Plant Cell Rep* 10: 81-84

Cheeseman JM (1988) Mechanism of salinity tolerance in plants. *Plant Physiol* 87: 547-550

Cheeseman JM (1994) Depressions of photosynthesis in mangrove canopies. In: photoinhibition of photosynthesis-From molecular mechanism to field(ed. NR Baker and JR Bowyer). Pp379-391. Bios Scientific Publishers, Oxford.

Cheeseman JM, Clough BF, Carter DR, Lovelock CE, Eong OJ and Sim RJ (1991) The analysis of photosynthetic performance in leaves under field conditions: a case study using *Bruguiera* mangroves. *Photosynth Res* 29: 11-2.

Cheeseman JM, Herendeen LB, Cheeseman AT and Clough BF (1997) Photosynthesis and photoprotection in mangroves under field conditions. *Plant Cell Environ* 20:579-588

Cherian, S, Reddy, MP, Pandya, JB (1999) Studies on salt tolerance in *Avicennia marina* (Forstk.) Vierh.: effect of NaCl salinity on growth, ion accumulation and enzyme activity. *Indian J. Plant Physiol.* 4: 266-270.

Claes B, Dekeeyser R, Villarroel R, Blucke MV, Bauw G, Montagu MV, Caplan A (1990) Characterization of a rice gene showing organ-specific expression in response to salt stress and drought. *Plant Cell* 2: 19-27

Clipson NJW, Tomos AD, Flowers TJ, Wyn Jones RG (1985) Salt tolerance in the halophyte *Suaeda maritima* L. Dum. The maintenance of turguor pressure and water potential gradients in plants growing at different salinities. 165: 392-396

Clough BF (1984) Growth and salt balance of the mangroves *Avicennia marina* (Forsk.) Vierh. and *Rhizophora stylosa* Griff. in relation to salinity. *Aust J Plant Physiol* 11: 419-430

Clough BF, Ong JE, Gong WK (1997) Estimating leaf area index and photosynthetic production in canopies of the mangrove *Rhizophora apiculata*. *Marine Ecol-Progres Series* 159:285-2922

Comba ME, Benavides MP, Tomaro ML (1998) Effect of salt stress on antioxidant defense system in soyabean root nodules. *Aust J Plant Physiol* 25: 665-671

Cowan IR (1986) Economics of carbon fixation in higher plantsIn: Givinish TJ (Ed.) on the economy of plant form and function. Cambridge University Press, Cambridge, 133-170

Cowan IR and Farquhar GD (1977) Stomatal function in relation to leaf metabolism and environment. In: Jennings DH (Ed) Integration of activity in the higher plant. Cambridge University Press, Cambridge, 471-505.

Crowe JH, Hoekstra FA, Crowe CM (1992) Anhydrobiosis. Annu Rev Plant Physiol 54: 579-599.

Cushman JC, Meyer G, Michalowski CB, Schmitt JM, Bohnert HJ (1989) Salt stress leads to differential expression of two isogenes of PEPCase during CAM Induction in the common ice plant. *Plant Cell* 1: 715-725

Das AB, Basak, UC, Das, P 1996. Karyotype analysis and 4C nuclear DNA estimation in three species of *Acanthus,* a mangrove associate from coastal Orissa. *Cytobios* 87: 151-159.

Das AB, Jena S, Sahoo P, Mohanty S. 2004. Karyotype, genome size and RAPD markers derived molecular phylogenetic relationship among five Indian mangroves associate legumes. *Genetica* (USA) (In Press).

Das AB, Parida A, Basak UC, Das P (2002) Studies on pigments, proteins and photosynthetic rates in some mangroves and mangrove associates from Bhirarkanika. *Orissa. Mar Biol* 141: 415-422.

Das P, Basak UC, Das AB (1997) Restoration of the mangrove vegetation in the Mahanadi Delta, Orissa, India. *Mangroves and Salt Marshes* 1:155-161

Das, A. B., Basak, U.C., Das, P. 1995. Karyotype diversity and genetic variability in some Indian tree mangroves. *Caryologia* 48(3-4): 319-328.

Das, A.B., Basak, U.C., Das, P. 1994. Karyotype diversity in three species of *Heritiera,* a common mangrove tree on the Orissa coast. *Cytobios* 80:71-78.

Das, A.B., Basak, U.C., Das, P. 1995. Chromosome number and karyotype diversity in the Rhizophoraceae found in mangrove forests of Orissa. *Cytobios* 81:27-35.

Das, A.B., Basak, U.C., Das, P. 1995. Variation in nuclear DNA contect and karyotype analysis in three species of *Avicennia,* a tree mangrove of coastal Orissa. *Cytobios* 84:93-102.

Das, A.B., Basak, U.C., Das, P. 1999. Genetic erosion of wetland biodiversity in Bhitarkanika forests of Orissa, India. *Biologia* 54(4):415-422.

Das, A.B., Mukherjee, A.K., Das, P. Molecular Phylogeny of *Heritiera* Aiton (Sterculiaceae), a tree mangrove: variations in RAPD markers and nuclear DNA content. *Botanical Journal of the Linnean Society* 136: 221-229.

Delphine S, Alvino A, Zacchini M, Loreto F (1998) Consequences of salt stress on conductance to $CO_2$ diffusion, Rubisco characteristics and anatomy of spinach leaves. *Aust J Plant Physiol* 25: 395-402

Demmig-Adams B, Wintwr K, Kruger A and Czygan F-C (1989) Zeaxanthin and the induction and relaxation kinetics of the dissipation of excitation energy in leaves in 2% $O_2$, 0% $CO_2$. *Plant Physiology* 90:887-893.

Ellison AM, Farnsworth EJ (1997) Simulated sea level change alters anatomy, physiology, growth and reproduction of red mangrove (*Rhizophora mangle* L.).

Elshintinawy F, Elshourbagy MN (2001) Alleviation of changes in protein metabolism in NaCl stressed wheat seedlings by thiamine. *Biologia Plant* 44: 541-545

Eusse A M, Aide TM (1999) Patterns of litter production across a salinity gradient in a *Pterocarpus officinalis* tropical wetland. *Plant Ecology* **145:** 307–315,

Farnsworth E (2000) The ecology and physiology of recalcitrant seeds. *Ann Rev Ecol and Systemat* 31:107-138

Farnworth EJ, Farrant JM (1998) Reductions in abscisic acid are linked with viviparous reproduction in mangroves. *Am J Bot* 85: 760-769.

Flowers TJ, Troke PF, Yeo AR (1977) Mechanism of salt tolerance in halophytes. *Annu Rev Plant Physiol* 128:89–121

Foyer CH, Noctor G (2000) Oxygen processing in photosynthesis: regulation and signaling. *New Phytol* 146: 359-388

Fukushima Y, Seasamoto H, Baba S, Ashihara H (1997) The effect of salt stress on the catabolism of sugars in leaves and roots of a mangrove plant *Avicennia marina. Z Naturfursch* 52c:187-192

GomezCadenas A, Tadeo FR, Primo Millo E, Talon M (1998) Involvement of abscisic acid and ethylene in the responses of citrus seedlings to salt shock. *Physiol Plant* 103: 475-484.

Gossett, DR, Millhollon EP, Lucas MC (1994) Antioxidant response to NaCl stress in salt tolerant and salt sensitive cultivars of cotton. *Crop Sci.* 34: 706-714

Greenway H. Munns R (1980) Mechanisms of salt-tolerance in non-halophytes. *Ann Rev Plant Physiol*, 31:149-190

Gunasekar M (1993) Changes in chlorophyll content, photosynthetic rate and saccharides level in developing *Rhizophora* hypocotyls. *Photosynthetica* 29: 635-638

Hagemann M, Erdmann N (1997) Environmental stresses. In: Rai AK (ed) Cyanobacterial nitrogen metabolism and environmental biotechnology. Springer, Heidelberg; Narosa Publishing House, New Delhi, India, pp 156-221.

Hamilton LS, Murphy DH (1988) Use and management of Nipa palm (*Nypa fructicans*, Arecaceae): a review. *Economic Bot* 42: 206-213

Hanagata N, Takemura T, Karube I and Dubinsky Z (1999) salt –water relationship in mangroves Israel. *Journal Plant Sciences* 47: 63-76

Harrington HM, Alm DM (1988) Interaction of heat and salt shock in cultured tobacco cells. *Plant Physiol* 88: 618-625

Hartzendorf T, Rolletschek H (2001) Effects of NaCl-salinity on amino acid and carbohydrate contents of *Phargmites australis. Aquat Bot* 69: 195-208

Hasegawa PM, Bressan RA, Zhu J-K, Bohnert HJ (2000) Plant cellular and molecular responses to high salinity. *Annu. Rev Plant Physiol Plant Mol Biol* 51: 463-499

Hassanein AM (1999) Alterations in protein and esterase patterns of peanut in response to salinity stress. *Biol Plant* 42: 241-248.

Hayashi H, Murata N (1998) Genetically engineered enhancement of tolerance in higher plants. In: Satoh K, Murata N (eds) Stress response of photosynthetic organisms: molecular mechanisms and molecular regulation. Elsevier, Amsterdam, pp 133-148

Hegemeyer J (1997) Salt. In: Prasad, MNV (ed.), Plant Ecophysiology. Wiley, New York, pp 173-206

Hernadez JA, Campillo A, Jimenez A, Alacon JJ, Sevilla F (1999) Response of antioxidant systems and leaf water relations to NaCl stress in pea plants. *New Phytol* 141: 241-251

Hernandez JA, Olmos E, Corpas FJ, Sevilla F, del Rio LA (1995). Salt-induced oxidative stress in chloroplasts of pea plants. *Plant Sci.* 105: 151-167

Heuer B (1991) Israel J Bot 40:275

Heuer B (1996) Photosynthetic carbon metabolism of crops under salt stress. In: Handbook of Photosynthesis, ed. by Pesserkali M., Marshal Dekar, Baten Rose , USA, pp: 897-909.

Hibinio T, Meng Y-L, Kawamitsu Y, Uehara N, Matsuda N, Tanaka Y, Ishikawa H, Baba S, Takabe T, Wada K, Ishii T and Takabe T (2001) Molecular cloning and functional characterization of two kinds of betaine-aldehyde dehydrogenase in betaine-accumulating mangrove, *Avicennia marina* (Forsk.)Vierh. *Plant Molecular Biology* 45:353-363.

Hoagrath P J (1999) The Biology of Mangroves. Academic Press, London.

Hong B, Barg R, Ho THD, (1992) Developmental and organ-specific expression of an ABA and stress-induced protein in barley. *Plant Mol Biol* 18: 663-674

Hurkman WJ, Tanaka CK (1988) Polypeptide changes induced by salt stress, water deficit and osmotic stress in barley roots, a comparison using two-dimensional gel electrophoresis. *Electrophoresis* 9: 781-787.

Hutchings P, Saenger P (1987) Ecology of mangroves. University of Queenseland Press, St Lucia.

Hwang YH, Chen SC (2001) Effects of ammonium, phosphate and salinity on growth, gas exchange characteristics and ionic contents of seedlings of mangrove *Kandelia candel* (L.) Druce. Bot *Bull Acad Sinica* 42: 131-139.

Ishitani M, Majumdar AL, Bornhouser A, Michealowski CB, Jensen RG, Bohnert HJ (1996) coordinate transcriptional induction of myoinositol metabolism during environmental stress *Plant J* 9: 537-548

Iyengar ERR, Reddy MP (1996) Photosynthesis in high salt tolerant plants. In: Pesserkali M (ed) Hand Book of Photosynthesis. Marshal Dekar. Baten Rose. USA, pp 56-65

Jana, S., Das, A.B. 2004. genetic variability in six ecotypes of mangrove, *Acanthus ilicifolius* L., as revealed by ganome size and RAPD markers. *Cytologia* 69(2): 131-139.

Jefferies RL (1981) *Biosciences* 31: 42

Jena, S., Das, A.B. 2003. Karyotype variation and genomic characterization in five monocotyledonous mangrove associate from Orissa coast. *Iran. Journ. Bot.* 10(1): 5-14.

Jena, S., Sahoo, P., Das, A. B. 2003. New reports of chromosome number and genome size in eight mangroves from coastal Orissa. *Caryologia* 56(3):353-358.

Jena, S., Sahoo, P., Mohnaty, S., Das A.B., Das, P. 2002. Karyotype variation and cytophotometric estimation of *in situ* DNA content in some minor associate mangroves of India. *Cytologia* 67:15-24.

Jones and Sutherland (1991) Stomatal control of xylem embolism. *Plant Cell Environment* 14:607-612

Joshi GV, Pimplaskar M, Bhosale LJ (1972) Physiological studies in germination of mangroves *Bot Mar* 15: 91-95

Kao WY and Tsai HC (1999) The photosynthesis and chlorophyll fluorescence in seedlings of *Kandelia candel* (L.) Druce grown under different nitrogen and NaCl controls. *Photosynthetica* 37:405-412

Kao WY, Chang KW (1998) Stable carbon isotope ratio and nutrient contents of the Kandelia candel mangrove populations of different growth forms. *Bot Bull Acad Sinica* 39:39-45

Kao WY, Tsai HC, Shih CN, Tsai TT, Handley LL (2002) Nutrient contents, delta C-13 and delta N-15 during leaf senescence in the manbgrove, *Kandelia candel* (L.) Druce. *Bot Bull Acad Sinica* 43: 277-282

Kao WY, Tsai HC, Tsai TT (2001) Effect of NaCl and nitrogen availability on growth and photosynthesis of seedlings of a mangrove species, *Kandelia candel* (L) Druce. *J Plant Physiol* 158: 841-846

Karimi SH (1984) Ecophysiological studies of *Atriplex triangularis* Willd. to environmental stress. Ph. D thesis, Ohio University, Athens, USA

Kathiresan K and Rajendran N (2002) growth of a mangrove *Rhizophora apiculata* seedling as influenced by GA(3), light and salinity. Revista de Biologia Tropical 50:525-530

Kawamitsu Y, Kitahara R, Nose A (1995) Effect of NaCl on leaf gas exchange rate and water potential in Okinawan mangroves. Ryukyu Nougakubu Gakujutsu Houkoku

Kerepesi I and Galiba G (2000) osmotic and salt-stress induced alteration in soluble carbohydrate content in wheat seedlings. *Crop Sci* 40: 482-487

Khan MA (2001) Experimental assessment of salinity tolerance of *Ceriops tagal* seedlings and saplings from the Indus delta, Pakistan. *Aquat Bot* 70: 259-268

Khavarinejad RA, Mostofi Y (1998) Effects of NaCl and $CaCl_2$ on photosynthesis and growth of alfalfa plants. *Photosynthetica* 35: 461-466

Knight H, Trewavas AJ, Knight MR (1997) Calcium signaling in *Arabidopsis thaliana* responding to drought and salinity. *Plant J* 12: 1067-1078

Kotmire SY, Bhosale LJ (1985) Photosynthesis in *Avicennia* and *Thespesia. Indian Bot Reporter* 4: 46-49

Kura-Hotta M, Mimura M, Tsujimura T, WashitaniNemoto S, Mimura T (2001) High salt treatment-induced $Na^+$ extrusion and low salt-treatment-induced $Na^+$ accumulation in suspension-cultured cells of the mangrove plant, *Bruguiera sexangula. Plant Cell Environ* 24: 1105-1112

Kurban H, Saneoka H, Nehira K, Adilla R, Premchandra GS, Fujita K (1999) Effect of salinity on growth, photosynthesis and mineral composition in leguminous plant *Alhagi pseudoalhagi* (Bieb.). *Soil Sci Plant Nutr* 45: 851-862

Kyle DJ (1984) The 32000 dalton $Q_B$ protein of phoptosystem II. *Photochem. Photobiol.* 41:107-116.

Lakhmi, M., Rajalakshmi, S., Parani, M., Anuratha, C. S., Parida, A. 1997. Molecular Phylogeny of mangrovesI. Use of molecular markers in assessing the intraspecific genetic variability in the mangrove species *Acanthus ilicifolius* Linn. (Acanthaceae). Theor. Appl. Genet. 94:1121-1127.

Lauchli A, Luttge U eds. (2000) Salinity: Environments-Plants-Molecules. Dodrecht, Netherlands: Kluwer.

Lawton JR, Todd A, Naidoo DK (1981) Preliminary investigations into the structure of the roots of the mangroves, *Avicennia marina* and *Bruguiera gymnorhiza,* in relation to ion uptake. *New Phytol* 88: 713-722

Levitt J (1980) Responses of plant to Environmental stress chilling, Freezing and high temperature stresses. New York: Academic, 2nd ed.

Liptz, Waisel Y (1974) *New Phytol* 1: 507

Loggini B, Scartazza A, Brugnoli E, Navari-Izzo F (1999) Antioxidative defense system, pigment composition and photosynthetic efficiency in two wheat cultivars subjected to drought. *Plant Physiol* 199: 1091-1099

Longstreth DJ, Nobel PS (1979) Salinity stress effects on leaf anatomy. *Plant Physiol* 63: 700-

Lovelock CE and Clough BF (1992) Influence of solar radiation and leaf angle on xanthophylls concentrations in mangroves. *Oecologia* 91:518-525.

Lovelock CE and De'ath G (1993) Influence of photosynthetic rate, solar radiation and leaf temperature on zeaxanthin concentrations in mangrove leaves. *Plant Cell, Environ*

Lu Congming, Qiu N, Wang B, Zhang J (2003) Salinity treatment shows no effects on photosystem II photochemistry, but increases the resistance of photosystem II to heat stress in halophyte *Suaeda salsa. Journal of Experimental Botany* 54: 851-860

Lugo AE, Sell M, Snedakar SC (1976) Mangrove ecosystem analysis. In: Systems analysis and simulation in ecology (ed. Patten C) pp 113-145, Academic Press, New York.

MacFarlane GR (2002) Leaf biochemical parameters in Avicennia marina (Dorsk.) Vierh as potential biomarkers of heavy metal stress in estuarine ecosystems. *Mar Pollun Bull* 44:244-256

MacFarlane GR and Burchett MD (1999) Zinc distribution and excretion in the leaves of the grey mangrove, *Avicennia marina* (Forsk.) Vierh. *Environ Exp Bot* 41:167-175

MacFarlane GR and Burchett MD (2000) Cellular distribution of copper, lead, zinc in the grey mangrove, *Avicennia marina* (Forsk.) Vierh. *Aquat Bot* 68:45-59

MacFarlane GR and Burchett MD (2001) Photosynthetic activity and peroxidase activity as indicators of heavy metal stress in the grey mangrove, *Avicennia marina* (Forsk.) Vierh. *Mar Pollut Bull* 42: 233-240

Mallery CH, Teas HJ (1984) The mineral ion relations of mangroves. I. Root cell compartments in a salt excluder and a salt secretor species at low salinities. *Plant Cell Physiol* 25:1123-1131

Mansour MM (2000) Nitrogen containing compounds and adaptation of plants to salinity stress. *Biol Plant* 43:491-500

Mauchamp A and Mesleard F (2001) Salt tolerance in *Phragmites australis* populations from coastal Mediterranean marshes. *Aquat. Bot.* 70:39-52

Meltcher PJ, Goldstein G, Meinzer FC, Yount DE, Jones TJ, Holbrook NM, Huang CX (2001) Water relations of coastal and estuarine *Rhizophora mangle*: xylem pressure potential and dynamics of embolism formation and repair. *Oecologia* 126: 182-192

Mishra S, Das AB (2004) Effect of NaCl on leaf salt secretion and antioxidative enzyme level in roots of a mangrove, *Aegiceras corniculatum. Ind J Exp Bot* 41: 160-166

Mitsuya S, Takeoka Y, Miyake H (2000) effects of sodium chloride on foliar ultrastructure of sweet potato (Ipomoea batatus Lam.) plantlets grown under light and dark conditions *in vitro. J Plant Physiol* 157: 661-667

Mittra B., Parida AK, Das AB, Das TK, Mohnaty P. 2004. Characterization and immunolocalization of a salt-sensitive SSP-23 protein in a mangrove *Bruguiera parviflora. Planta* (Electronic version realed).

Moon GJ, Clough BF, Peterson CA (1986) Apoplastic and symplastic pathways in *Avicennia marina* (Forsk.) Vierh roots revealed by fluorescent tracer dyes.

Moorthy and Kathiresan (1999) Photosynthetic efficiency in rhzophoracezn mangroves with reference to compartmentalization of photosynthetic pigments. *Revista de Biologia Tropical* 47: 21-25

Moorthy P, Kathiresan K (1997) Influence of ultraviolet-B radiation on photosynthetic and biochemical characteristics of mangrove *Rhizophora apiculata. Photosynthetica* 34:465-471

Moorthy P, Kathiresan K (1998a) Effects of UV-B radiation on biomass and uptake of nutrients in mangrove seedlings of *Rhizophora apiculata* (Rhizophorales: Rhizophoraceae). *Ind J Mar Sci* 27: 239-242

Moorthy P, Kathiresan K (1998b) UV-B induced alteration in composition of thylakoid membrane and amino acids in leaves of *Rhizophora apiculata* Blume. *Photosynthetica* 35: 321-328.

Mundy J, Chua NH (1990) Nuclear proteins bind conserved elements in the abscisic acid responsive promoter of a rice RAB gene. *Proc Natl Acad Sci* USA 87: 1406-1410

Munns R, Termatt A (1986) Whole plant responses to salinity. *Aust. J. Plant Physiol* 13:143-160.

Naidoo G, Rogalla H, vonWillert DJ (1997) Gas exchange responses of a mangrove species, *Avicennia marina*, to waterlogged and drained conditions. *Hydrobiologia* 352: 39-47

Naidoo G, Vonwillert DJ (1995) Diurnal gas exchange characteristics and water-use efficiency of 3 salt-secreting mangroves at low and high salinities. *Hydrobiologia* 295: 13-22

Nazaenko LV (1992) Effect of sodium chloride on ribulosebisphosphate carboxylase of *Euglena* cells. Fiziologiya Rastenii (Moscow) 39:748-752 (in Russian with English abstract)

Osborne DJ, Berjak P (1997) The making of mangroves: the remarkable pioneering role played by the seed of *Avicennia marina. Endeavour* 21:143-147

Pannier F, Fraino de Pannier R (1975) Physiology of vivipary in *Rhizophora mangle*. In: Proceedings of the international symposium on biology and management of mangroves. (Ed. Walsh GE, Snedakar SC, Teas HJ) pp 632-639 Institute of Food and Agricultural Sciences, University of Florida, Gainesville

Papageorgiou, GC Alygizaki-Zobra A, Ladas N, Murata N (!998) A method to probe the cytoplasmic osmolarity and osmotic water and solute fluxes across the cell membrane of cyanobacteria with Chl a fluorescence: experiments with *Synechococcus* Sp. PCC 7942 *Physiol Plat* 103: 215-224

Papageorgiou, GC, Govindjee (Eds.) Chlorophyll a Fluorescence: A Signature of Photosynthesis Series : *Advances in Photosynthesis and Respiration* , Vol. 19

Parani, M. Lakshmi, M., Ziegenhagen, B., Fladung, M., Senthilkumar, P., Parida, A. 2000. Molecular phylogeny of mangroves VI. PCR-RFLP of *trn*S-*psb*C and *rbc*L gene regions in 24 mangrove and mangrove-associate species. *Theor Appl Genet* 100:454-460.

Parani, M., Lakshmi, M., Elango, S., Ram, N., Anuratha, C.S., Parida, A. 1997a. Molecular phylogeny of mangroves II. Intra and inter-specific variation in *Avicennia* revealed by RAPD and RFLP markers. *Genome* 40:487-495.

Parani, M., Rao, C.S., Mathan, N., Anuratha, C.S., narayanan, K. K., Parida, A. 1997b. Molecular phylogeny of mangroves III. Parentage analysis of a *Rhizophora* hybrid using random amplified polymorphic DNA and restriction fragment length polymorphism markers. *Aquatic Botany* 58:167-172.

Parida A, Das AB, Das P (2002) NaCl stress causes changes in photosynthetic pigments, proteins and other metabolic components in the leaves of a true mangrove, *Bruguiera parviflora*, in hydroponic cultures. *J Plant Biol* 45: 28-36.

Parida AK, Das AB (2004) Effects of NaCl on nitrogen and phosphorus metasbolism in a true mangrove, *Bruguiera parviflora* grown under hydroponic culture. *J Plant Physiol*. 161: 921-928

Parida AK, Das AB, Mitra B, Mohnaty P. 2004. Salt stress induced alterations in protein profile and protease activity in the mangrove *Bruguiera parviflora*. *Z. Naturforsch* (Germany) 59c: 408-414.

Parida AK, Das AB, Mitra B (2004) Effects of salt on growth, ion accumulation, photosynthesis and leaf anatomy of the mangrove, *Bruguiera parviflora*. *Trees* 18: 167-174

Parida AK, Das AB, Mittra B (2003) Effects of NaCl stress on the structure, pigment complex composition and photosynthetic activity of mangrove *Bruguiera parviflora* chloroplasts. *Photosynthetica* 41: 191-200

Parida AK, Das AB, Mohanty P (2004) Defense potentials to NaCl in a mangrove, *Bruguiera parviflora*: Differential changes of isoforms of some antioxidative enzymes. *J. Plant Physiol.* 161: 531-542

Pelegri SP, Riveramonroy VH, Twilley RR (1997) A comparison of nitrogen fixation among three species of mangrove litter, sediments and pneumatophores in South Florida USA. *Hydrobiologia*, 356: 73-79

Peterson CA (1988) Exodermal casparian bands : their significance for ion uptake by roots. *Physiol Plant* 72:204-208

Plant AL, Cohen A, Moses MS, Bray EA (1991) Nucleotide sequence and spatial expression pattern of a drought and ABA-induced gene of tomato. *Plant Physiol* 97: 900-906

Popova LP, Stoinova ZG, Maslenkova LT (1995) Involvement of abscisic acid in photosynthetic process in *Hordeum vulgare* L. during salinity stress. *J Plant Growth Regul* 14: 211-218

Popp M (1984) Chemical composition of Australian mangroves. I. Inorganic ions and organic acids

Popp M (1984) Chemical composition of Australian mangroves. II. Low molecular weight carbohydrates. *Z. Pflanzenphysiol* 113: 411-421

Popp M (1994) Salt resistance in herbaceous halophytes and mangroves, In: Behnke H-D, *et al.*, (Eds.) Progress in Botany, Springer, Berlin, pp 416-429.

Popp M, Larther F, and Weigel P (1985) Osmotic adaptation in Australian mangroves. *Vegetatio* 61: 247-254.

Popp M, Larther F, Weigel P (1985) Osmotic adaptation in Australian mangroves. *Vegetatio* 61: 247-254.

Quintero fj, Garciadeblas B, Rodriguez-Navarro A (1996) The sal 1 gene of *Arabidopsis* encoding an enzyme with 3'(2') , 5' –Biosphere nucleotidase and inositol polyphosphate 1-phosphate activities, increases salt tolerance in yeast. *Plant Cell* 8: 529-537

Rada F, Goldstein G, Orozco A, Montilla M, Zabala O and Azocar A (1989) Osmotic and turgor relations of three mangrove ecosystem speies. *Aust J Plant Physiol* 16:477-486

Rawson HM and Munns R (1984) Leaf expansion in sunflower as influenced by salinity and

Reddy MP, Sanish S, Iyengar ERR (1992) Photosynthetic studies and compartmentation of ions in different tissues of *Salicornia brachiata* Roxb. Under saline conditions. *Photosynthetica* 26: 173-179

Redfield JA (1982) Trophic relationships in mangrove communities . In: Mangrove ecosystems in Australia: Structure, Function and Management (ed. BF Clough) pp 259-262. Austrlian National University Press, Canbera.

Reuveni M, Bennett AB, Bressan RA, Hasegawa PM (1990) Enhanced $H^+$ transport capacity and ATP hydrolysis activity of the tonoplast $H^+$–ATPase after NaCl adaptation. *Plant Physiol* 94: 524-530

Reviron MP, Vartanian N, Sallantin M, Huet JC, Pernollet JC, de Vienne D (1992) Characterization of a novel protein induced by progressive or rapid drought and salinity in *Brassica napus* leaves. *Plant Physiol* 100: 1486-1493

Robinson SP, Downton WJS, Millhouse JA (1983) Photosynthesis and ion content of leaves and isolated chloroplasts of salt-stressed spinach. *Plant Physiol* 73: 238-242.

Romeroaranda R, Soria T, Cuartero J (2001) Tomato plant-water uptake and plant-water relationships under saline growth conditions. *Plant Sci* 160: 265-272

Roth LC (1992) Hurricane and mangrove regeneration: effects of hurricane Joan. October 1988, on the vegetation of Isla del Venado, Bluefields, Niccaragua. *Biotropica* 24: 375-384.

Sadaka A, Himmelhoch S, Zamir A (1991) aA 150 kilodalton cell surface protein is induced by salt in the halotolerant green alga *Dunaliella salina. Plant Physiol* 95: 822-831

Sahoo, P., Mohnaty, S., Das, A. B. 1994. Interspecific genetic diversity in different ecotypes of *Bruguiera* L., a tree mangrove from mangrove forests of Orissa, revealed through karyotype, 4C DNA and RAPD analysis. *Proceedings of 8$^{th}$ Bigyan Congress* pp.121-130.

Saintilan N (1997) Above and below-ground biomasses of two species of mangrove on the Hawkesbury River estuary, New South Wales. *Mar Fresh Water Res* 48: 147-152

Santiago LS, Lau TS, Melcher PJ, Steele OC and Goldstein G (2000) morphological and physiological responses of *Hibiscus tiliaceus* populations to light and salinity International. *Journal Plant Sciences* 161: 99-106

Scholander PF (1968) How mangroves desalinate seawater. *Physiol Plant* 21:251-261.

Scholander PF, Bradstreet ED, Hammel HT, Hemmingsen EA (1966) Sap concentrations in halophytes and other plants. *Plant Physiol* 41: 529-532

Scholander PF, Bradstreet ED, Hammel HT, Hemmingsen EA, Garay W (1962) Salt balance in mangroves. *Plant Physiol* 37: 722-729

Scholander PF, Hammel HT, Hemmingsen EA, Bradstreet ED (1964) Hydrostatic pressure and osmotic potential in leaves of mangroves and some other plants. *Proceedings of National Academy of Sciences* USA 52:119-125.

Sehmer L, Alaui-Sosse B, Dizengremel P (1995) Effect of salt stress on growth and on the detoxifying pathway of pedunculate oak seedlings (*Quercus robur* L.). *J Plant Physiol* 147: 144-151

Singh NK, Bracker CA, Hasegawa PM, Handa AK, Buckel S, Hermodson MA, Pfancock E, Regnier FE, Bressan RA (1987) Characterization of osmotin. Ahumatin like protein associated with osmotic adaptation in plant cells. *Plant Physiol* 85:

Singh NK, Handa AK, Hasegawa PM, Bressan RA (1985) Proteins associated with adaptation of cultured tobacco cells to NaCl. *Plant Physiol* 79:126-137

Singh SK, Sharma HC, Goswami AM, Datta SP, Singh SP (2000) In vitro growth and leaf composition to grapevine cultivar as affected by sodium chloride. *Biol Plant* 43: 283-286

Smirnoff N, Cumbes QJ (1989) Hydroxy radical scavenging activity of compatible solutes. *Phytochem* 28: 1057-1060

Smith FA (1973) The internal control of nitrate uptake into excised barley roots with differing salt contents. *New Phytol* 72: 769-782

Smith JAC, Popp M, Luttge U, Cram WJ, Diaz M, Griffiths H, Lee HSJ, Medina E, Schafer C, Stimmel K-H and Thonke B (1989) Ecophysiology of xerophytic and halophytic vegetation of a coastal alluvial plain in northern Venezuela. VI. Water relations and gas exchange of mangroves. *New Phytologist* 111: 293-307.

Smith SM and Snedaker SC (2000) hypocotyls function in seedling development of the red mangrove, *Rhizophora mangle* (L.) *Biotropica* 32:677-685

Sobrado MA (1999) Leaf photosynthesis of the mangrove *Avicennia germinans* as affected by NaCl. *Photosynthetica* 36: 547-555

Sobrado MA (2000) Relation of water transport to leaf gas exchange properties in three mangrove species. *Trees - Struct and Funct* 14: 258-262

Sobrado MA (2001a) Effect of high external NaCl concentration on the osmolality of xylem sap, leaf tissue and leaf glands secretion of the mangrove *Avicennia germinans* (L.). *Flora* 196: 63-70

Sobrado MA (2001b) Hydraulic properties of *Avicennia marina* as affected by NaCl. *Biol Planta* 44:435-438

Sobrado MA and Greaves ED (2000) Leaf secretion composition of the mangrove species Avicennia germinans (L.) in relation to salinity: a case study by using total –reflection X-ray fluorescence analysis. *Plant Sci* 159:1-5

Soussi M, Liuch C, Ocana A (1998) Effect of salt stress on growth, photosynthesis and nitrogen fixation in chick pea (*Cicer arietinum* L.) cultivars under stress. *J Exp Bot* 50: 1701-1708

Sperry JS and Tyree MT (1988) Mechanism of water-stress induced xylem embolism. *Plant Physiol* 88:581-604.

Stelzer R, Kuo J, Koyro HW (1988) Substitution of $Na^+$ by $K^+$ in tissues and root vacuoles of barley (Hordeum vulgare L.cv. Aramir). *J Plant Physiol* 132: 671-677.

Stewart GR and Ahmed I (1983) Adaptations to salinity in angiosperm halophytes. In: Robbs DA, Pier point, WS (Eds.) Metals and Micronutrients. Academic Press, London, pp 33-50

Stewart GR, Lee JA, Orebamjo TO (1972) Nitrogen metabolism of halophytes I. Nitrate reductase activity in *Suaeda maritime. New Phytol* 71: 263-267

Suarez N, Sobrado MA (2000) Adjustments in leaf water relations of mangrove (*Avicennia germinans*) seedlings grown in a salinity gradient. *Tree Physiol* 20: 277-282

Suarez N, Sobrado MA, Medina E (1998) Salinity effects on the leaf water relations components and ion accumulation patterns in *Avicennia germinans* (L.) seedlings. *Oeccologia* 114: 299-304

Sugihara K, Hanagata N, Dubinsky Z, Baba S and Karube I (2000) Molecular characterization of cDNA encoding oxygen evolving enhancer protein 1 increased by salt treatment in the mangrove *Bruguiera gymnorhiza*. 41 (11): 1279-1285

Sussex K (1975) Growth and metabolism of the embryo and attached seedling of the viviparous mangrove *Rhizophora mangle. Am J Bot* 62: 948-953

Takemura T, Hanagata N, Dubinsky Z, Karube I (2002) Molecular characterization and response to salt stress of mRNAs encoding cytosolic Cu/Zn superoxide dismutase and catalase from *Bruguiera gymnorrhiza*. *Struct and Funct* 16(2-3): 94-99.

Takemura T, Hanagata N., Sugihara K., Baba S., Karube I., Dubinsky, Z. (2000) Physiological and biochemical responses to salt stress in the mangrove, *Bruguiera gymnorrhiza*. *Aquat Bot* 68: 15-28.

Tanaka S, Fukuda A, Nakamura A, Yamada A, Saito T, Ozeki Y, Mimura T (2000) Molecular cloning and characterization of mangrove $Na^+/H^+$ antiporter cDNA. *Plant Cell Physiol* 41 supplement 021 (1pB09)

Tanaka S, Ikeda K, Ono M and Miyasaka H (2002) Isolation of several antistress genes from a mangrove plant *Avicennia marina*. *World Journal of Microbiology and Biotechnology*, 18 (8): 801-804.

Taura T, Jwaikawa Y, Furumoto M, Katou K (1988) a model for radical water transport across plant roots. *Protoplasm* 144: 170-179

Theuri MM, Kinyamario JI, VanSpeybroeck D (1999) Photosynthesis and related physiological process in two mangrove species, *Rhizophora mucronata* and *Ceriops tagal* at Ghazi Bay, Kenya. *Afr J Ecol* 37:180193

Thiyagarajah M, Fry SC, Yeo AR (1996) In vitro salt tolerance of cell wall enzymes from halophytes and glycophytes. *J Exp Bot* 47: 1717-1724

Thomas JC, McElwain EF, Bohnert HJ (1992) Convergent induction of osmotic stress responses. *Plant Physiol* 100: 416-423

Tomlinson PB (1986) The Botany of Mangroves. Cambridge University Press, Cambridge.

Torres-Schumann S, Godoy JA, Pintor-Toro JA (1992) A probable lipid transfer protein gene is induced by NaCl stress in tomato plants. *Plant Mol Biol* 18: 749-757

Tuffers AV, Naidoo G and VonWillert DJ (1999) The contribution of leaf angle to photoprotection in the mangroves *Avicennia marina* (Forsk.) Vierh. and *Bruguiera gymnorhiza* (L.) Lam. Under field conditions in South Africa. *Flora* 194: 267-275

Vaidyanathan R, Kuruvilla S, Thomas G (1999) Characterization and expression pattern of an abscisic acid and osmotic stress responsive gene from rice. *Plant Sci* 140: 21-30

Vernon DM, Bohnert HJ (1992) A novel methyltransferase induced by osmotic stress in the facultative halophyte *Mesembryanthemum crystallinum*. *EMBO J* 11: 2077-2085

Vernon DM, Taraczynski MC, Jensen RG, Bohnert HJ (1993) Cyclitol production in transgenic tobacco. *Plant J* 4:199-205.

Wallbot V, Cullis CA (1985) Rapid genomic change in higher plants. *Ann Rev Plant Physiol Plant Mol Biol* 36:367-396

Walter H, Steiner M (1936) Die Okologie der on afrikanischenMangroven. *Z Botanik* 75:623-627

Wang Y, Nil N (2000) Changes in chlorophyll, ribulose bisphosphate carbooxylase-oxygenase, glycine betaine content, photosynthesis and transpiration in *Amaranthus tricolor* leaves during salt stress. *J Hortic Sci Biotech* 75 : 623-627

Werner A and Stelzer R (1990) Physiological responses in the mangrove *Rhizophora mangle* grown in the absence and presence of NaCl. *Plant Cell and Environment* 13:243-255

Winicov I (1998) New molecular approaches to improving salt tolerance in crop plants. *Annal Bot* 81: 703-710

Wu JL, Seliskar DM, Gallagher JL (1998) Stress tolerance in the marsh plane *Spartina patens:* Impact of NaCl on growth and root plasma membrane lipid composition. *Physiol Plant* 102: 307-317.

Yamada A, Saitoh T, Mimura T, Ozeki Y (2000) A novel resistant factor mangrin from a mangrove plant. *Plant Cell Physiol* 41, Supplement 201 (3aD11)

Yamada A, Sekiguchi M, Mimura T, Ozeki Y (2002) The role of plant CCT alpha in salt- and osmotic stress tolerance. *Plant Cell Physiol* 43: 1043-1048

Yancey P, Clark ME , Had SC, Bowlus RD, Somero GN (1982) Living with water stress: evolution of osmolyte system. *Science* 217: 1214-1222

Yasumoto E, Adachi K, Kato M, Sasamoto H, Baba S, Ashihara H (1999) Uptake of inorganic ions and compatible solutes in cultured mangrove cells during salt stress. *In vitro Cellular and Dev Biol-Plant* 35:82-85.

Yates EJ, Ashwath N, Midmore DJ (2002) Response to nitrogen, potassium and sodium chloride by three mangrove species in pot culture. *Trees- Structure and Function* 16: 120-125

Yen HE, Zhang DZ, Lin JH, Edwards GE, Ku MSB (1997) Salt-induced changes in protein composition in light-grown callus of *Mesembryanthemum crystallinum*. *Physiol Plant* 101:526-532

Yeo AR (1981) Salt tolerance in the halophyte *Suaeda maritima* L.Dum. : intracellular compartmentation of ions. *J Exp Bot* 32: 487-497

Yeo AR and Flower TJ (1986) Salinity resistance in rice (*Oryza sativa* L.) and a pyramiding approach to breeding varieties for saline soils. *Aust J Plant Physiol* 13:161-173

Youssef T, Saenger P (1998) Photosynthetic gas exchange and accumulation of phytotoxins in mangrove seedlings in response to soil physico-chemical characteristics associated with waterlogging. *Tree Physiol* 18:31-324.

Zheng WJ, Wang WQ and Lin P (1999) Dynamics of elemental contents during the development of hypocotyls and leaves of certain mangrove species. *J Exp Mar Biol Ecol* 233:247-257

Zhu J, Meinzer FC (1999) Efficiency of $C_4$ photosynthesis in *Atriplex lentiformis* under salinity stress. *Aust J Plant Physiol* 26:79-86

Zhu JK, Shi J, Singh U, Wyatt SE, Bressan RA, Hasegawa PM, Capita NC (1993) Enrichment of vitronectin and fibronectin like proteins in NaCl-adapted plant cells and evidence for their involvement in plasma-membrane-cell wall adhesion *Plant J* 3: 637-646

Zuberer DA, Silver WS (1975) Mangrove associated nitrogen fixation. In : Proceedings of the International Symposium on biology and management of mangroves. (ed. GE Walsh, SC Snedakar and HJ Teas), pp 643-653, University of Florida, Gainesville.

# APPENDICES

## APPENDIX I : TRUE MANGROVE SPECIES AVAILABLE IN INDIA

**Major Mangroves (20 species)**

1. Arecaceae

    1. *Nypa fruticana (Thumb)*

2. Avicenniaceae

    2. *Avicennia alba* Bl.
    3. Avicennia marina *(Forsk) Vahl*
    4. Avicennia officinalis *L.*

3. Combretaceae

    5. *Lumnizera racemosa* Willd.
    6. *Lumnizera littoralia* (Jack.) Voigt

4. Rhizophoraceae

    7. *Bruguiera cylindrica* (Linn.) Bl.
    8. *Bruguiera gymnorrhiza* (Linn.) Savign
    9. *Bruguiera sexangula* (Lour.)
    10. *Bruguiera parviflora* (Roxb.)
    11. *Ceriops decandra* (Griff.) Ding Hou
    12. *Ceriops tagal* (Perf.)
    13. *Kandelia candel* (L.) Druce
    14. *Rhizophora mucronata* L.
    15. *Rhizophora apiculata* Bc.
    16. *Rhizophora stylosa* Griff.

5. Sonneratiaceae

    17. *Sonneratia caseolaris* (Linn.) Engl.
    18. *Sonneratia apetala* Buch-Ham.
    19. *Sonneratia griffithi* Kurz.
    20. *Sonneratia alba* J. Sm.

**Minor Mangroves (10 species)**

6. Euphorbiaceae

    21. *Excoecaria agaocha* L.

7. Meliaceae

   22. *Xylocarpus granatum* Koen
   23. *Xylocarpus mekongenesis* Pierre.

8. Myrsinaceae

   24. *Aegiceras corniculatum* (Linn)

**9.Plumbaginaceae**

   ***25. Aegialitis rotundifilia* Roxb.**

10. Peteridacea

   26. *Acrostichum aureum* L.

11. Rubiaceae

   27. *Scyphiphora hydrophyllace* Gaerin f.

12. Sterculiaceae

   28. *Heritiera fomes* Buch-Ham.
   29. *Heritiera littoralis* Dryand. ex. Ait.
   30. *Heritiera macrophylla* Wallich ex. Kurz

## APPENDIX-II: MANGROVE ASSOCIATES (64 SPECIES)

13. Acanthaceae

   31. *Acanthus ilicifolius* L.
   32. *Acanthus volubilis* Wall.
   33. *Acanthus ebracteatus* Vahl.

14. Aizoaceae

   34. *Sesuvium portulacastrum* L.

15. Amaranthaceae

   35. *Alternanthera paronychoides* St. Hill

16. Amarylidaceae

   36. *Crinum defixum* Kar-Gawler

17. Annonaceae

   37. *Alphonsea madraspatana* Beddome

18. Apocynaceae

   38. *Cerbera manghas* L.

19. Araceae

   39. *Cryptocornye ciliata* (Roxb.) Schott.

20. Arecaceae

40. *Phoenix paludosa* (Roxb.)

21. Aristolociaceae

41. *Aristolochia indica* Linn

22. Asclepiadaceae

42. *Saccolobus carinatus* Wall.
43. *Saccolobus globosus* Wall.
44. *Tylophora tenuis* Bl.
45. *Pentatropis capensis* Lirtn.
46. *Finonsonia obovata* Wall.

23. Asteraceae

47. *Spaeranthus indicus* L.

24. Bignoniaceae

48. *Dolichondom spathecia* Linn. f.

25. Caesalpiniaceae

49. *Caesalpinia crista* (L.) Roxb.
50. *Caesalpinia bonduc* (Linn.) Roxb
51. *Cynometra ramiflora* L.
52. *Cynometra iripa* Kostel.

26. Chenopodiaceae

53.*Suaeda nudijlora* (Willd.) Moq.
54. *Suaeda maritime* (L.) Dumort
55. *Arthrocnemum indicum* (Willd.) Moq.
56. *Salicornia brachiata* Roxb.

27. Convulvulaceae

57. *Stictocardia tilifolia* (Desr.) Hall. F.
58. *Ipomea per-caprae* L.
59. *Ipomea tuba* (Schl.) G.Don.

28. Cyperaceae

60. *Fimbristylis ferrugina* (L.) Vahl.
61. *Scirpus littoralis* Schrad.

29. Fabaceae

62. *Dalbergia spinosa* Roxb.
63. *Derris heterophylla* Willd.
64. *Derris indica* (Lamk.) Bennet

65. *Derris scandens* Benth.
66. *Derris trifoliate* Lour
67. *Instia bijuga* (Colebr.) O. Kuntze
68. *Mucuna gigantica* (Willd.) D C.

30. Hippocratacea

69. *Salacia chinensis* L.

31. Lecythiadaceae

70. *Baringtonia racemosa* Roxb.

32. Loganiaceae

71. *Strychnos nux-vomica* Linn

33. Loranthaceae

72. *Viscum orientate* L.

34. Malvaceae

73. *Thespesia populnea* (L) Sol. ex Correa
74. *Thespesia populnoids* (Roxb.) Kostel.
75. *Hibiscus tiliaceus* L.
76. *Hibiscus tortuosus* Roxb.

35. Meliaceae

77. *Amoora cuculata* (Roxb.) Pellegrin

36. Pandanaceae

78. *Pandanus odoratimus* Linn. f.
79. *Pandanus tectorius* Sold. ex. Park.

37. Poaceae

80. *Aeluropus lagopoides* (L.) Trin.
81. *Spinifix littoreus* (Burm. t:) Merr.
82. *Portresia coarctata* (Roxb.) Tateoka
83. *Myriostachya wightiana* (Nees ex Steud.)
84. *Urochondra setulosa* (Trin.) Hubb.
85. *Phragmates karka* (Retz.) Stewd.

38. Rubiaceae

86. *Hydrophylax maritime* Linn.

39. Rutaceae

87. *Meropa angulata* (Willd.) Swingle

40. Salvadoraceae

88. *Salvadora persica* L.

41. Solanaceae

89. *Solanum trilobatum* L.

42. Tamariaceae

*90. Tamarix troupii* Hole
91. *Tamarix erecoides* Rottler and Willd.

43. Tiliaceae

*92. Brownlowia tersa* (Linn.) Kosterm
*93. Brownlowia lanceolata* (L.) Kosterm

44. Verbinaceae

94. *Clerondrom ineme* Gaertn.

## APPENDIX-III : SALT INDUCIBLE GENES IN GLYCOPHYTES

| Gene | Plants | Reference |
|---|---|---|
| Alfin1 | Alfalfa | Bastola *et al.* 1998a |
| ARSK1 | Arabidopsis | Hwang and Goodman, 1995 |
| ATCDPK1 | Arabidopsis | Urao *et al.* 1994 |
| ATCDPK2 | Arabidopsis | Urao *et al.* 1994 |
| Atmyb2 | Arabidopsis | Urao *et al.* 1993 |
| AtP5CS | Arabidopsis | Yoshiba *et al.* 1995 |
| AtPLC1 | Arabidopsis | Hirayama *et al.* 1995 |
| cor6.6 | Arabidopsis | Wang *et al.* 1995 |
| kin1 | Arabidopsis | Wang *et al.* 1995 |
| mlip15 | Maize | Kussano *et al.* 1995 |
| MsPRP2 | Alfalfa | Deutsch and Winicov, 1995 |
| OsBZB | Rice | Nakagawa, Ohmiya and Hattori, 1996 |
| PKABA1 | Wheat | Holappa and Walker-Simmons, 1995 |
| rd22 | Arabidopsis | Iwasaki,Yamaguchi-Shinozaki & Shinozaki, 1995 |
| rd29A(COR78) | Arabidopsis | Yamaguchi-Shinozaki and Shinozaki, 1994 |
| rd29B | Arabidopsis | Yamaguchi-Shinozaki and Shinozaki, 1994 |
| sal 1 | Arabidopsis | Quintero *et al.* 1996 |
| *Brassica napus* | Bnd 22 | Reviron *et al.* 1992 |
| *Citrus sinensis* | Saltassociated 23-25kDa | Bemn Hayyim *et al.* 1993 |

| | | |
|---|---|---|
| *Dunaliella salina* | protein<br>P 150 | Sadaka *et al.* 1991 |
| *Hordeum vulgare* | 26 and 27 kDa proteins (salt induced polypeptides SIP S1-S4)<br>hva 1 | Hurkman and Tanaka, 1988<br>Hong *et al.* 1992 |
| *Lycopersicum esculentum* | TAS-12<br>le-16 | Torres-Schumann *et cl.* 1992<br>Plant *et al.* 1991 |
| *Mesembryanthemun crystallinum* | ppc-1 and ppc-2 isogenes<br>lmt 1<br>lnps 1 | Cushman *et al.* 1989 |
| *Nicotiana tabacum* | 26 and 43 kDa polypeptides<br>30 kDa polypeptides<br>vitronectin and fibronectin like proteins<br>osmotin<br>RAB 21<br>SalT | Vernon and Bohnert 1992<br>Ishitani *et al.* 1996<br>Singh *et al.*, 1985<br>Harrington and Alm 1988<br>Zhu *et al.* 1993 |
| *Oryza sativa* | em | Singh *et al.* 1987<br>Mundy *et al.* 1990<br>Claes *et al.* 1990<br>Bostock and Quatrano 1992 |

## APPENDIX-IV: SALINITY EFFECT: HALOPHYTES VS. GLYCOPHYTES

Salt tolerance mechanism is a multigenic trait and therefore, biochemical pathways leading to products or processes that improve tolerance are likely to act additively and probably synergistically. So advantage of halophytes over glycophytes may be through more efficient performance of a new basic biochemical tolerance mechanism. Appendix-1V summarizes difference in halophytes and glycophytes with respect to salt tolerance mechanism.

| Halophytes | Glycophytes |
|---|---|
| 1.Salinity has a smaller effect upon the extent to which stomata limit photosynthesis | 1. Higher. |
| 2. High water use efficiency. | 2. Comparatively less. |
| 3. Low internal $CO_2$ concentration. | 3. Comparatively less. |
| 4. low light saturated photosynthetic rate. | 4. It is less. |
| 5. Efficient accumulating solutes. | 5. Less efficient. |
| 6. Low level of $Na^+$ and $Cl^-$ ions in the cytoplasm and chloroplast | 6. It is higher. |

## APPENDIX V : RESEARCH WORK ON SALT TOLERANCE IN MANGROVES

| Organism | Reference | Work |
|---|---|---|
| *Avicennia germinans*<br>*Rhizophora mangle*<br>*Laguncularia racemosa* | Sobrado (2000) | Gas exchange and hydraulic properties in relation to salinity tolerance |
| *Avicennia germinans* | Sobrado (2000) | Leaf secretion composition in relation to salinity |
| *Avicennia marina* | MacFarlane & Burchett (1999) | Zinc distribution amd excretion in the leaves |
| | MacFarlane & Burchett (2000) | Cellular distribution of copper, lead, zinc |
| | MacFarlane & Burchett (2001) | Photosynthetic activity and peroxidase activity as indicator of heavy metal stress |
| *Bruguiera gymnorhiza* | Takemura *et al.* (2000) | Physiological & biochemical responses to salt st.:ess |
| | Sugihara *et al.* (2000) | Molecular characterization of OEE protein 1 |
| | Hanagata *et al.* (2000) | Salt-water relation ship |
| *Aegiceras corniculatum*<br>*Avicennia marina* | Ball (1988) | Water use, carbon partitioning and salt balance |
| *Hibiscus tiliaceous* | Jaramillo & Bayona (2000) | Morphological and physiological responses to light and salinity |
| *Avicennia germinans* | Sobrado (1999) | NaCl stress on leaf photosynthesis |
| *Pterocarpus officinalis* | Eusse & Aide (1999) | Effect of salinity gradient on litter, flower and fruit production |
| *Rhizophora mucronata*<br>*Ceriops tagal* | Theuri *et al.* (1999) | Photosynthesis and related physiological processes |
| *Rhizophora apiculata*<br>*Rhizophora stylosa*<br>*Ceriops tagal*<br>*Aegiceras corniculatum*<br>*Avicennia marina*<br>*Kandelia candel* | Zheng *et al.* (1999) | Element content change during leaf development |
| *Avicennia germinans* | Sobrado (1999) | Drought effect under contrasting salinities |
| *Bruguiera gymnorhiza* | Yasumoto *et al.* (1999) | Inorganic ion and compatible solute uptake |
| *Bruguiera sexangula* | Yamada *et al.* (2002) | Salt tolerance genes |
| | Tanaka *et al.* (2002) | Anti-stress genes |
| *Kandelia candel* | Kao and Tsai (1999) | Photosynthesis and Chl a fluorescence under varied nitrogen & NaCl controls |
| *Bruguiera gymnorhiza*<br>*Avicennia marina* | Tuffers *et al.* (1999) | Contribution of leaf angle to photoprotection in mangroves |

| | | |
|---|---|---|
| *Avicennia marina* | Ball *et al.* (1987) | Atrazine-binding polypeptide of photosystem-II sensitive to salinity |
| *Bruguiera parviflora* | Parida *et al.* (2002) | Photosynthetic pigments, protins and other metabolic components in response to salt stress |
| *Aegiceras corniculatum*<br>*Aegialitis annulata*<br>*Avicennia marina* | Naidoo and Vonwillert (1995) | Gas-exchange characteristics and water-use efficiency in response to salinity |
| *Avicennia marina* | Sobrado and Greaves (2000) | Leaf secretion composition |
| *Avicennia germinans* | Suarez and Sobrado (2000) | Leaf –water relations |
| *Bruguiera gymnorrhiza* | Takemura *et al.* 2000 | Physiological and biochemical response to salt stress |
| *Avicennia marina* | MacFarlane and Burchett, (2000) | Cellular distribution of copper, lead and Zinc |
| | Farnsworth (2000) | Ecology and physiology of viviparous and recalcitrant seeds |
| *Hibiscus tiliaceus* | Santiago et al (2000) | Morphological and physiological responses |
| *Rhizophora mucronata* | Aziz and Khan (2001) | Effect of sea water on growth, ion content and water potential |
| *Suaeda maritima* | Thiyagarajah *et al.* (1996) | In vitro salt tolerance of cell wall enzymes |
| *Aegialitis annulata*<br>*Aegiceras corniculatum*<br>*Avicennia marina*<br>*Rhizophora stylosa*<br>*Sonneratia alba* | Bjorkman *et al.* (1988) | Response to high irradiance stress |
| *Bruguiera gymnorrhiza*<br>*Rhizophora stylosa*<br>*Xylocarpus granatum* | Lovelock *et al.* (1992) | Distribution and accumulation of ultraviolet-radiation –absorbing compounds |
| *Bruguiera gymnorrhiza*<br>*Rhizophora stylosa*<br>*Bruguiera parviflora* | Lovelock and Clough (1992) | Influence of solar radiation and leaf angle on leaf xanthophylls concentrations |
| *Bruguiera parviflora* | Cheesman *et al.* (1991) | Photosynthetic performance in leaves under field conditions |
| *Rhizophora stylosa*<br>*Rhizophora mangle* | Cheesman *et al.* (1997) | Photosynthesis and photoprotection |
| *Rhizophoracean mangroves* | Moorthy and Kathiresan (1999) | Photosynthetic efficiency with reference to photosynthetic pigments |
| *Kandelia candel* | Kao *et al.* (2002) | Nutreient contents, • $C^{13}$ and • $C^{15}$ during leaf senescence |
| *Rhizophora apiculatsa* | Kathiresan and Rajendran (2002) | growth of seedlings as influenced by $GA_3$, light and salinity |
| *Avicennia germinans* | Sobrado (2001) | Effect of high external NaCl concentration on the osmolality of xylem sap, leaf tissue and leaf glands secretion. |

| | | |
|---|---|---|
| *Avicennia marinas* | Aziz and Khan(2000) | Physiological adaptations to sea water concentrations |
| *Bruguiera sexangula* | KuraHotta *et al.* (2001) | Effect of high and low-salt treatment on sodium extrusion and accumulation in suspension cultured cells |
| *Rhizophora mangle* | Smith and Snedaker (2000) | Hypocotyls function in seedling development |
| | Melcher *et al.* (2001) | Water relations of coastal and estuarine plants |
| *Ceriops tagal* | Khan (2001) | Experimental assessment of salinity tolerance |
| *Ceriops tagal* | Aziz and Khan(2001) | Experimental assessment of salinity<br>Experimental assessment of salinity |
| *Kandelia candel* | Hwang and Chen (2001) | Effects of ammonium, phosphate and salinity on growth, gas exchange characteristics and ionic contents |
| *Rhizophora stylosa* | Andrews and Muller (1985) | Photosynthetic gas exchange |
| *Kandelia candel* | Kao *et al.* (2001) | Effect of NaCl and nitrogen availability on growth and photosynthesis of seedlings |
| *Avicennia marina* | Ashiara(2000) | Characterization of the mangrove metabolism |
| *Atriplex gmelini* | Matoh *et al.* (1987) | Sodium, potassium and betaine concentrations in isolated vacuoles |
| *Avicennia marina* | Dowton (1982) | Growth and osmotic relations |
| | Popp *et al.* (1985) | Osmotic adaptations in Australian mangroves |
| *Avicennia germinans*<br>*Aegialitis annulata*<br>*Aegiceras corniculatum* | Naidoo and Vonwillert | Diurnal gas-exchange characteristics and water-use efficiency |
| *Bruguiera sexangula* | Tanaka *et al.* (2000) | Molecular cloning and characterization of mangrove $Na^+/H^+$antiporter cDNA |
| | Yamada *et al.* (2000) | Salt-resistant factor mangrin |
| | Saitoh *et al.* (2000) | Expression of mangrin |
| | Yu-Ling *et al.* (2000) | Betaine aldehyde dehydrogenase |
| *Aegiceras corniculatum*<br>*Avicennia marina* | Ball and Farquahar (1984) | Photosynthetis and stomatal responses |
| *Phragmites australis* | Mauchamp and Mesleard (2001) | Salt tolerance Salt tolerance in *Phragmites australis* |
| *Phragmites australis* | Hartzendorf and Rolletschek (2001) | Effects of NaCl-salinity on amino acids and carbohydrate contents |
| *Phragmites communis* | Briens and Larher | osmoregulation |
| *Avicennia germinans* | Suarez *et al.* (1998) | Salinity effect on the leaf water relations |
| *Aegiceras corniculatum*<br>*Lumintzera racemosa etc.* | Popp (1984) | Chemical composition |
| *Atriplex nummularia* | Casas *et al.* (1991) | Cell growth and water relations |

| | | |
|---|---|---|
| *Rhizophora stylosa*<br>*Avicennia marina* | Clough (1984) | Growth and salt balance of the mangroves |
| *Suaeda maritima* | Yeo (1981) | Intercellular compartmentation of ions |
| *Suaeda maritima* | Clipson *et al.* (1985) | Maintenance of turguor pressure and water-potential gradients |
| *Avicennia marina* | Ball and Critchley (1982) | Photosynthetic responses to irradiance |
| *Salicornia fructicosa* | Abdulrahman and Williams (1981) | Temperature and salinity regulation of growth and gas exchange |
| *Avicennia marina* | Ball and Anderson (1986) | Sensitivity of photosystem II NaCl I relation to salinity tolerance |
| *Avicennia marina* | Moon *et al.* (1986) | Apoplastic and symplastic pathway |
| *Rhizophora mangle* | Werner and Stelzer (1990) | Physiological response in presence nad absence of NaCl |
| *Avicennia marina*<br>*Aegiceras corniculatum* | Burchett *et al.* (1989) | Growth and respiration at a range of salinity |
| *Avicennia marina* | Hibino *et al.* (2001) | Molecular cloning and characterization of two kinds of betaine-aldehyde dehydrogenase |
| *Suaeda maritima* | Ravindran *et al.* 2001 | Influence of UV-B supplementalra-diation on growth and pigment content |
| *Avicennia marina* | MacFarlane (2002) | Leaf biochemical parameters as potential biomarkers of heavy metal stress in estuarine ecosystems |
| *Bruguiera gymnorrhiza* | Banzai *et al.* (2002) | Identification and characterization of mRNA transcripts differentially expressed in response to high salinity by means of differential display |
| *Bruguiera gymnorrhiza* | Banzai *et al.* (2002) | Molecular cloning and characterization of genes encoding BURP domain-containing protein |
| *Bruguiera gymnorrhiza* | Takemura *et al.* (2002) | Molecular characterization and response to salt stress of mRNAs encoding cytosolic Cu/Zn superoxide dismutase and catalase |
| *Rhizophora apiculata* | Moorthy and Kathiresan (1998a) | Effects of UV-B radiation on biomass and uptake of nutrients in mangrove seedlings |
| | Moorthy and Kathiresan (1998b) | UV-B induced alteration in composition of thylakoid membrane and amino acids in leaves |
| | Moorthy and Kathiresan (1997) | Influence of ultraviolet-B radiation on photosynthetic and biochemical characteristics of mangrove |
| | Pelegri *et al.* 1997 | A comparison of nitrogen fixation among three species of mangrove litter, sediments and pneumatophores |
| | Youssef *et al.* 1998 | Photosynthetic gas exchange and accumulation of phytotoxins in |

| | | |
|---|---|---|
| | | mangrove seedlings in response to soil physico-chemical characteristics associated with waterlogging |
| *Kandelia candel* | Kao and Chang (1998) | Stable carbon isotope ratio and nutrient contents of the mangrove populations of different growth forms |
| *Avicennia marina* | Naidoo *et al.* (1997) | Gas exchange responses of a mangrove to waterlogged and drained condition |
| *Rhizophora mangle L* | Ellison and Farnsworth (1997) | Simulated sea level change alters anatomy, physiology, growth and reproduction of red mangrove |
| *Rhizophora apiculata* | Clough *et al.* 1997 | Estimating leaf area index and photosynthetic production in canopies of the mangrove |
| *of* | Ball (1998) | Mangrove species richness in relation to salinity and waterlogging: a case study along the Adelaide river floodplain, northern Australia |
| *Avicennia marina*<br>*Ceriops tagal*<br>*Rhizophora stylosa* | Yates *et al.* (2002) | Response to nitrogen, potassium and sodium chloride by three mangrove species in pot culture |
| *Avicennia marina* | Cherian, *et al.* (1999) | Effect of NaCl salinity on growth, ion accumulation and enzyme activity |
| *Ceriops mtagal*<br>*Ceriops decandra* | Ball (2002) | Interactive effects of salinity and irradiance on growth |

## APPENDIX – VI. SOMATIC CHROMOSOME NUMBER, KARYOTYPE, CHROMOSOME LENGTH, CHROMOSOME VOLUME AND 4C-DNA CONTENT IN DIFFERENT SPECIES OF MANGROVES

| Species/Family | Somatic chromosome number (2n) | Karyotype formulae | Total chromosome length (μm) | Total chromosome value (μm³) | 4C DNA content | TF% | References |
|---|---|---|---|---|---|---|---|
| **Rhizophoraceae** | | | | | | | |
| 1. *Rhizophora apiculata* | 36 | 28C+8D | 91.30 | 166.9 | - | 40.19 | |
| 2. *Rhizophora mucronata* | 36 | 2A+22C+12D | 58.4 | 24.2 | 15.21 | 35.07 | Das *et al.*, 1995a; Das *et al.*, 1995c, Das *et al.*, 1999 |
| 3. *Bruguiera parviflora* | 34 | 6A+20C+8D | 63.34 | 28.34 | 12.89 | 36.09 | Das *et al.*, 1995a, Das *et al.*, 1995c, Das *et al.*, 1999, Sahoo *et al.*, 2004, |
| 4. *Bruguiera gymnorhiza* | 34 | 4A+20C+6D | 49.10 | 19.84 | - | 34.96 | Sahoo *et al.*, 2004, |

| | | | | | | | |
|---|---|---|---|---|---|---|---|
| 5. *Bruguiera sexangula* | 34 | 4A+26C+2D+2E | 57.18 | 14.64 | - | 40.06 | Das *et al.*, 1995a, Das *et al.*, 1995c, Sahoo *et al.*, 2004, |
| 6. *Bruguiera cylindrical* | 34 | 2A+4B+24C+4D | 88.11 | 58.77 | 17.27 | 39.04 | Basak *et al.*, 1998, Sahoo *et al.*, 2004, |
| 7. *Ceriops decandra* | 36 | 4A+28C+4D | 55.62 | 18.28 | 11.12 | 41.25 | Das *et al.*, 1995a, Das *et al.*, 1995c, |
| 8. *Ceriops tagal* | 36 | 2A+24C+10D | 51.08 | 22.95 | 18.14 | 42.45 | Basak *et al.*, 1998, |
| 9. *Kandelia candel* | 38 | 4A+26C+8D | 65.4 | 27.34 | 19.23 | 39.40 | Das *et al.*, 1995a, Das *et al.*, 1995c, Das *et al.*, 1999 |
| **Avicenniaceae** | | | | | | | |
| 10. *Avicennia officinalis* | 64 | 4A+24C+36D | 72.25 | 49.92 | 51.24 | 34.43 | Das *et al.*, 1995b |
| 11. *Avicennia marina* | 62 | 6A+28C+28D | 65.22 | 47.48 | 50.35 | 40.52 | Das *et al.*, 1995b |
| 12. *Avicennia alba* | 66 | 2A+36C+28D | 82.5 | 58.26 | 56.24 | 42.55 | Das *et al.*, 1995b, Das *et al.*, 1999 |
| **Combretaceae** | | | | | | | |
| 13. *Lumnizera racemosa* | 24 | 14C+10D | 56.91 | 24.10 | 22.51 | 38.08 | Jena *et al.*, 2004 |
| **Meliaceae** | | | | | | | |
| 14. *Xylocarpus granayum* | 42 | 4B+16C+22D | 2.66 | 58.23 | 20.66 | 37.14 | Das *et al.*, 1999 |
| 15 *X. mekongenesis* | 44 | 2A+30C+12D | 102.56 | 78.08 | 25.20 | 38.38 | Basak *et al.*, 1998 |
| **Myrsinaceae** | | | | | | | |
| 16. *Aegiceras corniculatum* | 44 | 2B+36C+6D | 28.97 | 7.04 | 17.61 | 40.46 | Das *et al.*, 1995c; Das *et al.*, 1999 |
| **Euphorbiaceae** | | | | | | | |
| 17. *Excoecaria agallocha* | 130 | 106C+22D | 205.20 | 362.54 | 27.56 | - | Das *et al.* 1999 |
| **Plumbaginaceae** | | | | | | | |
| 18. *Aegialitis rotundifolia* | 34 | 2B+14C+18D | 134.44 | 108.33 | 35.72 | 34.86 | Basak *et al.*, 1998; Jena *et al.*, 2002 |
| **Sterculiaceae** | | | | | | | |
| 19. *Heritiera fomes* | 38 | 2A+2B+26C+8D | 74.26 | 46.06 | 28.66 | 37.07 | Das *et al.*, 1994; Das *et al.*, 1995c, Das *et al.*, 1999, Das *et al.*, 2001 |
| 20. *Heritiera littoralis* | 38 | 2A+28C+8D | 57.68 | 18.43 | 22.26 | 39.8 | Das *et al.*, 1994, Das *et al.*, 1995c, Das *et al.*, 2001 |
| 21. *Heritiera macrophylla* | 38 | 4A+20C+14D | 65.40 | 22.37 | 25.17 | 34.15 | Das *et al.*, 1994, Das *et al.*, 1995c, Das *et al.*, 2001 |
| **Acanthaceae** | | | | | | | |
| 22. *Acanthus ilicifolius* | 44 | 2A+18C+24D | 69.42 | 27.58 | 11.49 | 32.58 | Das *et al.*, 1996, Das *et al.*, 2004 |

| | | | | | | | |
|---|---|---|---|---|---|---|---|
| 23. *Acathus ebracteus* | 44 | 4A+28C+12D | 84.25 | 40.78 | - | 40.96 | Das *et al.*, 1996 |
| 24. *Acanthus volubilis* | 44 | 2A+10C+32D | 68.85 | 35.20 | 8.39 | 35.65 | Das *et al.*, 1996 |
| **Malvaceae** | | | | | | | |
| 25. *Thespesia populnia* | 26 | 2A+2B+12C+10D | 102.98 | 85.62 | 28.56 | 32.10 | Das *et al.*, 1995c, Das *et al.*, 1999, Jena *et al.*, 2002 |
| 26. *Hibiscus tiliaceus* | 86 | 2A+74C+10D | 114.20 | 95.13 | - | 43.73 | Jena *et al.*, 2003 |
| **Fabaceae** | | | | | | | |
| 27. *Caesalpinia crista* | 24 | 2A+12C+10D | 49.86 | 18.58 | 15.76 | 37.96 | Das *et al.*, 2005 |
| 28. *Cynometra ramiflora* | 26 | 2A+18C+6D | 54.00 | 25.54 | 8.97 | 38.71 | Das *et al.*, 1999, Das *et al.*, 2005 |
| 29. *Dalbergia spinosa* | 20 | 4B+12C+4D | 38.56 | 27.64 | 10.82 | 38.93 | Das *et al.*, 2005 |
| 30. *Derris heterophylla* | 24 | 4A+14C+6D | 59.16 | 27.64 | 16.96 | 39.33 | Das *et al.*, 2005 |
| 31. *Derris indica* | 22 | 4A+8C+10D | 67.40 | 74.90 | 28.73 | 35.38 | Das *et al.*, 2005 |
| **Apocynaceae** | | | | | | | |
| 32. *Cerbera manghas* | 40 | 38C+2D | 47.74 | 88.78 | 27.83 | 43.94 | Jena *et al.*, 2003 |
| **Bignoniaceae** | | | | | | | |
| 33. *Dolichondron spathecia* | 42 | 2A+32C+8D | 63.42 | 23.68 | - | 40.23 | Das *et al.*, 1995c |
| **Asclepiadaceae** | | | | | | | |
| 34. *Tylophora tenuis* | 24 | 22A+2D | 38.10 | 15.68 | - | 44.74 | Jena *et al.*, 2003 |
| 35. *Sarcolobus carinatus* | 22 | 20C+2D | 40.62 | 35.04 | 11.05 | 44.32 | Jena *et al.*, 2003 |
| **Asteraceae** | | | | | | | |
| 36. *Sphaeranthus indius* | 30 | 2B+8C+20D | 88.50 | 91.10 | - | 34.04 | Jena *et al.*, 2003 |
| **Chenopodiaceae** | | | | | | | |
| 37. *Sueada nudiflora* | 54 | 6A+2B+28C+18D | 233.00 | 228.30 | 29.05 | 36.29 | Das *et al.*,, 2002 |
| **Aizoaceae** | | | | | | | |
| 38 *Sesuvium portulacastrum* | 48 | 4A+32C+12D | 142.62 | 195.34 | - | 38.71 | Jena *et al.*, 2003 |
| **Amaryliadaceae** | | | | | | | |
| 39. *Crinum defixum* | 22 | 2A+4B+10C+6D | 93.86 | 335.39 | 47.15 | 35.32 | Jena & Das 2003 |
| **Arecaceae** | | | | | | | |
| 40. *Cryptocoryne ciliata* | 22 | 2B+\|12C+8D | 82.96 | - | 10.35 | 39.01 | Jena & Das 2003 |

| | | | | | | | |
|---|---|---|---|---|---|---|---|
| **Myrtaceae** | | | | | | | |
| 41. *Syzygium cumini* | 66 | 56C+10D | 45.43 | 18.56 | - | 43.83 | Jena *et al.*, 2003 |
| **Cyperaceae** | | | | | | | |
| 42. *Cyperus cephalotes* | 20 | 18C+2D | 45.50 | 91.46 | 6.31 | 42.64 | Jena & Das 2003 |
| **Asparagaceae** | | | | | | | |
| 43. *Asparagus racemosus var. javanica* | 20 | 2A+4B+10C+4D | 42.44 | 141.42 | 12.63 | 38.65 | Jena & Das 2003 |
| **Poaceae** | | | | | | | |
| 44. *Porteresia coarctata* | 48 | 4A+28C+16D | 91.64 | 74.86 | 16.63 | 38.97 | Jena & Das |

## APPENDIX – VII. MOLECULAR PHYLOGENETICAL STUDIES AND DNA MARKERS IN DIFFERENT SPECIES OF MANGROVES

| Species | Type of work | Type of DNA marker studied | Reference |
|---|---|---|---|
| *Acanthus ilicifolius* | Molecular phylogeny and genetic variability in ecotypic level | RAPD | Lakshmi *et al.*, 1997, Jena *et al.*, 2004 |
| *Heriteria fomes*<br>*H. litorallis*<br>*H. macrophylla* | RAPD markers and molecular phylogenetic studies | Cytology, Ctyphotometry, Genome size and RAPD | Das *et al.*, 2001 |
| *Avicennia marina*<br>*A. officinalis*<br>*A. alba* | Molecular phylogeny in interspecific and intra specific variation | RAPD, RFLP | Pariani *et al.*, 1997a |
| *Rhizophora apiculata*<br>*R. mucronata* | Molecular phylogenetic study and DNA polymorphism | RAPD, RFLP | Parani *et al.*, 1997b |
| *R. stylosa*<br>*Kandelia candel*<br>*Bruguiera cylindrical*<br>*B. gymnorrhiza*<br>*Ceriops decandra*<br>*Heriteria fomes*<br>*Xylocarpous granatum*<br>*Ammora cucullata*<br>*Sonneratia apetala*<br>*Avicennia alba*<br>*A. officinalis*<br>*Aegiceras corniculatum* | Molecular phylogeny and DNA polymorphism | RAPD, PCR-RFLP of *trans*S-*psb*C and *rbc*L genes | Parani *et al.*, 2000,<br>Jena *et al.*,, 2004<br>Das *et al.*, 2005 |

*Lumnitzera racemosa*
*Excoecaria agallocha*
*Acanthus ilicifolius*
*Saudea maritime*
*S. monoica*
*Sesuvium portulacastrum*
*Nypa fruticans*
*Porteresia coarctata*
*Caesalpinia crista*
*Cynometra ramiflora*
*Dalbergia spinosa*
*Derris heterophylla*
*Derris indica*

*Advances in Plant Physiology*, Vol. **9**
Ed. A. Hemantaranjan
Scientific Publishers (India), Jodhpur, 2006 pp. 155-167
E-mail: **info@scientificpub.com** www.scientificpub.com

# 7

# ADAPTIVE STRATEGIES OF COCONUT PALM UNDER STRESSFUL CONDITIONS

S. Naresh Kumar, V. Rajagopal and K.V. Kasturi Bai

Central Plantation Crops Research Institute, Kasaragod, 671 124
Email: nareshkumar.soora@gmail.com

## INTRODUCTION

Growth and productivity of coconut palms is influenced by the external factors such as rain fall, temperature, sun shine duration and relative humidity apart from the management practices like irrigation and nutrition. Further, the biotic factors also influence the growth and productivity (Fig. 1). The optimum weather conditions for good growth and nut yield in coconut are well distributed annual rainfall between 130 and 230 cm, mean annual temperature of 27°C, abundant sunlight ranging from 250 to 350 $Wm^{-2}$ with at least 120 hours per month of sun shine period (Child, 1974; Murray, 1977). The generally recommended levels of major nutrients like NPK is at 500N:320$P_2O_5$:1200 $K_2O$ per palm/year. The recommended irrigation levels are 200L/palm once in 4 days or @ 66% Eo as drip irrigation. Currently most of the information available is on drought stress tolerance in coconut. Any deviation from these optimal conditions causes the palms to experience the stress conditions. Added to these, the biotic factors like pests and diseases, cause considerable stress on coconut palms. In this chapter the response and adaptive strategies of coconut palms are discussed with respect to abiotic factors like drought and water stress. Coconut has adaptive strategies to withstand or overcome the stress conditions at biochemical, physiological, anatomical and morphological levels (Fig. 2). Obviously all these are either the manifestation of genetic setup of a cultivar alone or in interaction with the environment (G x E interaction).

The coconut palm experiences moisture stress when exposed to irradiation above 265 $Wm^{-2}$, temperature of 33°C and vapour pressure deficit of 26 m bar (Kasturi Bai *et*

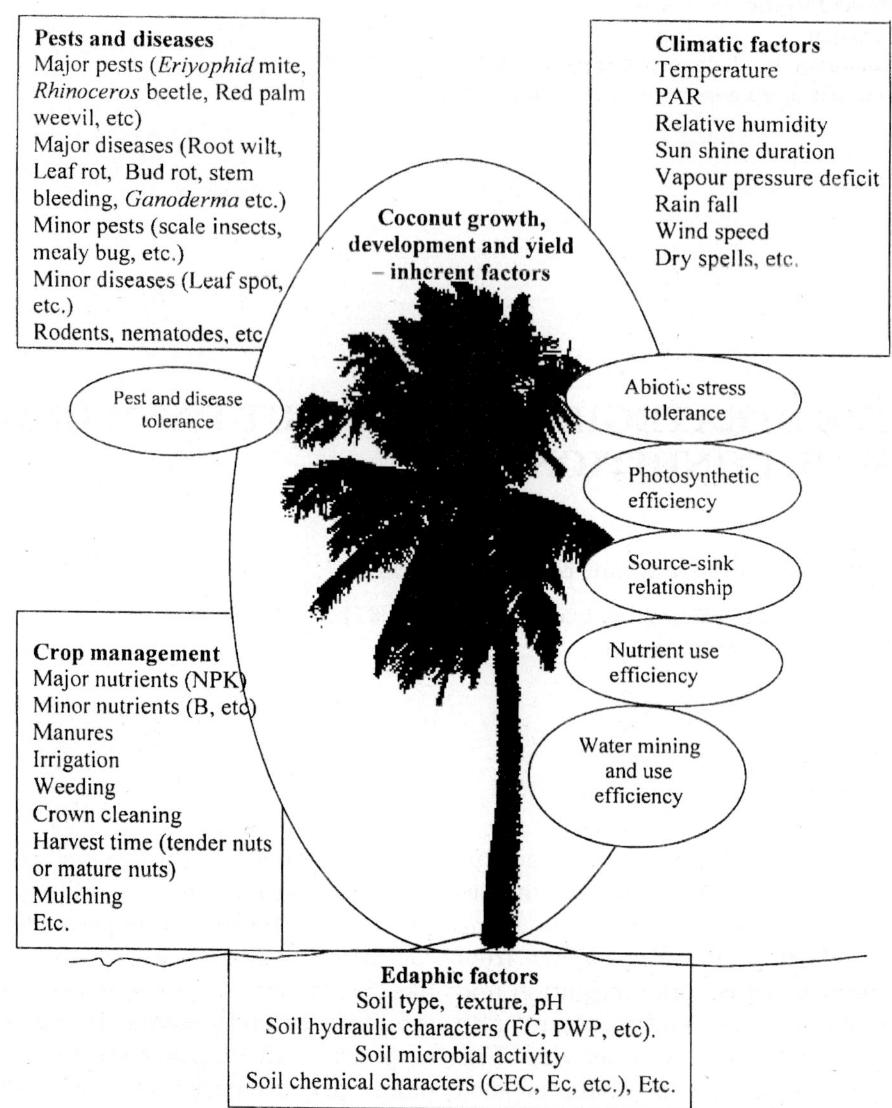


Fig 1. Factors influencing growth development and yield of coconut palm

*al.,* 1988), aggravated by soil water deficit during the period. A critical analysis of rainfall data in relation to growth stages of the inflorescence up to maturity of nuts, revealed that the duration of dry spell during initiation of inflorescence primordium, ovary development and button size nut stages, in that order have greater influence on nut yield than other stages (Rajagopal *et al.,* 1996).

## 2. RESPONSE OF COCONUT SEEDLINGS TO STRESSES

Coconut seedlings are sensitive to moisture stress and to high light intensity stresses. Studies indicated that the seedlings need to be supplied with adequate water and maintained under shade (Naresh Kumar, 2002-2003). The seedlings when exposed to

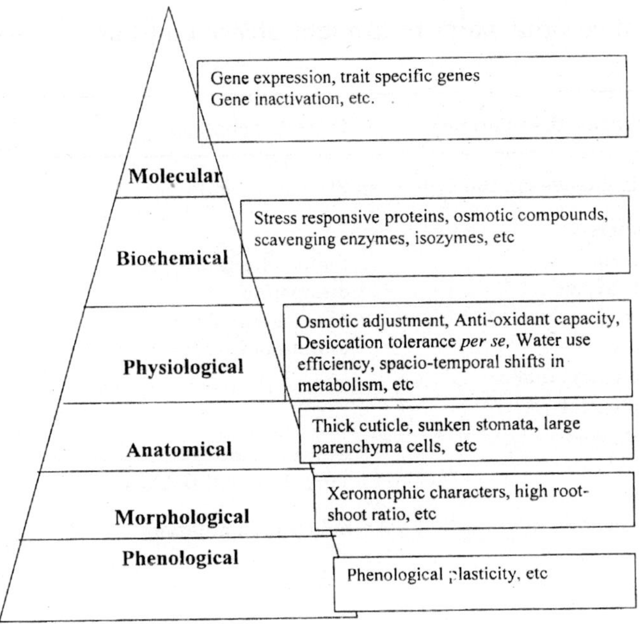


Fig 2. Adaptation levels of plants to abiotic stresses

high light intensities, experience photo-oxidative stress. Release of oxygen and super oxide radicals cause membrane damage thus causing the leaf scorching and in severe cases, seedling death (Naresh Kumar, 2002-2003). Even field-planted seedlings need to be protected from high sunlight intensities, may be by tying the leaves together or by providing a shield of dried coconut leaves around the coconut seedling canopy, which is tied. Further, field planted seedlings need to be irrigated for not only better establishment but also to protect them from moisture stress. Moisture stress in combination with high light intensity causes the seedling death.

## 2.1 Response of Coconut to Soil Moisture Levels

Depending on the soil type, the critical level of soil moisture, where coconut experiences water stress, varies. In sandy loam soil, water deficit of 110 mm is a critical level at which coconut suffers most as indicated by the stomatal closure (Rajagopal *et al.,* 1989). The photosynthetic rates, dry matter production and its partitioning are influenced by the soil water status (Kasturi Bai, 1993). High soil water deficit coupled with atmospheric evaporative demand during dry period *i.e.,* March-April, produced differential impact on coconut palms grown in red sandy loam and laterite soils (Voleti *et al.,* 1993). In general, palms suffered more in red sandy loam than in laterite soil as indicated by the stomatal resistance and leaf water potential components. Hybrids had higher stomatal resistance resulting in maintenance of higher water potentials during stress in laterite soil than in sandy loam. However the hybrid COD x WCT was found to be most sensitive to water stress in sandy loam.

## 2.2 Response and Adaptation of Coconut to Drought Stress

Coconut palm is influenced both by atmospheric and soil droughts. Under rainfed conditions, a prolonged dry spell extending

Table 1. Response of coconut palm to drought stress and factors responsible for drought tolerance

| Organ | Changes bringing about the tolerance | Impact of changes | Drought tolerance components |
|---|---|---|---|
| Root | • ABA/cytokinin changes in the xylem sap | Stomatal regulation | Uptake of water<br>Plant water balance |
| Leaf | • Leaf let thickness$^{+}$<br>• Cuticle thickness$^{+}$<br>• Sub stomatal cavity$^{+}$<br>• Palisade cell size$^{+}$<br>• Spongy cell size$^{+}$<br>• Stomatal frequency/ Index $^{-}$ | Light reflectance, Heat dissipation<br>Thermoregualation<br>Maintenance of water balance | Xeromorphic characters |
| | • Leaf water potential$^{+}$<br>• Stomatal resistance$^{+}$<br>• Transpirational rate$^{-}$<br>• Epicuticular wax$^{+}$ | Efficient metabolic functioning<br>Turgidity maintenance | Plant water balance |
| | • Total soluble sugar$^{+}$<br>• Amino acid$^{+}$ | Solute accumulation | Osmotic adjustment |
| | • Superoxide dismutase$^{+}$<br>• Peroxidase$^{+}$<br>• Catalase$^{+}$<br>• Polyphenol oxidase $^{-}$<br>• Acid phosphatase $^{-}$<br>• L aspartate 2- oxoglutarate amino transferase $^{-}$<br>• Nitrate reductase$^{+}$<br>• Lipid peroxidation $^{-}$ | Dissipation of super oxide, peroxide radicals,<br>Stable cells,<br>Efficient metabolic functioning<br>Cell membrane integrity | Oxidative stress resistance |
| | • Expression of stress proteins in Low, medium and high MW range | Stabilization of house keeping proteins/ enzymes<br>Stable metabolism | Dehydration tolerance |
| | • Net photosynthetic rate$^{+}$<br>• Water use efficiency$^{+}$<br>• Fv/Fm$^{+}$- chlorophyll fluorescence | Carbon skeloeton supply for the mainatance of metabolic and cell integrity,<br>Maintained/slightly reduced dry matter accumulation | Photosynthetic efficiency |
| Whole plant | • Total dry matter$^{+}$<br>• Vegetative dry matter $^{-}$<br>• Reproductive dry matter $^{+}$<br>• Harvest index $^{+}$ | Maintained/slightly reduced nut yield | Dry matter production and partitioning efficiency |

Position in drought tolerant types: + high/increased; - low/decreased

from 3 to 6 months affects the palm. The weekly aridity indices can be worked out using the formula Ia= WD/WNx 100, where WD is weekly water deficiency and WN is weekly water need and Ia is the aridity index for assessing drought (Rao, 1985). An aridity index of 100% for a prolonged period of 5 to 10 weeks drastically affects productivity of coconut palm.

Studies on coconut revealed adaptive strategies at different levels (Table 1). The drought tolerance mechanism in coconut is summarized in Figure 3.

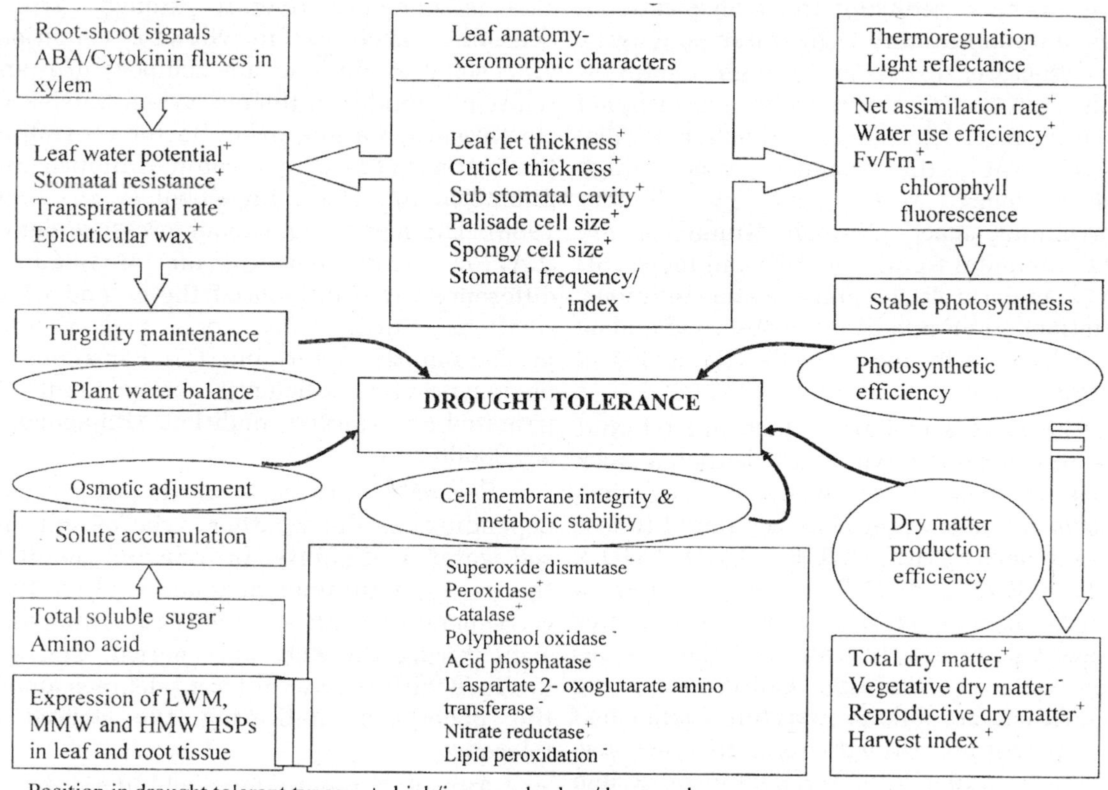


Fig 3. Functional relationship in mechanism of drought tolerance in coconut

### 2.2.1 Morphological symptoms and adaptations

The typical symptoms of coconut palm under drought stress are bending and breaking of dry leaves, poor spathe development and bunches with only one or two nuts. Drought Index, calculated based on the morphological symptoms, is related to nut yield.

Root activity and transpirational rates show marked variations among the tolerant and susceptible cultivars. The tolerant palms extract more soil moisture from the entire soil profile, but they conserve water in the tissues by reducing the transpiration rate through effective control of stomata. The root:shoot ratio in seedlings was higher in tolerant cultivars than in susceptible ones.

### 2.2.2 Leaflet anatomical adaptations in relation to drought tolerance

Anatomical basis of physiological efficiency for drought tolerance in coconut is delineated (Naresh Kumar *et al.,* 2000). The increase in leaflet thickness is mainly due to increase in parenchyma cell size. It is also associated with lowered stomatal frequency, an indication of adaptation to drought stress. Increased leaf thickness and thick cuticle are some of the xeromorphic characteristics as observed in WCT and FMS. Increase in thickness of leaflet causes decrease in the ratio of the external surface to its volume. Increased parenchyma cell size, as observed in tolerant types WCT and FMS, means less intercellular space/unit area. This may help in reducing the water conductance towards epidermis thus reducing the transpirational rates and maintaining high water potentials. The structure favorable to high photosynthetic rates (large palisade parenchyma tissue surface, i.e., small parenchyma cells) induces at the same time high transpirational rates due to higher intercellular space (Naresh Kumar *et al.*, 2000). Stomatal frequency (SF) and index (SI) play a major role in plant water relations. Variations in SF and SI were observed among the talls, dwarfs and hybrids (Rajagopal *et al.*, 1990)

The WCT and FMS, which are tolerant to water stress, have thick leaflets, thick cuticle on both sides, larger parenchyma, hypodermal and water cells compared to less tolerant ones (PHOT, WCT x COD, COD x WCT, GBGD and MYD) (Plate 1 and 2) (Naresh Kumar *et al.,* 2000). The water storage tissues supply water to other tissues when water is limiting. Cultivars having thick cuticle are able to maintain higher leaf water potentials. Drought tolerant types also have more scalariform thickening on xylem trachieds in vascular bundles and large sub-stomatal cavities. These traits are less favourable in moderately tolerant cultivars like PHOT and WCT x COD. The values for these traits are the least in susceptible types like COD x WCT, GBGD and MYD. However, the difference between moderately tolerant and susceptible cultivars for these traits is narrow. Certain parameters like epidermal cell size (upper and lower) and guard cell size are related to the drought tolerance characteristic of a cultivar. It is possible that cumulative effect of all these traits contributes to the adaptation to drought stress (Naresh Kumar *et al.,* 2000).

### *2.2.3 Physiological responses and adaptations to drought stress*

Coconut palm responds to drought stress in terms of stomatal regulation and epicuticular wax content to maintain leaf water potentials. The osmotic adjustment also is done to tolerate drought stress. Seasonal variations in climatic conditions viz., solar radiation, air temperature and relative humidity influence xylem tension by stomatal regulation, a key factor controlling the water balance in coconut (Milburn and Zimmermann, 1977; Rajagopal *et al.,* 1986; 1988). The leaf to air vapour pressure deficit (*LAVPD)* and leaf to air temperature difference ($\Delta T$ ) influenced the *gs* and water relations during day time and thereby predominantly determine the variations in photosynthetic efficiency of coconut in irrigated and rainfed conditions (Rajagopal *et al.,* 2000b).

Seasonal variations in the $\psi_{leaf}$ occurs depending on the weather, type of soil and soil water availability. In irrigated condition the $\psi_{leaf}$ is maintained at relatively high level corresponding with soil moisture availability even during the non-rainy period. The $\psi_{leaf}$ declined with time to different degrees among the genotypes, indicating the degree of tolerance.

Almost three to four fold increases in ECW during dry season was recorded in three coconut hybrids (Voleti and Rajagopal, 1991; Kurup *et al.,* 1993). The physiological age of palms and of leaves influence the formation of wax on leaf surface. Leaves of coconut seedlings have almost 50% less ECW than those from adult palms even at same degree of stress. TLC analysis revealed the qualitative difference in the composition of wax during different stages of palm growth. The alcohols could be identified only during the stress period, whereas fatty acids occurred only during post- stress.

The role of $K^+$ and $Cl^-$ nutrition in relation to drought tolerance in coconut has been explained on the basis of stomatal regulation. Unlike in most of the crops where malate serves as a balancing ion for $K^+$, in coconut the absence of chloroplasts in the guard cells deprives the availability of malate (Braconnier and D'Auzac, 1985). $Cl^-$ has been shown to increase water absorption and

Plate 1. Leaf breaking in drought affected palm

Plate 2. Button shedding and poor nut yield in drought affected palm

Plate 3. Drought coconut garden

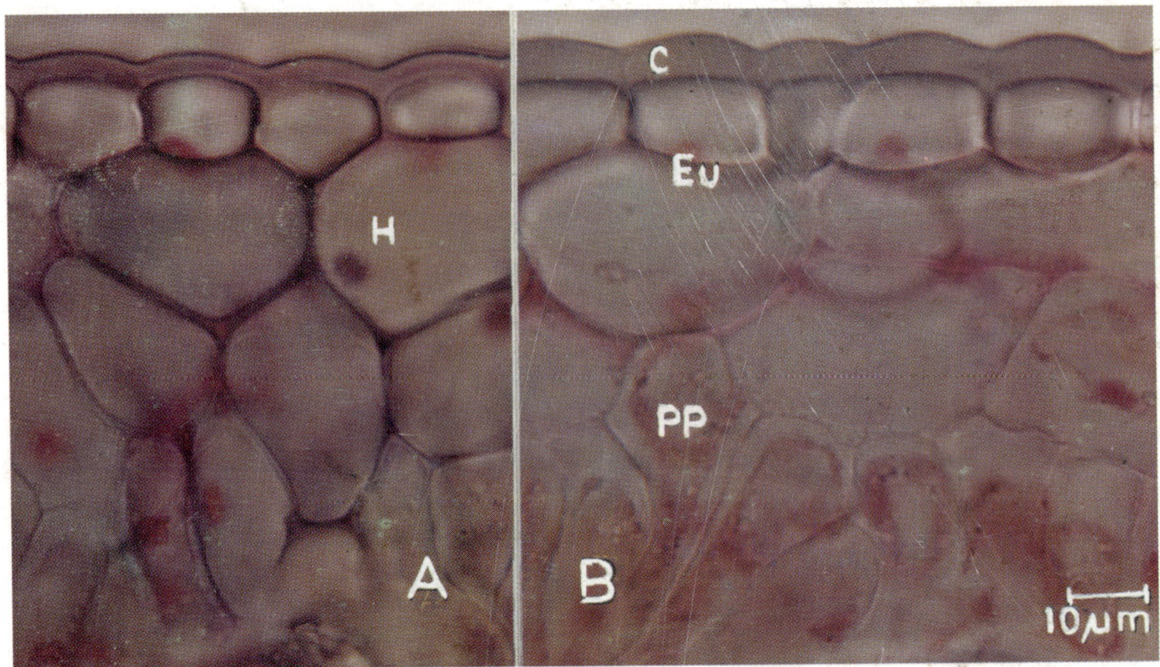


Plate 4. Variation in cuticle thickness in drought susceptible (A) and drought tolerant (B) cultivars of coconut (c- cuticle-adaxial; Eu-epidermis cells - upper; H- hypodermal cells; PP - palisade parenchyma)

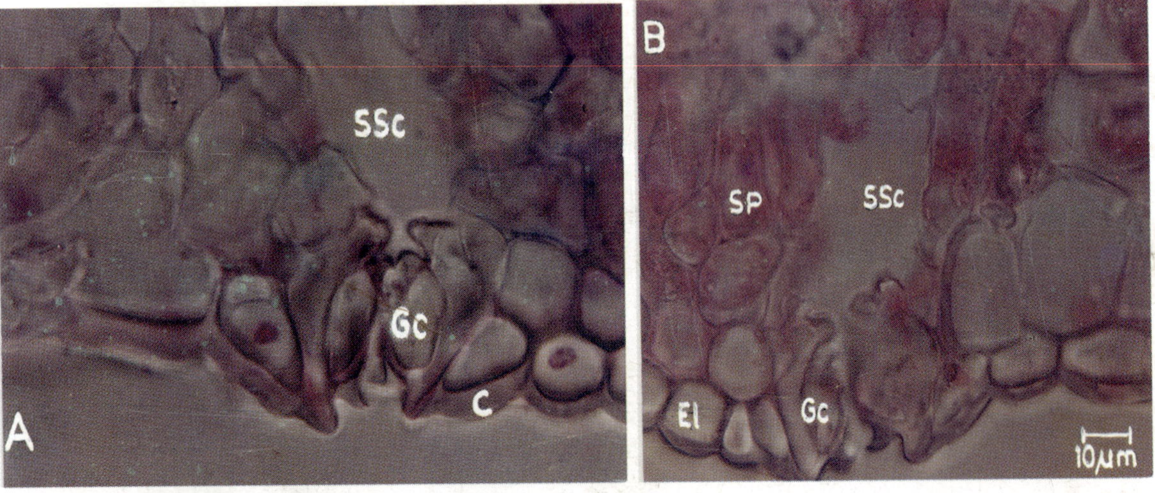


Plate 5. Variation in sub-stomatal cavity in drought susceptible (A) and drought tolerant (B) cultivars of coconut (c-cuticle-abaxial; El-epidermis cell-lower; Gc-guard cell; SP- spongy parenchyma)

reduce transpiration by stepping up osmotic pressure within the cells in coconut (Ollagnier *et al.,* 1983). Higher stomatal regulatory capacity in palms fertilized with higher levels of $K_2O$ under rainfed condition has been reported.

Coconut palms accumulate organic solutes such as sugars and amino acid during stress period. Accumulation of these solutes was found to be more in the tolerant types than in the susceptible types during severe stress condition (Kasturi Bai and Rajagopal, 2000). This implies that osmotic adjustment plays an important role in the drought tolerance mechanism in coconut.

Stomatal resistance (*rs*) differs during the period of adequate soil water availability *i.e.,* non-stress and during soil water deficit *i.e.,* stress among the cultivars and hybrids of coconut. Under stress conditions, where high evaporative demand in the atmosphere prevails, genotypes exhibit differential adaptability through stomatal regulation which is high in the hybrids followed by talls, whereas in dwarfs it is almost 50% less than that in hybrids. This indicates the higher transpiration loss of water in dwarfs than in talls and hybrids.

Transpiration rate (E) are inversely proportional to the degree of stomatal closure. The values of E are significantly lower in WCT x WCT, FMS, Java Giant, LCT x COD, LCT x GBGD, WCT x COD and WCT than in other cultivars and hybrids. This reflects the effective control of stomata on the water balance of palms. The sensitivity of stomata affects $\psi_{leaf}$ to a great extent. The low *rs* resulting in high *E* leads to lowering of $\psi_{leaf}$ in dwarfs, conversely high *rs* is associated with high $\psi_{leaf}$ among hybrids like WCT x WCT. Instantaneous and intrinsic WUE increased under mild water deficit conditions (Rajagopal *et al.,* 2000).

Hybrids viz., LCT x GBGD, LCT x COD, WCT x COD and the talls viz., JVGT, FMST, PHOT and CCNT show higher ECW content than the other cultivars including dwarfs. The *E* is inversely proportional to the content of ECW on the leaf surface.

### *2.2.4 Biochemical responses and adaptations to drought stress*

The biochemical responses of coconut palm to drought stress include upregulation or synthesis of scavenging enzymes to maintain cell membrane integrity thus enabling cells to tolerate stress. The metabolic status of the coconut palm gets affected during summer months when the available soil moisture falls below 20 mm during prolonged dry spell. Concomitant with the decrease in the leaf water status during drought stress, the activities of stress sensitive enzymes vary depending upon the nature and function of the enzyme in question. Drought stress was found to increase the activities of some of the stress sensitive enzymes, *viz.,* peroxidase (POD), polyphenol oxidase (PPO), superoxide dismutase (SOD), acid phosphatase (APH) and L-aspartate : 2-oxoglutarate amino transferase (AAT) in adult WCT palms, while activities of Malic dehydrogenase (MDH) and nitrate reductase (NR) were found to decrease (Shivasankar *et al.,* 1991, Kasturi Bai *et al.*, 1996b). Increase of APH and AAT activities in coconut leaves subjected to osmotic stress is correlated with decline in $\psi_{leaf}$ (Rajagopal *et al.,* 1988). Soluble protein levels also increase during stress indicating increased solubilization of enzyme protein or additional synthesis in response to stress. Being a hydrolytic enzyme, the sudden rise of APH activity during periods of water deficits indicates the degree of stress. (Shivashankar, 1990). Thermal stability of NR *in vivo* in several coconut cultivars was found to be correlated with drought tolerance (Shivashankar, 1992).

Many enzymes exist in multiple molecular forms called isozymes and changes in the activity or appearance of isozymes represent the relative tolerance of coconut cultivars to water stress. Two additional fast moving bands of PPO were located in the drought susceptible varieties under stress, while the drought tolerant cultivars showed no change (Shivashankar,1988). Increase in intensity of peroxidase bands could be considered as an adaptive mechanism. On the contrary, increased intensity of APH isozyme II shows the susceptibility of the genotype to water deficits since APH is a hydrolytic enzyme. In osmotically stressed coconut leaves peroxidase isozymes change quantitatively, while PPO (Shivashankar, 1988) and APH isozymes change both quantitatively and qualitatively. During stress one of the isozymes of APH (APH II) undergoes alterations in its kinetic parameters, especially in its Km value towards PNPP and other naturally occurring phosphate compounds like ATP, NADP, Glucose-6-phosphate etc. (Shivashankar and Nagaraja,1996). It is clear that the drought tolerant varieties are endowed with a biochemical mechanism to prevent the adverse effects of drought by appropriate regulation of enzyme activities.

At the cellular level, the impact of stress is generally seen on the integrity of membranes as the extent of solute leakage is regulated by the cell membrane stability. Solute leakage differs among the cultivars and with the maturity of leaf. Drought tolerant cultivar, WCT and hybrid WCT x COD had higher stability of membranes, while the susceptible hybrid COD x WCT had higher leakage. The membrane stability is also influenced by the potassium levels and phenol levels. Less $K^+$ leakage was seen in WCT x COD as compared with the susceptible hybrid The variations were found among different leaf positions and cultivars. Higher phenol content in leachate is also seen due to heat stress which varies among cultivars and leaf positions.

In coconut lipid peroxidation has been found to be high in drought susceptible cultivars as compared to tolerant ones (Chempakam *et al.,* 1993). Increase in the activities of peroxidase and SOD during stress appears to be essential for maintaining the integrity of the cell. Stress tolerance is thus characterized by higher activities of the protective enzymes like SOD, catalase and peroxidase and consequently coupled with lower levels of lipid peroxidation and higher membrane integrity (Shivashankar,1988, Chempakam *et al.,* 1993; Kasturi Bai *et al.,* 1996b). In seedlings a negative relationship between $\psi_{leaf}$ and peroxidation of cell wall lipids has been found. With a decrease of about 63 % in $\psi_{leaf}$, an increase of about 21% in lipid peroxidation has been observed during stress period as compared to non-stress period.

## 2.3 Molecular Studies

Studies on stress responsive proteins in coconut indicated expression of low, medium and high molecular weight stress proteins in seedlings exposed to drought stress (Naresh Kumar, 2001-2002). Recent studies using RAPD and ISSR analysis indicated correlation between leaf water potentials and molecular markers in 4 coconut cultivars (Manimekalai *et al.,* 2004). However, these studies need to be elaborated to other stress related traits before coming out with the molecular markers.

## 2.4 Screening for drought tolerance

Cell size and number, sub-stomatal cavity size, stomatal frequency, epicuticular wax content and thickness, leaf thickness, stomatal resistance, water potential components, cell membrane stability, water use efficiency, activity levels of scavenging enzymes are the essential anatomical and physiological traits for assessing moisture stress in plants (Rajagopal *et al.,* 1991; Kasturi Bai, 1993; Champakam *et al.,* 1993, Shivashankar *et al.,* 1991; Naresh Kumar *et al.,* 2000). Based on these, coconut germplasm collections comprising talls, dwarfs and hybrids were screened under field conditions for drought tolerance (Rajagopal *et al.*, 1990; Naresh Kumar *et al.*, 2000).

### *2.4.1 Ranking of cultivars for drought stress tolerance*

About 25 cultivars and hybrids of coconut were ranked for drought tolerance based on all stress sensitive parameters using parametric relationships. It is clear that all dwarfs performed badly with the ranks between 11 and 20, whereas all hybrids except COD x WCT and all talls except the SS Apricot, Andaman Ordinary and Laccadive Micro were ranked high. Based on anatomical adaptations characterized by thicker leaflets, thick cuticle on both sides, larger palisade and spongy parenchyma cells, larger hypodermal cells, water cells and sub-stomatal cavity, genotypes like WCT, FMS and PHOT and WCT x COD hybrid are identified as relatively tolerant to drought stress. Thus, coconut cultivars were identified for their degree of tolerance to drought stress based on the desirable traits, which reflect on the overall water relations of palms. Presence or absence of desirable traits impart higher degree of drought tolerance (eg. WCT x WCT) or drought susceptibility (eg. MYD) (Rajagopal *et al.*, 1990).

## 3. INFLUENCE OF DROUGHT STRESS ON PHOTOSYNTHETIC EFFICIENCY AND WATER USE EFFICIENCY IN COCONUT

The photosynthetic rates (*Pn*) get reduced by water stress mainly due to increase in stomatal or mesophyll resistance. Differences in the *Pn* characteristics have been observed between the stress period and non-stress period. Due to stress, more reduction in *Pn* is noticed in susceptible types than in tolerant types. The *gs* and *E* are lower in WCT x COD, LCT x COD and LCT x GBGD than in other cultivars during stress period (Kasturi Bai *et al.*, 1998). Drought tolerant hybrids such as WCT x COD, LCT x GBGD and LCT x COD exhibit higher increase in *Pn*/*gs* ratio than the susceptible types. The *Pn* rate is independent of $\psi_{leaf}$ until a threshold $\psi_{leaf}$ is reached. *Pn*/*E* (instantaneous WUE) decreases during stress period because of the lowered *gs* limiting *Pn* rate.

Significant differences in WUE has been observed between the cultivars and hybrids. Kasturi Bai *et al.*, (1996a) in their studies obtained a range in WUE between 28.8 and 69.3 gDM. mm water$^{-1}$ among the cultivars/hybrids. The three hybrids viz.; WCT x COD, LCT x GBGD and LCT x COD and WCT cultivar had higher WUE than the other cultivars and hybrids. WUE based on gas exchange measurement also followed similar trend with higher WUE in the above three hybrids than in others. Rajagopal *et. al.* (1989) recorded higher WUE in un-irrigated palms than in irrigated palms.

### 3.1 Day Time Fluctuations in Photosynthetic Efficiency

In coconut, the day time leaf surface to air VPD (LAVPD), ΔT influences the photosynthetic characteristics and water relations. Under mild stress conditions, WUE increases, whereas overall carbon assimilation efficiency is low in rainfed palms. The leaf surface temperature also was significantly higher in rainfed palms indicating that the coconut palm conserves water by closing the stomata, thus its leaflet temperatures are maintained higher than that of the ambient temperature. Thus increase in ΔT was significantly higher in rainfed palms (Rajagopal *et al.*, 2000).

## 4. INFLUENCE OF IRRIGATION AND SOIL MOISTURE CONSERVATION ON THE PHYSIOLOGICAL EFFICIENCY OF COCONUT

The studies on the extent of influence of irrigation on coconut palms grown on sandy and laterite soils with different levels of irrigation through drip {sandy-66, 100 and

133% of open pan evaporation (Eo); laterite - 33, 66 and 100% of Eo} and basin (100% of Eo) indicated that the source parameters viz., photosynthetic rate, $\Psi_{leaf}$ and PS II efficiency varied with the irrigation level and soil type. Palms receiving irrigation also showed marked improvement in female flower production and their retention to produce higher nut yields. Three types of physiological conditions in source and sink relationships were observed in the palms based on the type of irrigation treatment they were subjected to (Naresh Kumar *et al.*, 2002a). The drip irrigation provided conditions for better physiological efficiency of source and sink for high WUE and yield. WUE was found to increase at field and at plant and leaflet level (Naresh Kumar *et al.*, 2002a).

The net photosynthetic rates were higher in the palms grown on laterite soils compared to those grown on sandy soils and irrigation significantly increased the Pn rates and stomatal conductance. The chlorophyll fluorescence PS II efficiency parameter (Fv/Fm), an indicator of extent of physiological stress in leaf, has been found to be higher in irrigated palms compared to the rainfed palms (Naresh Kumar *et al.*, 2002a).

The female flower production and setting percentage normally increase with irrigation and the percent increase over rainfed palms in general was higher in sandy soils. The percentage of nut retention is an indicator of efficiency of conversion of female flowers into mature nuts. Percent increase in nut yield over rainfed palms was higher in sandy soils compared to laterite soils (Naresh Kumar *et al.*, 2002a).

The consequent increase in yield could be related to the increases in source (the Pn rates) and sink (female flower production) efficiency. The female flower production was higher in plants grown on laterite soils compared to those grown on sandy soils. Under rainfed conditions, the palms grown on sandy soil produced less number of female flowers compared to those grown on laterite soils. The physiological conditions of source and sink in palms grown under different systems of irrigation are defined (Naresh Kumar *et al.*, 2002a).

Soil moisture conservation also improved the yields by improving the souce-sink efficiencies (Naresh Kumar *et al.*, 2003). Results of a multi-location experiment indicated that the soil moisture conservation practices helped to retain and make soil moisture available for longer periods during summer. This helped to maintain higher the photosynthetic rates, water use efficiency thus helping to have improved nut retention and yield (Naresh Kumar *et al.*, 2003)

### 4.1 Identification and Characterization of *in situ* Tolerant Palms

The plants that can with stand the natural occurrence of drought and other stresses and still produce good yield are of premium value, as they may possess the desirable genes. Surveys in hotspot areas were conducted to identify the palms yielding very high compared to others in their vicinity. Palms at different agro-climatic regions were identified in farmers' plots with desirable canopy shape, leaf number with better yield. The physiological water use efficiency of these palms was also found to be high (Naresh Kumar *et al.*, 2002b). This type of *in situ* tolerant plants with desirable traits should be used in breeding programmes, which will help in reducing the time gap in breeding for drought tolerant cultivars in coconut (Naresh Kumar *et al.*, 2002b).

## CONCLUSIONS

To sum up, coconut responds to the stressful environments at morphological, anatomical, physiological and biochemical levels. Generally, drought coincides with high temperature and high light intensity stresses (Rajagopal *et al.*, 2000a; Rajagopal and Naresh Kumar, 2001; 2003). This makes the problem more complex. However, it is also important to note that the adaptive strategies of plants for different abiotic stresses are overlapped in most of the cases as the abiotic stresses themselves. Further studies should be focused to develop the molecular markers linked to desirable traits and to understand the inheritance patterns of these traits. The genotypes with desirable traits for tolerance to stress conditions can be used in breeding strategies for abiotic stress tolerance in future crop improvement strategies. Identification and characterization of field tolerant palms to abiotic stresses will be helpful in exploiting the natural tolerance to abiotic stresses.

## REFERENCES


Braconnier, S. and J.D. Auzec. 1985. Anatomical study and cytological demonstration of potassium and chlorine fluxes associated with oil palm and coconut stomatal opening. *Oleagineux*, **40**: 547-551.

Chempakam, B., K.V. Kasturi Bai. and V. Rajagopal. 1993. Lipid Peroxidation in Relation to Drought tolerance in Coconut (*Cocos nucifera* L.). *Pl. Physiol. & Biochem.*, **20**(1): 5-10.

Child, R. 1974. Coconut 2nd ed. Longman, London. Pp. 335.

Kasturi Bai, K.V. 1993. Evaluation of coconut germplasm for drought tolerance. Ph.D. Thesis, Mangalore University, Mangalore, India.

Kasturi Bai, K.V., S.R. Voleti. and V. Rajagopal.1988. Water relations of coconut palms as influenced by environmental variables. *Agri. Forest. Meteorol.*, **43**: 193-199.

Kasturi Bai, K.V., V. Rajagopal., C.D. Prabha., M.J. Ratnambal and M.V. George. 1996a. Evaluation of coconut cultivars and hybrids for dry matter production. *J. Plantn. Crops*, **24**: 23-28.

Kasturi Bai, K.V., V. Rajagopal., B. Chempakam. and C.D. Prabha 1996b. Assay of enzymes in coconut cultivars and hybrids under non-stress and stress conditions. *J. Plantn. Crops*, **24** (Supplement): 548-554.

Kasturi Bai, K.V., V. Rajagopal and D. Balasimha 1998. Variation in net carbon assimilation and related parameters in coconut (*Cocos nucifera* L.) under field conditions. *Plant Physol. Biochem.*, **25**(2): 163-166.

Kasturi Bai, K.V. and V. Rajagopal 2000. Osmotic adjustment as a mechanism for drought tolerance in coconut (*Cocos nucifera* L). *Indian J. Plant Physiol.*, **5** (4): 320-323.

Kurup, V.V.G.K., S.R. Voleti. and V. Rajagopal. 1993. Influence of weather variables on the content and composition of leaf surface wax in coconut. *J. Plantn. Crops*, **2**: 71-80.

Manimekalai, R., R. Nagarajan, B. Bharathi and S. Naresh Kumar. 2004. DNA polymorphism among coconut (*Cocos nucifera* L.) cultivars and reciprocal cross derivatives differing in drought tolerance. *J. Palntation Crops* (suppl.), 32: 117-122

Milburn, J.A. and M. H. Zimmermann. 1977. preliminary studies on sal flow in *Cocos nucifera*, L. 1 Water relations and xylem transport. *New Phytol.*, **79**: 535-541.

Murray, D.V. 1977. Coconut palm. Pp. 384- 407. In. Ecophysiology of Tropical Crops. (T.A. Alvim and T.T. Kozlowski, eds.). Academic Press, New York.

Naresh Kumar, S. 2001-02. CPCRI, Annual Report, 2001-2002 p 128. CPCRI Pub., Kasaragod, India pp.65.

Naresh Kumar, S. 2002-03. CPCRI, Annual Report, 2002-2003 p178. CPCRI Pub., Kasaragod, India, pp 90.

Naresh Kumar, S., V. Rajagopal. and Anitha Karun. 2000. Leaflet anatomical adaptations in coconut cultivars for drought tolerance. pp. 225-229. In: *Recent Advances in Plantation Crops Research*. (N. Muralidharan and R. Raj Kumar, eds). Allied Pub. Ltd. New Delhi.

Naresh Kumar, S., V. Rajagopal., R.H. Laxman and H.P. Maheswarappa 2002 a. Photosynthetic characteristics and water relations in coconut palm under drip irrigation on sandy and laterite soils. pp. 116-120. In: *Plantation Crops Research and Development in the New Millennium*. (P. Rethinam, H.H. Khan, V.M. Reddy, P.K. Mandal. and K. Suresh, eds.). Coconut Development Board, Pub, Kochi.

Naresh Kumar, S., V. Rajagopal., T. Siju Thomas, Vinu. K. Cherian., M. Hanumanthappa., B. Anil Kumar, Srinivasulu and D. Nagvekar. 2002 b. Identification and characterization of *in situ* drought tolerant coconut palms in farmers' fields in different agro-climatic zones. *PLACROSYM XV Proceeding, ISPC Pub., Kasaragod,* 335-339.

Naresh Kumar, S., V. Rajagopal., T. Siju Thomas. and Vinu K. Cherian. 2003. Influence of soil moisture conservation on the source-sink relationship in coconut (*Cocos nucifera* L.) under different agro-climatic conditions. Pp : 572. In Proceedings 2nd International Congress on Plant Physiology from 8 to 12 Jan. 2003 at IARI, New Delhi. Abs. S13-P25.

Ollagnier, M., R. Ochs., M. Pomier. and G. de Taffin. 1983. Effect of chlorine on the hybrid coconut PB 121- in the Ivory Coast and Indonesia - Growth, tolerance to drought, yield. *Oleagineux*, **38** (5): 309-321.

Rajagopal, V., Patil, K.D. and Sumathykutty Amma, B. 1986. Abnormal stomatal opening in coconut palms affected with root (Wilt) disease. *J. Exp. Bot.*, **37**: 1398-1405.

Rajagopal, V., S. Shivashankar; K.V. Kasturi Bai. and. S.R. Voleti. 1988. Leaf Water Potential as an Index of Drought Tolerance in Coconut (*Cocos nucifera* L.). *Pl. Physiol. and Biochem.*, **15**(1): 80-86.

Rajagopal, V., A. Ramadasan., K.V. Kasturi Bai. and D. Balasimha 1989. Influence of irrigation on leaf water relations and dry matter production in coconut palms. *Irrig. Sci.,* **10:** 73-81.

Rajagopal, V., K.V. Kasturi Bai. and S.R. Voleti 1990. Screening of coconut genotypes for drought tolerance. *Oleagineux,* **45** (5): 215-223.

Rajagopal, V., Voleti, S.R. Kasturi Bai, K.V. and Shivashankar, S. 1991. Physiological and Biochemical criteria for breeding for drought tolerance in coconut. In: Coconut Breeding and Management. Eg Silas, M. Aravindakshan and A.L. Jose (Eds.) Kerala Agricultural University, Vellanikkara, Trichur, pp 136-143.

Rajagopal, V., S. Shivashankar and Jacob Mathew. 1996. Impact of dry spells on the ontogeny of coconut fruits and it's relation to yield. *Plantation Research and Development*, **3**(4) : 251-255.

Rajagopal,V., K.V. Kasturi Bai and S. Naresh Kumar. 2000a. Drought management in plantation crops. Pp 30-35. In Plantation Crops Research and Development in the New Millennium. Proceedings PLACROSYM XIV, Hyderabad 2000 (P. Rathinam, HH Khan, VM Reddy, PK Mandal and K. Suresh eds.).

Rajagopl, V., S. Naresh Kumar; K.V. Kasturi Bai and R.H. Laxman. 2000 b. Day time fluctuations in juvenile coconut palms grown under rainfed and irrigated conditions. *J. Plant. Biol., 27*(1): 27-32.

Rajagopal, V. and S. Naresh Kumar. 2001. Avenues to improve productivity potential under drought condition- A case study on coconut. In: *Souvenir on National Seminar on Role of Plant Physiology for Sustaining Quantity and Quality of Food Production in Relation to Environment*, (Ed: M.B.CHETTI), 31-36

Rajagopal V. and S. Naresh Kumar, 2003. Management of abiotic stress in coconut. In: *Thematic Papers-national Seminar on Stress Management in Oil Seeds for Attaining Self-reliance in Vegetable Oils* (Mangala Rai, Harvir Singh and D.M. Hegde Eds). DOR pub. Hyderabad 113-136.

Rao, G.S.L.H.V.P. 1985. Drought and coconut palm. *Indian Cocon. J.*, **15** (12): 3-6.

Shivashankar, S. 1988. Polyphenoloxidase isozymes in coconut genotypes under water stress. *Pl. Physiol. and Biochem.*, **15**:87-92.

Shivashankar, S. 1990. Studies on soluble enzyme systems of coconut (*Cocos nucifera* L.) cultivars. Ph.D. thesis, Biochemistry, Mysore University, Mysore, India.

Shivashankar, S. 1992. Thermal stability of nitrate reductase in relation to drought tolerance in coconut. *J. Plantn. Crops*, **20** (Suppl.): 267-273.

Shivashankar, S., K.V. Kasturi Bai. and V. Rajagopal. 1991. Leaf water potential, stomatal resistance and activity of enzymes during the development of moisture stress in coconut palm. *Trop. Agric.*, **68** (2): 106-110.

Shivashankar, S. and K.V. Nagaraja. 1996. Water stress induces kinetic changes in the properties of an acid phosphatase isozyme from coconut leaves. *Pl. Physiol. and Biochem.*, **23** (1):21-26.

Voleti, S.R. and V. Rajagopal. 1991. Extraction and identification of epicuticular wax in coconut. *Pl. Physiol. and Biochem.*, **18**(2): 88-90.

Voleti, S.R., K.V Kasturi Bai., C.K.B. Nambiar and V. Rajagopal. 1993 a. Influence of soil type on the development of moisture stress in coconut (*Cocos nucifera* L.) *Oleagineux,* **46**: 505 - 509.

*Advances in Plant Physiology*, Vol. **9**
Ed. A. Hemantaranjan
Scientific Publishers (India), Jodhpur, 2006 pp. 169-192
E-mail: **info@scientificpub.com** www.scientificpub.com

# 8

# MOLECULAR PHYSIOLOGY OF HEAVY METAL STRESS IN PLANTS

S.K. Panda[1,2*], S. Choudhury[2] and H. Matsumoto[1]

[1]Research Institute for Bioresources, Okayama University, Kurashiki 7100046, Japan
[2]School of Life Sciences, Assam (Central) University, Silchar 788011, India

## INTRODUCTION

Heavy metals occur naturally and the plants for their normal physiological process require most of them. They are present in the environment with wide range of oxidation states and co-ordination number and these differences are closely associated with their toxicity (Pinto *et al.,* 2003). Metals that have density higher than 5g $cm^3$ are termed as heavy metals (Weast, 1984). Fifty three of ninety naturally occurring metals are heavy metals and classified as broadline heavy metals (As, Cd, Co, Cr, Cu, Ni, Sn, Zn) and class B heavy metals (Pb, Ag, Hg, Au, Ti). The toxicity level increase with the increase in class B character (Neiborer and Richardson, 1980). Environmental contamination with heavy metals has become as worldwide problem leading to severe agricultural losses and hazardous health effects. One of the fundamental factors that concerns over the persistence of metals in the environment is its non-biodegradability. The rise in the level of heavy metals in the environment is mainly due to extensive use of agricultural fertilizers, mining, sewage disposal and other related anthropogenic activities (Foy *et al.,* 1978; Dubey and Pessarakli, 1998). Chemically, heavy metals are transition metals with incompletely filled δ-orbital and under physiological conditions their redox state varies form –420 mV to +800 mV. Heavy metals of biological importance are divided into redox active and redox inactive metals (Schutzendubel and Polle, 2002).

The toxic effect generated by heavy metals depends on several factors like solubility of the metal, absorbability, transport, its chemical reactivity, pH of the medium and presence of other ions (Stohs and Bagchi, 1995; Das *et al.,* 1997). Heavy metals toxicity may result form the binding of the metal with the sulphadryl groups in the proteins leading to inhibition of enzyme activity of metabolic importance (Panda and Choudhury, 2005). The binding of heavy metals to the enzymes results in alteration of the catalytic function and inactivation (Van Assche and Clijsters, 1990). One of the common features of heavy metal toxicity is the increased permeability of the plasma membrane (Quartacci *et al.,* 2000; Choudhury and Panda, 2004a). Elevated levels of heavy metals induces ionic imbalance which is distinct than those observed in plants under salt, water, temperature stress and results the decrease of transpiration and relative water content (Barcelo and Poschenrieder, 1990). Uptake of essential metal ions can be affected under heavy metal stress and thus results in disturbed nutritional status in plants (Marshner, 1991). Elevated levels of heavy metals can result in expression of heat shock proteins (HSPs), without affecting the house keeping proteins (Wollgiehn and Neumann, 1995). One of the most important responses in plants to toxic concentrations of heavy metals is the disruption of the primary process of photosynthesis and chlorophyll biosynthesis. It has been observed that heavy metals like cadmium (Cd), lead (Pb), manganese (Mn), cobalt (Co) and nickel (Ni) enhances the activity of chlorophyll degrading enzyme chlorophyllase responsible for chlorophyll degradation and pigment loss (Nag *et al.,* 1981; Abdel Basset *et al.,* 1995).

The toxic effects of heavy metals are related to its ability to generate reactive oxygen species (ROS) resulting unbalanced cellular redox homeostasis (Dietz *et al.,* 1999; Panda, 2002). Oxidation and reduction reactions occur in both soil and aquatic environment.

## 2. PHYTOTOXICITY OF HEAVY METALS AND ITS PHYSIOLOGICAL EFFECTS ON PLANTS

### 2.1 Cadmium

Cadmium (Cd) is a non-essential heavy metal released into the environment by metal working industries, mining and agricultural mineral fertilizers (Nriagu and Pacyna, 1988). The background of Cd in agricultural soil is less than 1mg $kg^{-1}$ (Adriano, 2001), though higher levels have been observed in soils with long-term use of phosphate fertilizers and sewage sludge application (Chaney, 1980). It is a non-essential element that affects growth and has no essential biological function in plants. Under natural conditions, Cd occurs as a guest metal in Pb:Zn mineralization (Baker *et al.,* 1990). Though studies on potential toxicity of Cd on biological system have been conducted during last three decades, the mechanism underlying Cd toxicity is not clearly understood till date. Plants can withstand Cd loading into its tissue until it reaches a threshold level and turn toxic to plants. Under chronic Cd phytotoxicity, plants can withstand Cd and grow continuously with a low growth rate. However, under severe toxic concentrations, plants are unable to express their protective mechanism and suffer form Cd toxicity (Vassilev, 2003). Cd being a mobile metal has a long biological half life (Himly *et al.,* 1985) and the degree to which higher plants are able to take up Cd depends on its concentration in the soil, presence of organic matter, pH, redox potential and other ions. Cd penetrates the root through cortical tissue and as soon it enters the xylem either by apoplastic or symplastic pathway it is complexed by several ligands like organic acids and phytochelatins (Salt *et al.,* 1995). Normally Cd is generally retained in roots and small part is transported to the shoots. Higher level of Cd in soil can alter the uptake of essential ions by the plants either by affecting the availability of minerals in soil or by soil microbial population (Moreno *et al.,* 1999). Inhibition of stomatal opening, transpiration, photosynthesis and root Fe (III) reductase activity have been reported in plants under Cd toxicity (Alcantara *et al.,* 1994; Sanita di Toppi and Gibrielli, 1999).

Cd toxicity results in leaf roll, chlorosis and reduction in root and shoot growth. Cd can affect the growth rate of cells due to the irreversible inhibition exerted by Cd on proton pump responsible for the process (Aidid and Okamoto, 1993). It can inhibit the mitochondrial oxidative phosphorylation and increase the passive permeability of the $H^+$ of the mitochondrial inner membrane (Kessler and Brand, 1995). Several investigations have suggested the involvement of oxidative stress in plants under Cd toxicity via production of ROS leading to the inhibition or activation of the antioxidant defense mechanism (Somashekaraiah *et al.,* 1992; Gallego *et al.,* 1996; Sandalio *et al.,* 2001; Cho and Seo, 2004; Choudhury and Panda, 2004a). In soybean it has been observed that Cd treatment results in increased senescence leading to the generation of oxidative stress (Balestresse *et al.,* 2004). Choudhury and Panda (2004a) reported that elevated levels of Cd resulted in increased production on hydrogen peroxide ($H_2O_2$) and superoxide radical ($O_2^-$) in root cells resulting in loss of membrane integrity and lipid peroxidation. Moreover under elevated Cd concentrations on activity of several antioxidant enzymes viz., catalase (CAT), guaiacol peroxidase (GPx), superoxide dismutase (SOD) and glutathione reductase (GR) studied in root cells of rice showed decline in activity and the rate declined with increase in Cd in growth medium (Choudhury and Panda, 2004a). In Arabidopsis, high level of ROS like $H_2O_2$, $O_2^-$ and hydroxyl radical ($OH^-$) have been observed with simultaneous decline in antioxidant metabolism (Skorzynska-Polit *et al.,* 2003/4). On the other hand Cd has also been reported to induce antioxidant enzymes. In *Phaseolus vulgaris* and *Zea mays* Cd treatment increased the GPx activity (Chaoui *et al.,* 1997; Lagriffoul *et al.,* 1998). Similarly in Daucus carota no lipid peroxidation have been observed in root cells under Cd treatment (Sanita di Toppi *et al.,* 1998).

## 2.2 Chromium

Chromium (Cr) is the seventh most abundant metal on the earth crust and important environmental contaminant released into the environment due to industrial activities using Cr (Nriagu and Nieborer, 1988). In nature, Cr exists in two different stable oxidation states viz, the trivalent (Cr, III) and the hexavalent (Cr, VI). Both these forms of Cr differ in terms of mobility, bioavailability and toxicity (Panda and Patra, 1997; Panda and Choudhury, 2005a). Cr phytotoxicity results in inhibition of seed germination, pigment degradation, loss in photosynthesis, disturbances in the nutrient balance and alters the antioxidant enzymes (Poschenrieder *et al.,* 1991; Panda *et al.,* 2003; Choudhury and Panda, 2005; Panda and Choudhury, 2005a). The trivalent form of Cr is an essential nutrient for animals, however, at low concentrations causes toxic effects on plants. Cr (VI) on the other hand has no biological role and toxic to plants even at micromolar range of concentrations (Han *et al.,* 2004; Panda and Choudhury, 2005a; Choudhury and Panda, 2005). Cr (VI) toxicity results in growth retardation followed by reduction in number of palisade and spongy parenchyma cells in the leaves and increases the number of vacuoles and electron dense bodies along the wall of the xylem and phloem (Han *et al.,* 2004). Cr though thought to be a non-redox metal incapable to participate in Fenton type reactions, studies have revealed that Cr can indeed participate in such reaction proving its redox character (Shi and Dalal, 1989; Panda and Choudhury, 2005 and references therein).

The reactivity of Cr can be considered by its interaction with glutathione, NADH and $H_2O_2$ evolving $OH^-$ in cell free system (Shi and Dalal, 1989; Aiyar *et al.,* 1991). The oxidation state of Cr is most important as both trivalent and hexavalent forms differs in reactivity and toxicity (Carter, 1995). Manganese (Mn) oxidizes Cr (III) to Cr (VI), while FeS and organic matter present in the soil can convert Cr (VI) to least toxic Cr (III). Cr (III) and Cr (II) can enter Fenton reactions (Strile *et al.,* 2003). During radical oxidation, Cr (III), Cr (IV) and Cr (V) intermediates are thought to be involved. ESR studies at pH 5.8 and 7.1 for Cr (III) / $H_2O_2$ that the lower oxidation states of Cr can effectively participate in generating oxidizing species without generating Cr (VI) (Strile *et al.,* 2003). However, at higher pH of 8.9, the oxidation of Cr (III) is very rapid and although the concentration of Cr (VI) reaches a steady state the production of oxidizing species like Cr (IV) and Cr (V) proceeds at a slow rate. This indicates that both Cr (IV) and Cr (V) are catalytically active and possesses the ability to generate $OH^-$ (Strile *et al.,* 2003). However, for Cr (VI) this mechanism is not fully understood. Cr (VI) can generate ROS, though the exact mechanism is not properly understood, this phenomenon has been demonstrated in plants under Cr (VI) stress (Panda *et al.,* 2003; Panda and Choudhury, 2005a; Choudhury and Panda, 2005). Cr (VI) can also initiate lipid peroxidation. Treatment of plants with toxic concentrations of Cr (VI) can result in increment of malondialdehyde (MDA) content in cells and thereby resulting in increase in membrane permeability (Panda and Choudhury, 2005b; Choudhury and Panda, 2005). Cr (VI) induced distortion of cell membrane and chloroplast thylakoids have been reported in plants (Choudhury and Panda, 2005).

## 2.3 Aluminium

Aluminium (Al) toxicity is one of the most widespread metal toxicity encountered by plants. It is dependent on pH for its solubility and thus Al toxicity occurs in soil with pH values less than 5.5 (Vitorello *et al.,* 2005). Al toxicity is most severe in soil with low base saturation and poor in calcium (Ca) and magnesium (Mg). Al is the third most abundant metal on earth's crust, most of which is locked in the form of Al oxides or aluminosilicates. Acidification of soil makes Al soluble and potentially toxic to plants. Al toxicity is related to the changes in Ca homeostasis and membrane lipid composition in plants (Zhang and Krikham, 1994; Zhang *et al.,* 1997). Al also induces decline in cell wall extensibility due to cross-linking of the pectin carboxyl groups in the walls (Matsumoto *et al.,* 1976). It can inhibit root growth (Barcelo and Poschenrieder, 2002), though found to induce lateral root formation giving rise to compact root formation (Berckle, 1991). Roots are affected in the region of root cap, meristem, root hairs and branch initials (Foy *et al.,* 1978; Rengel, 1996). The root tip region is the most Al sensitive regions and detailed examination showed that the distal region of the transition zone (DTZ) is the most Al sensitive region of the roots (Sivaguru and Horst, 1998). The inhibition of root growth under Al toxicity is due to the decline in cell elongation in early developmental stages, while at latter stage cell division is arrested and resulting in reduced growth (Kochian, 1995; Barcelo and Poschenrieder, 2002; Ciamporova, 2002). Al can also affect shoots where cellular and ultrastructural modifications have been recorded. Beside these important toxic effects, Al may also result in decline in photosynthesis, chlorosis, foliar necrosis and disturbances in nutritional balance resulting in reduced plant biomass (Haug and Vitorello, 1996; Vitorello *et al.,* 2005). Al is soluble at lower pH (>5.5). The $H^+$ concentration has direct detrimental effect on roots (Koyama *et al.,* 2001). Low pH values have detrimental effect on plasma membrane by enhancing its permeability (Yan *et al.,* 1992; Koyama *et al.,* 2001). However, a positive correlation between $H^+$ and Al tolerance and also between $K^+$ efflux and Al uptake has been observed in wheat roots (Sasaki *et al.,* 1994; Souza, 1999). Extracellular acidity also increases the permeability of the plasma membrane to $H^+$ (Yan *et al.,* 1998). Both Al and $H^+$ toxicity are affected by ionic strength and cationic concentrations and the pattern of both Al and $H^+$ toxicity on roots are similar (Koyama *et al.,* 2001). For Al toxicity to be manifested in the plant its uptake is necessary and this has been demonstrated in cell suspensions and plant roots under Al stress (Yamamoto *et al.,* 1994; Kochian, 1995; Horst *et al.,* 1999; Matsumoto *et al.,* 2000). Al uptake is predominant in epidermis and outer cortex of the roots (Matsumoto *et al.,* 1996).

## 2.4 Lead

Lead (Pb) is considered as important environmental pollutant arising form mining, smelting activities and sewage sludge disposal (Eick *et al.,* 1999). Increase in Pb content in agricultural soils has been observed near industrial areas and significant amount of agricultural chemicals containing Pb are being used in agricultural soils that ultimately leads to Pb contamination. Till date there is no evidence of any potential biological role of Pb in cells and thus it is considered as non-essential metal. In various plant organs, the Pb content tends to decrease in the following order: roots>leaves>stem>infloresence>seeds (Sharma and Dubey, 2005). Physiological analysis of Pb toxicity in plants showed considerable changes in plants normal homeostasis (Verma and Dubey, 2003; Choudhury and Panda, 2004). It has been observed that treatment of Pb at toxic concentrations resulted in high accumulation of the metal in tissues (Verma and Dubey, 2003; Choudhury and Panda, 2004). Pb also induces the formation of ROS like superoxide radical ($O_2^-$) and hydrogen peroxide ($H_2O_2$) (Choudhury and Panda, 2004). Pb toxicity results in inhibition of seed germination and chlorotic symptoms in plants (Iqbal and Mustak, 1987). Pb is considered to affect the photosystems in plants (Becerril *et al.,* 1988). It mostly affects the photosystem II (PS II) and such inhibition was partly restored by using PS II specific electron donors such as hydroxylamine and manganese chloride (Rashid and Popovic, 1990). It has been demonstrated that Pb binds near the $Ca^{+2}$ and $Cl^-$ that acts as essential co-factor for oxygen evolution. They inhibit the key photosynthetic enzyme RUBP carboxylase (Moustakas *et al.,* 1994). The root cells are also severely affected by Pb showing retarded growth and altered branching (Shrama and Dubey 2005). At lower concentrations of Pb the development and extension of main root system is affected much higher than the lateral roots (Obroucheva *et al.,* 1998; Sharma and Dubey, 2005).

Activities of several enzymes have been reported to be affected by Pb. Pb showed increased activity of many enzymes like acid phosphatase, • -amylase, peroxidase in leaves (Lee *et al.,* 1976). Under Pb treatment, decline in nitrate reductase (NR) activity has been reported in plants. The inhibitory effect of Pb on the NR activity might be due to an effect on the intracellular mobilization of nitrate (Jain and Garde, 1997). In certain plants, Pb have induces accumulation of free amino acids (Shah and Dubey, 1998).

## 2.5 Copper

Copper (Cu) is an important transition metal involved in several important processes in plants. In the cellular system, Cu exists in multiple oxidation states participating in several regulatory processes (see Marschner,

1995 and references therein; Yruela, 2005). It acts as an important signalling element in protein trafficking machinery, transcription signal, iron mobilization and oxidative phosphorylation (Yruela 2005). However, high concentration of Cu in cells eventually turns toxic causing disruption of several cellular processes. Common visible symptoms of Cu toxicity involve chlorosis, necrosis, stunted growth and leaf color loss (Van Assche and Clijsters, 1990; Marschner, 1995). At the cellular level, Cu toxicity is manifested by its ability to inactivate enzymes of metabolic importance by binding to the sulfadryl groups of the protein, inducing nutrient deficiency, disrupting the cellular transport process and generating oxidative stress (Van Assche and Clijsters, 1990; Mehrag, 1994).

Presence of toxic levels of Cu results in the generation of reactive oxygen species (ROS), thus inducing oxidative load to the cells (Halliwell and Gutteridge, 1989). Induction of oxidative stress by Cu also leads to alterations in the antioxidant pathways. It has been observed that excess Cu results in alterations in the activity of antioxidant enzymes like glutathione reductase (GR), monodehydroascorbate reductase (MDHAR), superoxide dismutase (SOD) and guaiacol peroxidase (GPX) (De Vos *et al.,* 1992; Navari-Izzo *et al.,* 1998; Wang *et al.,* 2004). Moreover, lipid peroxidation also gets initiated by excess Cu in cells (Luna *et al.,* 1994). Cu toxicity also results in loss of biomass and pigments. Alterations in grana stacking, chloroplast lamellae, increase in size and number of plastoglobuli and formation of intrathylakoidal inclusion were reported under Cu stress in plants (Lindon and Henriques, 1991; Maksymiec *et al.,* 1994). Cu also leads to changes in the protein component of the photosynthetic membrane (Lindon and Henriques, 1991; Maksymiec *et al.,* 1994).

## 3. REACTIVE OXYGEN SPECIES AND OXIDATIVE STRESS: ROLE OF HEAVY METALS

Free radical reactions are the intrinsic features of plant senescence and induce oxidative stress and subsequently resulting in cell death (Thompson *et al.,* 1987). The atmospheric oxygen in ground state exists as biradical and is less reactive. During the univalent reduction of oxygen, ROS like $O_2^-$, $H_2O_2$, $HO^-$, etc are formed. These ROSs are detrimental to cells and acts as strong oxidizing agents, attacking almost all biomolecules (Charles and Halliwell, 1981; Elstner, 1982; Halliwell and Gutteridge, 1989; Panda, 2002). Under normal physiological conditions, the ROS and the antioxidants are in a balance with each other. Any imbalance in the level of ROS and antioxidants results in oxidative stress. Although ROS also acts as important signaling molecule, high concentrations can result in oxygen toxicity. Production of ROS poses severe threat to photosynthetic organisms due generation $O_2^-$ via electron transport chain (Krause, 1994; Kozaki and Takeba, 1998).

Plants exposed to wide range heavy metal generate oxidative stress via generation of toxic ROS and subsequently activating antioxidant response (Fig. 1) (Stohs and Bagchi, 1995). Redox metals like $Cu^{+2}$ or $Fe^{+2}$ can undergo Fenton reactions producing $OH^-$ from $H_2O_2$ and $O_2^-$ (Winterbourne, 1982). The $OH^-$ radical is most reactive of all known ROS in chemistry and initiate various irreversible chemical modifications in cells. The hydroperoxyl radical ($HO^-_2$) generated from $O_2^-$ can induce lipid peroxidation via protonation reaction (Mithofer *et al.,* 2004). Metals like $Pb^{+2}$, $Cd^{+2}$, $Hg^{+2}$, etc. are non – redox and are unable to replace $Cu^{+2}$ or $Fe^{+2}$ in Fenton reactions. Such metals can affect the pro – oxidant status by

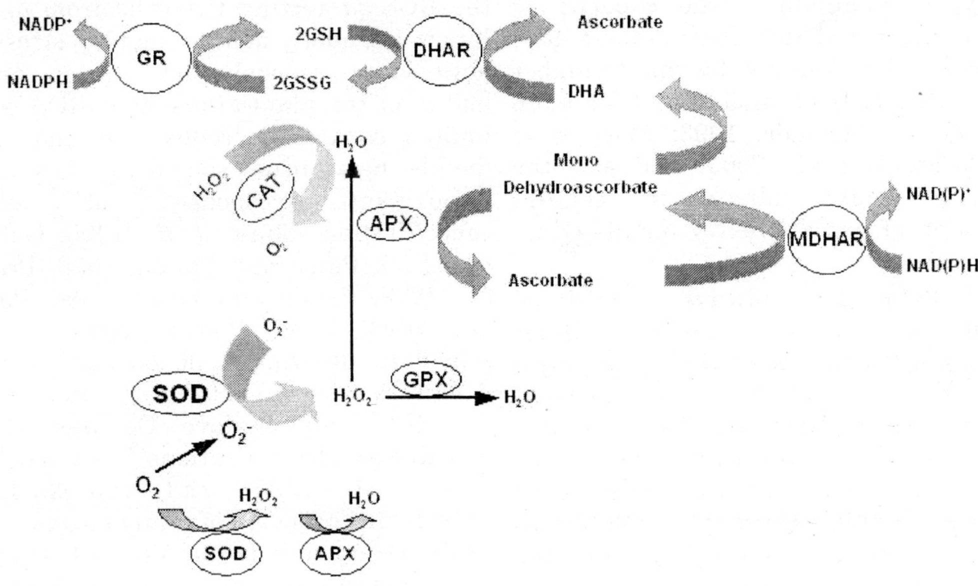


Fig. 1. Antioxidant response in plants to ROS.

depleting the antioxidant glutathione (GSH) pool and activates the calcium dependent system and affects redox metal mediated process (Pinto *et al.,* 2003).

The role of heavy metals in inducing oxidative stress has been well documented for many plants species. In *Pisum sativum,* exposure of 50 μM of $CdCl_2$ resulted in oxidative stress (Sandalio *et al.,* 2001). Inhibition of growth, reduction of photosynthetic rate and ultrastructural changes have been reported for *Pisum sativum* and *Phragmites australis* under Cd treatment (Sandalio *et al.,* 2001; Pietrini *et al.,* 2003). In *Brassica juncea,* toxic concentrations of $ZnCl_2$ showed increase in lipid peroxidation hypothesized to be due to ROS generation (Prasad *et al.,* 1999) and similar phenomenon has been reported for wheat under Zn toxicity (Panda *et al.,* 2003a). In rice, wheat and in moss *Taxithelium,* toxic concentrations of Cr have been reported to increase $H_2O_2$ production initiating membrane lipid peroxidation (Panda and Patra, 2000; Panda *et al.,* 2003 a, Choudhury and Panda, 2005). Irreversible changes in protein conformation by Cd via formation of metal thiolate bonds, which affects cell permeability by binding to the nucleophilic groups (Dafre *et al.,* 1996; Romas *et al.,* 2002). For Cd, it not clear not clear how ROS are generated in absence of Fenton type reactions, but is hypothesized that the normal photochemistry gets altered inducing cascade of $O_2^-$, $H_2O_2$, $OH^-$ and other ROS (Atal *et al.,* 1991; Pietrini *et al.,* 2003). $Cu^{+2}$ are though essential metal can turn toxic beyond certain threshold levels. Oxidative stress via lipid peroxidation is one of the important effects of Cu documented in plants. In many Cu has been reported to induce lipid peroxidation and can alter the membrane permeability by displacing various membrane stabilizing cations (Avery *et al.,* 1996; Demidchik *et al.,* 1997; Howlett and Avery, 1997; Murphy *et al.,* 1999; Quartacci *et al.,* 2001; Chen *et al.,* 2004). Thus, the type of reactivity of a particular heavy metal varies from system to system. In nature many plants can accumulate metal ions without showing any toxic symptoms and are referred to as hyperaccumulators. The capacity of survival under metal rich environment have been shown to be regulated by small number of genes with perhaps contribution from more modifier genes (Macnair, 1993; Macnair *et al.*, 2000; Schat *et al.*, 2000) and also the capacity of many plants in cellular detoxification of various heavy metals (Hall, 2002).

The ROS are always considered detrimental to cells and essential cellular processes, but it now known that ROS play a crucial role in redox regulation of important cellular processes and act as a broad-spectrum antibiotic, protecting the plants from invading pathogens via oxidative burst (Scandalios, 1997). However, deleterious effects of ROS are minimized by the coordinated action of enzymic and non – enzymic antioxidants. Metals are important component of many enzymes like SOD, which catalyses the reduction of $O_2^-$ to $H_2O_2$ evolving molecular oxygen (McCord and Fridovich, 1969). Metals like Mn, Cu and Fe are redox in behaviour and constitute important component of various enzymes. Cu acts as a co – factor for many enzymes catalyzing ROS detoxification reactions, but at too high concentrations Cu becomes toxic generating toxic ROS like $OH^{\cdot}$ via Fenton reactions. Antioxidants and Cu metabolism intersect in a diverse ways. Cu is essential for catalytic activity of Cu / Zn SOD that reduces the $O_2^-$ toxicity and play an important role in Cu buffering in cells (Galiazzo, *et al.*, 1988; Culotta *et al.*, 1995).

## 4. ANTIOXIDANT RESPONSE UNDER HEAVY STRESS

All types' stresses can disrupt the normal physiological process through generation of ROS. In order to control the ROS levels and maintain the cell homeostasis, the ROS need to be scavenged. Plants have developed complex antioxidant defense system by which they can scavenge the ROS protecting the cells from oxidative injury. Under heavy metal stress, the antioxidant response are varied and the ability of the plants to combat ROS becomes limited since both redox and non – redox metals have been shown to decrease the antioxidant response at elevated concentration (Shaw *et al.*, 1995; Gallego *et al.*, 1996; Patra and Panda, 1998; Prasad *et al.*, 1999; Panda and Patra, 2000; Panda *et al.*, 2003 a, b; Panda, 2003). A detail schematic description of various antioxidant defenses is shown in Fig. 1. Redox metals like $Fe^{+2}$ can produce $O_2^-$ and $H_2O_2$ at elevated concentrations (Caro and Puntraruol, 1996) and causes loss of antioxidant property of many enzymes (Polle, 1997). Heavy metals that do not belong to the transition group of metals can severely cause oxidative damage disrupting activities of many antioxidant enzymes (Gallego *et al.*, 1996; Prasad *et al.*, 1999; Panda *et al.*, 2003 a, b; Panda 2003, Choudhury and Panda 2004; Choudhury and Panda, 2005).

In *Brassica juncea*, however, toxic concentrations of $Zn^{+2}$ can induce enzyme specific increase in activities of several antioxidant enzymes like superoxide dismutase (SOD), catalase (CAT), guaiacol peroxidase (GPx), glutathione reductase (GR) (Prasad *et al.*, 1999). In case of wheat elevated Zn concentration inhibited the activity of important enzymes like SOD, CAT, GPx and GR with minor increase in early treatment days (Panda *et al.*, 2003a). However, in *Vigna radiata* $Al^{+2}$ treatments showed increase in activity of these antioxidant enzymes (Panda *et al.*, 2003 b).

The plant antioxidant defense system are either inhibited or activated by $Cd^{+2}$ (Gallego *et al.*, 1996; Sandalio *et al.*, 2001; Immelli *et al.*, 2002; Cagno *et al.*, 2001; Milone *et al.*, 2003). In sunflower treatment of $Cd^{+2}$ resulted in the inhibition of SOD, CAT and GPx, while increased response of GR to Cd was reported (Gallego *et al.*, 1996).

In pea plants $Cd^{+2}$ at 50 μM concentration range declined the SOD, CAT and peroxidase activities (Sandalio *et al.,* 2001). Inhibitions of these antioxidant enzymes have also been reported for heavy metals like $Zn^2$, $Cr^{+6}$, As and $Pb^{+3}$ (Panda *et al.,* 2003a, Choudhury and Panda, 2004; Choudhury and Panda, 2005). In many plants, however, heavy metals also induce antioxidant metabolism. A study on the antioxidant pattern in Phragmites australis under $Cd^{+2}$ stress showed highly positive antioxidant response (Immelli *et al.,* 2002). Ascorbate redox enzymes like ascorbate ,peroxidase have been reported to get activated by $Cd^{+2}$ in sunflower (Cagno *et al.,* 2001).

The non – enzymic antioxidants like ascorbate, glutathione, carotenoids and α - tocopherol are provital for protection against oxidative injury and has been widely studied in plants under Cu, Zn, Cd, Cr, Pb, Al and As (Prasad *et al.,* 1999; Sandalio *et al.,* 2001; Panda *et al.,* 2003 a,b; Panda, 2003; Choudhury and Panda, 2004; Choudhury and Panda 2005). Glutathione is the base unit for phytochelatins (PC) and are found in high concentration in cellular compartments (Foyer and Renenberg, 2000). With oxidized glutathione (GSSG), GSH form an important redox couple (Noctor *et al.,* 1998) and can interact with NADPH and NADH resulting in ATP generation via oxidative phosphorylation (May *et al.,* 1998; Hutchinson *et al.,* 2000). A study on Phragmites australis showed increase in glutathione pool under $Cd^{+2}$ and protects the photosynthetic enzymes, venerable to thiophilic binding of $Cd^{+2}$ (Pietrini *et al.,* 2003). Thus, for heavy metals like Cd, glutathione play an photosynthesis under Cd stress (Pietrini *et al.,* 2003). Ascorbate is an important antioxidant occurs usually in the reduced form (AsA) (Noctor and Foyer, 1998; Horemans *et al.,* 2000; Smrinoff, 2000). Under heavy metals, ascorbate shows differential pattern. In Brassica juncea, toxic concentrations of $Zn^{+2}$ showed enhancement in ascorbate content (Prasad *et al.,* 1999). $Cr^{+2}$ treatment to excised wheat leaves at various concentrations showed enhancement in the ascorbate level (Panda *et al.,* 2003a), proving ROS detoxifying role of this non – enzymic antioxidant. However, in case of sunflower, increase in ascorbate-based detoxification has been reported but found to be unable to provide complete detoxification of ROS under $Cd^{+2}$ (Cagno *et al.,* 2001).

Regulation and induction of antioxidant response to enhanced ROS levels is related to the expression of genes for the synthesis of various antioxidant enzymes (Bowler *et al.,* 1992; Foyer *et al.,* 1994, 1997). The induction of antioxidants is thus crucial in controlling the steady state level of ROS, specifically at sub cellular level, highly prone to oxidative injury.

## 5. MECHANISMS OF HEAVY METAL DETOXIFICATION IN PLANTS

Plants are able to synthesize certain compound that helps in sequestering heavy metal ions and in a way respond to heavy metals in different ways. Such mechanisms provide metal detoxification and repairs the damage induced by those metals. These mechanisms are primarily aimed to build up tolerance mechanism heavy metal stress (Hall, 2002). The strategies involve in detoxification process are diverse and such responses can be both in and out of the cell (see Fig. 2). Within the plants there exists a variety of potential mechanisms of heavy metal detoxification such as phytochelatins (PCs), metallothioneines (MTs), organic acids, metal compartmentalization and subsequent transport to the vacuoles. Plasmamembrane also plays a prominent role either by reducing the heavy metal uptake or stimulates efflux pumping of the metals that entered the cytosol (Hall, 2002). Mycorhiza, especially ectomycorhiza are also effective in reducing the effect of heavy metal in plants

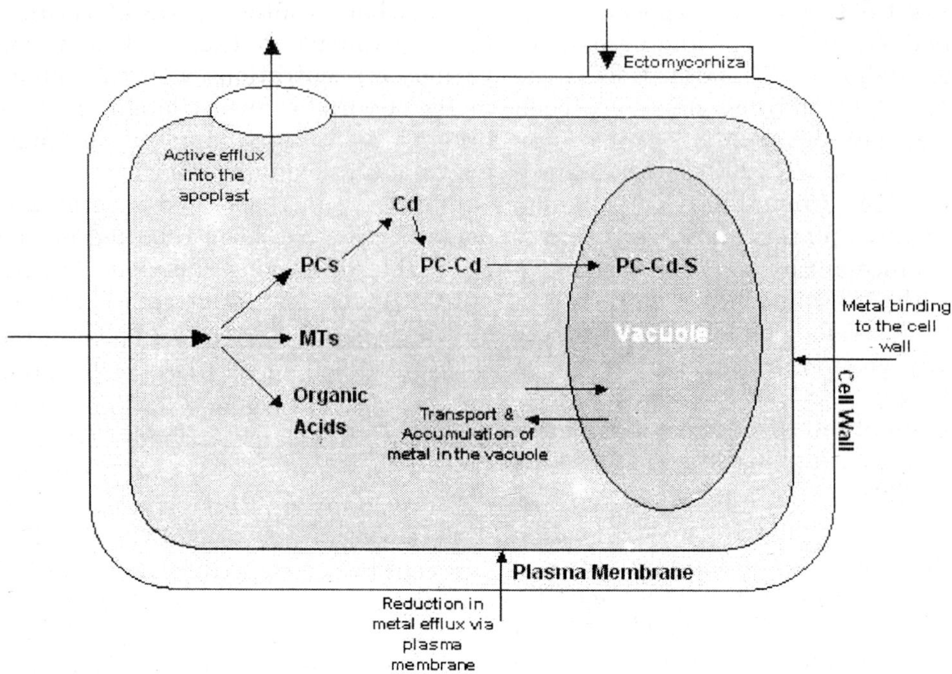


Fig. 2. The mechanisms involved in the bdetoxification of heavy metals in plants (after Hall, 2002).

(Marshner 1995; Huttermann *et al.,* 1999; Jentschke and Godbold 2000).

### 5.1 Metallothioneines (MT)

Metallothioneins (MTs) are the proteins products of mRNA translation. They are low molecular weight cystein – rich metal binding proteins (Kagi, 1993). The presence of large number of cystein residue in MTs binds a variety of heavy metals by mercaptide bonds (Cobbett and Goldsbrough, 2000). Several MT proteins have been identified in wheat and large numbers of MT genes were isolated from plants (Lane *et al.,* 1987; Cobbett and Goldsbrough, 2000). MTs were classified into four distinct categories basing on the amino acid sequence (Robinson *et al.,* 1993). These are Type 1, Type 2, Type 3 and Type 4 MTs. The Type 1 MT has six Cys-Xaa-Cys motifs that are distributed equally in two domains. The two domains are separated by 40 amino acids that include aromatic amino acids. The Type 2 contains two Cys rich domains separated by spacer of 40 amino acid residue, where a Cys-Gly-Gly-Cys motif is present in the N-terminal cystein rich domain. The overall sequence of the N-terminal domain in Type 2 is highly conserved. The Type 3 MTs contains only four Cys residue in the N-terminal domain while the Type 4 MTs differs by having three Cys rich domain each having five or six conserved Cys residue.

In plants, the specific role of PCs in cadmium tolerance is well established. It is, however, not clear that how MTs in plants function, though in animals MTs provide protection to Cd toxicity (Klaassen *et al.,* 1999). In plants, MTs are involved in copper tolerance and homeostasis. Certain plant MTs are functional copper binding proteins and expression of some MT genes are induced by Cu. It has also been observed that the MT gene expression in senescing leaves is coordinated with set of genes involved in copper homeostasis and level of expression of Type 2 MT gene has been closely co-related with Cu tolerance in Arabidopsis (Murphy and Taiz, 1995). The analysis of various defined null genotypes of plants can provide solution about the function of MTs in plants (Cobbett and Goldsbrough, 2000). For this T-DNA insertional mutagenesis is well developed, the objective still remains difficult, as MT genes present are very small for this approach (Krysan *et al.,* 1999; Cobbett and Goldsbrough, 2000). Suitable approaches other than this like targeted gene disruption strategies using transposable elements and RNA interference method are much efficient in determining the role of MTs in plants (Smith *et al.,* 1996).

MTs might have specific role in Cr detoxification in plants (Panda and Choudhury 2005a and references therein). This possible way of Cr detoxification in plants has been demonstrated in Sorghum, where MT – like proteins are expressed under Cr stress (Shanker *et al.,* 2004). The effect of Cr (VI) on MT3 gene expression in both Cr tolerant and susceptible variety of sorghum revealed a high intensity band matching the gene of interest in tolerant variety as compared to those in susceptible ones (Shanker *et al.,* 2004). It was suggested that the production of ROS and $H_2O_2$ as a result of Cr exposure might have triggered signals to induce MT mRNA transcription (Shanker *et al.,* 2004). However, the role of MTs in Cr tolerance in plants and possible detoxification is still not thoroughly understood.

## 5.2 Phytochelatins (PC)

Phytochelatins (PCs) are peptides that are synthesized enzymatically from reduced glutathione (GSH) by the enzyme phytochelatin synthase first characterized by Grill et al., in the year 1989. Structurally, they are the increasing repetition of γ-GluCys dipeptide followed by terminal Gly; $(\gamma\text{-GluCys})_n$-Gly, where n= 2-11. PCs are induced rapidly in plants subjected to heavy metal stress (Cobbett 2000; Goldsbrough, 2000). Though PC are thought to be involved in metal detoxification process, there are evidences that opposes this concept (Ernst *et al.,* 1992; Mehrag 1994; Zenk 1996). The presence of bonding between glutamate and cystein in PC suggest that they are not synthesized on ribosome and hence they are not direct gene product (Grill *et al.,* 1989). The presence of cystein residue confirms metal co-ordination via thiol group and high percentage of glutamic acid makes it highly water soluble (Bertrand *et al.,* 2002). In plant and also in *Saccharomyces pombe*, the PC-Cd complexes are formed to be sequestered in the vacuole. In *S. pombe*, this has been clearly demonstrated in the Cd sensitive mutant htm1 (Cobbett and Goldsbrough, 2002). In *S. pombe* there are two PC-Cd complexes, viz., HMW and LMW. The hmt1 mutant is unable to form HMW. The hmt1 gene encodes a ATP-binding cassette (ABC) membrane transport protein in the vacuolar membrane (Ortiz *et al.,* 1992). Both HMT1 and ATP are required for the transport of PC to the vacuolar membrane.

The gene for the enzyme PC-synthase was first identified genetically in *Arabidopsis*. Expression of *Arabidopsis* and wheat cDNA libraries in *S. cerevisae* were used to identify genes conferring Cd resistance, viz., AtPCS1 and TAPCS1 (Clemens *et al.,* 2001; Vatamaniuk *et al.,* 1999). Latter AtPCS1 was identified through positional cloning of CAD1 gene form Arabidopsis (Ha *et al.,* 1999). The activation of the enzyme PC synthase has direct interaction with metal ions as they are activated by wide range metals. In *Brassica juncea*, it has been observed that the accumulation of Cd is followed by synthesis of PC, sufficient enough to chalate Cd (Haag-Kerwer *et al.,* 1999). Treatment of Cu and Cd resulted in increase in transcription of genes responsible for glutathione synthesis (Xiang and Oliver, 1998). They suggested the response of Cu and Cd to be highly specific and resulted in the detoxification of metals.

The over expression of γ-glutamylcystein synthase gene form *E. coli* and *Brassica juncea* resulted elevated biosynthesis of GSH and PC resulted in increased tolerance to Cd (Zhu *et al.,* 1999). In order to compare the role of PC in heavy metal detoxification in *Arabidopsis* and *S. pombe*, sensitivity of Cad 1-3 were tested for the sensitivity of heavy metals (Ha *et al.,* 1999). It showed that PCs can detoxify Cd and arsenate efficiently, but have no role in detoxification of Zn, Ni and selenite. In S. pombe, the cad 1-3 showed very little sensitivity to Cu and Hg. Moreover, unlike Cd, Pb, etc, Cr is unable to induce PCs (Sanita di Toppi *et al.,* 2002).

### 5.3 Mycorhiza

Plant roots are associated with mycorhizal symbionts. These symbionts have specific role in modulating heavy metal stress in plants (Jentschke and Godbold, 2000; Schutzendubel and Polle, 2002). Among the mycorhizas, ectomycorhizas have characteristic property of heavy metal detoxification (Marshner, 1995; Hutterman *et al.,* 1999; Schutzendubel and Polle, 2002). Ectomycorhizal fungal species like *Paxillus involutus* can reduce $Zn^{+2}$ in pine roots by retaining most of the $Zn^{+2}$, while other species like *Thelephesa terrestris* cannot retain Zn (Colpaert and Van Assche, 1992). Many mycorhizal species other than these two have been shown to have varied responses for metals like Cu, Zn, Cd, etc (Tam *et al.,* 1995; Blaudez *et al.,* 2000; Hall, 2002). Mycorhizas usually adopt metal exclusion principle by which they restrict movement of metals into the roots (Hall, 2002). Such mechanisms involve absorption of metal by fungal hyphae, non–accessibility to apoplast and chelation (Goldbold, 2000). Variation in growth conditions also attribute to the detoxification role by the mycorhizas. For arbuscular Mycorhizas, heavy metals can either reduced form the host cell or they are incorporated via rapid absorption (Weissenhorn *et al.,* 1995). In many fungal species, metallothiones have been attributed for metal detoxification by mycorhizas (Mehra and Winge, 1991).

### 5.4 Organic Acids

Organic acids like amino acids and carboxylic acids have potential role in heavy metal detoxification in plants (Clemens, 2001). Though there is no strong evidence of the role of organic acids in heavy metal detoxification and tolerance in plants, it has been reported that about 36 fold increase in histidine content after exposure to Ni in *Alyssum lesbicum* (Kramer *et al.,* 1996). Moreover, it has also been observed that application of histidine to non Ni hyperaccumulating species greatly increase the Ni accumulation rate in that plant (Hall, 2002). In *Cassia tora* Al toxicity can be regulated by citrate efflux (Yang *et al.,* 2003). Citrate has been demonstrated to reduce $Ni^{+2}$ uptake and play significant role in $Ni^{+2}$ detoxification (Salt *et al.,* 2000).

### 5.5 Plasma Membrane

The plasma membrane of the cell acts as the first barrier for the entrance of the heavy metals inside the cell. However, its function may severely be affected by heavy metals e.g., increased leakage and lipid peroxidation resulting in loss of functionality and integrity (Choudhury and Panda, 2004, 2005; Panda and Choudhury 2005a, b).The tolerance mechanism involves the protection of the plasma membrane against heavy metals that would produce increased leakage of solutes form cells (Stange and Macnair 1991; DeVos *et al.,* 1991). In the metal tolerant plants there exists no defined mechanism of protection against ROS, but rather they possess a improved mechanism of metal homeostasis (Dietz *et al.,* 1999).

The role of plasma membrane in efflux transport of heavy metals is not clear, however, it has been demonstrated that there are several classes of metal transporters that play a key role in heavy metal tolerance. Such transporters include the heavy metal *CPx-ATPase, the Narmps, the CDF (cation diffusion facilitators)* and the *ZIP familiy* (Williams *et al.,* 2000; Guerinot 2000). Clearly more molecular evidence is essential to support any protective role against heavy metal stress.

## 6. MOLECULAR MECHANISMS OF PLANT AND HEAVY METAL INTERACTION

Living organisms live in habitats where many abiotic stress factors predominate and of which heavy metals is one such. One of the most important consequences of metal stress in plants is the generation of ROS that exerts oxidative damage to plant cells (Dietz *et al.,* 1999). In recent years, we have evidenced tremendous advances in our molecular understanding of metal stress in plants from experiments involving lower organisms like *Saccharomyces cerevisae* and *S. pombe*, which in many ways resembles molecular components of higher eukaryotes (Clemens, 2001). Studies on *S. cerevisae* and *S. pombe* have revealed the presence of metal transporters and complementation of *S. cerevisae* uptake mutants; cDNAs encoding these transporters have been cloned in many plants (Clemens, 2001). One such transporter is the *COPT1*, a Cu transporter (Kampfenkel *et al.,* 1995) and has been found to involve in Cu uptake in *Arabidopsis* (Clemens, 2001). Metals ions like $Fe^{+2}$ and $Zn^{+2}$ and taken up by plants via transporters belonging to ZIP family that includes *ZRT*, *IRT*, etc (Fox and Guerinot, 1998; Saier, 2000). *IRT*1 was first isolated from *Arabidopsis* and its transcription can be induced in roots via $Fe^{+2}$ deficiency. Besides involving in $Fe^{+2}$ transport, *IRT*1 can also regulate $Zn^{+2}$, $Cd^{+}$ and $Mn^{+2}$ uptakes (Korshunova *et al.,* 1999). Cloning of *IRT*1 has lead to the discovery of other related ZIP family like *ZRT*1 and *ZRT*2 that are identified on the sequence similarity with *IRT*1. The ZIP transporters (*ZIP*1 and *ZIP* 3) have a high affinity for $Zn^{+2}$, and in *Arabidopsis* they are found to confer $Zn^{+2}$ uptake activity (Grotz *et al.,* 1998; Guerinot and Eide, 1999). Both *ZIP* 1 and 3 are expressed in the roots and are activated only under $Zn^{+2}$ depletion conditions. Few plant metal transporters indentified till date are shown in Fig. 3.

In plants like fungi, small Cys – rich proteins generally known as metallothioneins are induced by heavy metals (Thiele, 1992). They are very useful for the organism and confer heavy metals tolerance in plants. However, many similar proteins and genes were identified in many plants, their function is still unknown (Zenk *et al.,* 1996; Foley *et al.,* 1997). Studies on metal sensitive strains of *Escherichia coli* showed that *TRX* – like genes could confer tolerance to heavy metals, suggesting that *TRX* gene are very important in conferring heavy metal tolerance in plants via protein repairing. In green alga *Chlamydomonas reinhardtii* it has been shown that the expression of *TRX* m and *TRX* h genes are regulated by heavy metals like Cd and Hg at m – RNA level (Lemaire *et al.,* 1999). *TRX* can thus be implicated to be involved in defense against heavy metals in photosynthetic organism (Lemaire *et al.,* 1999). In sea grass *Posidonia oceanica*, three DNA sequences *Pomt 2 a, Pomt 2b and Pomt 2c* putatively encoding metallothioneins (MTs) have been isolated, by PCR (Giordani *et al.,* 2000). *Pomt 2a and Pomt 2b* genes encodes for type II metallothioneins, while *Pomt 2c* is a pseudogene arises by the retrotransposition of *Pomt 2b* m RNA (Giordani *et al.,* 2000) and also observed in case of human genome (Hamer, 1986). However, the expression of MTs varies from plant species to plant species.

With the identification of various metal chaperons and transporters, especially for Cu in yeast and human, the intracellular metal trafficking is now becoming clear. In yeast, Cu – chaperons like *yCCS* that are encoded by *LYS7* are essential for CuSOD expression in yeast (Culotta *et al.,* 1997). In addition to

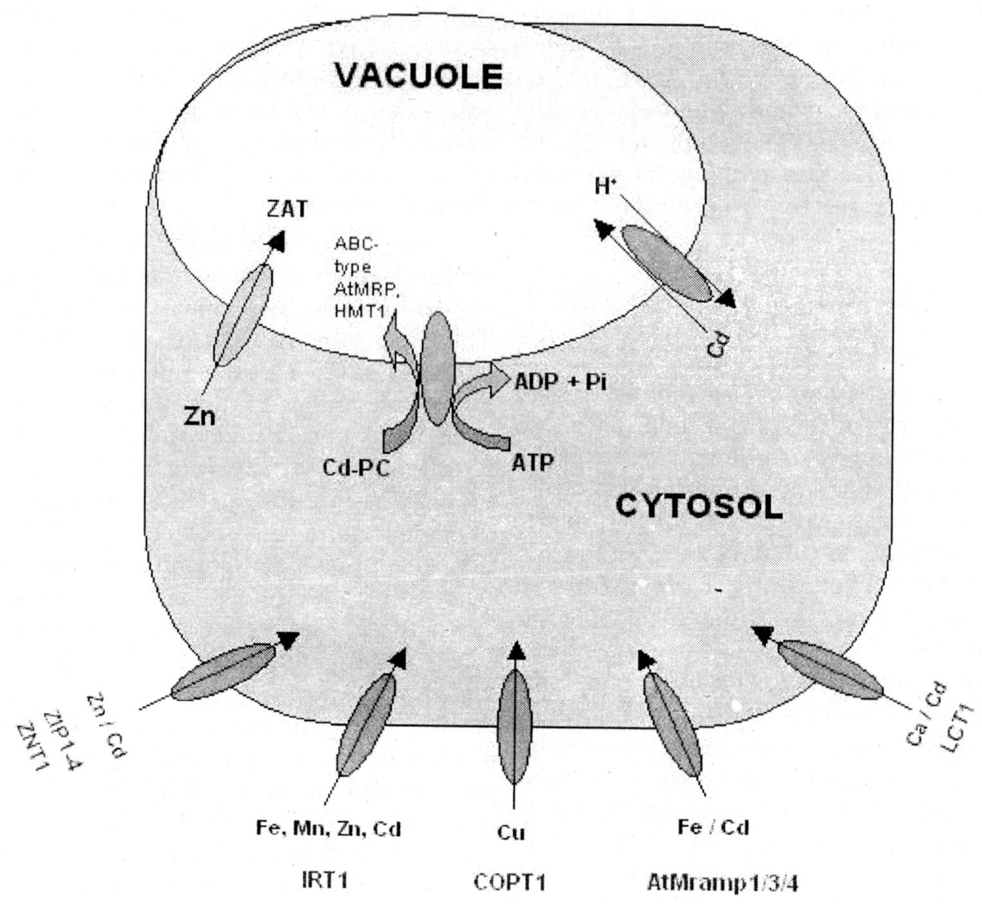


Fig. 3. Metal transporters identified involved in uptake, transport and sequestering of metals plants (modified after Clemens, 2001).

*yCCS*, other two types of cytosolic Cu chaperons *COX17* and *ATX1* have been identified (Glerum *et al.,* 1996). *COX17* is involved in incorporation of Cu to cytochrome c oxidase complex and *ATX1* is involved in Cu transfer to post golgi vesicles via Cu – pumping P – type ATPase called *CCC2* (Fu *et al.,* 1995; Pufahl *et al.,* 1997) and horologs of which have been identified in humans and higher plants (Klomp *et al.,* 1997; Himelblau *et al.,* 1999). In *Arabidopsis* the gene *RNA1* has been found to be associated with ethylene signaling (Hirayama *et al.,* 1999). It acts as a Cu – ATPase and resides in post golgi vesicles, involved in delivery of Cu into protein of the secreatory pathway (Rodriguez *et al.,* 1999). With the discovery and identification of RNA1 and *ATX 1* homologs in *Arabidopsis*, it now becoming clear that plants have Cu trafficking network analogous to those found in human and yeast (Himelblau and Amasio, 2000). Besides these many other chelators have been identified or await identification that is involved in metal trafficking in plants.

The primary driving force for transport processes is the electrochemical $H^+$ gradient, generated in both plasma membrane and vacuolar membrane. Many plant transporters like *IRT1*, *LCA1*, *CAX2* have a broad selectivity for ions. The *IRT1* transporter that was earlier selected as a $Fe^{+2}$

transporter found to transport $Mn^{+2}$ and $Zn^{+2}$ (Korshunova *et al.,* 1999). *LAC1* mediates uptake of $Ca^{+2}$ and $Cd^{+2}$ (Clemens *et al.,* 1999), while *CAX2* can transport $Ca^{+2}$, $Cd^{+2}$ and $Mn^{+2}$ in plants (Hrischi *et al.,* 2000). In *Arabidopsis*, increase in accumulation of metals like $Cd^{+2}$ and $Mg^{+2}$ is not proportional to increased expression of *CAX2*, but clearly suggested the broad ion selectivity of this transporter (Hrischi *et al.,* 2000).

In plants like *Triticum aestivum*, expression of a cDNA *TaPCS1* can induce dramatic increase in Cd tolerance (Clemens *et al.,* 1999). In *Arabidopsis* and fungi many genes families were identified, which can confer metal tolerance upon expression. Overexpression of bacterial citrate synthase gene in transgenic plant has been shown to confer Al tolerance in plants (de la Fuente *et al.,* 1997). In plants like wheat, tobacco and *Arabidopsis*, about 20 different genes were identified that are induced due to Al stress (Snowden and Gardner, 1993; Cruz – Ortega *et al.,* 1997; Delhaize *et al.,* 1999; Ezaki *et al.,* 1996, 1997; Sugimoto and Sakamoto, 1997; Richards *et al.,* 1998). Most of these genes are stress genes that are induced as a result of general stress responses like lowering in phosphate content or other metal associated toxicities or via oxidative stress (Ezaki *et al.,* 1995; Snowden *et al.,* 1995; Sugimoto and Sakamoto, 1997; Richards *et al.,* 1998) and many are induced as antioxidant enzymes (Ezaki *et al.,* 2000). It was thus proposed that there exist a common mechanism between gene induction and Al induced oxidative stress (Richards *et al.,* 1998). However, the exact biological importance of Al induced in plants is still unclear (Ezaki *et al.,* 2000). Many stress-induced genes however provide tolerance to metal stress in plants. The blue copper-binding gene (*AtBCB*), tobacco glutathione S transferase gene (*par B*), tobacco peroxidase gene (*NtPox*) and tobacco GDP – dissociation inhibitor gene (*NtGDI1*) has been found to confer Al tolerance in *Arabidopsis* and tobacco (Ezaki *et al.,* 2000). Both *AtBCB* and *par B* have been found to confer resistance to oxidative stress induced by metals and salts (Ezaki *et al.,* 2000). Plant genes that are induced by abiotic stresses are in general involved in protection against oxidative stress generated by that particular stressor (Camp *et al.,* 1996; Veena *et al.,* 1999). Conferring metal tolerance in crop plants is an important approach transgenic research. Construction of transgenic plants expressing metal induced genes can in a way provide metal tolerance and in a way allemiorate metal induced oxidative stress.

## 7. DEVELOPING MODEL SYSTEM TO UNDERSTAND HEAVY METAL AND PLANT INTERACTION

Understanding the plant and heavy metal interaction is important from the point of view of agriculture, human health and also the basic detoxification mechanism that can be used for phytoremedation. With the inclusion of *Arabidopsis thaliana*, as a model organism to study the molecular genetics of plant metal interaction, genes involved in metal metabolism have been identified using wide range of approaches. The elucidation of *Arabidopsis* genome sequence many gene families are now known that are involved in metal metabolism (Maser *et al.,* 2001). Many genes have now been identified with the mutants of yeasts and *E. coli* that are homologues to other higher organism known to be involved in metal metabolism in plants (Cobbett, 2003). One such family is the P – type ATPases that has been well studied in many prokaryotes, yeasts and mammals (Cobbett, 2003). Now, many plant genes are being studied with respect to their mutants that are involved in metal metabolism. With *Arabidopsis* as a model organism, mutants can be easily being identified solely basing on their phenotype; however, this aspect has been least used for metal metabolism and detoxification in plants (Cobbett, 2003). Non – plant model systems can also provide a greater understanding of metal metabolism in plants, for which unicellular models involving *Saccharomyces pombe* and *Chlamydomonas rheinhardtii* have been developed (Clemens and Simm, 2003; Hanikenne, 2003). *S. pombe* can express PCs that are involved in metal detoxification. *C. rheinhardtii* on the other hand is more advantageous than *S. pombe* because in addition PCs, photosynthesis can be traced with respect to essential metal ions (Cobbett, 2003).

Metal hyperaccumulation by many higher plant and cryptogams (bryophytes and lichens) has opened new ways of phytoremedation of metal polluted environment. Many higher plants like *Thlaspi caerulescens* can also be used as model organism in this regard. Now much attention has been paid on the basic mechanism of metal transport, metal accumulation and tolerance in plants and other organism to develop transgenic plants with immense phytoremedation capabilities. Expression of many bacterial genes involved in glutathione biosynthesis and arsenate reduction can significantly confer arsenic tolerance and arsenic accumulation (Dhankher *et al.*, 2002; Pilon – Smits and Pilon, 2002). Beside this, expression of arsenate reductase can also provide tolerance to Cd (Dhankher *et al.*, 2003). Identification and analysis of natural variation in the capacity of plants to accumulate and tolerate metals will provide immense information on the basic understanding of the process (Cobbett, 2003). In addition to this, application of molecular genetic tools to model organisms including the hyperaccumu-lating species and subsequent physiological studies can provide new ways to develop more new model organisms in this regard.

## CONCLUSIONS

Certain heavy metals like Cu, Zn and Fe are essential for various physiological processes. Other heavy metals like Cd, Al, Cr, Pb, Hg, As have no known biological role and potentially toxic to plants. Phytotoxicity of heavy metals results mainly in the inhibition of physiological processes leading to oxidative stress mainly due to free radicals or ROS. Plants have developed diverse antioxidant system involving enzymes like catalase, peroxidase, ascorbate peroxidase, glutathione reductase, mono dehydroascorbate reductase, superoxide dismutase and non – enzymes like $\alpha$ - tocopherol, ascorbic acid and glutathione to scavenge the deleterious effects of the ROS in cells. Antioxidants plays provital role to cope up the exceeding level of ROS in cells. Plants have developed various cellular and molecular mechanisms by which they detoxify metals. These include phytochelatins, organic acids, mycorhizas, efflux pumping, metal compartmentalization process by which the metals can be detoxified in plants. The transport and subsequent tolerance to particular metals depends on various factors. Plant genes are often induced by metal stress and with identification of many metals transporters in plants it is now possible to elucidate the exact mechanism involved in the process of metal detoxification, transport and tolerance in plants. Many antioxidant genes are induced by metal stress and their expressions under particular stressor determine the efficiency of the antioxidants that detoxify the ROS. It will thus be important to develop transgenic plants that can accumulate metals and in a way can be used for phytoremedation of metal polluted environment.

## REFERENCES

Angelone, M., Bini, C. 1992. Trace element concentration in soil and plants of Western Europe. – In: Adriano, D.C. (ed.): Biochemistry of trace metals. Pp. 19 – 60, Lewis Publisher, Boca Raton FL.

Atal, N., Pardha Saradhi, P., Mohanty, P. 1991. Inhibition of chloroplast photochemical reactions by treatment of wheat seedlings with low concentration of cadmium: analysis of electron transport activities and changes in fluorescent yield. *Plant Cell Physiol.*, **32**: 943 – 951.

Avery, S.V., Howlett, N.G., Radice, S. 1996. Copper toxicity towards *Saccharomyces cerevisae*: dependence on plasma fatty acid composition. *Appl. Environ. Microbiol.*, **62**: 3960 – 3966.

Balestrasse, K.B., Gallego, S.M., Tomaro, M.L. 2004. Cadmium induced senescence in nodules of soybean (*Glycine max* L.) plants. *Plant Soil*, **262**: 373 – 381.

Barcelo, J. Poschenrieder, C. 2002. Fast root growth responses, root exudates ans internal detoxification as clue to mechanism of aluminium toxicity and resistance. *Env. Exp. Bot.*, **48**: 75 – 92.

Blaudez, D., Botton, B., Chalot, M. 2000. Cadmium uptake and subcellular compartmentalization in ectomycorrhizal fungus *Paxillus involutus. Microbiology*, **146**: 1109 – 1117.

Blokhina, O., Virolainen, E., Fgrestedt, K.V. 2003. Antioxidant, oxidative damage and oxygen deprivation stress: A Review. *Annal. Bot.*, **91**: 179 – 194.

Bowler, C., van Montagu, M., Inze, D. 1992. Superoxide dismutase and stress tolerance – *Annu. Rev. Plant Physiol. Mol. Biol.*, **43**: 83 – 116.

Camp, W.V., Capiau, K.C., Van Montagu, M.V., Inze, D., Slooten, L. 1996. Enhancement of oxidative stress tolerance in transgenic tobacco plants overproducing Fe – superoxide dismutase in chloroplast. *Plant Physiol.*, **112**: 1703 – 1714.

Canesi, L., Caicci, C., Piccoli, G., Stocchi, V., Viarengo, A., Gallo, G. 1998. *In vitro* and *in vivo* effect of heavy metals on mussel digestive gland hexokinase activity: The role of glutathione. *Comp. Biochem. Physiol., C. Pharmacol. Toxicol., Endocrinol.*, **120**: 261 – 268.

Caro, A., Puntarulo, S. 1996. Effects of *in vivo* iron supplementation on oxygen radical production by soyabean roots. *Biochem. et. Biophys. Acta*, **1291**: 245 – 251.

Carter, D.E. 1995. Oxidation and reduction of metal ions. *Env.Health Presp.*, **103**: 17-19.

Charles, S.A., Halliwell, B. 1981. Light activation of fructose bisphosphate in isolated spinach chloroplast and activation by hydrogen peroxide. *Planta*, **151**: 242 – 246.

Chen, C.T., Chen, T.H., Lo, K.F., Chui, C.Y. 2004. Effect of proline on copper transport in rice seedlings under excess copper stress. *Plant Sci.*, **166**: 103 – 111.

Cho, U., Seo, N. 2004. Oxidative stress in *Arabidopsis thaliana* exposed to cadmium is due to hydrogen peroxide accumulation. *Plant Sci.*, **168**: 113-120.

Choui, A., Mazhoudi, S., Ghorbal, M.H., El Ferjani, E. 1997. Cadmium and zinc induction of lipid peroxidation and effects on antioxidant enzyme activities in bean (*Phaseolus vulgaris*). *Plant Sci.*, **127**: 139 – 147.

Choudhury, S., Panda, S.K. 2004.Induction of oxidative stress and ultrastructural changes in moss *Taxithelium nepalense* (Schwaegr.) Broth. under lead (Pb) and arsenic (As) phytotoxicity. *Curr. Sci.*, 87: 342-348.

Choudhury, S., Panda, S.K. 2005. Toxic effects, oxidative stress and ultrastructural changes in the moss *Taxithelium nepalense* (Schwaegr.) Broth. under lead and chromium toxicity.- Water, Air and Soil Pollut., (In Press).

Cho, U.H. and Park, J.O. 2000. Mercury – induced oxidative stress in tomato seedlings *Plant Sci.*, **156**: 1 – 9.

Chen, J., Zhau, J., Goldsbrough, P.B. 1997. Characterization of phytochelatin synthase from tomato. *Plant Physiol.*, **101**: 165 – 172.

Chrestensen, C.A., Strake, D.W., Mieyal, J.J. 2000. Acute cadmium exposure inactivate thiol transferase (glutaredoxin), inhibits intracellular reduction of protein – glutathionyl – mixed disulphides. *J. Biol. Chem.*, **275**: 26556 – 26565.

Ciamporova, M.: Morphological and structural responses of plants to aluminium at organ, tissue and cellular level. *Biol. Plant.*, **45**: 161 – 171.

Clemens, S., Kim, E.J., Neumann, D., Schroeder, J.I. 1999. Tolerance to toxic metals by a gene family of phytochelatin synthase from plant and yeast, *EMBO J.*, 18: 3325 – 3333.

Clemens, S. 2001. Molecular mechanism of plant metal tolerance and homeostasis. *Planta*, **212**: 475 – 486.

Clemens, S., Simm, C. 2003. *Schizosaccromyces pombe* as a model for metal homeostasis in plant cells: the phytochelatin dependent pathway in the main Cd detoxification mechanism. *New Phytol.*, **159**: 315 – 321.

Cobbett, C.S. 2000. Phytochelatins and their role in heavy metal detoxification. *Plant Physiol.*, **123**: 825 – 883.

Cobbett, C. 2003. Heavy metal and plant model system and hyperaccumulators. *New Phytol.*, **159**: 289 – 293,

Colpaert, J., van Assche, J. 1992. Zinc toxicity in ectomycorrhizal *Pinus sylvestris*. *Plant Soil*, **143**: 201 – 211.

Cruz – Ortega, R., Cushman, J.C., Ownby, J.D.: cDNA clones encoding 1, 3 - β - glucanase and fibrin like cytoskeletal protein are induced by Al toxicity in wheat roots. *Plant Physiol.*, **114:** 1453 – 1450.

Culotta, V.C., Klomp, L.W., Strain, J., Casareno, R.L., Krems, B., Gitlin, J.D. 1997. The copper chaperon for superoxide dismutase. *J. Biol. Chem.*, **272**: 23469 – 23472.

Dafre, A.L., Sies, H., Akerboom, T. 1996. Protein – S – thiolation and regulation of microsomal glutathione transferase activity by glutathione redox couple. *Arch Biochem. Biophys.*, **332**: 288 – 294.

de la Fuente, J.M., Ramirez – Rodriguez, V., Cabrera – Ponce, J.L., Herrera – Estrella, L. 1997. Aluminum tolerance in transgenic plants by alteration of citrate synthase. *Science*, **276**: 1566 – 1568.

de Vos, C.H.R., Schat, H., de Waal, M.A.M., Vooijs, R., Ernst, W.H.O. 1991. Increased resistance to copper induced damage of the root cell plasmalemma in copper tolerant *Silene cucubalus*. *Physiol. Plant.*, **82**: 523 – 528.

De, D.N.: Plant Cell vacuoles – CSIRO Publication, Collingwood, Australia, 2000.

Delhaize, E., Hebb, D.M., Richards, K.D., Lin, J.M., Ryan, P.R., Gardner, R.C. 1999. Cloning and expression of wheat (*Triticum aestivum* L.) phosphatidylserine synthase cDNA: overexpression in plants alters the composition of phsopholipids. *J. Biol. Chem.*, **274**: 7082 – 7088.

Demidchik, V., Sokolik, A., Yurin, V. 1997. The effect of Cu $^{+2}$ on ion transport system of plant cell plasmalemma. *Plant Physiol.*, **104**: 225 – 261.

Dhankher, O.P., Li, Y.J., Rosen, B.P., Fuhrman, M., Meagher, R.B. 2003. Increased cadmium tolerance and accumulation by plants expressing arsenate reductase. *New Phytol.*, **159**: 443 – 452.

Di Cagno, R., Zguidi, L., De Gara, L., Soldatini, G.F. 2001. Combined cadmium and ozone treatment affects photosynthesis and ascorbate - dependent defenses in sunflower. *New Phytol.*, **151**: 627 – 636,

Dietz, K – J., Baier, M., Kramer, U. 1999. Free radical and reactive oxygen species as mediator of heavy metal toxicity in plants – In: Prasad, M.N.V., Hagemeyer, J. (ed). Heavy metal stress in plants: From molecules to ecosystem. Pp. 73 – 97, Springer – Verlag, Berlin,

Dubey, R.S., Pessarakli, M.: Physiological mechanism of nitrogen absorption and assimilation in plants under stressful conditions. – In: Pessarakli, M. (eds). Handbook of plant and crop physiology. Pp. 605 – 625, Marcel Dakker Inc., New York, 1995.

Eick, M.J., Peak, J.D., Brady, P.V., Pesek, J.D. 1999. Kinetics of lead absorption / desorption in goethite: Residence time effect. *Soil Sci.*, **164**: 28 – 39.

Elstner, E.F. 1982. Oxygen activation and oxygen toxicity. *Ann. Rev. Plant Physiol.*, **33**: 73 – 96.

Ezaki, B., Tsugita, S., Matsumoto, H. 1996. Expression of moderately anionic peroxidase is induced by aluminum treatment in tobacco cells: possible involvement of peroxidase isoenzyme in aluminum ion stress. *Physiol. Plant.*, **96**: 21 – 28.

Ezaki, B., Koyanagi, M., Gardener, R.C., Matsumoto, H. 1997. Nucleotide sequence of cDNA for GDP dissociation inhibitor (GDI) which is induced by aluminum (Al) ion stress in tobacco cell culture (accession no. AF012823) (PGR 97 – 133). *Plant Physiol.*, 115: 314.

Foley, R.C., Liang, Z.M., Singh, K.B. 1997. Analysis of type 1 metallothionein cDNA in *Vicia faba*. *Plant Mol. Biol.*, **33**: 583 – 591.

Fu, D., Beeleer, T.J., Dunn, T.M. 1995. Sequence mapping and disruption of CCC2, a gene that cross complements the Ca (2+) – sensitive phenotype of csgl mutants and encodes a P – type ATPase belenging to the Cu (2+) – ATPase subfamily. *Yeast*, **11**: 283 – 292.

Fox, T.C., Guerinot, M.L. 1998. Molecular Biology of cation transport in plants. *Annu. Rev. Plant Physiol. Mol. Biol.*, **49**: 583 – 591.

Foy, C.D., Chaney, R.L., White, N.C. 1978. The physiology of metal toxicity in plants. *Annu. Rev. Plant Physiol.,* **29**:

Foyer, C.H., Renenberg, H. 1998. Regulation of glutathione synthesis and its role in abiotic and biotic stress – In: Renenberg, H., Burunold, C., De Kok, L.J., Stulen, I. (ed.). *Sulfur Nutrition and Sulfur Assimilation in Higher Plants*. Pp. 127 – 153, SPB Academic Publishing, The Huge, The Netherlands.

Foyer, C.H., Lelandais, M., Kunert, K.J. 1994. Photooxidative stress in plants. *Physiol. Plant.*, **92**: 696 – 717.

Foyer, C.H., Lopez – Delgrado, H., Dat, J.F., Scott, I.M. 1997. Hydrogen peroxide and glutathione – associated mechanism of acclamatory stress tolerance and signaling. *Physiol. Plant.*, **100**: 241 – 254.

Gallego, S.M., Benavioes, M.P., Tomaro, M.L. 1996. Effect of heavy metal ion excess on sunflower leaves – Evidence for involvement of oxidative stress. *Plant Sci.,* **121**: 151 – 159,

Glerum, D.M., Shtanko, A., Tazgoloff, A. 1996. Characterization of COX17, a yeast gene involved in copper metabolism and assembly of cytochrome oxidase. *J. Biol. Chem.*, **271**: 14504 – 14509.

Gollaoway, J.N., Thornton, J.D., Norton, S.A., Volcho, H.L., McLean R.A. 1982. Trace metals in the atmosphere deposition: a review and assessment. *Atmospheric Env.* **16**: 1677.

Goldbold, D.L., Polle, A. 2001. Cadmium – induced changes in antioxidant system, $H_2O_2$ content and differentiation in Pinus (*Pinus sylvestris*) roots. *Plant Physiol.*, **127**: 887 – 892.

Goldsbrough, P. 2000. Metal tolerance in plants: the role of phytochelatins and metallothiones – In: Terry, N., Banuelos, G. (ed) Phytoremedation of contaminated soil and water. Pp. 221 – 233, CRC Press, LLC.

Gortz, E., Fox, T., Connolly, E., Park, W., Guerinot, M.L., Eide, D. 1998. Identification of family of zinc transporter genes from *Arabidopsis* that respond to zinc deficiency. *Proc. Natl. Acda. Sci.* USA, **95**: 7220 – 7224.

Grill, E., Loffler, S., Winnacker, E – L., Zenk, M.H. 1989. Phytochelatins, the heavy metal binding peptides of plants are synthesized from glutathione by a specific $\gamma$ - glutamyl cystein dipeptidyl transpeptidase (phytochelatin synthase). *Proc. Nat. Acad. Sci.,* USA, **86**: 6838 – 6842.

Guerinot, M.L. 2000. The zip family of metal transporters. *Bichemica. et. Biophys. Acta*, **1465**: 190 – 198.

Guerinot, M.L., Eide, D. 1999. Zeroing in on zinc uptake in yeast and plants. *Curr. Opin. Plant Biol.,* **2**: 224 – 249.

Guo, Y., Marschner, H. 1995. Uptake, distribution and binding of cadmium and nickel in different plant species. *J. Plant Nutr.* **18** : 2691 – 2706.

Halliwell, B., Gutteridge, J.M.C.: Free radicals in biology and medicine – Clarendon Press, Oxford, Pp. 1 – 61, 1989.

Hanikenne, M. 2003. *Clamydomonas reinhardtii* as a model for studies of heavy metal homeostasis and tolerance in plants. *New Phytol.*, **159**: 331 – 350.

Hamer, D.H. 1986. Metallothionein. *Ann. Rev. Biochem.*, 55: 913 – 951.

Hanikenne, M. 2003. *Clamydomonas reinhardtii* as a model for studies of heavy metal homeostasis and tolerance in plants. *New Phytol.*, **159**: 331 – 350.

Han, F.X., Sridhar, B.B.M., Monts, D.L., Su, Y. 2003. Phytoavailability and toxicity of trivalent and hexavalent chromium to *Brassica juncea*. *New Phytol.*, **162**: 480-489.

Haug, A., Vitorello, V.: Aluminium co-ordination to calmodulin: Thermodynamic and kinetic aspects. *Cord. Chem. Rev.*, **149**: 113 – 124.

Heuttermann, A., Arduini, I., Goldbold, D.L. 1999.Metal pollution and forest decline – In: Prasad, M.N.V., Hagemeyer, J. (ed). Heavy metal stress in plants: From molecule to ecosystem. Pp. 253 – 272, Springer – Verlag, Berlin.

Himelblau, E., Mira, H., Lin, S.J., Culotta, V.C., Penarrubia, L., Amasino, R.M. 1998. Identification of functional homolog of the yeast copper homeostasis gene ATX1 from *Arabidopsis*. *Plant Physiol.*, **117**: 1227 – 1234.

Himelblau, E., Amasino, R.M.: Delivering copper with plants cells. 2000. *Curr. Opin. Plant Biol.*, **3**: 205 – 210.

Hirayama, T., Kieber, J.J., Hirayama, N., Kogan, M., Guzman, P., Nourizadeh, S., Alonso, J.M., Dailey, W.P., Dancis, A., Ecker, J.R. 1999. RESPONSIVE – TO – ANTAGONIST 1, a Menkes / Wilson disease – related copper transporter, is required for ethylene signaling in *Arabidopsis*. *Cell*, **97**: 383 – 393.

Horemans, N., Foyer, C.H., Asard, H. 2000. Transport and action of ascorbate in plasmamembrane. *Trends Plant Sci.*, **5**: 263 – 26.

Howlett, N.G., Avery, S.V.: Induction of lipid peroxidation during heavy metal stress in *Saccharomyces cerevisae* and influence of plasmamembrane fatty acid unsaturation. *Appl. Env. Microbiol.*, **63**: 2971 – 2976.

Hutchinson, R.S., Groom, Q., Ort, D.R. 2000. Differential effects of chilling induced photooxidation on redox regulation of photosynthesis enzyme. *Biochem.*, **39**: 6679 – 6688.

Iannelli, M.A., Pietrini, F., Fiore, L., Petrilli, L., Massacci, A. 2002. Antioxidant response to cadmium in *Phragmites austrslis* plants. *Plant Physiol. Biochem.*, **40**: 977 – 982.

Kampfenkel, K., Kushnir, S., Babiyhuk, E., Inze, D., Van Montagu, M. 1995. Molecular characterization of putative *Arabidopsis thaliana* copper transporter and its yeast homologue. *J. Biol. Chem.*, **270**: 28497 – 28486.

Kastori, R., Petrovic, M., Petrovic, N. 1992. Effect of excess lead, cadmium, copper and zinc on water relation in sunflower. *J. Plant Nutr.*, **15**: 2427 – 2439.

Kochian, L.V.1995 Cellular mechanism of aluminium toxicity and resistance in plants. *Ann. Rev. Plant Physiol. Mol. Biol.*, **46**: 237-260.

Korshunova, Y.O., Eide, D., Clark, W.G., Guerinot, M.L., Pakrasi, H.B. 1999. The IRT1 protein from *Arabidopsis thaliana* is a metal transporter with broad substrate range. *Plant Mol. Biol.*, **40**: 37-44.

Koyama, H., Toda, T., Hara, T. 2001. Brief exposure of low – pH stress causes irreversible damage to growing root in *Arabidopsis thaliana*: pectin – Ca interaction may play important role in proton rhizotoxicity. *J. Exp. Bot.* **52**: 361 – 368.

Klomp, L.W., Lin, S.J., Yuan, D.S., Lkausner, R.D., Culotta, V.C., Gitiin, J.D. 1997. Identification and functional expression of HAH1, a novel human gene involved in copper homeostasis. *J. Biol. Chem.*, **272**: 9221 – 9226.

Kramer, U., Cotter – Howells, J.D., Charnock, J.M., Baker, A.J.M., Smith, A.C.: Free histidine as a metal chelators in plants that accumulate nickel. *Nature,* 379: 635 – 638,

Lindon, F.C., Henriques, F.S. 1991. Limiting step in photosynthesis of rice plants treated with varying copper levels. *J. Plant Physiol.*, **138**: 115 – 118.

Macnair, M.R., Tilstone, G.H., Smith, S.E.: The genetics of metal tolerance and accumulation in higher plants – In: Terry, N., Banuelos, G. (ed.). Phytoremedation of contaminated soil and water. Pp. 235 – 250, CRC Press, LLC, 2000.

Maser, P., Thomine, S., Schroeder, J.I., Ward, J.M., Hirschi, K., Sze, H., Talke, I.N., Amtmann, A., Maathuis, F.J.M., Sanders, D., Harper, J.F., Tchieu, J., Gribskov, M., Persans, M.W., Salt, D.E., Kim, S.A., Guerinot, M.L. 2001. Phylogenetic relationship within cation transporter families of Arabidopsis. *Plant Physiol.*, **126**: 1646 – 1667.

Marschner, H.: Mineral nutrition in higher plants (2 nd ed). Academic Press, London, 1995.

Matsumoto, H., Hirasawa, E., Torikam H., Takahashi, E. 1976. Localization of absorbed aluminum in pea root and its binding to nucleic acids. *Plant Cell Physiol.*, **17**: 127 – 137.

Matsumoto, H., Senoo, Y., Kasai, M., Maeshima, M. 1996. Response of plant root to aluminium stress: Analysis of the inhibition of root elongation and changes in membrane function. *J. Plant Res.*, **109**: 99-105.

May, M.J., Vernouse, T., Leaver, C., van Montagu, M., Inze, D. 1998. Glutathione homeostasis in plants: implication of environmental sensing and plant development. *J. Exp. Bot.*, **49**: 649 – 667.

Mehra, R.K., Winge, D.R. 1991. Metal ion resistance – molecular mechanism and their regulated expression. *J. Cell Biochem.*, **45**: 30.

Mehrag, A.A. 1994. Integrated tolerance mechanism: constitutive and adaptive plant responses to elevated metal concentration in the environment. Plant Cell Env., **17**: 989 – 993.

Mehrag, A.A. 1993. The role of plasmamembrane in metal tolerance in angiosperm. *Physiol. Plant.*, **88**: 191 – 198.

Milone, M.T., Sgherri, C., Clijsters, H., Navari – Izzo, F. 2003. Antioxidative responses of wheat treated with realistic concentration of cadmium. *Env. Exp. Bot.,* **50**: 265 – 276.

Mithofer, A., Schulze, B., Boland, W. 2004. Biotic and heavy metal stress in plants: Evidence for common signals. *FEBS*, **566**: 1 – 5.

Murphy, A., Taiz, L. 1995. Comparison of metallothioneins gene expression and non – protein thiols in ten *Arabidopsis* ecotypes. Correlation with copper tolerance. *Plant Physiol.,* **113**: 1293 – 1301.

Milone, M.T., Sgherri, C., Clijsters, H., Navari – Izzo, F2003. Antioxidative responses of wheat treated with realistic concentration of cadmium. *Env. Exp. Bot.*, **50**: 265 – 276.

Neiboer, E., Richardson, D.H.S. 1980. The replacement of non – descript term << heavy metals >> by a biologically and chemically significant classification of metal ions. *Env. Pollut. Ser.,* **B1**: 3 – 26.

Nellesson, H., Fletcher, J.S. 1993. Assessment of published literature on the uptake, accumulation and translocation of heavy metals in vascular plants. *Chemosphere,* **9**: 1669 – 1680.

Noctor, G., Arisi, A., Jouanin, L., Kunert, K., Renenberg, H., Foyer, C. 1998. Glutathione biosynthesis, metabolism and relationship to stress tolerance explored in transformed plants. *J. Exp. Bot,.* **49**: 623 – 647.

Noctor, G., Foyer, C.H. 1998. Ascorbate and glutathione: keeping active oxygen under control. *Ann. Rev. Plant Physiol. Mol. Biol.,* **49**: 249 – 279.

Panda, S. K., Parta, H.K. 1997. Physiology of chromium toxicity in plants – A review. *Plant Physiol. Biochem.*, **24**(1): 10-17.

Panda, S.K., Patra, H.K. 2000. Dose chromium (III) produces oxidative damage in excised wheat leaves? *J. Plant Biol.,* **27**: 105 – 110.

Panda, S.K. 2002. The biology of oxidative stress in green cells. A Review – In: Panda, S.K. (ed.), *Advances in Stress Physiology of Plants*. Pp. 1 – 13. Scientific Publisher, Jhodpur, India.

Panda, S.K., Chaudhury, I., Khan, M.H. 2003a. Heavy metal induced lipid peroxidation and affects antioxidants in wheat leaves. *Biol. Plant.,* **42**(6): 289 – 294.

Panda, S.K., Singha, L.B., Khan, M.H. 2003b. Does aluminum phytotoxicity induces oxidative stress in green gram (*Vigna radiata*)? – *Bulg. J. Plant Physiol.*, **29**: 77 – 86.

Panda, S.K. 2003. Heavy metal phytotoxicity induces oxidative stress in moss *Taxithelium* sp. *Curr. Sci.*, **84**(5): 631 – 633.

Panda, S. K., Choudhury, S. 2005a. Chromium stress in plants. *Braz. J. Plant Physiol.* **17**(1): 95-102.

Panda, S.K., Choudhury, S. 2005b. Changes in nitrate reductase activity and oxidative stress in the moss *Polytrichum commune* subjected to chromium, zinc and copper phytotoxicity *Braz. J. Plant Physiol.*, **17**(2): 191-197.

Pietrini, F., Iannaelli, M.A., Pasqualini, S., Massacci, A. 2003. Interaction of cadmium with glutathione and photosynthesis in developing leaves and chloroplast of *Phragmites australis* (Cav.) Trin. ex. Steudel. *Plant Physiol.,* **133**: 829 – 837.

Pilon – Smits, E., Pilon, M. 2002. Phytoremedation of metals using transgenic plants. *Crit. Rev. Plant Sci.,* **21**: 539 – 566.

Pinto, E., Sigaud – Kutner, T.C., Leitao, M.A.S., Okamato, O.K., Morse, D., Colepicolo, P. 2003.Heavy metal induced oxidative stress in algae. *J. Phycol.*, **39**: 1008 – 1018.

Piqueras, A., Olmos, E., Martinez – Solano, J.R., Hellin, E. 1999. Cd induced oxidative burst in tobacco BY2 cells: time course subcellular localization and antiuoxidant response – *Free Rad. Res.*, **31**: 33 – 38.

Prasad, K.V.S.K., Pardha Saradhi, P., Sharmila, P. 1999. Concerted action of antioxidant enzymes and curtailed growth under zinc toxicity in *Brassica juncea*. *Env. Exp. Bot.* **42**: 1 – 10.

Pufahl, R.A., Singer, C.P., Peariso, K.L., Lin, S.J., Schmidt, P.J., Fahrni, C.J., Culotta, V.C., Penner – Hahn, J.E., O' Halloran, T.V. 1997. Metal ion chaperon function of soluble Cu (I) receptor ATX1. *Science,* **278**: 853 – 856.

Quartacci, M.F., Cosi, E., Navari – Izzo, F. 2001. Lipids and NADPH superoxide production in plasma membrane vesicles from roots of wheat grown under copper deficiency or excess. *J. Exp. Bot* ., **52**: 77 – 84.

Rangel, Z.: Uptake of aluminium by plant cells. *New Phytol.,* **134**: 389 – 406.

Rauser, W.E. 1999. Structure and function of metal chelators produced by plants: the case of organic acids, amino acids, phytin and metallothioneins. *Cell Biochem. Biophys.*, **31**: 19 – 48.

Richards, K.D., Schott, E.J., Sharma, Y.K., Davis, K.R., Gardener, R.C. 1998. Aluminum induces oxidative stress genes in *Arabidopsis thaliana*. *Plcnt Physiol.*, **116**: 409 – 418.

Romos, I., Esteban, E., Lucena, J.J., Garate, A. 2002. Cadmium uptake and subcellular distribution in plants of *Lacotica* sp. Cd – Mn interaction. *Plant Sci.,* **162**: 761 – 767.

Saier, M.H. 2000. A functional phylogenetic classification system for transmembrane solute transporter. *Microbiol. Mol. Biol. Rev.*, **107**: 1293 – 1301.

Salt, D.E., Blaylock, M., Kumar, P.B.N., Dushenkov, E., Ensley, B.B., Chet, I., Raskin, I. 1995. Phytoremedation: A noble strategy for the removal of toxic metals from the environment using plants. *Biotechnology,* **13**: 468 – 474.

Sandalio, L.M., Dalurozo, H.C., Gomez, M., Romero – Puertas, M.C., del Rio, L.A. 2001. Cadmium induced changes in growth and oxidative metabolism in pea plants. *J. Exp. Bot.* **52**: 2115 – 2126.

Santi di Toppi, L., Giabbrielli, R. 1999. Response to cadmium in higher plants. *Env. Exp. Bot.* **41**: 105 – 130.

Schat, H., Llugany, M., Bernhard, R. 2000. Metal specific patterns of tolerance, uptake and transport of heavy metals in hyperaccumulating and non – hyperaccumulating mettalophytes – In: Terry, N., Banuelos, G. (ed.). Phytoremedation of contaminated soil and water. Pp. 171 – 188, CRC Press LLC.

Schutzendubel, A., Polle, A. 2001. Plant responses to abiotic stress: Heavy metal induced oxidative and protection by mycorhization. *J. Exp. Bot.*, **53**(372): 1351 – 1365.

Schutzendubel, A., Schwanz, P., Teichmann, T., Gross, K., Langenfeld – Heyser, R., Stohs, S. and Bagchi, D. 1995. Oxidative mechanism in toxicity of metal ions. *Free Rad. Biol. Med.*, **18**: 321 – 336.

Sharma, P. and Dubey, R.S. 2005. Lead toxicity in plants. *Braz. J. Plant Physiol.*, **17**(1): 35-52.

Shi, X. and Dalal, N.S. 1990. Evidence for Fenton type mechanism for generation of OH radicals in the reduction of Cr (VI) in cellular media. *Arch. Biochem. Biophys.*, **281**: 90 – 95.

Shivaguru, M. and Horst, W.J.: The distal part of the transition zone is the most aluminium sensitive apical root zone of maize. *Plant Physiol.*, **116**: 155 – 163.

Silver, S. 1996. Bacterial resistance to toxic metal ions – a review. *Gene*, **179**: 9 – 19.

Smirnoff, N. 2000. Ascorbic acid: metabolism and function of a multi – facetted molecule – *Curr. Opion. Plant Biol.*, **3**: 229 – 235.

Somashekariah, B., Padmaja, K. and Prasad, A. 1992. Phytotoxicity of cadmium ions on germinating seedlings of mung beans (*Phaseolus vulgaris*) involvement of lipid peroxides in chlorophyll poverty. *Physiol. Plant.*, **85**: 85 – 89.

Stange, J. and Mcnair, M.R. 1991. Evidence for a role for the cell membrane in copper tolerance of *Mimulus guttatus* Fischer ex. DC – New Phytol. 119: 383 – 388.

Strile, M., Kolar, J., Selih, V.H., Kocar, D. and Pihlar, B.: A comparative study of several transition metals in Fenton like reaction system at circum – neutral pH. *Acta Chin. Slov.*, **50**: 619 – 632.

Sugimoto, M. and Sakamoto, W. 1997. Putative phospholipid hydroperoxide glutathione peroxidase gene from *Arabidopsis thaliana* induced by oxidative stress. *Genes Gent. Syst.*, **72**: 311: 316.

Tam, P.C.F. 1995. Heavy metal tolerance by ectomycorrhizal fungi and metal amelioration by *Pisolothus tincotrius. Mycorrhiza,* **5**: 181 – 187.

Thomine, S., Wang, R., Ward, J.M., Crawford, M.M. and Schroeder, J.I. 2000. Cadmium and iron transport by members of plant metal transporter family in *Arabidopsis* with homology to *Nramp* genes. *Proc. Nat. Acad. Sci.*, USA, **97**: 4997 – 4996.

Thompson, J.E., Ledge, R.L. and Barber, R.F. 1987. The role of free radical in senescence and wounding. *New Phytol.*, **105**: 317 – 344.

Van Assche, F. and Clijsters, H. 1986. Inhibition of photosynthesis in *Phaseolus vulgaris* by treatment with toxic concentrations of zinc effect on ribulose – 1,5 – bisphosphate carboxylase oxygenase. *J. Plant Physiol.*, **125**: 355 – 360.

Vassilev, A., Berova, M. and Zlatev, Z. 1998. Influence of Cd $^{+2}$ on growth, chlorophyll content and water relations in young barley plants. *Biol. Plant.*, **41**(4): 601 – 606.

Vassilev, A., Lidon, F.C., Ramalho, J.C., do Ceu Matos, M. and da Graca, M. 2004a. Shoot cadmium accumulation and photosynthetic performance of barley plants exposed to high cadmium concentration. *J. Plant Nutr.*, **27**(5): 773 – 793.

Veena, Reddy, V.S. and Sopory, S.K. Glyoxalase I from *Brassica juncea*: molecular cloning, regulation and its overexpression confer tolerance under transgenic tobacco under stress. *Plant J.*, **17**: 385 – 395.

Verma, S. and Dubey, R.S. 2003. Lead toxicity induces lipid peroxidation and alters the activities of antio-xidant enzymes in growing rice plants. *Plant Sci.*, **164**(4): 645 – 655.

Vitorello, V.A., Capaldi, F.R. and Stefunato, V.A.: Recent advances in aluminum toxicity and resistance in higher plants. *Braz. J. Plant Physiol.*, **17**(1): 129-143.

Weast, R.C. 1984. CRC handbook of chemistry and physics (64 th eds.), CRC Press, Boca Raton.

Weissenhorn, I., Leyval, C., Begly, G. and Berlthelin, J. 1995. Arbuscular mycorhizal contribution to heavy metal uptake by maize (*Zea mays* L.) in pot culture with contaminated soil. *Mycorrhiza,* **5**: 245 – 251.

Williams, L.E., Pittman, J.K. and Hall, J.L. 2000. Emerging mechanism of heavy metal transport in plants. *Bichemica. Biophys. Acta,* **77803**: 1 – 23.

Wilder, G.F. and Henkel, J. 1979. The effect of divalent cation ion on the activity of $Mg^{+2}$ – depleted ribulose – 1,5 – bisphosphate carboxylase oxygenase. *Planta,* **146**: 223 – 228.

Winterbourn, C.C.: Superoxide dependent formation of hydroxyl radicals in the presence of iron salts a feasible source of hydroxyl radical *in vivo*. *Biochem. J. Lett.*, **205**: 461 – 463.

Yamamoto, Y., Kobayashi, Y. and Mutsumoto, H. 2001. Lipid peroxidation is an early symptom triggered by aluminum, but not the primary cause of elongation inhibition in pea roots. *Plant Physiol.*, **125**: 199-208.

Yamamoto, Y., Rikiishi, S., Chang, Y.C., Ono, K., Kasai, M. and Matsumuto, M. 1994. Quantitative estimation of aluminium toxicity in cultured tobacco cells – correlation between aluminium uptake and growth inhibition. *Plant Cell Physiol.*, **35**: 575-583.

Yan, F., Schubert, S. and Mengel, K. 1992. Effect of low root membrane pH on net proton release, root respiration and root growth of corn (*Zea mays* L.) and broad bean (*Vicia faba* L.). *Plant Physiol.*, **99**: 415 – 421.

Yan, F., Feuerele, R., Schaffer, S., Frotmeier, H. and Schubert, S.: Adaptation of active proton pumping and plasma-lemma ATPase activity of corn roots to low root medium pH. – Plant Physiology., 117: 311-319.

Yreula, I. 2005. Copper in plants. *Braz. J. Plant Physiol.*, **17**(1): 145-156.

Zenk, M.H., Heavy metal detoxification in higher plants – a review. *Gene,* **179**: 21 – 30,

Zhang, J. and Kirkham, M.B. 1996. Drought stress induced changes in activities of superoxide dismutase, catalase and peroxidase in wheat species. *Plant Cell Physiol.*, **35**: 785-791.

Zhang, G., Slaski, J.J., Archambault, D.J. and Taylor, G.J. 1995. Alterations of the plasma membrane lipids in aluminium-resistance and aluminium-sensitive wheat genotypes in response to aluminium stress. *Physiol. Plant*, **99**: 302-308.

*Advances in Plant Physiology*, Vol. 9
Ed. A. Hemantaranjan
Scientific Publishers (India), Jodhpur, 2006 pp. 193-204
E-mail: **info@scientificpub.com** www.scientificpub.com

# 9

# PLANTS FOR HEAVY METAL TOXICITY ASSESSMENT: DUCKWEEDS (*Lemnaceae*)

Klaus-J. Appenroth[1*] and Kavita Shah[2]

[1]University of Jena, Institute of General Botany and Plant Physiology, Dornburger Str. 159, D-07743 Jena, Germany
[2]North Eastern Hill University, Department of Biochemistry, School of Life Sciences, Shillong, India

## INTRODUCTION

Increasing environmental pollution by heavy metals, especially from anthropogenic sources, requires specific and effective environmental control. Because of fluctuating contaminations, in some cases by unexpected toxic compounds, biomonitoring by plants is one way to solve this problem (Whitton and Kelly, 1995; Lewis and Wang, 1997; Prasad and Strazlka 2002). Wetland plants and water plants are in particular best suitable for environmental studies because they can be cultivated directly in wastewater streams containing pollutants (Mohan and Hosetti, 1999). In general most heavy metals induce more or less specific stress responses in plants that can be used to evaluate and monitor the quality of wastewater and the possible presence of heavy metals or other phytotoxic compounds. Several species of duckweeds have already been used for this task (Wang, 1990; Mohan and Hosetti, 1999; Michel *et al.*, 2004).

Duckweeds (*Lemnaceae*) represent a small group of water plants (Fig. 1) consisting of 38 different species (Landolt, 1986; Landolt and Kandeler, 1987; Les *et al.*, 2002). On the basis of recent molecular biological characterisation (Les *et al.*, 2002), duckweeds are now considered to be part of the plant family *Araceae*. Duckweeds posses several advantages that make them specially suitable for toxicity assessment (i) Water plants like duckweeds are not only in direct contact with the potentially contaminated wastewater but also the uptake of ions is carried out by the whole surface of the leafy organs, i.e. the whole plant, and not only by roots (Gorham 1941; Cedergreen and Madsen, 2002, 2003). Therefore, complications by long-distance transport from roots to leaves are avoided and the effect of toxic compounds on cell metabolism can immediately be investigated (ii) Duckweeds have a specific mechanism of propagation (Landolt, 1986). Most species flower rarely under normal physiological conditions. Propagation proceeds mainly through vegetative mode. Pairs of frond primordia being protected by internal pouches, emerge under favourable environmental conditions ("budding"). Each daughter frond (F1) emerging from pouches already contains two new generations of preformed daughter fronds (F2). Through this mechanism of clonal propagation, populations of genetically homogenous plants are formed. (iii) Propagation of duckweed fronds is very fast. Under laboratory conditions, the time for doubling the number of vegetative fronds may be as short as two days following an exponential law. Starting with 10 fronds of *Lemna minor* according to the standardised Lemna test (ISO/ CD 20079), more than 100 fronds are obtained at the end of the seven-day test period. This and other parameters can easily be used in routine laboratories to monitor the quality of water or wastewater. Even soil can be biomonitored this way (Youngman *et al.,* 1998; Schultz *et al.,* 2004). This makes duckweed immediately useful for bioassays.

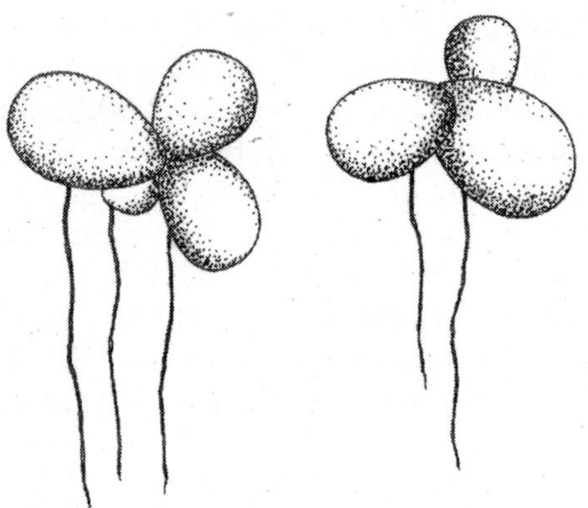


FIGURE: *Lemna minor* L. [Duckweed]

During the fast exponential growth, duckweeds take up mineral elements as well as heavy metals and other (partially toxic) compounds – as long as the extent of the toxic effects does not prevent further vegetative growth of the plants. These properties of duckweeds can be used for phytoremediation of water bodies (Smith and Moelyowati, 2000). In contrast to microorganisms, duckweeds can be removed simply by harvesting the plants.

To employ the above-mentioned advantages of duckweeds in future, more basic research concerning the phytotoxic properties of heavy metals (mechanisms of uptake and accumulation, metabolic effects and tolerance mechanisms) is required. At the same time, applied research concerning duckweed-based wastewater management like harvest technology, post-harvest management etc. is necessary (Smith and Moelyowati, 2000). The present review summarises the progress of heavy metal research using duckweeds during the past few years primarily since 1997.

## 2. PHYTOTOXICITY DUE TO SOME INDIVIDUAL HEAVY METALS

Depending upon the mechanisms of uptake of specific heavy metals, other compounds present simultaneously in wastewater may influence the rate of uptake. It is assumed that arsenate is taken up by a phosphate transporter as shown for *Lemna gibba* (Ullrich-Eberius and Novacky, 1984).

A high-affinity sulphate transporter takes up chromate and a clear competition with sulphate was demonstrated by Kaszycki *et al.,* (2004). Pre-cultivation of *Spirodela polyrhiza* at minimum sulphate concentration enhanced the rate of chromium accumulation. This effect was caused by the increased number of sulphate transporters which transport chromate into cells. A previously developed kinetic model (Appenroth *et al.,* 2000) was extended to include chromate transport and reduction in the presence of competing ions as sulphate (Kaszycki *et al.* 2004). The study concluded that (a) in absence of simultaneous application of sulphate, the transport of Cr(VI) from apoplast into plant cells and the reduction of Cr(VI) to Cr(V) within the apoplast are effectively competing processes with comparable rate constants. (b) The rate constant for reduction of Cr(V) to Cr(III) is much lower within cells than in apoplast showing that Cr(V) is stabilized in the symplast. (c) The rate of transport of Cr(VI) into plant cells is atleast one order of magnitude higher than that of Cr(V) or Cr(III) which might be the reason for the higher toxicity of chromate (cf. Dirilgen and Dogan, 2002). In these experiments, the reduction of Cr(VI) to Cr(V) was measured by L-band EPR spectroscopy. An interaction between the accumulation of chromate and the content of organic matter in wastewater has been demonstrated for *L. gibba* (Boniardi *et al.*, 1999).

The presence of chelators may be of high significance for the uptake of heavy metals. In *S. polyrhiza*, uptake of $Cd^{2+}$ (and most probably several other divalent cations) was inhibited by the chelator EDTA (Srivastava and Appenroth, 1995). Neither the free (i.e. non-chelated) nor the total heavy metal concentration of $Cd^{2+}$ determined the toxicity. Instead, the bio-availability actually depended on the equilibrium between chelator and heavy metal as dynamically influenced by the heavy metal uptake by plants. Analysis of the heavy metal content of *L. minor* grown in the Tiljala wetlands in West Bengal (India) demonstrated uptake and accumulation of $Pb^{2+}$, $Cd^{2+}$, Cr, $Zn^{2+}$, $Cu^{2+}$ and $Hg^{2+}$, most probably from industrial wastes (Mukherjee *et al.*, 2004). To investigate the toxicity of heavy metals of potential relevance to environment, a broad spectrum of duckweed species (Table 1) has been used, e.g. *L. minor* (Dirilgen and Dogan, 2002; Naumann *et al.*, 2004), *L. gibba* (Boniardi *et al.,* 1999; Dirilgen, 2001), *S. polyrhiza* (Appenroth *et al.*, 2003; Pandey *et al.,* 1999), *Lemna trisulca* (Prasad *et al.*, 2001, Maleva *et al.*, 2004), *Lemna valdiviana* (Dirilgen, 2001), *Lemna perpusilla*, *Lemna minuscula*, *Wolffia arrhiza* and *Landoltia punctata* (Xyländer and Augsten, 1992), and *Wolffia globosa* (Boonyapookana *et al.*, 2002). In most of these investigations, *L. minor* has been the species of choice. As a test parameter, inhibition of the multiplication rate (frond number) is most commonly used followed by fresh weight and dry weight (Srivastava and Pandey, 1999; Maleva *et al.*, 2004), and content of chlorophyll and carotenoids (Garnczarska and Ratajczak 2000a; Banaszak *et al.*, 2001). In most of these investigations, micromolar concentrations of most of the heavy metals used were effective to exhibit a stress response (Olkhovych and Smirnova, 1995; Naumann *et al.*, 2004). However, there are also more specific effects, partially related to toxicity and/ or partially related to the syndrome of adaptation and tolerance. $Ni^{2+}$ in the range of 1 to 100 μM not only inhibited growth of *L. paucicostata,* but also decreased the concentration of photosynthetic pigments, sugars and starch and increased the concentration of protein in the plant cells (Singh *et al.*, 1999). Prasad *et al.* (2001) compared the effcts of $Cd^{2+}$ and $Cu^{2+}$ using *L. trisulca* as a test system. $Cu^{2+}$ showed toxic effects at concentrations 1000 fold lower in comparison to $Cd^{2+}$. There occurred an inhibition of total gas exchange (respiration and photosynthesis) correlated with the decay of fluorescence from photosystem II caused due to $Cu^{2+}$ toxicity, however, $Cd^{2+}$ increased the impact of respiration in total gas exchange (Prasad *et al.*, 2001). $Cd^{2+}$ also induced the formation of a series of small molecular weight proteins, interestingly however, this effect was not observed with $Cu^{2+}$ (Prasad *et al.*, 2001). It has been suggested that these proteins represent stress proteins but further investigations are required for its demonstration.

TABLE 1. Broad spectrum of individual heavy metals based on studies by various research groups for metal toxicity assessment using water plants

| Water Plant | Metal | Reference |
|---|---|---|
| *Lemna gibba* | Nickel | Xylander & Augsten 1992 |
| *Lemna minuscula* | Nickel | Xylander & Augsten 1992 |
| *Wolffia arrhiza* | Nickel | Xylander & Augsten 1992 |
| *Landoltia punctata* | Nickel | Xylander & Augsten 1992 |
| *Spirodela polyrhiza* | Cadmium | Srivastava & Appenroth 1995 |
| *Lemna paucicostata* | Nickel | Singh *et al.* 1999 |
| *Lemna minor* | Selenium | Zayed *et al.* 1998 |
| *Lemna valdiviana* | Chromium | Diriligen 2001 |
| *Lemna trisulca* | Copper | Prasad *et al.* 2001 |
| *Wolffia globosa* | Cadmium, Chromium | Boonyapookana *et al.* 2002 |
| *Lemna minor* | Lead, Zinc, Mercury | Dirilgen & Dogan 2002 |
| *Spirodela polyrhiza* | Chromate | Kaszycki *et al.* 2004 |

For the toxic effects of chromate, reduction of Cr(VI) via Cr(V) finally to Cr(III), seems to be essential (Hunter *et al.*, 1997; Appenroth *et al.*, 2000). Analysis of plant cells by transmission electron microscopy and chemical analysis revealed the accumulation of starch grains in the chloroplasts following the application of chromate at low concentrations or for short periods (Appenroth *et al.*, 2003). An increase in the chromate concentration resulted in the disappearance of most of the starch grains. It has been suggested that this transient accumulation of starch is caused, first, by inhibition of the export of carbohydrates out of the plastids, and then by inhibition of photosynthesis. Chromate decreased the chlorophyll content and the chlorophyll $a/b$ ratio suggesting photosynthesis to be clearly an important target of chromate toxicity. The quantitative analysis of the chlorophyll-protein complexes showed that the photosystem II (core complex as well as connecting antenna) is more sensitive to chromate treatment than photosystem I and the outer light-harvesting complex of photosystem II (Appenroth *et al.*, 2003). This explains the previous results on time-resolved chlorophyll a fluorescence (Appenroth *et al.*, 2001). Electron microscopic analysis showed damage to several membrane systems as that of thylakoids, chloroplast envelope, plasmalemma and at higher concentrations, that of tonoplast and mitochondria. Thus, the membrane system is another target of chromate toxicity (Appenroth *et al.*, 2003).

Artetxe *et al.* (2002) were evidently surprised by the fact that *L. minor* showed lower toxicity effects when plants were cultivated in low light conditions. Perhaps, this depends on the formation of reactive oxygen intermediates under high light conditions. Formation of reactive oxygen species is a mechanism of heavy metal toxicity, e. g. induced by copper (Babu *et al.*, 2001). Several so-called antioxidant enzymes, capable of removing reactive oxygen intermediates or to regenerate other redox compounds, were induced like superoxide dismutase, peroxidase (Xu *et al.*, 2002), glutathione reductase, catalase, guaiacol peroxidase, pyrogallol peroxidase, and ascorbate peroxidase (Garnczarska and Ratajczak, 2000b;Teisseire and Guy, 2000). Inducibility of catalase and guaiacol peroxidase was highly significant with the application of $Cu^{2+}$ to *L. minor* and resulted in an increase of approximately 260% (Teisseire and Guy, 2000). The concentration of glutathione and ascorbate is often decreased and shown to be present rather in oxidised form (Teisseire and Vernet, 2000). The concentrations of thiol compounds were significantly enhanced in *S. polyrhiza* by $Ni^{2+}$ (Pandey *et al.*, 1999).

## 3. INTERACTION OF HEAVY METAL-INDUCED PHYTOTOXICITY WITH OTHER HEAVY METALS / TOXIC COMPOUNDS

When the toxicity of heavy metals has to be investigated under natural conditions, often several heavy metals together with other phytotoxic compounds are present. Possible interactions concerning uptake and toxicity have to be taken into consideration, both with respect to duckweed-based biomonitoring and for use of duckweeds in phytoremediation of wastewater. Several investigations were published which showed, unfortunately, to some degree contradictory results. This might be due to the intervals at which various concentrations were applied as well as by the test parameter used for evaluation.

Neither synergistic nor antagonistic effects were found upon investigating the uptake of $Pb^{2+}$ and $Ni^{2+}$ in *L. minor* (Axtell *et al.,* 2003). In *S. polyrhiza*, $Cu^{2+}$ had an antagonistic effect on $Cd^{2+}$ and reduced its absorption. $Cd^{2+}$, however, increased the absorption of $Cu^{2+}$ (Saadi *et al.,* 2002). A large number of morphological and biochemical parameters were used by Xu *et al.* (2002) to investigate the interaction of $Hg^{2+}$ and $Cd^{2+}$. According to the author's opinion, synergistic phytotoxic effects were detected. The number of different concentrations used in these experiments, however, was very small. The discharge of wastewater from a nuclear power plant in Ignalina, Lithuania induced investigations concerning possible interactions of heavy metals found, i.e. Cu, Cr, Cd, Ni, Mn, Zn and Pb (Marciulioniene *et al.,* 1998; Montvydiene *et al.,* 2000). A model mixture was produced and applied to several test organisms including *S. polyrhiza*. In all cases, single heavy metals were less toxic than those in the model mixture and their phytotoxic effects were found to be "more than additive". These results were found for several test organisms regardless of the phylogenetic level and ontogenesis (Marciulioniene *et al.,* 2002; Kazlauskiene *et al.,* 2003). A more advanced statistical approach was applied and concluded that the "additive toxicity mode" was more suitable than the "mixture toxicity index" to evaluate the experimental results (Montvydiene and Marciulioniene, 2004). Dirilgen and co-workers from the University of Bebek-Istanbul, Turkey presented most thorough investigations concerning the interaction of binary mixtures of heavy metals (Ince *et al.*, 1999; Dirilgen, 2001; Dirilgen and Dogan, 2002). Mainly *L. minor* was used to test the effects of $Cu^{2+}$, $Zn^{2+}$, $Co^{2+}$, Cr(III) and Cr(VI) and the data were subjected to careful statistical analysis. From all the data taken they concluded that in the majority of combinations of the two heavy metals, interactions were of antagonistic nature. Additive toxicity was the next frequent interaction, though much less often observed than antagonistic effects. Synergism was found to be totally unlikely in duckweeds within the test combinations used. Differences in conclusions from other groups might have appeared by the application of inadequate statistical methods.

Heavy metals and polycyclic aromatic hydrocarbons are often co-contaminants in industrial wastewater. Using *L. gibba* it has been shown that anthraquinone and $Cu^{2+}$ had synergistic effects on the photosystem II fluorescence, photosystem I activity and electron transport. The authors claimed that similar effects could be expected with other redox active heavy metals (Babu *et al.,* 2001). A fungicide, diethyldithiocarbamate, largely inhibited the absorption of $Cd^{2+}$ and $Cu^{2+}$ by *S. polyrhiza*. It has been concluded that the fungicide has a complexant effect on both heavy metals absorption and uptake and therefore reduce the toxicity due to metal by preventing the uptake process (Saadi *et al.,* 2002).

### 3.1 Biomonitoring

Principally, any of the above mentioned phytotoxic effects of heavy metals on any of the 38 duckweed species (cf. Xyländer and Augsten, 1992; Tsatsenko and Malyuga, 2000) could be used for biomonitoring heavy metals in potentially contaminated water. In fact, many duckweed-based bioassays have been described. Among these, there are vital staining of surviving cells (Tsasenko and Malyuga, 1998), conductometric measurements of the leakage from membrane system (Eich *et al.,* 1997), detection of esterase isoenzymes for heavy metal differentiation (Mukherjee *et al.,* 2004), induction of peroxidases (Garnczarska and Ratajczak, 2000b; Mohan and Hosetti, 1997; Teisseire and Guy, 2000), measurement of cytoplasmic malate dehydrogenase activity (Garnczarska and Ratajczak, 2000a), detecting concentration of thiol groups (Pandey *et al.*, 1999), and concentration of ascorbate and glutathione (Teisseire and Vernet, 2000). Li and Xiong (2004) reported a novel phytotoxic response to $Cu^{2+}$, $Cd^{2+}$, $Ni^{2+}$, $Zn^{2+}$, $Hg^{2+}$ and Cr, i.e. destruction of the stipes which connects mother frond to daughter frond resulting in separated fronds rather than frond colonies (described by the authors as "release daughter fronds from the mother fronds before maturity"). This process could also be used for biomonitoring.

Duckweed-based bioassays have already been successfully applied to demonstrate the sufficient high sensitivity to follow the process of composting oily waste (Juvonen *et al.*, 2000), to test eluate and leachate samples from smelter soils (Youngman *et al.*, 1998), or effluents from nuclear power plant (Montvydiene *et al.,* 2000). Investigation of the water quality of an illegal discharge of untreated wastewater, from the Taichung Industrial Park, Taiwan was carried out using duckweed toxicity test, which resulted in 60 to 100 % growth inhibition of *L. aequinoctialis* (Shyu *et al.*, 2002). Fomin *et al.* (2000) showed that duckweeds could be used to monitor extremely acidic mining leachates in Germany. Even soil can be investigated by the **Lemna test** following suitable methods of extraction (Schultz *et al.,* 2004). Although many of the duckweeds species have been applied for biomonitoring, *L. minor* and to a certain degree *S. polyrhiza* clearly dominate the field. The latter would have the advantages to possess additional stress reactions (accumulation of anthocyanin, turion formation, and root response) but its bigger size makes it slightly inferior to *L. minor* for routine experiments.

Besides the sensitivity of the test system, the bioassay should be simple and easy to carry out under routine conditions. In the last decade a standardization group has developed a detailed protocol how to use duckweeds for biomonitoring. In May 2004, a final ring test was organised on the basis of the ISO protocol in preparation (ISO/ CD 20079). *L. minor* has been selected as the species of choice. Although no specific clone was named to be used in bioassays the use of one defined clone is strongly recommended because of the known clonal differences in the heavy metal sensitivity (Kiss *et al.,* 2001). *Lemna minor* St is the oldest existing clone isolated in 1946 in Marburg, Germany (Pirson and Seidel, 1950). This clone has been already successfully used for several investigations. Further main conditions defined in the test protocol are (cf. Naumann *et al.,* 2004): temperature (25°C), time of treatment (7 days), light (100 $\mu$mol $m^{-2}$ $s^{-1}$ continuous white light) and the use of a specific nutrient medium, mainly on the basis of Steinberg (1946). As test parameter mainly, the multiplication rate (number of fronds) is suggested which also happens to be the parameter used by most research groups for duckweed-based bioassays. Other yield parameters that follow the exponential law like fresh weight, dry weight, chlorophyll content or carotenoid content can also therefore be used as additional parameters to increase the specificity of the assay (cf. Fomin *et al.*, 2000; Appenroth *et al.,* 2003; Naumann *et al.,* 2004). All these parameters are suitable to be used in routine experiments for biomonitoring. Moreover, Automatic Picture Analysis offers additional advantages and is now commercially available in its application for the so-called **Lemna test** (Lemnatec, Würselen, Germany). This permits the automatic measurement of number of fronds, individual frond area (which correlates to the fresh weight) and colour index (which correlates to the pigment content) and makes the statistical evaluation easier. An interesting alternative to the standardised Lemna test has been suggested by Olkhovych and Smirnova (1995). These authors showed that in some cases the chlorophyll content could be used as a parameter following a treatment with heavy metals for only 24 hours.

### 3.2 Phytoremediation

The biological basis for the use of duckweeds in phytoremediation is similar to that for biomonitoring: the fast exponential growth of the plants. Increasing the biomass results in uptake of nutritional elements from water, with almost unavoidable uptake of heavy metals. This might be because of limited specificity of the transporter. The most fundamental limitations under applied aspects are the temperature and the toxicity of heavy metal itself. The optimal temperature for the growth of vegetative fronds is between approximately 25°C and 30°C. This restricts the use of duckweeds for phytoremediation in the temperate climate zone to a certain vegetative period because in most cases growth ceases almost completely between 10°C and 15°C (Landolt and Kandeler, 1987). This restriction is without significance in warm countries. The toxicity of heavy metals itself restricts the use of duckweeds (as any other biological system). When the concentration of toxic compounds is too high, growth is inhibited and no further uptake is possible.

*L. minor* has been commonly used for phytoremediation of wastewater, e.g. for $Ni^{2+}$ (Axtell *et al.,* 2003; Kara *et al.,* 2003), chromate (Hasar and Obek, 2001), radioactive isotopes (Cecal *et al.,* 1999; Cecal and Popa, 2001), $Cd^{2+}$ (Hasar and Obek, 2001; Wang *et al.,* 2002) and $Pb^{2+}$ (Axtell *et al.,* 2003; Rahmani and Sternberg, 1999; Xin and Ma, 2003). However, also *S. polyrhiza* (Rai *et al.,* 1995; Sharma and Gaur, 1995) and other species like *L. gibba, L. paucicostata,* have been used in several experiments (Singh and Jaiswal, 1997; Boniardi *et al.,* 1999). The effluent of a paper mill, containing Cu, Zn, Mn and Fe was purified using *S. polyrhiza* by Srivastava and Pandey (1999). From the exogenously applied metal ~76% $Pb^{2+}$ and from ~82% $Ni^{2+}$ were removed by *L. minor* (Axtell *et al.,* 2003). The same species showed a factor of bio-enrichment for $Pb^{2+}$ of 858 in wastewater lagoons (Xin and Ma, 2003) and hyper accumulation of $Cd^{2+}$ (14 g $kg^{-1}$; Wang *et al.,* 2002). Zayed *et al.* (1998) measured accumulation factors in *L. minor* over a broad range of concentrations (0.1 to 10 mg $L^{-1}$) and obtained the following results (g $kg^{-1}$ duck-weed): 13 Cd, 4.3 Se, 3.4 Cu, 2.9 Cr, 1.8 Ni and 0.6 Pb. On the basis of these data, the authors concluded that duckweeds are good accumulators of Cd, Se, and Cu, moderate accumulators of Cr, and poor accumulators of Ni and Pb. However, the results of phytoremediation depend to a large degree on harvesting time and harvesting rate (Hasar and Obek, 2001), light condition, pH and the presence of chelating agents (Singh and Jaiswal, 1997), and the amount of organic matter dissolved in wastewater (Boniardi *et al.,* 1999). As there are known clonal differences (Kiss *et al.,* 2001), different species might result in different effectiveness. *L. aequinoctialis* and *S. polyrhiza* might be promising candidates beside *L. minor* owing to their robust and fast growth. Yamamoto *et al.* (2001) have shown that it is possible genetically to transform *L. minor* and *L. gibba*. Using genes that enhance heavy metal tolerance of duckweeds or increase the amount of accumulated heavy metals might be a promising way in the near future.

Besides scientific problems, practical problems like wastewater management and duckweed harvesting technology have to be taken into consideration (Smith and Moelyowati, 2000). The duckweeds harvested from eutrophic water without heavy metal contamination (e.g. municipal sewage) can be used to feed animals (or even as human diet) and it can also be used as green fertiliser. Both applications, however, are excluded if heavy metals are present. The high concentration factors of the heavy metals in duckweeds, however, make the duckweed-based water purification system even more attractive but in such cases the harvested plants have to be treated as heavy metal waste.

## REFERENCES


Appenroth, K-J., Bischoff, M., Gabrys, H., Stöckel, J., Swartz, H.M., Walczak, T., Winnefeld, K. 2000. Kinetics of chromium(V) formation and reduction in fronds of the duckweed *Spirodela polyrhiza* - a low-frequency EPR study. *Journal of Inorganic Biochemistry* **78**, 235-242.

Appenroth, K.-J., Keresztes, A., Sarvari, E., Jaglarz, A., Fischer, W. 2003. Multiple effects of chromate on *Spirodela polyrhiza*: Electron microscopy and biochemical investigations. *Plant Biology* **5**, 315-323.

Appenroth, K.-J., Stockel, J., Srivasatva, A., Strasser, R.J. 2001. Multiple effects of chromate on the photosynthetic apparatus of *Spirodela polyrhiza* as probed by direct, time-resolved chlorophyll a fluorescence measurements. *Environmental Pollution* **115**, 49-64.

Artetxe U., Garcia-Plazaola, J.I., Hernandez, A., Becerril, J.M. 2002. Low light grown duckweed plants are more protected against the toxicity induced by Zn and Cd. *Plant Physiology and Biochemistry* **40**, 859-863.

Axtell, N.R., Sternberg, S.P.K., Claussen, K. 2003. Lead and nickel removal using Microspora and *Lemna minor*. *Bioresource Technology* **89**, 41-48.

Babu, T.S., Marder, J.B., Tripuranthakam, S., Dixon, D.G., Greenberg, B.M. 2001. Synergistic effects of a photooxidized polycyclic aromatic hydro-carbon and copper on photosynthesis and plant growth: Evidence that *in vivo* formation of reactive oxygen species is a mechanism of copper toxicity. *Environmental Toxicology and Chemistry* **20**, 1351-1358.

Banaszak, A., Napieralska, A., Wozny, A. 2001. Inhibition of greening of *Lemna minor* by $Pb^{2+}$. *Biologia* **56**, 111-116.

Boniardi, N., Rota, R., Nano, G. 1999. Effect of dissolved metals on the organic load removal efficiency of *Lemna gibba*. *Water Resources* **33**, 530-538.

Boonyapookana, B., Upatham, E.S., Kruatrachue, M., Pokethitiyook, P., Singhakaew, S. 2002. Phytoaccumulation and phytotoxicity of cadmium and chromium in duckweed *Wolffia globosa*. *International Journal of Phytoremediation* **4**, 87-100.

Cecal, A., Palamaru, I., Popa, K., Caraus, I., Rudic, V., Gulea, A. 1999. Accumulation of Co-$60^{2+}$ and $UO_2^{2-}$ ions on hydrophyte plants. *Isotopes and Environmental Health Standards* **35**, 213-219.

Cecal, A.and Popa, K. 2001. Tl-(204)(+) ions adsorption from the low radioactivity solutions on *Lemna minor*. *Revista de Chimie* **52**, 382-385.

Cedergreen N. and Madsen TV. 2002 Nitrogen uptake by the floating macrophyte *Lemna minor*. *New Phytologist* **155**, 285-292.

Cedergreen N. and Madsen TV. 2003 Light regulation of root and leaf $NO_3^-$ uptake and reduction in the floating macrophyte *Lemna minor*. *New Phytologist* **161**, 449-457.

Dirilgen, N. 2001. Accumulation of heavy metals in freshwater organisms: Assessment of toxic interactions. *Turkish Journal of Chemistry* **25**, 173-179.

Dirilgen, N. and Dogan, F. 2002. Speciation of chromium in the presence of copper and zinc and their combined toxicity. *Ecotoxicology and Environmental Safety* **53**, 397-403.

Eich, J., Steger-Hartmann, T., Wagner, E. 1997. Use of conductivity tests for study of the toxicity of various membrane-active substances as well as wastewater testing. *Projekt "Angewandte Oekologie"* **22**, 257-268.

Fomin, A., Moser, H., Pickl, C. 2000. Ecotoxicological investigations of extremely acid mining lakes using bio-assays suitable for testing at low pH. *Toxicology and Environmental Chemistry* **76**, 237-254.

Garnczarska, M. and Ratajczak, L. 2000a. Metabolic responses of *Lemna minor* to lead ions – I. Growth, chlorophyll level and activity of fermentative enzymes. *Acta Physiologia Plantarum* **22**, 423-427.

Garnczarska, M. and Ratajczak, L. 2000b. Metabolic responses of *Lemna minor* to lead ions – II. Induction of antioxidant enzymes. *Acta Physiologia Plantarum* **22**, 429-432.

Gorham, P. R. 1941. Measurement of the response of *Lemna* to growth promoting substances. *American Journal of Botany* **28**, 98-101.

Hasar, H. and Obek, E. 2001. Removal of toxic metals from aqueous solution by duckweed (*Lemna minor* L.): Role of harvesting and adsorption isotherms. *Arabian Journal for Science and Engineering* **26**(2C), 47-54.

Hunter, D.B., Bertsch, P.M., Kemner, K.M., Clark, S.B. 1997. Distribution and chemical speciation of metals and metalloids in biotica collected from contaminated environments by spatially resolved XRF, XANES and EXAFS. *Journal de Physique IV*, **7**(C2, Vol. 2), 767-771.

Ince, N. H., Dirilgen, N., Apikyan, I. G., Tezcanli, G., Ustun, B. 1999. Assessment of toxic interactions of heavy metals in binary mixtures: a statistical approach. *Environmental Contamination and Toxicology* **36**, 365-372.

ISO/ CD 20079 Water quality – determination of the toxic effect of water constituents and wast water to duckweed (*Lemna minor*) – Duckweed growth inhibition test. *ISO TC* **147**/ SC 5/ WG 5, 2001

Juvonen, R., Martikainen, E., Schultz, E., Joutti, A., Ahtiainen, J., Lehtokari, M. 2000. A battery of toxicity tests as indicators of decontamination in composting oily waste. *Ecotoxicology and Environmental Safety* **47**, 156-166.

Kara, Y., Basaran, D., Kara, I., Zeytunluoglu, A., Genc, H. 2003. Bioaccumulation of nickel by aquatic macrophyta *Lemna minor* (Duckweed). *International Journal of Agriculture and Biology* **5**, 281-283.

P. Kaszycki, H. Gabry•, K.-J. Appenroth, A. Jaglarz, S. S'dziwy, T. Walczak & H. Koloczek 2004. Exogenously applied sulfate as a tool to investigate transport and reduction of chromate in the duckweed *Spirodela polyrhiza. Plant Cell Environment,* in press.

Kazlauskiene, N., Marciulioniene, D., Montvydiene, D., Svecevicius, G., Vosyliene, M.-Z. 2003. Comparative studies of the toxic effects of heavy metal model mixture on organisms of different phylogenetic level and ontogenesis. *Environmental and Chemical Physics* **25**, 116-122.

Kiss, I., Kovats, N., Szalay, T. 2001. Role of environmental factors on the reproducibility of *Lemna* test. *Acta Biologia Hungarian* **52**, 179-185.

Landolt E. 1986. The family of *Lemnaceae* - a monographic study. Vol. 1. Biosystematic Investigations in the Family of Duckweeds (*Lemnaceae*). Veröffentlichungen des Geobotanischen Instutes der ETH, Zürich.

Landolt E. and Kandeler R. 1987. The family of *Lemnaceae* - a monographic study. Vol. 2. Biosystematic Investigations in the Family of Duckweeds (*Lemnaceae*). Veröffentlichungen des Geobotanischen Instutes der ETH, Zürich.

Les, D.H., Crawford, D.J., Landolt, E., Gabel, J.D., Kimball, R.T. 2002. Phylogeny and systematics of Lemnaceae, the duckweed family. *Systematic Botany* **27**, 221-240.

Lewis, M. A. and Wang W. 1997. Water quality and aquatic plants. In: Plant for Environmental Studies (Wang, W., Gorsuch, J.W., Hughes J.S. eds.) CRC Lewis Publisher, Boca Raton, New York. pp 141-175.

Li, T. Y., Xiong, Z.T. 2004. A novel response of wild-type duckweed (*Lemna paucicostata* Hegelm.) to heavy metals. *Environmental Toxicology* **19**, 95-102.

Maleva, M.G., Nekrasova, G.F., Bezel, V.S. 2004. The response of hydrophytes to environmental pollution with heavy metals. *Russian Journal of Ecology (Ekaterinburg)* **35**, 230-235.

Marciulioniene, D., Lakacauskiene, R., Montvydieni, D., Kazlauskiene, N., Svecevicius, G., Burba, A. 1998. Effect of Ignalina nuclear power plant wastewater and model heavy-metal mixtures. *Ekologija* **4**, 53-59.

Marciulioniene, D., Montvydiene, D., Kazlauskiene, N., Svecevicius, G. 2002. Comparative analysis of the sensitivity of test-organisms of different phylogenetic level and life stages to heavy metals. *Environmental and Chemical Physics* **24**, 73-78.

Michel A., Johnson R. D., Duke S.O. and Scheffler B. E. 2004. Dose-response relationships between herbicides with different modes of action and growth of *Lemna paucicostata*: An improved ecotoxicological method. *Environmental Toxicology and Chemistry* **23**, 1074-1079.

Mohan, B.S. and Hosetti, B.B. 1997. Potential phytotoxicity of lead and cadmium to *Lemna minor* grown in sewage stabilization ponds. *Environmental Pollution* **98**, 233-238.

Mohan, B.S. and Hosetti, B.B. 1999. Aquatic plants for toxicity assessment. *Environmental Research Station A* **81**, 259-274.

Montvydiene, D., Lakacauskiene, R., Marciulioniene, D. 2000. Assessment of the toxicity of heavy metals in model and natural mixtures of higher plants. *Botanica Lithuanica* **6**, 281-287.

Montvydiene, D., Marciulioniene, D. 2004. Assessment of toxic interactions of heavy metals in a multicomponent mixture using *Lepidium sativum* and *Spirodela polyrrhiza*. *Environmental Toxicology* **19**, 351-358.

Mukherjee, S., Mukherjee, S., Bhattacharyya, P., Duttagupta, A. K. 2004. Heavy metal levels and esterase variations between metal-exposed and unexposed duckweed *Lemna minor*: field and laboratory studies. *Environment International* **30**, 811-814.

Naumann, B., Mattias, E., Appenroth, K.J. 2004. Heavy metal test assays using *Lemna minor* L., clone St on the basis of the ISO International Standard CD 20079 Water Quality. *Plant Cell Environment,* submitted.

Olkhovych, O.P. and Smirnova, N.N. 1995. The effect of heavy metals on the content of pigments in the higher aquatic plants. *Ukrains'kii Botanichnii Zhurnal* **52**, 213-219.

Pandey, S., Asthana, R. K., Kayastha, A. M., Singh N., Singh, S. P. 1999. Metal uptake and thiol production in *Spirodela polyrhiza* (L.) SP20. *Journal of Plant Physiology* **154**, 634-640.

Pirson, A. and Seidel F. 1950. Zell- und Stoffwechselphysiologische Untersuchungen an der Wurzel von *Lemna minor* L. unter besonderer Berücksichtigung von Kalium- und Kalziummmangel. *Planta* **38**, 431-473.

Prasad, M.N.W., Malec, P., Waloszek, A., Bojko, M., Stralka, K. 2001. Physiological responses of *Lemna trisulca* L. (duckweed) to copper bio-accumulation. *Plant Science* **161**, 881-889.

Prasad, M.N.V., and Strzalka, K. 2002. Physiology and Biochemistry of Metal Toxicity and Tolerance in Plants. Kluwer Academic Publisher, Dordrecht, Boston, London, 2002.

Rahmani, G.N.H., Sternberg, S.P.K. 1999. Bioremoval of lead from water using *Lemna minor*. *Bioresource Technology* **70**, 225-230.

Rai, U.N., Sinha, S., Tripathi, R.D., Chandra, P. 1995. Waste-water treatability potential of some aquatic macrophytes – Removal of heavy metals. *Ecological Engineering* **5**, 5-12.

Saadi, A., Guerbet, M., Garnier, J. 2002. Influence of diethyldithiocarbamate on cadmium and copper toxicity to freshwater macrophyte *Spirodela polyrrhiza*. *Water SA* **28**, 107-110.

Schultz, E., Joutti, A., Raisanen, M.L., Lintinen, P., Martikainen, E., Lehto, O. 2004. Extractability of metals and ecotoxicity of soils from two old wooden impregnation sites in Finland. *Science of the Total Environment* **326**, 71-84.

Sharma, S.S. and Gaur, J.P. 1995. Potential of Lemna polyrrhiza for removal of heavy metals. *Ecological Engineering* **4**, 37-43.

Shyu, T.H., Lee, Y.H., Chung, M.Y. 2002. Investigation of water quality in a ditch from the Taichung industrial park and evaluation of effluent toxicity using a duckweed toxicity test. *Zhiwu Baohu Xuehui Huikan* **44**, 157-169.

Singh, A. and Jaiswal, V.S. 1997. Uptake of nickel by *Lemna paucicostata*. *Indian Journal of Environmental Protection* **17**, 680-684.

Singh, A., Jaiswal, V.S., Singh, R. 1999. Toxicity of nickel to *Lemna paucicostata*. *Geobios (India)* **26**, 171-176.

Smith, M.D. and Moelyowati, I. 2000. Duckweed based wastewater treatment (DWWT): Design guidelines for hot climates. *Water Science and Technology* **43**, 291-299.

Srivastava, A. and Appenroth, K.J. 1995. Interaction of EDTA and iron on the accumulation of $Cd^{2+}$ in duckweeds (Lemnaceae). *Journal of Plant Physiology* **146**, 173-176.

Srivastava, P.K. and Pandey, G.C. 1999. Paper mill effluent induced toxicity in *Eichhornia crassipes* and *Spirodela polyrhiza*. *Journal of Environmental Biology* **20**, 317-320.

Steinberg, R. 1946. Mineral requirement of *Lemna minor*. *Plant Physiology* **21**, 42-48.

Teisseire, H. and Guy, V. 2000. Copper-induced changes in antioxidant enzymes activities in fronds of duckweed (*Lemna minor*). *Plant Science* **153**, 65-72.

Teisseire, H. and Vernet, G. 2000. Ascorbate and glutathione contents in duckweed, *Lemna minor*, as biomarker of the stress generated by copper, folpet and diuron. *Biomarkers* **5**, 263-273.

Tsatsenko, L.V. and Malyuga, N.G. 1998. Sensitivity of different tests for water pollution by heavy metals and pesticides based on the use of common Duckweed, *Lemna minor*. *Ekologiya (Ekaterinburg)* **5**, 407-409.

Tsatsenko, L.V. and Malyuga, N.G. 2000. Test system for evaluation of environmental contamination by heavy metals and pesticides. *Izvestiya Vysshikh Uchebnykh Zavedenii, Pishchevaya Tekhnologiya* **(2-3)**, 106-108.

Ullrich-Eberius, C.I. and Novacky, A. 1984. $H^+$/phosphate cotransport and its interaction with arsenate and vanadate in transport and metabolism in *Lemna gibba* G1. In: Membrane transport in plants (W. J. Cramed.ed) Symp. Abstr 433.

Wang, Q., Cui, Y., Dong, Y. 2002. Phytoremediation of polluted waters: potentials and prospects of wetland plants. *Acta Biotechnologica* **22**, 199-208.

Wang W. 1990. Literature review on duckweed toxicity testing. *Environmental Research* **52**, 7-22.

Whitton, B.A. and Kelly M.G. 1995. Use of algae and other plants for monitoring rivers. *Australian Journal of Ecology* **20**, 45-56.

Xin, X.Y. and Ma, X.D. 2003. Purification of wastewater in a lagoon by aquatic plants. *Shanxi Daxue Xuebao, Ziran Kexueban* **26**, 85-87.

Xu, N., Shi, G., Du, K., Zhang, X., Zeng, X., Zhou, H. 2002. Study on effect of Hg, Cd and their combined pollution in leaves of *Lemna minor* L.. *Nanjing Shida Xuebao, Ziran Kexueban* **25**, 109-115.

Xyländer, M., and Augsten, H. 1992. Different sensitivity of some Lemnaceae to nickel. *Beitraege Biologia Pflanzen* **67**, 89-99.

Yamamoto, Y.T., Rajbhadari, N., Lin, X., Bergmann, B.A., Nishimura, Y., Stomp, A.M. 2001. Genetic transformation of duckweed *Lemna gibba* and *Lemna minor*. *In vitro Cell Development Biology of Plants* **37**, 349-353.

Youngman, A.L., Williams, T.L., Lydy, M.J. 1998. Utilization of a duckweed bioassay to evaluate leaching of heavy metals in smelter contaminated soils. ASTM Special Technical Publication, STP 1333 (Environmental Toxicology and Risk Assessment: Seventh Volume), 227-236.

Zayed, A., Growthaman, S., Terry, N. 1998. Phytoaccumulation of trace elements by wetland plants: I. Duckweed. *Journal of Environmental Quality* **27**, 715-721.

*Advances in Plant Physiology*, Vol. 9
Ed. A. Hemantaranjan
Scientific Publishers (India), Jodhpur, 2006 pp. 205-225
E-mail: **info@scientificpub.com** www.scientificpub.com

# 10

# JASMONIC ACID - A STRESS RELIEVING PLANT CELL SIGNALING MOLECULE

B.K. Sarma

Department of Mycology and Plant Pathology, Institute of Agricultural Sciences, Banaras Hindu University, Varanasi 221005

## INTRODUCTION

A group of naturally occurring organic compounds that is active at very low concentration (e.g. <1 mM often 1 μM) is called hormones. Plant hormones control a diverse array of plant responses including growth, development and defense against biotic and abiotic stresses. A hormone, viz., indole-3-acetic acid, gibberellic acid, cytokinin, abscisic acid, brassinosteroids, jasmonic acid (JA), ethylene (ET) and salicylic acid (SA), is often formed in certain parts of the plant and then translocated to other sites where it evokes specific biochemical, physiological, and/or morphological responses. These organic compounds promote, inhibit, or qualitatively modify plant growth and development in the tissues where they are produced as well as in distant tissues to which they are translocated. This complex process requires a communication system that can operate over relatively long distances among different plant organs as well as different organelles within a single cell. Cells of different tissues and organs detect signals they receive from other parts of the plants, respond and transmit them in their own characteristic way in such a system.

## 2. HORMONE SIGNAL TRANSDUCTION PATHWAY

The induction of plant responses to any stimuli either exogenous or endogenous requires a perception by the plant via some specific signal molecules which may be chemical signals such as hormones and phytotoxins, and biotic signals such as fungal components known as elicitors, etc. JA is one among the signal molecules in plant cellular responses to different abiotic and biotic stresses including plant-herbivore as well as plant-pathogen interactions. It induces plant defense responses at a distance from the site of attack. Exogenous application of JA also induces defense responses at a distance. The first step in signaling involves high-affinity binding of the elicitor with a specific cellular recognition site of the plant called a receptor. The receptor then undergoes a conformational change that initiates a sequence of downstream events called signal transduction. Signal transduction defines a specific information pathway within a cell that translates an intra- or extra-cellular signal into a specific cellular response. After the signal transduction pathway is activated, the receptor may alter gene expression directly by acting as a transcription factor without transducing the activated signal to the pathway. Alternatively, the receptor may act as a molecular switch and pass the signal to the nucleus through a series of intermediary steps. The signaling components in the pathway are generally modified by phosphorylation or by the activation of low molecular weight GTP-binding proteins. Signaling cascade is often a very complex process and requires sensitivity and specificity of the transduction pathways that are coordinated and integrated with the specific signaling components. The stimulation of the receptor may be activated or inactivated by the relay components of the pathway, depending on the components of the pathway, through different types of cascading mechanism. The changes in signaling components not only permit a rapid response to the elicitor signal but also allow recycling of components of the signaling system to receive further signals. As a result, signal transduction not only modulates the enzyme activity in target cells, but also alters the biosynthetic rate of existing proteins or triggers the synthesis of new ones. So far, details of elicitor signaling are known only partially and there are still several intricacies that need to be revealed.

## 2.1 Jasmonic Acid Biosynthesis

Plants have evolved several complex mechanisms to resist attack by viruses, bacteria, fungi, nematodes, insects, and other pests and pathogens. Pests and pathogens in response to these resistance mechanisms have evolved alternate mechanisms to overcome plant resistance. New virulent strains of bacteria and fungi constantly place selection pressure on plants in a never-ending battle. In classical plant-pest interactions, the pathogen interacts with the plant, which is either resistant or susceptible to attack. If the plant is susceptible, then the reaction between the pest and plant is a compatible interaction and the pathogen is successful in its attack. If the plant is resistant then the resistance reaction is activated by a race-specific elicitor of the pathogen. The elicitor is then detected by a plant receptor, which triggers other genes and thereby helps to defeat the pathogen.

The most frequently occurring defense response of the plant is the hypersensitive response (HR), which occurs early during the infection process. The HR involves death of a few cells in the local region of pathogen attack to form a necrotic lesion that may contain antimicrobial compounds. Also, deposition of lignin and callose are triggered, as are hydroxyproline-rich glycoproteins and pathogenesis related (PR) proteins, including 1,3-$\alpha$-glucanases and chitinases. Other biochemical changes include the generation of reactive oxygen species (the oxygen burst) and the production of phenolics.

Several defense related pathways may be triggered which are required for the synthesis of JA and related compounds. These compounds act as triggers for the activation of other defense-related pathways and synthesis of defense-related chemicals. Physiologically JA in plant is known to act as a growth inhibitor and senescence-promoting substance. Jasmonates (JAs), such as jasmonic acid, are cyclopentanone derivatives synthesized from linolenic acid via the octadecanoic pathway. They are structurally related to prostaglandins, which are autacoidal hormones found in mammals. JA is synthesized from alpha-linolenic acid, which is a C18 poly-unsaturated fatty acid. It is synthesized from plant membranes by enzymes similar to lipase. Key JA biosynthetic enzymes have chloroplast transit peptides that directing their import into the chloroplast. Key enzymes involved in JA biosynthesis include lipoxygenase, allene oxide synthase, and allene oxide cyclase.

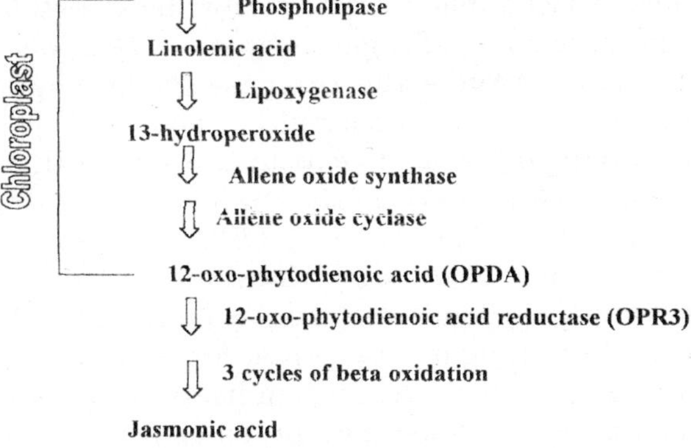


Fig. 1 Jasmonic acid biosynthetic pathway

JA can be conjugated to amino acids, ie. leucine, valine, isoleucine and the sugar, B-glucoside using UDP-glucose. (-)-JA and (-)-methyl jasmonate (MeJA) are major JAs in plants. JA induces vegetative storage proteins, osmotin, thionin (antifungal) and defensin. It induces the phytoalexin-related enzymes such as chalcone synthase, phenylalanine ammonia lyase (PAL), hydroxymethylglutaryl-COA reductase and protease inhibitors to control insects and PR-3, PR-4, PDF1.2 (defensin), chitinases (CHI-B), and hevein-like protein to control fungi. 12-oxo-phytodienoic acid is also a physiological signal for defense.

## 2.2 Genetic Evidence of Jasmonic Acid (JA) Signaling

Signals mediated by jasmonate plays a very crucial role in plant reproductive processes (Stintzi and Browse, 2000), protection from pests and pathogens (McConn *et al,* 1997; Staswick *et al.,* 1998; Engelberth *et al.,* 2004), as well as from abiotic stresses (Overmeyer *et al.,* 2000). Molecular and genetic analysis of JA biosynthesis or perception mutants revealed that JA is required for male fertility. For instance, external application of JA can restore fertility in male sterile mutants. However, it is important to note that application of JA does not complement a signaling mutant in contrast to a biosynthetic mutant. Therefore, for the development of the stamen and pollen, JA is required (Stintzi and Browse, 2000). JA is also required for development of the filament, development of pollen grains, and dehiscence of the anthers (Sanders *et al.,* 2000). However, it is not essentially true that male sterility is a general phenotype of JA mutants as some JA mutants are male fertile (Staswick *et al.,* 1998). Two possible explanations for this discrepancy have been put forward. First, the part of the signaling network that is affected in some JA mutants is not necessary for proper flower fertility. Second, some JA mutants that shows a less pronounced phenotype i.e. root growth and gene expression, than other JA mutants suggests that these mutants are weak alleles that allow JA perception and signaling up to some extent that is sufficient for plant reproduction. Another recent evidence in an *Arabidopsis* mutant *jar1* (*jasmonic acid resistant 1*) showed that it does not encode a signal transduction component, but rather an enzyme that biochemically modifies JA, suggesting that although required for aspects of JA response, this modification is apparently not necessary for pollen fertility (Staswick *et al.,* 2002).

Plant resistance against pathogens relies on the recognition of the pathogen by the plant and subsequent activation of effective defense. Resistance against specific races of a pathogen depends on the recognition of avirulence (AVR) gene products fromm the pathogen by resistance (R) gene products in the plant. In the absence of such a gene-for-gene recognition system, defenses seem to be elicited nonspecifically, similar to those observed in the innate immune response in animals. This type of defense is known as basal resistance, and restricts the development of the disease after pathogen attack. The resistance conferred by JA is also of similar fashion. Defects in JA response or interruptions in the JA biosynthetic pathway generally result in susceptibility of those plants to various pathogens and herbivores (Howe *et al.,* 1996; McConn *et al.,* 1997; Staswick *et al.,* 1998; Engelberth *et al.,* 2004). Recent evidence on the role of JAs providing basal resistance on plants came from genetic analyses of plant mutants and transgenics that are affected in the biosynthesis or perception of JAs. *Arabidopsis* mutant defective in JA-related processes, viz., both the *jar1* mutant with reduced sensitivity to MeJA and the *fad3fad7fad8* triple mutant, which is defective in JA biosynthesis exhibit susceptibility to the fungal pathogen *Pythium irregulare* (Staswick *et al.,* 1998). Increased susceptibility of *jar1* to *Fusarium oxysporum* and experiment of induced resistance against *Cucumber mosaic virus* in the triple mutant has also been reported recently (Berrocal-Lobo and Molina, 2004; Ryu *et al.,* 2004). The JA-insensitive mutant *coi1* (*coronative insensitivei*) shows enhanced susceptibility to the bacterial leaf pathogen *Erwinia carotovora* and the necrotrophic fungi *Alternaria brassicicola* and *Botrytis cinerea* (Norman-Setterblad *et al.,* 2000; Thomma et al., 1998). Accordingly, overexpression of a JA carboxyl methyl transferase increased endogenous levels of MeJA and resulted in higher resistance to *B. cinerea* (Seo *et al.,* 2001). Furthermore, constitutive activation of the JA signaling pathway in *Arabidopsis* resulted in enhanced resistance to the biotrophs *Erysiphe cichoracearum, Erysiphe orontii* and *Oidium lycopersicum* (Ellis *et al.,* 2002). In contrast to the role of JAs in resistance against pathogens with diverse life styles, JA has also been implicated in enhanced susceptibility to pathogen infection. For instance, *coi1* and the MAP kinase 4 mutant *mpk4,* which is impaired in JA-responsive gene expression show reduced susceptibility to the bacterial pathogen *Pseudomonas syringae* suggesting that in wild-type plants, JA-dependent responses promote susceptibility to this pathogen (Feys *et al.,* 1994; Kloek *et al.,* 2001).

The triple mutant of *Arabidopsis* is also more susceptible to attack by larvae of a saprophagous fungal gnat, *Bradysia impatiens.* Unlike the response of the triple mutant, the mutant plants for oxophytodienoic acid reductase 3 (*opr3*) that blocks JA biosynthesis beyond the JA biosynthetic precursor OPDA (12-oxo-phytodienoic acid), show the same resistance as wild types against *Bradysia* larvae as well as another fungal pathogen *A. brassicicola.* In summary, these results indicate that the regulation of resistance or susceptibility of the plant by JA-dependent signaling pathways is governed by the type of pathogen as well as the type of pathogenicity. The result in *opr3*, in relation to *Bradysia* larvae and the fungal pathogen *A. brassicicola* is particularly important because it shows that resistance to insect and fungal attack is possible even in the absence of JA. This suggests that JA and MeJA may not be essentially required for all jasmonate responses, and that the intermediate compound OPDA can signal defense responses against *Bradysia* larvae, as well as the fungal pathogen *A. brassicicola* in *Arabidopsis* (Stintzi *et al.,* 2001). Other intermediates of JA biosynthesis, dinor oxo-phytodienoic acid (dnOPDA), which is synthesized from hexadecatrienoic acid (16:3), and JA conjugates such as JA-amino acid and JA-glucosyl, may also act as important signaling molecules of the JA pathways. Furthermore, it is also becoming evident that the biochemical derivatives of JA may also be an important part of JA signaling (Staswick *et al.,* 2002). Current understanding of JA signaling and its interaction with other signaling pathways is not complete, and reveals the complexity of the process. However, new biochemical approaches and screens for additional mutants via insertional mutagenesis such as T-DNA and transposable elements may provide new opportunities to discover multiple control sites and to dissect the complexity of the pathway.

### 2.3 Pharmacological Evidence in Jasmonic Acid (JA) signaling

The role of JAs in plant defense has also been demonstrated by pharmacological evidence came from exogenous application of jasmonates. MeJA, a key compound in the JA signaling pathway and also regulates the JA biosynthesis pathway, showed that addition of MeJA to cell suspension cultures of different plant species induced defense related gene expression and elicited the accumulation of secondary metabolites (Gundalch *et al.,* 1992). Pretreatment of *Arabidopsis* with MeJA provides significant protection against *A. brassicicola* through induction of resistance *in planta* (Thomma *et al.,* 1998). Similarly, pretreatment with MeJA resulted in enhanced levels of resistance in potato and tomato against *Phytophthora infestans* (Cohen *et al.*, 1993), and in grapefruit against *Penicillium digitatum* (Droby *et al.,* 1999). In addition, MeJA treatment has been shown to be effective against *Pseudomonas syringae* in *Arabidopsis* and tomato (Pieterse *et al.*, 1998). However, MeJA application failed to induce resistance against the biotrophic pathogen *Peronospora parasitica* in *Arabidopsis* (Thomma *et al.*, 1998) or to *Blumeria graminis* in barley (Schweizer *et al.,* 1993) and thereby questioning its ability to induce resistance against biotrophic pathogens. Overall, the studies highlight the key role of JAs in basal and induced resistance to necrotrophic pathogens in different plant species.

## 3. JASMONIC ACID-MEDIATED INDUCED SYSTEMIC RESISTANCE (ISR) AGAINST ABIOTIC STRESS

Jasmonates occur in many plant species and are involved in various physiological processes (Creelman and Mullet, 1997). In barley (*Hordeum vulgare* L.) plant, changes in a number of photosynthetic parameters, such as a decrease in the rate of photosynthesis, $CO_2$ fixation and the activity of RuBP carboxylase have been achieved by exogenous treatment with JA or MeJA. Considerable increase in the rates of dark- and photo-respiration, $CO_2$ compensation point value, and stomatal resistance was observed in plants exogenously applied with the jasmonates (Popova *et al.*, 1988). A breakdown in the biosynthesis of Rubisco (Popova and Vaklinova, 1988), an inhibition of the Hill reaction activity and some changes in the kinetic characteristics of the flash-induced $O_2$ evolution (Maslenkova *et al.*, 1990) were occurred as a result of jasmonates application. Furthermore, jasmonates induced accumulation of a number of proteins (JIPs) in many plant species, including barley (Muller-Uri *et al.,* 1988). Induction of JIPs mainly belongs to the thylakoid-bounded polypeptides (Maslenkova *et al.,* 1992). Interestingly, most of the JA-induced polypeptides were identical to ABA- and NaCl-induced ones, leading to the assumption that externally applied jasmonates act as stress messengers. The intensive research on plant response to various environmental stresses recently has revealed the role of jasmonates as signaling molecules or stress-modulating compounds. They have also been involved in plant response to wound stress (Farmer and Ryan, 1992). Increased endogenous accumulation of jasmonates has also been found in plants suffering from drought and osmotic stress (Creelman and Mullet, 1997). In recent years jasmonates have been the focus of much attention because of their ability to protect planys from salinity stress (Tsonev *et al.*, 1998), UV irradiation (Mackerness *et al.*, 1999), or freezing stress in bromegrass (Wilen *et al.*, 1994), leading to the suggestion that jasmonates could mediate the defense response to various abiotic environmental stresses. The common link among such different stresses is that they all produce an oxidative burst. Chloroplasts, a major source of activated $O_2$ in plants (Foyer *et al.*, 1994; Iturbe-Ormaetxe *et al.*, 1998), and antioxidants, which may play an essential role in preventing oxidative damage, are greatly affected by such environmental stresses (Bowler *et al.*, 1994). Bipyridyl herbicides such as paraquat (Pq) and diquat are non-selective contact herbicides that act by intercepting electrons from the photosynthetic electron transport chain at PSI. Following this reaction there is production of bipyridyl radicals, which readily react with $O_2$ to produce superoxide and then, through a series of reactions, produce $H_2O_2$ and the hydroxyl radical. These toxic oxygen species cause extensive lipid peroxidation (Babbs *et al.*, 1989), chlorophyll breakdown (Shaaltiel *et al.*, 1988), loss of photosynthetic activity (Fedtke, 1982), leakage of electrolytes (Harris and Dodge, 1972), and loss in cell membrane integrity in plants (Kunert and Dodge, 1989). Cross tolerances were found between oxidant generating herbicides and environmental oxidants in different plant biotypes constitutively tolerant to paraquat, $SO_2$, $O_3$ (Shaaltiel *et al.*, 1988), to photoinhibition (Jansen *et al.*, 1989), or drought stress (Malan *et al.*, 1990). It can be concluded that pretreatment of barley plants with MeJA induced protection on photosynthesis against Pq. Pretreatment with MeJA causes protection of photosynthesis against Pq stress by decreasing in the levels of $H_2O_2$, lipid peroxidation and electrolyte leakage. It provides protection of membrane integrity but does not affect the activity of the ascorbate-gluthatione cycle. Most probably this protection is due to an involvement of the other antioxidative components like enzymes catalysing the detoxification of high $H_2O_2$ production, namely catalase and peroxidase, or lypoxigenase. There is a need to understand how wide a range of adverse environmental factors provoke the same changes in physiological processes, reflecting the changed expression of a common or overlapping set of biochemical reactions.

## 4. JASMONIC ACID-MEDIATED INDUCED SYSTEMIC RESISTANCE (ISR) AGAINST HERBIVORY

Plants provide a mean for food and shelter to microbes, plants and animals leading them to remain under a continuous threat. Despite this, plants have developed remarkable strategies to adapt to such stresses by using a range of constitutive or inducible biochemical and molecular mechanisms. Defensive strategies in plants initiate by "perceiving" the attack by pathogens and herbivore, followed by translating that "perception" into an appropriate adaptive response. Induced responses are the change that occurs as a defensive mechanism after herbivore attack. Plant induced responses to herbivory are initiated by elicitors when applied externally. Use of elicitors made it feasible to protect crop plants through induced systemic resistance (ISR) against herbivore, as inducing plant with biotic challenges for future protection from more damaging attackers is not generally feasible. Elicitors are a practical way to induce plant responses because they can be synthesized, cost effective due to its need in lesser quantity, and applied to large numbers of plants by conventional spray technologies. Manipulating plant-induced responses is an effective tool only if they benefit the plant by reducing the damage caused by herbivory attack. JA is observed to be induced in many plant species after caterpillar damage resulting in increased production of a chain of defense related compounds. External application of JA in tomatoes results in induction of proteinase inhibitors (*PIN*) and polyphenol oxidase and in a decrease in the preference, performance, and abundance of many common yield reducing herbivores in the field, including *Frankliniella occidentalis* Pergrande (thrips), *Spodoptera exigua* and *Trichoplusia ni* Hu¨bner (noctuid caterpillars), *Epitrix hirtipennis* Melsheimer (Pea beetles), and *Macrosiphum euphorbiae* Thomas, and *Myzus persicae* Sulzer (aphids) (Thaler *et al.,* 1996). Induced resistance to herbivores may however, only be beneficial if it is accompanied by increase in crop yield or economic value (Lyon *et al.*, 1995). Elicitors thus represent a potential and practical way to protect plants through induced resistance. It is commonly observed that the biochemical pathways that induce resistance against herbivores are highly conserved among different plant species, and thus shows the possibility of use of elicitors of these pathways as inducers in many crops (Meyer *et al.,* 1984). However, before application of elicitors in agricultural systems, one must understand the relationship between the chemicals that they induce, the effects these chemicals have on herbivores in the field, and the effects on plant value for the effective use of elicitors. Negative effects of induced resistance on herbivores do not necessarily translate into net benefits for the farmer. Isonicotinic acid (INA), for example, is an elicitor that induces strong negative effects on plant pathogens. It was considered as a pest management tool until its high level of phytotoxicity was recognized (Schneider *et al.,* 1996). Induced responses to herbivore may not benefit plants, even if they reduce herbivore numbers, if plants producing the induced chemicals have less energy to allocate to other economically important functions. In brief, application of JA induces resistance in plants against many herbivore in the field, and this induction has negative effects on several yield reducing herbivores. The induction of defenses by JA does not decrease plant performance, indicating that the costs of producing these defensive compounds can be easily compensated for. In environments with more severe pest problems or where plants are more susceptible to herbivore, induced resistance may be an important tool for pest management. With the use of chemical elicitors such as JA for plant resistance, it is possible to tailor the resistance profile of plants to suit the pest pattern of a particular region and a particular climate in which the crop is being grown.

Certain plant growth-promoting rhizobacteria (PGPR), which are a wide range of root colonizing bacteria with the capacity to enhance plant growth by increasing seed emergence, plant weight, crop yields and suppression of plant pathogens, also induces JA-mediated resistance in plants against herbivore attack. ISR inducer *Pseudomonas fluorescens* WCS417r when tested for its efficacy against the generalist herbivore *Spodoptera exigua* and the specialist herbivore *Pieris rapae*, it was found that performance of *S. exigua* larvae was significantly reduced on the ISR expressing plants, resulting in a slower gain in fresh weight. However, the performance of *P. rapae* was not affected on ISR-expressing plants. Upon investigation of the mechanism of the resistance observed against the generalist herbivore showed that *S. exigua* feeding induced the JA-responsive gene *PDF1.2* to a much higher level in ISR-expressing plants than in non-induced control plants. These suggest that priming for JA-responsive gene expression contributes resistance against some herbivore insects. Non-effectiveness of this priming after feeding by *P. rapae* suggests that this specialist somehow circumvents this defense response. In a recent study (De Vos *et al.*, 2005) showed that exposure of *Arabidopsis* to herbivorous insects with different modes of attack, viz., tissue-chewing caterpillars (*P. rapae*), cell-content-feeding thrips (*F. occidentalis*) and phloem-feeding aphids (*M. persicae*), showed that the kinetics of SA, JA and ET production varies greatly in both quantity and timing. A large number of *Arabidopsis* genes are recruited in response to *M. persicae* attack compared to the other two herbivores. Analysis of global gene expression profiles demonstrated that the signal signature characteristic of each *Arabidopsis*-attacker combination is orchestrated into a surprisingly complex set of transcriptional alterations. Comparison of the transcript profiles revealed that consistent changes induced by insects with very different modes of attack. Although, the two attackers, *P. rapae* and *F. occidentalis* stimulated JA biosynthesis, the majority of the changes in JA-responsive gene expression were attacker specific. All together these results show that SA, JA and ET play a primary role in the orchestration of the plant's defense response, but other regulatory mechanisms, such as pathway cross-talk or additional attacker-induced signals, eventually shape the highly complex attacker-specific defense response.

## 5. JASMONIC ACID-MEDIATED INDUCED SYSTEMIC RESISTANCE (ISR) AGAINST PATHOGENS

Plant innate immunity is based on a surprisingly complex response that is highly flexible in its capacity to recognize and respond to the invader encountered. When plants are challenged by pathogen attacks they exhibit both long- and short-term defense responses. However, the primary cause of crop loss generally represented by a cumulative effect of many stresses. Plant response originates from the perception of an extracellular signal, i.e. elicitors, and its transduction between and within plant cells. Specificity of the interactions between plants and pathogens is still an incomprehensible phenomenon with a complicated hierarchy of biological organization. Many resistance (R) genes, which confer resistance to various plant species against a wide range of pathogens, have been isolated. However, the key factors that switch on and switch off these genes during plant defense mechanisms remain poorly understood (Glazebrook, 2001).

Besides pathogenic interactions, beneficial rhizospheric bacteria (PGPR) specially fluorescent *Pseudomonas* spp. are among the most effective PGPR, which play important role in plant growth promotion and protection from pest and pathogen attack by various mechanisms (Van Loon *et al.*, 1998; Singh *et al.*, 2002; Singh *et al.*, 2003). Selected strains of rhizobacteria are able to induce a plant-mediated systemic resistance that is effective against a broad spectrum of pathogens. This phenomenon is known as rhizobacteria-mediated induced systemic resistance (ISR). *P. fluorescens* WCS417r-mediated ISR requires responsiveness to both JA and ET, but functions independently of SA (Pieterse *et al.*, 1998). *P. fluorescens* WCS417r-mediated ISR in *Arabidopsis* is predominantly effective against pathogens that in non-induced plants are resisted through JA/ET-dependent basal resistance, including *A. brassicicola*, *Xanthomonas campestris* and *P. syringae*. Therefore, ISR appears to be based on an enhancement of extant JA- and/or ET-dependent defense responses. Interestingly, the level of protection achieved by MeJA treatment in *Arabidopsis* against *A. brassicicola* and *P. syringae* was similar to that observed in ISR-expressing plants. Screening of *Arabidopsis* mutants with reduced basal resistance for their responsiveness to ISR-inducing rhizobacteria revealed that the inability of the mutants to mount ISR was associated with reduced sensitivity to MeJA (Ton *et al.,* 2002). It confirms that JA-dependent defense responses are essential for ISR. However, analysis of local and systemic levels of JA and ET in plants expressing ISR revealed that this type of induced resistance is not associated with detectable changes in their production (Pieterse *et al.,* 2000). Thus, rhizobacteria-mediated ISR is associated with an enhanced sensitivity of the induced tissues to these hormones, rather than an increase in their production. This phenomenon is known as priming. Priming has been implicated in several types of induced resistance. In most cases JA, SA or ET has been suggested to act as potentiation signals of defense-related gene expression. Pretreatment of parsley cell cultures with MeJA potentiates elicitor-induced accumulation of active oxygen species and elicitation of phenylpropanoid defense responses in these cells (Kauss *et al.,* 1994). In rice, JA potentiates the expression of the *PR-1* gene that is activated in response to fungal elicitors, from *Magnaporthe grisea* (Schweizer *et al.,* 1997). Priming is also an important mechanism in rhizobacteria-mediated ISR in *Arabidopsis*. The ISR-expressing *Arabidopsis* plants express the JA- and/or ET-responsive genes *VSP2*, *PDF1.2* and *HEL* to a higher level after subsequent elicitation. Microarray analyses showed that *P. fluorescens* WCS417r-induced ISR in *Arabidopsis* is not associated with detectable changes in gene expression in systemic tissues (Verhagen *et al.,* 2004). However, after challenge inoculation of WCS417r-induced plants with the bacterial leaf pathogen *P. syringae* pv. *tomato* DC3000, a large set of genes showed an augmented expression pattern in ISR-expressing leaves, suggesting that these genes were primed to respond faster and/or more strongly upon pathogen attack. The majority of the primed genes were predictably to be regulated by JA and/or ET signaling. Recently, Pozo *et al.* (2005) also demonstrated that ISR-expressing tissues are primed to respond faster and stronger to JA. Analysis of the promoter sequences of the JA-responsive, ISR-primed genes revealed that they are significantly enriched in binding sites for the transcription factor AtMYC2, a key regulator of the JA signaling pathway (Lorenzo *et al.,* 2004). Most research has focused on understanding how relevant genes are upregulated during pathogen-defense mechanism. However, plant responses to pathogen attack also involve the downregulation of several genes. In potato for example, the mRNA and protein levels of Rubisco are drastically reduced by pathogen infection or elicitor treatment. In persley, the expression of several genes involved in cell proliferation and cell-cycle regulation as well as flavanoid biosynthesis, are repressed to a large extent during defense responses (Somssich and Hahlbrock, 1998). Reports of Lorenzo *et al.,* (2003) confirmed the combination of positive and negative regulatory mechanisms of pathogen-acquired resistance in plants. The results support a central role for JA in priming phenomenon associated with rhizobacteria-mediated ISR. Priming of pathogen-induced genes allows the plant to react more effectively to the invader encountered, which might explain the broad-spectrum effectiveness of ISR.

Induction of a phosphorylation/ dephosphorylation cascade is necessary for the activation of R gene expression. The products of these genes are basically responsible for recognition of the products of pathogen-encoded Avr genes either directly or indirectly (Nimchuk *et al.,* 2001). Most of the R genes identified so far encode for intracellular proteins containing a predicted nucleotide-binding site (NBS) followed by a series of leucine-rich repeats (LRR) at their C termini, although some R genes are distinctly different (Xiao *et al.,* 2001). NBS-LRR resistance proteins generally contain one of two types of N-terminal domains, which have homology either with the Toll and

Interleukin-1 Receptor proteins (TIR) or a predicted coiled-coil domain (CC). The highly variable LRR domains are related to determine recognition of the pathogen Avr products (Dodds *et al.,* 2001; Jia *et al.,* 2000), whereas the more conserved TIR-NBS or CC-NBS regions are believed to propagate the perception signal (Tao *et al.,* 2000). Data on genetic basis postulates that Resistance (R) proteins i.e., products of R genes, act as receptors for pathogen-encoded Avr proteins to initiate a signal, which activates plant defense responses to arrest pathogen propagation. Binding of the tomato bacterial speck R gene product (Pto) to the corresponding AvrPto avirulence gene product of *Pseudomonas syringae* pv. *tomato* demonstrate a direct interaction between an R protein and an Avr protein at molecular level (Scofield *et al.,* 1996; Tang *et al.,* 1996). These findings led to development of genetically engineered transgenic plants by incorporation of R genes into susceptible plants. The R genes and genes encoding signal transduction proteins possess loci at their downstream sequences for production of pathogenesis-related proteins (PRs), enzymes involved in the generation of phytoalexins and protection of plants from oxidative stress, tissue repair, and lignification. Several pathogenesis-related protein families from different plant species have been characterized and found to play different protective roles against pathogens (Van Loon, 1997). The current revelation of R gene structure is not sufficient and much remains to be elucidated. For example, the NBS-LRR class of R genes represents only 1% of the *Arabidopsis* genome (Ellis *et al.,* 2000), which indicates the probable involvement of many other gene families in defense responses. In summary, different types of phytohormonal signaling pathways are involved in plant defense responses during pathogen attack. These results suggest that hormonal transcription factors are key elements in the integration of signal transduction for the regulation of defense-related genes.

## 6. CROSS TALK BETWEEN JASMONIC ACID (JA) AND OTHER HORMONE SIGNALING PATHWAYS

The diversity of plant response to deleterious as well as beneficial microorganisms relies on complex interplay between defense signaling pathways. Cross-talk between defense signaling pathways might help the plant to either prioritize the action of a particular pathway over another, or activate multiple pathways to achieve the most efficient response to the microorganism encountered. Global gene expression profiling studies strongly support the existence of substantial cross-talk between the SA, JA, and ET signaling pathways (Glazebrook *et al.,* 2003). Depending on the type of invader the interaction between JA, SA, and ET signaling pathways may be cooperative, synergistic or antagonistic in imparting resistance towards the invader (Rojo *et al.,* 2003, Pieterse and Van Loon, 2004).

### 6.1 Jasmonic Acid (JA) and Ethylene (ET) Signaling

A concerted action of two phytohormones ET and JA has been implicated in some host-pathogen interactions (McDowell and Dangl, 2000; Glazebrook, 2001; Thomma *et al.,* 2001). Depending on the type of pathogens, these two phytohormones mostly act synergistically to induce several defense genes including *PR1b, PR5* (osmotin), and *PDF1.2* (encodes a plant defensin with antimicrobial properties). Several findings confirmed that induction of *PDF1.2* is caused by a simultaneous activation of both the signaling pathways in *Arabidopsis* following infection with *A. brasicicola* (Penninckx et al., 1998). Mutations in genes of *Arabidopsis* that impair jasmonate signaling genes (*coi1* and *jar1*) or synthesis (fatty acid desaturase [*fad*]3-2, *fad7-2*, and *fad8*) have shown an increased susceptibility of *Arabidopsis* plants to fungal pathogens *B. cinerea, A. brassicicola, P. cucumerina,* and *Pythium* spp. (Staswick *et al.,* 1998; Thomma *et al.,* 2000; Vijayan *et al.,* 1998), or herbivore *Bradysia impatiens* (McConn *et al.,* 1997). Similarly, mutations leading to ethylene insensitivity has also been demonstrated to enhance the susceptibility of different plant species to several pathogenic fungi like *Septoria glycines, Rhizoctonia solani, Pythium* spp., *B. cinerea,* and *P. cucumerina* (Knoester *et al.,* 1998; Hoffman *et al.,* 1999; Thomma *et al.,* 1999), and to the soft-rot bacterium *E. carotovora* (Norman-Setterblad *et al.,* 2000). Moreover, in *jar1* and *etr1* (*ethylene resistant mutant1*) mutants impaired in their response to jasmonate and ethylene, respectively, are also blocked in ISR in *Arabidopsis* (Peiterse *et al.,* 1998). Increased *Arabidopsis* resistance to *B. cinerea* and *P. cucumerina* was correlated with the constitutive expression of *ERF1* (*ETHYLENE RESPONSE FACTOR1*), a downstream component of the ethylene-signaling pathway (Solano *et al.,* 1998, Berrocal-Lobo *et al.,* 2002). Exogenous application of MeJA to JA defective *Arabidopsis* plants induces resistance in such plants and reduces disease development from pathogenic fungi *A. brassicicola, B. cinerea,* and *P. cucumerina* in a dose dependent-manner (Thomma *et al.,* 2000). Recently, Lorenzo *et al.,* (2003) confirmed that transcription factor *ERF1* is a down-stream component of both ET and JA pathways. Its expression requires both signaling pathways simultaneously and is not expressed in plants defective in any of these two pathways. They also concluded that *ERF1* is an early ET- and JA-responsive gene and is responsible for the transcriptional activation of ET/JA-dependent defense-related genes. Transcription profiling of *ERF1*-overexpressing *Arabidopsis* plants also revealed that *ERF1* regulates a large number of genes that are responsive to both JA and ET.

ET and JA regulate different physiological processes in plants, including the activation of specific responses to different types of stress apart from their role in ISR. However, how the correct set of responses selected by the plants to a particular stress using the same two hormones is not clearly understood. It is believed that the type of responses that will be activated depends on the type of interaction that is established between these hormones after a given stress. Interactions between ET and JA may be antagonistic also. For instance, ET is required to suppress ozone-induced cell death in *Arabidopsis* and JA antagonizes this effect (Overmyer *et al.,* 2000). Similarly, ET and JA have antagonistic effects in the biosynthesis of the anti-herbivore compound nicotine in tobacco (Shoji *et al.,* 2000). Formation of the apical hook is another example of the participation of both hormones. While ethylene is required for hook development, the mutants that overproduce ethylene (*eto*) or that constitutively respond to the hormone (*ctr1*) develop an exaggerated apical hook (Wang *et al.,* 2002). Jasmonate antagonizes the effect of ethylene in this case also by preventing the formation of the apical hook even in *ctr1* (Ellis and Turner, 2001). However, the antagonistic effect of ET on JA was also exemplified in the case of the wound response in *Arabidopsis*. Wounding in *Arabidopsis* promotes the synthesis of both JA and ET. While JA induces expression of several wound-responsive genes systemically, locally synthesized ET at the wound site represses the expression of these genes. This antagonistic interaction between the two hormones ensures the correct spatial pattern of expression of systemically induced genes (Rojo *et al.,* 1999). In contrast to the negative interaction of these two hormones, a positive interaction between JA and ET is responsible for the induction of proteinase inhibitors (*PIN*) genes after wounding in tomato. ET alone is not able to induce *PIN* expression and therefore, it cooperates with JA to synergistically induce these genes (O'Donnell *et al.,* 1996).

Apart from ET and JA, phytohormone SA is also an important signaling molecule in plant defense to biotic stress. Plant defense responses are fine-tuned by an intricate signaling network involving all these three phytohormones. SA-dependent defense responses in *Arabidopsis thaliana* are mediated through both NPR1-dependent and -independent mechanisms. NPR1, also knaown as NIM1, is a regulatory protein that was identified in *Arabidopsis* through genetic screens for SAR compromised mutants (Shah *et al.,* 1997). Enhanced disease resistance through constitutive expression of the pathogenesis-related (PR) genes in the *Arabidopsis ssi1* mutant is observed following an NPR1 independent defense mechanism. Interestingly, the *ssi1* mutant also constitutively expresses the JA-responsive defensin gene *PDF1.2* mediated by SA, which also induce expression of PR genes. Hence, the *ssi1* mutant appears to target a step common to SA- and ET- or JA-regulated defense pathways. Recently, Nandi *et al.,* (2003) showed that in addition to SA, signaling of ET and JA are also required for the *ssi1*-conferred constitutive expression of *PDF1.2* and the NPR1-independent expression of PR-1. Enhanced susceptibility of *ssi1* mutants defective in *ein2* (ET-insensitive) and *jar1* was observed towards the fungus *B. cinerea*, but neither the ET- nor JA-signaling pathways compromise *ssi1*-conferred resistance to the bacterial pathogen *Pseudomonas synringae* pv. *maculicola* and the oomycete pathogen *Peronospora parasitica*. Interestingly, *ssi1* plants also exhibits a marginal increase in the levels of ET and JA, suggesting that low endogenous levels of these phytohormones are sufficient to activate expression of defense genes. Based on these findings it can be concluded that although *ssi1* renders expression of ET- or JA-responsive defense genes sensitive to SA and vice versa, it does not affect downstream signaling leading to resistance. The set of defenses activated in plants in response to different types of stress finally depends on the type of interaction (positive or negative) between these hormonal signaling pathways rather than on the independent contribution of each hormone. However, the knowledge on molecular mechanisms that underline these positive or negative interactions of phytohormones is not fully known. Thus, a thorough knowledge at the molecular level on the cross-regulation of ET and JA pathways has become essential to understand how plants activate the correct responses to a given stress, and thereby rationally improve these responses.

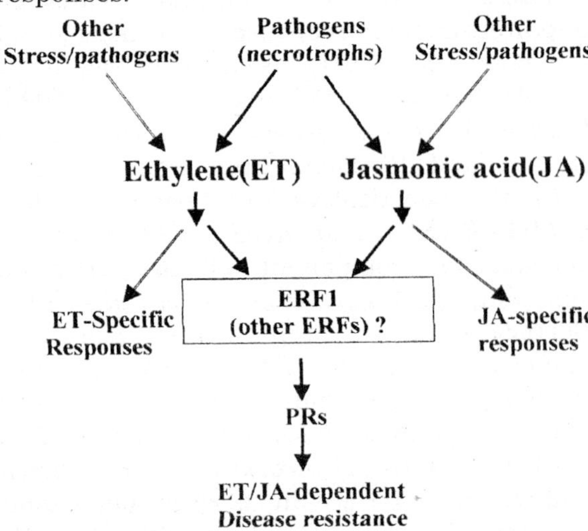


**Fig. 2** Scheme of the Ethylene/Jasmonate-dependent Pathway of the *Arabidopsis* Response to Pathogens (Lorenzo et al., 2003)

## 6.2 Jasmonic Acid (JA) and Salicylic Acid (SA) Signaling

Induced resistance may be expressed locally at the site of infection as well as systemically by the production of a signal at the infected site. It then translocates to other parts of the plant where it induces defense reactions known as systemic acquired resistance (SAR). Salicylic acid (SA) is a ubiquitous plant compound previously known to govern several plant growth and developmental factors as well as physiological processes (Pancheva *et al.,* 1996). As mentioned earlier, it also behaves as an endogenous regulatory signal transduction component in plant defense against pathogen attack (Durner *et al.,* 1997). It also accumulates in plants exposed to abiotic stresses like ozone or UV light (Sharma *et al.,* 1996) and plays a role in plants tolerance to salt and osmotic stresses (Borsani *et al.,* 2001). SA also has the ability to induce thermotolerance in mustard seedlings (Dat *et al.,* 1998) and to provide protection in maize plants against freezing stress (Janda *et al.,* 1999).

Plants are able to activate distinct defense responses that are effective especially against the invader (microbial pathogens and herbivorous insects) encountered. These induced defenses are initially expressed locally, which later transmits to the parts distant from the site of primary infection, and protect the plants systemically against subsequent attack. SA and JA functions as key signaling molecules in induced resistance, which is regulated by a network of interconnecting signal transduction pathways (Glazebrook, 2001; Thomma *et al.,* 2001). SA and JA accumulate in response to pathogen attack at the site of infection and activate a set of defense-related genes. Interestingly, mutant and transgenic plants that are affected in SA accumulation are often more susceptible to pathogen infection than wild-type plants. The blockage of the response to JA generally renders plants more susceptible to herbivore damage than necrotrophic pathogens (Howe *et al.,* 1996; McConn *et al.,* 1997; Thomma *et al.,* 2001; Wildermuth *et al.,* 2001). In general, interactions between SA and JA signaling are antagonistic as SA can antagonize the effect of JA signaling, a process called pathway cross-talk which allow the plant to fine-tune its defense response according to the type of invader.

Pathogen induced SAR is one of the most studied defense responses in plants. The onset of SAR is accompanied by a local and systemic increase in the endogenous levels of SA and induced responses including modifications of the cell wall, production of phytoalexins, and activation of programmed cell death, also called the hypersensitive reaction (HR). The concomitant upregulation of a large set of PR-genes that encode pathogenesis-related (PR) proteins is one among the most important manifestation of SAR (Van Loon and Van Strion, 1999; Ward *et al.,* 1991). Several PR proteins are antimicrobial and contribute to the state of resistance attained in the host plant. Transduction of the SA signal requires the function of NPR1 (Shah *et al.,* 1997). Analysis of the mutant *npr1*, which is expressed in SA signal transduction, revealed that SA-mediated down regulation of JA-responsive gene expression also requires the regulatory protein NPR1 (Pieterse and Van Loon, 2004; Spoel *et al.,* 2003). SA accumulates in those mutants at normal levels after pathogen infection but unable to express *PR* genes and to mount a SAR response. The *NPR1* gene encodes a protein with a BTB/BOZ domain and an ankyrin-repeat domain known to mediate protein-protein interactions and are present in proteins with diverse functions including the transcriptional regulator $I_kB$, which mediates animal innate immune responses (Aravind and Koonin, 1999; Baldwin, 1996). Upon induction of SAR, NPR1 is translocated to the nucleus where it interacts physically with members of the TGA/OBF subclass basic domain/Leu zipper (bZIP) transcription factors, which is required for the binding activity of these factors to promoter elements that play a crucial role in the SA-mediated activation of *PR* genes (Fan and Dong, 2002; Kinkema *et al.,* 2000; Niggeweg *et al.,* 2000). Further studies showed that nuclear localization of NPR1, which is essential for SAR, is not required for the suppression of JA signaling, indicating that cross-talk between SA and JA signaling pathways is modulated through a novel function of NPR1 in the cytosol.

Suppression of JA signaling in plants was observed in pathogen induced SAR which thereby prioritizes SA-dependent resistance to microbial pathogens over JA-dependent defense against insect herbivory (Felton and Korth, 2000). In addition, pharmacological and genetic experiments have shown that SA is a potent suppressor of JA-inducible gene expression (Harms *et al.,* 1998; Gupta *et al.,* 2000). In *Arabidopsis*, compared to wild-type plants, transgenic *NahG* (salicylate hydroxylase gene) plants showed enhanced expression of *LOX2* (*LIPOXYGENASE2*) and leading to higher levels of JA synthesis during pathogen infection. Moreover, the expression of JA-responsive genes *VSP* (encodes a vegetative storage protein) and *PDF1.2* was enhanced strongly in *NahG* plants, suggesting inhibition of JA signaling in wild-type plants by accumulating SA during infection. The antagonistic effect of SA on JA signaling shows a striking resemblance to the effect of the nonsteroidal anti-imflammatory drug acetylsalicylic acid (aspirin), a derivative of SA, on the formation of prostaglandins (structurally related to JA) in animal cells. JA and prostaglandins originate biosynthetically from linolenic acid and arachidonic acid, respectively, which are released from cell membranes upon phospholipid hydrolysis.

In plants, aspirin has been shown to inhibit the octadecanoid pathway by acetylating the key enzymes cyclooxygenase, allene oxide synthase, thereby affecting the formation of JA and the subsequent activation of stress-related gene expression (Pan *et al.,* 1998). Whereas aspirin is able to inhibit prostaglandin and JA biosynthetic enzymes by acetylating them, SA, which lacks the acetyl group, is ineffective in this respect. Rather no inhibitory effect of SA on allene oxide synthase activity was observed in *Arabidopsis* and flax (Harms *et al.,* 1998; Laudert and Weiler, 1998). Since acetylated form of SA does not occur naturally in plants (Pierpoint, 1997), it is unlikely that inhibition of the allene oxide synthase activity plays a major role in the cross-communication between SA and JA signaling in plants. In summary, it is concluded that SA is a strong negative regulator of JA-dependent cellular defense responses in plants (Gupta *et al.,* 2000).

Both aspirin and SA are able to reduce proinflammatory prostaglandin formation in animal cells by inhibiting the activity of the transcription factor NF-$_k$B that plays a key role in the transcriptional activation of many genes during the innate immune response (Baldwin, 1996; Hatada *et al.,* 2000), including the gene that encodes *CYCLOOXY-GENASE2*, which catalyzes a rate-limiting step in prostaglandin production (Newton *et al.,* 1997). In resting cells, NF-$_k$B is sequestered in the cytoplasm by combining with its inhibitory protein I$_k$B. In response to various cellular stress conditions, such as infection by microbial or viral pathogens, I$_k$B is ubiquitinated and degraded by the proteasome, releasing NF-$_k$B to migrate into the nucleus and activate gene expression (Baldwin, 1996; Hatada *et al.,* 2000). Both aspirin and SA block the activation of NF-$_k$B by inhibiting I$_k$B kinase, preventing the degradation of I$_k$B and retaining NF-$_k$B in the cytosol (Yin *et al.,* 1998). Interestingly, I$_k$B shares structural similarity with NPR1 in plants and believed to function similarly. In addition to the ankyrin-repeat domain, the phosphorylated Ser residues important for I$_k$B function also are conserved in the NPR1 protein (Ryals *et al.,* 1997).

The intriguing analogies between the actions of SA/aspirin, prostaglandin, and I$_k$B in animals and SA, JA, and NPR1 in plants, led the researchers to investigate further to check whether NPR1 plays a role in the SA-mediated negative regulation of JA signaling in *Arabidopsis*. In contrast to I$_k$B in animals cells, which functions in the cytosol, NPR1 function in the nucleus when acting as a positive regulator of SA-dependent, defense-related gene expression (Subramaniam et al., 2001). However, Spoel *et al.,* (2003) recently reported a novel function of NPR1 in the cytosol and provide evidence that cytosolic NPR1 plays a crucial role in cross-communication between SA- and JA-dependent plant defense responses. The striking parallels with processes involved in the animal innate immune response suggest that defense-signaling pathways in plants and animals are at least partly conserved. Although the mode of action of *NPR1* in the cytosol is unknown, it is speculated that it interferes with SCF$^{COI1}$ ubiquitin-ligase complex that regulates JA-responsive gene expression through targeted ubiquitination and subsequent proteasome-mediated degradation of a negative regulator of JA signaling (Pozo *et al.,* 2005). Further studies in this direction will make it more understandable how the cross-communication between SA- and JA-dependent pathways is regulated.

## REFERENCES


Aravind L, and Koonin EV (1999). Fold prediction and evolutionary analysis of the POZ domain: Structural and evolutionary relationship with the potassium channel tetramerization domain. *J. Mol. Biol.*, **285:**1353-1361.

Babbs CF, Pham, JA, and Coolbaugh, RC (1989). Lethal hydroxyl radical production in paraquat-treated plants. *Plant Physiol.*, **90**:1267–1270.

Baldwin AS (1996). The NF-$_k$B and I$_k$B proteins: New discoveries and insights. *Annu. Rev. Immunol,.* **14**:649-683.

Berrocal-Lobo M, and Molina A (2004). Ethylene response factor 1mediates *Arabidopsis* resistance to the soil borne fungus *Fusarium oxysporum*. *Mol. Plant-Microbe Interact.*, **17**:763-770.

Berrocal-Lobo M, Molina A, and Solano R (2002). Constitutive expression of ethylene-response-factor 1 in *Arabidopsis* confers resistance to several necrotrophic fungi. *Plant J.*, **29**:23-32.

Borsani O, Valpuesta V, and Botella MA (2001). Evideance for a role of salicylic acid in the oxidative damage genetared by NaCl and osmotic stress in *Arabidopsis* seedlings. *Plant Physiol.,* **126**:1024–1030.

Bowler C, Van Camp W, Van Montagu M, and Inze D (1994). Superoxide dismutase in plants. *Crit. Rev. Plant Sci.*, **13**:199–218.

Cohen Y, Gisi U, Niderman T (1993). Local and systemic protection against *Phytophthora infestans* induced in potato and tomato plants by jasmonic acid and jasmonic methyl ester. *Phytopathol.,* **83**:1054-1062.

Creelman RA, and Mullet JE (1997). Biosynthesis and action of jasmonates in plants. Annu. Rev. *Plant Physiol. Plant Mol. Biol.*, **48**:355–381.

Dat JF, Foyer CH, and Scott IM (1998). Changes in salicylic acid and antioxidants during induced thermotolerance in mustard seedlings. *Plant Physiol.*, **118**:1455–1461.

De Vos M, Van Oosten VR, van Poecke RMP, Van Pelt JA, Pozo MJ, Mueller MJ, Buchalla AJ, Metraux JP, Van Loon LC, Dicke M, and Pieterse CMJ (2005). Signal signature and transcriptome changes of *Arabidopsis* during pathogen and insect attack. *Mol. Plant-Microbe Interact.* (In Press)

Dodds PN, Lawrence GJ, and Ellis JG (2001). Six amino acid changes confined to the leucine-rich repeat ß-strand/ß-turn motif determine the difference between the P and P2 rust resistance specificities in flax. *Plant Cell,* **13**:495-506.

Droby S, Porat R, Cohen L, Weiss B, Shapiro B, Philosoph-Hadas S, and Meir S. (1999). Supressing green mold decay in grapefruit with postharvest jasmonate application. *J. Am. Soc. Hort. Sci.,* **124**:184-188.

Durner J, Shah J, and Klessig DF (1997). Salicylic acid and desease resistance in plants. Trends Plant Sci., 7:266–274.

Ellis J, Dodds P, and Pryor T (2000). Structure, function and evolution of plant disease resistance genes. *Curr. Opin. Plant Biol.,* **3**:278-284.

Ellis C, Karafyllidis L, and Turner JG (2002). Constitutive activation of jasmonate signaling in *Arabidopsis* mutant correlates with enhanced resistance to *Erysiphe cichoracearum*, *Pseudomonas syringae*, and *Myzus persicae*. *Mol. Plant-Microbe Interact.,* **15**:1025-1030.

Ellis C, and Turner JG (2001). The *Arabidopsis* mutant *cev1* has constitutively active jasmonate and ethylene signal pathways and enhanced resistance to pathogens. *Plant Cell,* **13**:1025-1033.

Engelberth J, Alborn HT, Scmelz EA, and Tumlinson JH (2004). Airborne signals prime plants against insect herbivore attack. *Proc. Natl. Acad. Sci.* USA, **101**:1781-1785.

Fan W, and Dong X (2002). In vivo interaction between NPR1 and transcription factor TGA2 leads to salicylic acid-mediated gene activation in *Arabidopsis. Plant Cell,* **14**:1377-1389.

Farmer EE, and Rayan CA (1992). Octadecanoid precursors of jasmonic acid activate the synthesis of wound-inducible proteinase inhibitors. *Plant Cell,* **4**:129–134.

Fedtke C (1982). Biochemistry and Physiology of Herbicide Action. Springer-Verlag, New York, pp. 65–69.

Felton GW, and Korth KL (2000). Trade-offs between pathogen and herbivore resistance. Curr. Opin. *Plant Biol.,* **3**:309-314.

Feys BJF, Beneditti CE, Penfold CN, and Turner JG (1994). *Arabidopsis* mutants selected for resistance to the phytotoxin coronatine are male sterile, insensitive to methyl jasmonate, and resistant to a bacterial pathogen. *Plant Cell,* **6:**751–759.

Foyer CH, Descourvieres P, and Kunert KJ (1994). Protection against oxygen radicals: an important defense mechanism studied in transgenic plants. *Plant Cell Environ.,* **17**:507–523.

Glazebrook J (2001). Genes controlling expression of defense responses in *Arabidopsis*-2001 status. *Curr. Opin. Plant Biol.,* **4**:301-308.

Glazebrook J, Chen WJ, Estes B, Chang HS, Nawarth C, Metraux JP, Zhu T, and Katagiri F (2003). Topology of the network integrating salicylate and jasmonate signal transduction derived from global expression phenotyping. *Plant J.,* **34**:217-228.

Gundalch H, Mueller MJ, Kutchan TM, and Zenk MH (1992). Jasmonic acid is a signal transducer in elicitor-induced plant cell cultures. *Proc. Natl. Acad. Sci.* USA, **89**:2389-2393.

Gupta V, Willits MG and Glazebrook J (2000). *Arabidopsis thaliana* EDS4 contributes to salicylic acid (SA)-dependent expression of defense responses: Evidemce for inhibition of jasmonic acid signaling by SA. *Mol. Plant-Microbe Interact.* **13**:503-511.

Harms K, Ramirez I, and Pena-Cortes H (1998). Inhibition of wound-induced accumulation of allene oxide synthase transcripts in flax leaves by aspirin and salicylic acid. *Plant Physiol.,* **118**:1057-1065.

Harris N, and Dodge AD (1972). The effect of paraquat on flax cotyledon leaves: physiological and biochemical changes. Planta 104:210–219.

Hatada EN, Krappmann D, and Scheidereit C (2000). NF-$_k$B and the innate immune response. *Curr. Opin. Immunol.,* **12**:52-58.

Hoffman T, Schmidt JS, Zheng X, and Bent A (1999). Isolation of ethylene-insensitive soybean mutants that are altered in pathogen susceptibility and gene-for-gene disease resistance. *Plant Physiol.*, **119**: 935-949.

Howe GA, Lightner J, Browse J, and Ryan CA (1996). An octadecanoid pathway mutant (*JL5*) of tomato is compromised in signaling for defense against insect attack. *Plant Cell,* **8**:2067-2077.

Iturbe-Ormaetxe I, Escuredo PR, Arrese-Igor C, and Becana M (1998). Oxidative damage on pea plants exposed to water deficit or paraquat. *Plant Physiol.*, **116**:173–181.

Janda T, Szalai G, Tari I, and Paldi E (1999). Hydroponic treatment with salicylic acid decreases the effects of chilling injury in maize (*Zea mays* L.) plants. *Planta,* **208**:175–180.

Jansen MAK, Shaaltiel Y, Kazzes D, Canaani O, Malkin S, and Gressel J (1989). Increased tolerance to photoinhibitory light in paraquat-resistant *Conyza bonariensis* measured by photoacoustic spectroscopy and 14 C02 - fixation. *Plant Physiol.*, **91**:1174–1178.

Jia Y, McAdams SA, Bryan GT, Hershey HP, and Valent B (2000). Direct interaction of resistance gene and avirulence gene products confers rice blast resistance. *EMBO J.*, **19**:4004-4014.

Kauss H, Jeblick W, Ziegler J, and Krabler W (1994). Pretreatment of parsley (*Petroselinum crispum* L.) suspension cultures with methyl jasmonate enhances elicitation of activated oxygen species. *Plant Physiol.*, **105**:89-104.

Kinkema M, Fan W, and Dong X (2000). Nuclear localization of NPR1 is required for activation of *PR* gene expression. *Plant Cell,* **12**:2339-2350.

Kloek AP, Verbsky ML, Sharma SB, Schoolz JE, Vogel J, Klessig DF, and Kunkel B.N. (2001). Resistance to *Pseudomonas syringae* conferred by an *Arabidopsis thaliana* coronative-insensitive (*coi1*) mutation occurs through two distinct mechanisms. *Plant J.,* **26**:509-522.

Knoester M, Van Loon LC, Van den Heuvel J, Henning J, Bol JF, and Linthorst H.J.M. (1998). Ethylene-insensitive tobacco lacks nonhost resistance against soil-borne fungi. *Proc. Natl. Acad. Sci.* USA, **95**:1933-1937.

Kunert KJ, and Dodge AD (1989). Herbicide-induced radical damage and antioxidative systems. In: Boger. P., Sandmann, G. (ed.): Target Sites of Herbicide Action. *CRC Press Inc: Boca Raton* Fl., 45–63.

Laudert D, and Weiler EW (1998). Allene oxide synthase: A major control point in *Arabidopsis thaliana* octadecanoid signaling. *Plant J.*, **15**:675-684.

Lorenzo O, Chico JM, Sanchez-Serrano JJ, and Solano R (2004). Jasmonate-insensitive1 encodes a MYC transcription factor essential to discriminate between different jasmonate-regulated defense responses in *Arabidopsis*. *Plant Cell,* **16:** 1938-1950.

Lorenzo O, Piqueras R, Sanchez-Serrano JJ, and Solano R (2003). Ethylene response factor 1 integrates signals from ethylene and jasmonate pathways in plant defense. *Plant Cell*, **15**:165-178.

Lyon GD, Reglinski T, and Newton, AC (1995). Novel disease resistance compounds: the potential to "immunize" plants against infection. *Plant Pathol.*, **44**: 407-427.

Mackerness SAH, Surplus SL, Blake P, John CF, Buchanan-Wollaston V, Jordan BR, and Thomas B (1999). Ultraviolet-B-induced stress and changes in gene expression in *Arabidopsis thaliana*: role of signalling pathways controlled by jasmonic acid, ethylene and reactive oxygen species. *Plant Cell Environ.*, **22**:1413–1423.

Malan G, Greyling MM, and Gressel J (1990). Correlation between CuZn superoxide dismutase and gluthatione reductase, and environmental and xenobiotic stress tolerance in maize inbreds. *Plant Sci.*, **69**:157–166.

Maslenkova LT, Miteva TS, and Popova LP (1992). Changes in the polypeptide patterns of barley seedlings exposed to jasmonic acid and salinity. *Plant Physiol.*, **98**:700–707.

Maslenkova LT, Zanev Y, and Popova LP (1990). Oxygen-evolving activity of thylakoids from barley plants cultivated on different concentrations of jasmonic acid. *Plant Physiol.*, **93**:1316–1320.

McConn J, Creelman RA, Bell E, Mullet JE, and Browse J (1997). Jasmonate is essential for insect defense in *Arabidopsis*. *Proc. Natl. Acad. Sci.* USA, **94**:5473-5477.

McDowell JM, and Dangl JL (2000). Signal transduction in the plant immune response. *Trends Biochem. Sci.*, **25**:79-82.

Meyer A, Miersch O, Buttner C, Dathe W, and Sem-bdner G (1984). Occurrence of the plant growth regulator jasmonic acid in plants. *J. Plant Growth Regul.*, **3**: 1-8.

Muller-Uri F, Parthier B, and Nover L (1988). Jasmonate-induced alteration of gene expression in barley leaf segments analyzed by in-vivo and in-vitro protein synthesis. *Planta,* **176**:241–247.

Nandi A, Kachroo P, Fukushige H, Hildebrand DF, Klessig DF, Shah J (2003). Ethylene and jasmonic acid signaling affect the NPR1-independent expression of defense genes without impacting resistance to *Pseudomonas syringae* and *Peronospora parasitica* in the *Arabidopsis* ssi1 mutant. *Mol Plant Microbe Interact.*, **16**:588-99

Newton R, Kuitert LME, Bergmann M, Adcock IM, and Barnes PJ (1997). Evidence for involvement of NF-$_k$B in the transcriptional control of *COX-2* gene expression by IL-1• . *Biochem. Biophys. Res. Commun.*, **237**:28-32.

Niggeweg R, Thurow C, Kegler C, and Gatz C (2000). Tobacco transcription factor TGA2.2 is the main component of *as-1*-binding factor ASF-1 and is involved in salicylic acid- and auxin-inducible expression of *as-1*-containing target promoters. *J. Biol. Chem.*, **275**: 19897-19905.

Nimchuk Z, Rohmer L, Chang JH, and Dangl JL (2001). Knowing the dancer from the dance: R-gene products and their interactions with other proteins from host and pathogen. *Curr. Opin. Plant Biol.,* **4**:288-294.

Norman-Setterblad C, Vidal S, and Palva ET (2000). Interacting signal pathways control defense gene expression in *Arabidopsis* in response to cell wall-degrading enzymes from *Erwinia carotovora*. *Mol. Plant-Microbe Interact.*, **13**:430-438.

O'Donnell PJ, Calvert C, Atzorn R, Wasternack C, Leyser HMO, and Bowles DJ (1996). Ethylene as a signal mediating the wound response of tomato plants. *Science,* **274**:1914-1917.

Overmayer K, Tuominen H, Kettunen R, Betz C, Langebartels C, Sandermann H Jr., and Kangasjarvi J (2000). Ozone-sensitive *Arabidopsis* rcd1 mutant reveals opposite roles for ethylene and jasmonate signaling pathways in regulating superoxide-dependent cell death. *Plant Cell,* **12**: 1849-1862.

Pan ZQ, Camara B, Gardner HW, and Backhaus RA (1998). Aspirin inhibition and acetylation of the cytochrome P450, allene oxide synthase, resembles that of animal prostaglandin endoperoxide H synthase. *J. Biol. Chem.*, **273**:18139-18145.

Pancheva TV, Popova, LP, and Uzunova AN (1996). Effects of salicylic acid on growth and photosynthesis in barley plants. *J Plant Physiol.*, **149**:57–63.

Penninckx IAMA, Thomma BPHJ, Buchala A, Metraux J-P, and Broekaert WF (1998). Concomitant activation of jasmonate and ethylene response pathways is required for induction of a plant defense. *Plant Cell,* **10**:2103-2113.

Pierpoint WS (1997). The natural history of salicylic acid: Plant product and mammalian medicine. *Interdiscip. Sci. Rev.*, **22**:45-52.

Pieterse, CMJ, and Van Loon LC (2004). NPR1: the spider in the web of induced resistance signaling pathways. *Curr. Opinion Plant Biol.*, **7**:456-464.

Pieterse, CMJ, Van Pelt JA, Ton J, Parchmann S, Mueller MJ, Buchala AJ, Metraux JP, and Van Loon LC (2000). Rhizobacteria-mediated induced systemic resistance (ISR) in *Arabidopsis* requires sensitivity to jasmonate and ethylene but is not accompanied by an increase in their production. *Physiol. Mol. Plant Pathol.*, **57**: 123:134.

Pieterse, CMJ, Van Wees SCM, Van Pelt JA, Knoester M, Laan R, Gerrits H, Weisbeek PJ, and Van Loon LC (1998). A novel signaling pathway controlling induced systemic resistance in *Arabidopsis. Plant Cell,* **10**:1571-1580.

Popova LP, Tsonev TD, and Vaklinova SG (1988). Changes in some photosynthetic and photorespiratory properties in barley leaves after treatment with jasmonic acid. *J. Plant Physiol.*, **132**:257–261.

Popova LP, and Vaklinova SG (1988). Effect of jasmonic acid on the synthesis of ribulose-1,5-bisphosphate carboxylase/oxygenase in barley. *J. Plant Physiol.*, **133**:210–215.

Pozo MJ, Van Loon LC, and Pieterse CMJ (2005). Jasmonates-signals in plant-microbe interactions. *J. Plant Growth Regul.*, **23**:211-222.

Rojo E, Leon J, and Sanchez-Serrano JJ (1999). Cross-talk between wound signaling pathways determines local versus systemic gene expression in *Arabidopsis thaliana. Plant J.*, **20**: 135-142.

Rojo E, Solano R, and Sanchez-Serrano JJ (2003). Interactions between signaling compounds involved in plant defense. *Plant Growth Regul.*, **22**:82-98.

Ryals J, Weymann K, Lawton K, Friedrich L, Ellis D, Steiner HY, Johnson J, Delany TP, Jesse T, Vos P, and Uknes S (1997). The *Arabidopsis* NIM1 protein shows homology to the mammalian transcription factor inhibitor IkB. *Plant Cell,* **9**:425-439.

Ryu CM, Murphy JF, Mysore KS and Kloepper JW (2004). Plant growth-promoting rhizobacteria systemically protect *Arabidopsis thaliana* against *Cucumber mosaic virus* by a salicylic acid and NPR1-dependent and jasmonic acid acid-dependent signaling pathway. *Plant J.*, **39**:381-392.

Sanders PM, Lee PY, Biesgen C, Boone JD, Beals TP, Weilerb EW, and Goldberg RB (2000). The *Arabidopsis DELAYED DEHISCENCE1* gene encodes an enzyme in the jasmonic acid synthesis pathway. *Plant Cell,* **12**:1041-1062.

Schneider M, Schweizer P, Meuwly P, and Metraux, JP (1996). Systemic acquired resistance in plants. *Int. Rev. Cytol.*, **168**: 303-339.

Schweizer P, Gees R, and Mosinger E. (1993). Effect of jasmonic acid on the interaction of barley (*Hordeum vulgare* L.) with the powdery mildew *Erysiphe graminis* f. sp. *hordei. Plant Physiol.*, **102**: 503-511.

Scofield SR, Tobias CM, Rathjen JP, Chang JH, Lavelle DT, Michelmore RW, and Staskawics BJ (1996). Molecular basis of gene-for-gene specificity in bacterial speck disease of tomato. *Science,* **274**:2063-2065.

Seo HS, Songh JT, Cheong JJ, Lee YH, LeeYW, Hwang I, Lee JS, and Chol YD (2001). Jasmonic acid carboxyl methyltransferase: A key enzyme for jasmonate-regulated plant responses. *Proc. Natl. Acad. Sci.* USA, **98**:4788-4793.

Shaaltiel Y, Glazer A, Bocion PF, and Gressel J (1988). Cross tolerance to herbicidal and environmental oxidants of plant biotypes tolerant to paraquat, sulfur dioxide and ozone. *Pest. Biochem. Physiol.*, **31**:13–23.

Shah J, Tsui F, and Klessig DF (1997). Characterization of a salicylic acid-insensitive mutant (*sai1*) of *Arabidopsis thaliana*, identified in a selective screen utilizing the SA-inducible expression of the *tms2* gene. *Mol. Plant-Microbe Interact.*, **10**: 69-78.

Sharma YK, Leon J, Raskin I, and Davis KR (1996). Ozone-induced responses in *Arabidopsis thaliana*-the role of salicylic acid in the accumulation of defense-related transcripts and induced resistance. *Proc Natl Acad Sci.* USA, **93**:5099–5104.

Shoji T, Nakajima K, and Hashimoto T (2000). Ethylene suppresses jasmonate-induced gene expression in nicotine biosynthesis. *Plant Cell Physiol.*, **41**:1072-1076.

Singh UP, Sarma BK, and Singh DP (2003). Effect of plant growth-promoting rhizobacteria and culture filtrate of *Sclerotium rolfsii* on phenolic and salicylic acid contents in chickpea (*Cicer arietinum* L.). *Curr. Microbiol.*, **46**: 131-140.

Singh UP, Sarma BK, Singh DP and Amar Bahadur (2002). Plant growth-promoting rhizobacteria-mediated induction of phenolics in pea (*Pisum sativum*) following infection with *Erysiphe pisi*. *Curr. Microbiol.*, **44**: 396-400.

Solano R, Stepanova A, Chao QM, and Ecker JR (1998). Nuclear events in ethylene signaling: A transcriptional cascade mediated by ethylene-insensitive 3 and ethylene-response-factor 1. *Gene Dev.*, **12**:3703-3714.

Somssich I, and Hahlbrock K (1998). Pathogen defense in plants: a paradigm of biological complexity. *Trends Plant Sci.*, **3**:86-90.

Spoel SH, Koornneef A, Claessens SMC, Korzelius JP, van Pelt JA, Mueller MJ, Buchala AJ, Metraux JP, Brown R, Kazan K, van Loon LC, Dong X, and Pieterse CMJ (2003). NPR1 Modulates cross-talk between salicylate- and jasmonate-dependent defense pathways through a novel function in the cytosol. *The Plant Cell,* **15**: 760-770.

Staswick PE, Tiryaki I, and Rowe M (2002). Jasmonate response locus *JAR1* and several related *Arabidopsis* genes encode enzymes of the firefly luciferase superfamily that show activity on jasmonic, salicylic, and indole-3-acetic acids in an assay for adenylation. *Plant Cell,* **14**:1405–1415.

Staswick PE, Yuen GY, and Lehman CC (1998). Jasmonate signaling mutants of *Arabidopsis* are susceptible to the soil fungus *Pythium irregulare*. *Plant J.*, **15**:747-754.

Stinzi A, and Browse J (2000). The *Arabidopsis* male-sterile mutant, opr3, lacks the 12-oxophytodienoic acid reductase required for jasmonate synthesis. *Proc. Natl. Acad. Sci.* USA, **97**:10625–10630.

Stintzi A, Weber H, Reymond P, Browse J, and Farmer EE (2001). Plant defense in the absence of jasmonic acid: The role of cyclopentenones. *Proc. Natl. Acad. Sci.* USA, **98**:12837-12842.

Subramaniam R, Desveaux D, Spickler C, Michnick SW, and Brisson N (2001). Direct visualization of protein interactions in plant cells. *Nat. Biotechnol.*, **19**:769-772.

Tao Y, Yuan F, Leister RT, Ausubel FM, and Katagiri F (2000). Mutational analysis of the *Arabidopsis* nucleotide binding site-leucine-rich repeat resistance gene RPS2. *Plant Cell,* **12**:2541-2554.

Tang X, Frederick RD, Zhou J, Halterman DA, Jia Y, and Martin GB (1996). Initiation of plant disease resistance by physical interaction of AvrPto and Pto kinase. *Science,* **274**:2060-2063.

Thaler JS, Stout MJ, Karban R, and Duffey SS (1996). Exogenous jasmonates simulate insect wounding in tomato plants (*Lycopersicon esculentum*) in the laboratory and field. *J. Chem. Ecol.*, **22**: 1767-1781.

Thomma BPHJ, Eggermont K, Broekaert WF, and Cammue BPA (2000). Disease development of several fungi on *Arabidopsis* can be reduced by treatment with methyl jasmonate. *Plant Physiol. Biochem.,* **38**:421-427.

Thomma BPHJ, Eggermont K, Penninckx IAMA, Mauch-Mani B, Vogelsang R, Cammue BPA, and Broekaert WF (1998). Separate jasmonate-dependent and salicylic-dependent defense response pathways in *Arabidopsis* are essential for resistance to distinct microbial pathogens. *Proc. Natl. Acad. Sci.* USA, **95**:15107-15111.

Thomma BPHJ, Eggermont K, Tierens KFMJ, and Broekaert WF (1999). Requirement of functional *Ethylene-Insensitive 2* gene for efficient resistance of *Arabidopsis* to infection by *Botrytis cinerea*. *Plant Physiol.*, **121**:1093-1101.

Thomma BPHJ, Penninckx IAMA, Broekaert WF and Cammue BPA (2001). The complexity of disease signaling in *Arabidopsis*. *Curr. Opin. Immunol.*, **13**:63-68.

Ton J, Van Pelt JA, Van Loon LC, and Pieterse CMJ. 2002. Differential effectiveness of salicylate-dependent and jasmonate/ethylene-dependent induced resistance in *Arabidopsis*. *Mol. Plant-Microbe Interact.*, **15**:27-34.

Tsonev TD, Lazova GN,. Stoinova Zh.G, and Popova LP (1998). A possible role for jasmonic acid in adaptation of barley seedlings to salinity stress. *J. Plant Growth Regul.*, **17**:153–159.

Van Loon LC (1997). Induced resistance in plants and role of pathogenesis-related proteins. *Eur. J. Plant Pathol.*, **103**:235-270.

Van Loon LC, Bakker PAHM, and Pieterse CMJ. 1998. Systemic resistance induced by rhizosphere bacteria. *Annu. Rev. Phytopathol.*, **36**:453-483.

Van Loon LC, and Van Strien EA (1999). The families of pathogenesis-related proteins, their activities, and comparative analysis of PR-1 type proteins. *Physiol. Mol. Plant Pathol.*, **55**:85-97.

Verhagen BWM, Glazebrook J, Zhu T, Chang HS, Van Loon LC, and Pieterse CMJ (2004). The transcriptome of rhizobacteria-induced systemic resistance in *Arabidopsis*. *Mol. Plant-Microbe Interact.*, **17**:895-908.

Vijayan P, Shockey J, Levesque CA, Cook RJ, and Browse J (1998). A role of jasmonate in pathogen defense of *Arabidopsis*. *Proc. Natl. Acad. Sci.* USA, **95**:7209-7214.

Wang KL, Li H, and Ecker JR (2002). Ethylene biosynthesis and signaling networks. *Plant Cell*, **14** (suppl.) S131-S151.

Ward ER, Uknes SJ, Williams SC, Dincher SS, Wiederhold DL, Alexander DC, Ahl-Goy P, Metraux JP, and Ryals JA (1991). Coordinate gene activity in response to agents that induce systemic acquired resistance. *Plant Cell*, **3**:1085-1094.

Wildermuth MC, Dewdney J, Wu G, and Ausubel FM (2001). Isochorismate synthase is required to synthesize salicylic acid for plant defense. *Nature*, **414**: 562-565.

Wilen RW, Bruce EE, and Gusta LV (1994). Interaction of abscisic acid and jasmonic acid on the inhibition of seed germination and the induction of freezing tolerance. *Can. J. Bot.*, **72**:1009–1017.

Xiao S, Ellwood S, Calis O, Patrick E, Li T, Coleman M, and Turmer JG (2001). Broad spectrum powdery mildew resistance in *Arabidopsis thaliana* mediated by RPW8. *Science*, **291**:118-120.

Yin MJ, Yamamoto Y, and Gaynor RB (1998). The anti-inflammatory agents aspirin and salicylate inhibit the activity of IkB kinase-•. *Nature*, **396**:415-426.

*SECTION — II*

## ***RECENT ADVANCES IN THE BIOSYNTHESIS AND FUNCTIONS OF CUTICULAR WAXES***

*Advances in Plant Physiology*, Vol. **9**
Ed. A. Hemantaranjan
Scientific Publishers (India), Jodhpur, 2006 pp. 229-251
E-mail: **info@scientificpub.com** www.scientificpub.com

# 11

# PLANT CUTICULAR WAXES: BIOSYNTHESIS AND FUNCTIONS

M. Sturaro* and M. Motto

CRA-Istituto Sperimentale per la Cerealicoltura, Sezione di Bergamo
Via Stezzano 24, 24126 Bergamo, ITALY.
e.mail: sturaro@iscbg.it

## INTRODUCTION

Plant science research is providing exciting advances in our understanding of plant growth and development including biosynthesis and physiological function of cuticular waxes (see Kunst and Samuels, 2003, for a recent review). Being at the boundary between the plant and its environment, cuticular waxes play important roles in limiting water loss and protecting the plant against several environmental stresses such as drought, high light radiation and pathogen attacks. In addition, they influence leaf absorption of chemicals including pesticides and foliar nutrients. Besides these roles in plant ecology, waxes have beneficial effects in human nutrition as cholesterol lowering compounds and have several industrial applications (Hargrove *et al.*, 2004). Therefore, it would be useful to control cuticular wax biosynthesis and secretion on leaf and accumulation of wax compounds in edible organs to increase plant adaptation to its environment and nutritional value of plant derived food.

Although biochemical and genetic analysis have clarified some steps in cuticular wax biosynthesis, a comprehensive view of the whole pathway has yet to be achieved. Several reviews have covered the topic of wax biology during the past years (von Wettsteins-Knowles, 1995; Post-Beittenmiller, 1996; Kolattukudi, 1996; Lemieux, 1996; Jenks and Ashworth, 1999; Kunst and Samuels, 2003). In this communication we will concentrate primarily on recent advances in the biosynthesis and functions of cuticular waxes, with emphasis on *Arabidopsis* and maize.

## 2. CUTICULAR WAX COMPOSITION

The cuticle is the lipid layer covering and protecting primary aerial organs of land plants against several environmental stresses (Post-Beittenmiller, 1996). This hydrophobic barrier is secreted by epidermal cells and consists of a polyester matrix, the cutin (Heredia, 2003), embedded and overlaid by amorphous waxes which are considered the main determinants of the cuticle functions. The outermost cuticular wax layer, properly referred to as epicuticular waxes, form the whitish or glaucous bloom on the plant surface which can be removed after a brief immersion in an organic (non polar) solvent. In addition, waxes are also associated with the triphyne layer of pollen grains and the suberin matrix in root and wound tissues as well as being components of seed storage lipids.

Cuticular waxes are a complex mixture of lipids, primarily very long chain fatty acids (VLCFAs: fatty acids with chain lengths of 20-32 carbons) and their derivatives: aldehydes, ketones, primary and secondary alcohols, alkanes and esters. Other minor components are triterpenes, sterols and flavonoids. Precursors of the wax biosynthetic pathway are C18 saturated fatty acyl moieties produced by *de novo* fatty acid biosynthesis in the chloroplast which are converted to VLCFAs and their derivatives in the cytosol. After synthesis, the different wax constituents are secreted on to the plant surface.

The amount and composition of the cuticular waxes that accumulate on plant surfaces has been extensively investigated by several authors (Avato *et al.*, 1990; Kolattukudy, 1996, Jenks and Ashworth, 1999). The main findings emerging from this study indicate that wax structure is highly variable; the coating may appear as dense or sparse platelets, rods, or tubes. In some cases wax structure has been related with wax chemistry. Cuticular wax composition can change between different species or ecotypes (Rashotte *et al.*, 1997; Cameron *et al.*, 2002) and within a single species according to tissue or developmental stage (Rhee *et al.*, 1998; Jenks *et al.*, 2001; Jetter *et al.*, 2001). For example, the surfaces of *Arabidopsis* stems and siliques are covered by a heavy wax bloom, whereas *Arabidopsis* leaves produce considerably less wax (Koornneef *et al.*, 1989; Hannoufa *et al.*, 1993; Jenks *et al.*, 1995). The major components of the waxes on *Arabidopsis* stems and siliques are alkanes, ketones and alcohols. In contrast, Arabidopsis leaves are almost devoid of ketons. Ontogenetic differences in wax composition are found in maize. In this species, juvenile leaves are covered with a heavy layer of wax bloom, but adult leaves appear glossy because of reduced level of wax accumulation and absence of epicuticular waxes. Although the waxes of juvenile leaves are composed mainly of alcohols, aldehydes and esters derived from C32 acyl moieties, the major constituents of wax from adult maize leaves are esters and alkanes with shorter chain lenghts (Bianchi *et al.*, 1985). Wax deposition is also regulated by environmental signals such as light, moisture, and temperature (Bengtson *et al.*, 1979; von Wettstein-Knowles *et al.*, 1979; Baker, 1980; Hadley, 1989; Bianchi *et al.*, 1985; von Wettstein-Knowles, 1993).

## 3. CUTICULAR WAX FUNCTIONS

The chemical, physical, and topographical properties of the leaf surface determine the relations of a leaf with its environment. Cuticular waxes are central to the function of the plant cuticle as a diffusional barrier for water and solutes and to its function in protecting the plant against abiotic and biotic stresses. Epicuticular waxes, influence the hydrophobic properties of the leaf surface and determine a self-cleaning mechanism which prevents water and contaminating particles from adhering to the leaf surface, the so called "Lotus effect" (Barthlott and Neinhuis, 1997) (Fig. 1). They also affect the retention and uptake of aqueous foliar sprays (Jenks and Ashworth, 1999).

Fig. 1. Wild type (left) and *glossy1* (right) seedling leaves sprayed with water. The epicuticular wax layer confers hydrorepellency and a glaucous appearance to wild type leaves compared to the shiny phenotype of the mutant seedling.

Plant surface waxes are essential for waterproofing the cuticle, thus limiting non-stomatal water loss and increasing plant tolerance to drought. How cuticular wax function in providing this water barrier is yet unclear. For example, little is known about the role that wax amount, composition, crystallization patterns, physico-chemical interaction between waxes and the cutin polymer of the cuticle membrane, or other wax properties play in plant water use (Premachandra *et al.*, 1991, 1994; Riederer and Schreiber, 2001; Vogg *et al.*, 2004). A complex association between intracuticular waxes and cutin meshwork is likely to define the water transport function of the cuticle membrane. Further studies are needed to investigate what role cuticle lipid structure and composition have in transpiration and plant drought tolerance. Wax mutants should provide a suitable model system to study these questions. In this respect, by investigating the wax profile of several *glossy* mutants of maize, Beattie and Marcell (2002) found that the physicochemical nature of cuticular waxes were probably responsible, either directly or indirectly, for the properties of several leaf surface traits.

Cuticular waxes also influence the reflective properties of leaves, and thus their exposure to light-driven temperature increases and other damages caused by excessive light absorption (Vogelmann, 1993; Barnes *et al.*, 1996).

The wax layer is also the site of the interaction of the plant with insects and micro-organisms acting as the primary defense against pathogens (Eigenbrode and Espelie, 1995; Juniper, 1995; Carver *et al.*, 1996; Beattie, 2001). For example, the prominent waxy bloom on leaves of *Brassica* and *Pisum sativum* (pea) reduces attachment by leaf feeding beetles and provides protection against other herbivorous insects (Stork, 1980; Stoner, 1990; Eigenbrode *et al.*, 1991; Bodnaryk, 1992; White and Eigenbrode, 2000). Moreover, studies in this field showed that plant wax chemical composition may influence host acceptance for oviposition by insect. These effects of the cuticular waxes have been demonstrated to depend on the amount of cuticular waxes produced, their composition, the presence of epicuticular crystalline waxes, and the size, distribution and orientation of these epicuticular crystalline waxes (Eigenbrode and Espelie, 1995; Jenks and Ashworth, 1999; Rashotte, 1999; Morris *et al.*, 2000; Beattie, 2001).

## 4. CUTICULAR WAX BIOSYNTHESIS

### 4.1. Methodological Approaches

The current model for wax biosynthesis was initially elaborated with biochemical approaches, including inhibitor studies, feeding experiments with radiolabeled substrates and assays of enzymatic activities in cell extracts, mainly in *Brassica oleracea*, *Pisum sativum,* and *Allium porrum* (Kolattukudy, 1967, 1971; Kolattukudy *et al.*, 1970, 1973; Cassagne and Lessire, 1978; Agrawal *et al.*, 1984 Lessire *et al.*, 1985; Cheesbrough and Kolattukudy, 1984; Cassagne *et al.*, 1994) However, direct biochemical analysis of wax biosynthesis is hampered by the scarcity of epidermis-specific proteins in extracts from whole organs and by the insolubility of many wax biosynthetic enzymes which are normally membrane bound.

Molecular genetic analysis of mutants impaired in wax metabolism has been an alternative approach used for dissecting the curticular wax biosynthetic pathway. Wax defective mutants, referred to as *glossy* (*gl*) or *eceriferum* (*cer*), have been described in several species including the monocots sorghum (*Sorghum bicolor*), barley (*Hordeum vulgare*), and maize (*Zea mays*), and the dicots cabbage (*Brassica oleracea*), rape (*Brassica napus*), pea (*Pisum sativum*), and *Arabidopsis thaliana* (Holloway *et al.*, 1977; Bianchi *et al.*, 1985, Lundqvist and Lundqvist, 1988; Eigenbrode *et al.*, 1991; Schnable *et al.*, 1994; Jianguo *et al.*, 1995; Jenks *et al.*, 1996, 2000).

The main species employed as model systems for the genetic analysis of wax biosynthesis are *Arabidopsis* and maize. The wax cover on the epidermis of aerial organs confers a typical glaucous appearance to stem and leaves. Therefore, identification of wax defective mutants was traditionally based on visual screens for shiny, bright green surfaces. However, only mutations in genes with a major role in wax biosynthesis or secretion are likely to be identified by visual screens. The *cer* and *gl* mutations identified so far do not saturate the wax biosynthetic pathway (Costaglioli *et al.*, 2005). Other genes with a minor role in wax metabolism have been isolated based on sequence homology.

*Arabidopsis* stems are more heavily covered with surface wax than leaves; therefore alteration in wax load or composition are more easily discerned on stems. Genetic studies indicate that in *Arabidopsis* there are currently 25 known *CER* loci, showing variable degrees of glossiness related to reduction of wax load (Koornneef *et al.*, 1989; McNevin *et al.*, 1993) (Fig. 2). Interestingly, the same shiny phenotype is conditioned by overaccumulation of epicuticular wax (Aharoni et *al.*, 2004). Another trait associated with defective wax biosynthesis in some Arabidopsis *cer* mutants is a conditional male sterility caused by strong reduction of the tryphine layer of pollen grains which impairs pollen hydration in dry environments. Additionally, some *cer* mutations cause slenderness and reduction of plant height (Koornneef *et al.*, 1989).

In maize at least 18 *GL* loci have been defined by mutations and most of them have been mapped (Schnable *et al.*, 1993, 1994; Neuffer, 1997). Maize is unique among cereals for the synthesis of a specific complement of cuticular waxes during the juvenile developmental phase (first five to seven leaves) which is absent from all subsequent adult leaves (Bianchi *et al.*, 1985; Moose and Sisco, 1994). These juvenile waxes confer a typical glaucous appearance to seedling leaves and allowed the identification of the *glossy* mutations which specifically impair their biosynthesis (Lorenzoni and Salamini, 1975) (Fig. 3). Comparisons of the wax composition of seedling leaves of each *gl* mutant to the wax of wild type seedlings were used to putatively identify the biochemical steps controlled by many of the *GL* genes. In view of these data, it has been suggested that cuticular wax biosynthesis occurs via two putative pathways (Bianchi *et al.*, 1985). One pathway is thought to be active throughout the plant life, with esters as the main end product, whereas the other is thought to be active only during the seedling

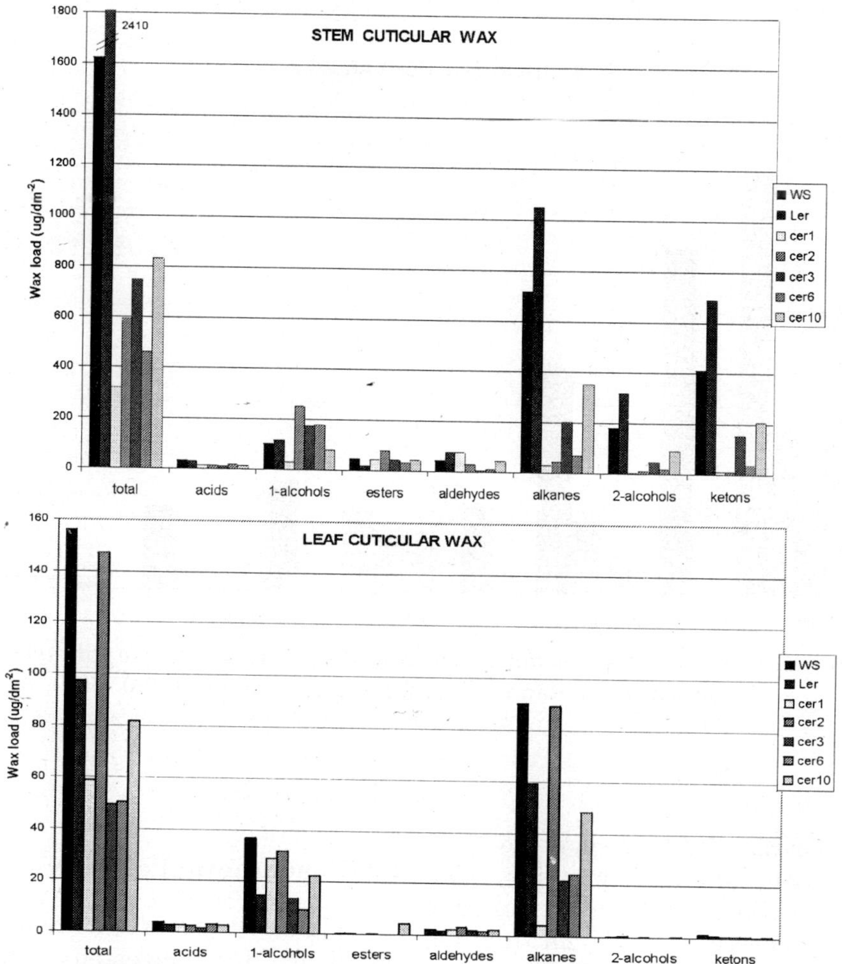


Fig. 2. Cuticular wax amount and composition on stems (A) and leaves (B) of two *Arabidopsis* ecotypes (WS: Wassilewskija; Ler: Landsberg *erecta*) and *eceriferum* (*cer*) mutants. (Data from: Jenks *et al.*, 1995; Rashotte *et al.*, 2000; Goodwin *et al.*, 2005)

stage and the end products of this pathway are mainly alcohols, aldehydes and alkanes.

Defining a role for the cloned *CER/GL* genes relies mainly on changes in wax composition induced by the corresponding mutation and on sequence homologies with proteins of known functions, like members of the chloroplast elongation complex (Jenks *et al.*, 1995; Rashotte, 2001). In few cases, evidence of gene function was deduced from expression in yeast or from functional complementation of yeast mutations. Spatial expression patterns of some wax related genes parallel differences in wax composition between different organs providing additional insights into gene function. However, the phenotype of *cer/gl* mutants is hardly ever conclusive in defining the precise role of the corresponding genes because of global changes in wax composition induced by most mutations. This is partly due to shunting of substrates upstream an enzymatic block into other branches of the biosynthetic pathway. Moreover, some CER/GL proteins (CER1,

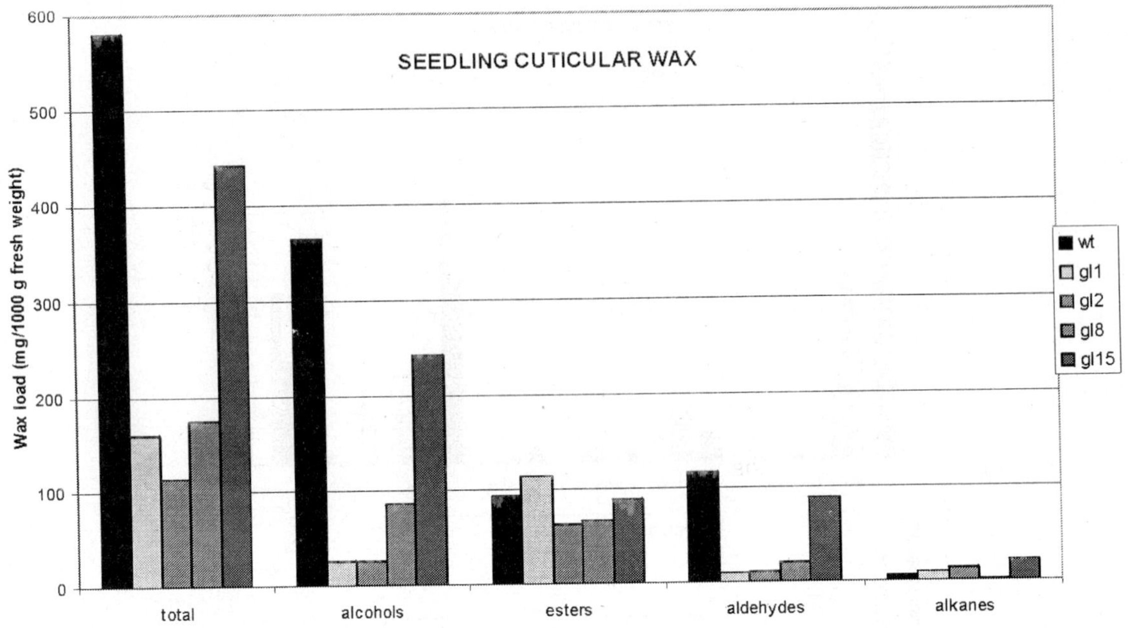


Fig. 3. Cuticular wax amount and composition on seedling leaves of maize wild type plants and *glossy* (*gl*) mutants. (Data from: Bianchi *et al.*, 1985)

GL1, WAX2, CER2, GL2) display no clear similarities with known proteins and contradictory attributions of gene function have been proposed based on limited sequence homology (see CER3). This explains why for most of the *CER* and *GL* genes the precise function in wax biosynthesis has not been determined. Other drawbacks have limited the elucidation of the wax biosynthetic pathway with molecular genetic approaches. None of the *cer* or *gl* mutants characterized so far are completely devoid of any single wax constituent suggesting extensive functional redundancy among structural genes. In addition, attempts to identify clear epistatic relations between *CER* or *GL* genes by means of double mutant analyses revealed complex gene interactions underlying cuticular wax production (Goodwin *et al.*, 2005).

## 4.2 Biosynthetic Pathway

Figure 4 gives a schematic representation of the suggested wax biosynthetic pathway in maize and *Arabidopsis* as deduced from biochemical and molecular genetic studies. It consists of an elongation pathway converting C18 acyl precursors into VLCFAs and two subsequent pathways, the decarbonylation and the acyl reduction pathway, which convert VLCFAs into the other wax constituents. An additional pathway, called the B-acyl-elongation pathway results in the production of B-diketones and their derivatives. These compounds are major components of waxes in barley and minor components in other species such as carnation and Brassicacee and will not discussed here since the topic has been extensively reviewed (Walton *et al.*, 1990).

This scheme may be an oversimplification of the actual process of wax biosynthesis since it is not clear whether different pools of precursors are channeled into the various branch pathways.

### *4.2.1 Acyl elongation*

In the epidermal cells, the C16 and C18 fatty acyl moieties produced by *de novo* fatty acid synthesis in the chloroplast are the precursors of three metabolic pathways originating glycerolipids, cutin and waxes, respectively. C18:0 fatty acyds are primarly channelled into wax biosynthesis while C16:0 and C18:1 fatty acids are mainly converted to membrane glycerolipids and cutin precursors. The first committed step in wax biosynthesis is the sequential elongation of the C 18:0 fatty acyl moieties by consecutive addition of two carbon units derived from malonyl-CoA until the acyl chain reaches C30-C34 carbons in length. These reactions are catalyzed by a multifunctional complex, the fatty acid elongase (FAE), consisting of four distinct enzyme activities: 3-ketoacyl-CoA synthase (KCS), 3-ketoacyl-CoA reductase (KCR), 3-hydroxyacyl-CoA dehydrase (HCD) and 2,3-enoyl-CoA reductase (ECR). The acyl chain elongation for wax synthesis resembles *de novo* fatty acid biosynthesis in several respects:

(a) the FAE complex operates through repeated cycles of condensation, reduction, dehydration and a second reduction step, similarly to the fatty acid synthase (FAS) complex in the chloroplast

(b) the elongating acyl chains remain esterified to acyl carriers during the whole process.

(c) NAD(P)H is the electron donor for the reduction reactions.

However, contrary to *de novo* fatty acid biosynthesis, acyl elongation for wax production takes place in the cytosol, involves microsomal enzymes and employs CoA instead of ACP as the acyl carrier.

In the chloroplast, three distinct FAS complexes are involved in *de novo* fatty acids biosynthesis. Instead it is not clear how many elongases participate in wax production. In *Arabidopsis*, inhibition studies, mutant characterization, and EST analysis (Newman *et al.*, 1994; Cooke *et al.*, 1996; Millar *et al.*, 1999; Todd *et al.*, 1999; Fiebig *et al.*, 2000) highlighted the presence of several distinct elongases with distinct substrate chain-length specificity, probably due to selectivity of the KCS enzyme (Millar and Kunst, 1997). The elongase KCSs are encoded by a multigene family some components of which display organ specific expression patterns. For example, the *FAE1* gene is specifically involved in erucic acid biosynthesis in seeds (James Jr. *et al*, 1995). The *Arabidopsis FAE1* and the jojoba (*Simmonshia chinensis*) *KCS1* genes have been cloned and their products characterized biochemically. Like FAE1, *KCS1* is involved in the production of VLCFAs in seeds where waxes are accumulated as the main storage lipids in jojoba (Lassner *et al.*, 1996). These enzymes have allowed the identification of homologous plant elongase KCSs active in wax biosynthesis. The additional components of the elongase complex remain largely uncharacterized.

### *4.2.2. Genes involved in acyl chain elongation*

It can be anticipated that mutations affecting the acyl elongation pathway would strongly impact cuticular wax composition altering both the decarbonylation and the acyl reduction pathway. Indeed, most of the wax defective mutants characterized so far are impaired in genes involved in acyl chain elongation.

Several genes encoding elongase KCSs have been identified. The *CER6* (*CUT1*) gene of *Arabidopsis* was isolated based on its sequence similarity with *KCS* genes (Millar *et al*., 1999) and with a positional cloning approach (Fiebig *et al*., 2000). Contrary to the *FAE1* and jojoba *KCS1* genes, *CER6* is specifically expressed in tissues active in wax biosynthesis (epidermis of aerial organs and tapetum cells) and its transcription is induced by environmental cues promoting cuticular wax accumulation (Kunst *et al*., 2000; Hooker *et al*., 2002). *cer6* mutants display a strong reduction of wax load to 20% of the wild type level and conditional male sterility. Products of both the decarbonylation and the acyl reduction pathway are affected, suggesting that in *Arabidopsis* the same pool of VLCFA is shared by the two branch pathways. The changes in wax composition conditioned by *cer6* mutations are consistent with a block in acyl chain elongation beyond C24. In fact, in contrast to the dramatic decrease of wax compounds longer than C26 which constitute >90% of total wax load in wild-type plants, levels of C24 wax compounds are increased in *cer6* mutants. In conclusion, sequence analysis and mutant wax phenotype indicate that CER6 is the main elongase KCS involved in acyl chain elongation beyond C24.

Sense suppression of *CER6* in transgenic *Arabidopsis* plant leads to an ever strong reduction of wax load than observed in *cer6* null mutants. This could be explained by the presence of additional *KCS* genes homolog to *CER6* and with overlapping functions. A possible candidate is *CER60* which shares 88% identity with *CER6* and shows a preferential expression in silk tissues but it is also active in stems, though to a lesser extent than *CER6* (Fiebig *et al*., 2000).

Another *CER6* homolog with high sequence similarities with known KCSs is the *Arabidopsis KCS1* gene (Todd *et al*., 1999). Direct evidence of KCS1 function was obtained expressing the corresponding cDNA in yeast. The enzyme converted saturated 18:0 CoA to 26:0 CoA thus confirming its elongation activity on the precursors of wax compounds. Furthermore, it was demonstrated that KCS1, like other fatty acid condensing enzyme, catalyzes multiple sequential reactions (Lassner *et al*., 1996; Kunst *et al*., 1992; Millar and Kunst, 1997). A knockout *kcs1* mutant displayed altered wax composition although total wax load was only slightly decreased compared to wild-type plants: the only visual alterations of *kcs1* mutants was reduced stem thickness. As in the case of *cer6* defective plants, no single wax component was completely absent from *kcs1* mutants, further suggesting the presence of multiple wax FAE complex in *Arabidopsis* with similar but not interchangeable functions.

An additional *Arabidopsis* gene with a putative role as a wax KCS enzyme is *FIDDLEHEAD* (*FDH*) (Lolle *et al*., 1997; Pruitt *et al*., 2000; Yephremov *et al*., 1999). This gene is mainly expressed in the epidermis of vegetative and floral organs and displays high homology with members of the FAE family and with other lipid condensing enzymes. However, besides sequence similarities, no other evidence supports the predicted function. *fdh* mutants have normal glaucous stems and although detailed characterization of wax composition has not been reported, analysis of crude cell wall fractions highlighted an increase in the amount of total long chain lipids condition by the *fdh* mutation. The mutant phenotype is complex and includes fusions of leaves and floral organs and increased epidermis permeability, both traits associated with defects in cutin synthesis (Sieber *et al*., 2000). Whether *FDH* plays a role in the metabolism of other lipids, perhaps those playing a role in cell-cell communication, has not been established.

A putative wax related KCR gene is the maize *GL8* (Xu *et al*., 1997, 2002). This genes is expressed in seedling leaves and, to a lesser extent, in other organs and at other developmental stages. The amino acid sequence exhibits significant sequence similarity to a group of enzymes from plants, eubacteria, and mammals that catalyzes the reduction of ketons, including plastidial 3' ketoacyl ACP reductases of *de novo* fatty acid biosynthesis. The Gl8 polypeptide is associated with the ER membranes and is immunodetected by antibodies raised against the leek epidermis acyl CoA elongase complex, strongly indicating its involvement in a homologous complex of maize. Moreover, anti-Gl8 antiserum inhibits total microsomal associated elongase activities from leek and maize, without significantly perturbing either of the other three activieties associated with 3-ketoacyl CoA reductase (KCS, HCD and ECR). Altogether, these findings and the chemical composition of the wax on *gl8* mutants suggest that GL8 is the 3 ketoacyl CoA reductase of an elongation complex involved in wax biosynthesis in juvenile leaves.

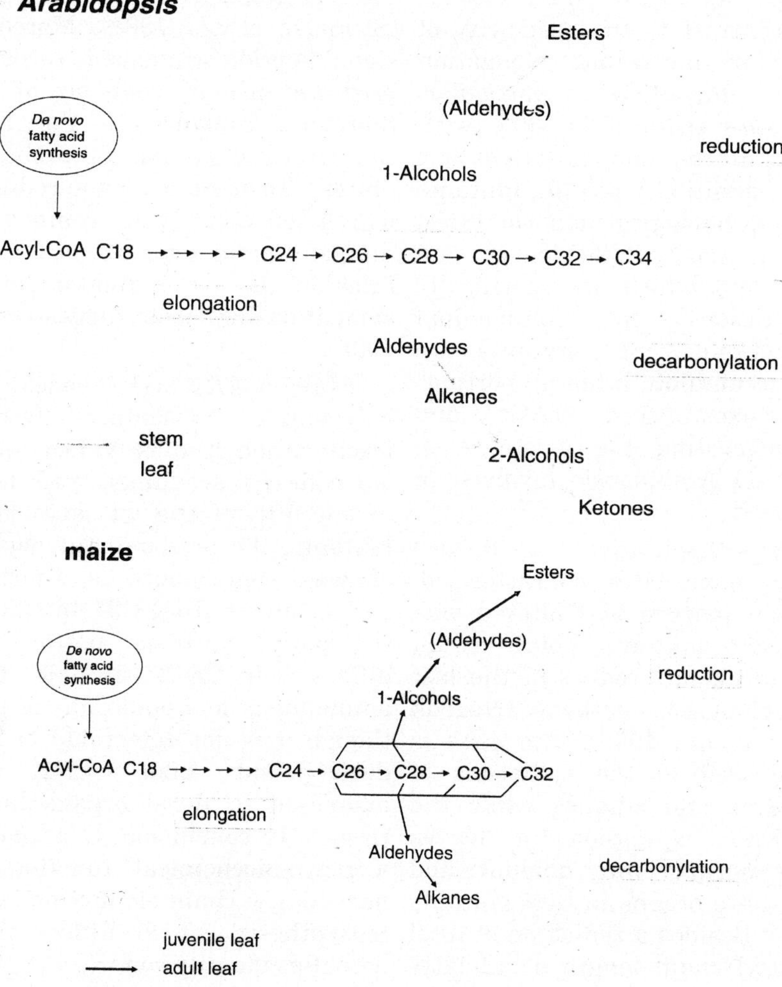


Fig. 4. Schematic representation of the proposed wax biosynthetic pathway for *Arabidopsis* stem and leaves (A) and maize juvenile and adult leaves (B). The main fluxes of VLCFAs into the reduction and the decarbonylation pathways are depicted with coloured arrows. The aldehyde intermediates in the reduction pathway (in brackets) are probably enzyme-bound and do not contribute to the pool of wax compounds. 1-Alcohols = primary alcohols; 2-Alcohols = secondary alcohols.

The *Arabidopsis CER10* gene was isolated as a gene encoding an enoyl-CoA reductase (ECR) involved in VLCFA biosynthesis based on sequence homologies and functional complementation of a yeast *ECR* gene (Kohlwein *et al.*, 2001; Gable *et al.*, 2004). Mutations of *CER10* cause reductions of all VLCFA containing lipids: cuticular waxes, seed storage triacylglycerols (TAGs) and sphingolipids, underlying an ubiquitous role of ECR in VLCFA biosynthesis in contrast to the specificity of KCS enzymes for individual elongation systems (Zheng *et al.*, 2005). In particular, cuticular waxes are reduced by 60% with large decreases of all the main classes of wax compounds. In addition, *cer-10* mutants display several morphological abnormalities, including smaller and crinkle leaves and reduced plant height which are specifically due to alterations in sphingolipid metabolism. Knockout *cer10* mutants are viable and still accumulate large proportions of cuticular waxes, seed TAGs and sphingolipids, suggesting the existence of additional ECRs in Arabidopsis involved in VLCFA biosynthesis.

Other genes with putative roles in acyl chain elongation have been characterized. Spatial expression pattern and altered wax phenotype in *cer2* mutants point to an involvement of the gene product in the last step of the acyl elongation pathway (Negruk *et al.*, 1996; Xia *et al.*, 1996). The gene is specifically expressed in the epidermis of stem, floral organs and siliques while the absence of *CER2* expression in leaves correlates with normal wax amount and composition on these organs in *cer2* mutants (Xia *et al.*, 1997). Besides a reduction of total wax load, the acyl chain length of all fatty acids derivatives turns out to be reduced by two carbon units in stems of *cer2* mutants compared to the wild type. The overall changes in wax composition in *cer2* defective stems closely resembles the changes conditioned by the *cer6* mutation, suggesting that the two genes play a role in the same biosynthetic steps. Nevertheless, the CER2 protein sequence has no homology to any FAE component. The only gene to which *CER2* is clearly related is the maize *GL2* (Tacke *et al.*, 1995) which plays an undefined function in wax metabolism (see below) and limited similarities are found with a class of CoA-dependent acyltransferases, although two highly conserved consensus sequences of these proteins are modified in CER2 and GL2 (St-Pierre *et al.*, 1998). Moreover, no clear signal peptide sequence is recognizable in the predicted primary sequence of CER2 which therefore, contrary to the FAE enzymes characterized so far, is probably a soluble protein. To account for the reduced acyl chain length of stem wax compounds and the unique protein sequence of CER2, a putative function as a regulator of acyl chain elongation has been suggested (Jenks *et al.*, 1995).

Like *CER2*, *GL2* codes for a soluble polypeptide, as deduced from subcellular fractionation studies (Velasco *et al.*, 2002). In *gl2* defective seedlings, wax load is reduced to one-fifth of the wt level and as in *cer2* mutants, the predominant acyl chain length of wax components is shortened by two carbon atoms from C32 to C30. As concerns the spatial expression pattern, *GL2* partially differs from *CER2* since its transciption is abundant in leaf epidermis of juvenile plants though it is not detectable at later stages of development. Like *CER2*, *GL2* is also expressed in floral organs but not in root tissues. In conclusion, *GL2* and *CER2* define a new biochemical function essential for normal acyl chain elongation in cuticular wax biosynthesis, which differs from the FAE activities identify so far.

*CER3* is another wax related gene with no clear homology with known proteins. Consistent with the broad expression pattern of the *CER3* gene, *cer3* mutants show an overall reduction of the various wax compounds compared to the wild type on both stems and leaves and conditional male sterility (Hannoufa *et al.*, 1996). Interestingly, in mutant *cer3* plants the homologous series of alkanes and stem primary alcohols are skewed towards longer chain lengths. Therefore, it was suggested the involvement of CER3 in the hydrolysis of fatty acyl-CoA into free fatty acids and CoA which stops their elongation and starts their conversion to other wax constituents (Jenks *et al.*, 1995). Anyway, the deduced aminoacid sequence of the large (90 kDa) CER3 protein displays no homology with members of the fatty acyl-ACP thioesterase gene family, which hydrolyze acyl-ACP in *de novo* fatty acid biosynthesis. Alternatively, it was suggested that CER3 acts as an acyl transporter involved in the regulation of the fatty acyl-CoA hydrolase activity. A further hypothesis implies an indirect control of this activity by *CER3* which would code for a transcriptional activator, given the presence of a putative nuclear localization sequence and two possible phosphorilation sites in the protein sequence, although CER3 subcellular localization has not been determined. Based on partial homology of CER3 with ubiquitin-like E3 protein ligase, a putative role in targeting wax related proteins for degradation was also suggested (Jenks *et al.*, 2002)

### *4.2.3. Reduction and decarbonylation pathways*

Both intermediate and end products of the acyl chain elongation pathway, that is acyl moieties from C24 to C34 are the initial substrates of two branch pathways, the acyl reduction pathway and the decarbonylation pathway which give rise to the different compounds constituting cuticular waxes in maize and *Arabidopsis*. It is not clear to what extent the VLCFA pool and the aldehydes derived from its initial reduction are shared by the two branch pathways. In *Arabidopsis*, shorter chain acyl-CoA precursors (up to C28) are channelled into the reduction pathway whereas longer acyl moieties are mainly substrates of the decarbonylation pathway.

*Acyl reduction pathway* - This pathway converts VLCFAs into aldehydes and primary alcohols by means of two reductive reactions catalyzed by one or two fatty acid reductase (FAR) enzymes. Subsequentially, wax synthase (WS) esterifies primary alcohols with VLCFAs of different chain lengths to produce wax esters.

At least in pea there are likely two different fatty acid reductases (FAR): the former reduces VLCFAs to primary alcohols in the reduction pathway without releasing the aldehyde intermediates, the latter converts VLCFAs into aldehydes in the decarbonylation pahway (Vioque and Kolattukudy, 1997). Thus wax aldehydes would derive exclusively from decarbonylation of VLCFA. Short chain acyl-CoA precursors (up to C28). A similar two-step reductase converting VLCFA into alcohols has been purified from jojoba seeds, which accumulate waxes as storage lipids. Expression of the corresponding cDNA in *E. coli* and *B. napus* resulted in the accumulation of fatty alcohols (Metz *et al.*, 2000). Transcripts related in sequence to the jojoba FAR have been found in several other species suggesting that these alcohol-generating reductases are widespread in plants but not in *Brassica oleracea* where two different enzyme catalyze the two reductive steps of the acyl reduction pathway (Kolattukudy, 1971).

Wax synthase (fatty acyl-CoA:fatty alcohols acyl transferase) has been characterized biochemically and partially purified from different sources (Kolattukudy, 1967). A cDNA encoding this activity has been cloned from developing jojoba embryos. Transgenic Arabidopsis plant expressing this WS clone together with the jojoba FAR and a KCS from *Lunaria annua*, accumulated high levels of wax esters in seeds, thus confirming the identity of the WS cDNA. (Lardizabel *et al.*, 2000).

*Decarbonylation pathway* - The first step in the decarbonylation pathway is again the reduction of VLCFAs to aldheydes which are then decarbonylated to odd-chain alkanes with release of carbon monoxide in a reaction catalyzed by aldehyde decarbonylase (ADC) enzymes. In *Arabidopsis* but not in maize, alkanes are further converted to secondary alcohols and ketones by means of enzymatic activities not yet characterized either biochemically or molecularly.

Aldehyde decarbonylases from pea and from the green alga *Botryococcus brownii* have been characterized to some extent but the corresponding genes have not been cloned (Dennis and Kolattukudy, 1992; Schneider-Belhadded and Kolattukudy, 2000). These plant and algal enzymes use fatty aldehydes as substrates, are membrane-bound and need metal cofactors for catalysis, Cu and Co respectively.

### 4.2.4. Genes involved in the reduction and decarbonylation pathways

Alkanes and their derivatives are the main constituents of stem wax in Arabidopsis. Therefore mutations in the aldehyde decarbonylase converting C30 aldehydes to alkanes in the acyl decarbonylation pathway should greatly impact total wax load and composition. In the *Arabidopsis cer1* mutant products of aldehyde decarbonylation, that is alkanes, secondary alcohols and ketones are strongly reduced while other wax constituents are less affected. Based on this wax phenotype, *CER1* was proposed to catalyze the conversion of fatty aldehydes to alkanes in the decarbonylation pathway of wax biosynthesis (Hannoufa *et al.*, 1993). The predicted *CER1* function is not bolstered by sequence homologies since no gene encoding a fatty aldehyde decarbonylase has been isolated so far. However, like the aldehyde decarbonylases from pea and from the green alga *Botryococcus brownii*, CER1 appears to be an integral membrane protein displaying three histidine-rich motifs proposed to bind non heme-iron ions (Aarts *et al.*, 1995). Similar Histidine-rich motifs are the signature sequence of a large family of membrane bound lipid processing enzymes which includes desaturases, alkane hydroxylases, xylene monooxygenases (Shanklin and Cahoon, 1998). No aldehyde decarbonylase has been identified among these activities and these enzymes have about half the size of CER1. Recently, based on global wax changes induced by the *cer1* mutation and the phenotype of double mutants, a role of *CER1* also in the reduction pathway was suggested (Goodwin *et al.*, 2005).

Sequences related to *CER1* are found in several plant species. Two of these have been characterized, the *WAX2* gene of Arabidopsis and its maize orthologue *GL1*.

*WAX2* shows high sequence homology with *CER1* (35% identity all over the protein sequence) and strong similarities between these two genes are also found in the gene structure since all the nine introns of *CER1* match exactly the position of nine introns of *WAX2* (Chen *et al.*, 2003). This implies that *WAX2* and *CER1* are likely paralogous genes. Important structural features of the CER1 polypeptide, like the three His-rich motifs and transmembrane domains are also conserved. Tissue specific expression of *WAX2* is similar to that reported for *CER1* except that *WAX2* is more strongly expressed in siliques and *CER1* in flowers, and mutations in both genes cause conditional male sterility (Ariizumi *et al.*, 2003). However, *WAX2* does not seem to code for an aldehyde decarbonylase. In *wax2* mutants, all products of the decarbonylation pathway are strongly decreased compared to the wild-type, including aldehydes whereas C30 primary alcohols are increased on stems. This was taken as evidence for the involvement of *WAX2* in the reduction of acyl-CoA to aldehydes in the decarbonylation pathway, although the same function has been proposed for at least four other *Arabidopsis* genes: *CER3*, *CER7*,- *CER8* and *CER13* (Post-Beittenmiller, 1998; Rashotte *et al.*, 2001). In addition, *wax2* mutations have a pleiotropic effect on epidermis development inducing alteration in cuticle membrane structure, post-genital organ fusion, increased epidermal permeability and abnormal trichome morphology (Chen *et al.*, 2003; Kurata *et al.*, 2003). This implies a complex role of *wax2* in both cuticular wax and cutin formation (Sieber *et al.*, 2000). It seems therefore that after divergence from a common ancestor gene, *CER1* and *WAX2* specialized towards quite different functions which have still to be clarified.

Although previously considered a putative membrane receptor (Hansen *et al.*, 1997), *GL1* appears to be the *WAX2* othologue of maize, as deduced from sequence comparison (68% identity at the protein level), and from overall wax changes and epidermal alterations conditioned by *gl1* mutations (Sturaro *et al.*, 2005). Expression of this gene is down regulated by environmental factors known to induce wax biosynthesis (light and drought) further implying a more complex role for *GL1* than wax biosynthesis (Sturaro *et al.*, unpublished data).

## 5. CUTICULAR WAX SECRETION

After biosynthesis, wax compounds are secreted to the epidermal surface. This topic has been recently reviewed (Kunst and Samuels, 2003). Because of its highly hydrophobic nature, there are some constrains to wax delivery from the ER to the plasma membrane. In plants there is no evidence of fatty acyl binding proteins like those interacting with VLCFAs in the cytosol of mammalian cells and mediating their movement in this compartment. Two hypothesis have been proposed to explain how wax transport to the PM could occur. The former predicts that a direct transfer of wax components could take place in regions where tight associations between cortical ER and plasma membrane occur. Such domains have been proposed to be site of active exchange of lipids between PM and ER for their recycling after exocytosis. The latter suggests that wax compounds could be traslocated via the secretory pathway by means of uncoated vescicles, like those delivering sphingolipids from the ER to the PM via the Golgi complex (Jenks *et al.*, 1994; Moreau *et al.*, 1998).

The subsequent movement of wax compounds from the PM to the apoplast is an active process, which could involve members of the "ABC transporter" protein family. These are widespread pumps with a broad substrate specificity localized on PM and internal membranes (Rea, 1998). On the PM of mammalian cells these transporters deliver a wide variety of molecules, including lipids, out of the cell. (Klein *et al.*, 1999). A recent breakthrough in wax export from PM to the apoplast has been the characterization of the *cer5* mutant of *Arabidopsis* (Pighin *et al.*, 2004). This mutant shows a reduction of wax load on the stem surface but total wax content in epidermal cells is comparable to the wild type level, thus suggesting an involvement of the gene product in wax transport. The CER5 protein is localized in the PM of epidermal cells and its structure resembles that of ABC transporters with a single ABC domain. In the cytosol of mutant epidermal cells, ultrastructural analysis revealed the presence of linear lipidic inclusions. Similar inclusions were observed in mammalian cells defective for an ABC transporter involved in translocation of VLCFAs into the peroxisome. Collectively these data indicate that CER5 is a component of the wax secretory pathway. However, *CER5* is not integral to the process of wax extrusion across the PM since the *cer5* phenotype is limited to stem and leaf while wax load on other aerial organs like flowers and silique is unaffected. Moreover, in a *cer5* null-mutant, residual surface waxes accumulate on stems suggesting that additional ABC-transporters or other transport mechanisms operate in wax extrusion to the cell wall.

After exported to the apoplast, wax compounds must move through the hydrophilic cell wall compartment to reach the cuticle. Non-specific lipid transfer protein (nsLTP) with broad substrate specificity for a variety of lipid molecules are abundant in plant wax and may be the candidate carriers for wax transport (Pyee *et al.*, 1994). However, from structural data and biochemical analysis of the maize nsLTP, it seems unlikely that these proteins could interact with acyl chains longer than 18 C (Zachowski *et al.*, 1998). Alternatively, wax could move in the cell wall through hydrophobic paths created by non polar regions of structural proteins, like glycine-rich proteins, or methyl esterified pectines. The last step in wax translocation in the apoplast is the movement of epicuticular waxes through the cuticle to the plant surface. There is no evidence of physical pores in the cuticle involved in wax extrusion, as initially postulated (Jeffree, 1996). Instead, Neinhuis *et al.* (2001) suggested a simple hypothesis to explain extrusion of epicuticular wax components which implies their movement in the water current permeating the cuticle.

## 6. GENETIC REGULATION OF WAX BIOSYNTHESIS

Most of the wax related genes characterized so far are structural genes involved in wax biosynthesis or secretion. Only three putative regulatory genes of wax biosynthesis have been identified: *WIN1/SHN1*, *WXP1* and *GL15*. All of them belong to the AP2/EREBP family of transcription factors and have a broader role than wax regulation. Therefore, regulatory genes specifically committed to wax biosynthesis, if any, have still to be characterized. Studies on these regulatory genes highlighted a complex regulatory network underlying wax biosyntheis, consistent with the genetically determined differences in wax amount and composition related to tissue and developmental stages.

The *WIN1/SHN1* gene of *Arabidopsis* was identified with both an activation tagging approach and a functional genomic approach (Aharoni *et al.*, 2004; Broun *et al.*, 2004). Transgenic *Arabidopsis* plants constitutively overexpressing the *WIN1/SHN1* gene displayed an up to six-fold increase of wax load on leaves and stem and altered wax composition. The concentration of all wax constituents was raised but the most remarkable increase was observed for products of the decarbonylation pathway, especially long chain alkanes. Plate-like wax crystal accumulated on leaf surfaces and TEM analysis revealed the presence of a thick internal lipid layer in the epidermis of transgenic plants. In addition, abnormal leaf morphology and epidermal cell differentiation were observed. Recovery after water stress was clearly enhanced in *WIN1/SHN1* overexpressing lines compared to wild type plants in spite of increased cuticle permeability. *WIN1/SHN1* belongs to a clade of three genes which are likely to perform similar functions since constitutive overexpression of either of them leads to the same phenotype. However their spatial expression patterns overlap only partially: *WIN1/SHN1* expression is detected in inflorescence, roots, and abscission zone of the silique but not in stem or mature leaves. *SHN2* transcription is limited to flowers, whereas *SHN3* is expressed in most organs and is induced upon wounding.

The *WIN1/SHN1* transcription level in transgenic plants appears correlated with the expression of some wax structural genes (*CER1, KCS1, CER2*), suggesting a role *WIN/SHN1* as a regulator of wax biosynthesis. However, microarray profiling in *WIN/SHN1* overexpressing lines showed that other genes involved in lipid metabolism turned out to be induced. The phenotype associated with *win1/shn1* null mutations was not reported. Anyway, spatial expression patterns of the three *WIN/SHN* genes and genomic analysis of gene transcription in overexpressing lines suggest that the *WIN/SHN* genes are functional in the regulation of lipid metabolism involved in plant protection against environmental stresses and upon abscission and dehiscence.

The *Medicago truncatula WXP1* gene is induced by cold, drought and ABA, all factors promoting wax accumulation (Zhang *et al.*, 2005). When constitutively overexpressed in *Medicago sativa* (alfalfa), wax loading on leaves was increased by about 1/3, mainly due to higher levels of C30 primary alcohols. This higher wax loading correlated with decreased transpirancy and increased drought tolerance. Consistent with the involvement of *WXP1* in the regulation of wax metabolism, three putative *FAE* genes were induced in the transgenic plants. However, the expression of other wax biosynthetic genes remained unaffected or even decreased. Similarly to *WIN1*, the *WXP1* regulatory functions are not limited to wax biosynthesis but extend to other lipid biosynthetic genes, as deduced from analysis of gene expression and lipid composition in overexpressing lines.

Comparison of the wax phenotype and gene expression patterns induced by *WIN1* and *WXP1* overexpression leads to the conclusion that regulation of wax biosynthesis is under the control of multiple regulatory genes. *WXP1* seems to be mainly involved in the control of the acyl reduction pathway while *WIN1* appears to regulate mainly the decarbonylation branch pathway. The direct activation of wax structural genes by *WIN1* and *WXP1* has not been proved. Therefore it can not be ruled out an indirect control of *WIN1* and *WXP1* on wax biosynthesis mediated by yet unidentified, committed transcription factors.

The maize *GL15* gene is involved in the juvenile-to-adult phase transition of maize seedlings (Evans *et al.*, 1994; Moose and Sisco, 1994, 1996). In particular, *GL15* regulates the juvenile identity of epidermal cells in leaves 3-7 where it is specifically expressed. In *gl15* mutants, epidermal cells of seedling leaves display morphological traits typical of adult leaves, including the absence of epicuticular wax and changes in intracuticular wax composition, the presence of bulliform cells and associated macrohairs, modifications of cell shape and cell wall structure.

While regulation of *GL15* by other developmental genes (*Corngrass1*, *Teopod1* and *Teopod2*) has been investigated, downstream target genes controlled by *GL15* have not been studied. Structural genes involved in juvenile wax biosynthesis, like *GL1*, *GL2* and *GL8* are some of the putative candidate. However the *GL* genes characterized so far are expressed also in leaves 1-2, where *GL15* transcription is barely detectable and its mutations have no phenotypic effect. Moreover, preliminary results underlined a different expression pattern of the structural *GL* genes in response to environmental cues inducing wax biosynthesis. Therefore, additional transcription factors are likely to control wax biosynthesis in juvenile leaves in response to both the normal developmental program and environmental signals.

## CONCLUSIONS

In this review we have attempted to cover recent research on the biosynthesis, secretion, regulation, and physiological function of cuticular waxes. This will provide a basis for genetic and metabolic engineering of crops for enhanced tolerance to biotic and/or abiotic stresses, and for commercially improving plants by altering the wax content of valuable organs or tissues such as seeds or fruits. In this respect, promising results have already been achieved:

- Overexpression of wax regulatory genes determined an increase of cuticular wax load and induced significant drought tolerance, thus providing evidence of potentiality of wax metabolic engineering (Aharoni *et al.*, 2004; Zhang *et al.*, 2005).
- *Arabidopsis* plants engineered to produce high levels of wax compounds in seed oil disclosed the possibility of producing high amount of waxes at low cost (Lardizabal et al., 2000).

As highlighted in the previous discussion, progress has been made in the identification of gene products involved in wax production and in cloning of a number of wax related genes. Despites of these advances, not all genes involved in the biosynthesis, regulation and transport of wax to the plant surface have been identified by genetic approaches and many gaps have to be filled before our understanding of wax metabolism is complete. In this respect, additional efforts should be focused on the nature and role of *trans*-acting factors specifically committed to the regulation of the wax biosynthetic pathway.

Advances in plant genetics and genomic technologies are contributing to the acceleration of gene discovery. The correlative analysis of extensive gene expression profiling and metabolic profiling is a powerful approach for the identification of candidate genes and enzymes (Brown and Botstein, 1999). The challenges are to identify new enzymes and to develop biochemical techniques that are suitable to large scale analyses (Fridman and Pichersky, 2005). Progress in these areas will allow the isolation of many new genes and the identification of regulatory control points that have a major impact on wax metabolism.

## Acknowledgements

We apologize to all those whose contributions we did not include in this review because of space constraints or simple oversight. Research in this laboratory is supported by Ministero per le Politiche Agricole e Forestali, Roma, special grant “Sviluppo di una cerealicoltura innovativa e competitiva”.

## REFERENCES

Aarts, M.G., C.J. Keijzer, W.J. Stiekema and A. Pereira. 1995. Molecular characterization of the CER1 gene of arabidopsis involved in epicuticular wax biosynthesis and pollen fertility. *Plant Cell*, **7**: 2115-2127.

Agrawal, V.P., R. Lessire and P.K. Stumpf. 1984. Biosynthesis of very long chain fatty acids in microsomes from epidermal cells of *Allium porrum* L. *Arch. Biochem. Biophys.*, **230**: 580-589.

Aharoni, A., S. Dixit, R. Jetter, E. Thoenes, G. van Arkel and A. Pereira. 2004. The SHINE clade of AP2 domain transcription factors activates wax biosynthesis, alters cuticle properties, and confers drought tolerance when overexpressed in Arabidopsis. *Plant Cell*, **16**: 2463-2480.

Ariizumi, T., K. Hatakeyama, K. Hinata, S. Sato, T. Kato, S. Tabata and K. Toriyama. 2003. A novel male-sterile mutant of Arabidopsis thaliana, faceless pollen-1, produces pollen with a smooth surface and an acetolysis-sensitive exine. *Plant Mol. Biol.*, **53**: 107-116.

Avato, P., G. Bianchi and N. Pogna. 1990. Chemosystematics of surface lipids from maize and some related species. *Phytochemistry*, **29**: 1571-1576.

Baker, E.A. 1980. Chemistry and morphology of plant cuticular waxes. In: *The Plant Cuticle* (eds. D.F. Cutler, K.L. Alvin and C.E. Price), pp. 139-165. Academic Press, New York.

Barnes, J.D., K.E. Percy, N. D. Paul, P. Jones, C.K. McLaughlin, P.M. Mullineaux, G. Creissen and A.. Wellburn. 1996. The influence of UV-B radiation on the physicochemical nature of tobacco (*Nicotiana tabacum* L.) leaf surfaces. *J. Exp. Botany,* **47**: 99-109.

Barthlott, W. and C. Neinhuis. 1997. Purity of the sacred lotus or escape from contamination in biological surfaces. *Planta*, **202**: 1-7.

Beattie, G.A. 2001. Leaf surface waxes and the process of leaf colonization by microorganisms. In: *Phyllosphere Microbiology* (eds. S.E. Lindow, E. Hecht-Poinar and V. Elliott) pp. 3-26. APS Press, Minneapolis, MN.

Beattie, G.A. and L.M. Marcell. 2002. Effect of alterations in cuticular wax biosynthesis on the physiochemical properties and topography of maize leaf surfaces. *Plant Cell Environ.*, **25**: 1-16.

Bengtson, C., S. Larsson and C. Liljenberg. 1979. Water stress, epicuticular wax and cuticular transpiration rate. In: *Advances in the Biochemistry and Physiology of Plant Lipids* (eds. L.A. Appelqvist and C. Liljenberg), pp. 269-274. Elsevier/North-Holland Biomedical Press, New York.

Bianchi, A., G. Bianchi, P. Avato and F. Salamini. 1985. Biosynthetic pathways of epicuticular wax of maize as assessed by mutation, light, plant age and inhibitor studies. *Maydica*, **30:** 179-198.

Bianchi, G. 1995. Plant waxes. In: *Waxes: Chemistry, Molecular Biology and Functions*. (ed. R.J. Hamilton) pp. 175–222. The Oily Press LTD, Dundee, Scotland.

Bodnaryk, R.P. 1992. Distinctive leaf feeding patterns on oilseed rapes and related Brassicaceae by flea beetles, *Phyllotreta cruiciferae* (Goeze) (Coleoptera: Chrysomelidae). *Can. J. Plant Sci.*, **72**: 575-581.

Broun, P., P. Poindexter, E. Osborne, C.Z. Jiang and J.L. Riechmann. 2004. WIN1, a transcriptional activator of epidermal wax accumulation in Arabidopsis. *Proc. Natl. Acad. Sci. USA*, **101**: 4706-4711.

Brown, P.O. and D. Botstein. 1999. Exploring the new world of the genome with DNA microarray. *Nat. Genet. Suppl.*, **21**: 33-37.

Cameron, K.D., M.A. Teece, E. Bevilacqua and L.B. Smart. 2002. Diversity of cuticular wax among Salix species and Populus species hybrids. *Phytochemistry*, **60**: 715-725.

Carver, T.L.W., S.M. Ingerson, and B.J. Thomas. 1996. Influences of host surface features on development of *Erysiphe graminis* and *Erysiphe pisi*. In: *Plant cuticles, an integrated and functional approach* (ed. G. Kersteins), pp. 33-82. Bios Scientific Publishers, Oxoford, UK.

Cassagne, C., R. Lessire, J.J. Bessoule, P. Moreau, A. Creach, F. Schneider and B. Sturbois. 1994. Biosynthesis of very long chain fatty acids in higher plants. *Prog. Lipid Res.*, **33**: 55-69.

Cassagne, C. and R. Lessire. 1978. Biosynthesis of saturated very long chain fatty acids by purified membrane fractions from leek epidermal cells. *Arch. Biochem. Biophys.*, **191**: 146-152.

Cheesbrough, T.M. and P.E. Kolattukudy. 1984. Alkane biosynthesis by decarbonylation of aldehydes catalyzed by a particulate preparation from Pisum sativum. Proc. Natl. Acad. Sci. USA, **81**: 6613-6617.

Chen, X., S.M. Goodwin, V.L. Boroff, X. Liu and M.A. Jenks. 2003. Cloning and characterization of the *WAX2* gene of Arabidopsis involved in cuticle membrane and wax production. *Plant Cell*, **15**: 1170-1185.

Cooke, R. 1996. Further progress toward a catalogue of all *Arabidopsis* genes: analysis of a set of 5000 non-redundant ESTs. *Plant J.*, **9**: 101-124.

Costaglioli, P., J. Joubes, C. Garcia, M. Stef, B. Arveiler, R. Lessire and B. Garbay. 2005. Profiling candidate genes involved in wax biosynthesis in *Arabidopsis thaliana* by microarray analysis. *Biochim. Biophys. Acta*, **1734**: 247-258.

Dennis, M. and P.E. Kolattukudy. 1992. A cobalt-porphyrin enzyme converts a fatty aldehyde to a hydrocarbon and CO. *Proc. Natl. Acad. Sci. USA*, **89**: 5306-5310.

Eigenbrode, S.D. and K.E. Espelie. 1995. Effects of plant epicuticular lipids on insect herbivores. *Ann. Rev. Entomol.*, **40**: 171–194.

Eigenbrode, S.D., K.A. Stoner, A.M. Shelton, and W.C. Kain. 1991. Characteristics of glossy leaf waxes associated with resistance to diamondback moth (Lepidoptera: Plutellidae) in *Brassica oleracea*. *J. Econ. Entomol.*, **84**: 1609-1618.

Evans, M., H.J. Passas and R.S. Poethig. 1994. Heterochronic effects of *glossy15* mutations on epidermal cell identity in maize. *Development*, **120**: 1971-1981.

Fiebig, A., J.A. Mayfield, N.L. Miley, S. Chau, R.L. Fischer, and D. Preuss. 2000. Alterations in CER6, a gene identical to *CUT1*, differentially affect long-chain lipid content on the surface of pollen and stems. *Plant Cell*, **12**: 2001-2008.

Fridman, E. and E. Pichersky. 2005. Metabolomics, genomics, proteomics, and the identification of enzymes and their substrates and products. *Plant Biol.*, **8**: 242-248.

Gable, K., S. Garton, J.A. Napier and T.M. Dunn. 2004. Functional characterization of the *Arabidopsis thaliana* orthologue of *Tsc13p*, the enoyl reductase of the yeast microsomal fatty acid elongating system. *J. Exp. Bot.*, **55**: 543-545.

Goodwin, S.M., A.M. Rashotte, M. Rahman, K.A. Feldmann and M.A. Jenks. 2005. Wax constituents on the inflorescence stems of double *eceriferum* mutants in *Arabidopsis* reveal complex gene interactions. *Phytochemistry*, **66**: 771-780.

Hadley, N.F. 1989. Lipid water barrier in biological systems. *Prog. Lipid Res.*, **28**: 1-33.

Hannoufa, A., J. McNevin and B. Lemieux. 1993. Epicuticular waxes of *eceriferum* mutants of *Arabidopsis thaliana*. *Phytochemistry* **33**: 851-855.

Hannoufa, A., V. Negruk, G. Eisner and B. Lemieux. 1996. The *CER3* gene of *Arabidopsis thaliana* is expressed in leaves, stems, roots, flowers and apical meristems. *Plant J.*, **10**: 459-467.

Hansen, J.D., J. Pyee, Y. Xia, T.J. Wen, D.S. Robertson, P.E. Kolattukudy, B.J. Nikolau and P.S. Schnable. 1997. The glossy1 locus of maize and an epidermis-specific cDNA from Kleinia odora define a class of receptor-like proteins required for the normal accumulation of cuticular waxes. *Plant Physiol.*, **113**: 1091-1100.

Hargrove, J.L., P. Greenspan and D.K. Hartle. 2004. Nutritional significance and metabolism of very long chain fatty alcohols and acids from dietary waxes. *Exp. Biol. Med.*, **229**: 215-226.

Heredia, A. 2003. Biophysical and biochemical characteristics of cutin, a plant barrier biopolymer. *Biochim. Biophys. Acta*, **1620**: 1-7.

Holloway, P.J., G.M. Hunt, E.A. Baker and M.J.K. Macey. 1977. Chemical composition and ultrastructure of the epicuticular wax in four mutants of *Pisum sativum* (L.). *Chem. Phys. Lipids* **20**: 141-155.

Hooker, T.S., A.A. Millar and L. Kunst. 2002. Significance of the expression of the CER6 condensing enzyme for cuticular wax production in Arabidopsis. *Plant Physiol.*, **129**: 1568-1580.

James, Jr. D.W., E. Lim, J. Keller, I. Plooy, E. Ralston and H.K. Dooner. 1995. Direct tagging of the Arabidopsis *FATTY ACID ELONGATION1* (*FAE1*) gene with the maize transposon Activator. *Plant Cell*, **7**: 309-319.

Jeffree, C.E. 1996. Structure and ontogeny of plant cuticles. In: *Plant cuticles an integrated functional approach* (ed. G. Kerstiens), pp. 33-82. Bios Scientific, Oxford, UK.

Jenks, M.A. and E.N. Ashworth. 1999. Plant epicuticular waxes: function, production, and genetics. *Hortic. Review* **23**: 1-68.

Jenks, M.A., P.J. Rich and E.N. Ashworth. 1994. Involvement of cork cells in the secretion of epicuticular wax filaments on *Sorghum bicolor* (L.) Moench. *Int. J. Plant Sci.*, **155**: 506-518.

Jenks, M.A., H.A. Tuttle, S.D. Eigenbrode and K.A. Feldmann. 1995. Leaf epicuticular waxes of the *eceriferum* mutants in *Arabidopsis*. *Plant Physiol.*, **108**: 369-377.

Jenks, M.A., A.M. Rashotte, H.A. Tuttle and K.A. Feldmann. 1996. Mutants in Arabidopsis thaliana altered in epicuticular wax and leaf morphology. *Plant Physiol.*, **110**: 377-385.

Jenks, M.A., P.J. Rich, D. Rhodes, E.N. Ashwort, J.D. Axtell and C.K. Din. 2000. Leaf sheath cuticular waxes on bloomless and sparse-bloom mutants of Sorghum bicolor. *Phytochemistry*, **54**: 577-584.

Jenks, M.A., L. Andersen, R.S. Teusink and M.H. Williams. 2001. Leaf cuticular waxes of potted rose cultivars as affected by plant development, drought and paclobutrazol treatments. *Physiol. Plant.*, **112**: 62-70.

Jenks, M.A., S. Eigenbrode and B. Lemieux. 2002. Cuticular waxes of *Arabidopsis*. In: *The Arabidopsis Book*. (eds. C. Somerville and E. Meyerowitz). American Society of Plant Biologists, Rockville, MD, USA.

Jetter, R. and S. Schaffer. 2001. Chemical composition of the Prunus laurocerasus leaf surface. Dynamic changes of the epicuticular wax film during leaf development. *Plant Physiol.*, **126**: 1725-1737.

Jianguo, M., L. Wanqu, Y. Qin and R.P. Bodnaryk. 1995. Inheritance of the waxless character of Brassica napus Nilla glossy. *Can. J. Plant Sci.*, **75**: 893-894.

Juniper, B.E. 1995. Waxes on plant surfaces and their interactions with insects. In: *Waxes: Chemistry, Molecular Biology and Functions*. (ed. R.J. Hamilton), pp. 157-175. The Oily Press LTD, Dundee, Scotland.

Klein, I., B. Sarkadi and A. Varadi. 1999. An inventory of the human ABC proteins. *Biochim. Biophys. Acta*. **1461**: 237-262.

Kohlwein, S.D., S. Eder, C.S. Oh, C.E. Martin, K. Gable, D. Bacikova and T. Dunn. 2001. Tsc13p is required for fatty acid elongation and localizes to a novel structure at the nuclear-vacuolar interface in *Saccharomyces cerevisiae*. *Mol. Cell Biol.*, **21**: 109-125.

Kolattukudy, P.E. 1967. Mechanisms of synthesis of waxy esters in broccoli (*Brassica oleracea*). *Biochemistry*, **6**: 2705-2717.

Kolattukudy, P.E. 1971. Enzymatic synthesis of fatty alcohols in *Brassica oleracea*. *Arch. Biochem. Biophys.*, **142**: 701-709.

Kolattukudy, P.E. 1996. Biosynthetic pathways of cutin and waxes and their sensitivity to environmental stresses. In: *Plant cuticles* (ed. G. Kerstiens), pp. 83-108. Oxford UK BIOS Scientific Publishers.

Kolattukudy, P.E. and T.Y. Liu. 1970. Direct evidence for biosynthetic relationships among hydrocarbons, secondary alcohols and ketones in *Brassica oleracea*. *Biochem. Biophys. Res. Commun.*, **41**: 1369-1374.

Kolattukudy, P.E., J.S. Buckner and T.Y. Liu. 1973. Biosynthesis of secondary alcohols and ketones from alkanes. *Arch. Biochem. Biophys.*, **156**: 613-620.

Koornneef, M., C.J. Hanhart, and F. Thiel. 1989. A genetic and phenotypic description of *eceriferum* (*cer*) mutants in *Arabidopsis thaliana*. *J. Hered.*, **80**: 118-122.

Kunst, L. and A.L. Samuels. 2003. Biosynthesis and secretion of plant cuticular wax. *Prog. Lipid Res.*, **42**: 51-80.

Kunst, L., S. Clemens and T. Hooker. 2000. Expression of the wax-specific condensing enzyme CUT1 in Arabidopsis. *Biochem. Soc. Trans.*, **28**: 651-654.

Kurata, T., C. Kawabata-Awai, E. Sakuradani, S. Shimizu, K. Okada and T. Wada. 2003. The *YORE-YORE* gene regulates multiple aspects of epidermal cell differentiation in Arabidopsis. *Plant J.*, **36**: 55-66.

Lardizabal, K.D., J.G. Metz, T. Sakamoto, W.C. Hutton, M.R. Pollard and M.W. Lassner. 2000. Purification of a jojoba embryo wax synthase, cloning of its cDNA, and production of high levels of wax in seeds of transgenic Arabidopsis. *Plant Physiol.*, **122**: 645-655.

Lassner, M.W., K. Lardizabal and J.G. Metz. 1996. A jojoba beta-Ketoacyl-CoA synthase cDNA complements the canola fatty acid elongation mutation in transgenic plants. *Plant Cell*, **8**: 281-292.

Lemieux, B. 1996. Molecular genetics of epicuticular wax biosynthesis. *Trends Plant Sci.* **1**: 312-318.

Lessire, R., H. Juguelin, P. Moreau and C. Cassagne. 1985. Nature of the reaction product of [1-14C]stearoyl-CoA elongation by etiolated leek seeding microsomes. *Arch. Biochem. Biophys.*, **239**: 260-269.

Lolle, S.J., G.P. Berlyn, E.M. Engstrom, K.A. Krolikowski, W.D. Reiter and R.E. Pruitt. 1997. Developmental regulation of cell interactions in the Arabidopsis fiddlehead-1 mutant: a role for the epidermal cell wall and cuticle. *Dev. Biol.*, **189**: 311-321.

Lorenzoni, C. and F. Salamini. 1975. *glossy* mutants of maize. V Morphology of the epicuticular waxes. *Maydica,* **20**: 5-19.

Lundqvist, U. and A. Lundqvist. 1988. Mutagen specificity in barley for1580 eceriferum mutants localized to 79 loci. *Hereditis*, **108**: 1-12.

McNevin, J.P., W. Woodward, A. Hannoufa, K.A. Feldmann and B. Lemieux. 1993. Isolation and characterization of *eceriferum* (*cer*) mutants induced by T-DNA insertions in *Arabidopsis thaliana*. *Genome*, **36**: 610-618.

Metz, J.G., M.R. Pollard, L. Anderson, T.R. Hayes and M.W. Lassner. 2000. Purification of a jojoba embryo fatty acyl-coenzyme A reductase and expression of its cDNA in high erucic acid rapeseed. *Plant Physiol.*, **122**: 635-644.

Millar, A.A., S. Clemens, S. Zachgo, E.M. Giblin, D.C. Taylor and L. Kunst. 1999. *CUT1*, an Arabidopsis gene required for cuticular wax biosynthesis and pollen fertility, encodes a very-long-chain fatty acid condensing enzyme. *Plant Cell*, **11**: 825-838.

Millar, A.A. and L. Kunst. 1997. Very-long-chain fatty acid biosynthesis is controlled through the expression and specificity of the condensing enzyme. *Plant J.*, **12**: 121-131.

Moreau, P., J.J. Bessoule, S. Mongrand, E. Testet, P. Vincent and C. Cassagne. 1998. Lipid trafficking in plant cells. *Prog. Lipid Res.*, **37**: 371-391.

Morris, B.D., S.P. Foster and M.O. Harris. 2000. Identification of 1-octacosanal and 6-methoxy-2-benzoxazolinone from wheat as ovipositional stimulants for Hessian fly, *Mayetiola destructor*. *J. Chem. Ecol.*, **26**: 859-867.

Moose, S.P. and P.H. Sisco. 1994. *Glossy15* controls the epidermal juvenile-to-adult phase transition in maize. *Plant Cell*, **6**: 1343-1355.

Moose, S.P. and P.H. Sisco. 1996. *Glossy15*, an *APETALA2*-like gene from maize that regulates leaf epidermal cell identity. *Genes Devel.*, **10**: 3018-3027.

Negruk, V., P. Yang, M. Subramanian, J.P. McNevin and B. Lemieux. 1996. Molecular cloning and characterization of the *CER2* gene of *Arabidopsis thaliana*. *Plant J.*, **9**: 137-145.

Neinhuis, C,, K. Koch and W. Barthlott. 2001. Movement and regeneration of epicuticular waxes through plant cuticles. *Planta*, **213**: 427-434.

Neuffer, M., E. Coe and S.R. Wessler. 1997. Mutants in Maize. Cold Spring Harbor Laboratory Press, Cold Spring Harbor, N.Y.

Newman, T., F.J. De Bruijn, P. Green, K. Keegstra, H. Kende, L. McIntosh, J. Ohlrogge, N. Raikhel, S. Somerville, M. Thomashow, E. Retzel and C. Somerville. 1994. Genes galore: a summary of methods for accessing results from large-scale partial sequencing of anonymous *Arabidopsis* cDNA clones. *Plant Physiol.*, **106**: 1241-1255.

Pighin, J.A., H. Zheng, L.J. Balakshin, I.P. Goodman, T.L. Western, R. Jetter, L. Kunst and A.L. Samuels. 2004. Plant cuticular lipid export requires an ABC transporter. *Science*, **306**: 702-704.

Post-Beittenmiller, D. 1996. Biochemistry and molecular biology of wax production in plants. *Annu. Rev. Plant Physiol. Plant Mol. Biol.*, **47**: 405-430.

Post-Beittenmiller, D. 1998. The cloned *Eceriferum* genes of *Arabidopsis* and the corresponding *Glossy* genes in maize. *Plant Physiol. Biochem.*, **36**: 157-166.

Premachandra, G.S., H. Saneoka, M. Kanaya and S. Ogata. 1991. Cell membrane stability and leaf surface wax content as affected by increasing water deficits in maize. *J. Exp. Botany,* **42**: 167-171.

Premachandra, G.S., D.T. Hahn, J.D. Axtell and R.J. Joly. 1994. Epicuticular wax load and water-use efficiency in bloomless and sparse-bloom mutants of *Sorghum bicolor* L. *Environ. Exp. Bot.*, **34**: 293-301.

Preuss, D., B. Lemieux, G. Yen and R.W. Davis. 1993. A conditional sterile mutation eliminates surface components from Arabidopsis pollen and disrupts cell signaling during fertilization. *Genes Dev.*, **7**: 974-985.

Pruitt, R.E., J.P. Vielle-Calzada, S.E. Ploense, U. Grossniklaus and S.J. Lolle. 2000. *FIDDLEHEAD*, a gene required to suppress epidermal cell interactions in *Arabidopsis*, encodes a putative lipid biosynthetic enzyme. *Proc. Natl. Acad. Sci. USA*, **97**: 1311-1316.

Pyee, J., H. Yu and P.E. Kolattukudy. 1994. Identification of a lipid transfer protein as the major protein in the surface wax of broccoli (*Brassica oleracea*) leaves. *Arch. Biochem. Biophys.*, **311**: 460-468.

Rashotte, A.M. 1999. Epicuticular wax in *Arabidopsis thaliana:* a study of the genetics, chemistry, structure, and interactions with insects. Ph.D. Dissertation, Plant Science. University of Arizona, Tucson.

Rashotte, A.M., M.A. Jenks, T.D. Nguyen and K.A. Feldmann. 1997. Epicuticular wax variation in ecotypes of Arabidopsis thaliana. *Phytochemistry*, **45**: 251-255.

Rashotte, A.M., M.A. Jenks and K.A. Feldmann. 2001. Cuticular waxes on eceriferum mutants of Arabidopsis thaliana. *Phytochemistry*, **57**: 115-123.

Rashotte, A.M., M.A. Jenks, A.S. Ross and K.A. Feldmann. 2004. Novel eceriferum mutants in Arabidopsis thaliana. *Planta*, **219**: 5-13.

Rea, P.A., Z-S. Li, Y-P. Lu and Y. Drozdowicz. 1998. From vacuolar GS-X pumps to multispecific ABC transporters. *Annu. Rev. Plant Physiol. Plant Mol. Biol.*, **49**: 727-760.

Rhee, Y., A. Hlousek-Radojcic, J. Ponsamuel, D. Liu and D. Post-Beittenmiller. 1998. Epicuticular wax accumulation and fatty acid elongation activities are induced during leaf development of leeks. *Plant Physiol.*, **116**: 901-911.

Riederer, M. and L. Schreiber. 2001. Protecting against water loss: analysis of the barrier properties of plant cuticles. *J. Exp. Bot.*, **52**: 2023-2032.

Schnable, P.S., B.J. Nikolau, P.S. Stinard, Y. Xia, J.D. Hansen, X. Xu, S. Heinen, M. Delledonne and M.E. Myszewski. 1993. Genetic approaches to isolating maize and *Arabidopsis* genes involved in cuticular wax biosynthesis. *Curr. Topics Plant Physiol.*, 9: 196-206.

Schnable, P.S., P.S. Stinard, T.-J. Wen, S. Heinen, D. Weber, L. Zhang, J.D. Hansen and B.J. Nikolau, 1994. The genetics of cuticular wax biosynthesis. *Maydica*, **39**: 279-287.

Schneider-Belhaddad, F. and P. Kolattukudy. 2000. Solubilization, partial purification, and characterization of a fatty aldehyde decarbonylase from a higher plant, Pisum sativum. *Arch. Biochem. Biophys.*, **377**: 341-349.

Shanklin, J. and E.B. Cahoon, 1998. Desaturation and related modifications of fatty acids. *Annu. Rev. Plant Physiol. Plant Mol. Biol.*, **49:** 611-641.

Sieber, P., M. Schorderet, U. Ryser, A. Buchala, P. Kolattukudy, J.P. Metraux and C. Nawrath. 2000. Transgenic Arabidopsis plants expressing a fungal cutinase show alterations in the structure and properties of the cuticle and postgenital organ fusions. *Plant Cell*, **12**: 721-738.

Stoner, K.A. 1990. Glossy leaf wax and host-plant resistance to insects in *Brassica oleracea* L. under natural infestation. *Environ. Entomol.*, **19**: 730-739.

Stork, N.E. 1980. Role of waxblooms in preventing attachment to brassicas by the mustard beetle, *Phaedon cochleariae*. *Entomol. Exp. Appl.*, **28**: 100-107.

St-Pierre, B., P. Laflamme, A.M. Alarco and V. De Luca. 1998. The terminal O-acetyltransferase involved in vindoline biosynthesis defines a new class of proteins responsible for coenzyme A-dependent acyl transfer. *Plant J.*, **14**: 703-713.

Sturaro, M., H. Hartings, E. Schmelzer, R. Velasco, F. Salamini and M. Motto. 2005. Cloning and characterization of *GLOSSY1*, a maize gene involved in cuticle membrane and wax production. *Plant Physiol.*, **138**: 478-489.

Tacke, E., C. Korfhage, D. Michel, M. Maddaloni, M. Motto, S. Lanzini, F. Salamini and H.-P. Döring. 1995. Transposon tagging of the maize *Glossy2* locus with the transposable element *En/Spm*. *Plant J.,* **8**: 907-917.

Todd, J., D. Post-Beittenmiller and J.G. Jaworski. 1999. *KCS1* encodes a fatty acid elongase 3-ketoacyl-CoA synthase affecting wax biosynthesis in *Arabidopsis thaliana*. *Plant J.*, **17**: 119-130.

Velasco, R., C. Korfhage, A. Salamini, E. Tacke, J. Schmitz, M. Motto, F. Salamini and H.-P. Döring. 2002. Expression of the *Glossy2* gene of maize during plant development. *Maydica*, **47**: 71-81.

Vioque, J. and P.E. Kolattukudy. 1997. Resolution and purification of an aldehyde-generating and an alcohol-generating fatty acyl-CoA reductase from pea leaves (*Pisum sativum* L.). *Arch. Biochem. Biophys.*, **340**: 64-72.

Vogelmann, T.C. 1993. Plant tissue optics. *Ann. Rev. Plant Physiol. Plant Mol. Biol.*, **44:** 231-251.

Vogg, G., S. Fischer, J. Leide, E. Emmanuel, R. Jetter, A.A. Levy and M. Riederer. 2004. Tomato fruit cuticular waxes and their effects on transpiration barrier properties: functional characterization of a mutant deficient in a very-long-chain fatty acid beta-ketoacyl-CoA synthase. *J. Exp. Bot.*, **55**: 1401-1410.

von Wettstein-Knowles, P. 1995. Biosynthesis and genetics of waxes: In: *Waxes: Chemistry, Molecular Biology and Functions*. (ed. R.J. Hamilton) pp. 91-129. The Oily Press LTD, Dundee, Scotland.

von Wettstein-Knowles, P., P. Avato and J.D. Mikkelsen. 1979. Light promotes synthesis of the very long chain fatty acyl chains in maize wax. In: *Biogenesis and Function of Plant Lipids* (eds. D. Mazliak, P. Benveniste, C. Costes and R. Douce), pp. 271-274. Elsevier/North-Holland Biomedical Press, New York.

Walton, T.J. 1990. Waxes, cutin and suberin. In: *Methods in Plant Biochemistry: Lipids, Membranes and Aspects of Photobiology*. (eds. J.L. Harwood and J.R. Bowyer) pp. 105-158. Academic Press, San Diego.

White, C. and S.D. Eigenbrode. 2000. Effects of surface wax variation in *Pisum sativum* on herbivorous and entomophagous insects in the field. *Environ. Entomol.*, **29**: 773-780.

Xia, Y., B.J. Nikolau and P.S. Schnable. 1997. Developmental and hormonal regulation of the arabidopsis *CER2* gene that codes for a nuclear-localized protein required for the normal accumulation of cuticular waxes. *Plant Physiol.*, **115**: 925-937.

Xia, Y., B.J. Nikolau and P.S. Schnable. 1996. Cloning and characterization of *CER2*, an *Arabidopsis* gene that affects cuticular wax accumulation. *Plant Cell*, **8**: 1291-1304.

Xu, X., C.R. Dietrich, M. Delledonne, Y. Xia, T.J. Wen, D.S. Robertson, B.J. Nikolau and P.S. Schnable. 1997. Sequence analysis of the cloned *glossy8* gene of maize suggests that it may code for a beta-ketoacyl reductase required for the biosynthesis of cuticular waxes. *Plant Physiol.*, **115**: 501-510.

Xu, X., C.R. Dietrich, R. Lessire, B.J. Nikolau and P.S. Schnable. 2002. The endoplasmic reticulum-associated maize GL8 protein is a component of the acyl-coenzyme A elongase involved in the production of cuticular waxes. *Plant Physiol.*, **128:** 924-934.

Yephremov, A., E. Wisman, P. Huijser, C. Huijser, K. Wellesen and H. Saedler. 1999. Characterization of the *FIDDLEHEAD* gene of Arabidopsis reveals a link between adhesion response and cell differentiation in the epidermis. *Plant Cell*, **11**: 2187-2201.

Zachowski, A., F. Guerbette, M. Grosbois, A. Jolliot-Croquin and J.C. Kader. 1998. Characterisation of acyl binding by a plant lipid-transfer protein. *Eur. J. Biochem.*, **257**: 443-448.

Zhang, J.Y., C.D. Broeckling, E.B. Blancaflor, M.K. Sledge, L.W. Sumner and Z.Y. Wang. 2005. Overexpression of WXP1, a putative Medicago truncatula AP2 domain-containing transcription factor gene, increases cuticular wax accumulation and enhances drought tolerance in transgenic alfalfa (*Medicago sativa*). *Plant J.*, **42**: 689-707.

Zheng, H., O. Rowland and L. Kunst. 2005. Disruptions of the *Arabidopsis* enoyl-CoA reductase gene reveal an essential role for very-long-chain fatty acid synthesis in cell expansion during plant morphogenesis. *Plant Cell*, **17**: 1467-1481.

*SECTION — III*

# ***PLANT BIOGENESIS***

*Advances in Plant Physiology*, Vol. 9
Ed. A. Hemantaranjan
Scientific Publishers (India), Jodhpur, 2006 pp. 255-294
E-mail: **info@scientificpub.com** www.scientificpub.com

# 12

# PROTEOLYSIS IN PLANT MITOCHONDRIA AND CHLOROPLASTS

P.F. Huesgen, H. Schuhmann and I. Adamska*

Department of Physiology and Plant Biochemistry, University of Konstanz, Universitätsstrasse 10, Postfach 5560 M601, D-78457 Konstanz, Germany
*Iwona.Adamska@uni-konstanz.de

## INTRODUCTION

Proteolysis is involved in a wide range of processes during biogenesis, maintenance of plant organelles and their senescence. Two types of proteolysis occur in plant organelles: protein processing (Gakh *et al.*, 2002; Zhang and Glaser, 2002; Jarvis and Robinson, 2004; Lister *et al.*, 2005) and destructive proteolysis (Andersson and Aro, 1997; Adam, 2000, 2001; Adam and Ostersetzer, 2001; Estelle, 2001; Yamamoto, 2001; Adam and Clarke, 2002). While protein processing is required for protein maturation, destructive proteolysis leads to a loss of particular proteins from various organellar compartments. Protein processing occurs during the import of nuclear-encoded protein precursors into organelles, the target of proteins to various organellar subcompartments or prior to the activation of latent forms of enzymes or receptors. Destructive proteolysis removes denaturated, unassembled and mistarget proteins or active proteins, when there is no further need for their physiological function.

It was proposed that proteins located in mitochondria and chloroplasts should contain distinct targeting sequences (Blobel and Dobberstein, 1975). The existence of processing enzymes in organelles became evident after the discovery that proteins imported into chloroplasts or mitochondria were synthesized in a higher molecular mass form (Dobberstein *et al.*, 1977; Maccecchini *et al.*, 1979 a,b). Some years later the first protein processing enzyme was partially purified from the stroma of pea chloroplasts (Robinson and Ellis, 1984). The plant mitochondrial processing enzyme was isolated much later and showed to be an integral part of the cytochrome (cyt.) bc1 complex of the respiratory chain in potato and spinach (Braun *et al.*, 1992; Eriksson *et al.*, 1996).

Until the beginning of 1980 our knowledge of the degradation of proteins in plant organelles was still in its infancy. At this time it was assumed that proteins are exported out of mitochondria or chloroplasts and that the vacuole is a place for the protein degradation in plants. It was shown for the first time in 1984 that isolated chloroplasts are able to degrade their own proteins (Liu and Jagendorf, 1984; Malek *et al.*, 1984) and ten years later a degradation activity was reported for spinach leaf and root mitochondria (Knorpp *et al.*, 1994). However, in spite of extensive biochemical attempts to isolate chloroplast and mitochondrial proteases only marginal success has been achieved in the following years. Genetic, biochemical and molecular approaches taken in recent years have led to the identification and characterization of several organellar proteases. A short overview of these enzymes is provided below. Although the identity of these proteases is known, only a very limited knowledge exists about the physiological processes that are mediated by these enzymes. Therefore, this review is not intended to summarize our knowledge on plant proteases but rather provide an update on recently identified endoproteolytic processes in chloroplasts and plant mitochondria.

## 2. PROTEOLYTIC SYSTEMS IN CHLOROPLASTS AND PLANT MITOCHONDRIA

### 2.1. Precursor Processing Proteases

This paragraph provides a summary of chloroplast and mitochondrial processing enzymes with experimentally confirmed identities. An extended list of putative processing enzymes identified in the genome of *Arabidopsis thaliana* is provided in Tables 1 and 2.

#### *2.1.1. Stomal processing peptidase (SPP)*

The SSP (summarized in Table 1 and Figure 1A) is an ATP-independent, soluble, monomeric Zn-metalloprotease related to the *Escherichia coli* pitrilysin, the insulysin from mammals and *Drosophila melanogaster*, the β-subunit of the mitochondrial processing protease (MPP) and the presequence-degrading peptidases (PrePs) (Oblong and Lamppa, 1992; Richter *et al.*, 2005). Pitrilysins are highly specific enzymes, which recognize their substrates without defined amino acid residues around the scissile bond, indicating that recognition is on the basis of higher-order structure rather than on the amino acid sequence (Anastasi *et al.*, 1993). Plant SPP has a conserved N-terminus containing an inverted Zn-binding motif (VanderVere *et al.*, 1995). Mutational analysis has demonstrated that this motif is crucial for SPP catalytic activity (Richter and Lamppa, 2003). The 3D crystal structure is available only for the pitrilysin enzyme from *E. coli* (PDB : 1Q2L).

#### *2.1.2. Thylakoid processing peptidase (TPP)*

The TPP (summarized in Table 1 and Figure 1A) is an ATP-independent, monomeric serine endopeptidase related to *E. coli* type I leader peptidase B (LepB) (Dalbey and von Heijne, 1992; Chaal *et al.*, 1998). Plant TPP is an integral thylakoid membrane protein with one predicted N-terminally-located transmembrane domain and a C-terminally-located catalytic domain, which protrudes into the thylakoid lumen (Kirwin *et al.*, 1988; Chaal *et al.*, 1998). The 3D structure is available for LepB from *E. coli* in complex with a lipopeptide inhibitor (PDB: 1T7D).

#### *2.1.3. Carboxyl-terminal processing (Ctp) peptidase*

Ctp proteases (summarized in Table 1 and Figure 1A) are ATP-independent serine endopeptidases closely related to the periplasmic carboxyl-terminal processing protease, the Tsp (tail-specific protease)/Prc (processing C-terminal) from *E. coli* (Silber *et al.*, 1992). The CtpA protease is a soluble enzyme that resides in the thylakoid lumen of higher plants (Inagaki *et al.*, 1996; Oelmüller *et al.*, 1996; Schubert *et al.*, 2002). While barley and spinach contain a single CtpA1 protease (Oelmüller *et al.*, 1996), three different Ctp isozymes were reported from the genome of *A. thaliana* (Sokolenko *et al.*, 2002). The X-ray structure of the CtpA1 enzyme from the green alga *Scenedesmus obliquus* (PDB : 1FC6) reveals that the enzyme is a monomer and is composed of three folding domains (Liao *et al.*, 2000). The middle domain is topologically homologous to a PDZ domain, which is known to play a role in substrate recognition and binding (Liao *et al.*, 2000; Sheng and Sala, 2001; Wilken *et al.*, 2004).

### 2.1.4. *Mitochondrial processing peptidase (MPP)*

The MPP (summarized in Table 2 and Figure 1B) is an ATP-independent, Zn-metalloendopeptidase related to enzymes from the pitrilysin/insulinase family (Glaser and Dessi, 1999). MPP forms a heterodimer composed of two structurally related subunits: the α-subunit (α-MPP) is responsible for peptide recognition and binding and the β-subunit (β-MPP) for catalysis (Braun and Schmitz, 1995). Although only the β-MPP contains the catalytic site consisting of an inverted Zn-binding motif, both subunits are required for enzymatic activity (Paetzel *et al.*, 2002). In contrast to yeast and mammals, where MPP exists as a soluble heterodimer in the mitochondrial matrix, the plant MPP is integrated into the membrane-bound cyt. bc1 complex, whereby the catalytic site is facing the mitochondrial matrix (Braun *et al.*, 1992; Braun and Schmitz, 1995; Eriksson *et al.*, 1996; Glaser and Dessi, 1999). Searches in the genome of *A. thaliana* revealed three *mpp* genes encoding two different α-MPP subunits and one β-MPP subunit. The expression of all three genes was confirmed experimentally by the analysis of the *A. thaliana* mitochondrial proteome (Kruft *et al.*, 2001; Millar *et al.*, 2001; Heazlewood *et al.*, 2004). 3D structures are available for β-MPP subunits from yeast without (PDB: 1HR6) or with bound a presequence peptide of the malate dehydrogenase (PDB: 1HR9).

### 2.1.5. *Presequence degrading protease (PreP)*

PrePs (summarized in Table 1 and 2 and Figure 1A and B) are ATP-independent, Zn-metalloendopeptidase related to enzymes from the pitrilysin/insulinase family (Ståhl *et al.*, 2002). Mutagenesis studies of the residues involved in metal binding confirmed two histidines and the proximal glutamate as essential residues for proteolytic activity (Moberg *et al.*, 2003). A dual location in the chloroplast stroma and the mitochondrial matrix was reported for these enzymes (Moberg *et al.*, 2003; Bhushan *et al.*, 2003, 2005). The 3D crystal structure is available only for the pitrilysin enzyme from *E. coli* (PDB: 1Q2L).

## 2.2. Proteases Involved in Protein Turnover

Several ATP-dependent proteases derived from bacterial/cyanobacterial ancestors have been shown to be involved in protein turnover in plant mitochondria and chloroplasts. These proteases belong to the families of ClpP (caseinolytic protease), Lon and FtsH (filamentation temperature sensitive) proteases with several representatives in both organelles (summarized in Tables 1 and 2 and Figure 1A and B). The three families of ATP-dependent proteases are also known as the AAA+ (ATPases Associated with various cellular Activities) proteases because they contain one or more copies of an ATPase domain, referred to as the AAA domain (Ogura and Wilkinson, 2001). Each AAA domain contains several conserved motifs, including those necessary for ATP binding and hydrolysis. It has been demonstrated that AAA domains act also as molecular chaperons during the assembly/ disassembly of proteins or protein complexes (Mogk and Bukau, 2004). In case of Lon and FtsH families, the AAA and proteolytic domains reside on the same polypeptide, whereas in the Clp family each of these domains is physically separated into different subunits of the complex (Table 1 and 2).

The ATP-independent proteases of bacterial/cyanobacterial origin participating in protein turnover in plant organelles are less characterized. This group consists of the chloroplast-located family of Deg (degradation of periplasmic proteins) proteases, the SppA1 (signal peptide peptidase A1) and the CND41 (chloroplast nucleoid DNA-binding) protease (Table 1 and Figure 1A). A short overview of the mentioned proteases is provided below.

### *2.2.1 Caseinolytic protease (Clp)*

By homology searches with a bacterial Clp sequence several Clp homologs were found in the chloroplast genome of tobacco, rice and the liverwort *Marchantia polymorpha* and their expression confirmed experimentally by immunoblotting (Maurizi *et al.*, 1990). Progressively, Clp homologs were characterized from *Pinus* (Clarke *et al.*, 1994), *A. thaliana* (Shanklin *et al.*, 1995; Sokolenko *et al.*, 1998), barley (Ostersetzer *et al.*, 1996), tomato (Sokolenko *et al.*, 1998) or pea (Halperin *et al.*, 2001a) chloroplasts. Clp proteases in plant mitochondria are less characterized and only few reports are available for this topic (Halperin *et al.*, 2001b; Peltier *et al.*, 2004).

ClpPs (summarized in Table 1 and 2 and Figure 1A and B) are ATP-dependent serine-endopeptidases (Wang *et al.*, 1997) which are present as high molecular mass complexes in the stroma of chloroplasts (Sokolenko *et al.*, 1998; Halperin *et al.*, 2001a; Peltier *et al.*, 2001, 2004) or in the matrix of plant mitochondria (Peltier *et al.*, 2004). However, a small portion of Clp complexes may also bind to stroma-exposed thylakoid membranes (Peltier *et al.*, 2004) and to the inner envelope membrane (Nielsen *et al.*, 1997). The Clp complexes in chloroplasts and plant mitochondria are composed of proteolytic and regulatory subunits. In higher plants the proteolytic subunits are encoded by the single copy chloroplast gene (*clpP1*) and the nuclear multigene family (*clpP2-6*), while the regulatory subunits ClpB3 and 4, ClpC1 and 2, ClpD, ClpS1 and 2, ClpT and ClpX1-3 are all nuclear-encoded (Adam *et al.*, 2001; Peltier *et al.*, 2001, 2004; Sokolenko *et al.*, 2002; Clarke *et al.*, 2005). The nuclear-encoded, plastid-located proteolytic subunits also exist in catalytically inactive versions, ClpR1-4 (Adam *et al.*, 2001; Sokolenko *et al.*, 2002; Peltier *et al.*, 2004). The regulatory subunits belong to two distinct classes of molecular chaperons (Porankiewicz *et al.*, 1999). Class I Clp/Hsp100 chaperons, consisting of ClpB, C and D subunits, contain two AAA domains that are essential for complex assembly and unfolding activity. Class II chaperons represented by mitochondrial ClpX subunits have only one AAA domain and serve as regulatory ATPases (Halperin *et al.*, 2001b; Peltier *et al.*, 2004). Recently, two novel ClpS1 and 2 regulatory subunits lacking AAA domains were identified from plastids of *A. thaliana* (Peltier *et al.*, 2001; Peltier *et al.*, 2004). Interestingly, ClpS subunits are unique for the land plants, being absent in algae or cyanobacteria (Peltier *et al.*, 2004). ClpS subunits share sequence similarity to the N-terminal domain of ClpC, thus it is expected that they bind to the ClpP/R core complex at the same site as ClpC, thereby blocking or interfering with the ClpC association. The plastid-located regulatory ClpT subunit (called in *E. coli* ClpS) was proposed to be an adaptor protein that interacts with ClpC and redirects its chaperone activity toward aggregated proteins as was proposed for bacterial ClpA homolog (Dougan *et al.*, 2002; Clarke *et al.*, 2005).

In photosynthetic and non-photosynthetic plastids the Clp protease forms a single mixed tetradecameric complex of 325-350 kDa that consists of 1-3 copies of five different proteolytic subunits (ClpP1, 3-6) and 4 non-proteolytic ClpR1-4 proteins with predicted stoichiometry of 3:3:2:1:1:1:1:1:1 for P4:P5:R4:P1:P3:P6:R1:R2:R3 (Peltier *et al.*, 2004). In addition, one copy of chaperone-like ClpS1 and 2 subunits is tightly associated with this core complex making 16 subunits in total per core (Peltier *et al.*, 2004). The non-proteolytic ClpR subunits have been suggested to influence the presentation of substrates to proteolytic ClpP subunits, to create lateral polar opening within the ring structure, enabling the diffusion of peptide fragments out and/or to influence chaperone association with the core (Peltier *et al.*, 2004). The chloroplast chaperone ClpC and ClpD subunits, but not the ClpB3 subunit, contain a conserved I(L)GF motif implicated in the interaction with the Clp core complex in *E. coli* (Kim *et al.*, 2001). The ClpC subunits were reported previously to form hexameric ring complexes interacting with the Clp core complex in the presence of ATP (Sokolenko *et al.*, 1998). However, more recent data indicate that ClpC subunits form a dimeric complex of 200 kDa in the chloroplast stroma that is composed mainly of ClpC1 subunits (Peltier *et al.*, 2004). Also the ClpB3 subunits were identified as a soluble ca. 200 kDa dimer in petal plastids (Peltier *et al.*, 2004). Whether and how these regulatory and chaperone subunits interact with the Clp core complex still remain to be elucidated.

In contrast, plant mitochondria contain a single 320 kDa homo-tetradecameric core complex consisting only of ClpP2 without association of ClpR or ClpS subunits (Peltier *et al.*, 2004). Three regulatory subunits ClpX1-3 were also detected in plant mitochondria (Halperin *et al.*, 2001b; Peltier *et al.*, 2004). All three ClpX1-3 subunits contain domains necessary for ClpP association and presumably form a complex with ClpP2.

Based on 3D homology modeling lateral exit gates for proteolysis products are proposed for the Clp complex in plants (Peltier *et al.*, 2004). The 3D crystal structure of the proteolytic ClpP subunit is available for *E. coli* (PDB: 1TYF) and *Streptococcus pneumoniae* (PDB: 1Y7O).

### 2.2.2. *Lon (La) protease*

Protease La (the product of the *lon* gene) was the first ATP-dependent protease identified in *E. coli* (Chung and Goldberg, 1981). This enzyme is now most commonly termed the Lon protease. Four Lon proteases are encoded by the genome of *A. thaliana* (Adam *et al.*, 2001; Janska, 2005) and two of them, Lon1 and Lon3 proteases are located in the mitochondrial matrix as confirmed by proteomics (Heazlewood *et al.*, 2004; Heazlewood and Millar, 2005). Lon4 protease is predicted to target to the chloroplast stroma. An experimental evidence for the presence of a Lon protease in the chloroplast stroma was provided also by immunoblot analysis (Adam *et al.*, 2001).

Lon proteases (summarized in Table 1 and 2 and Figure 1A and B) are soluble, ATP-dependent serine type proteases. Homo-heptameric ring complexes of about 700-800 kDa were reported for Lon proteases in the mitochondrial matrix of yeast (Stahlberg *et al.*, 1999) and maize (Mertova *et al.*, 2002). However, the Lon complex in maize was considerably less stable than the one in yeast or rat liver mitochondria (Stahlberg *et al.*, 1999; Mertova *et al.*, 2002). The 3D crystal structure of the catalytic domain is available for Lon homologs from Bacteria *E. coli* (PDB: 1RR9, 1RRE) and Archea *Methanocaldococcus jannaschii* (PDB: 1XHK) and *Archaeoglobus fulgidus* (PDB: 1Z0V, 1Z0W, 1Z0B, 1Z0C).

Table 1. Overview of experimentally confirmed chloroplast proteases in *Arabidopsis thaliana*.

| Name | Chromosomal Accession No. | Protein Size (aa) | Merops Clan, Family | Catalytic Motif | Energy Dependence | Location | Complex Formation | Reviewed |
|---|---|---|---|---|---|---|---|---|
| Processing Enzymes | | | | | | | | |
| SPP | At5g42390 | 1265-cTP54 | ME, M16X | HXXEH | No | Stroma | Monomer | 1-5 |
| TPP | At2g30440 | 340-cTP38 | SF, S24 | S-K | No | Thy-Mem | n.d. | 1-2, 5 |
| TPP-like | At3g24590 | 291-cTP67 | SF, S24 | S-K | No | Thy-Mem | n.d. | n.a. |
| CtpA1 | At3g57680 | 519-cTP79 | SK, S41 | S-K | No | Thy-Lumen | Monomer | 1 |
| CtpA2 | At4g17740 | 515-cTP85 | SK, S41 | S-K | No | Thy-Lumen | n.d. | 1 |
| CtpC | At5g46390 | 428-cTP12 | SK, S41 | S-K | No | Thy-Lumen | n.d. | 1 |
| PreP1 | At3g19170 | 1080-cTP56 | ME, M16C | HXXEH | No | Stroma | n.d. | 3 |
| PreP2 | At1g49630 | 1080-cTP68 | ME, M16C | HXXEH | No | Stroma | n.d. | 3 |
| Clp Complex | | | | | | | | |
| ClpP1 | AtCg00460 | 196 | SK, S14 | S-H-D | No | Stroma | Heterom. core | 1, 5-8 |
| ClpP3 | At1g66670 | 309-cTP71 | SK, S14 | S-H-D | No | Stroma | Heterom. core | 1, 5-8 |
| ClpP4 | At5g45390 | 292-cTP60 | SK, S14 | S-H-D | No | Stroma | Heterom. core | 1, 5-8 |
| ClpP5 | At1g02560 | 298-cTP62 | SK, S14 | S-H-D | No | Stroma | Heterom. core | 1, 5-8 |
| ClpP6 | At1g11750 | 271-cTP51 | SK, S14 | S-H-D | No | Stroma | Heterom. core | 1, 6-8 |
| ClpR1 | At1g49970 | 387-cTP41 | - | Missing | No | Stroma | Heterom. core | 1, 6-8 |
| ClpR2 | At1g12410 | 279-cTP54 | - | Missing | No | Stroma | Heterom. core | 1, 6-8 |
| ClpR3 | At1g09130 | 330-cTP43 | - | Missing | No | Stroma | Heterom. core | 1, 6-8 |
| ClpR4 | At4g17040 | 305-cTP68 | - | Missing | No | Stroma | Heterom. core | 1, 6-8 |
| ClpS1 | At4g25370 | 238-cTP63 | - | Regulatory | No | Stroma | Core assoc. | 1, 8 |
| ClpS2 | At4g12060 | 241-cTP58 | - | Regulatory | No | Stroma | Core assoc. | 1, 8 |

| | | | | | | | | |
|---|---|---|---|---|---|---|---|---|
| ClpB3 | At5g15450 | 968-cTP67 | - | Chaperone | ATP | Stroma | Dimer | 1, 8 |
| ClpC1 | At5g50920 | 929-cTP38 | - | Chaperone | ATP | Stroma | Dimer/Hexam. | 1, 5-6, 8 |
| ClpC2 | At3g48870 | 952-cTP45 | - | Chaperone | ATP | Stroma | Dimer/Hexam. | 1, 5-7, 8 |
| ClpD | At5g51070 | 945-cTP89 | - | Chaperone | ATP | Stroma | n.d. | 1, 5-8 |
| ClpT | At1g68660 | 159-cTP44 | - | Regulatory | No | Stroma | n.d. | 7 |
| Lon Family | | | | | | | | |
| Lon4 | At3g05790 | 942-cTP66 | SF, S16 | S-K | ATP | Stroma | n.d. | n.a. |
| *FtsH Family* | | | | | | | | |
| FtsH1 | At1g50250 | 716-cTP48 | MA, M41 | HEXXH | ATP | Thy-Mem | Heterohexam. | 1, 6, 8-9 |
| FtsH2 | At2g30950 | 695-cTP47 | MA, M41 | HEXXH | ATP | Thy-Mem | Heterohexam. | 1, 6, 8-9 |
| FtsH5 | At5g42270 | 704-cTP58 | MA, M41 | HEXXH | ATP | Thy-Mem | Heterohexam. | 1, 6, 8-9 |
| FtsH6 | At5g15250 | 687-cTP75 | MA, M41 | HEXXH | ATP | Thy-Mem | n.d. | 1, 6, 9 |
| FtsH7 | At3g47060 | 802-cTP55 | MA, M41 | HEXXH | ATP | Thy-Mem | n.d. | 1, 6, 9 |
| FtsH8 | At1g06430 | 685-cTP37 | MA, M41 | HEXXH | ATP | Thy-Mem | Heterohexam. | 1, 6, 9 |
| FtsH9 | At5g58870 | 806-cTP62 | MA, M41 | HEXXH | ATP | Thy-Mem | n.d. | 1, 8-9 |
| FtsH11 | At5g53170 | 806-cTP63 | MA, M41 | HEXXH | ATP | Thy-Mem | n.d. | 1, 9 |
| FtsH12 | At5g15250 | 687-cTP75 | MA, M41 | HEXXH | ATP | Thy-Mem | n.d. | 1, 8-9 |
| FtsHi1 | At4g23940 | 946-cTP53 | - | Chaperone | ATP | Thy-Mem | n.d. | 1 |
| FtsHi2 | At3g16290 | 876-cTP32 | - | Chaperone | ATP | Thy-Mem | n.d. | 1 |
| FtsHi3 | At3g02450 | 622-cTP42 | - | Chaperone | ATP | Thy-Mem | n.d. | 1 |
| FtsHi4 | At5g64580 | 855-cTP78 | - | Chaperone | ATP | Thy-Mem | n.d. | 1 |
| *Deg Family* | | | | | | | | |
| Deg1 | At3g27925 | 439-cTP101 | PA, S1 | H-D-S | No | Thy-Lumen | n.d. | 1, 5-6, 8, 10-11 |
| Deg2 | At2g47940 | 607-cTP69 | PA, S1 | H-D-S | No | Thy-Mem | n.d. | 1, 6, 8, 10-11 |

| | | | | | | | | |
|---|---|---|---|---|---|---|---|---|
| Deg5 | At4g18370 | 323-cTP28 | PA, S1 | H-D-S | No | Thy-Lumen | n.d. | 1, 6, 8, 10-11 |
| Deg8 | At5g39830 | 448-cTP26 | PA, S1 | H-D-S | No | Thy-Lumen | n.d. | 1, 6, 8, 10-11 |
| CND41 | At3g18490 | 500-cTP37 | AA, A1 | D-D | No | Envelops | Nucleoids | n.a. |
| SppA1 | At1g73990 | 677-cTP68 | SK, S49 | S-D | No | Thy-Mem | Homotetram. | 1, 8, 12 |

Protein precursor sizes are given in amino acids (aa) according to the TAIR database (http://www.arabidopsis.org/abrc). The sizes of chloroplast transit peptides (cTP) are according to the TargetP (http://www.cbs.dtu.dk/services/TargetP) prediction. The classification of proteases into Clans and Families is according to the Merops Database (http://merops.sanger.ac.uk). Abbreviations: Assoc., associated; AtCg, chloroplast-encoded; Heterom., heteromeric; Heterohexam., heterohexameric; Hexam., hexamer; Homotetram., homotetrameric; Mem, membrane; n.a., not available; n.d., no data; Thy, thylakoid.

Only recent review articles dealing with mentioned protease families are listed in this table, the original papers are quoted in the text. (1) Sokolenko *et al.*, 2002; (2) Paetzel *et al.*, 2002; (3) Schaller, 2004; (4) Richter *et al.*, 2005; (5) Adam, 2000; (6) Adam *et al.*, 2001; (7) Adam and Clarke, 2002; (8) Clarke *et al.*, 2005; (9) Adam *et al.*, 2005; (10) Kieselbach and Funk, 2003; (11) Huesgen *et al.*, 2005; (12) Sokolenko, 2005.

### 2.2.3. *Filamentation temperature sensitive (FtsH) protease*

The FtsH protease was originally identified in *E. coli* mutants that failed to septate when cultured at elevated temperatures (Santos and De Almeida, 1975). The first chloroplast FtsH protease was identified immunologically in spinach thylakoid membranes (Lindahl *et al.*, 1996). Mitochondrial FtsH proteases have been analyzed mainly in yeast. Only recently, the first experimental evidence was provided for the existence of a FtsH protease in pea mitochondria (Kolodziejczak *et al.*, 2002).

FtsH proteases (summarized in Table 1 and 2 and Figure 1A and B) are ATP-dependent Zn-metalloendopeptidases with the metal ion-binding motif characteristic of the thermolysin family. Of the 12 FtsH proteases identified in the genome of *A. thaliana* (Adam *et al.*, 2001; Sokolenko *et al.*, 2002; Adam *et al.*, 2005) nine (FtsH1, 2, 5-9, 11, 12) are targeted to the chloroplast and three (FtsH3, 4 and 10) to the mitochondria, as revealed by transient expression assay of green fluorescent protein (GFP)–fusion constructs in tobacco cell suspension cultures (Sakamoto *et al.*, 2003). In addition, four FtsHi (inactive FtsH) proteases with impaired Zn-binding domains were found in the genome of *A. thaliana* and a chaperone function was suggested for these proteins (Sokolenko *et al.*, 2002).

FtsH proteases are integral membrane proteins located in the thylakoid membrane of the chloroplast or in the inner mitochondrial membrane. In yeast mitochondria, two types of FtsH proteases with a different topology in the inner mitochondrial membrane were identified. The i-AAA proteases span the membrane once and expose their catalytic site toward the intermembrane space, while m-AAA proteases contain two transmembrane domains and act on the matrix side (Arnold and Langer, 2002).

In the chloroplast only the m-AAA type of FtsH proteases has been identified so far. It was shown that FtsH1, 2 and 5 proteases span the membrane with two transmembrane domains that are located at the N-terminus of the protein exposing ATP- and metal-binding sites to the stroma (Lindahl *et al.*, 1996; Chen

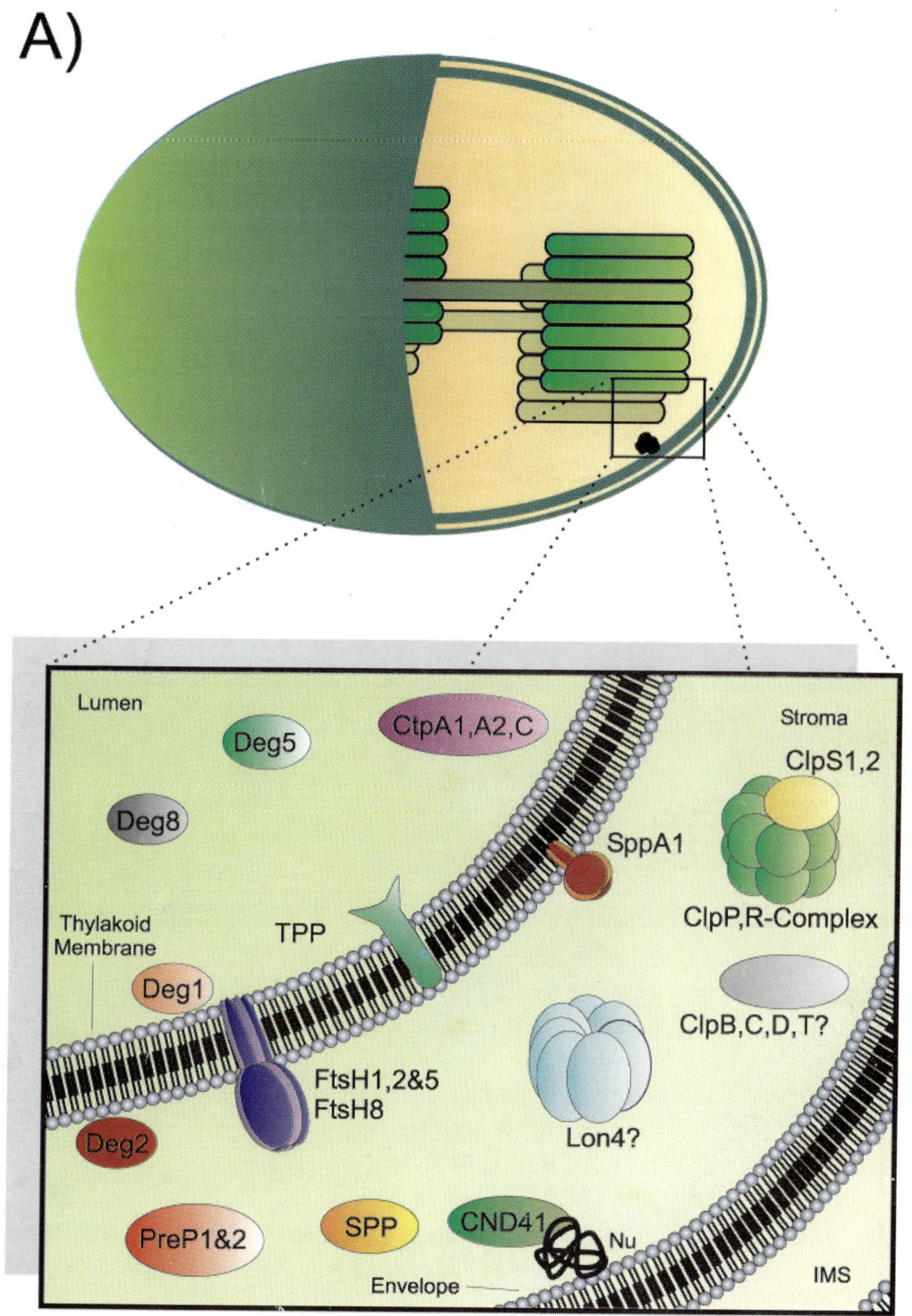


**Figure 1A. Overview of experimentally confirmed endoproteases in various chloroplast sub-compartments in *A. thaliana*.**

Abbreviations used: IMS, intramembrane space between two envelope membranes; Nu, nucleoids.

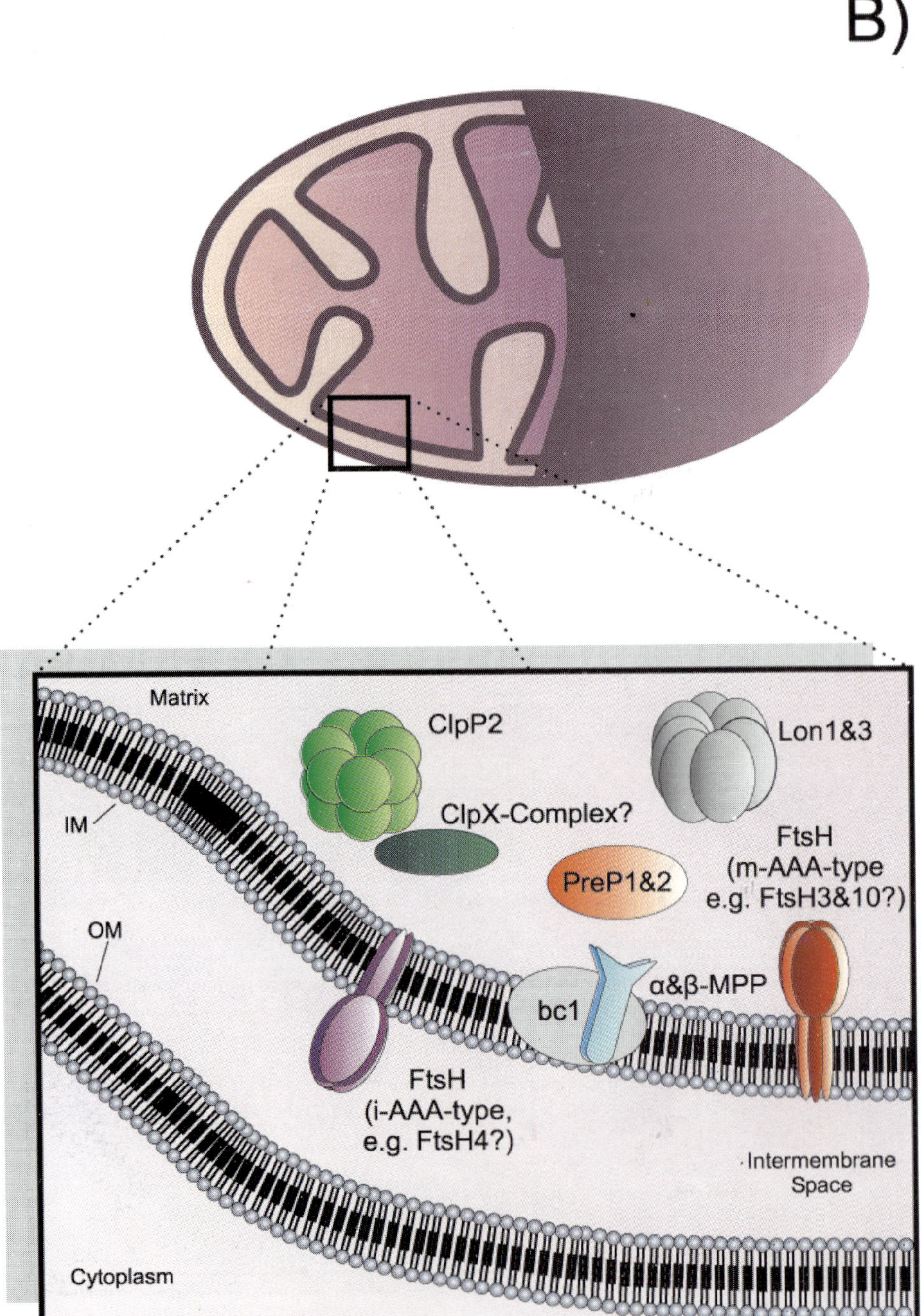


**Figure 2B. Overview of experimentally confirmed endopeptidases in various mitochondrial sub-compartments in *A. thaliana*.**

Abbreviations: IM, inner mitochondrial membrane; OM, outer mitochondrial membrane.

et al., 2000; Sakamoto et al., 2003). However, there are evidences that both types of FtsH proteases are present in mitochondria of higher plants. An m-AAA type of FtsH protease, which shares high sequence similarity with FtsH3 and 10 proteases from *A. thaliana* has been identified in pea mitochondria (Kolodziejczak *et al.*, 2002). The FtsH4 protease in the inner mitochondrial membrane of *A. thaliana* has a topology similar to i-AAA type proteases from yeast (Janska, 2005).

It was reported that mitochondrial FtsH proteases form large ring-like complexes of 1 MDa composed of identical (e.g. in the fungus *Neurospora crassa* or in the red alga *Cyanidioschyzon merolae*) or closely related (e.g. in yeast) subunits (Itoh *et al.*, 1999; Klanner *et al.*, 2001). Similar large oligomeric complexes are expected to be formed by FtsH proteases in mitochondria of higher plants. This assumption is supported by the fact that FtsH protease from pea could complement the yeast *ftsH* deletion mutants suggesting structural and functional relationship of these enzymes (Kolodziejczak *et al.*, 2002). The high degree of sequence identity of mitochondrial FtsH proteases suggests the hetero-oligomeric nature of such complexes in plants.

Similar to mitochondrial FtsH proteases, some chloroplast-located members of this family in *A. thaliana* also form 400-450 kDa complexes, most likely hexamers (Sakamoto *et al.*, 2003; Yu *et al.*, 2004). It was demonstrated that FtsH1, 2, 5 and 8 proteases are subunits of such heteromeric complexes, with FtsH2 and 5 proteases being the most abundant subunits (Sakamoto *et al.*, 2003; Sinvany-Villalobo *et al.*, 2004; Yu *et al.*, 2004, 2005). Furthermore, it was demonstrated that these complexes are formed by two closely related pairs of proteins, with FtsH2 and 8 proteases forming one pair and FtsH1 and 5 proteases the other (Sakamoto *et al.*, 2003; Yu *et al.*, 2004; 2005; Zaltsman *et al.*, 2005b). The 3D crystal structure of the ATPase domain of FtsH proteases is available for *E. coli* (PDB: 1LV7) and *Thermus thermophilus* with (PDB: 1IY1) or without (PDB: 1IXZ) bound ADP.

### *2.2.4. Degradation of periplasmic proteins/High temperature requirement (Deg/Htr) protease*

The first member of this family was identified in *E. coli* mutants, which were lacking the capability to degrade periplasmic proteins (Strauch and Beckwith, 1988) or were unable to grow at elevated temperatures (Lipinska *et al.*, 1989). The first plant Deg homolog was found in pea chloroplasts by cross-hybridization with an antibody raised against *E. coli* DegP enzyme (Itzhaki *et al.*, 1998).

Deg proteases (summarized in Table 1 and Figure 1A) are ATP-independent serine endopeptidases with the catalytic center of trypsin type and up to three PDZ domains (Pallen and Wren, 1997; Clausen *et al.*, 2002). Sixteen *deg* genes were found in *A. thaliana* (Adam *et al.*, 2001; Sokolenko *et al.*, 2002; Kieselbach and Funk, 2003; Huesgen *et al.*, 2005) and the chloroplast location was experimentally proven for four proteins. It was shown that Deg1, 5 and 8 is located in the thylakoid lumen (Itzhaki *et al.*, 1998; Peltier *et al.*, 2002; Schubert *et al.*, 2002) and Deg2 is attached to the stromal side of the thylakoid membrane (Haußühl *et al.*, 2001). Biochemical characterization of Deg1 protease from *A. thaliana* expressed in *E. coli* showed that this enzyme can form homo-hexameric complexes and was active both, as a complex and as a monomer (Chassin *et al.*, 2002). The formation of complexes by other Deg proteases is expected but not yet experimentally proven.

There is still no experimental evidence available for Deg proteases in plant mitochondria, even though up to six of the putative Deg proteases are predicted to target to this organelle in *A. thaliana* (Adam *et al.*, 2001; Kieselbach and Funk, 2003; Huesgen *et al.*, 2005). Circumstantial evidence came form proteomics projects, where Deg11 protease was found as a probable mitochondrial contamination in the thylakoid membrane fraction (Friso *et al.*, 2004).

Deg proteases from *E. coli* and *Homo sapiens* are the only Deg family members for which the whole structure has been studied on the molecular level. The 3D crystal structure of a proteolytic inactive variant of DegP/HtrA from *E. coli* (PDB: 1KY9B) shows a homotrimer as an essential unit. Two of these basic units dimerize to a cage-like hexamer, in which the protease domains form the top and the bottom and the PDZ domains the highly flexible side walls, which are believed to be the gate-keepers controlling access to the inner cavity of the hexamer (Krojer *et al.*, 2002). In contrast, HtrA2/Omi protease, a homolog of Deg protease from *H. sapiens* mitochondria (PDB: 1LCY) and DegS from *E. coli* crystallized without (PDB: 1TEO) or with the activation peptide (PDB: 1VCW) are homotrimers interacting through protease domains and leaving the PDZ domains free for interactions with substrate or regulatory molecules (Li *et al.*, 2002; Wilken *et al.*, 2004).

### *2.2.5. Chloroplast nucleoid DNA-binding (CND41) protease*

Nakano *et al.* (1997) isolated a 41 kDa DNA-binding protein from chloroplast nucleoids of cultured tobacco cells that contained an aspartyl protease active site motif of pepsin type (summarized in Table 1 and Figure 1A). Interestingly, the CND41 protease purified from tobacco has a strong proteolytic activity at acidic pH 2-4 (Murakami *et al.*, 2000). Furthermore, expression of a series of truncated CND41 protease constructs in *E. coli* revealed that this protease non-specifically binds chloroplast DNA and a lysine-rich region is essential for this binding (Nakano *et al.*, 1997). Expression studies showed that CND41 protease accumulates mainly in cells/tissues containing non-photosynthetic, actively growing plastids. A homolog of tobacco CND41 protease predicted to target to the chloroplast is present in the genome of *A. thaliana* (Table 1). The 3D crystal structure is not yet available for this enzyme.

### *2.2.6. Signal peptide peptidase A (SppA1)*

SppAs (called also protease IVand V) were initially purified from *E. coli* inner and outer cytoplasmic membranes and showed to degrade released signal peptides (Regnier, 1981; Pacaud, 1982). While two forms of this enzyme, the longer SppA1 form and the shorter SppA2 form, are present in *E. coli* and in the cyanobacterium *Synechocystis* sp. PCC6803, only the longer SppA1 form was found in the chloroplast of *A. thaliana* (Lensch *et al.*, 2001; Sokolenko, 2005).

The SppA1 (summarized in Table 1 and Figure 1A) is an ATP-independent serine protease containing two homolog proteolytic domains arranged in tandem (Lensch *et al.*, 2001; Sokolenko, 2005). However, only the C-terminally domain, located within the hydrophobic stretch, appears to be proteolytic active (Lensch *et al.*, 2001; Sokolenko, 2005). SppA1 in *A. thaliana* was found exclusively in the thylakoid membranes as a part of a 260 kDa, probably homotetrameric complex (Lensch *et al.*, 2001; Sokolenko, 2005). Since the predicted hydrophobic stretch of *A. thaliana* SppA1 is too short to span the entire membrane it was proposed that this protein integrates monotopically, i.e. into the stromal side of the lipid bilayer (Lensch *et al.*, 2001; Sokolenko, 2005). The 3D crystal structure is not yet available for these enzymes.

## 3. PROTEIN PROCESSING IN ORGANELLES

Chloroplast and mitochondria of higher plants contain as many as 3.600 (Baginsky and Gruissem, 2004) and 2.000-3.000 (Millar *et al.*, 2005) different proteins, respectively, out of which only a small portion is encoded by the organelle genome. The majority of these proteins are encoded in the nucleus and synthesized in the cytoplasm as precursor proteins with N-terminal transit peptides that target them to organelles. The transit peptides are proteolytically removed as proteins enter the organelle, either by a general stromal SPP or mitochondrial MPP, respectively. Both enzymes are closely related and evolved from a common ancestor (VanderVere *et al.*, 1995). The cleaved transit peptides in the stroma or the mitochondrial matrix are further degraded by PrePs. In addition, a subset of proteins targeted to the thylakoid lumen in chloroplasts or to various subcompartments in mitochondria contains two transit peptides organized in tandem. While the first signal targets the protein to the stroma or the mitochondrial matrix and is cleaved by SPP or MPP, respectively, the second signal determines the final destination. This second signal is cleaved by the thylakoid-located TPP after the translocation of proteins into the thylakoid lumen or by the putative plant mitochondrial proteases, such as the mitochondrial intermediate peptidase (MIP), the inner membrane peptidase (IMP) or the processing of cyt. c peroxidase protein 1 (Pcp1). An overview of these processing events is provided below.

## 3.1. Protein Processing in Plastids

### 3.1.1. General amino-terminal processing of nuclear-encoded proteins in the *chloroplast stroma*

It was reported nearly 20 years ago that pea chloroplast lysates contain a processing activity directed against the precursor of the small subunit (preSSU) of the ribulose-1,5-bisphosphate carboxylase / oxygenase (Rubisco) (Robinson and Ellis, 1984). This activity was further purified from pea chloroplasts and characterized in biochemical terms using the light-harvesting chlorophyll *a*/*b* precursor (preLhcp) as a substrate (Oblong and Lamppa, 1992). These studies lead to the identification of a stromal SPP. According to the existing knowledge a single SPP is responsible for the processing of most precursors that enter the stroma through the Toc (translocon of the outer chloroplast membrane)/Tic (translocon of the inner chloroplast membrane)-dependent import pathway (Vothknecht and Soll, 2005). The most prominent substrates of SPP are the stroma-located SSU, the thylakoid membrane-located Lhcp and ferredoxin or the thylakoid lumen-located subunits of the oxygen evolving complex (OEC) and plastocyanin (PC) (Richter and Lamppa, 1998; Richter *et al.*, 2005). Thus, the substrate range of SPP involves a great diversity of precursors destined for different compartments and biosynthetic pathways in various types of plastids, including chloroplasts, etioplasts and leucoplasts (Oblong and Lamppa, 1992; Richter and Lamppa, 1998; Richter *et al.*, 2005).

Some years later, it was proposed that a second stromal processing enzyme might exist in the chloroplast that differs in substrate specificity, apparent molecular mass and cleavage sites (Koussevitzky *et al.*, 1998). Similarly to SPP, this enzyme appears to be a metalloendopeptidase but in contrast to SPP it is inactive toward various preLhcps. Furthermore, it cleaves the precursor of the lumenal polyphenol oxidase immediately before the hydrophobic core of thylakoid processing sites for lumen proteins. The processing sites of other lumenal proteins determined so far were relatively distant from this domain (Peltier *et al.*, 2000). Thus, it is expected that several SPPs might exist in chloroplasts with preferences for different subsets of precursors (Koussevitzky *et al.*, 1998). No candidates for this processing activity were found in the genome of *A. thaliana*.

Table 2. Overview of experimentally confirmed mitochondrial proteases in *Arabidopsis thaliana*.

| Name | Chromosomal Accession No | Protein Size (aa) | Merops Clan, Family | Catalytic Motif | Energy Dependence | Location | Complex Formation | Reviewed |
|---|---|---|---|---|---|---|---|---|
| Processing Enzymes | | | | | | | | |
| MPP-α1 | At1g51980 | 503-mTP26 | - | Missing | No | Inner-Mem | Cyt. bc1 | 1-4 |
| MPP-α2 | At3g16480 | 499-mTP26 | - | Missing | No | n.d. | n.d. | n.a. |
| MPP-β | A3g02090 | 535-mTP26 | ME, M16C | HXXEH | No | Inner-Mem | Cyt. bc1 | 1-4 |
| MIP-like | At5g51540 | 860-mTP29 | MA, M3 | HEXXH | No | Matrix | n.d. | n.a. |
| IMP1-like | At1g53530 | 168-mTP15 | SF, S26 | S-K | No | Inner-Mem | n.d. | n.a. |
| IMP2-like | At1g06870 | 367-mTP24 | SF, S26 | S-K | No | Inner-Mem | n.d. | n.a. |
| Pcp1a-like | At1g18600 | 336-mTP47 | S-, S54 | N-S-H | No | Inner-Mem | n.d. | n.a. |
| Pcp1b-like | At1g74130 | 322-mTP52 | S-, S54 | N-S-H | No | Inner-Mem | n.d. | n.a. |
| PreP1 | At3g19170 | 1080-mTP56 | ME, M16C | HXXEH | No | Martix | n.d. | n.a. |
| PreP2 | At1g49630 | 1080-mTP68 | ME, M16C | HXXEH | No | Matrix | n.d. | n.a. |
| Clp Complex | | | | | | | | |
| ClpP2 | At5g23140 | 241-mTP30 | SK, S14 | S-H-D | No | Matrix | Homomer. core | 5-7 |
| ClpB4 | At2g25140 | 964-mTP39 | - | Chaperone | ATP | Matrix | n.d. | 5-7 |
| ClpX1 | At5g53350 | 579-mTP46 | - | Chaperone | ATP | Matrix | Core assoc. | 5-7 |
| ClpX2 | At5g49840 | 606-mTP66 | - | Chaperone | ATP | Matrix | Core assoc. | 5-7 |
| ClpX3 | At1g33360 | 656-mTP28 | - | Chaperone | ATP | Matrix | Core assoc. | 5, 7 |
| Lon Family | | | | | | | | |
| Lon1 | At5g26860 | 940-mTP61 | SF, S16 | S-K | ATP | Matrix | Homomer. | 6-7 |
| Lon3 | At3g05780 | 924-mTP64 | SF, S16 | S-K | ATP | Matrix | Homomer. | 7 |

| | | | | | | | | |
|---|---|---|---|---|---|---|---|---|
| FtsH Family | | | | | | | | |
| FtsH3 | At2g29080 | 809-mTP40 | MA, M41 | HEXXH | ATP | Inner-Mem | n.d. | 5-7 |
| FtsH4 | At2g26140 | 717-mTP15 | MA, M41 | HEXXH | ATP | Inner-Mem | n.d. | 6-7 |
| FtsH10 | At1g07510 | 813-mTP36 | MA, M41 | HEXXH | ATP | Inner-Mem | n.d. | 5, 7 |

Protein precursor sizes are given in amino acids (aa) according to the TAIR database (http://www.arabidopsis.org/abrc). The sizes of mitochondrial transit peptides (mTP) are according to the TargetP (http://www.cbs.dtu.dk/services/TargetP) prediction. The classification of proteases into Clans and Families is according to the Merops Database (http://merops.sanger.ac.uk). Abbreviations: Assoc., association; Homomer., homomeric; Mem, membrane; n.a., not available; n.d., no data. Only recent review articles dealing with mentioned protease families are listed in this table, the original papers are quoted in the text. (1) Braun and Schmitz, 1995; (2) Paetzel *et al.*, 2002; (3) Schaller, 2004; (4) Glaser and Dessi, 1999; (5) Sokolenko *et al.*, 2002; (6) Adam *et al.*, 2001; (7) Janska, 2005

The mechanism of the processing reaction is based on highly specific protein-protein interactions between SPP and substrates and proceeds in four steps: (a) recognition of the precursor by binding of SPP to the transit peptide (Richter and Lamppa, 1998); (b) processing reaction and the generation of a mature protein (Richter and Lamppa, 2003); (c) cleavage of the transit peptide and release of subfragments (Richter and Lamppa, 2002, 2003), and finally (d) degradation of released subfragments by PrePs (Ståhl *et al.*, 2002; Bhushan *et al.*, 2003; Moberg *et al.*, 2003; Bhushan *et al.*, 2005; Ståhl *et al.*, 2005).

How the SPP recognizes the transit peptides is still not known. Chloroplast transit peptides are highly divergent and vary substantially in length and do not share consensus sequences (Gavel and Heijne, 1990; Bruce, 2001). However, in general they contain three distinct regions, an uncharged N-terminal domain, a central domain lacking acidic residues and a C-terminal domain with the potential to form an amphiphilic β-strand. The secondary structure of targeting peptides is predicted to be mainly random coiled (Pilon *et al.*, 1992; Theg and Geske, 1992), although more recent data suggests that they can form α-helical structures in hydrophobic milieu (Bruce *et al.*, 2000).

Results from competition experiments (Richter and Lamppa, 2002) and mutant analysis (Clark and Lamppa, 1991; Archer and Keegstra, 1993; Bassham *et al.*, 1994) indicate that SPP interacts with the region of about 10-15 amino acids at the C-terminus of the transit peptide. It seems that this region alone mediates the binding, removal and internal cleavage catalyzed by SPP in a reaction that leads to the release of degradable transit peptide subfragments (Richter and Lamppa, 2002; Richter *et al.*, 2005). Furthermore, basic residues that are usually found in this region play a critical role for efficient processing (Clark and Lamppa, 1991; Archer and Keegstra, 1993; Bassham *et al.*, 1994). It has been showed by mapping of the cleavage sites that chloroplast targeting peptides contain a loosely conserved motif around the processing site, V/I-X-A/C ↓ A (Zhang and Glaser, 2002).

To investigate which regions of SPP are responsible for substrate recognition and binding various deletion constructs were engineered and tested (Richter and Lamppa, 2003). Deletion of the N-terminal half of the SPP sequence resulted in a loss of the substrate binding ability indicating that this part contains structural elements necessary for interaction with the transit peptide (Richter and Lamppa, 2003; Richter *et al.*,

2005). However, the N-terminal part of SPP alone could not reconstitute the optimal efficiency of substrate binding (Richter and Lamppa, 2003; Richter *et al.*, 2005).

The transit peptide binding by SPP and its cleavage are two separate steps of the overall processing reaction. The SPP precursor is synthesized as a latent form and needs to be processed for its activation (Richter and Lamppa, 2003). There is evidence that this activation of SPP occurs due to autoproteolysis (Richter and Lamppa, 2002). It was reported that SPP depends on the N-terminally located HXXEH zink-binding motif for catalytic activity although this motif is not involved in substrate recognition and binding (Richter and Lamppa, 2003). A mutated form of SPP with the substituted first histidine interacts with the transit peptide, but is not able to cleave it and convert it into subfragments. This suggests that the precursor processing by SPP and the transit peptide conversion depend on this motif (Richter and Lamppa, 2003). Furthermore, studies with SPP deletion constructs revealed that the structural integrity of the entire SPP polypeptide is required for an active conformation of this enzyme and the C-terminus of SPP confers an important feature for its activity (Richter and Lamppa, 2003).

The SPP removes the transit peptide in a single endoproteolytic step and whereas the mature protein is immediately released the transit peptide remains bound to SPP and is converted to subfragments (Richter and Lamppa, 1999). These subfragments are released and degraded by recently characterized PrePs (Ståhl *et al.*, 2002; Bhushan *et al.*, 2003; Moberg *et al.*, 2003; Bhushan *et al.*, 2005).

Antisense suppression of the *spp* gene in tobacco (Wan *et al.*, 1998) or *A. thaliana* (Zhong *et al.*, 2003) indicated that SPP is an essential component of the chloroplast Toc/Tic import machinery and necessary for chloroplast biogenesis and plant viability. Chloroplast import of transit peptides coupled to GFP was impaired in antisense lines, indicating that SPP-mediated maturation of chloroplast precursors in the stroma also affects earlier steps in the import pathways (Zhong *et al.*, 2003).

### *3.1.2. Amino-terminal processing of nuclear-encoded proteins imported into the thylakoid lumen*

Proteins located in the thylakoid lumen of the chloroplast have to pass three membrane systems and two soluble compartments to reach their final destination. Therefore, they are synthesized with bipartite transit peptides with two targeting signals arranged in tandem (Smeekens *et al.*, 1986). The first signal targets the protein through two envelope membranes into the stroma, where SPP removes this extension and generates intermediates ready for the transport into thylakoid membranes (Howe and Merchant, 1993). The second transit peptide directs intermediate forms across the thylakoid membrane into the lumen, where TPP removes the thylakoid target signal and generates a mature form of the protein (Hageman *et al.*, 1986). Although such a dual processing of lumen-located proteins has been known for very long time the gene encoding TPP in *A. thaliana* was identified and isolated relatively late (Chaal *et al.*, 1998).

Thylakoid lumen proteins known to be processed by TPP include prePC, preOEC33, 23 and 16 (Hageman *et al.*, 1986; James *et al.*, 1989). Interestingly, the substrate specificity of plant TPP is identical to that of the plasma membrane-located LepB from *E. coli* (Halpin *et al.*, 1989). *E. coli* LepB and plant TPP cleaved several thylakoid lumen precursors as well as proteins secreted or imported into the periplasmic space of *E. coli* at the same sites. However, one has to mention that structures of bacterial leader peptides for the periplasmic location and thylakoid-transfer domains have similar features. These include a positively charged N-terminal region, a hydrophobic central core and a polar C-terminal region, in which the –3 and –1 residues relative to the cleavage site are short uncharged amino acids (Shackleton and Robinson, 1991).

Recent analysis of the transit peptide cleavage sites of integral thylakoid membrane proteins suggests the presence of a second thylakoid processing protease that might recognize transit peptides of integral membrane proteins inserted through the spontaneous mechanism that is related to the secretory mechanism of Gram-negative bacteria (Gomez *et al.*, 2003). However, an enzyme involved in such processing is not yet identified. Several genes encoding TPP-like proteases related to bacterial LepB are present in the genome of *A. thaliana* and one of them is predicted to target to the chloroplast (Table 1). Furthermore, the TPP-like protein contains a predicted N-terminally located transmembrane domain suggesting its integral thylakoid membrane location, similar to that of TPP. Therefore, the TPP-like protein mentioned in Table 1 is a good candidate for an enzyme that might be involved in the processing of thylakoid membrane proteins.

### 3.1.3. *Carboxyl-terminal processing of chloroplast encoded thylakoid proteins*

The chloroplast-encoded D1 protein from the photosystem II (PSII) reaction center is synthesized as a precursor protein with a C-terminal extension (Reisfeld *et al.*, 1982). The lengths of this extension vary from as short as eight amino acid residues in the case of the green alga *Chlamydomonas reinhardtii* to nine residues for other eukaryotes and 16 residues for cyanobacteria (Nixon *et al.*, 1992). Proteolytic cleavage and removal of this extension in the thylakoid lumen is necessary for the integration of the water-splitting machinery into PSII complex and the oxygen evolution activity of PSII (Diner *et al.*, 1988; Nixon *et al.*, 1992; Roose and Pakrasi, 2004). Characterization of the *ctpA* deletion mutant of the cyanobacterium *Synechocystis* sp. PCC6803 revealed that despite of containing fully active PSII reaction center this mutant was not able to accept electron from water (Anbudurai *et al.*, 1994; Shestakov *et al.*, 1994). Immunoblot analysis showed that the D1 protein from PSII reaction center in this mutant was nearly 2 kDa larger than in wild type cells. Therefore, it was concluded that the deleted *ctpA* gene encodes a processing enzyme responsible for the maturation of preD1 protein. Subsequently, CtpA proteases were isolated and characterized in biochemical terms from the thylakoid lumen of *S. obliquus* (Trost *et al.*, 1997), spinach (Fujita *et al.*, 1995; Oelmüller *et al.*, 1996; Fabbri *et al.*, 2005) and barley (Oelmüller *et al.*, 1996). The PDZ motif of CtpA1 protease was proposed to be the site at which the substrate C-terminus binds (Liao *et al.*, 2000). The reminder of the substrate likely extends across the face of the enzyme, interacting at its scissile bond with the CtpA1 active site (Liao *et al.*, 2000).

Two *ctpA* and one *ctpC* genes are present in the genome of *A. thaliana* (Table 1) but it is still not known, whether all of them encode proteases involved in the C-terminal protein processing. Interestingly, a much broader function was reported for homolog Tsp/Prc proteases in heterotrophic or pathogenic bacteria. Tsp/Prc proteases were reported to be involved not only in the C-terminal maturation of cell envelope proteins in *Borrelia burgdorferi* (Ostberg *et al.*, 2004) or *Brucella suis* (Bandara *et al.*, 2005) but also to play a role in a quality control. In *E. coli* they are responsible for the degradation of mistranslated proteins after the addition of a C-terminal ssrA (small stable RNA) sequence-tags (Keiler *et al.*, 1996; Spiers *et al.*, 2002).

## 3.2. Protein Processing in Plant Mitochondria

### 3.2.1. General N-terminal processing of nuclear-encoded mitochondrial proteins

The nuclear-encoded mitochondrial precursor proteins are imported in unfolded state into the mitochondrial matrix across the double membrane through the Tom (translocase of outer mitochondrial membrane)/Tim (translocase of inner mitochondrial membrane) translocation complex (Ryan and Jensen, 1995; Schatz, 1996; Neupert, 1997; Pfanner and Meijer, 1997). After import, presequences are cleaved by a matrix-located MPP ((Braun *et al.*, 1992; Arretz *et al.*, 1994). MPP was originally identified from the mitochondria of non-photosynthetic organisms, such as *Saccharomyces cerevisiae* (Yang *et al.*, 1988) or rat (Ou *et al.*, 1989) and only some years later MPP homologs were reported from potato (Braun *et al.*, 1992), wheat (Braun and Schmitz, 1995) and spinach (Eriksson *et al.*, 1996).

MPP removes presequences in one proteolytical step, either from fully translocated proteins or from proteins that are in process of translocation (Glaser and Dessi, 1999). It has been shown that Tom/Tim translocation channel and MPP, which is an integral part of the cyt. bc1 complex, are located separately in the membrane of plant mitochondria. There is no correlation between electron transfer through the cyt. bc1 complex and protein processing by MPP (Eriksson *et al.*, 1996). Furthermore, using chimeric constructs consisting of a mitochondrial presequence fused to the mature dihydrofolate reductase it has been shown that the presequence is not processed directly upon exposure to the matrix, but the precursor must be translocated some distance beyond the cleavage site. This result implicates the lack of a direct link between precursor translocation and processing (Dessi *et al.*, 2000), as was shown for chloroplast SPP (see 3.1.1.).

Plants MPP forms a heterocomplex composed of α-MPP and β-MPP subunits. The catalytic site is conserved in β-MPP but degenerated in α-MPP (Glaser and Dessi, 1999). However, β-MPP in fungi has been shown to be unable to catalyze processing in the absence of α-MPP indicating an essential function of α-MPP in processing (Geli, 1993).

It was demonstrated that the whole catalytic region of yeast β-MPP is important for proper conformation of the active site and the mutation of the histidines and the glutamate in yeast and rat β-MPP abolished processing indicating that mutated residues are important for activity (Kitada *et al.*, 1995; Luciano *et al.*, 1997; Kitada *et al.*, 1998). Furthermore, the C-terminal region of β-MPP is also of importance for catalysis. The α-MPP can bind a precursor protein in the absence of β-MPP, suggesting that α-MPP is responsible for the binding of mitochondrial presequences (Luciano *et al.*, 1997). The truncation of the C-terminal 41 amino acids of α-MPP led to a loss of binding and processing activity (Shimokata *et al.*, 1998). Recent studies demonstrated that plant MPP binds presequences with high affinity only as a dimeric complex (Glaser and Dessi, 1999).

How does the MPP recognize mitochondrial presequences? Although different mitochondrial transit peptides do not share any sequence similarity they display some common features. A unique feature of plant mitochondrial presequences is that they have a high content of serine residues as compared to yeast or mammal presequences (Sjöling and Glaser, 1998). The C-terminal part of presequences most often contains a cleavage motif for MPP and a secondary structure in front of the cleavage site, which is compatible for processing (Klaus *et al.*, 1996; Sjöling and Glaser, 1998). This structure consists of a helix, or a helix followed by an extended stretch (Glaser and Dessi, 1999). The majority of mitochondrial presequences contain an arginine either at position −2 or −3 relative to the cleavage site, and have a loosely conserved motif around the cleavage site, R-X ↓ S-(S-T) T or R-X (F/Y) ↓ (A/S)(T/S/A). However, there are also presequences lacking the conserved arginine in the vicinity of the cleavage site that suggests that the arginine motifs are important but not essential for specific cleavage and additional higher order structural elements are required (Zhang and Glaser, 2002). Both NMR experiments and *in silico* analysis have shown that the N-terminal part of mitochondrial presequences has an ability to form amphiphilic α-helices (Chupin *et al.*, 1995). Circular dichroism studies and secondary structure predictions revealed that synthetic peptides missing the helical region had much lower affinity for spinach MPP and mutant peptides affected in the helical structure of the presequence were processed with lower efficiency (Sjöling *et al.*, 1996). These data indicate that the helical region is important for processing efficiency (Sjöling *et al.*, 1996). Thus, the known determinants for recognition of the processing site by MPP include proximal arginines, distal positively charged residues and secondary structure elements, such as a α-helix (Glaser and Dessi, 1999).

It was observed that in root mitochondria from spinach a part of the processing activity was located in the soluble fraction (Knorpp *et al.*, 1994). This suggested the presence of a second processing activity located in the mitochondrial matrix, in addition to the membrane-bound MPP. The presence of a matrix-located peptidase was confirmed by the processing of precursors of the alternative oxidase and the β–subunit of the ATP-synthase in mitochondrial matrix extracts isolated from soybean cotyledons and spinach leaves (Szigyarto *et al.*, 1998). The specificity of this processing reaction was confirmed by the absence of a general processing activity directed against other mitochondrial precursors incubated with the matrix. This additional processing activity was no further characterized in plants.

### *3.2.2 Processing of polyproteins*

The nuclear-encoded RPS14 (ribosomal protein S14) of rice mitochondria is synthesized in the cytoplasm as a polyprotein consisting of a large N-terminal domain comprising preSDHB (precursor of the succinate dehydrogenase B subunit) and the C-terminal RPS14 (Oshima *et al.*, 2005). After the preSDHB-RPS14 polyprotein is transported into mitochondrial matrix, it is processed into three peptides: the N-terminal presequence, the mature SDHB and the C-terminal mature RPS14. The MPP cleaves both, the N-terminal presequence and the connector region between two proteins (Oshima *et al.*, 2005). Mutational analysis around the cleavage site in the connector region suggested that MPP interacts with multiple sites in the region, possibly in a similar manner to the interaction with the N-terminal presequence. Crystal structure of a yeast MPP with bound presequence revealed that MPP incorporates presequence into the internal molecular cavity, possibly though a narrow cleft between the subunits and cleaves it in the hydrophilic environment (Taylor *et al.*, 2001). Since the cleft leading to the cavity of MPP is narrow, it seems likely that the folded polypeptide would have difficulty in entering the cavity. Thus, an unfolded state is essential for the polyprotein processing (Oshima *et al.*, 2005). Polyproteins are in an extended conformation during transport into mitochondria through the Tom/Tim complex, and mitochondrial chaperons keep proteins unfolded and prevent misfolding and protein aggregation (Ryan and Jensen, 1995; Schatz, 1996; Neupert, 1997; Pfanner and Meijer, 1997). Thus, it is expected that MPP may recognize the stretched polyprotein during passage the precursor through the Tom/Tim translocation apparatus and cleaves the connecting region between SDHB and RPS14 proteins even before the processing of the presequence (Oshima *et al.*, 2005).

### 3.2.3. *Processing of intermediate octapeptides*

A number of precursors targeted to the mitochondria are processed in two successive steps. Such precursor proteins contain a second targeting signal composed of eight amino acid residues located behind the presequence. Octapeptide targeting signals are typical of many mitochondrial proteins imported into the matrix and the inner membrane, primarily those involved in the assembly of Fe-S clusters (Nett and Trumpower, 1996). After cleavage of the matrix presequence by MPP, such octapeptides are removed by MIP, as was shown for rat and yeast mitochondria (Isaya *et al.*, 1992; Isaya *et al.*, 1994). MIPs are mitochondrial matrix-located Zn-metalloendopeptidases belonging to the family of thimet oligopeptidases (Table 2). MIP in rat mitochondria (Accession: NP_112314) is a monomeric enzyme of 75 kDa (Isaya *et al.*, 1992). Although a homolog of MIP is present in the genome of *A. thaliana*, the experimental evidence is still lacking, whether this type of protein processing exists in plant mitochondria.

### 3.2.4. *Processing of proteins exported into the intermembrane space*

The mitochondrial IMP in yeast cleaves certain mitochondrial proteins, e.g. the cyt. c oxidase subunit II, during their export from the matrix into the inner membrane or intermembrane space (Gakh *et al.*, 2002; Esser *et al.*, 2004). IMP in *S. cerevisiae* forms a heterocomplex composed of two catalytic protease subunits (Nunnari *et al.*, 1993; Jan *et al.*, 2000), IMP1 (NCBI Accession: AAB19704) and IMP2 (NCBI Accession: AAT930130) and a suppressor of IMP1, the SOM1 (NCBI Accession: AAB65036). The IMP1 and IMP2 subunits are serine endopeptidases with nonoverlapping substrate specificity spanning the inner mitochondrial membrane with one transmembrane domain (Nunnari *et al.*, 1993). Although IMPs were not characterized from plant mitochondria two potential candidates were identified in the genome of *A. thaliana* (Table 2).

### 3.2.5. *Processing of proteins of the carrier import pathway*

A number of polytopic proteins of the inner mitochondrial membrane, such as members of the mitochondrial metabolite carrier family, are imported via the carrier import pathway (Arco and Satrustegui, 2005). In yeast, all proteins imported through the carrier import pathway contain internal targeting signals and do not contain N-terminal cleavable extensions (Neupert, 1997; Pfanner and Geissler, 2001). In contrast, some inner membrane carrier proteins in plants contain an N-terminal transit peptide of 30 or more amino acid residues as reported for maize, potato, *A. thaliana* or rice (Winning *et al.*, 1992; Millar and Heazlewood, 2003; Murcha *et al.*, 2004; Picault *et al.*, 2004). Interestingly, this extension is not essential for the mitochondrial import but plays a role in increasing efficiency of insertion of the protein into the inner membrane (Zara *et al.*, 1992; Murcha *et al.*, 2005). The extension is removed in a two-step process. The first processing step is carried out by MPP, while the second step is mediated by not yet identified mitochondrial inner membrane protease of serine type (Murcha *et al.*, 2004). It was proposed that rhomboid-like Prc1 proteases might be involved in the second processing step of mitochondrial proteins in plants (Murcha *et al.*, 2004). Rhomboid proteases are polytopic intramembrane serine endopeptidases cleaving substrates within their membrane-spanning segments (Urban *et al.*, 2002; van der Bliek and Koehler, 2003). Sequence similarity searches in the genome of *A. thaliana* revealed the presence of 18 putative rhomboid proteases, two of which (Table 2) showed high sequence similarity to a Pcp1 protease from yeast (NCBI Accession: YGR101W) and both were predicted to be mitochondrial proteins (Murcha *et al.*, 2004).

### 3.3. Degradation of Cleaved Chloroplast and Mitochondrial Presequences

Despite the enormous amount of imported and processed precursors cleaved presequences have not been observed to accumulate. Mitochondrial presequences are potentially harmful because they can penetrate mitochondrial membranes, dissipate membrane potential and uncouple respiration (Nicolay *et al.*, 1994). Also chloroplast transit peptides have the ability to insert into lipid bilayers and cause severe damage to chloroplast membrane structure and functions (van't Hof *et al.*, 1993; van 't Hof and de Kruijff, 1995). To avoid the potential toxic effect of cleaved presequences these peptides should be removed by further degradation or export. Mitochondrial export of peptides has only been reported in yeast mitochondria (Young *et al.*, 2001). However, considering the low import efficiency proteolytic systems for the effective elimination of cleaved transit peptides have been postulated for chloroplasts and mitochondria. Recently, the PreP1 has been isolated from potato tuber mitochondria (Ståhl *et al.*, 2002) and later two homolog genes, *preP1* and *2*, were identified in the genome of *A. thaliana* (Bhushan *et al.*, 2003). Substrate specificity studies showed that PreP1 and 2 have the ability to degrade both, mitochondrial and chloroplast transit peptides (Bhushan *et al.*, 2003; Moberg *et al.*, 2003; Bhushan *et al.*, 2005). Immunological studies, measurements of proteolytic activity in isolated chloroplasts and mitochondria and a transient expression of PreP-GFP fusion constructs in tobacco protoplasts and *in planta* revealed the presence of PreP1 and 2 in both organelles (Bhushan *et al.*, 2003; Moberg *et al.*, 2003; Bhushan *et al.*, 2005).

PreP1 and 2 cleave peptides that are in the range of 10-65 amino acid residues, whereas folded or longer unfolded peptides or small proteins are not degraded (Ståhl *et al.*, 2005). The chloroplast and mitochondrial transit peptides do not share any amino acid consensus, although all of them are rich in hydrophobic, hydroxylated and positively charged amino acid residues and low in acidic residues (Peeters and Small, 2001). Serine residues are clearly over represented in chloroplast and plant mitochondria transit peptides. Both PreP1 and 2 showed preferences for basic amino acids in the P1 position and small, uncharged amino acids or serine residues in the P′1 position. Despite the high sequence identity between PreP1 and 2 and similarities in cleavage specificities, the cleavage site recognition differs for both enzymes. While PreP1 showed a preference for the N-terminal amphiphilic α-helix and positively charged amino acid residues, the PreP2 did not showed any positional preferences (Ståhl *et al.*, 2005).

## 4. PROTEIN TURNOVER IN ORGANELLES

In every cell proteins are continuously synthesized and degraded. The abundance of individual proteins is determined by a dynamic balance between the rates of synthesis and degradation. The protein turnover occurs in all organelles including chloroplasts and mitochondria and a variety of proteases are involved in these processes (Adam *et al.*, 2001). Proteases are employed in the removal of mistranslated or aberrant proteins, misfolded or unassembled subunits of multiprotein complexes and incorrectly targeted proteins (Adam, 2001). These major tasks are often summarized as protein quality control, clean-up proteolysis or housekeeping proteolysis.

The lifetimes of many metabolic enzymes and regulatory proteins need to be regulated as their optimal steady-state levels change during the life cycle of plants and during changing environmental conditions. Protein stability and susceptibility to damage also changes during adverse conditions (Andersson and Aro, 1997; Adam, 2001). Therefore, the integrity of proteins and protein complexes has to be continuously monitored and dynamically regulated. However, not all proteins that are misfolded or become denatured need to be degraded. Chaperones bind to non-native proteins, maintain their solubility thus preventing aggregation, promote correct protein folding to the native state and mediate the disassembly of protein aggregates and crosslinks. Many reported proteases or protease complexes combine proteolytic and chaperone-like activities (Suzuki *et al.*, 1997; Spiess *et al.*, 1999; Adam *et al.*, 2001). These appear ideally suited to fulfill the decisive task to discriminate between those proteins, which can be restored to functional integrity, and those, which need to be removed.

Below we attempt to illustrate the diverse and important functions of proteases in plant mitochondria and chloroplasts summarizing recently described examples.

### 4.1. Quality Control Proteolysis

When proteins are targeted to the wrong compartment, they are quickly degraded. This was shown by studies where the original bipartite transit peptide of the lumenal OEC33 was replaced by a target peptide of an integral thylakoid membrane protein, the Lhcb (Halperin and Adam, 1996). The chimeric OEC33 construct was targeted to the chloroplast stroma after *in vitro* import into intact chloroplasts and was rapidly degraded. The proteolytic activity responsible for the degradation of mistarget OEC33 was of serine type and resided mainly in the stroma fraction. Additionally, this activity was found to be stimulated by ATP (Halperin and Adam, 1996).

For many proteins, assembly with other proteins to form a complex is the next crucial step on their way to fulfill their physiological function. For instance, the major protein complexes in the thylakoid membrane, including those of PSII, cyt. b6f, photosystem I (PSI) and the ATP-synthase are all composed of multiple subunits. Inefficient assembly, the lack of a subunit or a damage of one of subunits usually leads to the degradation of other subunits of the complex (Choquet and Vallon, 2000; Adam, 2001). This was observed for mutated PsaC and PsbK proteins of *C. reinhardtii*, which resulted in destabilization and degradation of other subunits of PSI (Takahashi *et al.*, 1991) and PSII (Takahashi *et al.*, 1994) complexes, respectively. A recently reported example is the assembly of multimeric FtsH protease complexes in the thylakoid membranes of *A. thaliana*, which are formed by two pairs of subunits, FtsH1/FtsH5 and FtsH2/FtsH8 (Yu *et al.*, 2005; Zaltsman *et al.*, 2005b). The *ftsH5* knock-out mutants contain only a reduced amount of FtsH2 and 8 proteases (Sakamoto *et al.*, 2003; Yu *et al.*, 2004, 2005), while plants overexpressing FtsH1 protease in an *ftsH5* mutant background accumulated FtsH2 and 8 proteases in amounts comparable to wild type plants (Yu *et al.*, 2005). On the other hand, attempts to overexpress FtsH2 protease in the *ftsH5* mutant background failed (Yu *et al.*, 2005). This failure to accumulate higher amounts of FtsH2 protein was attributed to the lack of compatible partners of the FtsH1 and 5 protease pair. Furthermore, a posttranslational degradation of overexpressed FtsH2 protein was proposed since its mRNA accumulated in strongly increased amounts, but there is no information available for the protease(s) degrading this enzyme (Yu *et al.*, 2005). Interestingly, FtsH proteases have been also implicated to act as a general proteases involved in protein quality control (Adam *et al.*, 2005; Nixon *et al.*, 2005).

The unassembled thylakoid membrane Rieske Fe-S protein and its mutated soluble form were found to be rapidly degraded after *in vitro* import into isolated chloroplasts (Ostersetzer and Adam, 1997). The proteolytic activity was inhibited by the addition of antibodies raised against a native form of FtsH protease from *E. coli* and therefore it was postulated that proteases from the FtsH family might play a central role in the degradation of unassembled thylakoid membrane proteins (Ostersetzer and Adam, 1997).

Degradation of subunits of soluble protein complexes that are produced in excess was also observed for the chloroplast stroma and the thylakoid lumen (Adam, 2001). One of the first examples was the proteolytic removal of the nuclear-encoded SSU of Rubisco, when the synthesis of the chloroplast-encoded large subunits of Rubisco was impaired (Schmidt and Mishkind, 1983). Also a truncated form of OEC23 was quickly degraded in the thylakoid lumen after being correctly targeted and processed to its mature form because it failed to assemble with other subunits into the complex (Roffey and Theg, 1996). Degradation of proteins or protein complexes have also been observed in the absence of cofactors, which are crucial for correct protein folding and the assembly of complexes (Andersson and Aro, 1997; Adam, 2001).

Degradation of subunits is not limited to retain or restore the stoichiometry of multisubunit complexes, but can also occur in functional complexes after their damage. As all living matter, proteins are subjected to damage even during non-stressful conditions and hence have a limited lifetime. A recent effort characterized a proteolytic activity involved in the disassembly and degradation of PSI in thylakoid membranes isolated from *C. reinhardtii* (Henderson *et al.*, 2003). In the *in vitro* system used by the authors, the degradation of PSI consisted of multiple events and the proteolytic activity involved in these processes increased with temperature and was independent of ATP, GTP or stromal factors. The later steps of disassembly and proteolysis of PSI were dependent on the presence of divalent metals, such as $Zn^{2+}$ (Henderson *et al.*, 2003).

*In vitro* assays with the recombinant Deg1 protease from *A. thaliana* revealed that this enzyme degraded recombinant thylakoid lumen-located PC and OEC33 (Chassin *et al.*, 2002). This led to the conclusion that Deg1 protease might act as a housekeeping enzyme in the thylakoid lumen (Chassin *et al.*, 2002). However, these *in vitro* studies await confirmation by additional *in vivo* data.

In contrast to many reports about the involvement of proteases in quality control, e.g. in yeast mitochondria, only a limited number of proteolytic processes have been described for plant mitochondria. Though lacking identified natural substrates, the Clp protease complex has been proposed to play a housekeeping role in mitochondria of *A. thaliana* because of the constitutive expression and the constant level of its subunits under different environmental conditions (Zheng *et al.*, 2002). A chaperone-like activity was documented for the FtsH protease from pea mitochondria. It was demonstrated that this enzyme is involved in the correct assembly of the ATP-synthase complex (Kolodziejczak *et al.*, 2002).

Additionally, selective post-translational proteolysis of proteins synthesized from non-edited mRNA was observed in plant mitochondria (Binder and Brennicke, 2003). An interesting result about the post-translational gene regulation in mitochondria of higher plants has been published by Sarria *et al.* (1998). Accumulation of the *orf239* gene product in common bean, associated with cytoplasmic male sterility, was investigated in vegetative and reproductive tissues. Whereas the *orf239* gene was transcribed in both tissues, the protein accumulated only in vegetative tissues in the presence of a protease inhibitor. In contrast, in reproductive tissues the ORF239 protein was stable also in the absence of an inhibitor. A proteolytic activity involved in this process was identified as a mitochondrial Lon protease. Thus, the members of the Lon family in bean are responsible for a tissue-specific, post-translational gene regulation (Sarria *et al.*, 1998).

### 4.2. Proteolysis during Development/ Biogenesis of Organelles

Nearly all of our knowledge about proteases involved in the development/ biogenesis of the chloroplast was obtained from mutant studies. In *A. thaliana*, no viable homozygous lines have yet been obtained for any of the *clpP* genes, suggesting that the inactivation of any of them is lethal (Clarke *et al.*, 2005). If the content of various ClpP subunits was reduced by antisense repression, two distinct phenotypes were obtained, depending on which gene was targeted by the repression. Repression of the *clpP4* gene produced a "yellow heart" variegation phenotype, which showed severe chlorosis in young leaves, especially in the mid-vein region, but as the leaves matured, this chlorosis gradually disappeared with regreening along the edges of the leaf blade. In lines with lower repression, mature leaves were green like those from wild type plants. In contrast, lines with antisense repression of *clpP5* and *clpP6* genes also exhibited a chlorotic phenotype which was more severe in juvenile leaves, but this chlorosis also remained upon maturation (Clarke *et al.*, 2005). Evidence that ClpP protease is not crucial for general survival of the plant cell, but only for chloroplasts was obtained from the observation that several lines of non-photosynthetic maize suspension cultures do not carry *clpP* genes in their genome. Therefore, it was proposed that ClpP protease may only be essential for the development and function of plastids with the active gene expression (Cahoon *et al.*, 2003). This finding was also supported by the observation that heteroplasmic tobacco disruption mutants of the *clpP* gene exhibited two different types of chloroplasts, wild type chloroplasts and a population of smaller chloroplasts, which showed high chlorophyll fluorescence. Also the overexpression of the 5'-UTR (untranslated region) of the *clpP* gene produced a chlorotic phenotype (Shikanai *et al.*, 2001; Kuroda and Maliga, 2003).

Genetic studies have demonstrated that various regulatory/chaperone Clp subunits are also important for plastid development and function. Analysis of two *A. thaliana* T-DNA insertion mutants of the *clpC1* gene revealed their retarded growth, equally distributed leaf chlorosis, and reduced photosynthetic performance. This suggested that the ClpC1 subunit is likely involved in the biogenesis of photosystems, although no change in leaf cell anatomy or organelles ultrastructure was observed (Sjögren *et al.*, 2004). The ClpC2 subunit was proposed to be a suppressor of thylakoid biogenesis since a mutation in the *clpC2* gene does suppress the variegated leaf phenotype of the *ftsH2* mutant (see below) and the requirement for FtsH2 protease (Park and Rodermel, 2004).

The involvement of FtsH proteases in developmental processes of higher plants was demonstrated by analysis of variegated mutants of *A. thaliana* that were impaired in genes encoding FtsH2 (*var2* mutant) and FtsH5 (*var1* mutant) proteases (Chen *et al.*, 2000; Takechi *et al.*, 2000; Sakamoto *et al.*, 2002; Sakamoto, 2003; Sakamoto *et al.*, 2004). The *ftsH2* mutant had cotyledons like wild type plants; the first true leaf was mostly white and subsequent leaves variegated with decreasing ratio of white to green sectors. However, white sectors were present also in mature leaves. The green sectors contained morphologically normal chloroplasts, whereas in white sectors plastids with large vesicles and nearly no internal structure were reported to be accompanied by rare normal shaped chloroplasts (Chen *et al.*, 2000). All this data suggests that FtsH2 protease is involved in the early stages of chloroplast development, primarily in the biogenesis of thylakoids. The patchy nature of the phenotype suggests that the loss of FtsH2 protease could be compensated, at least in green sectors of the leaf, by other FtsH proteases. Furthermore, the compensation mechanism becomes more efficient with time or the role of FtsH2 protease becomes less important during development of subsequent leaves. The phenotype of the *ftsH5* mutant was similar to those of the *ftsH2* mutant described above, but less pronounced since first true leaves were already variegated instead of being white, and this variegation decreased in subsequent leaves until mature plants were totally green and not distinguishable from wild type plants (Sakamoto *et al.*, 2002). The phenotypes of both mutants did not change with altered light conditions, indicating a role of FtsH2 and 5 proteases in chloroplast development rather than in repair of photodamage (Zaltsman *et al.*, 2005a) as was proposed by some research groups (Nixon *et al.*, 2005).

The role of two other FtsH proteases, the FtsH1 and 8, during chloroplast development was not obvious from obtained null mutant phenotypes (Sakamoto *et al.*, 2003). However, a constitutive expression of FtsH1 protease in the *ftsH5* mutant background or FtsH8 protease in the *ftsH2* mutant background resulted in a rescue of the mutant phenotype, suggesting a similar physiological function of these proteases during chloroplast development (Yu *et al.*, 2004; Adam *et al.*, 2005). This was confirmed by studies using engineered double mutants (Zaltsman *et al.*, 2005b). A double T-DNA insertional mutant of *ftsH1* and *8* genes showed no obvious phenotype, whereas a mutation in *ftsH5* or *1* in the *ftsH2* background enhanced the variegated phenotype. In contrast, double mutants with interrupted *ftsH1* and *5* or *ftsH2* and *8* genes, respectively, showed a nearly lethal phenotype with white leaves and infertility. These data suggest a coordinative model, where the FtsH protease complex is assembled of two types of subunits, with FtsH1 and 5 proteases representing one type and FtsH2 and 8 proteases another. Within each subunit type, two proteases seem to be at least partially interchangeable (Zaltsman *et al.*, 2005b)

Much less biochemical data is available for the involvement of proteases in the development of chloroplasts and their structures. A light-inducible protease, mainly active against the Lhcb apoprotein, was found in bean etioplasts to be attached to the thylakoid membrane (Tziveleka and Argyroudi-Akoyunoglou, 1998). Purified Lhcb fraction contained a co-purified proteolytic activity directed against Lhcb, D1 and D2 proteins from PSII reaction center. This proteolytic activity was light-inducible during chloroplast biogenesis and classified as cysteine type by inhibitor studies (Georgakopoulos *et al.*, 2002).

To date, only very limited data is available about proteases involved in the biogenesis and development of plant mitochondria. Kolodziejczak *et al.* (1992) investigated a pea homolog of yeast m-AAA proteases. *In organello* experiments suggested that this protease is involved in the accumulation of the subunit 9 of the ATP-synthase (ATP9), since inactivation of the protease by inhibitors or temperature resulted in the reduction of the ATP9 protein level in the mitochondrial membrane (Kolodziejczak *et al.*, 2002).

### 4.3. Proteolysis during Stress or Adaptive Responses

Changes in the environment can lead to adverse conditions in organelles, which plants need to adapt to. Drought, salt stress, cold and heat as well as oxidative stresses caused by high light intensities, irradiation with UV light or $CO_2$ limitation can lead to increased damage, unfolding of proteins and/or their aggregation. The most intensively studied proteolytic process in the chloroplast is the turnover of the D1 protein from PSII reaction center. High intensity light causes photooxidative stress and leads to the damage of D1 protein (reviewed in (Andersson and Aro, 1997; Melis, 1999; Andersson and Aro, 2001; Yamamoto, 2001)). The damaged D1 protein is then subsequently degraded and replaced by a new copy. Extensive studies showed that D1 protein is initially cleaved in the stromal loop between its fourth and fifth membrane spanning helices (Andersson and Aro, 2001). Further biochemical studies characterized the proteolytic activity as that of serine type protease, which is associated with the stroma side of the thylakoid membrane and stimulated by GTP (Andersson and Aro, 1997, 2001). Some years later it was shown that the recombinant Deg2 protease from *A. thaliana* selectively cleaves the photodamaged D1 protein within the stromal loop generating two proteolytic fragments (Haußühl *et al.*, 2001). Deg2 protease was found to be associated with the stromal site of the thylakoid membrane and its activity against photodamaged D1 protein was stimulated by GTP (Haußühl *et al.*, 2001). The resulting proteolytic fragments, consisting of the N-terminal 23 kDa and the C-terminal 10 kDa polypeptides, were further degraded in an ATP-dependent manner and this process is stimulated by $Zn^{2+}$ (Spetea *et al.*, 1999; Andersson and Aro, 2001). *In vitro* experiments with recombinant FtsH1 protease from *A. thaliana* showed that this enzyme is involved in the degradation of the 23 kDa fragment (Lindahl *et al.*, 2000). Based on these findings, a two step model for the degradation of photodamaged D1 protein was proposed, where Deg2 protease performs the primary cleavage generating two types of proteolytic fragments, while FtsH protease(s) removes these fragments by secondary proteolysis (Andersson and Aro, 2001; Haußühl *et al.*, 2001). This hypothesis, which is predominantly based on *in vitro* data, is challenged by the fact that the variegated *ftsH2* mutant showed increased photoinhibition and a strongly reduced turnover of the D1 protein when exposed to high light intensities (Bailey *et al.*, 2002). According to an alternative model put forward in order to explain these findings, hetero-hexameric FtsH complexes are suggested to degrade photodamaged D1 protein without the help of other proteases *in vivo* (Nixon *et al.*, 2005). Adding to the already complex picture of D1 degradation, an increased accumulation of crosslinks, composed of D1 protein and other PSII subunits, such as the D2 protein, CP43 or the α-subunit of cyt. b559 have been observed under high light conditions *in vitro* and *in vivo* (Andersson and Aro, 2001; Yamamoto, 2001; Mizusawa *et al.*, 2003). These crosslinked products were removed by one or more still unidentified serine protease(s) located in the chloroplast stroma (Yamamoto, 2001; Mizusawa *et al.*, 2003). Therefore, it has been proposed that several pathways, involving various proteases, are utilized by plants for the efficient removal of photodamaged D1 protein and its crosslinks *in vivo* (Yamamoto, 2001; Mizusawa *et al.*, 2003). Furthermore, the proteolysis of photodamaged D1 protein is regulated through the phosphorylation of its N-terminus (Rintamäki and Aro, 2001) and the OEC33 protein in the thylakoid lumen has also been proposed to play a role in the regulation of D1 proteolysis (Yamamoto, 2001).

Plant defense reactions against high light intensities include the accumulation of stress-specific proteins from the Elip (Early Light Induced Protein) family in the thylakoid membrane (Adamska, 2001). Elips are stable during the exposure of plants to high light stress, but are rapidly degraded during the subsequent recovery at low intensity light or in darkness (Adamska, 2001). The protease involved in the degradation of Elip from pea was partially purified and characterized in biochemical terms. It revealed to be serine type endopeptidase with the apparent molecular mass of 65 kDa and the proteolytic activity depending on $Mg^{2+}$, but not on ATP hydrolysis (Adamska *et al.*, 1996).

Regulated degradation of proteins is often needed in order to adapt to new environmental conditions. When low light grown plants are transferred to higher light intensities for several days, Lhcb subunits of the PSII antenna system are degraded as the antenna size is reduced in order to lower light absorption and prevent photooxidative damage (Lindahl *et al.*, 1995; Andersson and Aro, 1997). The proteolytic activity involved in this process was initially characterized to be an ATP-dependent stromal cysteine or serine protease (Lindahl *et al.*, 1995). Furthermore, phosphorylation of Lhcb proteins prevented their degradation (Yang *et al.*, 1998). Another study identified two proteolytic activities in pea chloroplasts that degraded a recombinant N-terminal peptide of the Lhcb1 subunit (Forsberg *et al.*, 2005). One activity was ascribed to a 95 kDa stromal polypeptide that showed sequence similarity to a glutamyl endopeptidase purified from a crude extract of cucumber leaves. The second activity was located in the thylakoid membrane, inhibited by cysteine- or serine type inhibitors and stimulated by the addition of DTT (Forsberg *et al.*, 2005). Controversially, a recent study demonstrated that the thylakoid membrane FtsH6 protease is involved in the degradation of Lhcb proteins in *A. thaliana* (Zelisko *et al.*, 2005). The authors found that the *ftsH6* T-DNA insertion mutant was not able to degrade Lhcb1 or Lhcb3 subunits during high light acclimation. However, the biochemical properties of the FtsH6 protease have not yet been studied. The FtsH6 protease is further anticipated to accumulate only in residual amounts, as it has escaped detection in a recent study of the thylakoid proteome (Friso *et al.*, 2004).

The thylakoid membrane-bound SppA1 protease was also implicated in the acclimation of plants to high light intensities, since its protein level increased approximately four-fold when plants were subjected to the prolonged high light treatment (Lensch *et al.*, 2001). However, the exact function of SppA1 during acclimation to high light is still unknown.

Proteolysis is also needed during the adaptation to changing nutrient availability. The knock-down mutants of *C. reinhardtii*, which contained 25-40% of initial amounts of ClpP protease, showed normal growth rates under optimal growth conditions, but restricted ability to adapt to elevated CO2 levels (Majeran *et al.*, 2000). These mutants were also affected in the degradation rate of the cyt. b6f complex during nitrogen starvation. Therefore, the proteolytic disposal of fully or partially assembled cyt. b6f complex appears to be controlled by a Clp type protease (Majeran *et al.*, 2000).

Increases in mRNA and protein content for ClpD and several ClpP isomers were found during long-term high light and cold acclimation of *A. thaliana* plants. These results suggest that these proteins may also be involved in acclimation of plants to different physiological conditions (Zheng *et al.*, 2002).

Compared to the wealth of information available for chloroplasts, relatively little is known about proteolysis in plant mitochondria during stress conditions and acclimatory responses (Janska, 2005). It was reported that stress-responsive ATP-dependent proteases are responsible for the degradation of oxidatively modified proteins in plant mitochondria (Sweetlove *et al.*, 2002).

### 4.4. Proteolysis during Leaf Senescence

Senescence is the final stage in leaf development leading to death. Nutrients, especially from chloroplasts, are remobilized and relocated, e.g. to reproductive or storage organs (Matile, 2001). The main source of cellular nitrogen in a photosynthetic active plant cell is the chloroplast, thus proteolysis in the chloroplast represents an important process for nutrient remobilization (Matile, 2001; Hortensteiner and Feller, 2002). Although proteolytic processes during leaf senescence have been studied for long time (Matile, 2001) not much biochemical data is available for the specific proteases involved in the breakdown of chloroplast proteins during senescence.

In tobacco, the degradation of the Rubisco enzyme was linked to a DNA-binding aspartic protease, the CND41 (Kato *et al.*, 2004). Antisense CND41 plants showed a retarded senescence phenotype as compared to wild type controls, and no decrease of the Rubisco content was observed in older leaves upon nitrogen-depletion. *In vitro* studies revealed that the inactive Rubisco form was degraded by the purified CND41 protease, whereas the active form of this enzyme was not prone to the degradation by this enzyme (Kato *et al.*, 2004). Since no Rubisco degradation was observed in young and mature leaves of transgenic plants with an enhanced level of CND41 protease a posttranslational regulation was proposed to be involved in the control of CND41 activity during senescence (Kato *et al.*, 2005). The activation of CND41 by the N-terminal processing (Kato *et al.*, 2005) or by the reduced DNA content during senescence (Kato *et al.*, 2004) were discussed. The latter assumption is supported by the fact that DNA proved to be a competent inhibitor of CND41 protease activity (Kato *et al.*, 2004).

It was reported that an ATP-stimulated Zn-metalloprotease is responsible for the senescence-dependent degradation of the Lhcb3 subunit in barley (Zelisko and Jackowski, 2004). This protease was shown to be a thylakoid membrane-bound and constitutively active. Therefore, a regulation mechanism based on the substrate availability was proposed. Recently, the FtsH6 protease was identified to be responsible for the degradation of Lhcb3 in *A. thaliana* (Zelisko *et al.*, 2005). The authors demonstrated that the *ftsH6* T-DNA insertion mutant is unable to degrade Lhcb3 during senescence.

Surprisingly, no data is available about proteolytic events and the respective proteases involved in senescence of plant mitochondria.

## FUTURE SCOPES AND CONCLUSIONS

There are several problems that emerged in plant proteolysis field through the years. The most prominent were and still are: a lacking identity of organellar proteases, a missing link between observed proteolytic processes in organelles and proteases that are responsible for these events, confusing nomenclature and classification of organellar proteases and the absence of public databases summarizing the state of the art on plant enzymes.

In the past an approach was undertaken to identify organellar proteases through their homology to prokaryotic enzymes. This led to the successful identification of several prominent prokaryotic proteases, with Clp, Lon, FtsH or Deg proteases as the most prominent examples, in chloroplasts and/or mitochondria of various plant species. In recent years our knowledge on organellar proteases has dramatically increased due to functional genomics and proteomics studies. Multiple protease genes have been found in the genomes of completely sequenced plant species and their encoded products were detected in chloroplast or mitochondria proteomes. However, only a few of these proteases could be linked to the physiological processes in plants.

On the other hand many different proteolytic activities have been observed in chloroplasts but no proteolytic enzymes could be linked to these processes. In most cases, only a partial purification and characterization of the proteolytic activity could be achieved. The situation is even more dramatic for plant mitochondria, where proteolytic processes are hardly investigated, with an exception of protein processing. In recent years, a hypothesis-driven approach to identify putative target proteins has been helpful to link individual proteases to proteolytic processes.

Very little is known on the regulation of organellar proteases and proteolytic events. Examples show that conformational changes of substrates or their posttranslational modifications, e.g. the phosphorylation, might play important roles in the recognition of the substrate through the protease. Thus, the next important question is how the proteolytic events witnessed are regulated in organelles to proceed in an orderly manner?

The next serious problem in plant protease research is the existing nomenclature and classification of proteolytic enzymes, especially those located in plant organelles that are very confusing and do not meet our current needs. Standardized protease names within families and between various organisms should be introduced since the same homologs in prokaryota and eukaryota have often different names. Good examples here are Clp and Deg protease families in plants. Whereas two recently discovered Clp subunits, unique for land plants, were named ClpS1 and 2, the existing ClpS subunit in *E. coli* is actually homolog of the ClpT subunit in plants (see 2.2.1.). Also sixteen Deg family members in *A. thaliana* were originally named DegP1-16 because the first described Deg protease in plants was identified by a cross-reactivity with an antibody raised against *E. coli* DegP (Itzhaki *et al.*, 1998). Recently, it has been recognized that this nomenclature causes misunderstandings because it suggests that Deg enzymes from *A. thaliana* are more similar to bacterial DegP than to DegQ or DegS proteases, which is not the case (for details see: Huesgen *et al.*, 2005). Therefore, a standardized nomenclature is urgently required for recently discovered plant enzymes.

There is also urgency for the plant protease database that would be accessible to public. Due to the initiative of Neil D. Rawlings and Alain J. Barrett the MEROPS Protease Database was created and is available through the web site http://MEROPS.SANGER.AC.UK. Although this database summarizes the state of the art on prokaryotic and eukaryotic proteases, plant proteases are under represented in this database as compared with their bacterial or mammalian homologs. A link summarizing current data on protease families in *E. coli* (division into families, 3D structures) was created by Michael Ehrmann from Cardiff University, UK and is accessible through web link http://www.cf.ac.uk/biosi/staff/ ehrmann/ tools/ proteases.index.html. A comparable database, however, is still missing for plant enzymes.

## Acknowledgements

This work was supported by research grants from the Deutsche Forsuchungsgemeinschaft (AD 92/8-2 and SFB/TR11) and the University of Konstanz (547.71/15904703) to I. A.

## REFERENCES

Adam, Z. 2000. Chloroplast proteases: possible regulators of gene expression? *Biochimie,* **82**: 647-654.

Adam, Z. 2001. Chloroplast proteases and their role in photosynthesis regulation. In: *Regulation of Photosynthesis* (eds. E.-M. Aro and B. Andersson), vol. 11, pp. 265-276. Kluwer Academic Publishers. Dordrecht.

Adam, Z., Adamska, I., Nakabayashi, K., Ostersetzer, O., Haußühl, K., Manuell, A., Zheng, B., Vallon, O., Rodermel, S.R., Shinozaki, K. and A.K. Clarke. 2001. Chloroplast and mitochondrial proteases in *Arabidopsis*. A proposed nomenclature. *Plant Physiol.,* **125**: 1912-1918.

Adam, Z. and O. Ostersetzer. 2001. Degradation of unassembled and damaged thylakoid proteins. *Biochem. Soc. Trans,* **29**: 427-430.

Adam, Z. and A.K. Clarke. 2002. Cutting edge of chloroplast proteolysis. *Trends Plant Sci.,* **7**: 451-456.

Adam, Z., Zaltsman, A., Sinvany-Villalobo, G. and W. Sakamoto. 2005. FtsH proteases in chloroplasts and cyanobacteria. *Physiol. Plant.,* **123**: 386-390.

Adamska, I., Lindahl, M., Roobol-Boza, M. and B. Andersson. 1996. Degradation of the light-stress protein is mediated by an ATP-independent, serine-type protease under low-light conditions. *Eur. J. Biochem.,* **236**: 591-599.

Adamska, I. 2001. The Elip family of stress proteins in the thylakoid membranes of pro- and eukaryota. In: *Regulation of Photosynthesis* (eds. E.-M. Aro and B. Andersson), vol. 11, pp. 487-505. Kluwer Academic Publishers. Dordrecht.

Anastasi, A., Knight, C.G. and A.J. Barrett. 1993. Characterization of the bacterial metalloendopeptidase pitrilysin by use of a continuous fluorescence assay. *Biochem. J.,* **290**: 601-607.

Anbudurai, P.R., Mor, T.S., Ohad, I., Shestakov, S.V. and H.B. Pakrasi. 1994. The *ctpA* gene encodes the C-terminal processing protease for the D1 protein of the photosystem II reaction center complex. *Proc. Natl. Acad. Sci. USA,* **91**: 8082-8086.

Andersson, B. and E.-M. Aro. 1997. Proteolytic activities and proteases of plant chloroplasts. *Physiol. Plant.,* **100**: 780-793.

Andersson, B. and E.-M. Aro. 2001. Photodamage and D1 protein turnover in photosystem II. In: *Regulation of Photosynthesis* (eds. E.-M. Aro and B. Andersson), vol. 11, pp. 377-393. Kluwer Academic Publishers. Dordrecht.

Archer, E.K. and K. Keegstra. 1993. Analysis of chloroplast transit peptide function using mutations in the carboxyl-terminal region. *Plant Mol. Biol.,* **23**: 1105-1115.

Arco, A.D. and J. Satrustegui. 2005. New mitochondrial carriers: an overview. *Cell. Mol. Life Sci.,* **62**: 2204-2227.

Arnold, I. and T. Langer. 2002. Membrane protein degradation by AAA proteases in mitochondria. *Biochim. Biophys. Acta,* **1592**: 89-96.

Arretz, M., Schneider, H., Guiard, B., Brunner, M. and W. Neupert. 1994. Characterization of the mitochondrial processing peptidase of *Neurospora crassa*. *J. Biol. Chem.,* **269**: 4959-4967.

Baginsky, S. and W. Gruissem. 2004. Chloroplast proteomics: potentials and challenges. *J. Exp. Bot.,* **55**: 1213-1220.

Bailey, S., Thompson, E., Nixon, P.J., Horton, P., Mullineaux, C.W., Robinson, C. and N.H. Mann. 2002. A critical role for the *Var2* FtsH homologue of *Arabidopsis thaliana* in the photosystem II repair cycle *in vivo*. *J. Biol. Chem.,* **277**: 2006-2011.

Bandara, A.B., Sriranganathan, N., Schurig, G.G. and S.M. Boyl. 2005. Carboxyl-terminal protease regulates *Brucella suis* morphology in culture and persistence in macrophages and mice. *J. Bacteriol.,* **187**: 5767-5775.

Bassham, D.C., Creighton, A.M., Karnauchov, I., Herrmann, R.G., Klösgen, R.B. and C. Robinson. 1994. Mutations at the stromal processing peptidase cleavage site of a thylakoid lumen protein precursor affect the rate of processing but not the fidelity. *J. Biol. Chem.,* **269**: 16062-16066.

Bhushan, S., Lefebvre, B., Ståhl, A., Wright, S.J., Bruce, B.D., Boutry, M. and E. Glaser. 2003. Dual targeting and function of a protease in mitochondria and chloroplasts. *EMBO Rep.,* **4**: 1073-1078.

Bhushan, S., Ståhl, A., Nilsson, S., Lefebvre, B., Seki, M., Roth, C., McWilliam, D., Wright, S.J., Liberles, D.A., Shinozaki, K., Bruce, B.D., Boutry, M. and E. Glaser. 2005. Catalysis, subcellular localization, expression and evolution of the targeting peptides degrading protease, AtPreP2. *Plant Cell Physiol.,* **46**: 985-996.

Binder, S. and A. Brennicke. 2003. Gene expression in plant mitochondria: transcriptional and post-transcriptional control. *Philos. Trans. R. Soc. Lond. B Biol. Sci.,* **358**: 181-188.

Blobel, G. and B. Dobberstein. 1975. Transfer of proteins across membranes. I. Presence of proteolytically processed and unprocessed nascent immunoglobulin light chains on membrane-bound ribosomes of murine myeloma. *J. Cell Biol.,* **67**: 835-851.

Braun, H.P., Emmermann, M., Kruft, V. and U.K. Schmitz. 1992. The general mitochondrial processing peptidase from potato is an integral part of cytochrome c reductase of the respiratory chain. *EMBO J.,* **11**: 3219-3227.

Braun, H.P. and U.K. Schmitz. 1995. Are the 'core' proteins of the mitochondrial bc1 complex evolutionary relics of a processing protease? *Trends Biochem. Sci.,* **20**: 171-175.

Bruce, B.D. 2000. Chloroplast transit peptides: structure, function and evolution. Trends Cell Biol., **10**: 440-447.

Bruce, B.D. 2001. The paradox of plastid transit peptides: conservation of function despite divergence in primary structure. *Biochim. Biophys. Acta,* **1541**: 2-21.

Cahoon, A.B., Cunningham, K.A. and D.B. Stern. 2003. The plastid *clpP* gene may not be essential for plant cell viability. *Plant Cell Physiol.,* **44**: 93-95.

Chaal, B.K., Mould, R.M., Barbrook, A.C., Gray, J.C. and C.J. Howe. 1998. Characterization of a cDNA encoding the thylakoidal processing peptidase from *Arabidopsis thaliana*. Implications for the origin and catalytic mechanism of the enzyme. *J. Biol. Chem.,* **273**: 689-692.

Chassin, Y., Kapri-Pardes, E., Sinvany, G., Arad, T. and Z. Adam. 2002. Expression and characterization of the thylakoid lumen protease DegP1 from *Arabidopsis. Plant Physiol.,* **130**: 857-864.

Chen, M., Choi, Y., Voytas, D.F. and S. Rodermel. 2000. Mutations in the *Arabidopsis VAR2* locus cause leaf variegation due to the loss of a chloroplast FtsH protease. *Plant J.,* **22**: 303-313.

Choquet, Y. and O. Vallon. 2000. Synthesis, assembly and degradation of thylakoid membrane proteins. *Biochimie,* **82**: 615-634.

Chung, C.H. and A.L. Goldberg. 1981. The product of the *lon* (capR) gene in *Escherichia coli* is the ATP-dependent protease, protease La. *Proc. Natl. Acad. Sci. USA,* **78**: 4931-4935.

Chupin, V., Leenhouts, J.M., de Kroon, A.I. and B. de Kruijff. 1995. Cardiolipin modulates the secondary structure of the presequence peptide of cytochrome oxidase subunit IV: a 2D 1H-NMR study. *FEBS Lett.,* **373**: 239-244.

Clark, S.E. and G.K. Lamppa. 1991. Determinants for cleavage of the chlorophyll *a/b* binding protein precursor: a requirement for a basic residue that is not universal for chloroplast imported proteins. *J. Cell Biol.,* **114**: 681-688.

Clarke, A.K., Gustafsson, P. and J.A. Lidholm. 1994. Identification and expression of the chloroplast *clpP* gene in the conifer *Pinus contorta*. *Plant Mol. Biol.,* **26**: 851-862.

Clarke, A.K., MacDonald, T.M. and L.L.E. Sjögren. 2005. The ATP-dependent Clp protease in chloroplasts of higher plants. *Physiol. Plant.,* **123**: 406-412.

Clausen, T., Southan, C. and M. Ehrmann. 2002. The HtrA family of proteases: implications for protein composition and cell fate. *Mol. Cell,* **10**: 443-455.

Dalbey, R.E. and G. von Heijne. 1992. Signal peptidases in prokaryotes and eukaryotes--a new protease family. *Trends Biochem. Sci.,* **17**: 474-478.

Dessi, P., Rudhe, C. and E. Glaser. 2000. Studies on the topology of the protein import channel in relation to the plant mitochondrial processing peptidase integrated into the cytochrome bc1 complex. *Plant J.,* **24**: 637-644.

Diner, B.A., Ries, D.F., Cohen, B.N. and J.G. Metz. 1988. COOH-terminal processing of polypeptide D1 of the photosystem II reaction center of *Scenedesmus obliquus* is necessary for the assembly of the oxygen-evolving complex. *J. Biol. Chem.,* **263**: 8972-8980.

Dobberstein, B., Blobel, G. and N.H. Chua. 1977. *In vitro* synthesis and processing of a putative precursor for the small subunit of ribulose-1,5-bisphosphate carboxylase of *Chlamydomonas reinhardtii*. *Proc. Natl. Acad. Sci. USA,* **74**: 1082-1085.

Dougan, D.A., Reid, B.G., Horwich, A.L. and B. Bukau. 2002. ClpS, a substrate modulator of the ClpAP machine. *Mol. Cell,* **9**: 673-683.

Eriksson, A.C., Sjöling, S. and E. Glaser. 1996. Characterization of the bifunctional mitochondrial processing peptidase (MPP)/bc1 complex in *Spinacia oleracea*. *J Bioenerg. Biomembr.,* **28**: 285-292.

Esser, K., Jan, P.S., Pratje, E. and G. Michaelis. 2004. The mitochondrial IMP peptidase of yeast: functional analysis of domains and identification of Gut2 as a new natural substrate. *Mol. Genet. Genomics,* **271**: 616-626.

Estelle, M. 2001. Proteases and cellular regulation in plants. *Curr Opin Plant Biol* 4, 254-260.

Fabbri, B.J., Duff, S.M., Remsen, E.E., Chen, Y.C., Anderson, J.C. and C.A. Jacob. 2005. The carboxyterminal processing protease of D1 protein: expression, purification and enzymology of the recombinant and native spinach proteins. *Pest. Manag. Sci.,* **61**: 682-690.

Forsberg, J., Strom, J., Kieselbach, T., Larsson, H., Alexciev, K., Engstrom, A. and H.-E. Akerlund. 2005. Protease activities in the chloroplast capable of cleaving an LHCII N-terminal peptide. *Physiol. Plant.,* **123**: 21-29.

Friso, G., Giacomelli, L., Ytterberg, A.J., Peltier, J.B., Rudella, A., Sun, Q. and K.J. van Wijk. 2004. In-depth analysis of the thylakoid membrane proteome of *Arabidopsis thaliana* chloroplasts: new proteins, new functions, and a plastid proteome database. *Plant Cell,* **16**: 478-499.

Fujita, S., Inagaki, N., Yamamoto, Y., Taguchi, F., Matsumoto, A. and K. Satoh. 1995. Identification of the carboxyl-terminal processing protease for D1 precursor protein of photosystem II reaction center of spinach. *Plant Cell Physiol* 36, 1169-1177.

Gakh, O., Cavadini, P. and G. Isaya. 2002. Mitochondrial processing peptidases. *Biochim. Biophys. Acta,* **1592**: 63-77.

Gavel, Y. and G. von Heijne. 1990. A conserved cleavage-site motif in chloroplast transit peptides. *FEBS Lett.,* **261**: 455-458.

Geli, V. 1993. Functional reconstitution in *Escherichia coli* of the yeast mitochondrial matrix peptidase from its two inactive subunits. *Proc. Natl. Acad. Sci. USA,* **90**: 6247-6251.

Georgakopoulos, J.H., Sokolenko, A., Arkas, M., Sofou, G., Herrmann, R.G. and J.H. Argyroudi-Akoyunoglou. 2002. Proteolytic activity against the light-harvesting complex and the D1/D2 core proteins of photosystem II in close association to the light-harvesting complex II trimer. *Biochim. Biophys. Acta,* **1556**: 53-64.

Glaser, E. and P. Dessi. 1999. Integration of the mitochondrial-processing peptidase into the cytochrome bc1 complex in plants. *J. Bioenerg. Biomembr.,* **31**: 259-274.

Gomez, S.M., Bil, K.Y., Aguilera, R., Nishio, J.N., Faull, K.F. and J.P. Whitelegge. 2003. Transit peptide cleavage sites of integral thylakoid membrane proteins. *Mol. Cell. Proteomics,* **2**: 1068-1085.

Hageman, J., Robinson, C., Smeekens, S. and P. Weisbeek. 1986. A thylakoid processing protease is required for complete maturation of the lumen protein plastocyanin. *Nature,* **324**: 567-569.

Halperin, T. and Z. Adam. 1996. Degradation of mistargeted OEE33 in the chloroplast stroma. *Plant Mol. Biol.*, **30**; 925-933.

Halperin, T., Ostersetzer, O. and Z. Adam. 2001a. ATP-dependent association between subunits of Clp protease in pea chloroplasts. *Planta,* **213**: 614-619.

Halperin, T., Zheng, B., Itzhaki, H., Clarke, A.K. and Z. Adam. 2001b. Plant mitochondria contain proteolytic and regulatory subunits of the ATP-dependent Clp protease. *Plant Mol. Biol.,* **45**: 461-468.

Halpin, C., Elderfield, P.D., James, H.E., Zimmermann, R., Dunbar, B. and C. Robinson. 1989. The reaction specificities of the thylakoidal processing peptidase and *Escherichia coli* leader peptidase are identical. *EMBO J.*, **8**: 3917-3921.

Haußühl, K., Andersson, B. and I. Adamska. 2001. A chloroplast DegP2 protease performs the primary cleavage of the photodamaged D1 protein in plant photosystem II. *EMBO J.,* **20**: 713-722.

Heazlewood, J.L., Tonti-Filippini, J.S., Gout, A.M., Day, D.A., Whelan, J. and A.H. Millar. 2004. Experimental analysis of the *Arabidopsis* mitochondrial proteome highlights signaling and regulatory components, provides assessment of targeting prediction programs, and indicates plant-specific mitochondrial proteins. *Plant Cell,* **16**: 241-256.

Heazlewood, J.L. and A.H. Millar. 2005. AMPDB: the *Arabidopsis* Mitochondrial Protein Database. *Nucleic Acids Res.*, **33**: 605-610.

Henderson, J.N., Zhang, J., Evans, B.W. and K. Redding. 2003. Disassembly and degradation of photosystem I in an *in vitro* system are multievent, metal-dependent processes. *J. Biol. Chem.*, **278**: 39978-39986.

Hortensteiner, S. and U. Feller. 2002. Nitrogen metabolism and remobilization during senescence. *J. Exp. Bot.*, **53**: 927-937.

Howe, G. and S. Merchant. 1993. Maturation of thylakoid lumen proteins proceeds post-translationally through an intermediate *in vivo*. *Proc. Natl. Acad. Sci. USA,* **90**: 1862-1866.

Huesgen, P.F., Schuhmann, H. and I. Adamska. 2005. The family of Deg proteases in cyanobacteria and chloroplasts of higher plants. *Physiol. Plant.*, **123**: 413-420.

Inagaki, N., Yamamoto, Y., Mori, H. and K. Satoh. 1996. Carboxyl-terminal processing protease for the D1 precursor protein: cloning and sequencing of the spinach cDNA. *Plant Mol. Biol.,* **30**: 39-50.

Isaya, G., Kalousek, F. and L.E. Rosenberg. 1992. Sequence analysis of rat mitochondrial intermediate peptidase: similarity to zinc metallopeptidases and to a putative yeast homologue. *Proc. Natl. Acad. Sci. USA,* **89**: 8317-8321.

Isaya, G., Miklos, D. and R.A. Rollins. 1994. MIP1, a new yeast gene homologous to the rat mitochondrial intermediate peptidase gene, is required for oxidative metabolism in *Saccharomyces cerevisiae*. *Mol. Cell Biol.,* **14**: 5603-5616.

Itoh, R., Takano, H., Ohta, N., Miyagishima, S., Kuroiwa, H. and T. Kuroiwa. 1999. Two *ftsH*-family genes encoded in the nuclear and chloroplast genomes of the primitive red alga *Cyanidioschyzon merolae*. *Plant Mol. Biol.,* **41**: 321-337.

Itzhaki, H., Naveh, L., Lindahl, M., Cook, M. and Z. Adam. 1998. Identification and characterization of DegP, a serine protease associated with the luminal side of the thylakoid membrane. *J. Biol. Chem.,* **273**: 7094-7098.

James, H.E., Bartling, D., Musgrove, J.E., Kirwin, P.M., Herrmann, R.G. and C. Robinson. 1989. Transport of proteins into chloroplasts. Import and maturation of precursors to the 33-, 23-, and 16-kDa proteins of the photosynthetic oxygen-evolving complex. *J. Biol. Chem.* **264**: 19573-19576.

Jan, P.S., Esser, K., Pratje, E. and G. Michaelis. 2000. Som1, a third component of the yeast mitochondrial inner membrane peptidase complex that contains Imp1 and Imp2. *Mol. Gen. Genet.,* **263**: 483-491.

Janska, H. 2005. ATP-dependent proteases in plant mitochondria: What do we know about them today? *Physiol. Plant.,* **123**: 399-405.

Jarvis, P. and C. Robinson. 2004. Mechanisms of protein import and routing in chloroplasts. *Curr. Biol.,* **14**: 1064-1077.

Kato, Y., Murakami, S., Yamamoto, Y., Chatani, H., Kondo, Y., Nakano, T., Yokota, A. and F. Sato. 2004. The DNA-binding protease, CND41, and the degradation of ribulose-1,5-bisphosphate carboxylase/oxygenase in senescent leaves of tobacco. *Planta.* **220**: 97-104.

Kato, Y., Yamamoto, Y., Murakami, S. and F. Sato. 2005. Post-translational regulation of CND41 protease activity in senescent tobacco leaves. *Planta,* **222**: 1-9.

Keiler, K.C., Waller, P.R. and R.T. Sauer. 1996. Role of a peptide tagging system in degradation of proteins synthesized from damaged messenger RNA. *Science,* **271**: 990-993.

Kieselbach, T. and C. Funk. 2003. The family of Deg/HtrA proteases: from *Escherichia coli* to *Arabidopsis. Physiol. Plant.,* **119**: 337-346.

Kim, Y.I., Levchenko, I., Fraczkowska, K., Woodruff, R.V., Sauer, R.T. and T.A. Baker. 2001. Molecular determinants of complex formation between Clp/Hsp100 ATPases and the ClpP peptidase. *Nat. Struct. Biol.,* **8**: 230-233.

Kirwin, P.M., Elderfield, P.D., Williams, R.S. and C. Robinson.1988. Transport of proteins into chloroplasts. Organization, orientation, and lateral distribution of the plastocyanin processing peptidase in the thylakoid network. *J. Biol. Chem.,* **263**: 18128-18132.

Kitada, S., Shimokata, K., Niidome, T., Ogishima, T. and A. Ito. 1995. A putative metal-binding site in the beta-subunit of rat mitochondrial processing peptidase is essential for its catalytic activity. *J. Biochem. (Tokyo),* **117**: 1148-1150.

Kitada, S., Kojima, K., Shimokata, K., Ogishima, T. and A. Ito. 1998. Glutamate residues required for substrate binding and cleavage activity in mitochondrial processing peptidase. *J. Biol. Chem.,* **273**: 32547-32553.

Klanner, C., Prokisch, H. and T. Langer. 2001. MAP-1 and IAP-1, two novel AAA proteases with catalytic sites on opposite membrane surfaces in mitochondrial inner membrane of *Neurospora crassa. Mol. Biol. Cell,* **12**: 2858-2869.

Klaus, C., Guiard, B., Neupert, W. and M. Brunner. 1996. Determinants in the presequence of cytochrome b2 for import into mitochondria and for proteolytic processing. *Eur. J. Biochem.,* **236**: 856-861.

Knorpp, C., Hugosson, M., Sjöling, S., Eriksson, A.C. and E. Glaser. 1994. Tissue-specific differences of the mitochondrial protein import machinery: *in vitro* import, processing and degradation of the pre-F1 beta-subunit of the ATP synthase in spinach leaf and root mitochondria. *Plant Mol. Biol.,* **26**: 571-579.

Kolodziejczak, M., Kolaczkowska, A., Szczesny, B., Urantowka, A., Knorpp, C., Kieleczawa, J. and H. Janska. 2002. A higher plant mitochondrial homologue of the yeast m-AAA protease. Molecular cloning, localization, and putative function. *J. Biol. Chem.,* **277**: 43792-43798.

Koussevitzky, S., Ne'eman, E., Sommer, A., Steffens, J.C. and E. Harel. 1998. Purification and properties of a novel chloroplast stromal peptidase. Processing of polyphenol oxidase and other imported precursors. *J. Biol. Chem.,* **273**: 27064-27069.

Krojer, T., Garrido-Franco, M., Huber, R., Ehrmann, M. and T. Clausen. 2002. Crystal structure of DegP (HtrA) reveals a new protease-chaperone machine. *Nature,* **416**: 455-459.

Kruft, V., Eubel, H., Jansch, L., Werhahn, W. and H.P. Braun. 2001. Proteomic approach to identify novel mitochondrial proteins in *Arabidopsis. Plant Physiol.,* **127**: 1694-1710.

Kuroda, H. and P. Maliga. 2003. The plastid *clpP1* protease gene is essential for plant development. *Nature,* **425**: 86-89.

Lensch, M., Herrmann, R.G. andA. Sokolenko. 2001. Identification and characterization of SppA, a novel light-inducible chloroplast protease complex associated with thylakoid membranes. *J. Biol. Chem.,* **276**: 33645-33651.

Li, W., Srinivasula, S.M., Chai, J., Li, P., Wu, J.W., Zhang, Z., Alnemri, E.S. and Y. Shi. 2002. Structural insights into the pro-apoptotic function of mitochondrial serine protease HtrA2/Omi. *Nat. Struct. Biol.,* **9**: 436-441.

Liao, D.I., Qian, J., Chisholm, D.A., Jordan, D.B. and B.A. Diner. 2000. Crystal structures of the photosystem II D1 C-terminal processing protease. *Nat. Struct. Biol.,* **7**: 749-753.

Lindahl, M., Yang, D.H. and B. Andersson. 1995. Regulatory proteolysis of the major light-harvesting chlorophyll *a/b* protein of photosystem II by a light-induced membrane-associated enzymic system. *Eur. J. Biochem.,* **231**: 503-509.

Lindahl, M., Tabak, S., Cseke, L., Pichersky, E., Andersson, B. and Z. Adam. 1996. Identification, characterization, and molecular cloning of a homologue of the bacterial FtsH protease in chloroplasts of higher plants. *J. Biol. Chem.,* **271**: 29329-29334.

Lindahl, M., Spetea, C., Hundal, T., Oppenheim, A.B., Adam, Z. and B. Andersson. 2000. The thylakoid FtsH protease plays a role in the light-induced turnover of the photosystem II D1 protein. *Plant Cell,* **12**: 419-431.

Lipinska, B., Fayet, O., Baird, L. and C. Georgopoulos. 1989. Identification, characterization, and mapping of the *Escherichia coli htrA* gene, whose product is essential for bacterial growth only at elevated temperatures. *J. Bacteriol.,* **171**: 1574-1584.

Lister, R., Hulett, J.M., Lithgow, T. and J. Whelan. 2005. Protein import into mitochondria: origins and functions today. *Mol. Membr. Biol.,* **22**: 87-100.

Liu, X.-Q. and A.T. Jagendorf. 1984. ATP-dependent proteolysis in pea chloroplasts. *FEBS Lett.,* **166**: 248-252.

Luciano, P., Geoffroy, S., Brandt, A., Hernandez, J.F. and V. Geli. 1997. Functional cooperation of the mitochondrial processing peptidase subunits. *J. Mol. Biol.,* **272**: 213-225.

Maccecchini, M.L., Rudin, Y., Blobel, G. and G. Schatz. 1979a. Import of proteins into mitochondria: precursor forms of the extramitochondrially made F1-ATPase subunits in yeast. *Proc. Natl. Acad. Sci. USA,* **76**: 343-347.

Maccecchini, M.L., Rudin, Y. and G. Schatz. 1979b. Transport of proteins across the mitochondrial outer membrane. A precursor form of the cytoplasmically made intermembrane enzyme cytochrome c peroxidase. *J. Biol. Chem.,* **254**: 7468-7471.

Majeran, W., Wollman, F.A. and O. Vallon. 2000. Evidence for a role of ClpP in the degradation of the chloroplast cytochrome b(6)f complex. *Plant Cell,* **12**: 137-150.

Malek, L., Bogorad, L., Ayers, A.R. and A.L. Goldberg. 1984. Newly synthesized proteins are degraded by an ATP-stimulated proteolytic process in isolated pea chloroplasts. *FEBS Lett.,* **166**: 253-257.

Matile, P. 2001. Senescence and cell death in plant development: Chloroplast senescence and its regulation. In: *Regulation of Photosynthesis* (eds. E.-M. Aro and B. Andersson), vol. 11, pp. 277-296. Kluwer Academic Publishers. Dordrecht.

Maurizi, M.R., Clark, W.P., Kim, S.H. and S. Gottesman. 1990. Clp P represents a unique family of serine proteases. *J. Biol. Chem.,* **265**: 12546-12552.

Melis, A. 1999. Photosystem-II damage and repair cycle in chloroplasts: what modulates the rate of photodamage ? *Trends Plant Sci.,* **4**: 130-135.

Mertova, J., Almasiova, M., Perecko, D., Bilka, F., Benesova, M., Bezakova, L., Psenak, M. and E. Kutejova. 2002. ATP-dependent Lon protease from maize mitochondria - comparison with other Lon proteases. *Biologia Bratislava,* **57**: 739-745.

Millar, A.H., Sweetlove, L.J., Giege, P. and C.J. Leaver. 2001. Analysis of the *Arabidopsis* mitochondrial proteome. *Plant Physiol.,* **127**: 1711-1727.

Millar, A.H. and J.L. Heazlewood. 2003. Genomic and proteomic analysis of mitochondrial carrier proteins in *Arabidopsis*. *Plant Physiol.,* **131**: 443-453.

Millar, A.H., Heazlewood, J.L., Kristensen, B.K., Braun, H.P. and I.M. Moller. 2005. The plant mitochondrial proteome. *Trends Plant Sci.,* **10**: 36-43.

Mizusawa, N., Tomo, T., Satoh, K. and M. Miyao. 2003. Degradation of the D1 protein of photosystem II under illumination *in vivo*: two different pathways involving cleavage or intermolecular cross-linking. *Biochemistry,* **42**: 10034-10044.

Moberg, P., Ståhl, A., Bhushan, S., Wright, S.J., Eriksson, A., Bruce, B.D. and E. Glaser. 2003. Characterization of a novel zinc metalloprotease involved in degrading targeting peptides in mitochondria and chloroplasts. *Plant J.,* **36**: 616-628.

Mogk, A. and B. Bukau. 2004. Molecular chaperones: structure of a protein disaggregase. *Curr. Biol.,* **14**: 78-80.

Murakami, S., Kondo, Y., Nakano, T. and F. Sato. 2000. Protease activity of CND41, a chloroplast nucleoid DNA-binding protein, isolated from cultured tobacco cells. *FEBS Lett.,* **468**: 15-18.

Murcha, M.W., Elhafez, D., Millar, A.H. and J. Whelan. 2004. The N-terminal extension of plant mitochondrial carrier proteins is removed by two-step processing: the first cleavage is by the mitochondrial processing peptidase. *J. Mol. Biol.,* **344**: 443-454.

Murcha, M.W., Millar, A.H. and J. Whelan. 2005. The N-terminal cleavable extension of plant carrier proteins is responsible for efficient insertion into the inner mitochondrial membrane. *J. Mol. Biol.,* **351**: 16-25.

Nakano, T., Murakami, S., Shoji, T., Yoshida, S., Yamada, Y. and F. Sato. 1997. A novel protein with DNA binding activity from tobacco chloroplast nucleoids. *Plant Cell,* **9**: 1673-1682.

Nett, J.H. and B.L. Trumpower. 1996. Dissociation of import of the Rieske iron-sulfur protein into *Saccharomyces cerevisiae* mitochondria from proteolytic processing of the presequence. *J. Biol. Chem.,* **271**: 26713-26716.

Neupert, W. 1997. Protein import into mitochondria. *Annu. Rev. Biochem.,* **66**: 863-917.

Nicolay, K., Laterveer, F.D. and W.L. van Heerde. 1994. Effects of amphipathic peptides, including presequences, on the functional integrity of rat liver mitochondrial membranes. *J. Bioenerg. Biomembr.,* **26**: 327-334.

Nielsen, E., Akita, M., Davila-Aponte, J. and K. Keegstra. 1997. Stable association of chloroplastic precursors with protein translocation complexes that contain proteins from both envelope membranes and a stromal Hsp100 molecular chaperone. *EMBO J.,* **16**: 935-946.

Nixon, P.J., Trost, J.T. and B.A. Diner. 1992. Role of the carboxy terminus of polypeptide D1 in the assembly of a functional water-oxidizing manganese cluster in photosystem II of the cyanobacterium *Synechocystis sp.* PCC 6803: assembly requires a free carboxyl group at C-terminal position 344. *Biochemistry,* **31**: 10859-10871.

Nixon, P.J., Barker, M., Boehm, M., de Vries, R. and J. Komenda. 2005. FtsH-mediated repair of the photosystem II complex in response to light stress. *J. Exp. Bot.,* **56**: 357-363.

Nunnari, J., Fox, T.D. and P. Walter. 1993. A mitochondrial protease with two catalytic subunits of nonoverlapping specificities. *Science,* **262**: 1997-2004.

Oblong, J.E. and G.K. Lamppa. 1992. Identification of two structurally related proteins involved in proteolytic processing of precursors targeted to the chloroplast. *EMBO J.,* **11**: 4401-4409.

Oelmüller, R., Herrmann, R.G. and H.B. Pakrasi. 1996. Molecular studies of CtpA, the carboxyl-terminal processing protease for the D1 protein of the photosystem II reaction center in higher plants. *J. Biol. Chem.,* **271**: 21848-21852.

Ogura, T. and A.J. Wilkinson. 2001. AAA+ superfamily ATPases: common structure-diverse function. *Genes Cells,* **6**: 575-597.

Oshima, T., Yamasaki, E., Ogishima, T., Kadowaki, K., Ito, A. and S. Kitada. 2005. Recognition and processing of a nuclear-encoded polyprotein precursor by mitochondrial processing peptidase. *Biochem. J.,* **385**: 755-761.

Ostberg, Y., Carroll, J.A., Pinne, M., Krum, J.G., Rosa, P. and S. Bergstrom. 2004. Pleiotropic effects of inactivating a carboxyl-terminal protease, CtpA, in *Borrelia burgdorferi*. *J. Bacteriol.,* **186**: 2074-2084.

Ostersetzer, O., Tabak, S., Yarden, O., Shapira, R. and Z. Adam. 1996. Immunological detection of proteins similar to bacterial proteases in higher plant chloroplasts. *Eur. J. Biochem.,* **236**: 932-936.

Ostersetzer, O. and Z. Adam. 1997. Light-stimulated degradation of an unassembled Rieske FeS protein by a thylakoid-bound protease: the possible role of the FtsH protease. *Plant Cell,* **9**: 957-965.

Ou, W.J., Ito, A., Okazaki, H. and T. Omura. 1989. Purification and characterization of a processing protease from rat liver mitochondria. *EMBO J.,* **8**: 2605-2612.

Pacaud, M. 1982. Purification and characterization of two novel proteolytic enzymes in membranes of *Escherichia coli*. Protease IV and protease V. *J. Biol. Chem.,* **257**: 4333-4339.

Paetzel, M., Dalbey, R.E. and N.C. Strynadka. 2002. Crystal structure of a bacterial signal peptidase apoenzyme: implications for signal peptide binding and the Ser-Lys dyad mechanism. *J. Biol. Chem.,* **277**: 9512-9519.

Pallen, M.J. and B.W. Wren. 1997. The HtrA family of serine proteases. *Mol. Microbiol.,* **26**: 209-221.

Park, S. and S.R. Rodermel. 2004. Mutations in ClpC2/Hsp100 suppress the requirement for FtsH in thylakoid membrane biogenesis. *Proc. Natl. Acad. Sci. USA,* **101**: 12765-12770.

Peeters, N. and I. Small. 2001. Dual targeting to mitochondria and chloroplasts. *Biochim. Biophys. Acta,* **1541**: 54-63.

Peltier, J.B., Friso, G., Kalume, D.E., Roepstorff, P., Nilsson, F., Adamska, I. and K.J. van Wijk. 2000. Proteomics of the chloroplast: systematic identification and targeting analysis of lumenal and peripheral thylakoid proteins. *Plant Cell,* **12**: 319-341.

Peltier, J.B., Ytterberg, J., Liberles, D.A., Roepstorff, P. and K.J. van Wijk. 2001. Identification of a 350-kDa ClpP protease complex with 10 different Clp isoforms in chloroplasts of *Arabidopsis thaliana*. *J. Biol. Chem.,* **276**: 16318-16327.

Peltier, J.B., Emanuelsson, O., Kalume, D.E., Ytterberg, J., Friso, G., Rudella, A., Liberles, D.A., Soderberg, L., Roepstorff, P., von Heijne, G. and K.J. van Wijk. 2002. Central functions of the lumenal and peripheral thylakoid proteome of *Arabidopsis* determined by experimentation and genome-wide prediction. *Plant Cell,* **14**: 211-236.

Peltier, J.B., Ripoll, D.R., Friso, G., Rudella, A., Cai, Y., Ytterberg, J., Giacomelli, L., Pillardy, J. and K.J. van Wijk. 2004. Clp protease complexes from photosynthetic and non-photosynthetic plastids and mitochondria of plants, their predicted three-dimensional structures, and functional implications. *J. Biol. Chem.,* **279**: 4768-4781.

Pfanner, N. and M. Meijer. 1997. The Tom and Tim machine. *Curr. Biol.,* **7**: 100-103.

Pfanner, N. and A. Geissler. 2001. Versatility of the mitochondrial protein import machinery. *Nat. Rev. Mol. Cell. Biol.,* **2**: 339-349.

Picault, N., Hodges, M., Palmieri, L. and F. Palmieri. 2004. The growing family of mitochondrial carriers in *Arabidopsis*. *Trends Plant. Sci.,* **9**: 138-146.

Pilon, M., Rietveld, A.G., Weisbeek, P.J. and B. de Kruijff. 1992. Secondary structure and folding of a functional chloroplast precursor protein. *J. Biol. Chem.,* **267**: 19907-19913.

Porankiewicz, J., Wang, J. and A.K. Clarke. 1999. New insights into the ATP-dependent Clp protease: *Escherichia coli* and beyond. *Mol. Microbiol.,* **32**: 449-458.

Regnier, P. 1981. The purification of protease IV of *E. coli* and the demonstration that it is an endoproteolytic enzyme. *Biochem. Biophys. Res. Commun.,* **99**: 1369-1376.

Reisfeld, A., Mattoo, A.K. and M. Edelman. 1982. Processing of a chloroplast-translated membrane protein *in vivo*. Analysis of the rapidly synthesized 32000-dalton shield protein and its precursor in *Spivodefa oligorrhiza*. *Eur. J. Biochem.;* **124**: 125-129.

Richter, S. and G.K. Lamppa. 1998. A chloroplast processing enzyme functions as the general stromal processing peptidase. *Proc. Natl. Acad. Sci. USA,* **95**: 7463-7468.

Richter, S. and G.K. Lamppa. 1999. Stromal processing peptidase binds transit peptides and initiates their ATP-dependent turnover in chloroplasts. *J. Cell Biol.,* **147**: 33-44.

Richter, S. and G.K. Lamppa. 2002. Determinants for removal and degradation of transit peptides of chloroplast precursor proteins. *J. Biol. Chem.,* **277**: 43888-43894.

Richter, S. and G.K. Lamppa. 2003. Structural properties of the chloroplast stromal processing peptidase required for its function in transit peptide removal. *J. Biol. Chem.,* **278**: 39497-39502.

Richter, S., Zhong, R. and G.K. Lamppa. 2005. Function of the stromal processing peptidase in the chloroplast import pathway. *Physiol. Plant.,* **123**: 362-368.

Rintämäki, E. and E.-M. Aro. 2001. Phosphorylation of photosystem II proteins. In: *Regulation of Photosynthesis* (eds. E.-M. Aro and B. Andersson), vol. 11, pp. 395-418. Kluwer Academic Publishers. Dordrecht.

Robinson, C. and R.J. Ellis. 1984. Transport of proteins into chloroplasts. Partial purification of a chloroplast protease involved in the processing of important precursor polypeptides. *Eur. J. Biochem.,* **142**: 337-342.

Roffey, R.A. and S.M. Theg. 1996. Analysis of the import of carboxyl-terminal truncations of the 23-kilodalton subunit of the oxygen-evolving complex suggests that its structure is an important determinant for thylakoid transport. *Plant Physiol.,* **111**: 1329-1338.

Roose, J.L. and H.B. Pakrasi. 2004. Evidence that D1 processing is required for manganese binding and extrinsic protein assembly into photosystem II. *J. Biol. Chem.,* **279**: 45417-45422.

Ryan, K.R. and R.E. Jensen. 1995. Protein translocation across mitochondrial membranes: what a long, strange trip it is. *Cell,* **83**: 517-519.

Sakamoto, W., Tamura, T., Hanba-Tomita, Y. and M. Murata. 2002. The *VAR1* locus of *Arabidopsis* encodes a chloroplastic FtsH and is responsible for leaf variegation in the mutant alleles. *Genes Cells,* **7**: 769-780.

Sakamoto, W. 2003. Leaf-variegated mutations and their responsible genes in *Arabidopsis thaliana*. *Genes Genet. Syst.,* **78**: 1-9.

Sakamoto, W., Zaltsman, A., Adam, Z. and Y. Takahashi. 2003. Coordinated regulation and complex formation of yellow *variegated1* and yellow *variegated2*, chloroplastic FtsH metalloproteases involved in the repair cycle of photosystem II in *Arabidopsis* thylakoid membranes. *Plant Cell,* **15**: 2843-2855.

Sakamoto, W., Miura, E., Kaji, Y., Okuno, T., Nishizono, M. and T. Ogura. 2004. Allelic characterization of the leaf-variegated mutation *var2* identifies the conserved amino acid residues of FtsH that are important for ATP hydrolysis and proteolysis. *Plant Mol. Biol.,* **56**: 705-716.

Santos, D. and D.F. De Almeida. 1975. Isolation and characterization of a new temperature-sensitive cell division mutant of *Escherichia coli* K-12. *J Bacteriol* 124, 1502-1507.

Sarria, R., Lyznik, A., Vallejos, C.E. and S.A. Mackenzie. 1998. A cytoplasmic male sterility-associated mitochondrial peptide in common bean is post-translationally regulated. *Plant Cell,* **10**: 1217-1228.

Schaller, A. 2004. A cut above the rest: the regulatory function of plant proteases. *Planta*, **220**; 183-197.

Schatz, G. 1996. The protein import system of mitochondria. *J. Biol. Chem.*, **271**: 31763-31766.

Schmidt, G.W. and M.L. Mishkind. 1983. Rapid degradation of unassembled ribulose 1,5-bisphosphate carboxylase small subunits in chloroplasts. *Proc. Natl. Acad. Sci. USA*, **80**: 2632-2636.

Schubert, M., Petersson, U.A., Haas, B.J., Funk, C., Schroder, W.P. and T. Kieselbach. 2002. Proteome map of the chloroplast lumen of *Arabidopsis thaliana*. *J. Biol. Chem.*, **277**: 8354-8365.

Shackleton, J.B. and C. Robinson. 1991. Transport of proteins into chloroplasts. The thylakoidal processing peptidase is a signal-type peptidase with stringent substrate requirements at the -3 and -1 positions. *J. Biol. Chem.*, **266**: 12152-12156.

Shanklin, J., DeWitt, N.D. and J.M. Flanagan. 1995. The stroma of higher plant plastids contain ClpP and ClpC functional homologs of *Escherichia coli* ClpP and ClpA: an archetypal two-component ATP-dependent protease. *Plant Cell*, **7**: 1713-1722.

Sheng, M. and C. Sala. 2001. PDZ domains and the organization of supramolecular complexes. *Annu. Rev. Neurosci.*, **24**: 1-29.

Shestakov, S.V., Anbudurai, P.R., Stanbekova, G.E., Gadzhiev, A., Lind, L.K. and H.B. Pakrasi. 1994. Molecular cloning and characterization of the *ctpA* gene encoding a carboxyl-terminal processing protease. Analysis of a spontaneous photosystem II-deficient mutant strain of the cyanobacterium *Synechocystis sp.* PCC 6803. *J. Biol. Chem.*, **269**: 19354-19359.

Shikanai, T., Shimizu, K., Ueda, K., Nishimura, Y., Kuroiwa, T. and T. Hashimoto. 2001. The chloroplast *clpP* gene, encoding a proteolytic subunit of ATP-dependent protease, is indispensable for chloroplast development in tobacco. *Plant Cell Physiol.*, **42**: 264-273.

Shimokata, K., Kitada, S., Ogishima, T. and A. Ito. 1998. Role of alpha-subunit of mitochondrial processing peptidase in substrate recognition. *J. Biol. Chem.*, **273**: 25158-25163.

Silber, K.R., Keiler, K.C. and R.T. Sauer. 1992. Tsp: a tail-specific protease that selectively degrades proteins with nonpolar C termini. *Proc. Natl. Acad. Sci. USA*, **89**: 295-299.

Sinvany-Villalobo, G., Davydov, O., Ben-Ari, G., Zaltsman, A., Raskind, A. and Z. Adam. 2004. Expression in multigene families. Analysis of chloroplast and mitochondrial proteases. *Plant Physiol.*, **135**: 1336-1345.

Sjögren, L.L., MacDonald, T.M., Sutinen, S. and A.K. Clarke. 2004. Inactivation of the *clpC1* gene encoding a chloroplast Hsp100 molecular chaperone causes growth retardation, leaf chlorosis, lower photosynthetic activity, and a specific reduction in photosystem content. *Plant Physiol.*, **136**: 4114-4126.

Sjöling, S., Waltner, M., Kalousek, F., Glaser, E. and H. Weiner. 1996. Studies on protein processing for membrane-bound spinach leaf mitochondrial processing peptidase integrated into the cytochrome bc1 complex and the soluble rat liver matrix mitochondrial processing peptidase. *Eur. J. Biochem.*, **242**: 114-121.

Sjöling, S. and E. Glaser. 1998. Mitochondrial targeting peptides in plants. *Trends Plant Sci.*, **3**: 136-140.

Smeekens, S., Bauerle, C., Hageman, J., Keegstra, K. and P. Weisbeek. 1986. The role of the transit peptide in the routing of precursors toward different chloroplast compartments. *Cell*, **46**: 365-375.

Sokolenko, A., Lerbs-Mache, S., Altschmied, L. and R.G. Herrmann. 1998. Clp protease complexes and their diversity in chloroplasts. *Planta*, **207**: 286-295.

Sokolenko, A., Pojidaeva, E., Zinchenko, V., Panichkin, V., Glaser, V.M., Herrmann, R.G. and S.V. Shestakov. 2002. The gene complement for proteolysis in the cyanobacterium *Synechocystis sp.* PCC 6803 and *Arabidopsis thaliana* chloroplasts. *Curr. Genet.*, **41**: 291-310.

Sokolenko, A. 2005. SppA peptidases: family diversity from heterotrophic bacteria to photoautotrophic eukaryotes. *Physiol. Plant.*, **123**: 391-398.

Spetea, C., Hundal, T., Lohmann, F. and B. Andersson. 1999. GTP bound to chloroplast thylakoid membranes is required for light-induced, multienzyme degradation of the photosystem II D1 protein. *Proc. Natl. Acad. Sci. USA*, **96**: 6547-6552.

Spiers, A., Lamb, H.K., Cocklin, S., Wheeler, K.A., Budworth, J., Dodds, A.L., Pallen, M.J., Maskell, D.J., Charles, I.G. and A.R. Hawkins. 2002. PDZ domains facilitate binding of high temperature requirement protease A (HtrA) and tail-specific protease (Tsp) to heterologous substrates through recognition of the small stable RNA A (*ssrA*)-encoded peptide. *J. Biol. Chem.*, **277**: 39443-39449.

Spiess, C., Beil, A. and M. Ehrmann. 1999. A temperature-dependent switch from chaperone to protease in a widely conserved heat shock protein. *Cell*, **97**: 339-347.

Ståhl, A., Moberg, P., Ytterberg, J., Panfilov, O., Brockenhuus von Lowenhielm, H., Nilsson, F. and E. Glaser. 2002. Isolation and identification of a novel mitochondrial metalloprotease (PreP) that degrades targeting presequences in plants. *J. Biol. Chem.*, **277**: 41931-41939.

Ståhl, A., Nilsson, S., Lundberg, P., Bhushan, S., Biverstahl, H., Moberg, P., Morisset, M., Vener, A., Maler, L., Langel, U. and E. Glaser. 2005. Two novel targeting peptide degrading proteases, PrePs, in mitochondria and chloroplasts, so similar and still different. *J. Mol. Biol.*, **349**: 847-860.

Stahlberg, H., Kutejova, E., Suda, K., Wolpensinger, B., Lustig, A., Schatz, G., Engel, A. and C.K. Suzuki. 1999. Mitochondrial Lon of *Saccharomyces cerevisiae* is a ring-shaped protease with seven flexible subunits. *Proc. Natl. Acad. Sci. USA*, **96**: 6787-6790.

Strauch, K.L. and J. Beckwith. 1988. An *Escherichia coli* mutation preventing degradation of abnormal periplasmic proteins. *Proc. Natl. Acad. Sci. USA*, **85**: 1576-1580.

Suzuki, C.K., Rep, M., van Dijl, J.M., Suda, K., Grivell, L.A. and G. Schatz. 1997. ATP-dependent proteases that also chaperone protein biogenesis. *Trends Biochem. Sci.*, **22**: 118-123.

Sweetlove, L.J., Heazlewood, J.L., Herald, V., Holtzapffel, R., Day, D.A., Leaver, C.J. and A.H. Millar. 2002. The impact of oxidative stress on *Arabidopsis* mitochondria. *Plant J.*, **32**: 891-904.

Szigyarto, C., Dessi, P., Smith, M.K., Knorpp, C., Harmey, M.A., Day, D.A., Glaser, E. and J. Whelan. 1998. A matrix-located processing peptidase of plant mitochondria. *Plant Mol. Biol.*, **36**: 171-181.

Takahashi, Y., Goldschmidt-Clermont, M., Soen, S.Y., Franzen, L.G. and J.D. Rochaix. 1991. Directed chloroplast transformation in *Chlamydomonas reinhardtii*: insertional inactivation of the *psaC* gene encoding the iron sulfur protein destabilizes photosystem I. *EMBO J.*, **10**: 2033-2040.

Takahashi, Y., Matsumoto, H., Goldschmidt-Clermont, M. and J.D. Rochaix. 1994. Directed disruption of the *Chlamydomonas* chloroplast *psbK* gene destabilizes the photosystem II reaction center complex. *Plant Mol. Biol.*, **24**: 779-788.

Takechi, K., Sodmergen, Murata, M., Motoyoshi, F. and W. Sakamoto. 2000. The *YELLOW VARIEGATED* (*VAR2*) locus encodes a homologue of FtsH, an ATP-dependent protease in *Arabidopsis*. *Plant Cell Physiol.*, **41**: 1334-1346.

Taylor, A.B., Smith, B.S., Kitada, S., Kojima, K., Miyaura, H., Otwinowski, Z., Ito, A. and J. Deisenhofer. 2001. Crystal structures of mitochondrial processing peptidase reveal the mode for specific cleavage of import signal sequences. *Structure*, **9**: 615-625.

Theg, S.M. and F.J. Geske. 1992. Biophysical characterization of a transit peptide directing chloroplast protein import. *Biochemistry*, **31**: 5053-5060.

Trost, J.T., Chisholm, D.A., Jordan, D.B. and B.A. Diner. 1997. The D1 C-terminal processing protease of photosystem II from *Scenedesmus obliquus*. Protein purification and gene characterization in wild type and processing mutants. *J. Biol. Chem.*, **272**: 20348-20356.

Tziveleka, L.A., and J.H. Argyroudi-Akoyunoglou. 1998. Implications of a developmental-stage-dependent thylakoid-bound protease in the stabilization of the light-harvesting pigment-protein complex serving photosystem II during thylakoid biogenesis in red kidney bean. *Plant Physiol.*, **117**: 961-970.

Urban, S., Schlieper, D. and M. Freeman. 2002. Conservation of intramembrane proteolytic activity and substrate specificity in prokaryotic and eukaryotic rhomboids. *Curr. Biol.*, **12**: 1507-1512.

van der Bliek, A.M. and C.M. Koehler. 2003. A mitochondrial rhomboid protease. *Dev. Cell*, **4**: 769-770.

VanderVere, P.S., Bennett, T.M., Oblong, J.E. and G.K. Lamppa. 1995. A chloroplast processing enzyme involved in precursor maturation shares a zinc-binding motif with a recently recognized family of metalloendopeptidases. *Proc. Natl. Acad. Sci. USA*, **92**: 7177-7181.

van't Hof, R., van Klompenburg, W., Pilon, M., Kozubek, A., de Korte-Kool, G., Demel, R.A., Weisbeek, P.J. and B. de Kruijff. 1993. The transit sequence mediates the specific interaction of the precursor of ferredoxin with chloroplast envelope membrane lipids. *J. Biol. Chem.*, **268**: 4037-4042.

van 't Hof, R., and B. de Kruijff. 1995. Transit sequence-dependent binding of the chloroplast precursor protein ferredoxin to lipid vesicles and its implications for membrane stability. *FEBS Lett.*, **361**: 35-40.

Vothknecht, U.C. and J. Soll. 2005. Chloroplast membrane transport: interplay of prokaryotic and eukaryotic traits. *Gene*, **354**: 99-109.

Wan, J., Bringloe, D. and G.K. Lamppa. 1998. Disruption of chloroplast biogenesis and plant development upon down-regulation of a chloroplast processing enzyme involved in the import pathway. *Plant J.*, **15**: 459-468.

Wang, J., Hartling, J.A. and J.M. Flanagan. 1997. The structure of ClpP at 2.3 A resolution suggests a model for ATP-dependent proteolysis. *Cell*, **91**: 447-456.

Wilken, C., Kitzing, K., Kurzbauer, R., Ehrmann, M. and T. Clausen. 2004. Crystal structure of the DegS stress sensor: How a PDZ domain recognizes misfolded protein and activates a protease. *Cell*, **117**: 483-494.

Winning, B.M., Sarah, C.J., Purdue, P.E., Day, C.D. and C.J. Leaver. 1992. The adenine nucleotide translocator of higher plants is synthesized as a large precursor that is processed upon import into mitochondria. *Plant J.*, **2**: 763-773.

Yamamoto, Y. 2001. Quality control of photosystem II. *Plant Cell Physiol.*, **42**: 121-128.

Yang, D.H., Webster, J., Adam, Z., Lindahl, M. and B. Andersson. 1998. Induction of acclimative proteolysis of the light-harvesting chlorophyll *a/b* protein of photosystem II in response to elevated light intensities. *Plant Physiol.*, **118**: 827-834.

Yang, M., Jensen, R.E., Yaffe, M.P., Oppliger, W. and G. Schatz. 1988. Import of proteins into yeast mitochondria: the purified matrix processing protease contains two subunits which are encoded by the nuclear *MAS1* and *MAS2* genes. *EMBO J.*, **7**: 3857-3862.

Young, L., Leonhard, K., Tatsuta, T., Trowsdale, J. and T. Langer. 2001. Role of the ABC transporter Mdl1 in peptide export from mitochondria. *Science*, **291**: 2135-2138.

Yu, F., Park, S. and S.R. Rodermel. 2004. The *Arabidopsis* FtsH metalloprotease gene family: interchangeability of subunits in chloroplast oligomeric complexes. *Plant J.*, **37**: 864-876.

Yu, F., Park, S. and S.R. Rodermel. 2005. Functional redundancy of AtFtsH metalloproteases in thylakoid membrane complexes. *Plant Physiol.*, **138**: 1957-1966.

Zaltsman, A., Feder, A. and Z. Adam. 2005a. Developmental and light effects on the accumulation of FtsH protease in *Arabidopsis* chloroplasts--implications for thylakoid formation and photosystem II maintenance. *Plant J.*, **42**: 609-617.

Zaltsman, A., Ori, N. and Z. Adam. 2005b. Two types of FtsH protease subunits are required for chloroplast biogenesis and photosystem II repair in *Arabidopsis*. *Plant Cell*, **17**: 2782-2790.

Zara, V., Palmieri, F., Mahlke, K. and N. Pfanner. 1992. The cleavable presequence is not essential for import and assembly of the phosphate carrier of mammalian mitochondria but enhances the specificity and efficiency of import. *J. Biol. Chem.*, **267**: 12077-12081.

Zelisko, A. and G. Jackowski. 2004. Senescence-dependent degradation of Lhcb3 is mediated by a thylakoid membrane-bound protease. *J. Plant Physiol.*, **161**: 1157-1170.

Zelisko, A., Garcia-Lorenzo, M., Jackowski, G., Jansson, S. and C. Funk. 2005. AtFtsH6 is involved in the degradation of the light-harvesting complex II during high-light acclimation and senescence. *Proc. Natl. Acad. Sci. USA*, **102**: 13699-13704.

Zhang, X.P. and E. Glaser. 2002. Interaction of plant mitochondrial and chloroplast signal peptides with the Hsp70 molecular chaperone. *Trends Plant Sci.*, **7**: 14-21.

Zheng, B., Halperin, T., Hruskova-Heidingsfeldova, O., Adam, Z. and A.K. Clarke. 2002. Characterization of chloroplast Clp proteins in *Arabidopsis*: Localization, tissue specificity and stress responses. *Physiol. Plant.*, **114**: 92-101.

Zhong, R., Wan, J., Jin, R. and G.K. Lamppa. 2003. A pea antisense gene for the chloroplast stromal processing peptidase yields seedling lethals in *Arabidopsis*: survivors show defective GFP import *in vivo*. *Plant J.*, **34**: 802-812.

*SECTION — IV*

# *ROOT GROWTH REGULATION*

*Advances in Plant Physiology*, Vol. 9
Ed. A. Hemantaranjan
Scientific Publishers (India), Jodhpur, 2006 pp. 297-312
E-mail: **info@scientificpub.com** www.scientificpub.com

# 13

# IDENTIFICATION OF RICE ROOT PROTEINS REGULATED BY GIBBERELLIN USING PROTEOME ANALYSIS

Setsuko Komatsu* and Hirosato Konishi

National Institute of Agrobiological Sciences, Tsukuba 305-8602, Japan
* Correcpondence: Department of Molecular Genetics, National Institute of Agrobiological Sciences, 2-1-2 Kannondai, Tsukuba, Ibaraki 305-8602, Japan
Tel: 81-29-838-7446, Fax: 81-29-838-7408, E-mail: skomatsu@affrc.go.jp

## INTRODUCTION

Root architecture in plants is determined by interactions between their intrinsic developmental program and external biotic and abiotic stimuli (Schiefelbein and Benfey, 1991; Lynch, 1995). Root systems of plants growing in the field are marvelously successful at foraging for nutrients and water in a hostile, competitive environment where supplies of them are very limited, local and variable (McCully, 1995). The acquisition of soil resources by plant root systems is a subject of considerable interest in agriculture and ecology, as well as a complex and challenging problem in basic plant biology. Thus, root growth regulation is a critical function of terrestrial plants and is closely associated with the production, transport and response to plant hormones. To understand the mechanism by which plant hormones regulate the root growth and development of plants, it is necessary to identify and characterize proteins regulated by plant hormones.

Gibberellins (GAs) are essential endogenous regulators of plant growth and developmental processes (Jacobs, 1997; Kende and Zeevaart, 1997). Significant progress has been made in understanding the pathways involved in GA biosynthesis and on the mechanisms by which GA levels are regulated in plants (Hedden and Proebsting, 1999; Yamaguchi and Kamiya, 2000). The plant response to GA applied to intact root systems appears to be concentration-dependent, and can have rather variable effects. For example, a low concentration of exogenous GA plays a role in normal elongation of roots by maintaining the extensibility of the cell wall in GA-sensitive dwarf pea (Tanimoto, 1994). GA is a regulator of cell elongation in roots, and there is an apparent correlation between low levels of GA *in planta* and short roots (Dolan and Davies, 2004). Since root system architecture influences water and nutrient absorption, GA regulation of root growth is essential for plant survival. Plant cells with defects in vacuole expansion do not expand (Schumacher *et al.*, 1999) and the uptake of water by expanding vacuoles and rapid osmoregulation between the cytosol and vacuole are regulated by tonoplasts (Chaumont *et al.*, 1998). Vacuole expansion and cell elongation are dependent on several factors, such as rates of cell component biosynthesis, metabolite concentration and pH gradient across the tonoplast. GA is likely to be associated with tonoplast function.

Rice (*Oryza sativa* L.) is one of the most important crops in the world. It is the main staple food of more than half of the world's population (Sasaki and Burr, 2000). Since rice has a genome significantly smaller than other cereals, it is an ideal model plant for genetic and molecular studies, particularly among the monocots (Devos and Gale, 2000). Draft sequences of rice genomes have been reported for Subspecies *indica* (Yu *et al.*, 2002) and *japonica* (Goff *et al.*, 2002). Furthermore, the complete map-based genome sequences of chromosomes 1 (Sasaki *et al.*, 2002) and 4 (Feng *et al.*, 2002) for cultivar Nipponbare have been reported. The challenge ahead for the plant research community is to identify the functions, post-translational modifications, and regulation of proteins encoded by the plant's genes. Understanding the biological function of novel genes is a more difficult proposition than merely obtaining the nucleotide or peptide sequences. This is because the existing information on amino acid sequences of known proteins in the database is derived primarily from genetic (including 'reverse genetics') and biochemical studies, which are by nature focused and labor intensive; the use of a large number of plant species as experimental systems; and because of the extensive range of unique plant-produced secondary metabolites. Thus, the cumulative knowledge of functions of known and unidentified proteins does not match the wealth of nucleotide sequence information being generated through genome sequencing projects (Komatsu *et al.*, 2003). The analysis of proteins using high-resolution, two-dimensional polyacrylamide gel electrophoresis (2D-PAGE) is the most direct, first order approach to define gene function.

In this study, proteome analysis was used to investigate the effects of GA and aldolase increases in roots treated with $GA_3$ (Tanaka *et al.*, 2004a). Konishi *et al.* (2005a, on line) have reported that increases in expression of aldolase by exogenous $GA_3$ were reversed in roots treated with uniconazol, an inhibitor of GA biosynthesis, and that abscisic acid and aldolase levels were low in the roots of the GA-deficient mutant Tan-ginbozu. GA deficiency in this mutant is due to blocking the metabolic steps from *ent*-kaurene to *ent*-kaurenoic acid, catalyzed by *ent*-kaurene oxidase (Sakamoto *et al.*, 2004). Since the increase of aldolase caused by GA activates the glycolytic pathway, acceleration of root growth may be a direct result of these actions. Furthermore, antisense-transgenic rice plants were used to clarify the role of aldolase regulated by GA (Konishi *et al.*, 2005b, in press). In this review, the regulation of rice root proteins by gibberellin as revealed by proteome analysis is discussed.

## 2. PROTEOME ANALYSIS OF RICE ROOT BY TWO-DIMENSIONAL GEL ELECTROPHORESIS

### 2.1 Experimental

#### *2.1.1 Protein extraction from root*

A portion (400 mg) of the excised roots was homogenized with 1 ml phosphate buffer (Zhong *et al.*, 1997) using a glass mortar and pestle on ice. The homogenate was centrifuged at 15,000 rpm at 4°C for 10 min in an RA-50JS rotor (Kubota, Tokyo, Japan), and 50% trichloroacetic acid was added to the supernatant to a final concentration of 10%. The solution was placed on ice for 30 min and centrifuged at 15,000 rpm at 4°C for 10 min. The resultant precipitate was washed with 100 μl ice-cold ethanol, suspended in 100 μl lysis buffer containing 8 M urea, 2% Nonidet P-40, 2% Ampholine (pH 3.5-10.0; Amersham Biosciences, Piscataway, NJ, USA), 5% 2-mercaptoethanol and 5% polyvinyl pyrrolidone-40, and sonicated twice for 2 min each. The solution was centrifuged at 15,000 rpm for 5 min, and 90 μl of the supernatant was subjected to 2D-PAGE.

#### *2.1.2. Two-dimensional electrophoresis*

Prepared samples were separated in the first dimension by isoelectric focusing (IEF) or immobilized pH gradient (IPG) tube gel (Daiichi Pure Chemicals, Tokyo, Japan) and in the second dimension by SDS-PAGE. The IEF tube gel solution consisted of 8 M urea, 3.5% acrylamide, 2% NP-40, 2% Ampholines (pH 3.5-10.0 and pH 5.0-8.0), ammonium persulfate and TEMED (O'Farrell, 1975). Electrophoresis was carried out at 200 V for 30 min, followed by 400 V for 16 h and 600 V for 1 h. For IPG electrophoresis, samples were applied to the acidic side of gels and electrophoresed using IPG tube gels (pH 6.0-10.0) at 400 V for 1 h, followed by 1,000 V for 16 h and 2,000 V for 1 h. After IEF or IPG, SDS-PAGE in the second dimension was performed using 15% polyacrylamide gel. The gels were stained with Coomassie brilliant blue (CBB), and the individual protein positions were evaluated automatically using ImageMaster 2D Elite software (Amersham Biosciences). The isoelectric point (pI) and relative molecular mass ($M_r$) of each protein were determined using 2D-PAGE Markers (Bio-Rad, Richmond, CA, USA).

#### *2.1.3. Cleveland peptide mapping*

Following separation by 2D-PAGE, gel pieces containing protein spots were removed and the protein was electroeluted using an electrophoretic concentrator (ISCO, Lincoln, CA, USA) at 2 W constant power for 2 h. After electroelution, the protein solution was dialyzed against deionized water for 2 days and lyophilized. The purified protein was dissolved in 20 μl of SDS sample buffer (pH 6.8) and applied to a sample well in an SDS-PAGE gel and the sample solution was overlaid with 20 μl of 1/2 SDS sample buffer containing 50 ng/μl *Staphylococcus aureus* V8 protease (Pierce, Rockford, IL, USA). Electrophoresis was performed until the sample and protease were stacked in the stacking gel, interrupted for 30 min to allow digestion of the protein, and then continued (Cleveland *et al.*, 1977).

#### *2.1.4. N-terminal and internal amino acid sequencing*

Following 2D-PAGE or the Cleveland peptide method, the proteins were electroblotted onto a polyvinylidene difluoride (PVDF) membrane (Pall Bio Support Division, Port Washington, NY, USA) using a semidry transfer blotter (Nippon Eido, Tokyo, Japan), and detected by CBB staining. The stained protein spots were excised from the PVDF membrane and applied to a gas-phase protein sequencer Procise 494 (Applied Biosystems, Foster City, CA, USA). The amino acid sequences obtained were compared with those of known proteins in the

Swiss-Prot, PIR, Genpept and PDB databases with the Web-accessible search program FastA.

#### *2.1.5. Matrix-assisted laser desorption/ ionization time-of-flight mass spectrometry analysis*

The CBB-stained protein spots were excised from gels, washed with 25% methanol and 7% acetic acid for 12 h, and destained with 50 mM $NH_4HCO_3$ in 50% methanol for 1 h at 40°C. Proteins were reduced with 10 mM dithiothreitol (DTT) in 100 mM $NH_4HCO_3$ for 1 h at 60°C and incubated with 40 mM iodoacetamide in 100 mM $NH_4HCO_3$ for 30 min. The gel pieces were minced and allowed to dry and then rehydrated in 100 mM $NH_4HCO_3$ with 1 pmol trypsin (Sigma, St. Louis, MO, USA) at 37°C overnight. The digested peptides were extracted from the gel slices with three volumes of 0.1% trifluoroacetic acid in 50% acetonitrile/water. The peptide solution was dried and reconstituted with 30 μl of 0.1% trifluoroacetic acid in 5% acetonitrile/water, desalted with ZipTip C18™ pipette tips (Millipore, Bedford, MA, USA) and mixed with α-cyano-4-hydroxycinnamic acid. Matrix-assisted laser desorption/ionization time-of-flight mass spectrometry (MALDI-TOF MS) was performed using a Voyager TOF mass spectrometer (Applied Biosystems, Framingham, MA, USA). The mass spectra were subjected to a sequence database search using Mascot software (Matrix Science Ltd, London, UK).

### 2.2 Root Protein Identification Using Two-Dimensional Electrophoresis

Proteins were extracted from roots and separated by 2D-PAGE using IEF and IPG tube gels (Hirano *et al.* 2000) in the first dimension, with IEF in the low pI range (4.0 to 7.0) and IPG in the high pI range (6.0 to 10.0). After detection with CBB staining, protein spots were analyzed using Image-Master 2D Elite software. 2D maps of the low and high pI ranges overlapped at around pI 6.0. A total of 508 proteins from rice roots were detected in the 2D-PAGE patterns. Although clear images were obtained in the range of pI 3.5-6.0 from 2D-PAGE with the IEF tube gel, including ampholytes in the first dimension, it was difficult to effectively isolate the basic proteins. Therefore, electrophoresis using IPG tube gels (pH 6.0-10.0) was carried out for improved resolution of basic proteins. About 90% of the visible proteins detected by 2D-PAGE were present in the pI 4.0-7.5 range. Many proteins were detected in both the pI 4.0-6.0 and pI 6.0-7.5 ranges. Proteins separated under these conditions were reproducibly observed in their relative 2D map positions. Computer analyses using Image-Master 2D Elite software revealed 508 individual rice root protein spots. After 2D-PAGE, amino acid sequences were determined by MALDI-TOF and/or Edman sequencing.

Thirty-eight N-terminal sequences of the 94 root proteins separated by 2D-PAGE were determined by Edman sequencing. A further 35 separated proteins excised from gels were analyzed by MALDI-TOF MS. Using these approaches, at least some sequence information was determined for 73 individual proteins. The N-terminal amino acid sequences of 56 (59.6%) of the proteins could not be determined by the Edman technique likely due to a blocking group at the N-terminus. This result is consistent with a report in which 134 rice proteins were subjected to sequencing, of which 79 (59%) were found to have blocked N-termini (Tsugita *et al.* 1994).

### 2.3. Functional Classification of Root Proteins

Twenty-six percent of the identified proteins were involved in plant defense, indicating that the rice plant produces more disease-resistance and defense-related proteins than any other type. This functional category involves metallothionin, glutathione

S-transferases, chitinases, NBS-LRR-type resistant proteins, antifungal protein R, thaumatin-like proteins, superoxide dismutase [Cu-Zn] ([Cu/Zn]SOD), type-1 pathogenesis-related (PR-1) protein, and SalT protein. A root-specific protein, RCc3, was expressed in the elongation and maturation zones of primary and secondary roots, and in root caps (Xu *et al.*, 1995). These results may be consistent with marsh plants or helophytes such as rice, since they are genetically acclimatized to harsh soil environments and are tolerant of toxicity in the soil. In contrast, the root tip proteins of maize seedlings during hypoxic acclimation have been reported, and of 48 proteins analyzed by mass sprectrometry, 46 were identified (Chang *et al.*, 2000). Soluble metabolic enzymes were the most abundant, but no proteins involved in the defense category were identified (Chang *et al.*, 2000). These results reflect the differences in gene expression patterns at the protein level under conditions of stress.

Many root proteins were found in only one tissue in these assays, suggesting that their expression is tissue-specific. Since these proteins are related to the primary function of the specific tissue, they can provide valuable insights into the specialized physiological function of each tissue.

About 20% of root proteins were placed in the category of 'no assigned function'. The function of these proteins will need to be identified to provide information about their role in the physiology or anatomy of the tissue. This study presents the results of a direct differential display with 2D-PAGE for identification of proteins that undergo alterations in concentration, and potentially localization, under different physiological and developmental conditions in rice roots. This system readily visualizes changes in protein content, directly and rapidly extracts proteins of interest from gels, and analyzes the protein primary structure using MS and/or Edman sequencing. Harvesting the information stored in root tissues holds great promise in helping to predict the functions of many other proteins in response to environmental challenges and for expediting the molecular cloning of genes that are critical for both understanding plant function and practical application.

## 3. IDENTIFICATION OF ROOT PROTEINS REGULATED BY GIBBERELLIN IN RICE

### 3.1. Gibberellin Regulation of Seedling Root Growth

Although GAs are well-characterized, essential endogenous regulators of root growth and developmental processes (Tanimoto 1994), the precise role of GA in root growth is unclear. To assess the elongation of seminal roots, germinated seeds were grown for 72 h in soil amended with$GA_3$. Shoots were measured to test the effects of $GA_3$ on germinated seeds and seedlings. Shoot elongation increased with the concentration of $GA_3$. Roots grown in 0.1 μM $GA_3$ were 3.8% longer than controls.

Root lengths regulated by GA were analyzed by comparing a normal sized cultivar, Nipponbare, with a semi-dwarf cultivar, Tan-ginbozu, and by treatment with uniconazole, an effective inhibitor of GA biosynthesis (Izumi *et al.*, 1985). Germinated seeds of Nipponbare, Ginbozu and Tan-ginbozu with or without 0.1 μM $GA_3$, 10 μM uniconazole and 10 μM abscisic acid (ABA) were grown for 72 h in soil. Seedling root lengths were measured and compared to the control (Nipponbare). Ginbozu roots were slightly longer (0.5%) and Tan-ginbozu roots were 69.8% shorter than Nipponbare. Treatment with 10 μM uniconazole decreased root growth by 44.6% and 10 μM ABA caused a 79.2% decrease in root length. Root elongation was normal in Tan-ginbozu plants treated with 0.1 μM $GA_3$ and Nipponbare treated with 10 μM uniconazole or 0.1 μM $GA_3$. Since root elongation was inhibited in both Tan-ginbozu and uniconazole-treated

Nipponbare, and could be recovered by $GA_3$ treatment, the proteins whose accumulation was affected by these treatments may be involved in the mechanisms for GA-induced root growth. It is well established that GA and ABA act antagonistically in many aspects of plant development (Gilroy and Jones, 1992), as illustrated by the reduction in seedling root length when seeds were germinated in 10 $\mu$M ABA.

### 3.2. Differential Increase of Four Proteins in Seedling Roots Treated with Gibberellin

To clarify the relationship between root elongation and GA, root proteins differentially accumulated in Tan-ginbozu and in Nipponbare roots treated with uniconazole or ABA were investigated using a proteomics approach. Extracted proteins were separated in the first dimension by IEF or IPG, and in the second dimension by SDS-PAGE. Images of 2D-PAGE from 0.1 $\mu$M $GA_3$-treated and control extracts were synthesized and the positions of individual proteins were evaluated automatically using ImageMaster 2D Elite software. Since the same samples were subjected to IEF and IPG, additional equipment for electrophoresis followed the same procedure. Protein patterns resulting from triplicate protein extractions and triplicate 2D gels were compared to ensure reproducibility. Using protein differential display analysis, it became apparent that 4 proteins were up-regulated due to treatment with $GA_3$. Following separation by 2D-PAGE or by the Cleveland method, proteins were electroblotted onto a PVDF membrane. The CBB-stained protein spots were excised from the PVDF membrane and subjected to gas-phase protein sequencing. Amino acid sequences were compared with those of known proteins using the Web-accessible search program FASTA. All but one of the sequences of differentially accumulated proteins were found to be identical to proteins previously reported. The other three proteins were homologous to the beta 5 subunit of 20S proteasome, aldolase and glyceraldehyde 3-phosphate dehydrogenase.

Aldolase is a widely distributed enzyme and a key constituent of both the glycolytic/gluconeogenic pathways and the reductive pentose phosphate cycle in plants. This enzyme plays a vital role in carbohydrate metabolism and in the production of triose phosphates in signal transduction (Schaeffer, *et al.*, 1997). Glyceraldehyde 3-phosphate dehydrogenase is also a ubiquitous enzyme involved in glycolysis and gluconeogenesis (Duée *et al.*, 1996). Two of the four proteins up-regulated by GA treatment are key enzymes in the glycolytic pathway, suggesting that GA signaling and root growth may be mediated by the glycolytic production of ATP.

Another identified protein, the 20S proteasome, is a proteolytic complex that is involved in removing abnormal proteins (Sassa *et al.*, 2000). The 26S proteasome is composed of a 20S core proteasome with 19S regulatory complexes on either side. GA-insensitive dwarf 2 (GID2) is a positive regulator of GA signaling and slender rice1 protein is degraded via the ubiquitin/26S proteasome pathway mediated by the $SCF^{GID2}$ complex (Sasaki *et al.*, 2003). Increases in the beta 5 subunit of the 20S proteasome by $GA_3$ may be related to activation of the $SCF^{GID2}$-proteasome pathway.

### 3.3. Differential Expression of Proteins from Roots Treated with Uniconazole and Abscisic Acid

There were thirteen proteins identified that were expressed in significantly different amounts in Tan-ginbozu and Nipponbare and/or when treated with with uniconazole or ABA treatment that had amino acid sequence similarity to other plant proteins. A scheme summarizing the regulation of the characterized proteins is shown in Figure 1. PR1 protein accumulated in large amounts in Tan-ginbozu. GSH-dependent dehydroascorbate reductase 1 and [Mn]SOD accumulated in large amounts in Tan-ginbozu and also increased with uniconazole treatment. Antifungal protein-R, nicotianamine synthase 2, and SalT protein increased after treatment with ABA. Cyclophilin 2 accumulated in smaller amounts in Tan-ginbozu than controls. Bowman-Birk trypsin inhibitor accumulated in small amounts in Tan-ginbozu and decreased with uniconazole treatment. Oleosin decreased with ABA treatment. Aldolase accumulated to low levels in Tan-ginbozu and decreased after treatment with uniconazole or ABA.

We propose that all of the proteins placed in the "up-regulation" category of figure 1 are down-regulated by $GA_3$. Plant dehydroascorbate reductase reduces oxidized ascorbate to maintain an appropriate level of ascorbate in plant cells. Dehydroascorbate reductase activity in rice seedlings is stimulated at high temperature (Urano *et al.*, 2000). SOD catalytically scavenges superoxide radicals, providing a defense against reactive oxygen species (Fridovich, 1975). GA is likely to suppress the expression of GSH-dependent dehydroascorbate reductase and SOD. PR-1 protein, which plays a role in defense against stress, accumulated in large amounts in Tan-ginbozu. However, the correlation between PR-1 protein and root growth caused by GA is not clear. The action of ABA is unlikely to be limited strictly to antagonism of GA, and can be expected to have independent effects on protein expression. Antifungal protein R, nicotianamine synthase 2 and SalT protein increased as a result of treatment with 10 μM ABA in comparison with the controls, indicating that ABA induced the synthesis of these proteins. PR proteins constitute lines of defense against pathogen or herbivory, and antifungal protein R is a thaumatin-like (PR-5) protein from barley (Hejgaard *et al.*, 1991). Nicotianamine synthase catalyzes the trimerization of S-adenosylmethionine to form one molecule of nicotianamine, an intermediate in the biosynthetic pathway of the mugineic acid family of phytosiderophores. Iron deficiency in rice induces nicotianamine synthase gene expression in roots (Higuchi *et al.*, 2001). Salt stress causes water deficit responses and a defense response reminiscent of aspects of the plant's response to wounding and pathogen attack. Jasmonic acid, ABA and NaCl induce accumulation of the *salT* transcript in roots (Moons *et al.*, 1997).

Proteins in the "down-regulation" category of figure 1 may be up-regulated by $GA_3$. Bowman-Birk proteinase inhibitor is believed to play a role in defense: its induced gene expression follows a kinase-signaling cascade in rice (Rakwal *et al.*, 2001). Cyclophilin is a specific cytosolic binding protein responsible for the accumulation of the immunosuppressant cyclosporin A by lymphoid cells, and *Cyp2*, a rice cyclophilin, may be preferentially translated during stress conditions (Buchholz *et al.*, 1994). Bowman-Birk trypsin inhibitor and cyclophilins are likely to be related to the action triggered by GA. On the other hand, oleosin decreased in roots treated with ABA. Oil bodies are lipid storage organelles in which plants store energy as polysaccharides or lipids to support periods of active metabolism (Murphy and Vance, 1999). The most abundant oil body-associated proteins are oleosins, and much of the oil body surface may be covered by oleosin (Frandsen *et al.*, 2001). Stored energy for active metabolism may be regulated by the balance between the action of ABA, decreasing the amount of stored energy, and GA exerting the opposite effect.

### 3.4. Relative Aldolase mRNA and Protein Levels

The relationship between transcriptional regulation and translational expression not always strictly proportional, but is a key factor in understanding the complex interactions that affect tissue growth and development. To compare the expression of aldolase protein and mRNA, root protein extracts from plants treated as above were probed with anti-aldolase antibody in an immunoblot assay. Fructose-bisphosphate aldolase accumulated in small amounts in Tan-ginbozu and levels decreased with uniconazole or ABA treatment. The enzyme accumulated at normal levels in Tan-ginbozu treated with $GA_3$ and in Nipponbare treated simultaneously with uniconazole and $GA_3$.

Northern blot analysis was carried out with an aldolase DNA probe. Relative amounts of mRNA were proportional to the accumulation of aldolase protein (data not shown). Aldolase protein increased in $GA_3$-treated roots, was at low levels in Tan-ginbozu roots, and levels further diminished in roots treated with uniconazole or ABA suggesting that an adequate amount of aldolase in rice roots is required for normal root growth. The low level of aldolase protein observed in untreated Tan-ginbozu, likely due to phenotypic selection for semi-dwarfism, was increased by treatment with uniconazole or ABA. Accumulation of aldolase was at 'normal' levels in Tan-ginbozu treated with $GA_3$, and in Nipponbare treated simultaneously with uniconazole and $GA_3$. Furthermore, levels of aldolase mRNA and protein were directly proportional, indicating that aldolase may act as a mediator between GA signaling and root growth.

## 4. CHARACTERIZATION OF ALDOLASE REGULATED BY GIBBERELLIN IN RICE ROOTS

### 4.1. Time-Course and Dose-dependent Changes in mRNA and Protein Levels of Aldolase in Response to Gibberellin

The growth of rice seedling roots is markedly stimulated by the addition of $GA_3$ and is correlated with a significant increase in aldolase accumulation. In rice roots, 14 proteins were up-regulated by $GA_3$ (Tanaka *et al.*, 2004a). Aldolase was shown to be at a low level in Tan-ginbozu roots, and decreased in Nipponbare roots treated with uniconazole or ABA (Konishi *et al.*, 2005a, in press). In the root, GA is a regulator of cell elongation that is dependent on several factors, including the pH gradient across the tonoplast. Thus, we focused our attention on the glycolysis enzyme aldolase and its quantitative and functional regulation by $GA_3$, because stimulation of aldolase accumulation in $GA_3$-treated roots is thought to modulate cell elongation and proliferation by controlling the rate of metabolism and tonoplast proton pumps.

The exogenous application of GA to rice roots was shown to affect the level of aldolase (Tanaka *et al.*, 2004a). To examine how $GA_3$ changes the level of aldolase in rice roots, proteins were extracted from roots treated with $GA_3$ at various concentrations for 0 to 72 h and subjected to Western blot analysis with an anti-aldolase antibody. Measurable increases in aldolase accumulation were noted after 24 h exposure to $GA_3$. By 48 h there was a pronounced increase in aldolase protein that continued until 72 h. Aldolase levels increased in a dose-dependent manner, and treatment with 0.1 μM $GA_3$ was the optimum concentration for root elongation. Changes in aldolase mRNA accumulation during treatment with $GA_3$ were also examined for 24 h by Northern blot analysis. The transcript level of aldolase was extremely low at 0 h, and reached a maximum at 12 h. Transcript levels also increased in a dose-dependent manner and were coincident with aldolase protein levels. Root elongation of rice seedlings is accelerated by treatment with 0.1 μM $GA_3$ (Konishi *et al.*, 2005a, in press). Treatment with $GA_3$ for 48 h for aldolase protein, or 12 h for transcripts was optimal at 0.1mM. Therefore, rice plant tissues were treated with 0.1 μM $GA_3$ for 48 h for protein or 12 h for mRNA in the following experiments.

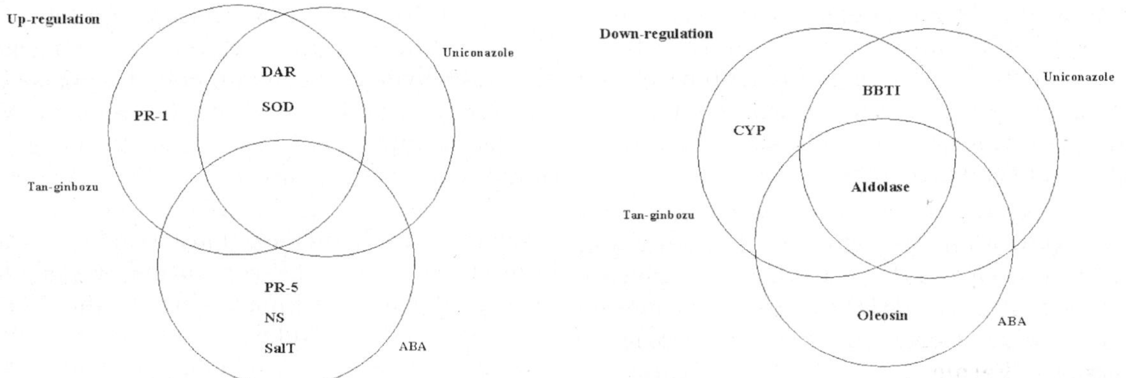


Fig. 1. Venn diagrams of the effects of the Tan-ginbozu semi-dwarfism, uniconazole-P, and ABA on protein accumulation. Tan-ginbozu, uniconazole, and ABA indicate protein expression was increased or decreased in Tan-ginbozu and by uniconazole or ABA. PR-1, type-1 pathogenesis-related protein; DAR, dehydroascorbate reductase; SOD, superoxide dismutase; PR-5, antifungal protein-R; NS, nicotianamine synthase; SalT, SalT protein; CYP, cyclophilin; BBTI, Bowman-Birk tripsin inhibitor.

### 4.2. Changes in Aldolase Transcript and Protein Levels in Response to Gibberellin in Plant Organs

To examine the organ-specific accumulation of aldolase in roots in response to $GA_3$, proteins were extracted from organs of roots treated with 0.1 μM $GA_3$ for 48 h and subjected to Western blot analysis with anti-aldolase antibody. Aldolase predominantly accumulated in roots and was detectable in leaf sheaths in the absence of $GA_3$. Treatment with $GA_3$ increased aldolase mRNA and protein levels in roots and leaf sheaths. However, aldolase protein and transcript levels in leaf blades were very low.

Furthermore, to examine the zone-specific accumulation of aldolase in roots in response to $GA_3$, proteins extracted from the basal, medial and apical regions of the root were subjected to Western blot analysis. Aldolase protein and transcripts accumulated prominently in the apical region, and treatment with $GA_3$ led to an increase in aldolase protein accumulation in all three regions.

These findings support the hypothesis that aldolase is implicated in the tissue growth and differentiation triggered by $GA_3$. The accumulation of aldolase induced by $GA_3$ was most notable in the apical region, which includes the root meristem (Ranathunge *et al.*, 2004). Exogenous $GA_3$ increased aldolase expression in all regions of the roots, although the response varied with the region. Increased levels of aldolase may stimulate the growth of all of the root cells in the roots by causing an increase in energy production. Also, metabolic activation in the roots could cause increased water uptake.

### 4.3. Repression of Root Growth in Aldolase-Antisense Transgenic Rice

Aldolase-antisense transgenic rice plants were constructed to determine the effects of aldolase function on rice root growth. An aldolase cDNA clone was introduced into rice cells in an antisense orientation under the control of the CaMV 35S promoter in a pIG121-Hm binary vector by *Agrobacterium*-mediated transformation. Transgenic plants selected on media containing hygromycin were transplanted to soil and grown to maturity. Following a second round of selection on hygromycin, seeds were harvested and used in the following experiments. In a root elongation assay, roots of vector-only control rice plants grew to 55 mm by day 6, reaching the bottom of the container, whereas the growth of aldolase-antisense transgenic rice lines was only about half of the wild type rate.

The glycolytic pathway, which catalyzes the oxidization of glucose to pyruvate, produces two ATP molecules per glucose molecule. It is essential for providing energy and a wide range of other physiological functions (Fernie *et al.*, 2004). Activation of glycolytic enzyme expression by increasing aldolase may promote root growth. Addition of glucose to cells growing on a nonfermentable carbon source results in major changes in enzymatic activities and gene expression (Rolland *et al.*, 2001). However, the induction of glycolysis occurs via a glucose-independent mechanism in potato tubers (Trethewey *et al.*, 2001). The up-regulation of glycolysis may accelerate cell growth. The present study clearly demonstrated the physiological importance of aldolase in root growth using the aldolase-antisense transgenic rice plants. Aldolase activity increased with exogenous $GA_3$ treatment in control plant roots but the increase was marginal in aldolase-antisense transgenic plant roots.

An aldolase assay was performed to confirm whether aldolase activity in roots is affected by $GA_3$ in aldolase-antisense transgenic plants. In roots of vector-only control plants, treatment with 0.1 μM $GA_3$ caused an increase in aldolase activity of 25%. In the roots of aldolase-antisense transgenic plants, aldolase activity was 43% lower than in vector control plants and increased very little with 0.1 μM $GA_3$ treatment.

### 4.4. Co-Immunoprecipitation of Aldolase with V-ATPase

In yeast cells deficient in aldolase, the peripheral $V_1$ domain of V-ATPase dissociates from the integral membrane $V_0$ domain, indicating that there is a coupling of glycolysis to the proton pump (Lu *et al.*, 2001). Since the glycolytic enzyme glyceraldehyde 3-phosphate dehydrogenase is physically associated with the aldolase-V-ATPase complex and the glycolytic enzyme complex is directly coupled to the V-ATPase proton pump (Lu *et al.*, 2004), association of aldolase with V-ATPase may result in the location of the ATP-generating glycolytic enzyme P-glycerate kinase close to the ATP-utilizing ATPase. Aldolase deletion mutant cells display a growth phenotype similar to that observed in V-ATPase subunit deletion mutants and the V-ATPase abnormalities shown in aldolase deletion mutant cells can be restored to normal levels by aldolase complementation (Lu *et al.*, 2004). The binding of aldolase to V-ATPase should provide the cells with a means for localized ATP generation by glycolysis. Furthermore, cytoplasmic aldolase has been identified in the vacuolar membrane fraction isolated from rice (Tanaka *et al.*, 2004b), suggesting that aldolase may associate with the vacuolar membrane protein.

Immunoprecipitation analysis was performed to examine whether aldolase interacts with V-ATPase in rice roots. V-ATPase is a multi-subunit complex and two antibodies were prepared that are specific either to the 100-kDa subunit or to the subunit A of V-ATPase. The antibodies were separately added to 1% TritonX-100-extracted proteins from roots. The immunoprecipitated proteins were isolated using Protein A-Sepharose and incubated with anti-aldolase antibody. IgG was stained with CBB, and aldolase was detected in samples co-immunoprecipitated with anti-V-ATPase 100-kDa subunit and subunit A antibodies.

The V-ATPase complex co-precipitated with aldolase, thus indicating that they are physically associated in rice roots. Co-immunoprecipitation was mediated by both antibodies raised against V-ATPase 100-kDa subunit and subunit A. The accumulation of osmotically active ions accompanies vacuole expansion, and tonoplast proton pumps generate the trans-tonoplast electrical and proton gradient that drives the uptake of many solutes (Dolan and Davies, 2004). Blocking the expression of V-ATPase subunit A causes the inhibition of root elongation, confirming the importance of this proton pump in cell expansion (Gogarten *et al.*, 1992). Vacuolar solute accumulators that are energized by the $H^+$ electrochemical potential gradient also influence the expansion of root cells (Cheng *et al.*, 2003). Vacuolar expansion is important for cell elongation, and thus may also be implicated in root cell growth mediated by aldolase activity.

### 4.5. Effect of $Ca^{2+}$ on Aldolase Accumulation in the Roots of OsCDPK13-Antisense Transgenic Rice

In plants, the role of cytosolic $Ca^{2+}$ concentration in the coupling of stimulus and response is not completely clear. However, $GA_3$ induces a sustained increase in cytosolic $Ca^{2+}$ concentration (Gilroy and Jones, 1992) and the locus, HvCDPK1, that encodes a $Ca^{2+}$-dependent kinase, regulates vacuolar function during the $GA_3$ response in barley aleurone (McCubbin *et al.*, 2004). Moreover, OsCDPK13 expression is increased in rice leaf sheath segments treated with $GA_3$ and is highly constant in leaf sheath and root as compared with leaf blade (Yang *et al.*, 2003). To investigate the correlation between CDPK and aldolase accumulation, OsCDPK13-antisense transgenic rice (Abbasi *et al.*, 2004) was used. To assess the possible involvement of $Ca^{2+}$ in aldolase accumulation, rice seedlings were treated with $CaCl_2$ and the $Ca^{2+}$ channel blocker $LaCl_3$. In roots of vector-only control seedlings, aldolase accumulation was stimulated when 5mM $CaCl_2$ and 5 μM calcium ionophore A23187 were added together to excised roots for 48 h. Under similar conditions, aldolase accumulation was dramatically inhibited by 5mM $LaCl_3$. In roots of OsCDPK13-antisense transgenic seedlings, the accumulation of aldolase was markedly less than in control seedlings. Aldolase levels were slightly enhanced by treatment with 5mM $CaCl_2$ and 5 μM A23187. Aldolase was undetectable in 5mM $LaCl_3$-treatment roots.

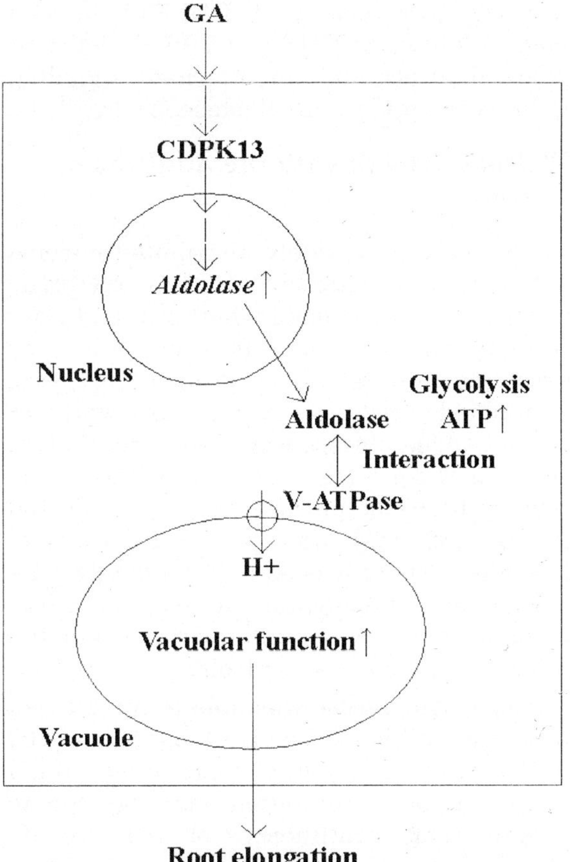


Fig. 2. Schematic summary of the signal transduction between GA and root elongation. $GA_3$-induced aldolase enhances the metabolic rate of glycolysis in rice roots and aldolase activates V-ATPase through physical interactions. As a result, the rate of cell growth in seedling roots may be significantly enhanced. Also, $GA_3$ signaling for root growth promotion may be mediated by CDPK13.

Accumulation of aldolase in the roots of OsCDPK13-antisense transgenic seedlings (Abbasi *et al.*, 2004) was markedly decreased compared with vector-only control transformant seedlings but was increased by $Ca^{2+}$.

CDPK mediates aldolase expression regulated by $GA_3$, and the cytosolic $Ca^{2+}$ concentration should correlate with this signaling pathway. These results indicate that $GA_3$ up-regulates the aldolase level through cytosolic $Ca^{2+}$ concentrations and CDPK13 in rice roots. Addition of $Ca^{2+}$ to CDPK13-deficient plants may activate an alternate signaling pathway to up-regulate aldolase levels.

### 4.6. Root Growth with the Addition of Glucose

Aldolase is a widely distributed enzyme and a key constituent of the glycolytic pathway. To investigate whether the glucose consumption due to ATP generation by glycolysis affects root growth in rice, germinated seeds of Nipponbare with or without added glucose were grown for 72 h in soil. Root growth was influenced by concentrations of glucose ranging from 100 nM to 100 mM and was concentration-dependent. There was an approximately 9.1% increase in root elongation at 100 μM glucose, but root growth was inhibited at 10 and 100 mM as compared with controls.

Exogenous glucose was able to inhibit root elongation at high concentrations (10 and 100 mM), and promote root elongation at 100 μM glucose. Growth inhibition may be due to harmful osmotic influences on root growth. Glucose is a universal nutrient preferred by most organisms, and plants use hexokinase as a glucose sensor to interrelate hormone signaling networks for controlling growth and development (Moore *et al.*, 2003). The present study suggests that fructose-bisphosphate aldolase is involved in GA-stimulated root growth through activation of the glycolytic pathway.

## Conclusion

In conclusion, the present results suggest that $GA_3$-induced aldolase enhances the metabolic rate of glycolysis in rice roots. The present work suggests that aldolase activates V-ATPase through physical interaction. As a result, the rate of cell growth of seedling roots may be efficiently enhanced. Also, $GA_3$ signaling for promotion of the root growth may be mediated by CDPK13 (Figure 2).

A number of preliminary experiments have suggested that a number of modes of action may be involved in linking gibberellin, aldolase and root growth. More refined experiments will be needed to provide detail about the exact relationships between the components of this regulation. The final result may be that the growth of roots, and other complex plant organs, may be regulated by a sensitive matrix of interacting components, of which GA and its antagonists may play the most central role, but that they act through primary metabolic intermediaries like aldolase and CDPK.

## Acknowledgements

This work was supported by a grant from the Program for Promotion of Basic Research Activities for Innovative Biosciences and a MAFF Rice Genome Project grant (Japan).

## REFERENCES

Abbasi, F., Onodera, H., Toki, S., Tanaka, H. and Komatsu, S. 2004. OsCDPK13, a calcium-dependent protein kinase gene from rice, is induced by cold and gibberellin in rice leaf sheath. *Plant Mol. Biol.*, **55**:541-552.

Buchholz, W.G., Harris-Haller, L., DeRose, R.T. and Hall T.C. 1994. Cyclophilins are encoded by a small gene family in rice. *Plant Mol. Biol.*, **25**: 837-843.

Chang, W.W.P., Huang, L., Shen, M., Webster, C., Burlingame, A.L. and Roberts, J.K.M. 2000. Patternes of protein synthesis and tolerance of anoxia in root tips of maize seedlings acclimated to a low-oxygen environment, and identification of proteins by mass spectrometry. *Plant Physiol.*, **122**: 295-317.

Chaumont, F., Barrieu, F., Herman, E.M. and Chrispeels, M.J. 1998. Characterization of a maize tonoplast aquaporin expressed in zones of cell division and elongation. *Plant Physiol.*, **117**: 1143-1152.

Cheng, N.H., Pitman J.K., Barkla, B.J., Shigaki, T. and Hirchi, K.D. 2003. The *Arabidopsis cax1* mutant exhibits impaired ion homeostasis, development and hormonal responses and reveals interplay among vacuolar transporters. *Plant Cell,* **15**: 347-364.

Cleveland, D.W., Fischer, S.G., Kirschner, M.W. and Laemmli, U.K. 1977. Peptide mapping by limited proteolysis in sodium dodecyl sulphate and analysis by gel electrophoresis. *J. Biol. Chem.*, **252**: 1102-1106.

Devos, M.K. and Gale, D.M. 2000. Genome relationships: the grass model in current research. *Plant Cell,* **12:** 637-646.

Dolan, L. and Davies, J. 2004. Cell expansion in roots. *Curr. Opin. Plant Biol.*, **7**: 33-39.

Duée, E., Olivier-Deyris, L., Fanchon, E., Corbier, C., Branlant, G. and Dideberg, O. 1996. Comparison of the structures of wild-type and a N313T mutant of Escherichia coli glyceraldehyde 3-phosphate dehydrogenases: implication for NAD binding and cooperativity. *J. Mol. Biol.*, **257**: 814-838.

Feng, Q., Zhang, Y., Hao, P., Wang, S., Fu, G., Huang, Y., Li, Y., Zhu, J., Liu, Y., Hu, X., Jia, P., Zhang, Y., Zhao, Q., Ying, K., Yu, S., Tang, Y., Weng, Q., Zhang, L., Lu, Y., Mu, J., Lu, Y., Zhang, L.S., Yu, Z., Fan, D., Liu, X., Lu, T., Li, C., Wu, Y., Sun, T., Lei, H., Li, T., Hu, H., Guan, J., Wu, M., Zhang, R., Zhou, B., Chen, Z., Chen, L., Jin, Z., Wang, R., Yin, H., Cai, Z., Ren, S., Lv, G., Gu, W., Zhu, G., Tu, Y., Jia, J., Zhang, Y., Chen, J., Kang, H., Chen, X., Shao, C., Sun, Y., Hu, Q., Zhang, X., Zhang, W., Wang, L., Ding, C., Sheng, H., Gu, J., Chen, S., Ni, L., Zhu, F., Chen, W., Lan, L., Lai, Y., Cheng, Z., Gu, M., Jiang, J., Li, J., Hong, G.., Xue, Y. and Han, B. 2002. Sequence and analysis of rice chromosome 4. *Nature*, **420**: 316-320.

Fernie, A.R., Carrari, F. and Sweetlove, L.J. 2004. Respiratory metabolism: glycolysis, the TCA cycle and mitochondrial electron transport. *Curr. Opin. Plant Biol.*, **7**: 254-261.

Fridovich, I. 1975. Superoxide dismutases. *Ann. Rev.Biochem.*, **4**: 47-159.

Gilroy, S. and Jones, R.L. 1992. Gibberellic acid and abscisic acid coordinately regulate cytoplasmic calcium and secretory activity in barley aleurone protoplasts. *Proc. Natl. Acad. Sci.* USA, **89**: 3591-3595.

Goff, S.A., Ricke, D., Lan, T.-H., Presting, G., Wang, R., Dunn, M., Glazebrook, J., Sessions, A., Oeller, P., Varma, H., Hadley, D., Hutchison, D., Martin, C., Katagiri, F., Lange, B.M., Moughamer, T., Xia, Y., Budworth, P., Zhong, J., Miguel, T., Paszkowski, U., Zhang, S., Colbert, M., Sun ,W.-I., Chen, L., Cooper, B., Park, S., Wood, T.C., Mao, L., Quail, P., Wing, R., Dean, R., Yu, Y., Zharkikh, A., Shen, R., Sahasrabudhe, S., Thomas, A., Cannings, R., Gutin, A., Pruss, D., Reid, J., Tavtigian, S., Mitchell, J., Eldregde, G., Scholl, T., Miller, R.M., Bhatnagar, S., Adey, N., Rubano, T., Tusneem, N., Robinson, R., Feldhaus, J., Macalma, T., Oliphant, A. and Briggs, S. 2002. A draft sequence of the rice genome (*Oryza sativa* L. ssp. *japonica*). *Science*, **296**: 92-100.

Gogarten, J.P., Fichmann, J., Braun, Y., Morgan, L., Styles, P., Taiz, S.L., de Lapp, K. and Taiz, L. 1992. The use of antisense mRNA to inhibit the tonoplast $H^+$-ATPase in carrot. *Plant Cell,* **4**: 851-864.

Hedden, P. and Proebsting, W.M. 1999. Genetic analysis of gibberellin biosynthesis. *Plant Physiol.*, **119**: 365-370.

Hejgaard, J., Jacobsen, S. and Svendsen, I. 1991. Two antifungal thaumatin-like proteins from barley grain. *FEBS Lett.*, **291**: 127-131.

Higuchi, K., Watanabe, S., Takahashi, M., Kawasaki, S., Nakanishi, H., Nishizawa, N.K. and Mori, S. 2001, Nicotianamine synthase gene expression differs in barley and rice under Fe-deficient conditions. *Plant J.,* **25**: 159-167.

Hirano, H., Kawasaki, H. and Sassa, H. 2000. Two-dimensional gel electrophoresis using immobilized pH gradient tube gels. *Electrophoresis*, **21**: 440-445.

Izumi, K., Kamiya, Y., Sakurai, A., Oshio, H. and Takahashi, N. 1985. Studies of sites of action of a new plant growth retardant (*E*)-1-(4-chlorophenyl)-4,4-dimethyl-2-(1,2,4-triazole-1-yl)-1-penten-3-ol (S-3307) and comparative effects of its stereoisomers in a cell-free system from *Cucurbita maxima. Plant Cell Physiol.*, **26**: 821-827.

Jacobs, T. 1997. Why do plant cells divide? *Plant Cell*, **9**: 1021-1029.

Kende, H. and Zeevaart, J.A.D. 1997. The five "classical" plant hormones. *Plant Cell*, **9:** 1197-1210.

Komatsu, S., Konishi, H., Shen, S. and Yang, G. 2003. A step toward functional analysis of the rice genome. *Mol. Cell. Proteomics*, **2**: 2-10.

Konishi, H., Kitano, H. and Komatsu, S. 2005a Identification of rice root proteins regulated by gibberellin using proteome analysis. *Plant Cell Environ.* (in press)

Konishi, H.,Yamane, H., Maeshima, M. and Komatsu, S. 2005b. Characterization of fructose-bisphoaphate aldolase regelated by gibberellin in roots of rice seedling. *Plant Mol. Biol.* (in press)

Lu, M., Sautin, Y.Y., Holliday, L.S. and Gluck, S.L. 2004. The glycolytic enzyme aldolase mediates assembly, expression, and activity of vacuolar $H^+$- ATPase. *J. Biol. Chem.,* **279:** 8732-8739.

Lynch, J. 1995. Root architecture and plant productivity. *Plant Physiol.*, **109**: 7-13.

McCubbin, A.G., Ritchie, S.M., Swanson, S.J. and Gilroy, S. 2004. The calcium-dependent protein kinase HvCDPK1 mediates the gibberellic acid response of the barley aleurone through regulation of vacuolar function. Plant J. 39: 206-218.

McCully, M. 1995. How do real roots work? *Plant Physiol.*, **109**: 1-6.

Moons, A., Prinsen, E., Bauw, G. and van Montagu, M. 1997. Antagonistic effects of abscisic acid and jasmonates on salt stress-inducible transcripts in rice roots. *Plant Cell,* **9**: 2243-2259.

Moore, B., Zhou, L., Rolland, F., Hal,l Q., Cheng, W.H., Liu, Y.X., Hwang, I., Jones, T. and Sheen, J. 2003. Role of the *Arabidopsis* glucose sensor HXK1 in nutrient, light, and hormonal signaling. *Science,* **300**: 332-336.

O'Farrell, P.H. 1975. High resolution two-dimensional electrophoresis of proteins. *J. Biol. Chem.*, **250**: 4007-4021.

Rakwal, R., Agrawal, G.K. and Jwa, N.S. 2001. Characterization of a rice (*Oryza sativa* L.) Bowman-Birk proteinase inhibitor: tightly light regulated induction in response to cut, jasmonic acid, ethylene and protein phosphatase 2A inhibitors. *Gene*, **263**: 189-198.

Ranathunge, K., Kotula, L., Steudle, E. and Lafitte, R. 2004. Water permeability and reflection coefficient of the outer part of young rice roots are differently affected by closure of water channels (aquaporins) or blockage of apoplastic pores. *J. Exp. Bot.,* **55**: 433-447.

Rolland, F., Winderickx, J. and Thevelein, J.M. 2001. Glucose-sensing mechanisms in eukaryotic cells. *Trends Biochem. Sci.,* **26**: 310-317.

Sasaki, T. and Burr, B. 2000. Internal rice genome sequencing project: the effort to completely sequence the rice genome. Curr. Opin. Plant Biol., 3: 138-141.

Sasaki, T., Matsumoto, T., Yamamoto, K., Sakata, K., Baba, T., Katayose, Y., Wu, J., Niimura, Y., Cheng, Z., Nagamura, Y., Antonio, B.A., Kanamori, H., Hosokawa, S., Masukawa, M., Arikawa, K., Chiden, Y., Hayashi, M., Okamoto, M., Ando, T., Aoki, H., Arita, K., Hamada, M., Harada, C., Hijishita, S., Honda, M., Ichikawa, Y., Idonuma, A., Iijima, M., Ikeda, M., Ikeno, M., Ito, S., Ito, T., Ito, Y., Ito, Y., Iwabuchi, A., Kamiya, K., Karasawa, W., Katagiri, S., Kikuta, A., Kobayashi, N.,

Kono, I., Machita, K., Maehara, T., Mizuno, H., Mizubayashi, T., Mukai, Y., Nagasaki, H., Nakashima, M., Nakama, Y., Nakamichi, Y., Nakamura, M., Namiki, N., Negishi, M., Ohta, I., Ono, N., Saji, S., Sakai, K., Shibata, M., Shimokawa, T., Shomura, A., Song, J., Takazaki, Y., Terasawa, K., Tsuji, K., Waki, K., Yamagata, H., Yamane, H., Yoshiki, S., Yoshihara, R., Yukawa, K., Zhong, H., Iwama, H., Endo, T., Ito, H., Hahn, J., Kim, H.-L., Eun, M.-Y., Yano, M., Jiang, J. and Gojobori T. 2002. The genome sequence and structure of rice chromosome 1. *Nature,* **420**: 312-316.

Sasaki, A., Itoh, H., Gomi, K., Ueguchi-Tanaka, M., Ishiyama, K., Kobayashi, M., Jeong, D.H., An, G., Kitano, H., Ashikari, M. and Matsuoka, M. 2003. Accumulation of phosphorylated repressor for gibberellin signaling in an F-box mutant. *Science,* **299**: 1896-1898.

Sassa, H., Oguchi, S., Inoue, T. and Hirano, H. 2000. Primary structural features of the 20S proteasome subunits of rice (*Oryza sativa*). *Gene*, **250**: 61-66.

Schaeffer, G.W., Sharpe, F.T. and Sicher, R.C. 1997. Fructose 1,6-bisphosphate aldolase activity in leaves of a rice mutant selected for enhanced lysine. *Phytochem.*, **46:** 1335-1338.

Schiefelbein, J.W. and Benfey, P.N. 1991. The development of plant roots: new approaches to underground problems. *Plant Cell*, **3**: 1147-1154.

Schumacher, K., Vafeados, D., McCarthy, M., Sze, H., Wilkins, T. and Chory, J. 1999. The *Arabidopsis det3* mutant reveals a central role for the vacuolar $H^+$-ATPase in plant growth and development. *Genes. Dev.*, **13**: 3259-3270.

Tanaka, N., Konishi, H., Khan, M.M.K. and Komatsu, S. 2004a. Proteome analysis of rice tissues by two-dimensional electrophoresis: an approach to the investigation of gibberellin regulated proteins. *Mol. Gen. Genomics*, **270**: 485-496.

Tanimoto, E. 1994. Interaction of gibberellin $A_3$ and ancymidol in the growth and cell-wall extensibility of dwarf pea roots. *Plant Cell Physiol.*, **35**: 1019-1028.

Trethewey, R.N., Fernie, A.R., Bachmann, A., Fleischer-Notter, H., Geigenberger, P. and Willmitzer, L. 2001. Expression of a bacterial sucrose phosphorylase in potato tubers results in a glucose-independent induction of glycolysis. *Plant Cell Environ.*, **24**: 357-365.

Tsugita, A., Kawakami, T., Uchiyama, Y., Kamo, M., Miyatake, N. and Nozu, Y. 1994. Separation and characterization of rice proteins. *Electrophoresis*, **15**: 708-720.

Urano, J., Nakagawa, T., Maki, Y., Masumura, T., Tanaka, K., Murata, N. and Ushimaru, T. 2000. Molecular cloning and characterization of a rice dehydroascorbate reductase. *FEBS Lett.*, **466**: 107-111.

Xu, Y., Buchholz, W.G., DeRose, R.T., Hall, T.C. 1995. Characterization of a rice gene family encoding root-specific proteins. *Plant Mol. Biol.*, **27**: 237-248.

Yamaguchi, S. and Kamiya, Y. 2000. Gibberellin biosynthesis: Its regulation by endogenous and environmental signals. *Plant Cell Physiol.*, **41**: 251-257.

Yang, G., Shen, S., Yang, S. and Komatsu, S. 2003. *OsCDPK13*, a calcium-dependent protein kinase gene from rice, is induced in response to cold and gibberellin. *Plant Physiol. Biochem.,* **41**: 369-374.

Yu, J., Hu, S., Wnag, J., Wong, G.K.-S., Li, S., Liu, B., Deng, Y., Dai, L., Zhou, Y., Zhang, X., Cao, M., Liu, J., Sun, J., Tang, J., Chen, Y., Huang, X., Lin, W., Ye, C., Tong, W., Cong, L., Geng, J., Han, Y., Li, J., Liu, Z., Li, L., Liu, J., Qi, Q., Liu, J., Li, L., Li, T., Wang, X., Lu, H., Wu, T., Zhu, M., Ni, P., Han, H., Dong, W., Ren, X., Feng, X., Cui, P., Li, X., Wang, H., Xu, X., Zhai, W., Xu, Z., Zhang, J., He, S., Zhang, J., Xu, J., Zhang, K., Zheng, X., Dong, J., Zeng, W., Tao, L., Ye, J., Tan, J., Ren, X., Chen, X., He, J., Liu, D., Tian, W., Tian, C., Xia, H., Bao, Q., Li, G., Gao, H., Cao, T., Wang, J., Zhao, W., Li, P., Chen, W., Wang, X., Zhang, Y., Hu, J., Wang, J., Liu, S., Yang, J., Zhang, G., Xiong, Y., Li, Z., Mao, L., Zhou, C., Zhu, Z., Chen, R., Hao, B., Zheng, W., Chen, S., Guo, W., Li, G.,

Liu, S., Tao, M., Wang, J., Zhu, L., Yuan, L. and Yang, H. 2002. A draft sequence of the rice genome (*Oryza sativa* L. ssp. *indica*). *Science*, **296:** 79-92.

Zhong, B., Karibe, H., Komatsu, S., Ichimura, H., Nagamura, Y., Sasaki T. and Hirano, H. 1997. Screening of rice genes from a cDNA catalog based on the sequence data-file of proteins separated by two-dimensional electrophoresis. *Breed. Sci.*, **47**: 245-251.

*SECTION — V*

## ***CROP MODELING***

*Advances in Plant Physiology*, Vol. 9
Ed. A. Hemantaranjan
Scientific Publishers (India), Jodhpur, 2006 pp. 315-323
E-mail: **info@scientificpub.com** www.scientificpub.com

# 14

# CROP SIMULATION MODELING: IMPLICATIONS AND POTENTIALS FOR RUBBER

S. K. Dey

Rubber Research Institute of India, Regional Research Station, Agartala-799 006 , India

## INTRODUCTION

A model is any representation of a real system. Simulation means that a model mimics like a real crop e.g. gradually growing leaves, stem, roots etc. during the season. It does not just predict some final state such as biomass or yield. All simulation models therefore, contain same mechanism, but there is still a wide range in their capabilities. Crop simulation models are gaining more importance now a days because they behave like real crop in the field. Extensive work is going on all over the world to build crop simulation models. Computer based simulation modeling of agriculture ecosystem is rapidly expanding scientific methodology for two reasons. On the one hand the volume of scientific knowledge on agricultural production systems in relevant scientific disciplines is expanding rapidly. Secondly, computer based simulation and modeling is an efficient tool to integrate this knowledge and to apply it to practical problems (Jand *et al.,* 1995). Moreover, there are related disciplines, such as hydrology and meteorology, where model use is routine. System analysis and simulation modeling can be applied to a wide range of research questions related to improved crop production and to more efficient resource management. It can provide quantitative insight in to the interactions between the crop, its environment and its management. Crop simulation models can be divided into two groups: those that aspire to improve our understanding of the physiology and environmental interactions of crops, and those that aspire to provide sound management advice to farmers or sound predictions to policy makers.

### 1.1 Early History

The first steps for crop modeling were models developed to estimate light interception and photosynthesis in crop canopies (Loomis and Williams, 1963; de Wit, 1965; Duncan *et al.*, 1967). These models calculate the light profile in a canopy and made it possible to assess the sensitivity of crop photosynthetic rates to solar radiation, leaf angle distribution and the latitudinal position of the crop. These were relatively simple models, but they opened the way to quantitative, mechanistic estimates of maximum attainable growth rates. Crop growth and potential yield became quantitatively and demonstrably linked via biochemical and biophysical mechanisms to the amount of solar energy and biomass by plants. The complexity of crop models increased as the various details of crop mass accumulation and an accounting of the factors that alter plant growth were incorporated in to the models (Sinclair and Seligman, 1996). Important advances in describing various sub-components of carbon assimilation in particular were also made during this period. The significance of stomatal conductance in regulating leaf gas exchange was quantitatively described (Cowan, 1977). The fate of photo assimilates in respiratory pathways was carefully analyzed (Penning de Vries, 1975). Crop modeling, the computerized simulation of dynamic crop systems, was born about 40 years ago, when systems analysis and modern computers presented a new technique to crop scientists. From about 1970, when computers became easily available to help us deal with the complexity of crops, the craft of crop simulation modeling developed rapidly. Two distinct types of model emerged: one was essentially practical, and combined a few rules of thumb to predict the behaviour of crops. The other was seemingly scientific in spirit, and sought to represent the biological and physiological processes thought to occur in plants and their environments (Passioura, 1973).

## 1.2 The Structure and Development of Different Simulation Models

A scientific model is the numerical quantification of the relationship in the form of single equation or a set of equations. A mathematical model is an equation or set of equations, which represents the behavior of a system. Mathematics is merely a tool enabling the biologist to express his biological hypothesis so that quantitative prediction is possible, which can then be compared with the real world. A complete crop model is a simple representative of a crop. A crop model can be defined as a quantitative scheme for predicting the growth, development and yield of a crop, given a set of genetic coefficients and relevant environmental variables. It is conventional to distinguish between mechanistic crop models, in which all quantified processes have a sound physical or physiological basis, and empirical models consisting of functions that are chosen, often arbitrarily, to fit measurements from field or laboratory (Montheith, 1996). In practice, however, most models represent a compromise between rigor and utility. At one end of the spectrum are models of plants and crops based almost exclusively on long-established and robust principles from physics and chemistry. It used to study growth and compute growth responses to the environment. Dynamic simulation crop model can predict the changes in crop status with time a, function of exogenous parameters. Phenological models are broad class models that predict crop development from one growth stage to another. Physiologically and physically based simulation models are those mechanistic models where plant or soil processes can be physiologically, physically or chemically described. In other words model is an organized and refined expression of knowledge with regard to factors interacting with each other. They help (i) interpreting experimental results, (ii) as an agronomic research tool and (3) as an agronomic grower tools (Boote *et al.*, 1996). Crop models are being formulated with the goal of their ultimate use by the growers.

The models are broadly classified by Jand *et al.* (1995) based on testing as (1) Relatively untested models (Hypotheses), (2). Partially tested models (Theories) and (3) Well tested models (Laws). The fundamental difficulty is that all models are basically a collection of hypotheses and not a single falsifiable hypothesis, so they inherently cannot be validated (Oreskes *et al.* 1994). Not only can other collections of hypotheses approximate the experimental results equally well, but the validation data themselves are flawed by substantial experimental and observational error. Calibration means adjusting certain model parameters or relationships to make the model work for a site or sites. Finally, the crop models must be validated. Validation is determining whether the model works with totally independent data sets; that is, does it accurately predict growth, yield, and processes? Attempts to validate models can show only how well a model performs in a particular circumstance. They cannot guarantee the performance of the model under any other environmental condition especially when the model has been calibrated to fit a specific circumstance or set of circumstances (Oreskes *et al.*, 1994). Validation exercises are probably useful only when a model is to be moved to an applications role. The users of the model need to be given some notion of situations in which the model has proven useful. An essential requirement for a good working crop model is that it be able to predict with reasonable accuracy through out the range of variables over which it was calibrated (Passioura, 1996).

## 2. GENE BASED APPROACHES TO CROP SIMULATION

The current research aims to significantly improve the crop knowledge and technology by linking research activities at different levels of integration, from gene to plant to ecosystem. The genome maps characterize vertical traits that have implications at the crop level. Attention is also drawn to characterize quantitative links between genome characteristics, physiological traits and crop production possibilities in a given environment. A gene based model, Gene Gro, was developed to simulate in common bean. Modeling gene action through linear estimates of effects on model parameters has shown promise in the common bean model Gene Gro. Over the next decade, genetic information probably will have most contribution in under standing temporal and tissue level variation in genetic control of specific processes and for more applied modeling, in improving the representation of cultivar differences (White and Hoogenboom, 2003).

## 3. MODELING BY USING REMOTE SENSING TECHNOLOGY

In remote sensing, spectral reflectance's, thermal radiations or other electromagnetic spectra from fields are measured by sensors installed on satellites, aircrafts etc. To estimate crop growth from remotely sensed data different models can be used i.e. empirical simulation model, the biomass production model as a function of absorbed or intercepted solar radiation etc. Different areas like soil survey for identification of suitable areas for rubber cultivation and identification of different rubber clones have been attempted by this technology (Rao, 2003).

### 3. 1 Some Aspects of Modeling in Rubber

Rubber tree (*Hevea brasiliensis*) is a natural forest tree originated from Amazonian basin of South America. This species is now grown in the tropical regions of Asia, Africa and America (Dey, 2004). Around 9.2 million hector of area is now under cultivation of rubber in 15 countries of the world (Table 1). Rubber is a long perennial plant having a life span of 33 years out of which nearly 20 years is the productive phase. The crop harvesting system of rubber is also interesting and different from other plants. The economic yield (dry rubber), latex is being extracted by wounding the tree on alternate day. The yield is highly variable under different environmental conditions (Dey *et al.*, 2004). Even though, it is difficult to bring the rubber plant physiological functions to mathematical equations, however, several researchers have attempted for modeling of rubber plant to predict growth and yield.

Table 1. Area under rubber in main producing countries in the world

| Territory | End of | Total (ha) |
|---|---|---|
| Indonesia | 2002 | 3318000 |
| Thailand | 2002 | 1973000 |
| Malaysia | 2000 | 1431000 |
| China | 1998 | 618000 |
| India | 2002 | 569670 |
| Vietnam* | 2002 | 420000 |
| Sri lanka | 2002 | 157000 |
| Nigeria | 1999 | 150000 |
| Liberia | 1999 | 109000 |
| Myanmar | 1995 | 105000 |
| Brazil | 2001 | 101000 |
| Cote d Iovire | 1998 | 96000 |
| Philippines | 1999 | 92000 |
| Cameroon | 1997 | 42000 |
| D.R. of Congo** | 1999 | 35000 |

*Estimate ** Formerly Zaire

### 3.2 Determination of Biomass in Different Stages

Shorrocks *et al.* (1965) has developed a linear equation (weight in kg, w = 0.002604 $G^{2.7826}$) for estimation of shoot biomass in rubber. Clone specific (RRIM 600) and location specific regression (in kg, w = 0.0202 $G^{2.249}$) for mature rubber tree biomass estimation was developed subsequently (Dey *et al.*, 1996).

To estimate the biomass of immature tree, Choudhury *et al.* (1995) has developed the regression equation of two clones in North East region of India. The equation is given as biomass in g, (w) = 2.2784 X 2.2683 (where, x is the girth at 15 cm height from bud union). A comparative differences were shown by regression equations in seasonal relative growth of rubber during immature period with different climatic parameters in humid and dry sub-humid region of India (Dey *et al.*, 1998). Different equations for 25 clones of rubber were developed to predict immature growth in traditional region of India, where cumulative rainfall and sunshine are the prime data inputs (Dey and Vijayakumar, 2002).

### 3.3 Estimation of Yield

The first attempt was made by Sethuraj (1981) to express yield through different yield components by a simple equation like y = (F.l.Cr)/P), where Y = yield, F = initial flow, l = length of tapping cut; Cr = dry rubber content and P = plugging index. Subsequently, the effect of vapour pressure, maximum and minimum temperature in humid and dry sub-humid climatic conditions was incorporated to these components (Dey *et al.*, 1999).

### 3.4 Empirical Regression Models for Rubber

Homogeneous rainfall areas conducive for rubber cultivation in the North East India have been identified by Kriging interpolatory methods (Raj and Dey, 2003). Multiple regression equation was developed to estimate latex yield of clone RRII 105 on a tapping day, with different antecedent weather conditions for traditional region in India (Rao *et al.*, 1998). In this model, the weekly weather conditions associated with optimum yields were maximum temperature of 30.4°C, minimum temperature of 22.8°C, 5.9 h sunshine and 72 mm of rainfall. Any deviation from these meteorological conditions up to 6 months elevation may significantly influence latex production. Raj *et al.* (1998) has developed regression equations for five clones in North East Indian climatic condition, where yield could be predicted by antecedent environmental parameters under alternate day tapping system. Recently, Alam *et al.* (2003) developed an equation for prediction of yield, where biochemical parameters like latex sugar, thiol, inorganic phosphorus, total solid content were incorporated along with climatic parameters like maximum and minimum temperature and soil moisture in the North East Indian climatic condition.

### 3.5 Simulation Rubber Model

For simulation of growth and productivity of rubber, the EMB-RUBBER model is in use. This model was originally built to simulate rubber growth and production, to assess the relative importance of alternative research priorities and generate hypotheses in theoretical studies (Castro, 1988, Brummer, 1992). Later, this model was refined by Bernardes *et al.* (1994) and written in FST (Rappldt and Kraalingen, 1996). The model consists of carbon and water balance and simulates dry matter accumulation, girth of tree, leaf area index, rubber yield and other variables with one day time interval. It was adapted from the level II basic crop growth simulator (MACROS), developed by Penning de Vries *et al.* (1989) and originally written in the Continuous System Modeling Programme (CSMP), with some subroutines and functions written in FORTRAN 77. Some of the abbreviations used in this model were taken from Simple and Universal CROP growth Simulator termed as SUCROS2. The relational diagram of EMB – RUBBER model is presented in Fig. 1. This model can predict growth and yield of rubber by using the weather data of a particular location.

## 4. PRACTICAL APPLICATIONS OF MODELS

Crop models can be applied for predicting the crop performance for climates when the crop has not been grown before or not grown under optimal conditions. The yield can be predicted some time before harvest by using expected weather data which is important for trade or planning and post harvest operation, supply/demand and market rate etc. The dynamism of crop simulation models in terms of predicting responses to different environments and limiting factors lead to their wider application in agro-climatic zonation and studying the impact of infestation of pests and diseases, water and nutrient shortages, natural damage and various agronomic practices. Teng *et al.* (1991) used a crop model to predict the incidence of blast epidemic in rice in Thailand. Prediction of yield loss in rice due to expected intensity of drought in Tarlac province of Philippines by using crop models have also been reported (Woperes *et al.*, 1993). Agarwal *et al.* (1994) used a crop model for agro-climatic zonation for wheat production in India. Dey *et al.* (2002) first attempted to use EMB RUBBER model for agro-ecological zonation at different rubber growing regions of India for productivity of rubber. The yield of clone RRIM 600 was predicted in 27 locations spread all over the traditional and non-traditional rubber cultivating regions of India. On the basis of the predicted yield, the non-traditional areas have been classified into high (above 1000 $kg^{-1}$ $ha^{-1}$), medium (700 to 1000 kg $ha^{-1}$) and low (less than 700 kg) productivity zones (Fig. 2 ).

### 4.1 Limitations in Models

The regression models have many shortcomings due to errors in assumptions. As the relation between crop yield and predictive variables is derived from historical data set, the predictive ability of these

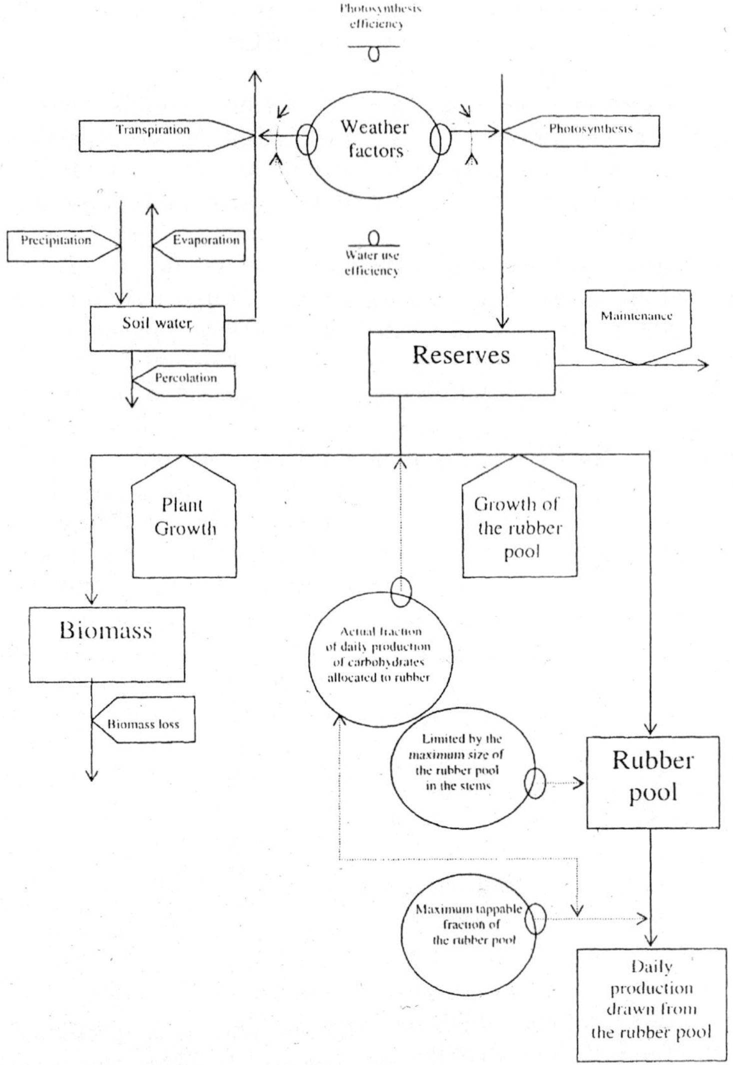


Fig. 1. The relation diagram of EMB-RUBBER model

models is poor out side the scope of data set range in both time and space domains. However, different researchers have attempted to forecast yield of *Hevea* by empirical regression models.

## 5. FUTURE SCOPE FOR MODELING IN RUBBER

Many crop growth models have been developed during the last three decades. Initially, the focus of most of these efforts was on understanding processes and interpreting experimental data. In order to this, most models became very complex and no one person other than the developer could understand them limiting their role in problem solving. Lately, several models have been developed which have a relatively simple structure. Models are now available for all major cereals such as wheat, rice, maize, sorghum and millet. These models are well developed and have been reported satisfactory in their performance at several places in the world (Aggarwal, 2003). However, there are enough scope for modeling in rubber. Several potential areas like identification of suitable new areas for rubber cultivation, prediction of growth and yield where rubber is not grown, identification of suitable clone specific crop harvesting (exploitation) system are deserving attention for modeling in rubber. There are some progress in remote sensing work, however needs more effort in areas like impact of seasonal drought on productivity of rubber, disease and pest forecasting based on weather condition etc. Advances in genomic research in rubber suggest the possibility of using information on gene action to improve simulation models.

## Conclusion

Crop models cannot produce all the answers to crop production problems. However, crop growth modeling is a potentially valuable tool in research, crop management and policy decisions, when model is constructed at a reasonable level and validated. Research can make use of crop growth models to conduct hypothetical studies and compare with experimental data. A model allows the synthesis and mobilization of existing knowledge about individual processes leading to yield and help pin point deficiencies of knowledge. There is much that we do not know about the rubber plant and their interaction with their environment. We believe that use of rubber models will play an increasingly important role in research understanding, crop management and policy questions. Modeling in rubber needs expertise of scientists from plant physiology, agronomy, breeding, statistics, plant protection, molecular biologist and computer application to better assess and harness the data resulting from their research.

## REFERENCES

Aggrawal, P. K. 2003. Crop modeling : a tool for accelerating utilization of plant physiology knowledge. Paper presented at National Seminar on "Physiological interactive for improved crop productivity and quality : opportunities and constraints". December 2003 at Tirupati, India.

Aggrawal, P. K., Karla Naveen and Sankeran, V. M. 1994. Simulating the effect of climatic factor and genotype on productivity of wheat, pp. 71-88. In: *simulating the effect of climatic factors genotype and management on productivity of wheat in India*, ICAR, India.

Alam, B., Das, G., Raj, S., Roy, S., Pal, T. K. and Dey, S. K. 2003. Studies on yield and biochemical subcomponents of latex of rubber trees (*Hevea brasiliensis*) with a special reference to the impact of low temperature in the non-optimal environment. *Journal of Rubber Research,* **6** (4) : 241 – 257.

Bernardes, M. S. and Goudriaan, J. L. 1994. Evaluation of the "EMB-RUBBER" Model for estimating growth and yield.In: *Proceedings of 3 rd ESA Congress*, Alano – Padova.

Brummer, B. M. 1992. A modeling and simulation study of planting density and tapping systems in rubber (*Heave brasiliensis*) ESALQ/USP, Piracicaba, Brazil.

Boote, K. J., Janes, J. W and Pickering, N. B. 1996. Potential uses and limitations of crop models. *Journal of Agronomy,* 88 : 704-716.

Castro, A. M. G. 1988. A system approach to determining priorities for natural rubber research in Brazil (Ph.D. Thesis) 292p.

Choudhuri, D., Vinod, K. K., Potty, S.N., Sethuraj, M. R., Pothen, J. and Reddy, Y.A.N. 1995. Estimation of biomas in *Hevea* clones by regression method : 1. Relation between girth and biomass. *Indian Journal of Natural Rubber Research,* **8**(2) : 113-116.

Cowan, I. R. 1977. Stomatal behaviour and environment. *Adv. Bot. Res.* 4: 117-228.

De wit., T. 1965. Photosynthesis of leaf canopies. Inst. Biol. Chem. Res. Field Crops Herb. Agric. Res. Rep. 663. Wageningen, The Netherlands.

Dey, S. K., Choudhuri, D., Vinod, K. K., Pothen, J. and Sethuraj, M. R. 1996. Estimation of biomass in *Hevea* clones by regression method : 2. Relation of girth and biomass for nmature tree of clone RRIM 600. *Indian Journal of Natural Rubber Research*, **9**(1) : 40-43.

Dey., S. K., Chandrashekar, T. R., Nair, D. B., Vijayakumar, K. R., Jacob, J. and Sethuraj, M. R. 1998. Effect of some agroclomatic factors on the growth of rubber (*Heave brasiliensis*) in a humid and a dry sub-humid location. *Indian Journal of Natural Rubber Research* , 11 (1& 2) : 104-109.

Dey, S. K., Vijayakumar, K. R., Nair, D. B. and Subramanian, P. 1999. Effect of temperature and vapour pressure on major yield components of rubber in Humid and dry sub-humid climatic regions. *Indian Journal of Natural Rubber Research,* **12** (1& 2) : 69-76.

Dey, S. K., Joseph, T., Barnardes, M.S. and Penning De Vries F.W.T. 2002. Agroecological zoning for yield prediction in rubber plantations in India. *Indian Journal of Natural Rubber Research,* **15** (2): 158-164

Dey, S.K. and Vijayakumar, K.R. 2002. Selection of vigorous clones of *Hevea* through regression method. *Indian Journal of Natural Rubber Research,* **15** (1):76-87.

Dey. S.K. 2004) Eco-friendliness and sustainability of rubber plantations. *Planters' Chronicle* 100 (4) 23-29

Dey, S.K., Das, G., Alam, B. and Sarma, A.C. 2004. Path analysis of yield components of *Hevea* clone RRIM 600 under different environmental conditions in North Eastern region of India. 17(1) *Natural Rubber Research,* (In Press).

Duncan, W.G., Loomis, R.S., Williams, W.A. and Hanau, R. 1967. A model for simulating photosynthesis in plant communities. *Hilgardia*, **38** : 181-205.

Jand, V. Bains, G. S. and Mavi, H. S. 1995. Modeling plant growth an over view. In *Agro's Ann. Plant Physiology* (BandA) II. (Ed. S. S. Purohit) Agro botanical Publisher (India) Bikaner. pp 1-12.

Loomis, R. S. and Williams, W. A. 1963. Maximum crop productivity : An estimate. *Crop Science*, **3**: 67-72.

Monteith, J.L. 1996. The quest for balance in crop modeling. *Journal of Agronomy,* **88** : 695-697

Oreskes, N., Shrader – Frechette, K. and Belitz, K. 1994. Verification, validation and confirmation of numerical models in earth sciences. *Science* (Washington, DC), **263** : 641-646.

Passioura, J.B. 1973. Sense and nonsense in crop simulation. *Journal of Aust. Inst. Agric. Sci.,* **39** : 181-183.

Passioura, J.B. 1996. Simulation models : Science, Snake Oil, Education or Engineering ? *Journal of Agronomy*, **88** : 690-694.

Penning De Vries, F.W.T., Jansen, D. M., Ten Berge, H.F.M. and Bakema, A. H. 1989. Simulation of eco-physiological progress in several annual crops. Simulation Monographs. PUDOC, Wageningen, The Netherlands. 271 p.

Penning De Vries, F.W.T. 1975. Use of assimilates in higher plants. P. 459-480. In photosynthesis and productivity in different environments (Ed. J.P. Cooper). Cambridge University Press, Cambridge.

Raj, S. and Dey, S. K. 2003. Spatial analysis of rainfall variability over the North East with respect to rubber cultivation. Paper presented at National seminar on Agrometeorology in the new millennium prospectives and challenges held at PAU, Ludhiana on October 29-31, 2003. Abstract p. 39.

Raj, S., Das, G., Dey, S.K. and Sethuraj, M.R. 1998. Relationship of yield with antecedent environmental parameters and its predictive potential in *Hevea brasiliensis*. National Symposium on Current Trends in Plant Physiology and Biochemistry, 1998. University of Hyderabad, Hyderabad, India, p. 168.

Rao, N. 2003. Possible applications for remote sensing in rubber. *Planter's chronicle,* **12** : 26-35.

Rao, P. S., Saraswathyamma, C.K. and Sethuraj, M. R. 1998. Studies on the relationship between yield and meteorological parameters of para rubber tree (*Hevea brasiliensis*). *Agricultural and Forest Meteorology*. **90** : 235-245.

Rappoldt, C. and Kraalingen van., D.W.G. 1996. The Fortran simulation translator manual, QASA Report. 5, AB-DLO, Wageningen. P.178

Shorrocks, V. M., Templetor, T. K. and Iyer, G. C. 1965. Mineral nutrition growth and nutrient cycle of *Hevea brasiliensis* : 3. Relationship between girth and shoot dry weight. *Journal of Rubber Research Institute of Malaysia*, **19** : 85092.

Sethuraj, M. R. 1981. Yield components in *Hevea brasiliensis* – theoretical considerations. *Plant, Cell and Environment*, **4** : 81-83.

Sinclair, T. R. and Seligman, N.G. 1996. Crop modeling: From infancy to maturity. *Journal of Agronomy,* **88** : 698-704.

Teng, P. S., Calvero, S., Pinnschmidt, H. 1991. Sumulation of rice patho systems and disease losses. Pages 27-33. In: *systems simulation at IRRI. JRRI Res.paper* Ser. 151. (Eds. Penning de Vries F. W. T., Kropff, M. J. Teng, P. S.and Krik G.J.D.)

White, J. W. and Hoogenboom, G. 2003. Gene based approaches to crop simulation : past experiences and future opportunities. *Agronomy Journal,* **95** : 52 – 64.

Wopereis, M.C.S., Kropff, M.J., Hunt, E.D., Sanidad, W. Bouma, J. 1993. Case study on regional application of crop growth simulation models to predict rainfed rice yield. Tarlac province. Pages 27-46 (Ed. Bouman B.A.M., Van Laar, H.H., Wang Zhaoguian).In: *Agro ecology of rice – based system. SARP Research Proceedings.* October 1993. (AB-DLO, TPE-WAU and IRRI), ABO-Wageningen.

*SECTION — VI*

## *ACTINORHIZAL AND LEGUMES NITROGEN FIXING SYMBIOSES*

*Advances in Plant Physiology*, Vol. 9
Ed. A. Hemantaranjan
Scientific Publishers (India), Jodhpur, 2006 pp. 327-344.
E-mail: **info@scientificpub.com** www.scientificpub.com

# 15

# MAKE YOUR WAY TO NODULES. EARLY EVENTS IN ACTINORHIZAL AND LEGUMES NITROGEN FIXING SYMBIOSES

Daphné Autran, Laurent Laplaze, Valérie Hocher, Florence Auguy, Mame Oureye-Sy, Mariana Obertello, Benjamin Péret, Claudine Franche and Didier Bogusz*

Rhizogenesis lab. IRD (Institut de Recherche pour le Développement). 911 Avenue Agropolis, BP 64501, 34394 Montpellier Cedex 5, FRANCE.
*corresponding author : bogusz@mpl.ird.fr

## INTRODUCTION

With the exception of water, nitrogen availability constitutes the main limiting factor of plant growth. The amounts of available nitrogen for plants in soils are relatively low. The atmosphere, being constituted of 78% of di-nitrogen, potentially represents an infinite source of nitrogen, but the only existing organisms able to exploit this ressource to grow are the diazotroph bacteria which possess the enzymatic complex nitrogenase. Some higher plants are able to associate with nitrogen fixing bacteria, creating within their roots a favorable environnement for the bacteria, in the form of a new organ : the nodule. The bacteria living in the nodules in turn provide fixed nitrogen to the host plant. Only two groups of higher plants form these so-called nitrogen fixing nodule symbiosis : plants of the Legumes familly – as well as one species member of *Ulmaceae* family, *Parasponia andersonii* - interact with bacteria of the family *Rhizobiaceae,* and actinorhizal plants establish symbiosis with bacteria of the genus *Frankia*.

Legumes, *Parasponia* and actinorhizal plants are flowering plants encompassing families all found within the Eurosid I clade of the Eudicots (Soltis *et al.,* 1995). Legumes are within the order Fabales and are represented by a single family, the *Fabaceae* (formerly the Leguminosae); however, more than 650 genera in the family contain species that can form rhizobial root nodules.

*Parasponia* is one of the 18 genera of the *Ulmaceae* (order Rosales), and is the only genus outside the legumes known to establish an $N_2$-fixing symbiosis with rhizobia (Akkermans *et al.*, 1978). In contrast, actinorhizal plants are more taxonomically diverse. They are simply defined as a group by their ability to be nodulated by *Frankia* bacteria. They represent approximately 200 species, encompassing 25 genera, in eight different families, in three different orders: the *Betulaceae, Casuarinaceae* and *Myricaceae* of the order Fagales; the *Rosaceae*, *Rhamnaceae* and *Elaeagnaceae* of the order Rosales; and the *Coriariaceae* and *Datiscaceae* of the order Cucurbitales (Vessey *et al.*, 2004; Gualtieri and Bisseling, 2000). Actinorhizal plants are mainly woody shrubs and trees and are predominantly found in temperate climes but also extending into the tropics, especially the *Casuarinaceae*. Only one specie, *Datisca glomerata*, is a herbaceous plant.

Like the wild legumes, actinorhizal plants have the capacity to grow in poorly fertile soils, and many are early successional plants, by their ability of colonizing poor soils and allowing growth of secondary species (Benson and Silvester, 1993). Actinorhizal plants play extremely important roles in the N cycle of forests and in the revegetation of various landscapes. Rates of $N_2$ fixation in the range of 90 kg N ha-1 year-1 are possible, as it has been estimated for *Coriria aborea* plantations in alluvial flood terraces in New Zealand over a 20 year period (Sprent and Parsons, 2000). For comparison, the amount of $N_2$ fixed by legumes can be in the range of 300-400 kg N ha-1 year-1 (Peoples et al., 2002). Actinorhizal plants have been used in erosion control, soil reclamation, agroforestry and dune stabilization, as well as in fuel wood production. *Casuarinaceae* are utilized in stabilizing desert and coastal dunes (i.e., in shelter belts), and in the reclamation of salt-affected soil (Diem and Dommergues, 1990), as well as in revegetation of limestone mines, for instance in Kenya with *Casuarina equisitifolia* plantations (Sprent and Parsons, 2000). Thus, the understanding of actinorhizal symbiosis is of great importance for the implementation of agronomy and agroforestery strategies, especially in tropical zones.

The properties of actinorhizal plants and wild Legumes as pionneer species are linked mainly with the ablity of their root to develop original organs, the nodules, containing the symbiotic N-fixing bacteria, in response to low nitrogen conditions. In both symbiosis, the interaction between micro- (bacteria) and macrosymbiont (plant) starts with a molecular dialogue at the root-bacteria interface which determine if the partners are compatible or not. If so, recognition of the bacteria trigger a number of differenciation responses in the plant root, in distinct tissues. The epidermis is the site of entry of the bacteria, which could occur either intracellularly *via* root hairs or intercellularly. Infection-threads of proliferating bacteria are formed and progress in a controled manner in the plant root. A cascade of signal transductionis induced in the epidermal cells, and lead to the reprogramming of inner root tissues, either cortex or pericycle, depending of the host plant. These cells will re-enter the mitotic cycle to form nodule primordia which will be colonized by the bacteria once the infection threads reach the primordium. Nitrogen fixation is then established by the bacteria hosted in the mature nodules.

Our knowledge on the establishment of nitrogen-fixing nodule symbiosis mainly rely on current numerous studies on Legumes-*Rhizobia* interaction. Two species have been choosen as models in Legumes, the temperate *Medicago trunctula* and the tropical specie *Lotus japonicus*. These plants are amenable for large scale molecular and genetics studies and provided important recent breakthroughs in the study of the mechanisms controling the establisment of the symbiosis with *Rhizobia* (Riely *et al.*, 2004). In contrast, actinorhizal symbioses are still poorly understood at the mechanistic level. So far, model plants have not been identified in actinorhizal plants. The majority of actinorhizal plants are trees or woody shrubs with long generation times, thus no genetic systems for the study for these symbioses are available. The only herbaceous actinorhizal specie, *Datisca*, have not been studied deeply so far because its specific microsymbiont strains are not determined (Vessey *et al.*, 2004). Molecular approaches have progressed during the last decade mainly for the temperate specie *Alnus glutinosa* and the tropical trees of the *Casuarinaceae* family (Laplaze *et al.*, 2003, Obertello *et al.*, 2003). For the latter, transgenics plants have been obtained for two species, *Casuarina glauca* and *Allocasuarina verticillata* (Franche *et al.*, 1998).

Thus, actinorhizal symbioses emerged recently as simple and original systems of root nodules initiation, and they offer the opportunity to explore developmental strategies significantly different from Legumes to form nitrogen fixing nodules. In this chapter, we summarize the current knowledge in Legumes - *Rhizobia* symbiosis, and present the recent approaches developped in actinorhizal plants. We focus on the early steps necesary to establish symbiosis, from the recognition mechanisms between rhizobacteria and their host plants, to the cellular and morphological events in different root tissues associated with nodule primordium formation. The study of later stages of symbiosis, *i.e.* the mechanisms of nodule functionning (metabolism, maintenance) and nodule senescence have been described recently in different reviews (see for instance, Vessey *et al.*, 2004; Pawloski and Sirrenberg, 2003) and are not adressed here.

## 2. INITIATION OF SYMBIOTIC INTERACTION : RECOGNITION DIALOGUES BETWEEN RHIZOBACTERIA AND THE HOST PLANT

To initiate the formation of the nitrogen-fixing nodule inside root tissues, the bacteria must be attracted by the root and recognized by their host plant. The importance of the determination of the correct host is shown by the complex mechanisms of host specificity that have been developped by both bacteria and plants during evolution, especially in the case of *Rhizobia*-Legumes symbiosis. As symbiosis has a high energetical cost for the plant which supply carbohydrates to the hosted bacteria, host specificity systems would allow the plant to maintain a selection of the « ideal » microsymbiont among the welth of bacteria present in the rhizosphere, ensuring efficient nitrogen fixation, and as welll minimizing the chances of pathogenic infections (Perret *et al.,* 2000).

### 2.1 Rhizobia-Legumes Symbiosis : NOD Factors Based- Host Specificity Mechanisms

Host specificity is a relative concept as, depending on the legume, a single nodule may contain several strains of *Rhizobia*, but in the majority of cases a nodule hosts a single strain of *Rhizobia* (Martinez-Romero, 2003).

#### *2.1.1 Induction of bacterial response at the root surface*

The establishment of N-fixing symbioses is initiated by the exudation of flavonoids by the host plant which acts as chemo-attractants and growth promoter to the bacteria. Flavonoids production is enhanced in N limiting nitrogen conditions (Coronado *et al.*, 1998), and many differents types of flavonoids (flavonols, flavones, flavanones, isoflavonoids...) are produced according to the Legumes specie considered (and presumably *Parasponia*), with different effects on

*Rhizobia*, therefore constituting a first level of host specificty determination (2000). Other substances such as nutrients (sugars, amino acids) and secondary metabolites have also been implicated in regulating cell surface interactions and growth of the bacteria (Miklashevichs *et al.*, 2001). However, if isoflavonoids are specific to Legumes, other non symbiotic plants are also source of flavonoids and nutrients, therefore the cocktail of compounds produced by the plant is not sufficient to ensure proper recognition.

### *2.1.2 NOD factors synthesis*

A second level of host specificity determination lies in the fact that flavonoids act as well as inducers of *Rhizobial* genes required for nodulation, the *nod* genes, via the activation of NodD proteins. NodD proteins respond to specific flavonoids, and act as transcriptional activators that bind the promoters of the nod genes (Perret *et al.,* 2000). *Nod* genes products catalyze the synthesis of the so-called NOD factors molecules involved in recognition of the bacteria by the host plant (see below).

NOD factors are lipo-chitooligosaccharides (LCOs) consisting of a core of ß-1-4 N-acetylglucosamine with terminal residues substituted with acyl chains. The number and nature of the substitutions (or decorations) are specific of each *Rhizobia* species (Geurts and Bisseling, 2002). The lenght of NOD factor backbone and the decorations have been shown to be crucial determinants of host specificity.as shown by *Rhizobia* mutants affected in the different genes encoding *nod* enzymes (Perret *et al.,* 2000 ). NOD factors level produced by the bacteria is limitant for host specificty, and different NOD factors could be produced by a given *Rhizobia* strain, and that NOD factor could cooperate with different molecules (chitin derivatives) to induce nodulation (Perret *et al.,* 2000), complicating again the puzzle of the recognition system.

### *2.1.3 NOD factors perception and signaling in epidermal cells*

Perception of Nod factor by the host legume results in numerous responses involved in infection and nodule formation : root hair cellular responses such as membrane depolarization and characteristic oscillations of intracellular calcium ($Ca^{++}$ spiking), associated with root hair deformation (swelling and curling) to entrap bacteria, development of pre-infection threads, cortical cell divisions, and induction of nodulation-specific genes expressed early in nodule development, the ENOD genes (Guerts and Bisseling, 2002).

As NOD factors elicit plant response with a high affinity and specificity, it has been postulated that they are perceived by the plant *via* receptors expressed in the root hair cells. Genetic analysis in crop and models legumes led recently to important breakthroughs in NOD factor perception and signaling (Figure 2, reviewed in Riely *et al.*, 2004). Notable results are the identification of new plant genes required for nodulation that encodes receptor-like kinases (RLKs) with extracellular LysM domains, known for their peptidoglucan binding properties in procaryotes. Mutants of the *NFR1* and *NFR5* RLKs genes identified in *Lotus japonicus* are impaired the earliest cellular response to NOD factor (membrane depolarization, occuring within seconds after NOD factor application) (Radutoiu *et al.*, 2003, Madsen *et al.*, 2003). Because *NFR1* and *NFR5* genes products are similar, as well as their mutants phenotypes, it has been suggested that they could directly interact to perceive NOD factors (Radutoiu *et al.*, 2003). A closely related gene, LYK3, was identified in *Medicago truncatula* (Limpens *et al.*, 2003) which could be involved as well in NOD factor perception or represent a NFR1 ortholog (Riely *et al.*, 2004, Parniske and Downie, 2003). It was shown that NFR1 and 5 genes act upstream of another distinct RLK

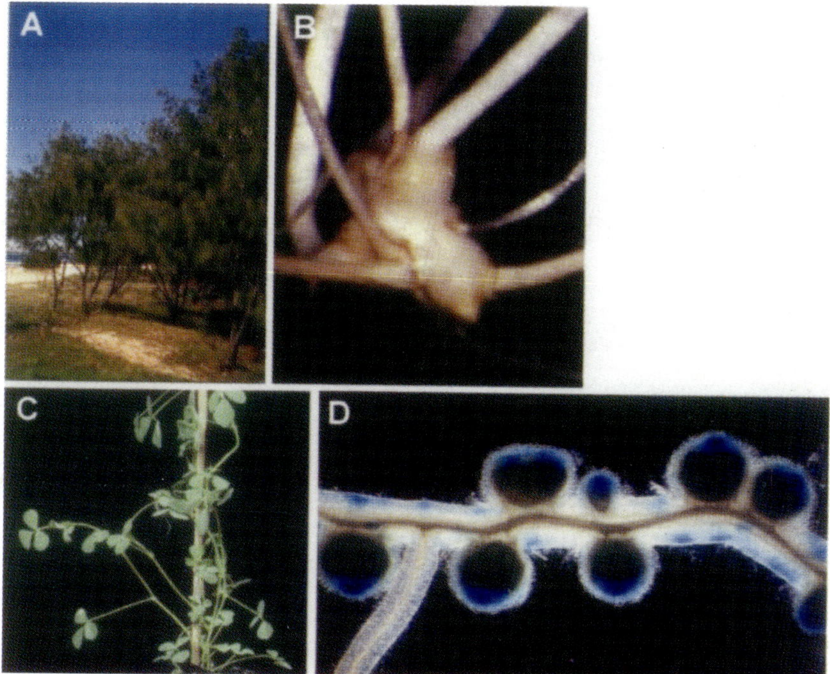


**Figure 1: Nitrogen fixing symbioses. Examples of actinorhizal plants and Legumes and nitrogen fixing root nodules.**

**A:** *Casuarina equisitifolia*. Tropical trees of the *Casuarinaceae* family are among the model species of actinorhizal plants. **B:** Nodules of *Casuarina glauca*. The typical multilobed structure and the nodular roots emerging from each nodular lobes are shown. **C:** *Medicago truncatula*, one of the model plants of the Legumes family. **D:** Nodules of *Medicago truncatula*. The nodules are unilobed and do not show nodular roots.

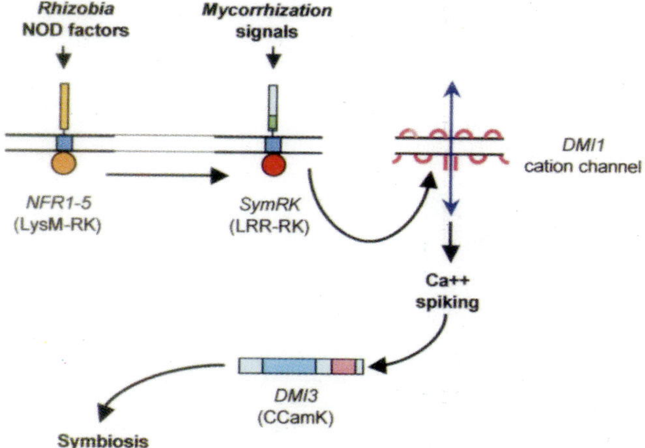


**Figure 2: NOD factor signaling pathway in Legumes.**

During *Rhizobia*-Legumes symbiotic interaction, bacterial NOD factors induce a signal transduction cascade in root epidermal cells. The *NFR1* and *NFR5* LysM receptors kinase (LysM-RK) are the putative initial point NOD factor perception. They act upstream of the *SymRK* leucine-rich-repeat receptor kinase (LRR-RK) and a cation channel encoded by the *DMI1* gene. Intracellular calcium oscillations (Ca++ spiking) are induced by NOD factors. Transduction of the calcium signal by the *DMI3* Calcium-Calmodulin kinase (CCamK) is required for the establishment of dowstream symbiotic responses.

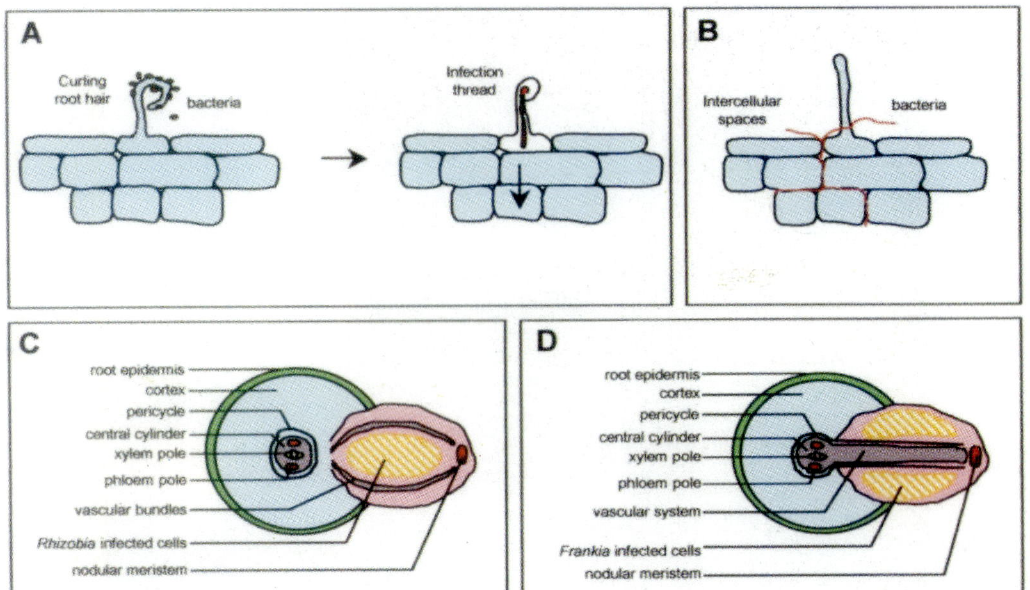


**Figure 3: Infection routes and nodules structures in Legumes and Actinorhizal plants.**
**A:** Intracellular infection pathway. **B:** Intercellular infection pathway. **C:** Structure of Legumes nodule. **D:** Structure of Actinorhizal plants nodule.

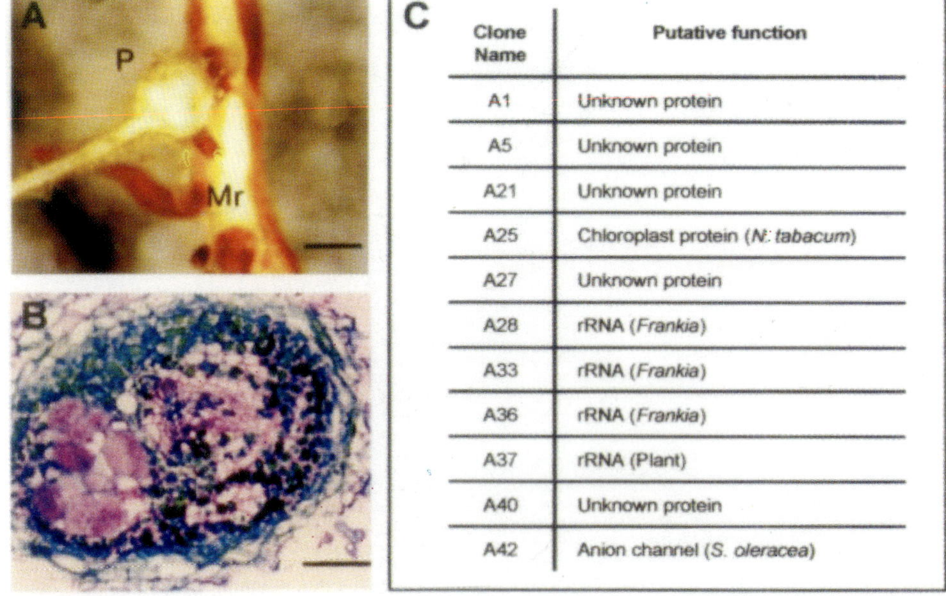


| Clone Name | Putative function |
|---|---|
| A1 | Unknown protein |
| A5 | Unknown protein |
| A21 | Unknown protein |
| A25 | Chloroplast protein (*N. tabacum*) |
| A27 | Unknown protein |
| A28 | rRNA (*Frankia*) |
| A33 | rRNA (*Frankia*) |
| A36 | rRNA (*Frankia*) |
| A37 | rRNA (Plant) |
| A40 | Unknown protein |
| A42 | Anion channel (*S. oleracea*) |

**Figure 4. The actinorhizal prenodule.**
**A:** Prenodule in the actinorhizal tree *Casuarina glauca*. **B:** Semi-thin section through a *C. glauca* prenodule. *Frankia* infected cells appear as large purple stained cells. **C:** Molecular analysis of prenodules : example of genes identified in *Alnus glutinosa* prenodules *via* a *Differential Display* RT-PCR approach. The sequence homologies found are shown. *P:*Prenodule. *Mr:* Main root.

gene, the *Lotus japonicus SYMRK* gene (also known as *NORK/DMI2/NN1008/SYM19* in different Legumes), which codes for a kinase with an extracellular leucine-rich repeat domain. SYMRK is not necessary for early membrane depolarization, but is required for downstream NOD factor responses such as calcium spiking, root hair curling and infection thread formation (Endre *et al.*, 2002; Stracke *et al.*, 2002). The same phenotypes are observed when *Medicago truncatula DMI1* gene is mutated. *DMI1* encodes a putative ligand-gated ion channel that could act closely to SYMRK to regulate ions efflux or calcium movements at the plasma membrane (Ané *et al.*, 2003). The *DMI3* gene, coding for a calcium- calmodulin-dependant kinase, is a potential negative regulator of calcium spiking response to NOD factor, as *dmi3* mutants showed an hypersentivity to NOD factor in this response (Lévy *et al.*, 2004 ; Mitra *et al.*, 2004). In addition to these importants regulators recently identified through genetics, transduction of the NOD factor signal in epidermal cells possibly involves as well other signaling components, such as small G-protein, phosphatidic acid and reactive oxygen species, as shown by pharmocological and cellular studies (reviewed in Riely *et al.*, 2004).

## 2.2 Specificity in Frankia-Actinorhizal Plants Interaction

Actinorhizal plants' partner, *Frankia*, is a filamentous, branching, Gram-positive actinomycete (Benson and Sylvester, 1993). The biology and genetics of *Frankia* are still at their infancy, mainly due to their slow growth rate, poor spore germination, and the subsequent difficulty to isolate pure culture of specific strains. As a consequence, important tools such as an efficient transformation system are still awaited. However, recent progress have been made, with the identification of *Frankia* mutants resistant to antibiotics, usefull in genetic approaches (Myers and Tisa, 2004) and, more importantly, with the sequencing of *Frankia* genome. Estimation of genome size is between 12000 and 8700 kbp according to the strains (Benson and Sylvester, 1993), and several plasmids ranging from 7kb to 190kb have been detected (Lavire et Cournoyer, 2003). To date, about 293 Kbp have been analyzed, which correspond to rRNA and tRNA genes, nitrogen fixation and ammonium assimilation genes, recombination genes and proteolyse genes (Lavire et Cournoyer, 2003). Recently, the complete sequence of three strains of *Frankia* was obtained an international consortium (www.web.uconn.edu/mcbstaff/benson/Frankia/FrankiaGenome.htm). Notably, so far, no DNA sequences homologous to the *Rhizobia nod* genes have been reported, suggesting that *Frankia* evolved a different host determination system than *Rhizobia* (Cérémonie *et al.*, 1998; Lavire and Cournoyer, 2003).

### 2.2.1 Phylogeny of Frankia and host-specificity groups in actinorhizal symbiosis

Two third of the *Frankia* DNA sequence data were generated to study the phylogeny of the genus. As many *Frankia* strains have not been purified and other is difficult to grow, it is difficult to evaluate the diversity of *Frankia* strains living in root nodules of actinorhizal plants. Several studies have use 16S RNA sequences to group strains for phylogenetic analysis (Normand *et al.*, 1996; Clawson and Benson, 1999). Four hetero-genous groups have been distinguished : *Frankia* strains of group I nodulate actinorhizal plants from the hamamelid families *Betulaceae*, *Casuarinaceae* and Myricaceae; group II include strains interacting with the families *Elaeagnaceae* and *Rhamnaceae* and some hamamelid genera; group III *Frankia* strains occupy nodules of the families *Coriariaceae*, *Datiscaceae* and from *Ceanothus* of the

*Rhamnaceae,* and group IV strains represent a loosely associated group of non nodulating actinomycetes. Isolates that represent groups I, II and IV have been obtained, but not from group III (Clawson and Benson, 1999).

These studies support the idea that actinorhizal symbiosis, like their *Rhizobia*-Legumes counterparts, is established in a host specific fashion. Nethertheless, several reports suggest that *Frankia*-actinorhizal plants interactions might be more heterogenous than *Rhizobia*-Legumes ones (Wessey *et al.*, 2004). The level of diversity of the strains hosted in the actinorhizal nodules appear to vary greatly from specie to specie, even within a same family. Early work by Benson and Hanna (1982) using protein profiling indicated that more than one genetically distinct *Frankia* were present within the same actinorhizal nodule. In the promiscuous specie, *Myrica pensylvanica*, *Frankia* from groups I, II, and IV were found in nodules (Clawson and Benson, 1999). In contrast, hybridization based on a 23S rRNA insertion target which allowed for the analysis of *Frankia* populations in nodules of various *Alnus* species revealed the presence of only one *Frankia* population in every nodule homogenate (Zepp *et al.*, 1997). Cross inoculation experiments in the greenhouse have shown for instance that group I of *Frankia* strains encompass two host-specificity groups (HSG) : the *Alnus-Comptonia* (*Betulaceae*) group and the *Casuarina* group (Baker, 1987). From greenhouse experiments, *Myricaceae* were considered has promiscuous hosts, i.e. tolerating strains from all *Frankia* HSGs, which could be explained by the fact that they are considered as the more ancient genus of actinorhizal plants. This suggested that actinorhizal symbiosis has evolved toward more specificity (Maggia and Bousquet, 1994). However, recent field experiments have shown that certain *Myrica* species (*Myrica gale*) are nodulated only by *Frankia* strains of the *Alnus* HSG in their natural habitat, while other are more promiscious (Clawson and Benson, 1999, Huguet *et al.*, 2005). Moreover, *Frankia* strains nodulating European *Myrica gale* were shown to be disctinct genetically from strains effective on American *Myrica gale*, suggesting that both symbionts have diverged in parallel (Huguet *et al.*, 2004). The same is true for the *Casuarinceae*, where species from the *Gymnostoma* gendus show different host specificities than species from the *Casuarina* and *Allocasuarina* genera (Navarro *et al.*, 1997). Taken together these findings suggest that host-specificity in actinorhizal plants is probably the result of processes of coevolution between the plant and the bacteria, more than a strict evolution toward specificity (Huguet *et al.*, 2005). The diversity of effective strains for each host plant may reflect the nature of their inhabitats : environments with high constraints (as exemplified by the acidic soils colonized by *Myrica gale* – Clawson and Benson, 1999) would increase selection pressure on microsymbiont populations, leading to a reduction of diversity. Another determinant for host specificity in actinorhizal symbiosis could be the mode of infection by the bacteria (see below), as the first two phyletic groups of *Frankia* strains I and II are characterized by different modes of infection, intracellular and intercellular, respectively (Normand *et al.*, 1996).

### 2.2.2 Regulations at bacteria-root interface in actinorhizal symbiosis

In actinorhizal plants, substances similar to Legumes inducing chemo-attraction and proliferation of *Frankia* in the rhizosphere have been identified. Flavonols (e.g., quercetin and kaempferol) can enhance the level of nodulation of Alnus *glutinosa* (Hughes *et al.*, 1999). Some microelements such as boron, which is required for *Frankia* growth especially in nitrogen free conditions, might be necessary for both partners to establish actinorhizal symbiosis as examplified in *Discaria Trinervis* (Bolanos *et al.*, 2002). It was also shown that *Frankia* produces plant hormones or hormones-like subtances that could play a role in the symbiotic interaction. *Frankia* produces cytokinins (Stevens and Berry, 1988), and pseudoactinorhiza (i.e. empty nodules) can be induced by cytokinins in some cases (Rodriguez-Barrueco and de Castro 1973; C. Santi, C. Franche and E. Duhoux, unpublished results), however there is no evidence that cytokinins synthesized by *Frankia* are required for the formation of functional nodules. Similarly, phenyl acetic acid (PAA), known as an auxinomimetic subtance; exhibiting the same effects on plant growth as indole acetic acid (IAA); is produced by *Frankia* strains, and exogenous PAA added to *Alnus glutinosa* roots resulted in the formation of short lateral roots which resembled actinorhizal nodules. It was suggested that PAA could be a way for the bacteria to limit the induction of defense response while contacting their host plant (Hammad *et al.*, 2003).

### *2.2.3 Looking for Frankia equivalent of NOD factors ?*

If the initial recognition dialogue between *Frankia* and their host plant seems to require many different molecules that are being discovered, even more elusive appear the mechanisms by which the microsymbiont is recognized by the plant and how bacterial signals are propragated to prepare symbiotic differenciation.

Albeit many attempts undertaken during the last decade, equivalents of *Rhizobia*l NOD factors have not been identified for the *Frankia* genus. Supernatants of *Frankia* culture were shown to induce root hair deformation (Van Ghelue *et al.*, 1997), however this response might not be specific to *Frankia* symbiosis (Knowlton *et al.*, 1980). Attempts to purify *Frankia* symbiotic factors using the protocol developed for *Rhizobia* NOD factors (Lerouge *et al.*, 1990) has failed, indicating chemical differences between both types of molecules (Cérémonie *et al.*, 1999). So far, functionnal complementations of *Rhizobia* mutants affected in *nod* genes with *Frankia* DNA have also been unsuccessfull (Cérémonie *et al.*, 1998), probably due to the fact that most *Frankia* promoters are not active in *Rhizobia* (Lavire and Cournoyer, 2003).

The current progress of *Frankia* genetics and the isolation of antibiotic resistant mutants (mentionned above) should allow the beginning of mutagenesis approaches in *Frankia* to screen for the bacterial factors required for symbiosis initiation.

A crucial tool in such approaches would be a reliable bioassay to assess the trigerring of early symbiotic plant response. Molecular markers would be the tool of choice to identify and follow successfull infection events, ie plant roots genes showing a differential expression upon symbiosis initiation. Such markers could be vizualized either by molecular quantitative techniques or spatially within plant tissues using reporter gene fusions. Both approaches are undertaken in our group using the actinorhizal tropical trees *Casuarina glauca* and *Allocasuarina verticillata*.

To identify genes differentially expressed during symbiosis with *Frankia*, cDNA libraries of *Casuarina glauca* have been constructed from uninoculated roots, roots 4 days after inoculation (early stages) and from mature nodules. Corresponding ESTs have been sequenced, constituting a database of about two thousands sequences and representing the largest genomic effort for actinorhizal plants (Hocher, Auguy and Bogusz, unpublished datas). Differential screening of SSH- generated early stages libraries allowed the identification of candidate *Casuarina glauca* genes potentially overespressed in the root in response to *Frankia* inoculation. Quantitative confirmation of the time course of expression of these candidates is underway and could provide usefull early markers of nodulation (Hocher, Auguy and Bogusz, unpublished results). Moreover these approaches woud allow, at the era of comparative genomics, comparisons with global transcriptional changes occuring at different steps of *Rhizobia*-Legumes symbiosis (Colebacth *et al.*, 2004, El Yayahoui *et al.*, 2004).

The development of genetic transformation techniques for the *Casuarinaceae* (Diouf *et al.*, 1995, Franche *et al.*, 1997, Franche *et al.*, 1998, Smouni *et al.*, 2000) opened the door to the use of reporter gene fusion in actinorhizal plants. Both endogenous *C. glauca* and heterologous Legumes promoters have been tested for early expression upon *Frankia* inoculation in *Casuarinaceae*.

The earliest endogenous actinorhizal gene reported so far is the subtilase Cg12 (Laplaze *et al.*, 2001, Svistoonoff *et al.*, 2003), homologue of *Alnus* Ag12 (Ribeiro *et al.*, 1995), which is specifically expressed in *Frankia* infected cells of the plant. Cg12 expression starts simutaneously with infection of root hairs cells, and later on continues in cortex and nodules cells (Svistoonoff *et al.*, 2003). However, Cg12 promoter is not yet active during the very early steps of plant-*Frankia* recognition, before bacteria entry.

The promoters of Legumes ENOD genes have been choosen as potential candidates for early symbiosis markers in the context of the actinorhizal plants *C.glauca* and *A.verticillata*. The gene ENOD40 is involved in nodule morphogenesis in *Medicago sativa* (Charon *et al.*, 1999) and induced within 1 hour after *Rhizobia* (or NOD factor) inoculation, in the pericyle and cortex cells (Campalans *et al.*, 2001). The CgENOD40 (for *Casuarina glauca* ENOD40) gene is unique in *C.glauca* genome and its promoter showed a non-symbiotic expression pattern similar to a Legume ENOD40 reporter fusion introduced in *Casuarinaceae*, however both promoters were not induced by *Frankia* and in mature actinorhizal nodules. In addition, CgENOD40 was not either induced by NOD factor when expressed in the Legume *M.truncatula* (Santi *et al.*, 2003). These results suggest that the element(s) of the NOD factor transduction pathway necessary for ENOD40 induction are absent in *Casuarinaceae* roots. Other ENOD promoter is currently under investigations in our group such as *Pisum sativum* ENOD12 promoter (Oureye Sy, Franche and Bogusz. unpublished results). Legumes ENOD12 genes are expressed in root epidermis, within 2-3 hours upon either NOD factor or Rhizobia application (Pichon *et al.*, 1992; Jourmet *et al.*, 1994; .Vijn *et al.*, 1995). Moreover, the induction of ENOD12 transcription is dependent on the same host-specificty determinants that govern the Legumes-Rhizobia interaction, as it was shown in *M. truncatula*, where ENOD12 expression is not activated by a *nodH Rhizobia* mutant which synthetizes non sulfated NOD factors, ineffective in this host specie (Jourmet *et al.*, 1994). The results of ENOD12 activation in *Casuarinaceae* will allow exploring further the possibility of conservation of early Legumes response in actinorhizal plants. Another homolog of Legumes ENOD genes has been reported in an actinorhizal plant, the dg93 gene common to soybean nodules and *Datisca glomerata*; however its early symbiotic expression has not been examined (Okubara *et al.*, 2000). Dg93 represents therefore another interesting candidate to be tested.

### 2.2.4 Reverse genetics to explore early actinorhizal symbiosis signaling

As mentionned earlier, because the majority of actinorhizal plants display long generation times, no genetic systems for the study for these symbioses have been established. The only actinorhizal plant identified so far that would offer the potential for genetic analysis is *Datisca glomerata*, which is a herbaceous specie with a generation time of 6 months and a small genome (Pawlowski and Bisseling, 1996).

However the isolation of pure culture of compatible *Frankia* strains have been shown to be highly difficult, impairing further development of the study of *Datisca-Frankia* interaction. Therefore, possibilities of functional studies of early symbiosis genes in actinorhizal plants currently rely mostly on reverse genetics approaches. Transformation techniques are available in two *Casuarinaceae* species either for the generation of whole transgenic plants (Franche *et al.*, 1997, Smouni *et al.*, 2000), or composite plants where only the root is transformed thanks to the properties of the bacteria *Agrobacterium rhizogenes* (Diouf *et al.*, 1995). The expression pattern of a 35S-GUS reporter gene is similar in the transgenic roots from both systems, and *A.rhizogenes* generated roots can be efficiently nodulated (Diouf *et al.*, 1995; Franche *et al.*, 1998). This later techniques allow rapid generation of transgenics roots and can possibly be applied to other actinorhizal plants species, in the same way it is more and more used in different Legumes species (Endre *et al.*, 2002 ; Limpens *et al.*, 2003; 2004, Boisson-Dernier *et al.*, 2003). Coupled to targeted inactivation strategies such as RNA interference and misexpression experiments, transgenic approaches should allow the examination of the consequences of loss or gain of function of candidate genes on actinorhizal nodules formation.

A first class of interesting candidate genes for functional analysis are those identified in differential expression screens at early stages of actinorhizal symbiosis. Interestingly, in our screen in *C. glauca* at 4 days after *Frankia* inoculation, homologues of Legumes NOD factors signaling regulators were encountered; suggesting shared molecular components in both systems. For instance, *C.glauca* transcripts showing significant homologies to NFR1/LYK3 Legumes LysM receptor kinase are expressed at stage 4 days after inoculation (Auguy and Bogusz, unpublished results). Although the specific induction of this NFR1-like gene upon inoculation still must be confirmed in *C.glauca*, because NFR1 is not induced by NOD factor (or Rhizobia) in *M.truncatula* but rather expressed at the same level in uninoculated and inoculated roots (Radutoiu *et al.*, 2003), it could represents an interesting difference in the regulation of signaling genes between the two types of symbioses.

A second class of candidate genes are therefore homologues of symbiotic regulators, identified by genetic analysis in models Legumes. We identified a SYMRK homologue in *C. glauca* roots, called CgSYMRK, showing 72% overall similarity with the Legumes SYMRK (Autran, Giczey, Parniske and Bogusz, unpublished results). All the characteristic domains found in Legumes SYMRK receptor kinase are also present in CgSYMRK, suggesting that these genes could have similar function. To test this hypothesis, inactivation of the *CgSYMRK* gene by RNAi in *Casuarinaceae* using the *A. rhizogenes* root transformation system and heterologous complementation of *Lotus japonicus symrk* mutants are currently underway. This strategy, coupled to the study of the regulation of *CgSYMRK,* should provide usefull informations on the convergences and/or diverging points in the symbiotic signaling pathways triggered in the host plants by *Rhizobia* and *Frankia*.

## 3. CONTROL OF INFECTION AND NODULE PRIMORDIUM FORMATION : REPROGRAMMING OF INNER ROOT TISSUES

The signaling cascades induced by bacteria in the root epidermis participate in the entry of the symbiotic bacteria and are the inducers of inner cells responses. Bacteria can enter root cells either intracellularly *via* root hairs or intercellularly *via* epidermis « craks » or intercellular spaces, and both routes of infections are found in *Rhizobia*-Legumes and actinorhizal symbiosis (Figure 3. A,B). Simultaneously to bacteria entry, divisions are reactivated in the cortex, which form nodule primordium in Legumes and an original structure, the prenodule, in actinorhizal plants where nodule primordium will formed from pericycle cells. Infection threads progress through root tissues until they reach the nodule primodium (or the prenodule). All infection and nodule primordium steps are tighly controled by the host plant and although a great diversity of modes of infection and nodule structure exist in nitrogen fixing symbioses (reviewed in Vessey *et al.*, 2004), common mechanisms emerged, with the notable characteristic of the recruitment or « hijacking » of existing non-symbiotic developmental programs to built the specialized symbiotic organs (Szcyglowski and Amyot, 2003).

### 3.1 Infection Routes : Microsymbionts on the Road

Intracellular infection begins when bacteria enter the root *via* curled root hairs (Figure 3A), creating a localized hydrolysis of the cell wall in the pocket curl, while the plasma membrane invaginates and a new cell wall is formed around the bacteria. Infection threads growth within plant cells is very similar to the pollen tube growth within ovules. Indeed, the two kind of structures might be controled in a similar way by the plant, since the same cell-wall modifying enzymes are expressed in *Medicago* infection threads and pollen tubes (Rodriguez-Llorente, 2004). Construction of infection thread by the plant involves as well extensive cytoskeleton activity, the synthesis of which is induced by bacterial NOD factor (Lhuissier *et al.*, 2001).

The mechanisms by which the host plant controls infection thread propagation and potential release of the bacteria in cortex cells (which occur in most of the symbiotic Legumes species to form the symbiosome) depend again on interactions with bacterial factors, as in the process of NOD factor host-specificity control. Secreted bacterial proteins co-regulated with NOD factor by the NodD gene, collectively called Nops proteins (Marie *et al.*, 2004), have been proposed has a second « code » allowing the bacteria to stimulate infection propagation and symbiosome formation in the nodule primordium. *Rhizobia* secrete different types of polysaccharides that are important to « inform » the plant to proceed with the symbiotic interaction and develop a functionnal nodule, as it was shown recently for lipolysaccharides of the microsymbiont *Azorhizobium caulinodans* in the tropical Legume *Sesbania rostrata* (Mathis *et al.*, 2005). In addition, receptors involved in the epidermis NOD factor transduction pathway could play a second role (maybe with different affinities for their ligands) in infection thread and symbiosome formation (Parniske and Downie, 2003). Indeed, incomplete inactivation of either SYMRK or LYK3 receptors induce an arrest of the nodulation process but at a later stage than null mutants (it has to be noted that LYK3 knock out were not published, but probably correspond to *nrf1* mutants – Limpens *et al.*, 2003; Parniske and Downie, 2003). In such incomplete inactivation situation, infection threads are formed but arrest prematurly and absence of bacterial release in the primordium nodule is observed (Holsters *et al.*, 2004). Consistenly with this hypothesis, SYMRK orthologue of *M.truncatula*, DMI2, is expressed in root epidermis and in nodule primordium (Besoult *et al.*, 2004). This level of control may as well involve bacterial ligands. It is therefore proposed that the infection of the root cells by *Rhizobia* involve « multiple perceptions » systems (Ardourel, 1994; Limpens and Bisseling, 2003, Parniske and Downie, 2003).

In actinorrhizal plants, intracellular infection takes place in the *Betulaceae*, *Casuarinacae* and *Myricaceae*, and proceeds through similar steps as in Legumes. However, *Frankia* hyphae are embedded directly within a cell wall-like material or 'interfacial matrix' (Berg, 1999), which is the equivalent of the infection thread wall in Legume nodules, and there is no space between infection thread wall and bacteria as in Legumes root. The actinorhizal subtilisin-like Cg12/Ag12 could be involved in the control of the infection process, as they are expressed in all *Frankia* infected cells, before the onset of nitrogen fixation (Laplaze *et al.*, 2000a, Svistoonoff *et al.*, 2003). These proteases could act on cell wall remodeling and allow infection thread to pass from one cell to another, or be part of a signaling cascade in response to *Frankia* infection (Svistoonoff *et al.*, 2003). A similar expression pattern has been described for agNt84/ag164, two genes that were isolated from *Alnus glutinosa* (Pawlowski *et al.*, 1997). These genes encode respectively for a glycine and histine rich proteins, with extracellular targeting signals which presumably positionned the proteins at the interface between the two symbionts. AgNt84 produced in *E. coli* was shown to have nickel binding properties, suggesting a role in metal ions signal regulation (Pawlowski *et al.*, 1997).

The intercellular infection route is believed to be more primitive because less control could be exerted by the plant on the spread of infection when bacteria progress exclusively through the apoplast (Figure 3.B, Pawlowski and Bisseling, 1996). Indeed, in some legumes infected *via* the intercellular pathway, pathogen-like behaviour are observed for the symbiotic *Rhizobia* which provoque the collapse of cortex cells, forming an « infection strand » to reach the nodule primordium (Alazard and Duhoux, 1990). In Legumes, intercellular entry of *Rhizobia* is observed in a few species that form stem-nodules as *Sesbania* and *Aeschymomene*, and take place at the sites of lateral or adventitious roots, taking adavantage of gaps in the epidermis (Pawlowski and Bisseling, 1996). Moreover intercellular infection could represent an alternative route to root hair-based bacteria entry, as in certain genetic context such as mutants impaired in root hair formation; nodule formation is still possible and occurs *via* intercellular infection (Karas *et al.*, 2005). In the actinorhizal group, intercellular infection is more widespread, and takes place in host plants of the *Rhamnaceae, Elaeagnaceae* and *Rosaceae* families. *Frankia* hyphae enter the root between epidermal cells as, in contrast to *Rhizobia*, *Frankia* does not depend on gaps in the root epidermis for entering the root (Valverde and Wall, 1999, Racette and Torrey, 1989). In host plants of the actinorhizal Cucurbitales (*Datisca* and *Coriaria*), the infection mechanism has not been examined yet, but since no prenodules or infection threads are found in these plants, infection is assumed to follow the intercellular pathway (Vessey *et al.*, 2004).

### 3.2 Formation of Nodule Primordia in Legumes and Actinorhizal Plants Root Tissues

Upon infection, *Rhizobia* and *Frankia* induce the formation of the new symbiotic organs, the nodules. The morphology and ontology of nodules vary according to the host plant (Figure 3).

#### *3.2.1 Mitotic reactivation of cortex cells and nodules positionning in Legumes*

For Legumes nodule primordium formation, synchronous with infection thread development, starts the mitotic reactivation of the cortical cells, induced by the NOD factor pathway. In determinate nodules (lacking a meristem) mostly found in tropical and subtropical legumes, divisions are induced in the outer cortex. In contrast, in temperate Legumes producing indeterminate nodules (showing a meristem responsible for continuous growth of the nodule), reactivated cells to form a primordium are located in the inner cortex. In either cases, nodule primordia are initiated opposite to protoxylem poles. Many of the early nodulin genes (ENOD) are associated with induction of cortical cell divisions, and cell-wall modifications. Cell cycle genes are as well re-expressed upon NOD factor activation of cortex cells (Foucher and Kondorosi, 2003). *Lotus japonicus nin* mutants are blocked in the mitotic activation of the cortex, whereas several epidermal responses are not affected. The NIN gene, which was the first Legume gene cloned using gene tagging, encodes a putative leucine-zipper transcrition factor (Schauser *et al.*, 1999; Catoira *et al.*, 2000). Nodule primordia, like lateral roots, are initiated opposite to protoxylem poles. This suggests that the stele provide positional informations controling the cells to be reactivated to form nodule primordia. Ethylene is likely to be produced by the stele cells opposite the phloem pole which highly express ethylene synthesis enzymes, and because ethylene insensitive mutant *sickle* (*skl*) forms more nodule primordia abberantly positioned at the phloem pole, it is proposed that ethylene act as repressor of phloem pole cortical cells activation (Heidstra *et al.,* 1997; Guert and Bisseling, 2002). The cytokinin/auxin balance in the inner root tissues also probably acts to determine mitotic reactivation and where nodule primordia are formed (Crespi and Galvez, 2003).

### 3.2.2 « Root-like » nodules of actinorhizal plants

The actinorhizal nodules are characterized by a « root-like » topology (with central vascular bundles) differing from the « stem-like » topology (with peripheral vascular systems) of legume nodules (Figure 3. C, D). Moreover, the actinorhizal nodules develop from primordia that are induced *de novo* in the pericycle, like lateral roots, and not from cortex cells. Thus, actinorhizal nodule primordia formation may involve common mechanisms to the lateral root development programm. In this regard, the growing number of genes shown to be important for lateral root development in the model plant *Arabidopsis* (Casimiro *et al.*, 2003) represent interesting candidate mechanisms to elucidate the specificty of actinorhizal nodules development. In our group, comparison of the role of auxin transport in lateral root development (in *Arabidopsis*) and in actinorhizal nodules of the *Frankia* is underway (Peret and Laplaze, unpublished results). *Arabidopsis* is also used to study the regulation and function of *Casuarina* genes potentially involved in the pathway.

Interestingly, a "root-like" topology is also characteristic of the nodules of the only non Legume *Rhizobia* symbiont, *Parasponia*. The nodules are formed from the pericycle cells, and *Rhizobia* remain located within infection threads as in actinorhizal symbioses, and do not undergo endocytosis to form symbiosome as in Legumes (Hadri *et al.*, 1998). Thus, *Parasponia* nodule development and structure in some ways can be seen as an intermediate between legume and actinorhizal nodules, but it also show unique features (reviewed in Vessey *et al.*, 2004). Despite early observations that inoculation affected root hair growth and morphology, it is shown that Parasponia *Rhizobia* do not infect the plant through root hairs (Becking, 1992). Instead, *Rhizobia* erode the root epidermis below an area of bacterial colonization on the root surface. The subsurface swelling zone is refered as a prenodule, comparable with the prenodule development that occurs in actinorhizal plants. However, Becking (1992) argues that these swellings are not always observed and are not morphogenetically comparable to prenodules of actinorhizal plants (see below). The key example of *Parasponia* nodules suggest that even if *Rhizobia* and *Frankia* induced nodules appear distinct in their organisation and ontogeny, they have evolved from common developmental programs. Other examples of flexibility in the type of nodule formed include some papilionoids (*Arachis*, *Stylosanthe*, and *Aeschynomeneae*), where nodules emerge not from primordia induced *de novo* but directly from lateral root primordia. In these simplified nodules, infection threads are not formed and bacteria are endocytosed from intercellular spaces (Provorov *et al.*, 2002).

### *3.2.3 Role of actinorhizal prenodule*

As mentionned above, another important feature of actinorhizal nodules primordia formation (probably shared with *Parasponia*) is the induction of cortical cell divisions giving rise to a small protuberance, called the prenodule (Figure 4, A,B). This structure is induced concomitantly to root-hair infection, and the infection thread grows to the dividing cortical cells, and infects some of them by branching of *Frankia* hyphae within the cells (Wall and Berry, 2000). The prenodule is an obligatory steps of intracellular infection but is not directly involved in nodule formation. Depending on the host plant species, more than one nodule primordium can be formed *per* prenodule (Torrey and Callaham, 1979).

Prenodule function has been adressed in *C.glauca* by investigating the differenciation of *Frankia* and plant cells in the prenodule (Laplaze *et al.*, 2000b). It was shown that *Frankia* can fix nitrogen within the prenodule as demonstrated by the expression of the nitrogenase *nifH* gene. Moreover, prenodule cells differentiate to allow nitrogen fixation, starting to express a symbiotic hemoglobin gene and by intensive cell wall lignification. The prenodule also contain non infected cells that accumulate starch, in the same way as uninfected cells in nodules. Taken together, these results suggest that prenodule ressemble a simple version of nodules. It might be a rest of the evolution of nitrogen-fixing endosymbioses in plants (Laplaze *et al.*, 2000b).

To explore further the function of prenodules, differential screening experiments have been performed in the actinorhizal system *Alnus glutinosa* (Laplaze and Bogusz, unpublished results). Transcripts differentially expressed in root containing prenodules as compared to uninoculated roots were identified by the sensitive technique DDRT-PCR. This technique was shown to be usefull in the characterization of plant-microorganisms interactions (Taylor and Harrier, 2000). The expressions of about 5200 genes have been analyzed, and 42 of these displayed prenodule specific expression. The corresponding fragments were cloned and sequenced. Among the potential prenodule sequences identified, most do not show any homology to known sequences in the databases and could represent novel actinorhizal genes (Figure 4C). Interestingly, a sequence showing homologies to chloroplastic tobacco sequences was identified in this prenodule screen, which is not expressed in uninoculated roots and neither in mature nodules (Laplaze and Bogusz, unpublished results). Recently, two highly homologous genes from *Lotus japonicus,* CASTOR and POLLUX, that are indispensable for *Rhizobia* infection and act upstream of intracellular calcium spiking, have been shown to be localized in the plastids of root cells, indicating a previously unrecognized role of this ancient endosymbiont in controlling intracellular symbioses that evolved more recently (Imaizumi-Anraku *et al.*, 2005). Therefore, it is possible that plastidial protein is also involved in bacteria-plant cells interaction in the prenodule. Nethertheless, once the prelimary data from this approach being comfirmed by quantitative expression analysis during nodule development, and subsequent functional analysis, they could represent usefull molecular markers for early inner root cells infection by *Frankia*, and provide insight for the elucidation of the molecular mechanisms governing prenodule development.

In actinorhizal species infected by the intercellular route, no prenodule is observed. During the colonization of the cortex, the root cortical cells secrete an electron-dense pectin- and protein-rich material into the intercellular spaces, and the formation of a nodule primordium is induced in the root pericycle. *Frankia* hyphae infect directly primordium cells from the apoplast (Valverde and Wall, 1999).

## Conclusions

In this review we adressed how much we know and the approaches developped in the initiation of the symbiotic interaction in actinorhizal plants, and compared these original symbiotic systems with the well known Legumes-*Rhizobia* symbiosis.

In the last decade, molecular mechanisms of actinorhizal symbiosis have started to unravel. Genomic approaches as well as reverse genetics strategies allowed by the progress in transformation systems in actinorhizal plant species will greatly enhance functional studies. Implementation of *Frankia* genetic systems and elucidation of *Frankia* genomic sequences in different strains are awaited to explore further the specificities of actinorhizal symbiotic interactions.

Many of the NOD factor signaling regulators identified in Legumes are as well necessary for the establisment of fungal endosymbiosis. With a few exceptions like the *nfr1* and *nfr5* mutants which do form arbuscular mychoriza and represent therefore a NOD factor specific pathway, most *nod-* mutants (impaired in nodulation) identified to date show also a *myc-* phenotype (impaired in mycorhization). This indicates a partial overlapping between the regulatory circuits controling the two type of symbiosis, and suggests that bacterial nitrogen fixing nodules symbiosis have recruited pre-existing cellular mechanisms established by fungal symbiosis which appeared very early during evolution (Kirstner and Parniske, 2002; Szcyglowski and Amyot, 2003). The comparison with other symbiotic systems is thus of crucial importance to test this hypothesis. In actinorrhizal plants, the promoter of subtilase Cg12, expressed after *Frankia* entry in root cells in *C.glauca*, is as well activated in Legumes upon *Rhizobia* infection, but does not respond to fungal signals (Svistoonoff *et al.,* 2004). These results suggest that the transcriptional activation pathways triggered by bacterial infection are at least partially conserved between *Rhizobia* and *Frankia* induced symbioses, and confirmed that specific bacterial symbiotic pathways exist. If the majority of the flowering plants can form symbioses with fungi, only species from the Eurosid I class of the dicotyledons are able to interact with nitrogen-fixing bacteria, in the form of root nodules. Legumes and actinorhizal plants could share a common ancestor characterized by a « predisposition » to form symbiotic nodules. Interestingly, actinorhizal symbioses show features related to « infection thread symbiosis », a hypothethical missing link between arbuscular mycorrhiza and legumes root nodules symbiosis, notably with the absence of bacterial release in the nodules cells (Kistner and Parniske, 2002). Therefore, the elucidation of actinorhizal symbiosis mechanisms could shade light on the evolution of endosymbioses.

## Aknowledgements

We thank Dr D. Barker'group (LIPM, INRA-CNRS, Castanet-Tolosan, France) for the kind gift of *Medicago truncatula* pictures in Figure 1. We apologize to our colleagues whose work we could not cite in this chapter. Our work on actinorhizal symbiosis is funded by IRD (Institut de Recherche pour le Développement).

## REFERENCES


Akkermans ADL, Abdulkadir S and Trinick MJ.1978. N2-fixing root nodules in *Ulmaceae*: *Parasponia* or (and) *Trema* spp? *Plant Soil,* **49** :711-715.

Alazard D and Duhoux E. 1990. Development of stem nodules in a tropical forage legume, *Aeschynomene afraspera. J.Exp.Bot.*, **41**:1199-1206.

Ane JM, Kiss GB, Riely BK, Penmetsa RV, Oldroyd GE, Ayax C, Levy J, Debelle F, Baek JM, Kalo P, Rosenberg C, Roe BA, Long SR, Denarie J and Cook DR. 2004. *Medicago truncatula DMI1* required for bacterial and fungal symbioses in legumes. *Science,* **27**:1364-7.

Ardourel M, Demont N, Debelle F, Maillet F, de Billy F, Prome JC, Denarie J. and Truchet G. 1994. *Rhizobium meliloti* lipooligosaccharide nodulation factors: different structural requirements for bacterial entry into target root hair cells and induction of plant symbiotic developmental responses. *Plant Cell*, **6**:1357-74.

Baker DD and Mullin BC. 1992. Actinorhizal symbiosis. *In* Biological Nitrogen Fixation. Eds. G Stacey, H Evans and R Burris. pp. 259–292. Chapman and Hall, New York.

Becking J H 1992. The *Rhizobium* symbiosis of the nonlegume *Parasponia*. *In* Biological Nitrogen Fixation: Achievements and Objectives. Eds. G. Stacey, R H Burris and H J Evans. pp. 497–559. Chapman and Hall, New York.

Benson DR and Silvester WB 1993. Biology of *Frankia* strains, actinomycete symbionts of actinorhizal plants. *Microbiol Rev.*, **57**:293-319

Berg RH. 1999. *Frankia* forms infection threads. *Can. J. Bot.*, **77** :1327–1333.

Besoult A, Camut S. and Cullimore, JV. 2004. Characterization and regulation of the expression of the DMI2 gene in the model legume *Medicago truncatula. 6th conference on European Nitrogen Fixing Symbiosis.* Toulouse, France.

Bolanos L, Redondo-Nieto M, Bonilla I and Wall LG. 2002. Boron requirement in the *Discaria trinervis* (*Rhamnaceae*) and *Frankia* symbiotic relationship. Its essentiality for *Frankia* BCU110501 growth and nitrogen fixation. *Physiol Plant.*, **115**:563-570.

Callaham D and Torrey JG. 1977. Prenodule formation and primary nodule development in roots of *Comptonia* (*Myricaceae*). *Can J Bot.*, **51**:2306-2318.

Ceremonie H, Cournoyer B, Maillet F, Normand P and Fernandez MP. 1998. Genetic complementation of rhizobial *nod* mutants with *Frankia* DNA: artifact or reality? *Mol Gen Genet.*,**260**:115-119.

Ceremonie H, Debelle F and Fernandez MP. 1999. Structural and functional comparison of *Frankia* root hair deforming factor and rhizobia Nod factor. *Can J.Bot.,* **77**:1293-1301.

Clawson ML and Benson DR. 1999. Natural diversity of *Frankia* strains in actinorhizal root nodules from promiscuous hosts in the family *Myricaceae*. *Appl Environ Microbiol.*, **65**:4521-4527.

Colebatch G, Desbrosses G, Ott T, Krusell L, Montanari O, Kloska S, Kopka J and Udvardi MK. 2004. Global changes in transcription orchestrate metabolic differentiation during symbiotic nitrogen fixation in *Lotus japonicus*. *Plant J.,* **39**:487-512.

Coronado C, Zuanazzi J, Sallaud C, Quirion JC, Esnault R, Husson HP, Kondorosi A and Ratet P. 1998. Alfalfa root flavonoid production is nitrogen regulated. *Plant Physiol.*, **108** :533-542.

Diouf D, Gherbi H, Prin Y, Franche C, Duhoux E and Bogusz D. 1995. Hairy root nodulation of *Casuarina glauca*: a system for the study of symbiotic gene expression in an actinorhizal tree. *Mol Plant Microbe Interact.,* **8**:532-537.

Downie JA and Parniske M. 2002. Plant biology: fixation with regulation. *Nature,* **28**:369-370.

El Yahyaoui F, Kuster H, Ben Amor B, Hohnjec N, Puhler A, Becker A, Gouzy J, Vernie T, Gough C, Niebel A, Godiard L and Gamas P. 2004. Expression profiling in *Medicago truncatula* identifies more than 750 genes differentially expressed during nodulation, including many potential regulators of the symbiotic program. *Plant Physiol.*, **136**: 3159-3176.

Endre G, Kereszt A, Kevei Z, Mihacea S, Kalo P and Kiss GB. 2002. A receptor kinase gene regulating symbiotic nodule development. *Nature*, **27**:962-966.

Franche C, Laplaze L, Duhoux E and Bogusz D. 1998. Actinorhizal symbioses: Recent advances in plant molecular and genetic transformation studies. *Crit. Rev. Plant Sci.,* **17** :1-28.

Foucher F and Kondorosi E. 2000. Cell cycle regulation in the course of nodule organogenesis in *Medicago. Plant Mol Biol.*, **43**:773-786.

Geurts R and Bisseling T. 2002. *Rhizobium* nod factor perception and signalling. *Plant Cell*,**14** S :239-249.

Gualtieri G and Bisseling T. 2000. The evolution of nodulation. *Plant Mol. Biol.,* **42** :181-194.

Hammad Y, Marechal J, Cournoyer B, Normand P and Domenach AM. 2001. Modification of the protein expression pattern induced in the nitrogen-fixing actinomycete *Frankia* sp. strain ACN14a-tsr by root exudates of its symbiotic host *Alnus glutinosa* and cloning of the *sodF* gene. *Can J Microbiol.*, **47**:541-547.

Hughes M, Donnelly C, Crozier A and Wheeler CT. 1999. Effects of the exposure of roots of *Alnus glutinosa* to light on flavonoids and nodulation. *Can. J. Bot.,* **77** :1311–1315.

Huguet V, Mergeay M, Cervantes E and Fernandez MP. 2004. Diversity of *Frankia* strains associated to *Myrica gale* in Western Europe: impact of host plant (*Myrica* vs. *Alnus*) and of edaphic factors. *Environ Microbiol.,* **6**:1032-1041.

Huguet V, Gouy M, Normand P, Zimpfer JF and Fernandez MP. 2005. Molecular phylogeny of *Myricaceae*: a reexamination of host-symbiont specificity. *Mol. Phylogenet. Evol.,* **34**:557-568

Kistner C, Parniske M. 2002. Evolution of signal transduction in intracellular symbiosis. *Trends Plant Sci.,* **7**:511-518.

Knowlton S, Berry A and Torrey JG. 1980. Evidence that associated soil bacteria may influence root hair infection of actinorhizal plants by Frankia. Can. J. Microbiol., **26:** 971-977.

Karas B, Murray J, Gorzelak M, Smith A, Sato S, Tabata S and Szczyglowski K. 2005. Invasion of *Lotus japonicus root hairless 1* by *Mesorhizobium loti* involves the nodulation factor-dependent induction of root hairs. *Plant Physiol.,* **137**:1331-1344.

Laplaze L, Ribeiro A, Franche C, Duhoux E, uy F, Bogusz D and Pawlowski K. 2000. Characterization of a *Casuarina glauca* nodule-specific subtilisin-like protease gene, a homolog of *Alnus glutinosa* ag12. *Mol Plant Microbe Interact.*, **13**:113-117.

Laplaze L, Duhoux E, Franche C, Frutz T, Svistoonoff S, Bisseling T, Bogusz D and Pawlowski K. 2000. *Casuarina* glauca prenodule cells display the same differentiation as the corresponding nodule cells. *Mol Plant Microbe Interact.*, **13**:107-112.

Laplaze L., Svistoonoff S., Santi C., Auguy F., Franche C and Bogusz D. 2003. Molecular biology of actinorhizal symbioses. In "Nitrogen fixation research: origins and progress" Vol. VI: *Actinorhizal symbioses.* W.E. Newton (ed). Kluwer Academic Publishers. Norwell.

Cournoyer B and Lavire C. 1999. Analysis of *Frankia* evolutionary radiation using *glnII* sequences. *FEMS Microbiol Lett.*, **1**:29-34.

Lerouge P, Roche P, Faucher C, Maillet F, Truchet G, Prome JC and Denarie J.1990. Symbiotic host-specificity of *Rhizobium meliloti* is determined by a sulphated and acylated glucosamine oligosaccharide signal. *Nature,* **19**:781-784.

Levy J, Bres C, Geurts R, Chalhoub B, Kulikova O, Duc G, Journet EP, Ane JM, Lauber E, Bisseling T, Denarie J, Rosenberg C and Debelle F. 2004. A putative Ca2+ and calmodulin-dependent protein kinase required for bacterial and fungal symbioses. *Science,* **27**:1361-1364.

Limpens E, Franken C, Smit P, Willemse J, Bisseling T and Geurts R. 2003. LysM domain receptor kinases regulating rhizobial Nod factor-induced infection. Science, **24:**630-633.

Limpens E and Bisseling T. 2003. Signaling in symbiosis. *Curr. Opin. Plant Biol.,* **6**:343-650.

Limpens E, Ramos J, Franken C, Raz V, Compaan B, Franssen H, Bisseling T and Geurts R. 2004. RNA interference in *Agrobacterium rhizogenes*-transformed roots of *Arabidopsis* and *Medicago truncatula*. *J. Exp. Bot.,* **55**:983-992.

Madsen EB, Madsen LH, Radutoiu S, Olbryt M, Rakwalska M, Szczyglowski K, Sato S, Kaneko T, Tabata S, Sandal N and Stougaard J. 2003. A receptor kinase gene of the LysM type is involved in legume perception of rhizobial signals. *Nature,* **9**:637-640.

Maggia L, and Bousquet J. 1994. Molecular phylogeny of the actinorhizal *Hamamelidae* and host specificity relationship towards *Frankia. Mol. Ecol.*, **3**:459-457.

Mathis R, Van Gijsegem F, De Rycke R, D'Haeze W, Van Maelsaeke E, Anthonio E, Van Montagu M, Holsters M and Vereecke D. 2005. Lipopolysaccharides as a communication signal for progression of legume endosymbiosis. *Proc. Natl. Acad. Sci. USA,* **15**:2655-2660.

Mitra RM, Shaw SL, Long SR 2004. Six nonnodulating plant mutants defective for Nod factor-induced transcriptional changes associated with the legume-rhizobia symbiosis. Proc. Natl. Acad. Sci. USA, **6:**10217-10222.

Mitra RM, Gleason CA, Edwards A, Hadfield J, Downie JA, Oldroyd GE and Long SR. 2004. A Ca2+/calmodulin-dependent protein kinase required for symbiotic nodule development: Gene identification by transcript-based cloning. *Proc. Natl. Acad. Sci. USA*, **30**:4701-4705.

Myers AK and Tisa LS. 2004. Isolation of antibiotic-resistant and antimetabolite-resistant mutants of *Frankia* strains EuI1c and Cc1.17. *Can. J. Microbiol.*, **50**:261-267.

Navarro E, Nalin R, Gauthier D and Normand P. 1997. The nodular microsymbionts of *Gymnostoma* spp. are *Elaeagnus*-infective *Frankia* strains. *Appl. Environ. Microbiol.*, **63**:1610-1616.

Normand P, Orso S, Cournoyer B, Jeannin P, Chapelon C, Dawson J, Evtushenko L and Misra AK. 1996. Molecular phylogeny of the genus *Frankia* and genera and emendation of the family *Frankia*ceae. *Int. J. Syst. Bacteriol.*, **46**:1-9.

Obertello M., Sy M.O., Laplaze L., Santi C., Svistoonoff S., Auguy F., Bogusz D. and Franche C. 2003. Actinorhizal nitrogen fixing nodules: infection process, molecular biology and genomics. *African Journal of Biotechnology,* **2**:528-538.

Okubara PA, Fujishige NA, Hirsch A and Berry AM. 2000. *Dg93*, a nodule-abundant mRNA of *Datisca glomerata* with homology to a soybean early nodulin gene. *Plant Physiol.*, **122**: 1073-1079.

Parniske M and Downie JA. 2003. Plant biology: locks, keys and symbioses. Nature, **9:**569-570.

Pawlowski K and Bisseling T. 1996. Rhizobial and Actinorhizal Symbioses: What Are the Shared Features? *Plant Cell,* **8**:1899-1913.

Pawlowski K, Twigg P, Dobritsa S, Guan C and Mullin BC. 1997. A nodule-specific gene family from *Alnus glutinosa* encodes glycine- and histidine-rich proteins expressed in the early stages of actinorhizal nodule development. *Mol. Plant Microbe Interact.*, **10**:656-664.

Pawlowski K and Sirrenberg A. 2003. Symbiosis between *Frankia* and actinorhizal plants: root nodules of non-legumes. *Indian J. Exp. Biol.*, **41**:1165-1183.

Perret X, Staehelin C. and Broughton WJ. 2000. Molecular basis of symbiotic promiscuity. *Microb. Mol. Biol. Rev.,* **64**:180-201.

Provorov NA, Borisov AY and Tikhoich IA. 2002. Developmental genetics and evolution of symbiotic structures in nitrogen-fixing nodules and arbuscular mycorrhiza. *J. Theor. Biol.*, **21**:215-232.

Racette S and Torrey JG. 1989. Root nodule initiation in *Gymnostoma* (*Casuarinaceae*) and *Shepherdia* (*Elaeagnaceae*) induced by *Frankia* strain HFPGpI1. *Can. J. Bot.*, **67**:2873–2879.

Radutoiu S, Madsen LH, Madsen EB, Felle HH, Umehara Y, Gronlund M, Sato S, Nakamura Y, Tabata S, Sandal N and Stougaard J. 2003. Plant recognition of symbiotic bacteria requires two LysM receptor-like kinases. *Nature*, **9**:585-592.

Ribeiro A, Akkermans AD, van Kammen A, Bisseling T and Pawlowski K. 1995. A nodule-specific gene encoding a subtilisin-like protease is expressed in early stages of actinorhizal nodule development. *Plant Cell,* **7**:785-794.

Riely BK, Ane JM, Penmetsa RV and Cook DR. 2004. Genetic and genomic analysis in model legumes bring Nod-factor signaling to center stage. *Curr. Opin. Plant Biol.*, **7**:408-413.

Santi C, von Groll U, Ribeiro A, Chiurazzi M, Auguy F, Bogusz D, Franche C and Pawlowski K. 2003. Comparison of nodule induction in legume and actinorhizal symbioses: the induction of actinorhizal nodules does not involve ENOD40. *Mol. Plant Microbe Interact.*, **16**:808-816.

Soltis DE, Soltis PS, Morgan DR, Swensen SM, Mullin BC, Dowd JM and Martin PG. 1995. Chloroplast gene sequence data suggest a single origin of the predisposition for symbiotic nitrogen fixation in angiosperms. Proc. Natl. Acad. Sci. USA, **28:**2647-2651.

Stracke S, Kistner C, Yoshida S, Mulder L, Sato S, Kaneko T, Tabata S, Sandal N, Stougaard J, Szczyglowski K and Parniske M. 2002. A plant receptor-like kinase required for both bacterial and fungal symbiosis. *Nature,* **27**:959-962.

Svistoonoff S, Laplaze L, uy F, Runions J, Duponnois R, Haseloff J, Franche C. and Bogusz D. 2003. cg12 expression is specifically linked to infection of root hairs and cortical cells during *Casuarina glauca* and *Allocasuarina verticillata* actinorhizal nodule development. *Mol. Plant Microbe Interact.*, **16**:600-607.

Svistoonoff S, Laplaze L, Liang J, Ribeiro A, Gouveia MC, uy F, Fevereiro P, Franche C and Bogusz D. 2004. Infection- activation of the cg12 promoter is conserved between actinorhizal and legume-rhizobia root nodule symbiosis. Plant Physiol., 136:3191-3197.

Szczyglowski K and Amyot L. 2003. Symbiosis, inventiveness by recruitment? *Plant Physiol.*, **131**:935-940.

Torrey JG and Callaham D. 1979. Early nodule development in *Myrica gale*. Soil actinomycete causing nodulation. *Bot. Gaz.*, **140** :S10–S14.

Torrey JG. 1976. Initiation and development of root nodules of *Casuarina* (*Casuarinaceae*). *Amer. J. Bot.,* **63**:335-345.

Valverde C and Wall LG. 1999. Time course of nodule development in the *Discaria trinervis* (*Rhamnaceae*)-*Frankia* symbiosis. *New Phytol.*, **141**:345–354.

Van Ghelue M, Lovaas E, Ringo E and Solheim B. 1997. Early interaction between *Alnus glutinosa* and *Frankia* strain ArI3. Production and specificity of root hair deformation factor(s). *Physiol Plant.,* **99**:579-587.

Vessey JK, Pawlowski K and Bergman B. 2004. Root-based N2-fixing symbioses: Legumes, actinorhizal plants, *Parasponia* sp. and cycads. *Plant and Soil,* **1**:1-26.

Wais RJ, Keating DH and Long SR. 2002. Structure-function analysis of nod factor-induced root hair calcium spiking in *Rhizobium*-legume symbiosis. *Plant Physiol.*, **129**:211-224.

Wang HY and Berry AM. 1996. Plant regeneration from leaf segments of *Datisca glomerata*. *Acta Bot. Gallica,* **143**:609-612.

Zepp K, Hahn D and Zeyer J. 1997. Evaluation of a 23S rRNA insertion as target for the analysis of uncultured *Frankia* populations in root nodules of alders by whole cell hybridization. *Syst. Appl. Microbiol.*, **20**:124–132.

*SECTION — VII*

# ***MOLECULAR BASIS OF METABOLISM***

*Advances in Plant Physiology*, Vol. 9
Ed. A. Hemantaranjan
Scientific Publishers (India), Jodhpur, 2006 pp. 347-367
E-mail: **info@scientificpub.com** www.scientificpub.com

# 16

# FRUCTAN EXOHYDROLASES (FEHS) IN FRUCTAN AND NON-FRUCTAN PLANTS

Wim Van den Ende and André Van Laere

Laboratory for Molecular Plant Physiology, Kasteelpark Arenberg 31,
B-3001 Leuven (Heverlee), Belgium
Wim.VanDenEnde@bio.kuleuven.ac.be

## INTRODUCTION

Fructans are fructose-based oligo-and polysaccharides known to accumulate in about 15% of flowering plants (Hendry, 1993), in many bacterial and in a few fungal species in the genera *Aspergillus, Fusarium, Penicillium* and *Xanthophyllomyces* (Kritzinger *et al.*, 2003). The number of organisms capable of fructan biosynthesis is probably underestimated since numerous species and even genera have not yet been investigated. Fructans generally, but not necessarily, contain a terminal glucose (Glc) molecule (Lewis, 1993). Neokestose-type fructans contain an internal Glc (Vijn and Smeekens, 1999) while reducing inulo-*n*-oses and levan-*n*-oses contain no Glc at all (Van den Ende *et al.*, 1996a).

Like Raffinose-family oligosaccharides (RFO; Tapernoux-Luthi *et al.*, 2003), fructans may be considered as a polymeric but still soluble extension of vacuolar sucrose (Suc) in plants (Frehner *et al.*, 1984; Darwen and John, 1989; Wiemken *et al.*, 1995). However, significant fructan concentrations were also found in the apoplast of oat crowns during second-phase cold hardening (Livingston and Henson, 1998), suggesting that two different pools of fructans might exist in plants. Fructans were also found in the phloem (Wang and Nobel, 1998) and the fructan trisaccharide 6-kestose can be loaded into the phloem and transported to sink tissues (Zuther *et al.*, 2003). Since fructans are not degraded by most invertases, they can temporarily withhold hexoses from mainstream metabolism and form a carbohydrate reserve, different from starch in chemical nature (fructose versus glucose), cellular localization (vacuole versus plastids) and physical state (dissolved versus crystalline-precipitated) (Van Laere and Van den Ende, 2002). Based on work with transgenic plants, (apoplastic) fructans were proposed to protect plants against drought and freezing (Pilon-Smits *et al.*, 1995, Hisano *et al.*, 2004, Parvanova *et al.*, 2004) most probably by stabilizing plant plasma membranes (Vereyken *et al.*, 2001; Hincha *et al.*, 2000, 2002). It can be speculated that fructans are exchanged between vacuole and apoplast by exo- or endocytotic process, as observed for Suc (Echeverria, 2000, Etxeberria *et al.*, 2005). An inspection of the carbohydrate-active enzyme database (http://afmb.cnrs-mrs.fr/CAZY) shows that so far no •-fructosidase enzymes have been reported in humans and animals. Consequently, fructans are excellent prebiotics which are now widely applied in functional food and feed. They selectively promote growth of beneficial intestinal bacteria such as lactobacilli and bifidobacteria (Roberfroid and Delzenne, 1998). Moreover, they promote $Ca^{2+}$ uptake in the colon and are possibly anti-carcinogenic (Pool-Zobel *et al.*, 2002). Although *Helianthus tuberosus* tubers have been mentioned (Barta and Patkai, 2000), up to now only chicory (*Cichorium intybus* L.) roots are widely used for extraction of inulin (Kaur and Gupta, 2002; De Leenheer, 2004) as a prebiotic and health improving compound. Chicory yields are satisfactory and there is a vast agricultural experience with this crop (Wilson *et al.,* 2004). A major drawback of the use of chicory, however, is the induction of hydrolytic enzymes (fructan 1-exohydrolases or inulinases) in cold temperatures, lowering both the quality and quantity of the inulin at the end of the growth period and during storage (Van Laere and Van den Ende, 2002; Desprez *et al.*, 2004). To further improve inulin quality and extractability, an extensive study of chicory fructan 1-exohydrolases was started including gene cloning (see below).

Fructans are biosynthesized by transferring β (2,1) and/or β (2,6) linked fructofuranosyl units to the primary hydroxyl groups of Suc (Han, 1990; Vijn and Smeekens, 1999; Van Laere and Van den Ende, 2002). Linear β (2,1)-type fructans (inulins) mainly occur in dicot plant families (Asteraceae, Boraginaceae, Campanulaceae) but also in a few micro-organisms like *Aspergillus sydowi* (Heyer and Wendenburg, 2001), *Streptococcus mutans* (Rossell and Birkhed, 1974) and *Lactobacillus reuteri* (Van Hijum *et al.*, 2002). Linear β (2,6)-type fructans (levans) or (2,6 and 2,1) branched fructans (graminans) are common in monocots (Pollock, 1986; Chatterton *et al.*, 1990). Most fructan-producing bacteria accumulate linear levans with very high degree of polymerization (DP), although species-specific (2,1)-branches might occur (Iizuka *et al.*, 1993). Neokestose-type fructans (with an internal Glc) accumulate in Asparagales such as onion and Asparagus (Shiomi, 1992; Shiomi *et al.*, 1997; Ernst *et al.*, 1998) and in Poaceae such as *Avena sativa* (Livingston *et al.*, 1993) and *Lolium perenne* (Pavis *et al.*, 2001).

Two distinct enzymes are needed for inulin biosynthesis in plants (Edelman and Jefford, 1968; Lüscher *et al.*, 1996; Koops and Jonker, 1996; Van den Ende and Van Laere, 1996a). In the first step, 1-kestose and Glc are synthesized from two Suc molecules by sucrose:sucrose 1-fructosyltransferase (1-SST). In a second step, fructan:fructan 1-fructosyltransferase (1-FFT) catalyzes chain elongation by transferring a fructosyl residue from one inulin molecule to an other. 1-FFT can also use fructose (Fru) as an acceptor for the production of inulo-*n*-oses (Van den Ende *et al.*, 1996a). Some Asteracean 1-FFT enzymes (e.g. *Cynara scolymus* and *Echinops ritro*) are capable of producing higher DP inulins than others (Hellwege *et al.*, 2000; Vergauwen *et al.*, 2003). 1-SST and sucrose: fructan 6-fructosyltransferase (6-SFT) are key enzymes for fructan biosynthesis in cereals but the role of fructan: fructan 1-fructosyltransferase (1-FFT) is less clear (Wei *et al.*, 2002; Kawakami and Yoshida, 2002; Nagaraj *et al.*, 2004). 1-SST and fructan:fructan 6G-fructosyl transferase (6G-FFT) are key enzymes for neokestose-type fructan synthesis in *Allium cepa*. Onion 6G-FFT has an intrinsic 1-FFT activity (Ritsema *et al.*, 2003; Fujishima *et al.*, 2005) while a distinct 1-FFT besides 1-SST and 6G-FFT is believed to occur in *Asparagus officinalis* (Ueno *et al.*, 2005). In contrast to plants, where two or three enzymes are involved in fructan biosynthesis, bacteria use a single enzyme for this purpose: levansucrase or inulosucrase (Chambert *et al.*, 1974; Van Hijum *et al.*, 2002). Depending on conditions, these bacterial enzymes can hydrolyze fructans and sucrose or produce oligo- or polysaccharides (Hernandez *et al.*, 1995).

This review focuses on the enzymes catalyzing fructan breakdown in plants: Fructan ExoHydrolases (FEHs; $G\text{-}F_n + H_2O \Rightarrow G\text{-}F_{(n-1)} + F$ with $n>1$). Both FEHs and fructan biosynthetic enzymes are believed to originate from invertases. All these enzymes belong to glycosyl hydrolase family 32. A summary of enzyme specificities within the group is presented in Table I.

Both exo- and endo-type fructan hydrolases occur in bacteria and fungi (Van Damme and Derycke, 1983). Although the presence of reducing fructans in e.g. chicory (Van den Ende *et al.*, 1996a) suggests inulin degradation by endo-inulinase, so far no evidence for such enzyme could be obtained in plants. Most likely, plants only contain FEHs releasing terminal Fru units from fructan substrates. The recent discovery of FEHs in plants that do not accumulate fructan (so called non-fructan plants) suggest (a) clearly different function(s) for these enzymes (Van den Ende *et al.*, 2004a). Therefore, FEHs from non-fructan plant will be treated in a different section.

## 2. FEHs IN FRUCTAN PLANTS

### 2.1 Properties of Dicot 1-FEHs

Depending on the linkage type attacked, 1-FEH (inulinase, degradation of β 2,1 linkages) and 6-FEH (levanase, degradation of • 2,6 linkages) type enzymes can be differentiated. Within the *Asteraceae*, apparently only inulin-type fructans accumulate which are subject to degradation by 1-FEHs. Although two 1-FEH isoforms A and B were partially purified from tubers of *Helianthus tuberosus* (Edelman and Jefford, 1964) and roots of *Taraxacum officinale* (Rutherford and Deacon, 1972), the first electrophoretically pure 1-FEH reported was the enzyme from chicory roots (Claessens *et al.*, 1990). This enzyme was termed 1-FEH I since later a second 1-FEH (1-FEH II) was discovered in chicory. 1-FEH I was described as a 70 kDa glycoprotein and showed a neutral pI and an acidic pH optimum. The enzyme was unable to degrade Suc or bacterial levan and was not inhibited by Suc. Marx *et al.* (1997) purified a 1-FEH from *Helianthus tuberosus* with similar properties. Since no short fructans or Suc are produced as long as longer fructan chain are available, it was concluded that 1-FEHs work via a multichain attack mechanism. The hydrolysis of inulin oligomers followed normal saturation kinetics: $K_m$ values for 1,1-kestotetraose and 1,1,1-kestopentaose were 8.3 and 12 mM, respectively. De Roover *et al.* (1999) purified and characterized a second 1-FEH (1-FEH IIa) from chicory roots. Although the enzyme had similar properties as 1-FEH I, it had an acid pI and was very avidly and specifically inhibited by Suc and not by Suc analogues. After cloning of the cDNAs (see further), a further distinction was made between 1-FEH IIa and IIb type isoenzymes (Van den Ende *et al.*, 2001). The detailed substrate specificities of chicory and *Helianthus* 1-FEHs are shown in Table II.

### 2.2 Properties of Monocot FEHs

Many monocots contain fructans with both β (2,1) and β (2,6) linkages. It is conceivable that they contain a mixture of either 1-FEH and 6-FEH enzymes or alternatively enzymes capable of breaking both linkage types. Orchardgrass (*Dactylis glomerata*) mainly accumulates linear levan-type fructans. A 6-FEH was partially purified from stems of orchardgrass after defoliation (Yamamoto and Mino, 1985) and a 1-FEH was partially purified from wheat (Jeong, 1991). Many other 1-FEH and 6-FEH activities were reported from grasses (as reviewed in Simpson and Bonnett, 1993; Bonnett and Simpson, 1993, 1995) but none of them was purified up to electrophoretic purity. The first electrophoretically pure monocot 6-FEH enzymes were obtained from *Avena sativa* (Henson and Livingston, 1996) and *Lolium perenne* (Marx *et al.*, 1997b). Their detailed substrate specificities are listed in Table II. Other FEHs include an enzyme capable of attacking both β (2,1) and β (2, 6) linkages from barley (Henson and Livingston, 1998) and two 1-FEH isoforms from stems of wheat that are strongly inhibited by Suc (Van den Ende *et al.*, 2003a). Interestingly, the two isoforms 1-FEH w1 and w2 from wheat differed in their capacity to degrade several graminan-type fructans (Van den Ende *et al.*, 2003a). Besides a 6-FEH that breaks down bacterial levan (Van Riet *et al.*, 2004), a peculiar FEH was recently discovered in wheat (Van den Ende *et al.*, 2004b; Van den Ende *et al.*, in press). This FEH was termed 6-kestosidase or 6-kestose exohydrolase (6-KEH) since it preferentially degrades 6-kestose. 6-KEH fails to degrade Suc and a number of other fructan substrates. A similar enzyme was reported in *Asparagus officinalis*. (Ueno *et al.*, 2004). Interestingly, this species contains no endogenous 6-kestose.

Overall, monocot FEHs showed properties similar to dicot 1-FEHs: monomeric glycoproteins, an acidic pH optimum and the occurrence of forms that are inhibited by Suc and others that are not. Although Simpson and Bonnett (1993) proposed that β-fructofuranosidases (enzymes capable of degrading both fructans and Suc) occur in grasses, so far no electrophoretically pure plant enzyme was obtained that shows both FEH and invertase activities to a considerable extent. Therefore, the described plant β-fructofuranosidases might have been a mixture of both FEH and invertase enzymes. Assuming this, it might be concluded that probably *all* plant FEHs are unable to degrade Suc, in contrast to microbial exo-hydrolases that generally can also degrade sucrose and are better designated as β-fructosidases. It can be concluded that the EC number 3.2.1.80 (enzymes showing activity against both Suc and fructans) is inappropriate for plant FEHs. Therefore, a new EC number should be appointed for this class of enzymes.

### 2.3 Molecular Characterization of FEHs

Many plant fructosyl transferases (Ritsema and Smeekens, 2003) and both endo- and exo-type microbial FEHs (see references in Michiels, 2003) were cloned in the nineties. The delay in cloning of plant FEHs can probably be explained by the fact that many groups directly focused on the DNA level and used PCR primers based on conserved sequences in vacuolar invertases. Indeed, it was expected that FEHs, reported tot be vacuolar enzymes (Wagner and Wiemken, 1986), would resemble vacuolar invertases. Furthermore, at that time it was already clear that plant fructosyl transferases were closely related to vacuole invertases, suggesting a common evolutionary origin. However, a breakthrough in cloning plant FEHs was only realized by work at the protein level. Peptide sequences were generated from the purified chicory 1-FEH I and this information led to the cloning of a first plant FEH (Van den Ende *et al.*, 2000). Functional expression in potato tubers convincingly demonstrated the 1-FEH nature of the protein coded by the isolated cDNA and the complete absence of invertase activity. Most surprisingly, it was found that 1-FEH I was more homologous to cell-wall invertases than to vacuolar invertases, despite its alleged vacuolar localization. It was suggested that plant FEHs evolved from ancestral cell wall type invertase genes by capturing a vacuolar targeting signal and loosing the high pI for interaction with acidic cell wall components, typical for cell wall enzymes. Without a functional analysis at the protein level, the 1-FEH I cDNA would simply have been classified as a cell wall invertase. This clearly indicates the weakness of relying only on sequence information to predict the function and localization of a protein.

Using essentially the same approach, chicory 1-FEH IIa and IIb were subsequently cloned (Van den Ende *et al.*, 2002). An unrooted phylogenetic tree shows that the three FEH cDNAs from chicory cluster in a distinct subgroup of cell wall type invertases (Fig. 1, C I), together with a putative cell wall invertase from *Arabidopsis thaliana* (AtcwINV6, revised nomenclature of Sherson *et al.*, 2003). Later, Michiels *et al.* (2004) succeeded in cloning the complete 1-FEH IIa gene including its promoter by using a different PCR approach (Michiels *et al.*, 2003). Interestingly, the 1-FEH IIa gene showed an intron/exon organisation that was more similar to vacuolar invertases than to cellwall invertases.

Van den Ende *et al.* (2003a; 2005, in press) reported cloning of the first monocot FEHs. Two highly similar 1-FEH cDNAs (1-FEH w1 and w2) and 6-KEH cDNAs (6-KEH w1 and w2) cluster together in one subgroup (Fig. 1, C II) in a phylogenetic tree, quite close to the dicot subgroup C I. The functionality of the cDNAs was demonstrated by heterologous expression in *Pichia pastoris* (Fig. 2). This expression technique is now widely used (Cregg *et al.*, 2000) and proved a valuable system to judge on the functionality of plant fructosyl transferases (Kawakami and Yoshida, 2002), FEHs, invertases (Van den Ende *et al.*, 2003b; Verhaest *et al.*, 2004; De Coninck *et al.*, 2005) and bacterial levansucrases (Trujillo *et al.*, 2002). The substrate specificities of a number of recombinant cell wall type GH 32 enzymes from plants are compared in Fig. 2, convincingly demonstrating that all plant FEHs are unable to degrade Suc.

## 2.4. Functions of FEHs (and Fructans)

### *2.4.1 Breakdown of fructan reserves*

The main function of fructans in underground storage organs of Asteraceae is their use as a long-term storage compound (Van Laere and Van den Ende, 2002). However, also short-term fructan storage can be induced in chicory seedlings by incubation in Suc solutions (Vijn *et al.*, 1997) or by growth inhibition induced by drought or N-shortage. Addition of water or nitrogen to the latter plants induced fructan remobilization by 1-FEH induction (Van den Ende *et al.*, 1999; De Roover *et al.*, 1999). FEHs are necessary to provide hexoses for leaf re-growth, sprouting, flowering and forcing processes (Yamamoto and Mino, 1985, 1999; Solhaug and Aares, 1994; Van den Ende and Van Laere, 1996b, Van den Ende *et al.*, 1996b; De Roover *et al.*, 1999; Morvan-Bertrand *et al.*, 2001; Asega *et al.*, 2004). In grasses, fructans merely accumulate in the vegetative tissues and are usually stored for a shorter period of time (Pollock, 1996). Fructans temporarily accumulate in stems of wheat. Subsequently they are degraded by FEHs (Van den Ende *et al.*, 2003a, Yang *et al.*, 2004) and the carbohydrate skeletons are subsequently transported (as Suc) to the wheat kernels to fuel the grain-filling process. Although 1-FEH activities were found to be considerably higher during the fructan breakdown phase in wheat stems, it was possible to purify substantial amounts of 1-FEH from actively fructan biosynthesizing stems suggesting that 1-FEH might play a role as a β (2,1) trimmer throughout the period of active graminan biosynthesis.

Therefore, the species and developmental stage-specific complex fructan patterns found in monocots might be determined by the relative proportions and specificities of both fructan biosynthetic and breakdown enzymes (Van den Ende *et al.*, 2003a). .

One of the most detailed studies on fructan breakdown by FEHs is the remobilization of fructan reserves after grazing or cutting in *Lolium perenne* (Morvan-Bertrand *et al.*, 2001). FEH activity peaks in the growth zone after defoliation suggesting that fructans stored in the leaf growth zone were hydrolyzed and recycled in that zone to sustain the leaf regrowth.

### 2.4.2 Breakdown of fructan for stress protection

Chicory root fructans are partly depolymerised in autumn (fist frost) and this is accompanied by an increase in Fru content, the appearance of inulo-*n*-ose type fructans and a strongly increased 1-FEH activity (Van den Ende *et al.*, 1996b; Van den Ende *et al.*, 1996b). Both 1-FEH I and II enzymes are involved in this process (Van Laere and Van den Ende, 2002) and cold is probably the most essential trigger to enhance FEH gene transcription (Michiels *et al.*, 2004). In part, the degradation process may be needed in terms of energy provision (see 2.4.1), but the huge increase of Fru in the roots and the absence of visual re-growth for several months suggest that the fructan depolymerization might have an additional or other role. A similar fructan depolymerization process and Fru accumulation can be observed during second-phase cold-hardening (< 0°C) of young winter wheat and oat plants (Yukawa *et al.*, 1994; Yoshida *et al.*, 1998; Livingston and Henson, 1998). It can be speculated that the accumulation of (apoplastic) fructans is a specific adaptation that helps plants to survive freezing temperatures. On the contrary, fructan biosynthesis is often induced during first-phase cold-hardening (> 0°C) in grasses. It was shown that fructan biosynthetic enzymes are less sensitive to inactivation at low temperatures than starch synthesizing enzymes. Fructan synthesis might prevent sugar-induced feedback inhibition of photosynthesis, an advantage during winter growth (Pollock, 1986). When temperatures slowly rise in spring, it can be expected that the soluble fructans can be mobilized much faster (either as such, or after breakdown by FEHs) than the insoluble starch. Consequently, fructans might be very useful compounds in environments that require frequent and rapid changes in resource allocation especially at low temperatures.

The relationship between drought stress and fructan accumulation is more complex, possibly because of two distinct roles of fructans as membrane protectants on the one hand and carbohydrate stores on the other. Fructans and 1-SST increase in chicory seedlings under drought stress (De Roover *et al.*, 2000) but, on the other hand, chicory feeder roots already containing fructans depolymerize them under drought stress (Van den Ende *et al.*, 1998). The outcome however is the same: accumulation of fructans of low degree of polymerization (DP). Therefore, it can be speculated that a mixture of oligofructans, together with Suc and hexoses, might be optimal for membrane protection and associated stress tolerance.

### 2.4.2 Breakdown of fructans for osmoregulation

The possibility that fructans act as osmoregulators has been suggested by many studies (reviewed in Hendry, 1993). However, the most clear-cut example to date of fructans serving an osmotic function is provided by studies on flower opening in *Hemerocallis* (Bieleski, 1993) and *Campanula* (Vergauwen *et al.*, 2000). Like in *Asteraceae*, inulin-type fructans occur in the *Campanulaceae*. It was found that inulins are present in all organs and developmental stages of *Campanula rapunculoides*. Strikingly, inulins accumulate high concentrations in petals of closed flowers. However, just before flower opening a rapid breakdown of inulin to Fru was observed in these petals. The results suggest that the inulin-to-Fru conversion strongly reduces the water potential and drives petal expansion by massive water inflow. It was demonstrated (Bieleski, 1993) that cycloheximide halted flower opening suggesting that *de novo* FEH synthesis is required in the process. The degradation of inulin to Fru is accomplished by a 1-FEH that is induced prior to flower opening in *Campanula rapunculoides* petals (unpublished results).

## 2.5 Regulation of FEHs

Studies on the regulation of fructan metabolism are still in their infancy. Moreover, these studies mainly focused on fructan anabolism (Van den Ende *et al.*, 2002; Nagaraj *et al.*, 2004; Morcuende *et al.*, 2004; Ritsema *et al.*, 2004). The most prominent outcome of this work is the strong induction of fructosyl transferases by Suc at the transcriptional level.

The isolation of the promoter of chicory 1-FEH IIa revealed a complex array of regulatory motifs (Michiels *et al.*, 2004). Cold appears to be the major trigger for gene induction but etiolation and plant hormones might also play a role (Michiels, 2003). The data correlate well with the expression of the FEH IIa gene throughout the chicory growing season and after cold induction (Van Laere and Van den Ende, 2002). Overall, northern blot profiles of chicory 1-SST, 1-FFT, 1-FEH IIa and 1-FEH I were fully consistent with enzymatic activity measurements and carbohydrate profiles (Van Laere and Van den Ende, 2002). Therefore it appears that all chicory fructan enzymes are regulated at the transcriptional level.

Glu, Fru and Suc were all able to partially inhibit the increases in FEH activity after defoliation of *Dactylis glomerata* and *Lolium perenne* (Yamamoto and Mino, 1987; Lothier *et al.*, 2004). Similarly, Suc was able to stop the expression of 1-FEH IIa at the transcriptional level (Michiels, 2003). Additionally and perhaps even more importantly, Suc is a strong competitive inhibitor of many FEHs including chicory 1-FEH IIa and IIb and wheat 1-FEH w1 and w2 (Van den Ende *et al.*, 2001, Van den Ende *et al.*, 2003b). It was suggested that these FEHs are effectively inhibited by normal physiological Suc concentrations. However, these enzymes could be activated immediately when Suc concentrations fall (by damage, wounding, cutting, grazing or other stresses).

Table 1. A list of GH 32 family enzymes occurring in plants. Preferential fructosyl donor and acceptor substrates are indicated.

| Enzyme | Fructosyl donor | Fructosyl acceptor |
|---|---|---|
| Invertase | sucrose | water |
| 1-FEH (Fructan 1-Exo Hydrolase) | inulin | water |
| 6-FEH (Fructan 6-ExoHydrolase) | levan | water |
| 6-KEH (6-kestose Exo Hydrolase) | 6-kestose | water |
| 6&1-FEH (Fructan Exohydrolase) | fructan | water |
| 1-SST (Sucrose: Sucrose 1-fructosyl Transferase) | sucrose | sucrose |
| 6-SFT (Sucrose: Fructan 6-fructosyl Transferase) | sucrose | 1-kestose |
| 1-FFT (Fructan: Fructan 1-fructosyl Transferase) | 1-kestose/ inulin | inulin/ sucrose |
| 6G-FFT (Fructan: Fructan 6-fructosyl transferase) | 1-kestose/ fructan | 1-kestose/ sucrose |

Yamamoto and Mino (1998) found that exogenously applied gibberellic acid, cytokinin, 8-bromoadenosine-3,5-cyclic monophosphate and theophylline increased 6-FEH activity after defoliation of orchardgrass. Induction was suppressed by continuous treatment with ABA. *Lolium perenne* FEH activity was found to be strongly inhibited by uniconazole, an inhibitor of gibberellin biosynthesis. Inhibition could be overcome by a subsequent treatment with gibberellic acids (Morvan *et al.*, 1997).

### 2.6 Structure of FEH

The different plant fructan-enzymes (see Table 1) all essentially transfer a Fru moiety from a donor to an acceptor substrate, but each enzyme has a preferred or even unique substrate as donor (fructan, sucrose) and/or acceptor (fructan, sucrose, or water). They all belong to the glycosyl hydrolase (GH) family 32 in the carbohydrate-active enzyme database (http://afmb.cnrs-mrs.fr/CAZY), a classification based on overall amino acid sequence homologies (Henrissat and Davies, 1997). GH family 32 is combined in a superfamily together with GH family 68 containing microbial levansucrases and β-fructosidases. Recently, the 3D-structures of three microbial enzymes have been resolved within this superfamily (Table III): a GH 68 levansucrase from *Bacillus subtilis* (Meng and Fütterer, 2003), a GH 32 •-fructosidase from *Thermotoga maritima* (Alberto *et al.*, 2004) and a GH 32 inulinase from *Aspergillus awamori* (Nagem *et al.*, 2004). Chicory 1-FEH IIa is now the first higher plant GH 32 enzyme of which the 3D structure has been resolved (Verhaest *et al.*, 2004; Verhaest *et al.*, 2005). The structure of 1-FEH IIa was described at a resolution of 2.35 Å and consists of an N-terminal five-fold β-propeller domain and a second domain containing two C-terminal β-sheets (Fig. 3). The putative active site is located entirely in the β-propeller domain and is formed by amino acids highly conserved within the GH family 32 (Fig. 3). Preliminary investigations indicate that the substrate binds in a cleft between the two subdomains (Le Roy *et al.*, 2004). In contrast to the multi-functional microbial enzymes, FEH IIa is a monofunctional enzyme (Table III). The structure provides an ideal template for other plant members of the family GH 32 and will prove an important tool to understand what molecular determinants are important for the different substrate specificities within the group. These fine molecular insights will lead to the development of superior fructosyl transferases capable of producing designer fructans with beneficial characteristics for specific agronomical and/or industrial applications.

## 3. FEHs IN NON-FRUCTAN PLANTS

### 3.1 Discovery

The complete absence of FEHs in plants that are unable to accumulate fructans (so-called non-fructan plants) has long been considered as trivial. To overcome the problem of endogenous FEH induction and associated inulin breakdown in chicory roots during autumn, molecular biologists introduced several plant and bacterial fructosyl transferases into non-fructan plants like sugar beet and potato (Hellwege *et al.*, 2000; Sévenier *et al.*, 1998; Cairns, 2003; Weyens *et al.*, 2004;) with the intended purpose of obtaining a large scale industrial production of tailor-made fructans. Although the approach was successful for plant enzymes mainly biosynthesizing β 2,1 linkages, more difficulties arose for enzymes creating β 2,6 linkages. Transgenic plants harbouring bacterial levansucrases accumulated only very low amounts of levan which were apparently not present in the vacuole. Intriguingly, toxic effects of levans were found when bacterial levansucrases were overexpressed in non-fructan plants (Cairns, 2003).

When chicory 1-FEH I was heterologously expressed in potato (Van den Ende *et al.*, 2000) and Fru production from inulin was measured in these transgenic plants, some Fru production was also detected in wild-type potatoes that were used as controls in these experiments. However, this Fru production was always considered as a side-activity of

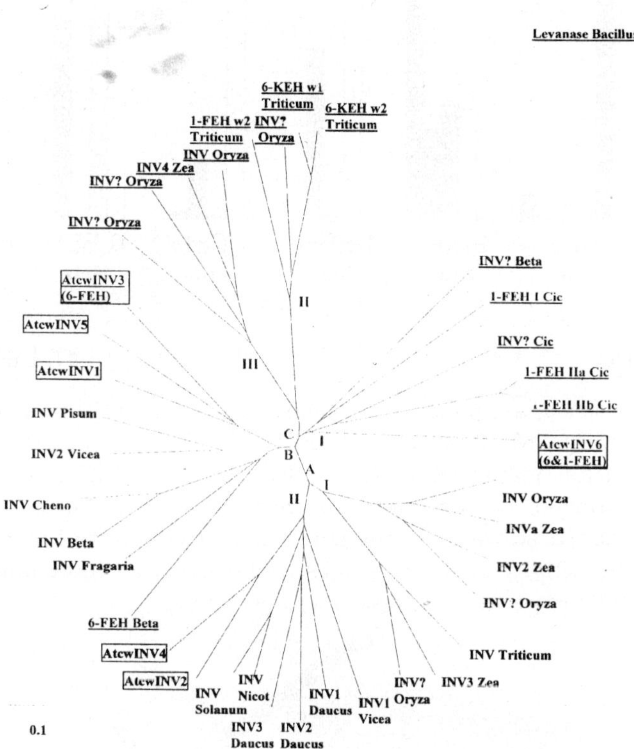


Fig. 1. Unrooted phylogenic tree of cDNA derived amino acid sequence of plant cell wall invertases and FEHs.

Three main groups can be discerned. Within the first group (A) a monocot (I) and dicot (II) subgroup are discerned. A I: *Zea mays* INV1, 2 and 3; two putative (indicated with ?) and one effective *Oryza sativa* INV; *Triticum aestivum* INV. A II: *Vicia faba* INV1; *Daucus carota* INV 1, 2 and 3; *Solanum tuberosum* INVF; *Nicotiana tabacum* INV; *Arabidopsis thaliana* INV2 and 4; Second group (B): *Beta vulgaris* 6-FEH and INV; *Arabidopsis thaliana* INV1, 5 and 6-FEH (AtcwINV3); *Vicia faba* INV2; *Pisum sativum* INV; *Chenopodium rubrum* INV1; and *Fragaria* × *ananassa* putative INV. The third group (C) splits out in three subgroups (I-III). C I: putative INV from *Cichorium intybus* and *Beta vulgaris*; *Cichorium intybus* 1-FEH I, IIa and IIb and a 6&1 FEH from *Arabidopsis thaliana* (AtcwINV6). C II: 1-FEH w2 from *Triticum aestivum*, a putative invertase from *Oryza sativa* and *Triticum aestivum* 6-KEH w1 and w2. C III: *Zea mays* INV4; two putative and one effective INV from *Oryza sativa*. Enzymes with acid iso-electric points are underlined. FEHs are in red. Arabidopsis enzymes are boxed. *Aspargillus awamori* inulinase and *Bacillus subtilis* levanase are included as outliers. The scale bar indicates a distance value of 0.1. INV: cell wall invertase. Revised nomenclature was used for *Arabidopsis thaliana* invertases according to Sherson *et al.*, 2003.

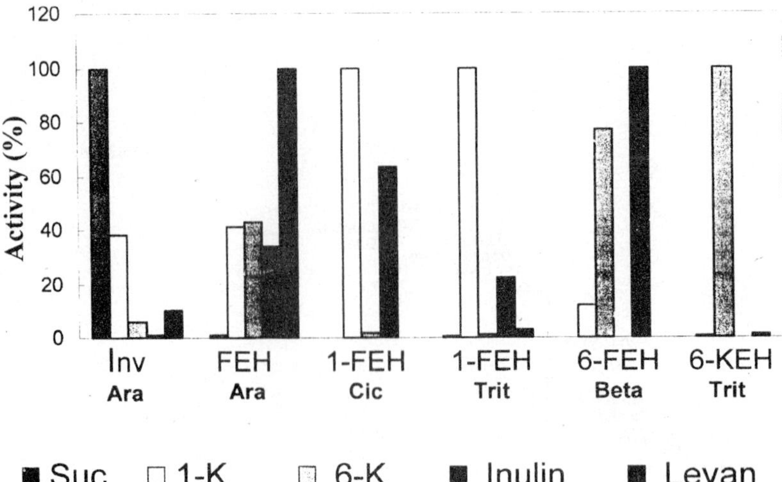


Fig. 2. Substrate specificities of a number of recombinant FEHs. A selection of *Pichia pastoris* derived FEHs from fructan (*Cichorium intybus* 1-FEH IIa, *Triticum aestivum* 1-FEH w2 and 6-KEH w1) and non-fructan plants (*Arabidopsis thaliana* (6&1)-FEH, *Beta vulgaris* 6-FEH) were tested with equimolar concentrations of Suc, 1-kestose, 6-kestose, inulin and levan as substrates and compared with recombinant Arabidopsis cell wall invertase 1 (AtcwINV1). For each enzyme, the activity of the best substrate was set at 100% and the activities obtained with the other substrates were expressed as a percentage.

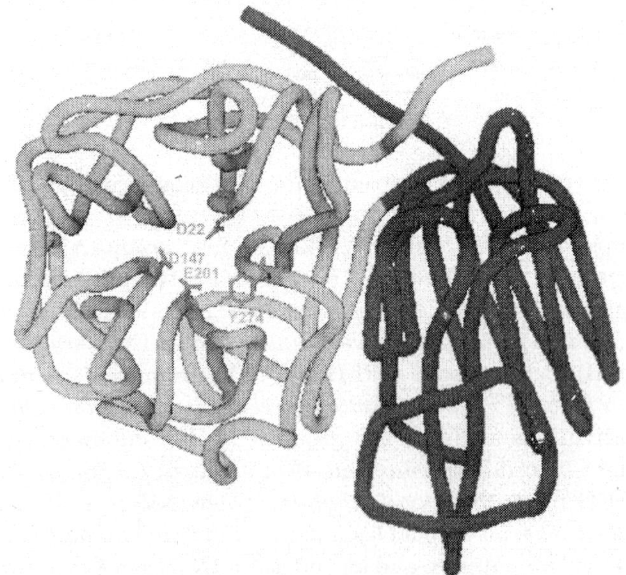


Fig. 3. Schematic presentation of the 3D stucture of chicory 1-FEH IIa. Two domains can be differentiated: a 5-bladed β-propeller (green) containing the active site and a β sandwich module (blue). The presumptive catalytic amino acids (red) are D22 (nucleophile), E201 (acid-base catalyst), D147 (transition state stabilizer) and Y274 putative pKa modulator. For details see Verhaest *et al.*, 2005.

invertases on inulin. The detection of high levels of FEH activity together with the complete absence of fructan in etiolated endive leaves (Van den Ende *et al.*, 2001) were strongly suggestive for the occurrence of FEH enzymes in non-fructan accumulating tissues. Additionally, it was noticed that one putative cell wall invertase from *Arabidopsis thaliana* (AtcwINV6) clusters together with FEHs from fructan plants in a phylogenetic tree (Fig. 1, group C I), suggesting that it might be a FEH rather than an invertase. All these findings inspired us to look more in detail for FEH activities in non-fructan plants.

A selection of non-fructan plants belonging to a variety of plant families were tested for 6-FEH and 1-FEH activities. Surprisingly, considerable 6-FEH activities could be detected in about 85% of the species tested, suggesting that 6-FEHs are rather widely distributed in plants. The highest 6-FEH enzyme activity was found in sugar beet tap roots (*Beta vulgaris* L.). Sugar beet roots accumulate Suc to a high extent and therefore these tissues show no or only very low invertase activities. These findings clearly indicated that the 6-FEH activity could not be considered as a side-activity of invertases. Still, we remained very cautious about these data and suspected that the 6-FEH activity might originate from microbes present within the plant material. However, sugar beet seedlings grown under sterile conditions also showed this 6-FEH activity. Therefore, a purification of the sugar beet 6-FEH enzyme was attempted.

### 3.2 Purification, Characterization and Cloning of FEHs

Sugar beet 6-FEH was purified from tap roots by a combination of ammonium sulphate precipitation, affinity chromatography (Concanavalin A), anion- and cathion exchange chromatography, gel filtration and electrophoresis (Van den Ende *et al.*, 2003b). Overall, its properties were very similar to FEHs purified from fructan plants (glycoprotein, 70 kDa, low pI and acidic pH optimum). However, the enzyme appeared as a heterotrimer after SDS-PAGE. Like all other plant FEHs, the 6-FEH enzyme showed no invertase activity. Instead, the enzyme turned out to preferentially hydrolyze β 2,6 linkages occurring in bacterial and plant levans, neokestose and levanbiose (Table II). To further prove that the enzyme did not originate from endophytic or pathogenic bacteria or fungi, the corresponding cDNA was cloned by a PCR based approach and the functionality of the 6-FEH cDNA was further demonstrated by heterologous expression in *Pichia pastoris* (Fig. 2). Like for FEHs from fructan plants, a high sequence homology to plant cell wall invertases was found, excluding a microbial origin. Surprisingly, the sugar beet 6-FEH falls in another subgroup (Fig. 1, B) than the FEHs from fructan plants (Fig. 1, C I and II). This polyphyletic profile suggests that FEHs might have evolved more than once from different cell-wall invertases, and perhaps only a limited number of mutations might determine the functional differences between invertases and FEHs on the one hand and between 1-FEHs and 6-FEHs on the other hand.

The presence of 6-FEH in sugar beet further stimulated us to investigate whether some of the six putative invertases in *Arabidopsis thaliana* (Sherson *et al.*, 2003), another non-fructan plant widely used in plant molecular biology, could in fact be FEHs. FEH activities could indeed be detected in crude extracts of Arabidopsis. AtcwINV1, 3 and 6 cDNAs (see Fig. 1) were chosen for heterologous expression in *Pichia pastoris* (De Coninck *et al.*, in press). Not surprisingly, AtcwINV1 encoded an invertase (Fig. 2) but AtcwINV3 and 6 recombinant proteins (Fig. 2) had no invertase activities. Instead, AtcwINV3 proved to be a 6-FEH with almost identical substrate specificities as the sugar beet 6-FEH (De Coninck *et al.*, in press). Both 6-FEHs fall in the same

Table 2. Substrate specificities of a number of native FEHs derived from fructan and non-fructan(*) plants. The activity with routine substrates (bold, different for each case) were set at 100% and the activities with other substrates were expressed as a percentage.

| Substrate | DP | 6-FEH *Avena* 1996 | 6-FEH *Lolium* 1997 | 1-FEH *Helianthus* 1997 | 1-FEH IIa *Cichorium* 1999 | 1-FEH w2 *Triticum* 2003 | 6-FEH *Beta** 2003 | 6-FEH *Arabidopsis** 2005 |
|---|---|---|---|---|---|---|---|---|
| sucrose | 2 | 0 | 3 | <1 | <1 | <1 | 0 | 0 |
| inulobiose | 2 | | | 20 | | | | |
| levanbiose | 2 | | | | 18 | | 63 | 76 |
| 1-kestose | 3 | 0 | 21 | 34 | **100** | **100** | 12 | 23 |
| 6-kestose | 3 | | 103 | 4 | | 1 | 77 | 77 |
| neokestose | 3 | 29 | 21 | 7 | | 3 | 93 | 151 |
| melizitose | 3 | | 0 | <1 | | | | |
| raffinose | 3 | | | 9 | | | | |
| stachyose | 4 | | | 6 | | | | |
| inulotetraose | 4 | | | | 210 | | | |
| 1,1 nystose | 4 | | 24 | **100** | 75 | 83 | 0 | 12 |
| 6,6 nystose | 4 | | **100** | 6 | | | | |
| bifurcose | 4 | | 93 | 20 | | | | |
| $6^G$, 6-kestotetraose | 4 | **100** | | | | | | |
| 1&$6^G$ kestotetraose | 4 | 8 | | | | | | |
| 1,1,1 kestopentaose | 5 | | | 180 | | | | |
| $6^G$, 1&6 kestopentaose | 5 | 60 | | | | | | |
| phlein | 4-12 | | | | | | 84 | 156 |
| wheat fructan | >7 | | 41 | | | | | |
| inulin | >10 | | | 133 | 83 | 22 | 0 | 4 |
| levan | nd | | | | 20 | 3 | **100** | **100** |
| neokestin | | | | | | | 25 | |

Table 3. Summary of fructosidase enzymes of which the 3D structure has been resolved and indication of their transferase, invertase, 1-FEH and 6-FEH activities.

| Enzyme | Organism | Family | Transferase | Invertase | 1-FEH | 6-FEH |
|---|---|---|---|---|---|---|
| Levansucrase | *Bacillus subtilis* | GH68 | +++ | ++ | — | + |
| Invertase | *Thermotoga maritima* | GH32 | — | +++ | ++ | — |
| Inulinase | *Aspergillus awamori* | GH32 | — | + | +++ | ++ |
| 1-FEH | *Cichorium intybus* | GH32 | +/— | — | +++ | — |

subgroup in the tree (Fig. 1, group B) suggesting that these genes are homologs. The recombinant AtcwINV6 however, was able to degrade both 2,1 and 2,6 linkage types in linear and branched fructans. The term (6&1)-FEH was proposed for this enzyme. To verify that recombinant 6-FEH activity was not an artifact from the *Pichia pastoris* expression system, the native 6-FEH enzyme was purified and indeed showed only 6-FEH and no invertase activity (Table II).

The position of the Arabidopsis (6&1)-FEH in the phylogenetic tree (Fig. 1) and its rather aspecific substrate specificity profile towards fructans (Fig. 2) make it a kind of evolutionary intermediate between aspecific microbial β-fructofuranosidases and specific plant FEHs. The latter lost their ability to degrade Suc and further specialized to degrade a single linkage type (e.g. 1-FEH) or even a single fructan (e.g. 6-KEH). Indeed, inclusion of two microbial •-fructosidases into the tree (Fig. 1) shows that they cluster into the low pI group C. The splitting of •-fructo-sidase into FEH and invertase that probably occurred in plants allows independent control of sucrose and fructan concentrations.

### 3.3 Functions of FEHs in Non-Fructan Plants

The occurrence of plant enzymes in tissues lacking their substrates was described before. A good example is the presence of chitinases in higher plants. These enzymes probably attack fungal cell-walls and play an important role in plant defence (Graham and Sticklen, 1994). The occurrence of functional FEHs in non-fructan plants suggests novel physiological roles (Van den Ende *et al.*, 2004a). Two main putative functions can be postulated:

1. The most plausible function of plant 6-FEHs would be to degrade or even prevent the formation of exogenous β (2,6) linked levans from bacterial origin. Since bacterial levans always show β (2,1) side chains, (6&1)-FEHs might be useful to degrade branch-points. 6-FEHs could participate directly in hydrolysis of levan-containing slimes surrounding phytopathogenic bacteria such as *Pseudomonas* and *Erwinia*. Levan is considered to play a crucial role during the early phase of infection by creating a separating layer between bacteria and the plant cell wall, preventing pathogen recognition and the plant defense response (Kasapis *et al.*, 1994; Hettwer *et al.*, 1995; Bereswill *et al.*, 1997). *Erwinia amylovora* strains deficient in levansucrase showed reduced virulence (Tharaud *et al.*, 1994; Bereswill *et al.*, 1997). It can be speculated that 6-FEHs may counteract bacterial levansucrases and detoxify the products they form inside the plant. Toxic effects of levan were often found as a result of bacterial levansucrase overexpression in non-fructan plants (Cairns, 2003). Moreover, 6-FEH might limit sequestration of apoplastic sucrose through bacterial levan synthesis. Remarkably and in severe contrast, levans accumulate to a high extent in the apoplast of sugar cane and in this case no toxic effects or enhanced plant defense responses are noticed (Hernandez *et al.*, 1995). These levans are produced by the nitrogen-fixing symbiotic endophyte *Gluconacetobacter diazotrophicus*. It can be speculated that FEHs could be involved in stabilizing the symbiosis between plants and fructan-producing endophytic bacteria. Since levan structures might be highly species-specific, unique branched oligofructan structures might be released by plant FEHs acting on levans in bacterial slimes. Structures derived from the symbiont might suppress the plants defense system while other structures might have the opposite effect.
2. 6-FEHs might be involved in signaling events. Invertases have the capacity to produce the three kestose trisaccharides 1-kestose, 6-kestose and neokestose *in vitro* from high Suc concentrations (Cairns and Ashton, 1991). It cannot be excluded therefore that the production of these trisaccharides in plants lacking fructosyl transferases is of physiological importance. It is conceivable that Suc derivatives or homologues like e.g. kestoses (and trehalose) fulfil crucial

roles in plant carbohydrate partitioning and can act as signal molecules captured by specific sensors (Van den Ende *et al.*, 2004a). Assuming such a role for kestoses, ideally the signal should be destroyed after the sensing process by a highly specific FEH that could also be involved in sensing these signals. The discovery of extremely specific 6-KEH enzymes in plants might in itself be indicative for such a role. Unfortunately, so far the highly specific 6-KEH enzymes have only been described in fructan plants (see higher).

## Conclusion

In fructan plants, FEHs are involved in many fundamental processes such as reserve breakdown for remobilization, osmoregulation and stress tolerance. The unraveling of the first 3D structure of a plant FEH will boost structure-function research on FEHs, invertases and fructosyl transferases and this will open the way to promising agronomical and industrial applications. The discovery that FEHs are not restricted to fructan accumulating plants but occur in non-fructan plants opens a completely new research area. Novel physiologically important functions in defence and signalling processes are now proposed but wait further confirmation. The unexpected appearance of FEHs among cell wall invertases highlights the strict necessity of a full functional characterization in the group.

## REFERENCES


Alberto, F., C. Bignon, G. Sulzenbacher, B. Henrissat and M. Czjzek. 2004. The three-dimensional structure of invertase (beta -fructosidase) from *Thermotoga maritima* reveals a bimodular arrangement and an evolutionary relationship between retaining and inverting glycosidases. *J. Biol. Chem.*, **279**: 18903-18910.

Asega, A.F. and M.A.M de Carvalho. 2004. Fructan metabolising enzymes in rhizophores of *Vernonia herbacea* upon excision of aerial organs. *Plant Physiol. Biochem.*, **42**: 313-319.

Barta, J. and G. Pátkai. 2000. Food Products from Jerusalem artichoke. Proc. 8th seminar on inulin. A. Fuchs Ed. European Fructan Association, pp. 122-124.

Bereswill, S., S. Jock, P. Aldridge, J.D. Janse and K. Geider. 1997. Molecular characterization of natural *Erwinia amylovora* strains deficient in levan synthesis. *Phys. Mol. Plant Path.*, **51**: 215-225.

Bieleski, R. 1993. Fructan hydrolysis during petal expansion in the ephemeral dailily flower. *Plant Physiol.*, **109**: 557-565.

Bonnett, G.D. and R.J. Simpson. 1993. Fructan-hydrolyzing activities from *Lolium rigidum* Gaudin. *New Phytol.*, **123**: 443-451.

Bonnett G.D. and R.J. Simpson. 1995. Fructan exohydrolase activities from *Lolium rigidum* that hydrolase • 2,1 glycosidic and • 2,6 glycosidic linkages at different rates. *New Phytol.*, **131**: 199-209.

Cairns, A.J. 2003. Fructan biosynthesis in transgenic plants. J. Exp. Bot., 54: 549-567.

Cairns, A.J. and J.E. Ashton. 1991. The interpretation of in vitro measurements of fructosyl transferase activity: an analysis of patterns of fructosyl transfer by fungal invertase. New Phytol., 118: 23-24.

Chambert, R., G. Treboul and R. Dedonder. 1974. Kinetic studies of levansucrase of *Bacillus subtilis*. Eur. J. Biochem., 41: 285-300.

Chatterton, N.J., P.A. Harrison, W.R. Thornley and E.A. Draper. 1990. Oligosaccharides in foliage of *Agropyron, Bromus, Dactylis, Festuca, Lolium and Phleum*. New Phytol., 114: 167-171.

Claessens G., A. Van Laere A and M. De Proft. 1990. Purification and properties of an inulinase from chicory roots (*Cichorium intybus* L.). J. Plant Physiol., **136:** 35-39.

Cregg, J.M., J.L. Cereghino, J.Y. Shi and D.R. Higgins. 2000. Recombinant protein expression in *Pichia pastoris.* Mol. Biotech., 16: 23-52 .

Darwen C.W.E. and P. John. 1989. Localisation of the enzymes of fructan metabolism in vacuoles isolated by a mechanical method from tubers of Jerusalem artichoke (*Helianthus tuberosus* L.). Plant Physiol., 89: 658-663.

De Coninck, B., K. Le Roy, I. Francis, S. Clerens, R. Vergauwen, A. Halliday, S.M. Smith, A. Van Laere and W. Van den Ende. 2005. Arabidopsis AtcwINV3 and 6 are not invertases but are fructan exohydrolases (FEHs) with different substrate specificities. Plant Cell Env., In press.

De Leenheer, L. Industrial chicory production and processing. Fifth International Fructan Symposium, Havana, Cuba, December 5-9, 2004, pp. 136.

De Roover J., M. De Winter, A. Van Laere, J.W. Timmermans and W. Van den Ende. 1999. Purification and properties of a second fructan exohydrolase from the roots of *Cichorium intybus* L. Physiol. Plant., 106: 28-34.

De Roover, J., A. Van Laere and W. Van den Ende. 1999. Effect of defoliation on fructan pattern and fructan metabolizing enzymes in chicory roots (*Cichorium intybus*). Physiol. Plant., 106: 158-163.

De Roover, J., K. Van den Branden, A. Van Laere and W. Van den Ende. 2000. Drought induces fructan synthesis and 1-SST (sucrose: sucrose 1-fructosyl transferase) in roots and leaves of chicory seedlings (*Cichorium intybus* L). Planta, 210: 808-814.

Desprez B., R.G. Wilson, L. Delesalle and J. Lepeltier. 2005. Chicory breeding for the production of high quality inulin. Abstract at the Fifth International Fructan Symposium. December 5-9, 2004, Havana, Cuba. pp. 135.

Echeverria, E. 2000. Vesicle-mediated solute transport between the vacuole and the plasma membrane. Plant Physiol., 123: 1217-1226.

Edelman, J. and T.G. Jefford. 1964. The metabolism of fructose polymers in plants. 4. •-fructofuranosidases of tubers of *Helianthus tuberosus*. Biochem. J., 93: 148-161.

Edelman, J. and T.G. Jefford. 1968. The mechanism of fructosan metabolism in higher plants as exemplified in *Helianthus tuberosus*. New Phytol., 67: 517-531.

Etxeberria E., E. Baroja-Fernandez, F.J. Munoz and J. Pozueta-Romero. 2005. Sucrose inducible endocytosis as a mechanism fror nutrient uptake in heterotrophic plant cells. Plant and Cell Physiol., In press.

Ernst, M.K., N.J. Chatterton, P.A. Harrison and G. Matitschka. 1998. Characterization of fructan oligomers from species of the genus *Allium* L. J. Plant Physiol., 153: 53-60.

Frehner M., F. Keller and A. Wiemken. 1984. Localisation of fructan metabolism in the vacuoles isolated from protoplasts of Jerusalem artichoke. J. Plant Physiol., 116: 197-208.

Fujishima, M., H. Sakai, K. Ueno, N. Takahashi, S. Onodera, N. Benkeblia and N. Shiomi. 2005. Purification and characterization of a fructosyltransferase from onion bulbs and its key role in the synthesis of fructo-oligosaccharides *in vivo.* New Phytol., 165: 513-524.

Graham, L.S. and M.B. Sticklen. 1994. Plant chitinases. Can. J. Bot., 72: 1057-1083.

Han, Y.W. 1990. Microbial levan. Adv. Appl. Microbiol., 35: 171-194.

Hellwege, E.M., S. Czapla, A. Jahnke, L. Willmitzer and A.G. Heyer. 2000. Transgenic potato (*Solanum tuberosum*) tubers synthesize the full spectrum of inulin molecules naturally occurring in globe artichoke (*Cynara scolymus*) roots. Proc. Natl. Acad. Sci. *USA,* 97: 8699-8704.

Hendry, G.A.F. 1993. Evolutionary origins and natural functions of fructans - a climatological, biogeographic and mechanistic appraisal. New Phytol., 123: 3-14.

Henrissat, B. and G. Davies. 1997. Structural and sequence-based classification of glycoside hydrolases. Curr. Op. Struct. Biol., 7: 637-644.

Henson, C.A. and D.P. Livingston. 1996. Purification and characterization of an oat fructan exohydrolase that preferentially hydrolyzes β-2,6 fructans. Plant Physiol., 110: 639-644.

Henson, C.A. and D.P. Livingston. 1998. Characterization of a fructan exohydrolase purified from barley stems that hydrolyzes multiple fructofuranosidic linkages. Plant Physiol. Biochem. Paris, 36: 715-720.

Hernandez, L., J. Arrieta, C. Menendez, R. Vazquez, A. Coego, V. Suarez, G. Selman, M.F. PetitGlatron and R. Chambert. 1995. Isolation and enzymatic properties of levansucrase secreted by *Acetobacter diazotrophicus* SRT4, a bacterium associated with sugar cane. Biochem. J., 309: 113-118.

Hettwer, U., M. Gross and K. Rudolph. 1995. Purification and characterization of an extracellular levansucrase from *Pseudomonas syringae* pv Phaseolicola. J. Bact., 177: 2834-2839.

Heyer, A.G. and R. Wendenburg. 2001. Gene cloning and functional characterization by heterologous expression of the fructosyltransferase of *Aspergillus sydowi* IAM 2544 Appl. Env. Microbiol., 67: 363-370.

Hincha, D.K., E.M. Hellwege, A.G. Heyer and J.H. Crowe. 2000. Plant fructans stabilize phosphatidylcholine liposomes during freeze-drying. Eur. J. Biochem., 267: 535-540.

Hincha, D.K., E. Zuther, E.M. Hellwege and A.G. Heyer. 2002. Specific effects of fructo-and gluco-oligosaccharides in the preservation of liposomes during drying. Glycobiology 12: 103-110.

Hisano, H., A. Kanazawa, A. Kawakami, M. Yoshida, Y. Shimamoto and T. Yamada. 2004. Transgenic perennial ryegrass plants expressing wheat fructosyltransferase gene accumulate increased amounts of fructan and acquire increased trance on a cellular level of freezing. Plant Science, 167: 861-868.

Iizuka, M., H. Yamaguchi, S. Ono and N. Minamiura. 1993. Production and isolation of levan by use of levansucrase immobilized on the ceramic support SM-10. Biosci. Biotechnol. Biochem., 57: 322-324.

Jeong, B-R. 1991. Fructan metabolism in wheat (*Triticum aestivum* L.) Ph. D. thesis, Perdue University, USA.

Kasapis, S., E.R. Morris, M. Gross and K. Rudolph. 1994. Solution properties of levan polysaccharide from *Pseudomonas syringae* pv *phaseolicola*, and its possible primary role as a blocker of recognition during pathogenesis. Carbohyd. Polymers, 23: 55-64.

Kaur, N. and A.K. Gupta. 2002. Applications of inulin and oligofructose in health and nutrition. J. Biosci., 27: 703-714.

Kawakami, A. and M. Yoshida. 2002. Molecular characterization of sucrose:sucrose 1-fructosyl transferase and sucrose:fructan 6-fructosyl transferase associated with fructan accumulation in winter wheat during cold hardening. Biosci. Biotechnol. Biochem., 66: 297-2305.

Koops, A. and H. Jonker. 1996. Purification and characterisation of the enzyme of fructan biosynthesis in tubers of *Helianthus tuberosus* Colombia. 2. Purification of sucrose:sucrose 1-fructosyltransferase and reconstitution of fructan synthesis *in vitro* with purified sucrose:sucrose 1-fructosyltransferase and fructan:fructan 1-fructosyltransferase. Plant Physiol., 100: 1167-1175.

Kritzinger, S.M., S.G. Kilian, M.A. Potgieter and J.C. du Preez. 2003. The effect of production parameters on the synthesis of the prebiotic trisaccharide, neokestose, by *Xanthophyllomyces dendrorhous* (*Phaffia rhodozyma*). Enzyme and microbial technology, 32: 728-737.

Le Roy K., A. Van Laere and W. Van den Ende. Glycosylation affects substrate specificity in 1-FEH. Fifth International Fructan Symposium, Havana, Cuba, December 5-9, 2004, pp. 107.

Lewis D.H. 1993. Nomenclature and diagrammatic representation of oligomeric fructans - a paper for discussion. New Phytol., 124: 583-594.

Livingston, D.P., N.J. Chatterton and P.A. Harrison. 1993. Structure and quantity of fructan oligomers in oat (*Avena* spp.). New Phytol., 123: 725-734.

Livingston, D.P. and C.A. Henson. 1998. Apoplastic sugars, fructans, fructan exohydrolase, and invertase in winter oat: Responses to second-phase cold hardening. Plant Physiol., 116: 403-408.

Lothier, J., A.S. Miclot, A. Ernstsen, V. Amiard, M.P. Prud'homme and A. Morvan-Bertrand. 2004. Regulation of fructan exohydrolase activity by sugars in *Lolium perenne* after defoliation. Fifth International Fructan Symposium, Havana, Cuba, December 5-9, 2004. pp. 85.

Lüscher, M., C. Erdin, N. Sprenger, U. Hochstrasser, T. Boller and A. Wiemken. 1996. Inulin synthesis by a combination of purified fructosyltransferases from tubers of *Helianthus tuberosus*. FEBS Lett., 385: 39-42.

Marx, S.P., J. Nösberger and M. Frehner. 1997a. Seasonal variation of fructan-β-fructosidase (FEH) activity and characterisation of a β-(2,1)-linkage specific FEH. New Phytol., 135: 267-277.

Marx, S.P., J. Nösberger and M. Frehner. 1997b. Hydrolysis of fructan in grasses: A β-(2,6)-linkage specific fructan-β-fructosidase from stubble of *Lolium perenne*. New Phytol., 135, 279-290.

Meng, G. and K. Fütterer. 2003. Structural framework of fructosyl transfer in *Bacillus subtilis* levansucrase. Nat. Struct. Biol., 10: 935-941.

Michiels, A. 2003. Molecular cloning and functional characterization of fructan 1-exohydrolases in *Cichorium intybus* L. var. foliosum Hegi. Ph. D. Thesis. K.U.Leuven.

Michiels, A., M. Tucker, W. Van den Ende and A. Van Laere. 2003. Chromosomal walking of flanking regions from short known sequences in GC rich plant genomic DNA. Plant Mol. Biol. Rep., 21: 1-8.

Michiels A., A. Van Laere, W. Van den Ende and M. Tucker. 2004. Expression analysis of a chicory fructan 1-exohydrolase gene reveals complex regulation by cold. J. Exp. Bot., 55: 1325-1333.

Morcuende, R., S. Kostadinova, P. Perez, I.M.M del Molino and R. Martinez-Carrasco. 2004. Nitrate is a negative signal for fructan synthesis, and the fructosyltransferase-inducing trehalose inhibits nitrogen and carbon assimilation in excised barley leaves. New Phytol., 161: 749-759.

Morvan, A., G. Challe, M.P. Prud'homme, J. LeSaos and J. Boucaud. 1997. Rise of fructan exohydrolase activity in stubble of *Lolium perenne* after defoliation is decreased by uniconazole, an inhibitor of the biosynthesis of gibberellins. New Phytol., 136: 81-88.

Morvan-Bertrand, A., J. Boucaud, J. Le Saos and M.P. Prud'homme. 2001. Roles of the fructans from leaf sheaths and from the elongating leaf bases in the regrowth following defoliation of *Lolium perenne* L. Planta, 213: 109-120.

Nagaraj, V.J., D. Altenbach, V. Galati, M. Luscher, A.D. Meyer, T. Boller, and A. Wiemken. 2004. Distinct regulation of sucrose: sucrose-1-fructosyltransferase (1-SST) and sucrose: fructan-6-fructosyl transferase (6-SFT), the key enzymes of fructan synthesis in barley leaves: 1-SST as the pacemaker. New Phytol., 161: 735-748.

Nagem, R.A.P., A.L. Rojas, A.M. Golubev, O.S. Korneeva, E.V. Eneyskaya, A.A. Kulminskaya, K.N. Neustroev and I. Polikarpov. 2004. Crystal structure of exo-inulinase from *Aspergillus awamori*: the enzyme fold and structural determinants of substrate recognition. J. Mol. Biol., 344: 471-480.

Pavis, N., J. Boucaud and M.P. Prud'homme. 2001. Fructans and fructan-metabolizing enzymes in leaves of *Lolium perenne*. New Phytol., 150: 97-109.

Parvanova, D., S. Ivanov, T. Konstantinova, E. Karanov, A. Atanassov, T. Tsvetkov, V. Alexieva and D. Djilianov. 2004. Transgenic tobacco plants accumulating osmolytes show reduced oxidative damage under freezing stress. Plant Physiol. Biochem., 42, 57-63.

Pilon-Smits, E.A.H, M.J.M. Ebskamp, M.J.W. Jeuken, P.J. Weisbeek and S.C.M Smeekens. 1995. Improved performance of transgenic fructan-accumulating tobacco under drought stress. Plant Physiol., 107: 125-130.

Pollock, C.J. 1986. Fructans and the metabolism of sucrose in vascular plants. New Phytol. 104: 1-24.

Pool-Zobel, B., J. van Loo, I. Rowland and M.B. Roberfroid. 2002. Experimental evidences on the potential of prebiotic fructans to reduce the risk of colon cancer. Br. J. Nutr., 87: S273-81.

Ritsema, T. and S. Smeekens. 2003. Fructans: beneficial for plants and humans. Curr. Op. Plant. Biol., 6: 223-230.

Ritsema, T., J. Joling and S. Smeekens. 2003. Patterns of fructan synthesized by onion fructan: fructan 6G-fructosyl transferase expressed in tobacco BY2 cells - Is fructan: fructan 1-fructosyl transferase needed in onion? *New Phytol.,* 160: 61-67.

Ritsema, T., J. Tuynman, S. Diks, C. Bos, D. Brodmann, D., V. Nagaraj, M. Peppelenbosch, T. Boller and A. Wiemken. The promoter of sucrose: fructan 6-fructosyltransferase (SFT) and its regulation by sucrose. Fifth International Fructan Symposium, Havana, Cuba, December 5-9, 2004. pp. 69.

Roberfroid, M. B. and N.M. Delzenne. 1998. Dietary fructans. Annu. Rev. Nutr., 18: 117-143.

Rossell, K.G. and D. Birkhed. 1974. Acta Chem. Scand., B 28: 589.

Rutherford, P.P. and A.C. Deacon. 1972. Fructofuranosidases from roots of dandelion (*Taraxacum officinale* Weber). Biochem J., 126: 569-573.

Sévenier, R., R.D. Hall, I.M. van der Meer, H.J.C. Hakkert, A.J. van Tunen A.J. and A.J. Koops. 1998. High level fructan accumulation in a transgenic sugar beet. Nature Biotech., 16: 843-846.

Sherson, S.M., H.L. Alford, S.M. Forbes, G. Wallace and S.M. Smith. 2003. Roles of cell wall invertases and monosaccharide transporters in the growth and development of *Arabidopsis*. J. Exp. Bot., 54: 525-531.

Shiomi, N. 1992. Content of carbohydrate and activities of fructosyltransferase and invertase in Asparagus roots during the fructo-oligosaccharide-accumulating and fructo-polysaccharide-accumulating season. New Phytol., 122: 421-432.

Shiomi, N., S. Onodera and H. Sakai. 1997. Fructo-oligosaccharide content and fructosyltransferase activity during growth of onion bulbs. New Phytol., 136: 105-113.

Simpson, R.J. and G.D. Bonnett. 1993. Fructan exohydrolase from grasses. New Phytol., 123: 453-469.

Solhaug, K.A. and E. Aares. 1994. Remobilisation of fructans in *Phippsia algida* during rapid inflorescence development. Phys. Plant., 91: 219-225.

Tapernoux-Luthi, E.M., A. Bohm and F. Keller. 2004. Cloning, functional expression, and characterization of the raffinose oligosaccharide chain elongation enzyme, galactan: galactan galactosyltransferase, from common bugle leaves. Plant Physiol., 134: 1377-1387.

Tharaud, M., M. Menggad, J.P. Paulin and J. Laurent. 1994. Virulence, growth, and surface characteristics of *Erwinia amylovora* mutants with altered pathogenicity. Microbiol.-UK, 140: 659-669.

Trujillo, L.E., A. Banguela, J. Pais, Y. Tambara, J.G. Arrieta, M. Sotolongo and L. Hernandez. 2002. Constitutive expression of enzymatically active *Gluconacetobacter diazotrophicus* levansucrase in the methylothrophic yeast *Pichia pastoris*. Afinidad, 59: 365-370.

Ueno, K., S. Onodera, A. Kawakami, M. Yoshida and N. Shiomi. 2004. Molecular characterization and expression of a cDNA encoding unique • -fructofuranosidase from asparagus (*Asparagus officinalis* L.). Fifth International Fructan Symposium, Havana, Cuba, December 5-9, 2004. pp. 73.

Ueno, K., S. Onodera, A. Kawakami, M. Yoshida and N. Shiomi. 2005. Molecular characterization and expression of a cDNA encoding fructan: fructan 6G-fructosyltransferase from asparagus *(Asparagus officinalis)*. New Phytol., In press.

Van Damme, E.J. and D.G. Derycke. 1983. Microbial inulinases. Fermentation process, properties and applications. Adv. Appl. Microbiol., 29: 139-176.

Van den Ende, W. and A. Van Laere. 1996a. *De novo* synthesis of fructans from sucrose *in vitro* by a combination of two purified enzymes (SST and FFT) from chicory roots (*Cichorium intybus* L.). Planta, 200: 335-342.

Van den Ende, W. and A. Van Laere. 1996b. Fructan synthesizing and degrading activities in chicory roots (*Cichorium intybus*) during field-growth, storage and forcing. J. Plant Physiol., 149: 43-50.

Van den Ende, W., J. De Roover J and A. Van Laere. 1996a. *In vitro* synthesis of fructo-oligosaccharides from inulin and fructose by purified chicory root fructan:fructan fructosyltransferase. Phys. Plant., 97: 346-352.

Van den Ende, W., A. Mintiens, H. Speleers, A.A. Onuoha and A. Van Laere. 1996b. The metabolism of fructans in roots of *Cichorium intybus* L. during growth, storage and forcing. New Phytol., 132: 555-563.

Van den Ende, W., G. Van Hoenacker , S. Moors and A. Van Laere. 1998. Effect of osmolytes on the fructan pattern in feeder roots produced during forcing of chicory (*Cichorium intybus* L.). J. Plant Physiol., 153: 290-298.

Van den Ende, W., J. De Roover and A. Van Laere. 1999. Effect of nitrogen concentration on fructan and fructan metabolizing enzymes in young chicory plants (*Cichorium intybus*). Physiol. Plant., 105: 2-8.

Van den Ende, W., A. Michiels, J. De Roover, P. Verhaert and A. Van Laere. 2000. Cloning and functional analysis of chicory root fructan 1-exohydrolase I (1-FEH I): a vacuolar enzyme derived from a cell wall invertase ancestor? Mass fingerprint of the 1-FEH I enzyme. Plant J., 24: 447-456.

Van den Ende, W., A. Michiels, D. Van Wonterghem, S. Clerens, J. De Roover and A. Van Laere. 2001. Defoliation induces 1-FEH II (fructan 1-exohydrolase II) in witloof chicory roots. Cloning and purification of two isoforms (1-FEH IIa and 1-FEH IIb). Mass fingerprint of the 1-FEH II enzymes. Plant Physiol., 126: 1186-1195.

Van den Ende, W., A. Michiels, J. De Roover and A. Van Laere. 2002. Fructan biosynthetic and breakdown enzymes in dicots evolved from different invertases. Expression of fructan genes throughout chicory development. The ScientificWorldJOURNAL, 2: 1273-1287.

Van den Ende, W., S. Clerens, R. Vergauwen, L. Van Riet, A. Van Laere, M. Yoshida and A. Kawakami. 2003a. Fructan 1-exohydrolase: ß(2,1) trimmers during graminan biosynthesis in stems of wheat (*Triticum aestivum* L.)? Purification, characterization, mass mapping and cloning of two 1-FEH isoforms. Plant Physiol., 131: 621-631.

Van den Ende, W., B. De Coninck, S. Clerens, R. Vergauwen and A. Van Laere 2003b. Unexpected presence of fructan 6-exohydrolases (6-FEHs) in non-fructan plants: characterization, cloning, mass mapping and functional analysis of a novel 'cell wall invertase-like' specific 6-FEH from sugar beet (*Beta vulgaris* L.). Plant J., 36: 697-710.

Van den Ende, W., B. De Coninck and A. Van Laere. 2004a. Plant fructan exohydrolases: a role in signaling and defense? *TRENDS Plant Sci.*, 9: 523-528.

Van den Ende, W., M. Yoshida, S. Clerens, R. Vergauwen and A. Kawakami. 2004b. Novel 6-kestose exohydrolases (6-KEHs) in wheat (Triticum aestivum L.). Fifth International Fructan Symposium, Havana, Cuba, December 5-9, 2004. pp.82.

Van den Ende, W., M. Yoshida, S. Clerens, R. Vergauwen and A. Kawakami. 2005. Cloning, characterization and functional analysis of novel 6-kestose exohydrolases (6-KEHs) from wheat (*Triticum aestivum* L.) New Phytol., In press.

Van Hijum, S.A.F.T., G.H. van Geel-Schutten, H. Rahaoui, M.J. van der Maarel, and L. Dijkhuizen. 2002. Characterization of a novel fructosyltransferase from *Lactobacillus reuteri* that synthesizes high-molecular-weight inulin and inulin oligosaccharides. Appl. Env. Microbiol., 68: 4390-4398.

Van Laere, A. and W. Van den Ende. 2002. Inulin metabolism in dicots: chicory as a model system. Plant, Cell Env., 25: 803-815.

Van Riet, L., W. Van den Ende, A. Wiemken and A. Van Laere. Purification, cloning and properties of a 6-FEH from wheat (*Triticum aestivum*). Fifth International Fructan Symposium, Havana, Cuba, December 5-9, 2004. pp. 72.

Vergauwen, R., W. Van den Ende and A. Van Laere. 2000. The role of fructans in flowering of *Campanula rapunculoides*. *J. Exp. Bot.,* 51: 1261-1266.

Vergauwen, R., Van Laere, A. and W. Van den Ende. 2003. Properties of Fructan: Fructan 1-fructosyltransferase (1-FFT) from *Cichorium intybus* L. and *Echinops ritro* L., two Asteracean Plants Storing Greatly different types of inulin. Plant Physiol., 133: 391-401.

Verhaest, M., W. Van den Ende, M. Yoshida, K. Le Roy, Y. Peeraer, S. Sansen, C.J. De Ranter, A. Van Laere and A. Rabijns. 2004. Crystallization and preliminary X-ray diffraction study of fructan 1-exohydrolase IIa from *Cichorium intybus*. Acta Cryst. D., 60: 553-554.

Verhaest, M., W. Van den Ende, K. Le Roy, C.A De Ranter, A. Van Laere and A. Rabijns. 2005. X-Ray Diffraction Structure of a plant Glycosyl Hydrolase family 32 protein: Fructan 1-Exohydrolase IIa of *Cichorium intybus*. Plant J. 41: 400-411.

Vereyken, I.J., V. Chupin, R.A. Demel, S.C.M. Smeekens and B. De Kruijff. 2001. Fructans insert between the headgroups of phospholipids. BBA Biomem., 1510: 307-320.

Vijn, I., A. van Dijken, N. Sprenger, K. van Dun, P. Weisbeek, A. Wiemken and S. Smeekens. 1997. Fructan of the inulin neoseries is synthesized in transgenic chicory plants (*Cichorium intybus* L.) harbouring onion (*Allium cepa* L.) fructan: fructan $6^G$ fructosyltransferase. Plant J., 11: 387-398.

Vijn, I. and S.C.M Smeekens. 1999. Fructan: more than a reserve carbohydrate? Plant Physiol., 120: 351-359.

Wagner, W. and A. Wiemken. 1986. Properties and subcellular localization of fructan hydrolase in the leaves of barley (*Hordeum vulgare* L. cv. Gerbel). *J. Plant Physiol.* 123: 429-439.

Wang N. and P.S. Nobel. 1998. Phloem transport of fructans in the Crassulacean acid metabolism species *Agave deserti*. *Plant Physiol.,* 116, 709-712.

Wei, J.Z., N.J. Chatterton, P.A. Harrison, R.R.C. Wang and S.R. Larson. 2002. Characterization of fructan biosynthesis in big bluegrass (*Poa secunda*). J. Plant Physiol., 159: 705-715.

Weyens, G., T. Ritsema, K. Van Dun, D. Meyer, M Lommel, J. Lathouwers, I Rosquin, P. Denys, A. Tossens, M. Nijs, S. Turk, N. Gerrits, S. Bink, B. Walraven, M. Lefebvre and S.C.M. Smeekens. 2004. Plant Biotech. J., 2: 321-327.

Wiemken, A., N. Sprenger and T. Boller. 1995. Fructans - an extension of Sucrose by Sucrose. In HG Pontis, GL Salerno, EJ Echeverria, eds, Current Topics in Plant Physiology Volume 14. ASPP, Rockville, pp 179-189.

Wilson, R.G., J.A. Smith and C.D. Yonts. 2004. Chicory root yield and carbohydrate composition is influenced by cultivar selection, planting, and harvest date Crop Sci., 44: 748-752.

Yamamoto, S. and Y. Mino. 1985. Partial purification and properties of phleinase induced in stem base of orchardgrass after defoliation. Plant Physiol., 78: 591-595.

Yamamoto, S. and Y. Mino. 1987. Effect of sugar level on phleinase induction in stem base of orchardgrass. Physiol. Plant., 69: 456-460.

Yamamoto, S. and Y. Mino. 1998. Effect of exogenous plant bioactive substances on phleinase induction in stem base of orchardgrass. Grassland Sci., 43: 380-384.

Yamamoto, S., S. Amano and Y. Mino. 1999. Carbohydrate metabolism in the stem base of timothy and orchardgrass in winter. Grassland Sci., 44: 315-319.

Yang, J., J. Zhang, Z. Wang, Q. Zhu and L. Liu. 2004. Activities of fructan- and sucrose-metabolizing enzymes in wheat stems subjected to water stress during grain filling. Planta, 220: 331-343.

Yoshida, M. J. Abe, M. Moriyama and T. Kuwabara. 1998. Carbohydrate levels among winter wheat cultivars varying in freezing tolerance and snow mold resistance during autumn and winter. Physiol. Plant., 103: 8-16.

Yukawa, T., Y. Watanabe and S. Yamatoto. 1994. Studies on fructan accumulation in wheat (*Triticum aestivum L.*) 2. Changes in degree of polymerization of fructan under treatment at 1 degrees °C in dark. Jpn. J. Crop. Sci., 63: 430-435.

Zuther, E., M. Kwart, L. Willmitzer and A.G. Heyer. 2003. Expression of a yeast-derived invertase in companion cells results in long-distance transport of a trisaccharide in an apoplastic loader and influences sucrose transport. Planta., 218: 759-766.

*Advances in Plant Physiology*, Vol. 9
Ed. A. Hemantaranjan
Scientific Publishers (India), Jodhpur, 2006 pp. 369-375
E-mail: **info@scientificpub.com** www.scientificpub.com

# 17

# 4-CHLOROINDOLE-3-ACETIC ACID : METABOLISM AND BIOACTIVITY IN PLANTS

B. Ali, S. Hayat, Q. Fariduddin and A. Ahmad*

Deptt. of Botany, Aligarh Muslim University, ALIGARH-202002, India
*Corresponding author E-mail.aqil_ahmad@lycos.com

## INTRODUCTION

It is well recognized that indole-3-acetic acid (IAA) is the major and ubiquitous auxin in plants, that plays a key role in the regulation of plant growth and development (Moore, 1989; Davies, 1995). The other naturally occurring auxins include 4-chloroindole-3-acetic acid (4-Cl-IAA), indole-3-butyric acid (IBA) and phenyl acetic acid (PAA) (Normanly *et al.*, 1995; Normanly, 1997). 4-Cl-IAA occurs in plants, belonging to the family of Fabaceae (Engvild, 1975, 1980; Engvild *et al.*, 1978, 1980; Hofinger and Böttger, 1979; Katayama *et al.*, 1987) and also in the seeds of *Pinus sylvestris* (Ernstsen and Sandberg, 1986). The precursor of 4-Cl-IAA is identified as (4-Cl-Trp) 4-chlorotryptophan (Manabe *et al.*, 1999). A number of bioassays have demonstrated high degree of biological activity of 4-Cl-IAA. It includes, pea stem split curvature growth, pea stem straight growth, oat and wheat coleoptile growth, mungbean hypocotyl growth and increased root initiation and ethylene evolution in pea cuttings (Porter and Thimann, 1965; Marumo *et al.* 1974; Böttger *et al.*, 1978; Katekar and Geissler, 1983; Ahmad *et al.*, 1987). Moreover, 4-Cl-IAA also induced the synthesis and/or activation of specific enzymes (Hirasawa, 1989; Ahmad and Hayat, 1999; Ahmad *et al.*, 2001a) and also activates the rate of photosynthesis and seed production in *Brassica juncea* (Ahmad *et al.*, 2001b).

## 2. OCCURRENCE

4-Cl-IAA and its methyl ester were, for the first time noticed in immature seeds of *Pisum sativum* (Gandar and Nitsch, 1967; Marumo *et al.*, 1968a) and later on in mature pea seeds (Marumo *et al.*, 1968b; Schneider *et al.*, 1985), leaves and both mature and immature seeds of *Vicia faba* (Pless *et al.*, 1984; Hofinger and Böttger, 1979; Engvild et al., 1980), immature seeds of *Lens culinaris, Vicia sativa, Lathyrus maritinus, Lathyrus sativus and Lathyrus orodatus* (Engvild *et al.*, 1981), *Vicia amurensis* (Katayama *et al.*, 1987) and *Lathyrus lotifolius* (Engvild, 1980). The seeds of *Pinus sylvestris* also possess 4-Cl-IAA or its methyl ester (Ernstsen and Sandberg, 1986). However, in some species of vicae tribe namely, *Glycine max, Vigna catiang, Dolichos lablab, Glycine soja* and *Cicer arietinum* do not possess 4-Cl-IAA/methyl ester (Hofinger and Böttger, 1979; Katayama *et al.*, 1987, Engvild, 1994). Moreover, Engvild (1975) tested 14 plant types (*Hordeum vulgare, Avena sativa, Triticum aestivum, Secale cereale, Zea mays, Allium schoenoprasum, Phaseolus vulgaris, Glycine max, Lapidium sativum, Brassica napus, Linum usitatissimum, Nicotiana tabacum, Lycopersicum esculentum* and *Halianthus annuus)* and observed that none of these synthesized detectable quantities of either 4-Cl-IAA/ its methyl ester but incorporated chloride into unidentified compounds. However, the identification of 4-Cl-IAA in *Pinus sylvestris* (Ernstsen and Sandberg, 1986) provided a clue for the wider destribution of 4-Cl-IAA. A concrete conclusion may, therefore, be drawn after quantifying its level in other plant species, by adopting a most sensitive method.

## 3. METABOLISM

The metabolic pathways for biosynthesis and catabolism of 4-Cl-IAA in plants have been studied a little, due to non-availability of labeled 4-Cl-IAA and its precursors (Reinecke, 1999).However, protocols for $^{14}$C-4-Cl-IAA synthesis have been described (Engvild, 1994; Baldi *et al.*, 1985), but radiolabeled 4-Cl-IAA has so far been used only as a tracer in the quantification of endogenous 4-Cl-IAA (Pless *et al.*, 1984; Magnus *et al.*, 1997).

Watering of pea plants, at preflowering stage, with radiolabeled chlorine ($^{36}$Cl) resulted in the accumulation of radiolabeled 4-Cl-IAA and its methyl ester, in immature seeds (Engvild, 1975). L-4-Cl-tryptophan was identified in two week old plants of *Vicia faba* and these plants also incorporated deuterium labeled indole to 4-chloroindole. Moreover, deuterium labeled L- and D-4-Cl-tryptophan were synthesized and supplied to *Vicia* plants. L-4-Cl-tryptophan was incorporated to 4-Cl-IAA but the incorporation ratio was very low (Manabe *et al.*, 1999). Natural occurrence of L-4-Cl-tryptophan has also been reported in *Vicia faba* seeds and young plants (Fock *et al.*, 1992; Higashi *et al.*, 1998). On the basis of these observations, it was concluded that L-4-Cl-tryptophan is the precursor of 4-Cl-IAA, against the earlier belief where D-tryptophan was considered as the precursor of 4-Cl-IAA (Marumo and Hattori, 1970).

Tryptophan is probably not the only source for auxin synthesis as tryptophan deficient mutants of maize (Wright *et al.*, 1991) and *Arabidopsis* (Fink, 1991) synthesize a sufficient quantity of IAA. Hence, the question of biosynthesis of IAA and that of 4-Cl-IAA is completely open (Engvild, 1994). Therefore, the biosynthesis of 4-Cl-IAA by a non-4-Cl-tryptophan pathway is not ruled out (Reinecke, 1999).

The metabolism of 4-Cl-IAA to 4-Cl-dioxindole in *Bradyrhizobium japonicum* (Jensen *et al.*, 1995) is likely mediated by 4-Cl-dioxindole-3-acetic acid, since oxidases in *B. japonicum* converted 5-Cl-IAA to 5-Cl-dioxindole-3-acetic acid and 5-Cl-dioxindole (Reinecke, 1999). This shows a synergism with the catabolism of IAA (Bandurski *et al.*, 1995). Similarly, methyl ester and amide conjugates of 4-Cl-IAA have been identified in pea seeds and pericarp (Marumo *et al.*, 1968; Hattori and Marumo, 1972) and quantified, by following base hydrolysis (Magnus *et al.*, 1997). The methyl ester of 4-Cl-IAA has also been identified in immature seeds of *Vicia faba* (Hofinger and Böttger, 1979). Moreover, 4-Cl-IAA-aspartate has also been reported to be highly active in the *Phaseolus mungo* hypocotyl swelling assay (Hattori and Marumo, 1972). Apart all this, much more work is required before the emergence of a clear cut understanding of the metabolism of chloroindole auxins.

## 4. BIOACTIVITY

The biological activity of 4-Cl-IAA has been tested in a number of bioassays where it is some times 50 fold more than IAA (Engvild, 1985). 4-Cl-IAA is highly active in *Avena* coleoptile curvature assay (Gerhold and Muir, 1989), the classical auxin bioassay. Similarly, *Avena* coleophile elongation (Muir and Hansch, 1953; Marumo *et al.*, 1974; Böttger *et al.*, 1978; Hatano *et al.*, 1987), wheat coleoptile elongation (Marumo *et al.*, 1974; Katekar and Geissler, 1983; Za• ímalová and Kutá• ek, 1985), pea stem elongation (Katekar and Geissler, 1982), split pea curvature (Porter and Thimann, 1965), mungbean hypocotyl elongation (Marumo *et al.*, 1974), mungbean hypocotyl swelling (Marumo *et al.*, 1968a,b, 1974; Hatano *et al.*, 1987) exhibited highly active response to 4-Cl-IAA. Moreover, 4-Cl-IAA also induced lateral root initiation and their subsequent growth in pea cuttings (Ahmad *et al.*, 1987) and wheat and cucumber seedlings (Stenlid and Engvild, 1987). 4-Cl-IAA also enhanced the root primordia formation in pea, although, a few of them developed into root proper (Wightman *et al.*, 1986).

In maize, which is not a natural source of chlorinated auxins (Hofinger and Böttger, 1979) the exogenous application of 4-Cl-IAA activated its growth, over and above that of IAA (Fischer *et al.*, 1992; Karcz and Burdach, 1995, 2002; Karcz *et al.*, 1999). Similarly, 4-Cl-IAA favoured growth in maize coleoptile segments that was 10 times faster over that of 1-Naphthylactic acid (1-NAA) but the binding affinities of 4-Cl-IAA to auxin binding site-I (ABS-I) and site-II were 100 times lower, compared with 1-NAA (Hertel, 1994, 1995; Stotz and Hertel, 1994). This discrepancy, between the observed biological activity of 4-Cl-IAA and its binding efficiency to ABS-I and II, was argued by these authors by stating that auxin binding sites (I and II) are not the true auxin receptors. However, at a later stage (Rescher *et al.*, 1996) a correlation was observed between the growth promoting effect of 4-Cl-IAA and its binding affinity to ABS-I. Furthermore, the value of membrane potential, in response to 4-Cl-IAA, was two times more than that of IAA, suggesting a specific signal pathway, generated by 4-Cl-IAA in maize coleoptile (Karcz and Burdach, 2002).

Germinating seeds, maintain GA metabolism which is considered to be indispensable for pericarp growth (Sponsel, 1982; Ozga *et al.*, 1992). However, 4-Cl-IAA restored pericarp growth in deseeded ovaries of pea (Reinecke *et al.*, 1995) by favouring the conversion of $GA_{19}$ to $GA_{20}$ (Ozga and Brenner, 1992; Huizen *et al.*, 1995).

4-Cl-IAA strongly favours the production of ethylene in cucumber and wheat seedlings (Stenlid and Engvild, 1987) and pea leafy cuttings, where it is almost irreversible, for a certain duration, causing the loss of activity at apical meristem and the origin of laterals is favoured (Ahmad *et al.*, 1987). Like auxin, in deseeded bean pods, 4-Cl-IAA causes leaf senescence (Tamas *et al.*, 1981). It also inhibited the level of the enzyme (9-cytokinin-alanin synthase) that metabolizes cytokinin (Parker *et al.*, 1988).

Out of various chloroindole auxins tested, seed germination gave maximum response to 4-Cl-IAA (Ahmad *et al.*, 2001a) possibly by mediating the effect on hydrolases (Hirsawa, 1989; Ahmad *et al.*, 2001a). The level of the other enzyme, nitrate reductase (NR) increased in response to 4-Cl-IAA in 12 days old pea seedlings (Ahmad and Hayat, 1999) and 60 day old mustard plants (Ahmad *et al.*, 2001b). Moreover, leaf-applied

4-Cl-IAA also enhanced the activity of carbonic anhydrase, chlorophyll content, the rate of photosynthesis in the plants of *Brassica juncea* and seed yield, at harvest (Ahmad *et al.*, 2001b).

## Conclusion

A measurable quantity of chloroindole auxin and its methyl ester is reported in a limited number of plant species (e.g. *Pisum sativum, Vicia faba, Lens culinaris, Vicia sativa, Vicia amurensis, Lathyrus lotifolius, L. maritinus, L. orodatus, L. sativus and Pinus sylvestris*). However, in others, it may be a matter of time when the use of more sensitive methodologies may quantify them. Strong auxin like activity of 4-Cl-IAA makes it the most potential phytohormone to generate beneficial exploitation of the genetic potential of the plants to result in higher biological yield. It may also be used as a tool to understand the sequence of biological events and the path of signal transduction in various plant responses as it resists getting metabolised.

## REFERENCES

Ahmad, A. and S. Hayat. 1999. Response of nitrate reductase to substituted indole acetic acids in pea seedlings. In : Srivastava, G.C., K. Singh and M. Pal (ed.) : *Plant Physiology for Sustainable Agriculture*. P. 252-259. Pointer Publishers, Jaipur.

Ahmad, A., A.S. Andersen and K. Engvild. 1987. Rooting, growth and ethylene evolution of pea cuttings in resonse to chloroindole auxins. *Physiol. Plant.*, **69**: 137-140.

Ahmad, A., S. Hayat, Q. Fariduddin and I. Ahmad. 2001b. Photosynthetic efficiency of plants of *Brassica juncea,* treated with chlorosubstituted auxins. *Phytosynthetica*, **39**: 565-568.

Ahmad, A., S. Hayat, Q. Fariduddin and S. Alvi. 2001a. Germination and α-amylase activity in the grains of wheat, treated with chloroindole acetic acids. *Seed Technol.*, **23**: 88-91.

Baldi, B.G., J.P. Slovin and J.D. Cohen. 1985. Synthesis of $^{14}C$-labeled halogen substituted indole-3-acetic acid. *J. Label Comp. Radiopharm.*, **22**: 279-285.

Bandurski, R.S., J.D. Cohen, J.P. Slovin and D.M. Reinecke. 1995. Auxin biosynthesis and metabolism. In : Davies, P.J. (ed.) *Plant Hormones. Physiology, Biochemistry and Molecular Biology,* 2nd edn. Dordrecht : Kluwer Academic Publishers, pp. 649-670.

Böttger, M., K.C. Engvild and H. Soll. 1978. Growth of *Avena* coleoptile and pH drop of protoplast suspensions induced by chlorinated indole acetic acids. *Planta*, **140**: 89-92.

Davies, P.J. 1995. The Plant Hormones : Their nature, occurrence and functions. In: Davies, P.J. (ed) *Plant Hormones, Physiology, Biochemistry and Molecular Biology*. 2nd edn. Dordrecht : Kluwer Academic Publishers, pp. 1-12.

Engvild, K.C. 1975. Natural chlorinated auxins labeled with radioactive chlorine in immature seeds. *Physiol. Plant.*, **34**: 286-287.

Engvild, K.C. 1980. Simple identification of the neutral chlorinated auxin in pea by thin layer chromatography. *Physiol. Plant.*, **48**: 435-437.

Engvild, K.C. 1985. The chloroindole auxins of pea and related species. In: Purohit, S.S. (ed) *Hormonal Regulation of Plant Growth and Development,* Vol. II. Dordrecht : Martinus Nijhoff, pp. 221-234.

Engvild, K.C. 1994. The chloroindole auxins of pea, strong plant growth hormones or Endogenous Herbicides? D.Sc. Thesis, Denmark : Riso National Laboratory.

Engvild, K.C., H. Egsgaard and E. Larsen. 1978. Gas chromatographic mass spectrometric identification of 4-chloroindolyl-3-acetic acid methyl ester in immature green peas. *Physiol. Plant.*, **42**: 365-368.

Engvild, K.C., Egsgaard, H., and E. Larsen. 1980. Determination of 4-chloroindole-3-acetic acid methyl ester in *Lathyrus, Vicia* and *Pisum* by gas chromatography mass spectrometry. *Physiol. Plant.* **84**: 499-503.

Engvild, K.C., H. Egsgaard and E. Larsen. 1981. Determination of 4-cloroindole acetic acid methyl ester in Vicieae species by gas chromatography mass spectrometry. *Physiol. Plant.*, **53**: 79-81.

Ernstsen, A. and G. Sandberg. 1986. Identification of 4-chloroindole-3-acetic acid and indole-3-aldehyde in seeds of *Pinus sylvestris. Physiol. Plant.*, **68**: 511-518.

Fink, G.I. 1991. Plant molecular biology blossoms in the desert. Meeting Report. *Plant Cell*, **3**: 1259-1260.

Fischer, C., H. Luthen, M. Böttger and R. Hertel. 1992. Initial transient growth inhibition in maize coleoptile following auxin application. *J. Plant Physiol.*, **141**: 88-92.

Fock, A., J. Kettner and M. Böttger. 1992. The occurrence of 4-chlorotryptophan in *Vicia faba. Phytochem.*, **31**: 2327-2328.

Gandar, J.C. and C. Nitsch. 1967. Isolement de l'ester methylique d'un acide chloro-3-indolylacetique a partir de graines immatures de pois, *Pisum sativum* L. *CR Acad. Sci.*, Paris, Series D. **265**: 1795-1798.

Gerhold, L.S. and R.M. Muir. 1989. Structural activity differences of halogenated indole acetic acids in the growth response of the *Avena* coleoptile. *Crop Physiol. Abstr.*, **15:** 252.

Hatano, T., M. Katayama and S. Marumo. 1987. 5,6-Dichloroindole-3-acetic acid as a potent auxin : its synthesis and biological activity. *Experimentia*, **43** : 1237-1239.

Hattori, H. and S. Marumo. 1972. Monomethyl-4-chloroindolyl-3-acetyl-L-aspartate and absence of indolyl-3-acetic acid in immature seeds of *Pisum sativum. Planta.*, **102**: 85-90.

Hertel, R. 1994. A critical review on proposed hormone action : the example of auxin. In : Smith, C.J. *et al.*, (eds.) *Biochemical mechanisms involved in plant growth regulation.* Oxford : Clarendon Press, 1-15.

Hertel, R. 1995. Auxin binding protein I is a red herring. *J. Exp. Bot.*, **46**: 461-462.

Higashi, M., M. Hatori, K. Manabe and Y. Sakagami. 1998. Biosynthesis studies of chlorinated indole in *Vicia faba. Plant Growth Regul.*, **27**: 15-19.

Hofinger, M. and M. Böttger. 1979. Identification by GC-MS of 4-chloroindolyl acetic acid and its methyl ester in immature *Vicia faba* seeds. *Phytochem.*, **18**: 653-654.

Hirasawa, E. 1989. Auxins induce α-amylase activity in pea cotyledons. *Plant Physiol.*, **91**: 484-486.

Huizen, R., J.A. Ozga, D.M. Reinecke, B. Twitchin and L.N. Mandez. 1995. Seed and 4-chloroindole-3-acetic acid regulation of gibberellin metabolism in pea pericarp. *Plant Physiol.*, **109**: 1213-1217.

Jensen, J.B., H. Egsgaard and O.H. Van. 1995. Catabolism of indole-3-acetic acid and 4- and 5-chloroindole-3-acetic acid in *Bradyrhizobium japonicum. J. Bact.*, **177:** 5762-5766.

Karcz, W. and Z. Burdach. 1995. Effect of temperature on IAA and 4-Cl-IAA and FC-induced growth and $H^+$-extrusion in *Zea mays* L. coleoptile segments. In : *10th International Workshop on Plant Membrane Biology,* Ragensburg, August 6-11, Abstract S16.

Karcz, W. and Z. Burdach. 2002. A comparison of the effect of IAA and 4-Cl-IAA on growth, proton secretion and membrane potential in maize coleoptile segments. *J. Exp. Bot.*, **53**: 1089-1098.

Karcz, W., H. Luthen and M. Böttger. 1999. Effect of IAA and 4-Cl-IAA on growth rate in maize coleophile segments. *Acta Physiol. Plant.*, **21**: 133-139.

Katayama, M., S.V. Thiruvikraman and S. Marumo. 1987. Identification of 4-chloroindole-3-acetic acid and its methyl ester in immature seeds of *Vicia amurensis* (the tribe Vicieae) and their absence from three species of *phaseolae. Plant Cell Physiol.*, **28**: 383-386.

Katekar, G.F. and A.E. Geissler. 1982. Auxins II : The effect of chlorinated indolylacetic acids on pea stems. *Phytochem.*, **21**: 257-260.

Katekar, G.F. and A.E. Geissler. 1983. Structure activity differences between indoleacetic acid auxins on pea and wheat. *Phytochem.*, **22**: 27-31.

Magnus, V., J.A. Ozga, D.M. Reinecke, G.L. Pierson, T.A. Larue, J.D. Cohen and M.L. Brenner. 1997. 4-chloro-indole-3-acetic acid and indole-3-acetic acid in *Pisum sativum*. *Phytochem.*, **46**: 675-681.

Manabe, K., M. Higashi, M. Hatori and Y. Sakagami. 1999. Biosynthetic studies on chlorinated indoles in *Vicia faba*. *Plant Growth Regul.*, **27**: 15-19.

Marumo, S. and H. Hattori. 1970. Isolation of D-4-chlorotryptophan derivatives as auxin related metabolites from immature seeds of *Pisum sativum*. *Nature*. **219**: 959-960.

Marumo, S., H. Abe, H. Hattori and K. Munakata. 1968a. Isolation of noval auxin, methyl 4-chloroindole acetic acid from immature seeds of *Pisum sativum*. *Agri. Biol. Chem.*, **32**: 117-118.

Marumo, S., H. Hattori, H. Abe and K. Munakata. 1968b. Isolation of 4-chloroindolyl-3-acetic acid from immature seeds of *Pisum sativum*. *Nature*, **219**: 959-960.

Marumo, S., H. Hattori and A. Yamamoto. 1974. Biological activity of 4-chloroindolyl-3-acetic acid. In : *Plant Growth Substances* 1973. Tokyo: Hirokawa Publishing Co., pp. 419-428.

Moore, T.C. 1989. Biochemistry and physiology of plant hormones, 2nd edn. New York : Springer-Verlag Inc., 28-85.

Muir, R.M. and C. Hansch. 1953. On the mechanism of action of growth regulators. *Plant Physiol.*, **28**: 218-235.

Normanly, J. 1997. Auxin metabolism. *Physiol. Plant.*, **100**: 431-442.

Normanly, J., J.P. Slovin and J.D. Cohen. 1995. Rethinking auxin biosynthesis and metabolism. *Physiol. Plant.*, **107**: 323-329.

Ozga, J.A. and M.L. Brenner. 1992. The effect of 4-Cl-IAA on growth and GA metabolism in deseeded pea pericarp (abstract No. 12). *Plant Physiol.*, **99**: S-2.

Ozga, J.A., M.L. Brenner and D.M. Reinecke. 1992. Seed effects on gibberellin metabolism in pea pericarp. *Plant Physiol.*, **100**: 88-94.

Parker, C.W., B. Entsch and D.S. Letham. 1986. Inhibitors of two enzymes which metabolize cytokinins. *Phytochem.*, **25**: 303-310.

Pless, T., M. Böttger, P. Hedden and J. Graebe. 1984. Occurrence of 4-chloroindoleacetic acid in broad beans and correlation of its levels with seed development. *Plant Physiol.*, **74**: 320-323.

Porter, W.L. and K.V. Thimann. 1965. Molecular requirements for auxin action-I. Halogenated indoles and indoleacetic acids. *Phytochem.*, **4**: 229-243.

Reinecke, D.M. 1999. 4-chloroindole-3-acetic acid and plant growth. *Plant Growth Regul.*, **27**: 3-13.

Reinecke, D.M., J.A. Ozga and V. Magnus. 1995. Effect of halogen substitution of indole-3-acetic acid on biological activity of pea fruit. *Phytochem.*, **40**: 1361-1366.

Rescher, U., A. Walther, C. Schiebl and D. Klämbt. 1996. *In vitro* binding affinities of 4-chloro-, 2-methyl-, 4-methyl- and 4-ethylindoleacetic acid to auxin binding protein I (ABPI) correlate with their growth stimulating activities. *J. Plant Growth Regul.*, **15**: 1-3.

Schneider, E.A., C.W. Kazakoff and F. Wightman. 1985. Gas chromatography – mass spectrometry evidence for several endogenous auxins in pea seedling organs. *Planta*, **165**: 232-241.

Sponsel, V.M. 1982. Effects of applied gibberellins and naphthylacetic acid on pod development in fruits of *Pisum sativum* L. cv. Progress No. 9. *J. Plant Growth Regul.*, **1:** 147-152.

Stenlid, G. and K.C. Engvild. 1987. Effects of chlorosubstituted indoleacetic acids on root growth, ethylene and ATP formation. *Physiol. Plant.*, **70**: 109-113.

Stotz, H.U. and R. Hertel. 1994. Reevaluation of the role of auxin binding site II. *J. Plant Growth Regul.*, **13**: 79-85.

Tamas, I.A., C.J. Engels, S.L. Kaplan, J.L. Ozbun and D.H. Wallace. 1981. Role of indoleacetic acid and abscissic acid in the correlative control by fruits of axillary bud development and leaf senescence. *Plant Physiol.*, **68**: 476-481.

Wightman, F., E.A. Schneider and K.V. Thimann. 1980. Hormonal factors controlling the initiation and development of lateral roots. II. Effects of exogenous growth factors on lateral root formation in pea roots. *Physiol. Plant.*, **49**: 304-314.

Wright, A.D., M.B. Sampson, M.G. Neuffer, L. Michalczuk, J.P. Slovin and J.D. Cohen. 1991. Indole-3-acetic acid biosynthesis in the mutant maize orange pericarp, a tryptophan auxotroph. *Science*, **254**: 998-1000.

Za•ímalová, E. and M. Kutá•ek, 1985. Auxin binding site in wheat shoots : interactions between indole-3-ylacetic acid and its halogenated derivatives. *Biol. Plant.*, **27**: 114-118..

*SECTION — VIII*

## *CROP PHYSIOLOGY & BIOCHEMISTRY*

*Advances in Plant Physiology*, Vol. 9
Ed. A. Hemantaranjan
Scientific Publishers (India), Jodhpur, 2006 pp. 379-448
E-mail: **info@scientificpub.com** www.scientificpub.com

# 18

# A STUDY INTO GRAIN GROWTH OF *TRITICUM AESTIVUM* L. WHEATS AS STEERED BY ANATOMICAL, BIOCHEMICAL AND PHYSIOLOGICAL PARAMETERS

I. S. Dua and Davood Eradatmand Asli*

Professor in Department of Botany, Panjab University, Chandigarh, 160 014
*Research Scholar, Department of Botany, Panjab University, Chandigarh, 160 014

## INTRODUCTION

The development of the cereal grain as a system constitutes a group of diverse biological phenomena related to the rise and fall of metabolites, enzyme activities associated with differential gene expression and/or its regulation. Its capacity to grow and gather photosynthetic assimilates is considered to be an important parameter governing economic yield. An attempt has been made to review the researches pertaining to role played by diverse physiological parameters in grain growth and development and the relationship that exists between the yield of grains and its steering attributes have shown the involvement of diverse factors like;

1. Physiological basis of yield.
2. Photosynthetic contributions of different organs to grain yield.
3. Mobilization, transport and regulation of flow of assimilate.
4. Plant growth regulators and their role in determining sink efficiency.
5. Physiological and biochemical changes during grain development.
6. Role of enzymes in regulation of sink efficiency.
7. Role of vascular system of inflorescence on grain formation.
8. Effect of alternative oxidative pathway on sink efficiency.

## 2. PHYSIOLOGICAL BASIS OF YIELD

The physiological basis of yield in the wheat has been the subject of investigation for many years. These investigations are directed towards the identity what determines the basic components of the yield, the number of inflorescence per unit area and the number and weight of grain per inflorescence. Research has been devoted to both ends of yield system, the source of assimilates derived mainly from the flag leaf and the ear (Thorne, 1966; Murchie *et al.*, 2002). However, it has been found that the storage organs (sink) have an important regulatory effect on the yield formation. As yield is the most important attribute of crop the prime purpose is to find what are the factors of the plant which control the production of useful dry matter? The development of cereal grain as a system constitutes some of diverse biological processes related to the rise and fall of metabolites, enzyme activities associated with differential gene expression and/or its regulation. Growing capacity and gather photosynthetic assimilates to be an important factor governing economic yield which including of the interaction of a variety of primary characters of the plant with each other and with environment.

## 2.1 Analysis of Yield Components

In cereals, its life history is divided to pre- and post-anthesis phases (Austin, 1982). According to Hunt *et al.* (1991) grain growth in wheat consist of three phases; each of them is influenced by the environment, with little increase in grain dry weight during lag phase, lasting for 4-6 days at 18°C, immediately after anthesis. There is a rapid increase in grain dry weight during the grain filling period until a maximum dry weight is attained, after which it remains approximately stable, while the grain dries. The end of the grain filling period has been termed as mass maturity (Ellis and PietaFilho, 1992; Talbert *et al.*, 2001). According to Gallagher *et al.* (1976) grain dry weight is quantified as product of grain filling. Rate and duration of grain filling determine final grain dry weight, a component of grain yield (Darroch and Baker, 1995; Talbert *et al.*, 2001). Data from the genotypes studied showed that it is possible to produce a high grain dry weight by decreasing the duration of the grain filling period and increasing the rate of grain filling (Talbert *et al.*, 2001). Wheat may initiate as many as 10-12 florets per spikelets but rarely more than 4 or 5 of these produce grains (Thorne, 1982). Okawa *et al.* (2003) characterized the process of partitioning and remobilization of carbon assimilated at different stages during the early phase of grain filling under controlled environmental conditions. The results indicated that a shift of the major sink from the culm to the spike occurs at around 10 days after anthesis.

Understanding variation in kernel number per unit area is a major importance in understanding yield and in identifying opportunities to increase yield potential (Bindraban *et al.*, 1998). Many studies have suggested that changes in grain number per spike and per $m^2$ are closely associated with changes in biomass allocated to spikes during spike development. The number of grains per ear is determined by the number of unifloral per spikelet in barley and the products of spikelet per ear and fertile florets per spikelet in wheat (Mathews and Thomson, 1984). Not all spikelets develop at the same pace, at the first time the mid ear ones developing earlier, more slowly and more fully than the progressively later, basal and apical one. Generally use of growth retardants (Mathews *et al.*, 1982) may slow down development, although the synchronized production of fewer, evened and more competitive tillers may be a concomitant factor. Fisher and Stockman (1980); Stockman *et al.* (1983) reported that floret competency was closely related to spike dry weight at anthesis. Fisher (1985) concluded that the period just prior to anthesis is of major importance in determining kernel set and grain yield. Genotype influences kernel number as semidwarf cultivars tended to produce more kernels per unit area than tall cultivars while not differing in overall dry matter production (Brooking and Kirby, 1981).

In a broader sense green revolution related to shorter cultivar with higher harvest index coupled with growth regulators and fungicide use. Capacity of tillering varies between cultivars and rapidly compensates for wide variation in plant population. In isolation wheat plant may produce 100 tillers rarely produce more than two to five ear bearing shoots (Reilly, 1984). Tillering is intrinsically a part of early stage of vegetative growth. Optimization of tillering is favoured by a few early and even tillers per plant, thus avoiding subsequent redundancy and vegetative floral competition (Mathews and Thomson, 1984).

Cereal shoot rarely carries more than three to four fully functional leaves with little pre-storage of assimilates for grain production. Siddique and Whan (1993) showed that weight (ear:stem ratio in wheat) may give a better indication of potential yield and harvest index and concluded that selection for high ear:stem ratio at anthesis may result in further improvement in grain yield of wheat. According to Goudriaan and Laar (1994) crop growth related to potential dry matter production in early stages of crop development, when canopy is still open to the adequate simulation of leaf area. Seeds of cereals such as barley (Pieta Fillho and Ellis, 1991) and wheat (Ellis and Pieta Filho, 1992) increase in seed quality between the end of seed filling phase and harvest maturity, when cereal seed crops have dried naturally to about 15 percent moisture content. Grain dry weight increases as linear function of time during grain filling phase (Gallagher *et al.*, 1976; Brock-Lehurst, 1977). The dry matter in winter wheat grains is mainly derived from both reserve photosynthate in the sheath and stem stored prior to anthesis and current photosynthate produced after anthesis (Evans *et al.*, 1975). Yu (1990); Jin (1996) suggested that 2/3 to 3/4 of dry matter in winter wheat grains originates from photosynthate produce after anthesis. Kruk *et al.* (1997) suggested that during post-anthesis, the number and the capacity of the grains demanding assimilates (i.e. sink strength) is more important in determining yield potential than the supply of assimilates to the growing grains (i.e. source strength).

## 2.2 Factors Limiting Grain Yield

The potential number of grains is influenced both by environmental as well as physiological factors. Thorne (1965) showed that effect of temperatures on growth and grain yield depended on the stage of development of plant. According to Thorne *et al.* (1968) large increase in tiller number of wheat kept at 15°C instead of 20°C for whole period increased yield by 21 per cent. Soil temperature of 21+1.5°C and soil moisture of 100-15 per cent is suitable for growth of wheat (Dubey, 1990). Wardlaw (1994) showed the effect of high temperature on kernel development and suggested that mean temperature greater than 15-18°C following anthesis could result in decrease in kernel weight at maturity, while breeding for high temperature tolerance during kernel filling could prove a significant increase in yield. Warmer temperature decreased carbohydrate supply by hastening the natural senescence of the leaf (Thorne *et al.*, 1968; Wardlaw, 1970). Wardlaw and Moncur (1995) reported that high temperature resulted in a considerable drop in kernel dry weight at maturity, ranging from 30 to 60 per cent decrease, those cultivars most tolerant were those where the rate of kernel filling was most enhanced by high temperature, i.e. the increased rate compensated for the reduced duration of kernel filling. Water stress during early grain filling has a marked effect on grain yield through reduced endosperm cell number and thus sinks capacity to accumulate dry matter (Ahmadi and Baker, 2001).

Pre-emergence shading treatment decreased 35 per cent of grain yield by decreasing grain size and number of grain per ear (Welbank *et al.*, 1968). High temperature reduced final grain size in all floret position, which suggests that it is directly affecting the synthetic processes in grain rather than availability of assimilates. Shortage of assimilate supply due to inhibition of photosynthetic processes is one of the major factors determining grain filling (Matsushima and Wada, 1958; Yoshida, 1981; Evans, 1996; Egli, 1998). Inhibition of photosynthesis during the grain filling period due to environmental stresses such as shading or water deficit can result in a major reduction in grain dry matter (Kobata and Moriwaki, 1990). The studies showed clearly that potential grain growth was not realized when available assimilate failed to meet the assimilate requirement. Other studies suggest that potential grain growth rate (i.e. ability of the grain to fill, or sink strength) is influenced by change in environmental conditions (e.g., radiation, temperature, and fertilizer application) during the early phase of grain filling (Tanaka and Matsushima, 1963; Yoshida, 1981; Egli, 1998). In both wheat and rice, reductions in grain weight or filling percentage (decline in potential of grain growth) from shortage of assimilate supply are believed to be due to reduction in the number or size of cells in the endosperm (Singh and Jenner, 1984; Nakamura *et al.*, 1992; Horie *et al.*, 1997). In the first 10 days after fertilization, cell division and expansion in the endosperm of most grains ends and starch deposition begins (Egli, 1998).

It was suggested by Shukla *et al.* (1995) that refixation of respired carbon dioxide by green pericarp and remobilization of carbon back to endosperm by seed coat amylase activity contributes in sustaining the grain weight under high temperature in tolerant wheat genotype. Wheeler *et al.* (1996) showed that plants grown at the same temperature resulted in heavier grains, under increased $CO_2$ concentrations. Sild *et al.* (1999) found that the above ground biomass and grain yield of wheat increased mainly in $CO_2$ concentration range between 400 and 500 μ mol mol$^{-1}$. Some authors have observed a slight advancement of floral initiation (Marc and Gifford, 1984) and anthesis (Sionit *et al.*, 1980), while others found no significant effects (Krenzer and Moss, 1975; Schonfeld *et al.*, 1989; Slafer and Rawson, 1997) of $CO_2$ elevated on wheat. Mare and Gifford (1984) observed a slight advance of floral initiation (approximately 3days) in wheat grown under elevated $CO_2$ compared with ambient $CO_2$, but the plants under different $CO_2$ environments reached anthesis simultaneously, indicating that those plants under elevated $CO_2$ experienced a slight extension of the period from floral initiation to anthesis.

Li *et al.* (2000) reported that the number of grain per main shoot and grain weight did not increase with $CO_2$ enrichment unless these exceptionally low in ambient $CO_2$ conditions. Further, there was a small (7 percent) but significant decrease in nitrogen content of wheat grain in $CO_2$ enrichment at ambient temperature. Pinter *et al.* (1996) observed $CO_2$ enrichment in wheat caused a 21 percent yield increase in limited water and 8 percent increase in well-watered treatments. Kimball *et al.* (1995); Pinter *et al.* (1996) observed $CO_2$ elevated increased whole canopy photosynthesis by about 20 percent in the well watered conditions, thereby structural and non-structural carbohydrate accumulation increased (Havelka *et al.*, 1984; Estiarte *et al.*, 1999). According to Wheeler *et al.* (1996) $CO_2$ enrichment to 680 μ mol mol$^{-1}$ has been shown to increase grain-filling rate by 21 to 24 percent per spike over the temperature range of 14 to 20°C, respectively, linear grain-filling duration was unaffected by $CO_2$ enrichment.

According to Sofield *et al.* (1977) grain filling duration is determined mainly by temperature and is shorter at warmer temperature. Studies of Pandey *et al.* (1997)

also showed that application of 100 kg nitrogen per hectare and seed flow mulching with farmyard manure (50 quintal/hectare) recorded significantly higher grain yield and net return. The early sown wheat varieties favoured all the yield contributing characters as well as the overall performance of crop grain and straw yield (Kumar *et al.*, 1998). The studies of Zhao *et al.* (1997) showed that apart from effects on yield, the sulphur nutrition of a crop significantly improved the bread making quality of field growth wheat, as well as the net protein content of flour. Hasan and Kamal (1998) showed that the applications of urea and organic manure increased grain and straw yields.

## 3. PHOTOSYNTHETIC CONTRIBUTION OF DIFFERENT ORGAN TO GRAIN YIELD

Grain growth has been shown to depend mainly on current supply of photosynthates (Kumar and Singh, 1980) and thus yield is strongly dependent on the total amount of photosynthesis during the growing season. In cereals, flag leaf and the inflorescence have a major role in affecting the grain yield (Boonstra, 1937; Thorne, 1966; Rawson and Hofstra, 1969; Murchie *et al.*, 2002), although other parts of plant contribute to some extent towards the grain yield (Blum, 1985). Distribution of photoassimilates between organs may change during grain filling in relation to the part of the plant. Source or sink organs that exert control on the partitioning at a given moment (Marcelis and De-Koning, 1995). Some authors (Evans and Rawson, 1970; Richards, 1996) reported that photosynthesis of the spike and flag leaf blade alone could meet the requirements of the grains at all times during grain filling.

Boonstra (1929) recorded for first time 27 per cent of total grain weight were derived from the photosynthesis of ear. According to Smith (1933) ear assimilates contributed 41 percent of total dry weight, this represented the dry weight of grain produced by ear alone. Asana and Mani (1948, 1950) reported that when plants were defoliated and shaded shoots the contribution of 11-59 percent, 20-42 percent and 20-42 percent by ear, leaves and stem respectively. Asana and Basu (1963); Larsson and Hensen (1992) considered that greater potentiality for the growth of grain in cereals might be conducive to higher yield in wheat. According to Watson *et al.* (1958) about 25.5 percent of the dry matter in barley grain came from ears (10 percent from awns), 50 percent from flag leaf Lamina, stem and sheath and 15 percent from shoot below the flag leaf. Thorne (1963) is of the view that in barley the shoot provides 40-50 per cent of carbohydrates to the grains, whereas 40 per cent by photosynthesis in the ear.

Radio carbon techniques showed that flag leaf and parts above it constituted the main source of assimilates for developing grains (Stoy, 1965; Wardlaw, 1968; Rawson and Hofstra, 1969). The photosynthates from penultimate leaf and below it are used for rest of plant (Quinlan and Sagar, 1962; Stoy, 1963; Lupton, 1966). Simmons (1967) and Laver *et al.* (1983) by $^{14}C$-labelling reported that inflorescence of barley received photo assimilates from top, two to three leaves only on the plant. Relative contribution of assimilates by the ear to the cereal grain yield has been estimated to range any where from 13 percent (Biscoe *et al.*, 1975) to 76 percent (Evans and Rawson, 1970).

Welbank *et al.* (1968) emphasized that high yields are associated with prolonged duration of leaf area available for photosynthates from anthesis. Wicke *et al.* (1972) associated final grain yield with leaf area duration. Dhiman *et al.* (1980) found positive correlation between flag leaf area and grain yield in wheat. Sayed and Gadallah (1983) pointed out existence of relation between grain yield and the photosynthetic activity above flag leaf node. Rao *et al.* (1984) suggested that grain size is not related to grain yield during grain filling. Batten and

Wardlaw (1987) have also proved by the labelling experiments in wheat flag leaf that maximum translocation of assimilates occurred towards ear.

Ellen (1990) while working on winter wheat anthesis differing in harvest index observed that total dry matter production was not correlatable to grain yield. Davidson and Chevalier (1992) observed that grain yield of wheat depended in part on carbohydrate reserves available in the stem. Takahashi *et al.* (1994) studied effects of shading on dry matter accumulation in grain and culm was shown to prevent wheat canopy from producing any photoassimilate and affected increase in weight by 50-60 percent of the control. Pre-anthesis reserve contributed 11-29 percent to the total mass of carbon in grains at maturity; pre-anthesis C accumulation in grains was dependent on both the mass of pre-anthesis of C mobilized in above ground vegetative parts and pre-anthesis C deposited in grain. Multiple regression analysis indicated that C from mobilized pre-anthesis water soluble carbohydrates may be used more efficiently in grain filling than C present in proteins at anthesis (Gebbing *et* al., 1999). Correlation coefficient among yield and physiological traits showed that leaf photosynthetic rate, total dry matter production, specific leaf weight, absolute growth rate and harvest index had significant positive correlation with leaf photosynthetic rate, grain yield and harvest index (Kumar *et al.*, 1998).

The awns of the ear occupy a favourable position for providing photosynthates to growth kernels. The photosynthetic capacity of individual awned ears is more than awnless ones (Evans and Rawson, 1970; Evans *et al.*, 1972; Teare *et al.*, 1972). In the most literature yield of awned genotypes are higher than the awnless counter parts. Also the kernel weight and test weight is higher in most of the cases (Grundbacher, 1963). The mean contribution of awns to total carbon exchange rate (CER) was roughly in the range of 40-80 percent in different species of wheat (Blum, 1985). The stem contributed 5-10 percent (Wardlaw and Porter, 1967) and 2.7-12.2 percent (Rawson and Evans, 1971) of assimilates to the grains of wheat. Leaf area is frequently correlated with growth and yield, and this is undoubtedly often a causal relationship. According to Watson (1946) large leaf area, particularly the flag leaf, is conducive to high grain yields. Wicke *et al.* (1972); Murchie *et al.* (2002) found a relation between the flag leaf area index and the grain yield. Berdahl *et al.* (1972) found that large leaf area in barley also favoured higher kernel weights. Evidence from our most intensively breed crops such as wheat and maize suggests that, except for a period just before anthesis, photosynthate production is not limiting and crops have a reserve capacity for photosynthesis under favourable conditions. This may account for the near doubling of the genetic yield potential during this century without any change in the rate of photosynthesis per unit leaf area (Richards, 2000).

## 4. MOBILISATION, TRANSPORT AND REGULATION OF FLOW OF ASSIMILATES

Though plants are photoautotrophic they are composed of heterotrophic tissue systems that must import sugars and amino acids in a process called assimilate partitioning. In the higher plants leaf is so called "source" tissue, containing photosynthetic machinery which transforms light energy into useful biological energy and much of captured energy is transported from leaf in the form of sucrose and amino acids in order to satisfy the biochemical needs of heterotrophic cells.

The efficiency of the source-sink association in producing yield is measurable in the harvest index (i.e. the fraction of total biomass recovered as grain). Harvest index reflects the efficiency with which yield constituents like C and N are transferred from vegetative to the reproductive system (Reilly, 1990). Studies of Gamalei (1998) showed that structural development of transport communications in ontogeny is accompanied by the formation of source-sink functional relationship. Their study showed the correlation of apoplast and endoplast structure with the dynamics of photosynthesis and photosynthate. In general, C and N particularly have principal contributors to the grain (Cregan and Berkum, 1984). The effect of sink size measured by grains per unit area is positive, but proportionally less than leaf area enhancement. Protein deposition as a terminal process related to leaf senescence and gradually accelerated during grain development. In one survey grain protein content were associated with the level of protein degrading enzymes in the senescing vegetation (Abrol *et al.*, 1984). In unfavourable conditions (high temperature) both grain protein content and senescence may be simultaneously prompted often at the expense of protein and grain yield (Murchie *et al.*, 2002).

### 4.1 Source-Sink Relationship and Yield

Mason and Maskell (1928) while working on the physiology of cotton coined the terms 'sink'. It was defined as the region where carbohydrates were used for further growth. Some of researches (Stoy, 1979; Paul and Foyer, 2001) indicate that 80 to 90 percent of grain carbohydrate is derived from post anthesis photosynthetic activity. The results of investigations show that grain yield is directly associated with the leaf area duration after flowering (Evans *et al.*, 1975) and it is possible that a single day delay in flag leaf senescence may result in an extra 200 kg $ha^{-1}$ grain dry matter (De-Vos, 1979). The flag and penultimate leaf are the principal contributors, but ears and especially awns may contribute substantially (Wojcieska *et al.*, 1984).

It seems to be little doubt that sinks phenomena regulates the partitioning in the post anthesis period (Austin, 1982; Paul and Foyer, 2001). The strength of demand dictating the rate of assimilation in the source leaves (Watson and Casper, 1984). Source-sink manipulation could regulate the net photosynthetic rate of winter wheat after anthesis; however the direction and magnitude of regulation varied with time of the anthesis. Source and sink relation markedly effected photosynthetic rate during grain filling. The photosynthetic rate was increased by source reduction and decreased by sink reduction, significantly, which indicated that photosynthetic rate was closely associated with change of source or sink size (Wang *et al.*, 1998). Source-sink relationship between a leaf and other organs is complex. It is a function of growth stage of the plant and position of leaf. King *et al.* (1967) showed that in wheat two weeks after anthesis, 45 percent of flag leaf assimiiates were transferred to the developing ear, which was itself photosynthesizing.

Demand for assimilates can also influence the rate, velocity, and translocation pattern in wheat (Rawson and Evans, 1970) and presumably in other plants. Another subtle relationship that has been recognized for many years (King *et al.*, 1967) is that the size of activity in the sink may have an influence on photosynthetic rate in the source leaves. If assimilates are not transported to sinks, photosynthetic rate is depressed, and if new sinks are provided, photosynthetic rate is increased. In wheat, when the ear is removed, the photosynthetic rate of the flag leaf, which is the major source of the ear, decreases rapidly because the major sink is impaired. When the lower leaves of a plant from which the ear has been removed are darkened, the photosynthetic rate of the flag leaf recovers considerably because those leaves and roots become the sink of the flag leaf (King *et al.*, 1967). Blum *et al.* (1988) found that decreasing the sink size of two wheat cultivars with different kernel size caused an increase in the weight of remaining kernels on the small kernel cultivar and no change in kernel size of large kernel cultivar.

### *4.1.1. Source characteristics*

Until the flowering when canopy completion the crop growth rate of cereals is proportional to intercepted light radiation (Marshall and Biscoe, 1982) thereafter photosynthetic productivity drops rapidly, coupled to an increasing demand for development. Some reports and literatures (Evans, 1975; Spiertz *et al.*, 1984; Richards, 2000) indicate that leaf area duration after anthesis is a major and prevailing determinant of grain yield. Over a ten years period grain yield and hours of sunshine during the period of gamete formation, fertilization and grain filling (Borojevic and William, 1982) were highly correlated with a strong suggestion that yield improvement came from the telescoping of the earlier vegetative period to facilitate more prolonged grain development (Watson and Casper, 1984; Richards, 2000).

A long grain filling period result in improve yield by increased kernel number and greater seed size. The production and distribution of major grain photo assimilate such as C and N are intimately linked (Cregan and Berkum, 1984; Paul and Foyer, 2001) thus the rate of reduction in gross photosynthesis in flag leaves is closely related to the total N already exported by them (Marshall and Biscoe, 1982).

### *4.1.2 Sink characteristics*

Nearly 85 percent of cereal grain is derived from a short burst (10-14 days) of cellular activity immediately following fertilization which produces about 100 to 150 thousand endosperm cells (Cochrane and Duffus, 1982). In the early cellular phase plant hormones such as cytokinin with perhaps auxins and gibberellins involved in cell enlargement (Gale, 1979; Banowetz *et al.*, 1999a). Rate and duration of endospermic activity in wheat is affected by increasing temperature (Cochrane and Duffus, 1982). The difference in rate and duration of grain filling resulted in genetic variation of ultimate grain weight (Spiertz, 1979; Talbert *et al.*, 2001). Richards (2000); Paul and Foyer (2001) have shown that the numbers of cells in the sink organ may be a good measure of sink size because cell division activity is crucial in attracting photoassimilate to sink organs in early stages of development. Gleadow *et al.* (1982) also reported that final grain dry matter is potentially determined by the numbers of cells in the endosperm of wheat grains. During post-anthesis, the number and the capacity of the grains demanding assimilates (i.e. sink strength) is more important in determining yield potential than the supply of assimilates to the growing grains (i.e. source strength) (Bruckner and Frohberg, 1991).

According to Evans *et al.* (1975) temperature has a significant effect on the duration of grain filling which takes longer at lower temperature, but produces a higher grain yield. Correlative influence within individual ears and spikelets resulted in limiting grain fill. In a single spikelets that have a common vascular system exposing them to competitive and hormonal effect.

Until maturity development of the vascular system of individual spikelets along the ears is irregular (Hutley-Bull and Schwabe, 1982). Thus generally grain weight is larger in spikelets supplied by an unbranched vascular trace. Application of ABA to wheat ear shortly after anthesis can inhibit grain set in third and higher florets in each spikelet (Radley, 1982).

### *4.1.3 Sink strength and competition*

Competition between sink organ results in availability of mobile assimilates to a particular sink. Venkateswarlu and Visperas

(1987b) suggested that sink strength is the product of sink size and unit activity, which can also be stated as the product of sink weight and relative growth rate. According to Ho and Baker (1982); Ho (1988) sink size reflected the physical constraint while sink activity reflects the physiological constraint upon a sink organ's assimilating import. When dry matter production in flowering tomato plant was reduced, the limited amount of mobile assimilate was imported by the developing young leaves to sustain the growth, and the initiating inflorescence were aborted (Kinet, 1977). Development of the nodule of soybean because of high demand by the pods restricted by inadequate supply of assimilates (Lawn and Brun, 1974; Latimore *et al.*, 1977).

In situations where several sinks are competing for a limited supply of assimilates, the relative magnitude of the sinks may be of overriding importance in partitioning, with a pronounced bias in favour of the largest sink. According to Wardlaw (1990) sink strength may increase with the increasing sink sizes because of an associated increase in size of surface (membrane) across which metabolites are transferred from the vascular system to the zone of utilization (unloading area). Seeds are the high priority sink getting their requirement of photosynthate despite the fluctuation in supply. The priority rank order is summarized as seeds>fleshy fruit parts= shoot spices and leaves>cambium>roots> storage (Wardlaw, 1990). Jenner and Hawker (1993) suggested the activity of single enzymes like sucrose synthase, invertase and ADPG-phosphorylase as suitable measure of sink strength. According to Liang *et al.* (2001) dry matter accumulated by grains was linearly related to the increases in endosperm cell numbers and the activities of sucrose synthase (SS) and ADP-glucose pyrophosphorylase (AGPase) in rice.

Plant growth regulators have also been related to sink strength (Kuiper, 1993). The availability of mobile assimilate to the particular sink may be determined by the competition between sink organs. In the primitive wheat, roots, tillers, stem and grains are various sinks, where the assimilates are partitioned more or less evenly among all of them. In contrast, the modern wheat at this stage shows all the assimilates going towards the grains (Evans, 1975) but there also exists a tendency for differential sink efficiency resulting in different size of grains. Similarly, in maize, though several cobs may differentiate, but only the upper most 1 or 2 may fully develop in modern cultivars, whereas primitive maize had many smaller cobs (Evans, 1975) and coblets (Mishra, 1990).

In terms of assimilate transport, the ability of sink organs to import assimilate depends on the sink strength. Although actual sink strength would be affected by the availability of assimilate supply and the proximity of the sink to the source, the most critical determinant is the intrinsic ability of the sink to receive or attract assimilate (Cook and Oparka, 1983). Due to increase in spikelet number in rice, serious competition for carbohydrates would take place between the spikelet, and weak spikelets in the lower parts of the panicle would fail to be fertilized, or would abort the florets (Wardlaw, 1968).

Ho (1988) suggested that the intrinsic ability of sink to attract assimilates acts as critical determinant of organ growth. The source-sink ratio expressed as the ratio of total plant growth to total potential growth rate is used as a measure for the level of assimilates available for growth (Bertin, 1995). Hence increase in sink size is accompanied by a proportionally greater decline in sink strength, i.e., import rate. Therefore, apart from the availability of leaf assimilates and competition with other sinks, the rate of import is not entirely determined by potential sink size but also is regulated by the metabolic activity of the sink organ during development (Ho, 1988).

## 4.2 Translocation of Assimilates from Source to Sink

One of the major factor limiting yield is translocation of assimilates from source to sink. Stoy (1963) studied translocation of metabolites from flag leaf to the ear in three wheat cultivars and found that a positive correlation between rate of translocation and differences in grain yield. Sink activity is the inherent capacity of the sink to create a translocation gradient from the source to center of accumulation. In the reproductive structures, sink becomes more active and shifts to a higher potential only after fertilization. Some hydrolytic changes being after fertilization and create a translocation gradient (Venkateswarlu *et al.*, 1986).

Chikov *et al.* (1984) found that partial removal of an ear decreased $^{14}C$ assimilate translocation from leaves and increased assimilate entry into roots. Carbohydrates stored in the stem of cereal crop plants is a temporary form of stored photosynthetic product which would be translocated eventually to the grains and contribute to the grain yield along with the current photosynthetic products (Bell and Incoll, 1990). Carbohydrates translocated to the grain could originate from two sources; a 'pre-anthesis' and a 'post-anthesis' source, which is translocated directly to the grain. Contribution of pre-anthesis source was estimated to be as much as 30 percent of the grain yield in rice (Cock and Yoshida, 1972). Wardlaw (1968) reported that translocation of assimilates to grain depends on the number of grains in an ear. The consequences of plant growth and morphogenesis of variations in photosynthesis depend on the efficiency of the conversion of triose-p into succinate and of phloem loading at the sites of succinate production and unloading in growing sinks (Masle, 2000).

The transport system comprising of assimilates from source to sink organs via the phloem, has been discussed under the following sections:

### *4.2.1 Phloem loading in the source*

The surplus of assimilates produced by photosynthesis in the leaves is exported during both day and night through the vascular system to the sink organ. Phloem loading is a process where photo assimilates constantly leak away and retrieved along the transport path (Minchin and Thorpe, 1984, 1996). Roeckl (1949) provided evidence for phloem loading by asserting that osmotic pressure of the sieve sap was considerably higher than that of surrounding leaf tissue. Only a few compounds are translocated in large amounts in the phloem, e.g. several sugars, some amino acids and several inorganic ions (Pate *et al.*, 1974). Phloem loading is capable of accumulating solutes in the minor veins of the phloem to a level above that in surrounding tissues (Geiger *et al.*, 1973). High osmotic pressure in the minor vein causes water to enter and produces a high hydrostatic pressure in the phloem. Entry of solutes by active transport seems to produce a gradient of osmotic pressure in the phloem from source to sink site.

Pate and Gunning (1972) have shown that phloem loading of sucrose take place mainly in the minor veins and transfer cells playing major role. Geiger (1975) showed that though symplastic transport is important but in mesophyll tissue apoplastic transport is dominant. Transport studies with isolated cells and plasma membrane vesicles have demonstrated the presence of carrier mediated apoplastic transport process (Bush, 1990; Williams *et al.*, 1992).

Lalonde *et al.* (1999) found that inter connected network of sieve elements forms a supra cellular compartment for the transport of nutrients, phytohormones and macromolecules from site of nutrient assimilation (sources) to site of nutrient utilization (sinks). Partitioning of phloem-delivered nutrients between competing sinks is governed by their relative ability to unload major osmotic species from the importing phloem sieve elements (Patrick, 1997). This process depends upon a set of inter cellular (post-sive element) transport events which are integrated with growth or storage function of the recipient sink tissues.

#### *4.2.2 Phloem unloading in the sink*

The process of transfer of solutes from the sieve tube into the surrounding tissue at the sink sites is not well understood. Indirect evidence supports the involvement of an active process in phloem unloading (Jenner, 1980b; Bel and Van Patrick, 1985). Unloading of sucrose affected by phytohormone such as ABA (Tanner, 1980; Schussler *et al.*, 1984) and even the low concentration of ABA increases the rate of sucrose efflux from the phloem tissue (Vreughenhil, 1983). Unloading of sucrose into the apoplast of sink tissue is facilitated by the concentration gradient created by the consumption of sucrose during growth and storage of sucrose, either as sucrose in sugar beet or as starch in cereal grains. Invertase enzyme is involved in the phloem unloading of sucrose in stems of sugarcane (Hawker and Hatch, 1985), in apical zones of growing roots and other growing sinks (Eschrich, 1984). Similarly, in storage sinks accumulating starch, the unloading of sucrose accelerated by high degree of sucrose consumption and a correspondingly steep gradient in sucrose concentration from phloem to the apoplast of storage cells (Jenner, 1980b; Wang and Fisher, 1994).

It is shown that the sieve element transport in the developing seed of both cereals and grain legumes is characterized by extensive symplastic routes, with exchange to and from the seed apoplast, being confronted to cells located at or near the material filial interface (Patrick and Offler, 1995). Sieve elements in seeds of grain legumes and cereals are symplasmically interconnected with the surrounding ground tissues (Wang and Fisher, 1994; Patrick *et al.*, 1995; Tegeder *et al.*, 1999). Circumstantial evidence suggests that these symplasmic routes allow sieve elements to unload water (Murphy, 1989; Wang and Fisher, 1994), sucrose (Offler and Patrick, 1993; Wang and Fisher, 1994), amino acids (Hsu *et al.*, 1984; Fisher and Macnicol, 1986), certain mineral ions e.g. iron (Marentes and Grusuk, 1998) and macromolecules (Fisher *et al.*, 1992). Even release of photoassimilates along seed vascular systems (Offler *et al.*, 1989; Ugalde and Jenner, 1990a,b) is consistent with phloem sap being unloaded under high pressures by bulk flow through a lateral path with an hydraulic conductance lower than that for axial flow (Patrick and Offler, 2001).

It is not always true that uptake of sugars by the sink cells is a selective process. However, in the older starch storage sinks such as grains of wheat (Jenner, 1972) and barley (Felker *et al.*, 1984), imported sucrose was not hydrolyzed prior to the uptake by sink cells. Apoplastic sucrose inversion could set up an osmotic gradient in a direction assisting water removal from the sieve tubes. Whether unloading into the apoplast be active or passive, there must be some feedback control over the process related to the sugar concentration in the recipient free space and to the needs of sink growth.

#### *4.2.3 Partitioning and regulation of flow of assimilates*

Partitioning of assimilated carbon among sink organs largely determines the rate and pattern of plant growth. Only a minor fraction of the products of photosynthesis remains at the site of production in fully expanded leaves, most of them are translocated to other organs where in they are used as building blocks for various cell constituents or deposited as storage products.

Translocation of photosynthetics follows a definite distributive pattern (Brouwer, 1962; Wardlaw, 1968). This pattern of movement changes continuously (qualitatively and quantitatively) during the growth of the plant and exerts a profound effect on morphological and yielding properties of the plant. Very young leaf imports photosynthetic products from other parts of the plant to build up its own structure but very soon it becomes self-supporting and in short time starts exporting assimilates (Shiroya *et al.*, 1961; Joy, 1964). As long as the plant is young, this export is mainly directed towards the centers of active growth, e.g. developing leaves, root tips or shoot apex, but later on much of the assimilate transport is diverted to the storage organs, such as fruits, grains or tubers. The sink primarily regulates dry matter partitioning among sink organs themselves (Evans, 1975; Ho, 1988; Paul and Foyer, 2001). Amounts of carbon fixed, its partition and redistribution in plants are studied with help of pulse-chase labelling involving $^{14}CO_2$ (Rawson and Hofstra, 1969; Hume and Criswell, 1973; Bidinger *et al.*, 1977). Mitcham *et al.* (1989) employed $^{14}C$ sucrose to study photoassimilate partitioning. Tracing carbon with steady state $^{14}CO_2$ labelling over a 24 hours period gave data for carbon partitioning among various sinks during that photoperiod and for current value of daily relative growth rate for each sink part (Geiger *et al.*, 1989; Sung *et al.*, 1994).

Distribution or transport of assimilates from the sites of synthesis (sources) to sites of utilization (sinks) is partitioning. It involves phloem loading at source site, translocation in phloem and phloem unloading at sink site. The work of Quinlan and Sagar (1962); Doodson *et al.* (1964); Lupton and Pinthus (1969); Rawson and Hofstra (1969) gave information regarding partitioning in the cereals. The pattern of distribution changes as plant grows and develops new leaves and sinks in wheat (Rawson and Hofstra, 1969). Rate and pattern of plant growth is result of translocation of assimilated carbon along alternative pathways that supply metabolism, accumulation or incorporation in an array of sinks (Farrar and Williams, 1991). Translocation under various environmental conditions and at different stages of plant development shows a remarkable degree of regulation of partitioning and export of material derived from photosynthesis. Control of initiation of export and regulation of partitioning of the metabolites during plant development are fairly known (Fellow and Geiger, 1974). The ability of sink to attract photosynthate is determined by sink strength and by their priority rank ordering for supply in the presence of reduced availability of photosynthate (Minchin and Thorpe, 1996).

Yin *et al.* (1998) showed that effect of source sink changes on grain mass in wheat was in order upper>basal>middle spikelets on spike and the effect was found mainly in the grain mass at the positions three and four from base of the spikelet. Gladun *et al.* (1998) studied the change of source-sink relation during rice grain formation and filling period and suggested the participation of phytohormones in the regulation of photoassimilate flow.

### 4.3 Source-Sink Manipulation and Yield

In striving to increase the yield potential of cultivated plants it is important to determine the physiological factors limiting grain yield. The first step towards this is to assess whether the growth of harvested organs is limited by the availability of substrates (source limited) or by the capacity of the organ to assimilate and utilize the available substances for growth (sink limited) (Patrick, 1988).

Grain filling and ripening are affected by many environmental factors, including water, temperature, radiation, and soil nutritional condition (Yoshida, 1981). Grain dry matter increase is supported by available assimilates, as defined by C assimilation during the grain filling period plus assimilates reserve stored in the straw (Cock and Yoshida, 1972; Weng *et al.*, 1982).

Shortage of assimilate supply due to inhibition of photosynthetic processes is one of the major factors determining grain filling (Matsushima and Wada, 1958; Yoshida, 1981; Evans, 1996; Egli, 1998). Attempts to identify physiological factors limiting yield must integrate source-sink interactions both spatially and temporally (Patrick, 1988). Results from the response of kernel weight and grain set to source-sink manipulations suggested yield limitation by both the sink and the source, depending on the seasons, time of day, etc. (Evans and Wardlaw, 1996). However, the time courses of source and sink control have not been well documented. Some authors (Evans and Rawson, 1970; Richards, 1996) suggested that grain yield in irrigated spring wheat may not be limited by the supply of carbon at any time during grain filling. However, there are data showing significant increases in mass per grain associated with reductions in grain number (Fisher and HilleRisLambers, 1978; Ledent and Stoy, 1985; Koshkin and Tararina, 1989), implying source limitation at least on some occasion after anthesis. Source limiting situations are more important in high temperature environments (Fisher, 1983; Blum *et al.*, 1994). This suggests that physiological processes limiting yield in temperate area can not be directly extrapolated to subtropical or tropical environment.

Analysis of source-sink interactions should also consider the role of alternative sinks in the plant (Schnyder, 1993). In wheat particularly, special attention should be given to stems since competition exists between the growing upper internodes and reproductive organs in the weeks before anthesis (Wardlaw, 1968; Bingham, 1972; Patrick, 1972; Brooking and Kirby, 1981; Siddique *et al.*, 1989), the out come of which depends on both genotype and environment. Besides, there is good evidence that temporary storage is very important under stress conditions (Blum *et al.*, 1994; Wardlaw and Willenbrink, 1994). Also genotypic differences in the patterns of allocation and mobilization of dry matter in the stem of wheat have been demonstrated (Blum *et al.*, 1994). Therefore, the inclusion of alternative carbohydrate storage pools in the analysis of source-sink interactions should be of major importance under tropical environments. Grain yield in wheat is the end products of a complexity of assimilate producing and consuming pathways. It depends upon the balance between assimilation by sources and consumption by sinks, and may be limited by either (Fisher and HilleRisLambers, 1978). Sink or yield components of wheat include spike numbers, spikelet numbers per spike, kernel numbers per spikelet, and kernel weight. Final number of spikelets per spike appears to be affected by environmental conditions such as photoperiod and temperature prior to initiation of the terminal spikelet, while fertile floret numbers per spikelet depend upon environmental conditions from double ridge to after anthesis (Simmons, 1987). Floret competency is also affected by the position of the floret within the spikelet and the spikelet position upon the spike (Evans *et al.*, 1972).

In crop plants, the physiological basis of dry matter production is dependent on the source-sink concept, where the source is the potential capacity for photosynthesis and the sink is the potential capacity to utilize the photosynthetic products. If the sink is small, the yield cannot be high, and even if the sink is large, the yield cannot be high if the source capacity is limited. Several crop plants including wheat and rice have shown changes in source and sink where the leaf (source) is short and erectofile to capture greater solar radiation and the sink is expanded by increased number and size of grain. The continued growth of any plant depends primarily on photosynthetic activity in the leaves and the transport of organic compounds from the leaves to heterotrophic cells. The source supplies assimilates (C-compounds) to the sink. The sink accepts and consumes assimilates for its own growth or accumulates them for a certain period. The sink and source functions of any plant part may change depending on its developmental stage. For example, in wheat, before flowering the leaf sheath and culm accumulate sugars and starch, i.e., act as sink (Yoshida and Ahn, 1968). After flowering, most of the accumulated carbohydrates move into the spikelets (Cock and Yoshida, 1973).

The size of an organ that functions as a sink depends on two factors; including capacity, which can be expressed by dry weight, and intensity, which can be expressed by specific activity. Young growing organs such as very young trusses and the stem can have very high intensity but small capacity (Tanaka and Fujita, 1974). The intensity of very young leaves as sinks for the photosynthates of other leaves is not very high, because most probably, they start to photosynthesize from an early stage of development. The intensity becomes extremely low when the leaves are old. However, their intensity as sinks for their own photosynthates is maintained for a long period. The intensity of grains or reproductive organs as sinks is high at very early stages of their development and becomes relatively low with growth. However, since capacity increases with growth, the sink size of grains becomes large.

In cereals, particularly in wheat and rice, sink size is increased through appropriate combinations of cultural, nutritional, and water management factors. However, the potential has a limit beyond which it is almost impossible to increase sink size in terms of number. In cereals like wheat, potential sink size is based largely on spikelet number, while the realized sink in terms of filled grains is invariably lower. For understanding the yield of physiology of wheat and determining the limiting factors for different conditions and cultivars for obtaining suitable management and doing genetical improvement in order to eliminating or decreases limiting factors of yield, manipulation of source and sink have used under the following heads:

### 4.3.1 Source increasing

Enrichment of $CO_2$ concentration is one of the methods of experimentally changing the assimilatory capacity that use for the study of source-sink relationships of the plants and determining the limiting factors for different conditions. Elevated of $CO_2$ concentration at different stages of plant growth is postulated to increase assimilate availability for grain filling, biomass and grain yield (Kimball, 1983; Ravi *et al.*, 2001; Sharma-Natu *et al.*, 2004).

There has been an increasing trend in the atmospheric $CO_2$ concentration since preindustrial period due to burning of fossil fuels and deforestation (Clark *et al.*, 1982). Presently the $CO_2$ concentration in the atmosphere is around 370 ppm, which is expected to become double of pre-industrial $CO_2$ concentration by the middle of present century (Watson *et al.*, 1990; Olszyk and Wise, 1997). Different crops respond differently to elevated $CO_2$ concentration (Ghildiyal and Sharma – Natu, 2000). In general, $C_3$ plants species respond favourably to elevated $CO_2$ compared to $C_4$ species. Elevated $CO_2$ can cause an increase in biomass and yield ranging between 10 and 40 percent in many $C_3$ crops (Kimball, 1983; Weigel *et al.*, 1994; Ravi *et al.*, 2001; Sharma-Natu *et al.*, 2004). Such a variation in the performance of $C_3$ species under elevated $CO_2$ results from the variations in photosynthesis rate and sinks strength of the species (Biswas *et al.*, 1996; Ulman *et al.*, 2000; Sharma-Natu *et al.*, 2004).

Li *et al.* (2000) found elevated $CO_2$ increased final grain weights of the upper main stem spike section by 10 percent and the lower main spike section by 24 percent under drought stress conditions, under well-watered conditions, elevated $CO_2$

concentration increased final grain weights in lower main stem spike section by 14 percent, compared with ambient $CO_2$ treatment. Grain weight increase under elevated $CO_2$ in this experiment was due to sufficient source of assimilate and a faster rate of grain filling. These observation indicate that a spike intrinsic regulation mechanism probably exists which allows assimilate distribution to the lower and middle sections first, then to the upper section during later grain filling stage. Photosynthetic acclimation has intrigued researches in the area of plant responses to elevated $CO_2$ over the last decade, as evidenced by two recent reviews (Stitt and Krapp, 1999). Acclimation, which occurs only after prolonged exposure of plant to elevated $CO_2$, is characterized by lower leaf photosynthetic capacity (Drake *et al.*, 1997). Jitla *et al.* (1997) observed an increase in $CO_2$ assimilation rate at elevated $CO_2$ after seeding emergence, high $CO_2$ concentration accelerates development in the shoot apex and enhances leaf area growth. These changes in $CO_2$ assimilation rate, after exposure to high $CO_2$ are reflected in greater net assimilation and relative growth rate at the whole plant level, and are the forerunners to greater dry mass in the vegetative and reproductive organs later in development.

The grain weight increase of a spike caused by elevated $CO_2$ was due to an enhanced grain-filling rate and possibly increased number of grain per spike (Li *et al.*, 2000), also greater assimilate supply is the most likely explanation for the higher grain weight and faster grain-filling rate with elevated $CO_2$ (Wheeler *et al.*, 1996). Garcia *et al.* (1998); Wall *et al.* (1999) found elevated $CO_2$ increased both pre-and post-assimilate supply in wheat because of enhanced flag leaf and spike photosynthesis. As a $C_3$ plant, rice also showed a positive response to elevated $CO_2$ concentration (Baker *et al.*, 1990a;Ziska and Teramura, 1992; Ziska *et al.*, 1996; Hamid *et al.*, 2003). At elevated $CO_2$, rice leaf N concentrations are generally reduced during vegetative and reproductive phases due to the greater carboxylation efficiency of Rubisco (Makino *et al.*, 1994). Hamid *et al.* (2003) found that elevated $CO_2$ enhanced biomass accumulation resulting in nearly 38 percent higher dry matter than ambient grown rice, compared to the plants grown in open top chamber at ambient $CO_2$, elevated $CO_2$ treated plants had more favorable partitioning of dry matter into panicles. In this experiment plants with elevated $CO_2$ produced higher seed yield (about 24 percent) than the ambient and field grown rice.

According to Smart *et al.* (1994) elevated $CO_2$ increased the accumulation of starch, fructans, sucrose and total non-structural carbohydrates (TNC) in the leaves from the upper layer of the wheat canopy, while only starch and fructans accumulated in the lower canopy layers. Tuba *et al.* (1994) found that the concentration of starch in leaves and roots of wheat responded positively to elevated $CO_2$. Blumenthal *et al.* (1996) observed that elevated $CO_2$ not only increased the grain yield, but also decreased the grain quality of wheat by decreasing grain protein content. No significant effects of elevated $CO_2$ on either total starch content or starch structure in the grains were evident in that study.

Baker *et al.* (1992) observed carbon dioxide enrichment at 28/21°C increased both biomass accumulation and tillering and increased grain yield by 60 percent, with increasing temperature to 37/30°C grain yield decreased from 10.4 to 1.0 $Mgha^{-1}$ in rice plants, these results indicate that while future increases in atmospheric $CO_2$ are likely to be beneficial to rice growth and yield, potentially large negative effects on rice yield are possible if air temperatures also rise. Carbon dioxide enrichment has been shown to increase growth by stimulating tillering in both wheat (Sionit *et al.*, 1980) and rice (Baker *et* al., 1990b).

Estiarte *et al.* (1999) found $CO_2$ enriched in wheat plants had higher flavonoid concentrations by 14 percent, higher TNC concentrations and lower N concentrations in upper canopy leaves throughout the growth cycle. Lieffering *et al.* (2004) studied the impact of elevated $CO_2$ on the elemental concentrations of field-grown rice grains. They found no changes in the concentrations of any of the microelements analysed. In this study elevated $CO_2$ increased biomass and grain production and decreased grain N concentrations.

Recent work has identified a gene HIC (HIgh Carbon dioxide) whose disruption leads to large increases in the number of stomata initiated in response to $CO_2$ enrichment. This response contrasts with the typical, but not universal, decrease in stomatal initiation under $CO_2$ enrichment. The HIC gene encodes an enzyme involved in the synthesis of those long-chain fatty acids that are typically found in the cuticle of leaves. Changes in these fatty acids may influence the cell-to-cell signaling of stomatal development (Gray *et al.*, 2000).

Most of studies are on the response of plant to concentration below 1200 $\mu$ mol $mol^{-1}$ $CO_2$ in air. The majority of the cases which examined $CO_2$ enrichment up to 1200 $\mu$ mol $mol^{-1}$, found a positive effect on plants in terms of increased photosynthetic rate, enhanced crop yield, and elevated vegetative dry weight biomass (Egli *et al.*, 1970; Fischer and Aguilar, 1978; Cloux *et al.*, 1989). Grotenhuis and Bugbee (1996) studied the effect of 350, 1200 and 2500 $\mu$ mol $mol^{-1}$ $CO_2$ on growth and yield of two cultivars of wheat grown hydroponically in growth chamber. They found that the peak of vegetative biomass and seed yield was obtained at 1200 $\mu$ mol $mol^{-1}$ $CO_2$, further elevation of $CO_2$ to 2500 $\mu$ mol $mol^{-1}$ reduced seed yield by 22 percent in cv. Veery-10, and by 15 percent in cv. Usu-Apogee. They also found that exposing wheat plant to 2500 $\mu$ mol $mol^{-1}$ 2 weeks before and after anthesis, mimicked the results of a constant high $CO_2$ exposure treatment. The reduction in photosynthetic capacity after growth at high $CO_2$ is attributed to lower concentrations of Rubisco, and is more pronounced at low N supplies (Rogers *et al.*, 1996). Whether the reduction in Rubisco concentration is caused by accumulation of soluble carbohydrates is still a matter of debate (Stitt and Krapp, 1999).

Plant thinning is one of the methods of experimentally changing the assimilatory capacity of the crop canopy. Thinning at anthesis is postulated to increase the light available to remaining plants and thus increase assimilates availability for grain filling (Fischer and Laing, 1976). Thinning at anthesis may increase grain set (Nerson, 1980) and either have no effect (Willey and Holliday, 1971) or increase kernel weight (Fischer and Laing, 1976).

Jedel and Hunt (1990) observed in a post anthesis thinning treatment of wheat plants, in which stem number was reduced by 50 percent at anthesis time, thinning increased grain yield of plants with increased kernel weight. Increased assimilate availability due to thinning at anthesis may be expected to increase kernel weights. According to the definitions of sink and source limitations developed by Fischer and HilleRisLambers (1978) therefore, the results indicated that the wheat crops in this experiment were source limited.

Fischer and Laing (1976) have used thinning as a technique for increasing photosynthate supply for developing kernels that increased kernel weight. Blade and Baker (1991) found that thinning treatments in wheat, either at the six-leaf stage or at heading, resulted in increased number of kernel per spikelet/spike. But thinning at the six leaf stage caused a significant reduction in kernel weight, this was likely due to the increase in number of spike per plant and reflects the effect of thinning at this stage on the tillering process. Use of thinning as a method for increasing the source of photosynthate seems highly dependent on environmental condition.

#### 4.3.2 Source decreasing

Light, a critical natural resource, essentially controls morphogenesis and production in crop plants. Its role in photosynthesis, photoperiodism, and photonasty are well known. Crop plants in a sense harvest solar radiation with their complex biological machinery to produce human food. Manipulation of source and sink by shading has been used to understand the yield physiology and limiting factors of wheat. Developmental stages when shading is imposed, as well as shading duration, have varying impacts on kernel characteristics and grain yield (Venkateswarlu and Visperas, 1987; Grabau *et al.*, 1990; Pararajasingham and Hunt, 1991; Kiniry, 1993; Kobata *et al.*, 2000; Okawa *et al.*, 2003; Wang *et al.*, 2003).

Grain yield in wheat is dependent on photosynthate production and allocation. Factors such as light intensity affect photosynthesis and photosynthate allocation and influence grain yield and yield component in a number of ways (Jin, 1996). Various authors have demonstrated that pre anthesis shading primarily reduces spike density, spikelet number and grain number per spike, that post-anthesis shading mainly reduces grain weight and the grain number (Willey and Holliday, 1971; Kemp and Whingwiri 1980; Fischer, 1985; Mcmaster *et al.*, 1987; Jedel and Hunt, 1990; Slafer *et al.*, 1994; Kobata, *et al.*, 2000; Okawa *et al.*, 2003; Wang *et al.*, 2003).

Wang *et al.* (2003) suggested that the adaptability of the small-spike, small-grain cultivar to either pre-or post-anthesis shading is much greater than that of the large-spike, large-grain cultivar in terms of many characteristics closely related to grain yield. They suggested that, in areas where low light intensity often occurs, the small-spike, small-grain cultivar would be more likely to produce high, stable grain yields. Low irradiance reduces the photosynthetic rate, resulting in a reduction in the amount of carbon available for grain development (Tanaka and Matsushima, 1971). During the early phase of grain filling in rice as in other gramineous plants, the assimilated carbon is temporarily stored as non-structural carbohydrates in the culm and leaf sheath (Evans and Wardlaw, 1976). A highly positive correlation is obtained between the amount of stored non-structural carbohydrate available per grain in the initial 10 days of grain filling and the relative growth rate of grain in field-grown rice (Chan and Bhargawa, 1998). This period corresponds to the reported time in which endosperm cell number is determined. Furthermore, the effect of shading on dry matter accumulation in the spikelets depends on the position of the spikelet within the spike (Oshima, 1996). The above-cited research suggests that changes in the partitioning of assimilated carbon at the whole plant level and within the spikelets occur under low light conditions during the early phase of grain filling.

Light intensity and quality have pronounced effects on photosynthate production and allocation. Previous work on wheat has shown both curvilinear and linear responses of production and yield to light intensity, with the response partly dependent on the duration, timing, and intensity of shading (Pendleton and Weibel, 1964; Willey and Holliday, 1971; Kemp and Whingwiri, 1980; Stockman *et al.*, 1983; Fischer, 1985). McMaster *et al.* (1987) found that shading of winter wheat from booting to one-week after anthesis affected tiller mortality, spike morphology, kernel weights and ultimately grain yield. Respiration significantly influences the carbon balance of a crop. In wheat, biomass equivalent to between 40 and 75 percent final grain mass can lost through shoot respiration during grain filling. Pararajasingham and Hunt (1991) examined the relationship between changes in biomass and respiration of the above ground plant parts of shaded and unshaded wheat during grain filling. They observed the $CO_2$ efflux rate for unshaded plants declined significantly from anthesis through maturity. Imposition of shade resulted in significantly lower $CO_2$ efflux rates compared to the unshaded plants. It was concluded, therefore, that shading affects total respiration through its impact on growth, but exerts no direct effect on the basic pattern of change in maintaining respiration during grain filling.

A factorial field experiment was executed with three bread wheat cultivars and two shading treatments (shading to 50 percent of the incoming solar radiation from the beginning of stem elongation to heading or no shading). In this experiment shading reduced both the number of spikes per $m^2$ and the number of grains per spike, preanthesis shading reduced number of grain per $m^2$ on main shoots less (45 percent) than on tillers (65 percent) (Slafer *et al.*, 1994; Abbate *et al.*, 1997). Demotes-Mainard and Jeuffroy (2004) observed shading reduced spike dry matter but increased spike N concentration, it reduced spike N content whenever there was a large decrease in incident photosynthetically active radiation (PAR).

Okawa *et al.* (2003) quantified the effect of low irradiance on the partitioning of assimilated carbon during the early phase of grain filling in rice by using $^{13}C$ as a tracer, they suggested that under low irradiance at around 10 days after heading, the priority of the panicle and the superior spikelets of the panicle in the partitioning of previously and recently assimilated carbon is intensified by the compensational decrease of assimilates into the culm and the leaf sheath and into the inferior spikelets in the rice. This phenomenon is thought to be an important strategy, to achieve a certain proportion of ripened grains even under conditions of low irradiance. Wang *et al.* (2003) observed both pre- and post-anthesis shading caused a decrease in both dry matter (DM) accumulation and allocation to grain, the decreases in grain number or weight per spikelet in pre-and post-anthesis shading took place mainly in the upper and basal spikelets.

Shading probably affects not only assimilate supply to the grain, but may also influence other growth processes, such as nutrient absorption or leaf senescence, leading to combined growth after effects (Fujiwara and Ishida, 1963). In a study of wheat grown under shaded conditions, Takashi and Kanazawa (1996) found that final grain weights where affected by the size of starch granules formed after completion of endosperm structural development, rather than by the number of endosperm cells. At around 10 days after heading the enzyme activities related to starch synthesis in the endosperm are enhanced at this stage (Liang *et al.*, 2001).

Leaf removal treatment on cereal crops have caused significant reductions in grain yield and seed weight and used for determining of source-sink limitation (Nagato and Chaudhry, 1979; Rajewski and Francis, 1991). The magnitude of yield reduction appears to be affected by the time and severity of leaf loss. In several crop plants, and particularly in wheat, there is a division of function among the leaves on a culm after the elongation of internodes. The sink for photosynthates from the upper two to three leaves is the spike that for the lower leaves is the root system (Tanaka, 1961). Leaves in an intermediate position may send assimilates in either or both directions (Yoshida, 1972, Venkateswarlu and Visperas, 1987b).

The flag leaf which reaches full expansion after anthesis and senescence during grain filling, supplies a large proportion of photosynthate to the grain. Flag leaf and spike photosynthesis are the primary contributors of C assimilates during grain filling (Austin and Edrich, 1975; Bidinger *et al.*, 1977; Rawson *et al.*, 1983). When photosynthesis during grain filling is reduced (artificially or from stress), more carbohydrates are mobilized from stem reserves for grain development (Gallagher *et al.*, 1976; Bidinger *et al.*, 1977).

Leaf clipping and organ shading studies (Venkateshwarlu, 1976; Venkateswarlu and Visperas, 1987a) have indicated that the leaf is a major determinant of the nature of grain filling. In wheat, the leaves contribute about 51 percent of the grain yield, while non-structural carbohydrate contributes 15 percent, the stem up to 18 percent, the panicles up to 16 percent and the ear contributes 10-49 percent (Kriedemann, 1966). There is hardly any contribution of reserves in wheat under normal conditions, while under stress conditions accumulated reserves are utilized to varied degrees. Wareing *et al.* (1968) showed that the stimulatory effect of partial defoliation on remaining leaves caused increased levels of protein and particularly of carboxylating enzymes. They suggested that a corresponding increase in photosynthetic capacity, rather than increased relative demand by the sink, was responsible for the higher photosynthetic rate of surviving leaves, and they favoured a dynamic hormonal interaction between sink and source.

The flag leaf is the major source in crop plants, removal of the flag leaf or a portion of it has been used as a method to reduce the amount of photosynthate available to developing kernels. Blade and Baker (1991) observed removal of the flag leaf blade at anthesis or 8 days after anthesis resulted in significant decrease in kernel weight. Removal at 18 days after anthesis resulted in a small but insignificant decrease in kernel weight. The flag leaf blade is the principal source of photoassimilates imported by grains during grain filling (Rawson *et al.*, 1976; Murchie *et al.*, 2002). In spite of this, removal of the flag leaf may lead, in some circumstances, to enhancement of the photosynthetic activity of other leaves and green parts (Koch, 1996) and remobilization of stored carbohydrates (Schnyder, 1993). These mechanisms avoid the restriction of grain filling in such a manner that often in source limitation occurs (Richards, 1996). However, it has been pointed out that there is a higher degree of source limitation in warmer environments due to earlier senescence of green parts (Fischer, 1983). Cruz - Aguado *et al.* (1999) observed the removal of the flag leaf reduced the mass per grain, this reduction was greatest in one of cultivar with largest flag leaf area.

King *et al.* (1967) showed that in wheat, 2 weeks after anthesis, 45 percent of the flag leaf assimilates were transferred to the developing ear, which was itself photosynthesizing. Removal of the ear resulted in a 50 percent reduction in photosynthetic rate of the flag leaf within 15 hours. Darkening of the other leaves resulted in recovery of photosynthetic rate of the flag leaf, but the assimilates being diverted to roots and shoot (Venkateswarlu and Visperas, 1987a). Wang *et al.* (1997) found that source reduction by partial defoliation increased leaf net photosynthetic rate and sink reduction decreased photosynthetic rate of irrigated wheat. Source reduction enhanced photosynthate translocation into grain and decreased the partitioning of photosynthates into the upper parts including grains of plant.

### *4.3.3 Sink decreasing*

High yield potentials in wheat have been achieved through increases in partitioning of photo assimilates into economic sinks without adversely affecting light interception or the hardiness of the crop (Gifford *et al.*, 1984). Decreasing the reproductive sink size by removal of wheat spikelets has provided a way to study source-sink relationship. Such manipulations enhanced the availability of assimilates to wheat spikes (Jenner, 1980).

Generally, partial degraining increased the growth rates and final weight of the remaining individual grains (Bingham, 1967; Cook and Evans; 1978; Fischer and

HillRisLambers, 1978; Simmons *et al.*, 1982; Cruz-Aguado *et al.*, 1999). For some wheat cultivars, partial degraining had no effect on the growth of remaining kernels (Fischer and HilleRisLambers, 1978; Jenner, 1980; Atsmon *et al.*, 1986; Blum *et al.*, 1988; Mackown *et al.*, 1989). Blum *et al.* (1988); Ma *et al.* (1990) found that decreasing the sink size of two wheat cultivars with different average kernel size caused an increase in the weight of remaining kernels on the small-kernel cultivar and no change in kernel size of the large kernel cultivar. They hypothesized that compensatory kernel growth in spikes with reduced kernel number is related to average kernel size of the cultivar. These results contrast with those of Blade and Baker (1991) who observed that large-seeded cultivars tended to show the greatest response to spikelet removal.

Kernel size is dependent on endosperm cell number, which is established during the early developmental stages of the kernel (Brock-Lehurst 1977). Partial degraning increases the endosperm cell number of kernels (Brock-Lehurst, 1977). Apparently, any effect of enhanced assimilate supply to degrained spikes caused a proportionate increase in the size and number of cells in all remaining kernels. Limitation of yield by source or sink is subjected to both pre- and post-anthesis conditions (Wardlaw, 1994), and varies from environment to environment (Stapper and Fischer, 1990; Slafer and Miralles, 1992). Modification of sink size could inhibit photosynthesis in such a manner that even the filling of the remaining grains and whole plant dry matter grain could be affected by the availability of photoassimilates i.e. the rate of photosynthesis would be diminished to a greater extent than the demand. The principal cause of such inhibition of photosynthesis might be the degree of reduction in size of the sink or the position of the trimmed grains or both.

Richards (1996) suggested that grain yield of irrigated spring wheat may not be limited by the supply of carbon. However, there are data in the literature showing a significant increase in mass of grains associated with reductions in grain number after anthesis (Fisher and HillRisLambers, 1978; Ledent and Stoy, 1985; Koshkin and Tararina, 1989) implying source limitation, at least some time during grain filling. On the other hand, the lack of difference in grain mass between control and degraining treatments could be a consequence of source limitation at early stages of grain formation, when the number of endosperm cells is fixed (Brock-Lehurst, 1977; Wardlaw, 1994). These conflicting results suggest a strong environmental and genotypic interaction in the response of grain mass to modifications in source-sink relationships (Blade and Baker, 1991). Manipulation of source-sink ratios by artificial reduction in grain number per inflorescence has been used in several cereal grain species to estimate potential kernel weight and study the grain fill process (Fischer and HilleRisLambers, 1978; Simmons *et al.*, 1982). Actual kernel weight is less than potential kernel weight because of competition among kernels for available assimilate and interplant competition for light, water and nutrients (Peterson, 1983). Potential Kernel weight is obtained when Kernel number is reduced to a point at which competition among Kernels for assimilate no longer exists (Ficsher and HilleRisLambers, 1978; Peterson, 1983). Fischer and HilleRisLambers (1978) defined potential kernel weight as the kernel weight resulting from an 80 percent reduction in kernel number per spike (imposed at anthesis by surgical removal of spikelets) and estimated postanthesis source limitation by comparing potential to actual kernel weights.

Cruz-Aguado *et al.* (1999) manipulated the source-sink ratio to examine the performance of source-sink interactions after anthesis and the factors(s) limiting grain filling in tropical conditions. Plants of three wheat cultivars were artificially degrained to give different source-sink ratio and partitioning of $^{14}C$ photoassimilates was measured after anthesis. Modification of sink led to different patterns of allocation of dry matter between cultivars. The reduction in sink size was enhanced mass per grain in two cultivars. Removing all spikelets except the four-central spikelets of the spike in one of the treatment declined both mass per grain and translocation of $^{14}C$ photoassimilates, apparently due to feed back inhibition of photosynthesis. The pattern of partitioning of dry matter observed in plants in this investigation suggests a source limitation.

Zamaski and Grunberger (1995) suggested that of the two high- yielding wheat cultivars that usually give identical yields, the one with long multi-spikelet ears and lighter grains has significantly greater unexpressed potential for high yield than the cultivar with shorter ears and heavier grains. In this experiment removing one third of the ear's distal end, increased the number of grains on the untouched basal spikelets of the decapitated ears in both cultivars. Thus, the total yield (number x mass) achieved by the untouched portion of the decapitated ear was 100 percent in the long ears and only 80 percent in the short ears type.

Strong competition between the vegetative and reproductive organs may also develop during the transition to flowering. This is critical in cereals because the potential floret number is established before anthesis (Evans *et al.*, 1975) when vegetative growth, especially in the upper internodes of the stem, still persists (Care and Wardlaw, 1965; Rawson and Evans, 1971; Bonnett and Incoll, 1993a,b). Therefore, removal or sterilization of florets at an early stage of development results in increased growth of the remaining florets (Solansky, 1979) or even in the formation of grains, which would otherwise not have occurred (Rawson and Evans, 1970). In wheat, anthesis (followed by grain formation and filling) usually begins just above the middle of the ear, and proceeds upwards and downwards (Rawson and Evans 1970; Zadoks *et al.*, 1975). Moreover, Spikelets in the middle of the ear are more fertile than those at the base or tip (Hanif and Longer, 1972). In wheat, differences in grain weight were found within a particular spikelet (Bremner and Rawson, 1978) and were mostly due to differences in grain filling rates (Rawson and Ruwali, 1972; Simmons and Crookston, 1979). After a sink reduction treatment, similar weight increases in individual grains were found regardless of their positions within the spikelet (Simmons *et al.*, 1982). It has been reported that sink reduction may delay leaf senescence in the plant (Patterson and Brun, 1980; Gwathmey *et al.*, 1992) although this is not always obvious (Labrana and Avaus, 1991).

Voltas *et al.* (1998) decreased sink strength of main stems of barley by half at anthesis by sterilizing central and lateral florets, they have found that a 50 percent reduced sink, increased significantly grain weight of both central and lateral spikelets by simulating grain filling rate in all trials, and also by lengthening grain filling duration.

Perez *et al.* (1989) found that removal of the top halves of ears of winter wheat at various times after anthesis increased the nitrogen content in the grains of lower half of the ear. The increase was greater with early than late removal. Nitrogen uptake of shoots after anthesis decreased with halving. The decrease in kernel number of partially degrained spikes would reduce the demand for assimilates and consequently may have reduced the relative ability of degrained spikes to compete with other sinks for assimilates (Cook and Evans, 1978; Ma, *et al.*, 1990). This effect should be more pronounced in cultivars that obtain a large portion of their assimilates from leaf photosynthesis and mobilized vegetative reserves.

Photosynthetic rate responds to the demand for assimilates. In wheat, during grain filling, when most of assimilates from the flag leaf are translocated to the ear, removal of the ear leads to an accumulation of assimilates in the flag leaf and to a fall in its photosynthetic rate by about half within hours. If the lower leaves are shaded to that the flag leaf has to support the rest of the plant, assimilates are exported at a high rate, and the photosynthetic rate rises to its original level (King *et al.*, 1967). Such pronounced feedback effects have not always been found, probably because alternate sinks for assimilates, such as young tillers, were present (Venkateswarlu and Visperas, 1987a).

## 5. PLANT GROWTH REGULATORS AND THEIR ROLE IN DETERMINING SINK EFFICIENCY

The history of plant growth regulators (PGRs) goes back long before the time of Jesus, when it was a common practice to place a drop of olive oil to develop figs in the Middle East (Nickell, 1986). It is now known that heat caused the breakdown of the oil, releasing ethylene, which in turn affected the development of the fig. Tropism in plants, especially the way in which differential cell extension was controlled, led to the discovery of very first plant hormone. Darwin's study of coleoptile phototropism provided the initial stimulus.

Plant growth regulators have apparently been used owing to their probable beneficial effect on growth and maturity of plants. According to Nickell (1979) PGRs are expected to play an important role in rectifying the hurdles on the manifestation of biological productivity. The complexities in searching for right PGRs for a given crop have been amply studied and presented by Archer and his colleagues (1982). This study clearly demonstrates the need for knowledge of life cycle of crop involved and the need to determine the rate limiting factors and their interactions in order to determine when a PGR must be applied to have its maximum effect. Due to higher biostability *in vitro*, synthetic analogues of plant hormones, synonymously also known as PGRs, are generally preferred for horticultural and agricultural uses (Bruinsma, 1985). Five major classes of PGRs have been recognized viz., auxins, gibberellins, cytokinins, abscisic acid, ethylene (Morgan, 1979; Bruinsme, 1985; Nickell, 1985; Symons and Reid, 2003; Gibson, 2004) and many others are awaiting their grouping as plant hormones, since they are of plant origin and active at extremely lower concentration, e.g. triacaptanol of polyamines etc. (Mishra, 1990). Growth hormones have been reported to be involved in grain filling in determining yielding ability and in partitioning of assimilates between ear and shoot or ear and rest of the plant (Stahli *et al.*, 1995; Peltonen-Sainio *et al.*, 2003).

One of the oldest uses of PGRs has been to initiate and/or accelerate the rooting of cuttings by auxins, of which indole butyric acid was commonly used. The malic hydrazide came into use very extensively to prevent sprouting of potatoes and onions and to inhibit the growth of grasses in lawns, on golf courses, in parks and along the highways. One of the most widely used regulators was CCC (chlormequat, cycocel) which shortened the height of wheat without changing the size and quality of spikes. This decrease in height prevented lodging of wheat plant after rain and wind.

Large scale, commercial and extensive agricultural utilization came by using auxins for weed control and raising the cereal yields by as much as 30 percent. Of late, they were used as tool of new biological warfare in the form of defoliant. Of all the commercial growth regulators, gibberellins, cytokinins and auxins are true growth promoters, and many of their applications involve a modification of development, rather than a simple effect on growth rate or final size. Plant hormones or phytohormones are plant produced growth regulators and, therefore, are naturally occurring plant substances, PGRs, on the other hand, are organic compounds other than nutrients, which are either synthetic or natural compounds and when applied directly to a plant, promote, inhibit or, otherwise, modify any physiological process in plants. The various endogenous growth substances probably play an important role in determination of yield components and their interaction. In recent years it has been suggested that PGRs might play an important role in the determination of grain number and size (Peltonen-Sainio *et al.*, 2003). The involvement of growth hormones in determining yielding ability and partitioning of assimilate between ear and shoot or ear and rest of the plant has been well explained.

### 5.1 Auxins

The first plant hormone, auxin, was discovered by a group of workers at California Institute of Technology (Vanderhoff and Kosuge, 1984). Auxins are a group of plant hormones, which promote growth along the long axis when applied in low concentration to the shoots of plant, freed as far as possible, from their own inherent growth promoting substances (Dharmasiri *et al.*, 2003). Free IAA is chemically clusterable and is rapidly degraded by plant enzymes (Reinecke and Bandurski, 1981) and by light (Symons and Reid, 2003). In contrast, conjugates of IAA are very stable compounds and are not destroyed by plant degradatory enzymes such as peroxidases (Cohen and Bandurski, 1978).

It was Choloduy (1936) who was the first to claim an increase in growth and yield of wheat and oats by treating the seeds with indole acetic acid (IAA). Vamuda *et al.* (1963) demonstrated the presence of substantial amounts of free auxin in oat endosperm, noticed its disappearance during germination and provided evidence for the movement of this auxin to coleoptile tip. Thus, the auxin present in the endosperm was transported to the tip and the tip itself does not synthesize any auxin. Microinjection of labelled IAA into endosperm results in the later appearance of labelled IAA in the coleoptile tip (Sheldrake, 1973). Naryanan *et al.* (1981) are of the view that developing ovules and immature fruits are the excellent sources of auxin. Mishra (1990) observed auxin caused higher grain yield in rice due to increase in number and length of panicles and grain numbers. Bhardwaj and Verma (1987) reported that the developing grains of wheat produce high level of auxin.

According to Asana and Bagga (1966) IAA affects the growth and yield of wheat and its application at time of onset of reproductive phase increased the grain number per ear as well as 1000 grain weight. Leopold (1949) have reported that the auxin content in bareiy plant was 78 percent higher under long day conditions than in the short days. This may be one of the causes for lower number of tillers under long-day conditions. According to Leopold (1949) auxin decreased the number of tillers in grasses by 60 percent. Further, evidence that auxin controls tiller formation comes from the work of Saini *et al.* (1975), where in they have shown that the endogenous auxin content decreased when the apex of grass plants was treated with x-rays or some inhibitory substances, e.g. TIBA and coumarin. This treatment led to the increase in tiller number. The major portion of auxin in the grain was located in the accumulating endosperm (Rademacher, 1978). Wheeler (1972) have also shown that the auxin content is at maximum during development in wheat. Developing wheat grains contain high concentrations of endogenous auxins at the time of rapid photoassimilate accumulation (Wheeler, 1972; Bangerth *et al.*, 1985). Therefore, it would be reasonable to expect that photoassimilate import by developing wheat grains may, at least in part, be attributable to their auxin contents. This conclusion is supported by the finding that grains of ears treated with IAA exhibited enhanced import of $^{14}$C-labelled photo- assimilates exported from the flag leaves (Prochazka, 1978).

There are some reports describing hormonal effects on photoassimilate transport in maternal tissues of developing legume seeds (e.g.Gibson, 2004). Based on considerations of the cellular pathway of photoassimilate transport within developing wheat grains (Patrick and Offler, 1995), auxin could act on the symplasmic and membrane, transport of photoassimilates. According to Sacher and Berry (1984) auxin modulates the enzyme activity, either by affecting the de novo synthesis or by activation of preformed, otherwise, dormant enzymes. They were also of the view that application of phytohormones to intact plant tissues is known to activate preformed enzyme molecules. Developing wheat grains contain high concentrations of endogenous auxins at the time of rapid photoassimilate accumulation (Darussalam *et al.*, 1998). Therefore, it would be reasonable to expect that photoassimilate import by developing wheat grains may, at least in part, be attributable to their auxin contents. This conclusion is supported by the finding that grains of ears treated with IAA exhibited enhanced import of 14C-labelled photoassimilates exported from the flag leaves (Darussalam *et al.*, 1998; Jianchang *et al.*, 2000).

Singh (1963) has reported an increase in the number of filled grains, weight of grains per panicle and yield of total grains of IAA and kinetin treatment. Concurrently, Normanly (1997) also showed that the developing grains, capable of producing higher levels of auxins, could mobilize a greater proportion of assimilates from flag leaf or elsewhere. Contemporary studies directed to establish the hormonal regulation of grain growth of their development indicate that the capacity of grains to grow and accumulate photosynthates is dependent on the levels of the distribution of a plethora of substances with cytokinins, auxins and gibberellins occupying prime positions. Somehow, the role of inhibitory substances in balancing or operating to counter the diverse promotery substances has not been given a befitting stature. A similar trend of a specific correlation between the grain weight with that of endogenous levels of hormones have been shown by Dua *et al.* (1990) in *Fagopyrum esculentum* thus hinting at the possibility of pivotal underlying mechanism which operates at a broader spectrum especially wherever the precipitation of photosynthetase is involved by growing sink.

Dua and Sehgal (1981) found that middle grains in spike which had significantly higher weight than the two peripheral grains, had more auxins. In the same direction, Patrick and Offler (1995), suggested the interpositional differences in the levels of endogenous auxins, amongst the spikelets (apical, central, basal) and florets to be the major factor responsible for determining the difference in the grain weight. According to Singh (1963) IAA level in wheat grains increase dramatically during period of most dry matter accumulation.

### 5.2 Gibberellins

The first isolation of a gibberellin from a higher plant was achieved by MacMillan and Suter in 1958, who found gibberellic acid in immature seeds of *Phaseolus coccineus L.* Subsequently many gibberellins were reported and at present 125 naturally occurring gibberellins have been chemically characterized to date (Taiz and Zeiger, 2002). The effects of gibberellins on stem extension and growth are well known (Brian and Hemming, 1955). Bioactive gibberellins (GAs) affect a number of processes during plant development including seed germination, leaf expansion, stem elongation, flower initiation and flower and fruit development (Sun, 2000).

Gibberellic acid (GA) is known to influence many physiological processes in growth and development of wheat. According to Wheeler (1972) GA content reaches to peak activity in the period of grain filling. Radley (1976) has reported that at higher temperature GA content of ear increased rapidly. An increase in the dry weight and height of plant by $GA_3$ treatment but decline in tiller number was observed, whereas an increase in yield could be seen only at post flowering stage in rice (Gogol and Baruah, 2000).

Schliemann and Schneider (1993) reported that a combination of cycocel (CCC) and $GA_3$ pretreatment followed by $GA_3$ application was quite influential in increasing the growth and yield component. However, the beneficial effects decreased with the increase in the number of pretreatment cycles. Biswas and Choudhuri (1977) reported an increase in grain yield per ear by GA treatment although there was no improvement in harvest index. Schliemann and Schneider (1993) observed that the polymeric gibberellin forms (Soluble and insoluble) displays a high activity during the growth of wheat seedlings and that there was a certain relationship between activity of polymeric forms and structure and molecular weight of polymer.

In addition to their role in internode elongation, gibberellins causes bolting and flowering of long day plants kept under short-day conditions (Lang and Reinhard, 1961). The greater rate of GA accumulation was associated with a greater rate of dry matter increase. Correlation between endogenous levels of GA and grain number has been reported (Bhardwaj and Dua, 1975). Rahman *et al.* (1981) observed an increase in yield due to increased grain number in normal height wheat varieties whereas semi-dwarf varieties showed the opposite effect. Application of $GA_3$ to the spike at heading markedly depressed grain number per plant because $GA_3$ application have been known to cause abnormal anther development and hence male sterility in barley (James and Lund, 1960; Rahman *et al.*, 1981).

Gibberellin has been implicated as a senescence deterring hormone by exogenous applications which delay senescence (Itoh *et al.*, 2003). Spraying of $GA_3$ at post flowering stage in rice, i.e. 96 DAS could increase the Chlorophyll content (Biswas and Chaudhari, 1977). Under the influence of $GA_3$ Chlorophyll content in barley leaves decreased while arise in photochemical activity of the chloroplasts was observed (Popova *et al.*, 1982). $GA_3$ treatment on the whole plant after anthesis in wheat increased the Chlorophyll and protein levels of flag leaf. An increase in the leaf area of rice plant was observed by $GA_3$ treatments during flowering stages (Angrish *et al.*, 2001). Sankhla and Huber (1974) observed an inhibitory effect of $GA_3$ on Rubisco activity in pennisetum whereas Popova *et al.* (1982) reported an increased activity in mature bean leaves and barley seedlings respectively. There are very few reports on gibberellin induced photosynthate translocation to the active sink (i.e. ear). $GA_3$ application at 100 days in rice (Ray and Chaudhuri, 1980) has also shown increased translocation of photosynthate in different varieties of rice. Chatterjee *et al.* (1976); Gogol and Baruah (2000) observed higher endogenous gibberellin content in heavier grains of rice and further reported an increase in plant dry matter by its exogenous application.

Stahi *et al.* (1995) studied the effect of $GA_3$ on phospholipid metabolism and α-amylase production in the aleurone tissue in wheat (*Triticum aestivum* L.). They reported the $GA_3$ promoted phospholipid break down and terminated the independent turnover of choline N methyl group in phosphatidyl choline and promoted turn over of the whole choline breakdown. Angrish *et al.* (2001) reported that $GA_3$ treatment results in the synthesis and the secretion of hydrolytic enzymes, notably α-amylase by the cascades signals released from aleurone layers and the protoplast. Gibberellins have been shown to induce expression of some α-amylase (Gibson, 2004). Sun (2000) reported that $GA_3$ (5 and 30 μm) increased fresh weight, dry weight and protein content of cultured cells but effect of $GA_3$ (50 μm) was not significantly different. $GA_3$ at 5 μm increased starch content while at 50 μm starch accumulation was reduced.

Guoping (1997) showed that GA3 weakened the effect of paclobutrazol on tillering and shortening the plant, increased the dry matter accumulation of different tillers, improved plant nitrogen metabolism by favoring nitrogen translocation in to tillers at late growth stage and stimulated tiller development. All of these reduced the difference in spike weight among tillers and the main stem. Miyoshi and Sato (1997) reported the stimulatory effects of gibberellin on the germination of rice seeds under aerobic and anaerobic conditions. Applications of GA3 to tall wheat plants stimulate the effect of long-day plant, e.g. by accelerating the initiation of primordia and spikelet defferntition at the shoot apex, whereas growth retardants have the opposite effect, which can be relieved by subsequent treatment with GA3 (Evans *et al.*, 1995; Symons and Reid, 2003).

### 5.3 Cytokinins

Cytokinins play an important role in the regulation of plant growth and development. One of the consequences of cytokinin over production could be beneficial for modern agriculture is delaying of leaf senescence in plants and extend their photosynthate productivity (Heyl and Schmulling, 2003; Dua and Eradatmand-Asli, 2005).Cytokinins modulate gene expression in a wide variety of plant tissues and cell types over a variety of response time(Schmulling *et* al.,1997). Letham (1963) isolated zeatin from immature kernels of *Zea mays*. Subsequently, zeatin and zeatin riboside have been identified from a range of plants and assumed to be present in many others (Kende, 1971). Cytokinins are generally associated with the phenomena of cell division and enlargement (Fosket *et al.*, 1977), and they apparently play an important role in nutrient mobilization within plants (Mothes *et al.*, 1961). Both these processes are of the most importance in seed development and germination. The general picture which emerges from the literature is that cytokinin levels in the vegetative apices are low prior to and at anthesis, but that these levels increase greatly after fertilization and during periods when there is rapid seed growth. This is particularly noticeable when cell division and enlargement are occurring (Davey and Van Staden, 1977).. These levels again decrease when the food reserves have accumulated and the plant mature.

Besides regulating the rate of cell division and direction of cell elongation, cytokinins are also assumed to play a characteristic role in the formation of new attraction sites for assimilates in certain tissues. Cytokinins have also been implicated with other plant hormones in a sink-source phenomenon. By changing the sink capacity these substances influence the intensity and direction of assimilate flow (Saha *et al.*, 1986). In rice and wheat, grain yield is influenced not only by the production of photosynthates in leaves during the filling period but also by the efficiency of translocation and absorbing capacity of the grain (Yoshida, 1972). In cereal, increase in number of tillers by cytokinins is observed (Langer *et al.*, 1973). Further Ray *et al.* (1983) have also shown that kinetin treatment significantly increased the number of spikelets per panicle, number of panicles and the percentage of filled grains. Biswas and Chaudhuri (1977) have reported an increase in grain weight per ear with a slight increase in harvest index in wheat. Growth substances may be involved in the determination of yield capacity and distribution of assimilates between panicle and shoot (Herzog, 1982). In wheat chloroplasts contain a wide spectrum of cytokinins and a relatively high activity of cytokinin oxidase, and they react to darkness with specific changes in the level of various cytokinin metabolites (Benkova *et al.*, 1999). In general, cytokinin and chlorophyll quantities were correlated; i.e. varieties which showed the most rapid decline in chlorophyll also contained the lowest levels of cytokinins (Banowetz, 1997).

In general the high free cytokinin content is correlated with active overall metabolism of the normal plant (Herzog, 1982). Dua and Kalsi (1988) studied the physiological factors influencing the differential yield capacity of the same ear of wheat under *in vitro* conditions and their result revealed that smaller grains remained small and bolder grains retained its boldness even under acute stress. The incorporation of auxin and kinetin in liquid media in which sample grains were cultured enhanced the dry matter of small grains than the bold grains. This behavior was due to the fact that the small grain possessed fewer amounts of these substances *in vivo* as compared to the bold grains. Biswas and Chaudhri (1977); Banowetz (1997) have also shown an increase in both chlorophyll and protein content of flag leaf of wheat by kinetin treatment. Feierabend (1970); Treharne *et al.* (1970); Feierabend and Bore (1978) have reported the sensitivity of Rubisco to cytokinin treatment. Similarly, Srivastava and Goswami (1988) have also found an increased enzyme activity in sunflower by BA treatment.

One of the most important physiological effects produced by cytokinins is the direct transport of solutes towards the site of its application. Application of kinetin at a particular site made amino acids move towards these sites. Subsequently it was shown that other solutes also moved in a similar manner (Mothes *et al.*, 1961). Micheal and Seller-Keilbitsch (1972) observed a correlation between grain weight and level of endogenous cytokinins in barley and wheat. Bhardwaj and Verma (1985) found that bolder grain size of middle spikelet was related to higher levels of cytokinins. Studies showed that cereal endosperm cell proliferation is correlated with kernel cytokinin accumulation and that this accumulation is thermosensitive (Cheikh and Jones, 1994; Banowetz *et al.*, 1999a,b). The observations that maximum kernel cytokinin accumulation precedes or occurs simultaneously with endosperm cell proliferation suggest that cytokinins are a critical regulatory component of kernel development (Banowetz *et al.*, 1999b). Micheal and Beringer (1970) suggested that cytokinin enhances seed size by the establishment of an increased endosperm cell number, providing larger storage capacity. Endogenous cytokinins from a variety of cereal tissues have been identified and changes in cytokinin content have been correlated with developmental or growth processes (Banowetz, 1992, 1997).

Kaminek *et al.* (2000) reported that cytokinins zeatin and zeatin riboside reached their maximum amount by 4-11 days after anthesis and the auxin content was inversely proportional to the total cytokinin level and reached maximum during filling stage of grain development. The studies of Chen (1997) suggested that the concentration of cytokinin in tissue capable of generative development in winter wheat should be on a level that occurred in inflorescence cell lives in comparison to immature embryo all time. Also, different levels of hormone may be induced to the developmental stages of ears of wheat to flower, in winter wheat plants regenerated *in vitro*. Dua *et al.* (1982); Normanly (1997), were able to inter-link the endogenous production potential of auxins and cytokinins or their distribution to the inherent differentiality with regard to the weight of individual grains. Concrete evidence to this postulation was lent when their exogenous applications, to a specific pre-determined cultivar, led to an enhancement in its harvest index (Dietrich *et al.*, 1995).

### 5.4 Abscisic Acid

Liu and Carns (1961) isolated a crystalline material, called abscisin-I from mature cotton fruit bolls. A different abscission speeding substance was isolated from young cotton fruits by a different group of workers (Addicott *et al.*, 1968). This substance initially called abscisin-II came later to be known as Abscisic Acid (ABA) (Addicott *et al.*, 1968).

Abscisic acid (ABA), generally as an inhibitory growth hormone (Tanner, 1980), increases markedly in leaves (Ober and Setter, 1990), floral organs (Saini and Aspinall, 1982) and developing grains (Goldbach and Goldbach, 1977) of water stressed plants. Water stress induced reduction in grain set in wheat (Morgan, 1979) and decreased rates of endosperm cell division in water stressed maize have been attributed to elevated levels of ABA. However observed effects of ABA on the capacity of the sink to accumulate dry matter are not consistent, both inhibition and stimulation have been reported. The increase in ABA levels towards the end of grain filling and its rapid fall during maturation have raised questions about the role of ABA in controlling dry matter accumulation (King, 1982; Ahmadi and Baker, 1999). Indeed in several cases the application of ABA to the medium has enhanced accumulation of reserves, particularly storage proteins in legumes, and production of mRNA. The studies of Radi *et al.* (2002) revealed that ABA treatment resulted in non significant changes in dry matter production of stressed seedlings, the fresh matter production at most salinization levels tend to increase in wheat seedlings.

ABA has been implicated in the assimilate transport, since application of ABA to developing wheat grains enhanced import of assimilates (Dewdney and McMwha, 1979). Stimulatory effects of ABA on assimilate unloading (Tanner, 1980; Schussler *et al.*, 1984) and *in vitro* sucrose uptake (Schussler *et al.*, 1984) has also been reported. Nevertheless, these promoting effects are not always observed, and depending on the concentration and the timing of ABA application.

ABA levels in grains play a role on the regulation of maturity and may influence grain size and yield (Ma *et al.*, 1990). There is considerable accumulation of ABA during growth in wheat till maximum grain dry weight is obtained. ABA activity in cereal grain increases during grain growth, more or less, continuously (Goldbach and Micheal, 1976). ABA is transported from the mature leaves to the shoot tops (Zeevart, 1977) or to young leaves (and apical parts of plant), which are temporarily richer in ABA than older ones (Goldbach *et al.*, 1975). According to Waters *et al.* (1984) ABA caused reduction in the grain yield by reduction in grain number and decreasing uptake of sucrose through decreasing stomatal conductance and transpiration. Hall (1973) found that the daily application of ABA to growing wheat plants initially inhibited growth, but after a short lag resulted in an increase in the number of leaves and tillers after 84 days.

### 5.5 Role of Growth Regulators in Sink Development

A number of explanations have been advanced to explain the mechanism controlling the efficiency of sink. Sweet and Wareing (1966) assumed that levels of growth substances in the sink might affect its growth rate and thereby inducing changes in photosynthetic rate affected by assimilate accumulation. Contrary to the above King *et al.* (1967) with the help of defoliation studies in wheat reported that photosynthesis by the flag leaf was regulated directly by the demand for assimilates. The rate of photosynthesis by flag leaf could be extensively, rapidly and reversibly changed by varying the demand (Stahli *et al.*, 1995).

Wareing *et al.* (1968) reported an increase in the net rates of photosynthesis of bean (*phaseolus vulgaris*) leaves by resorting to partial defoliation and advocated that this increase was related to the higher levels of carboxylating enzymes per unit leaf area. It was thought that the roots supplied higher amounts of cytokinins which, in turn, increased the levels of above enzymes. This postulation was confirmed by supplying cytokinins and by removal of roots.

Evans (1972) have suggested that detailed analysis of the ways to increase storage capacity of cereals ears would be of significance in the analysis of processes limiting grain yield. It appeared that the mechanism was more likely to be correlative and hormonal in nature rather than a competition for assimilates. Plant growth regulating substances have apparently been used owing to their probable beneficial effects on growth and productivity of plants. Bhardwaj and Dua (1975) working on the effect of 2,4- D, NAA and IAA on tall wheat pbc-591 noticed sufficient increases in the yield of grain by pre-treating the seeds with the above substances. The increase in grain yield could only be attributed to the increased vigor of the main shoot, as indicated by significant increases in plant height and 1000 grain weight. Asana and Bagga (1966) studying the effect of spraying the foliage with IAA and NAA also noticed stimulation in growth and yield of wheat. The beneficial response was linked to the higher survival rate of tillers as well as increase in grain number and 1000 grain weight.

Bhardwaj (1962) observed that protein content of the wheat grains was higher in the hormone treated plants. It is possible that hormonal treatment might lead to the differences in the accumulation of proteins and carbohydrates in the grain, which might be related to the impeding of carbohydrates translocation and enhanced protein synthesis in the developing grains (Ma *et al.*, 1990). Das (1963) reported that in rice application of NAA (Nepthalene Acetic Acid) increased the yield which was attributed to higher number of tillers and mean number of grains per panicle. In same line Yamuda *et al.* (1963) showed that spraying of rice plants with IAA or NAA resulted in reduced tillering and panicle height but hastened the emergence of panicles.

Hayashi (1961) reported an increase of 10 to 15 percent in photosynthetic activity of whole plant by the application of $GA_3$ and it was shown by Taylor *et al.* (1989)that this increase was due to the increase in efficiency of photosynthetic systems rather than due to enhanced rate of translocation. Singh (1963) treated the rice plants with $GA_3$ and it was seen that the number of spikelet per plant, combined length of all ears, weight of stem and number of grains per plant were significantly increased. The higher rate of GA accumulation was associated with a greater rate of dry matter increase. Correlation between endogenous level of GA and grain number in wheat has been reported (Bhardwaj and Dua, 1975). Michael *et al.* (1970) observed a correlation between grain weight and level of endogenous cytokinins in barley and wheat. Bhardwaj and Verma (1985) found that bolder grain size of peripheral spikelet was related to high levels of cytokinin. Michael and Beringer (1980) suggested that cytokinin enhanced seed size by the establishment of an increased cell number providing larger storage capacity. Jianchang *et al.* (2000) suggested that cytokinins in the grains and roots during the early phase of grain development play an important role in regulating grain filling pattern and consequently influence grain filling percentage in rice plants.

From the above it is clear that applications of growth regulating substances in various ways have a profound effect on growth and yield though quite often contradictory results have been reported. The beneficial effects have generally been explained on the basis of improvement in number of tillers, number of grains per ear and/or 1000-grain weight.

## 6. PHYSIOLOGICAL AND BIOCHEMICAL CHANGES DURING GRAIN DEVELOPMENT

Growth of grain from initiation to terminal ripening stage follows a complex pattern of growth comprising a vegetative growth, flower and grain development. The most important period is from anthesis to maturity as during this yield of grain in ear is determined by the number of grain which develop and the size they attain. To know about the yield components and various sources of biochemical products, researches has been conducted on the biochemical aspect of wheat and their time course developments in grain, which are captioned under the following heads:

### 6.1 Starch

It is the major constituent of wheat grain and determines its yield and yield is a measure of the activity of processes contributing to deposition of starch in the grain (Jenner *et al.*, 1991; Chinnusamy and Khanna-Chopra, 2003). Sucrose, the primary product of photosynthesis and assimilates carbon, acts as the main raw material for starch formation (Porter and May, 1955). Berry *et al.* (1971) isolated, studied and compared the starches of triticales, rye hard red spring (HRS) and durum wheats. They reported that the greatest percentage of starch occurred on the triticale flour samples as compared to wheat and rye. Jenner (1980b) correlated the starch synthesis in endosperm to concentration of sucrose in tissue. Starch accumulation appeared to cease during development in the oat prior to the total dry weight accumulation, while soluble carbohydrates primarily sucrose and glucose remained at high level (Koch and Peterson, 1991). Black *et al.* (1996) studied in embroys of wheat that starch accumulated in the axis and scutella from about 20 days post anthesis (dpa) to reach a maximum at approximately 35 dpa, after wards it declined to a very low value in late maturation. The concentration of starch, sugars and proteins in rice decreased under both high and low concentration of Cu (Nautiyal *et al.*, 1999).

According to Jenner *et al.* (1991) starch content was initially low but increased rapidly during maturation while reducing sugars, sucrose and fructose decreased in two spring wheat cultivars. Jenner and Rathjen (1972) concluded that magnitude of photosynthesis is adequate to maintain the flow of sucrose into the wheat grain and thus flow of sucrose is limited by the capacity of the processes transporting the sugars in final stages of its passages into grains. Koch and Peterson (1991) showed that active starch synthesis started from 14 days after anthesis onwards and continued until 35 days after anthesis.

Pena (1982) stated that starch content of mature grains varied with extent of grain shriveling in secondary hexaploid culture. Metabolic pathway of starch biosynthesis in the developing wheat grains was studied with the help of labelled sugar and NMR spectrometry (Keeling *et al.*, 1987).

### 6.2 Sugars

Many plant developmental, physiological and metabolic processes are regulated, at least in part, by nutrient availability. In particular, alteration in the availability of soluble sugars, such as glucose and sucrose, help regulate a diverse array of processes (Gibson, 2004). The accumulation of sugars in the storage cells is crucial for the size of grain. According to Sterans (1970) sugars constituted nearly 10 percent of dry matter of wheat aleurone cells. Paul *et al.* (1971)

reported that in developing seeds of rice reducing sugars were gradually converted into non-reducing forms during development. Sterans (1970) reported that in hard red spring wheat, the contents of reducing and non-reducing sugars decreased as grain matured and the contents of non-reducing sugars was higher than reducing at maturity stages. Daul *et al.* (1971) reported that the concentration of glucose and fructose decreased while raffinose which appeared at later stages increased with maturity. Duffus and Rosie (1973) found changes in soluble reducing sugars and studied that reducing sugars remained low throughout development. Caputo and Barneix (1999) observed that glucose, fructose and glucofructose contents reached maximum before the phase of rapid starch synthesis but then decreased as the kernel matured in case of spring wheat. According to Singh and Juliano (1977) level of soluble carbohydrates remained high in caryopsis and also in milled rice after starch accumulation suggesting that supply of sugar precursors does not limit starch accumulation in rice grain because of high level of reducing sugars.

Singh and Singh (1982) noticed that total soluble sugars decrease during course of grain development in rice (cv. Jaya) treated with IAA, $GA_3$ and kinetin. Black *et al.* (1996) investigated the relationships between starch sucrose and raffinose with onset of dessication tolerance and found that starch accumulated from about 20 days post anthesis (dpa) in axes and scutella and reached maximum at approximately 35 days after anthesis and then declined to a very low value in late maturation whereas starch sucrose and raffinose appeared 26 dpa and sucrose rise was paralleled with accumulation of starch in grain. Caputo and Barneix (1999) have studied the relationship between amino acid and sugar export to the phloem in wheat plants. Study showed that the sugar concentration in the phloem exudates was increase by higher light intensities, but there was no difference in the amino acid concentration of the phloem exudates and thus the amino acid: sugar ratio in the phloem decreased in the high-light plants. The present results suggest that amino acids can be exported to the phloem independently of the export of sugars.

In addition to affecting a number of developmental processes, sugars have been implicated in the regulation of a large number of genes (Koch, 1996). The α-amylase gene family provides a particularly well-characterized example of sugar-regulated gene expressing. Given their biological function, it is, perhaps, not surprising that the expression of many α-amylase genes has been shown to be repressed by soluble sugars, such as glucose and sucrose, thus providing at mechanism by which starch breakdown may be regulated to provide and adequate supply of soluble sugars (Gibson, 2004). The regulation of α-amylase expression by sugars is complicated, occurring at multiple levels and via multiple response pathways that can be dependent on sugar concentration or flux. For example α-amylase expression is regulated by sugars at both the transcriptional and posttranscriptional levels (Gibson, 2004)

### 6.3 Proteins

Proteins of seed, either enzymatic or storage reserves are synthesized during seed development and maturation and are deposited within membrane bound protein bodies (Pernollet, 1978; Matile, 1987). At the onset of germination, the protein bodies are degraded inside these organelles and products are transported to the growing parts of plant for use in the biosynthesis (Ashton, 1976; Mikkonen, 1986). Wheat grain proteins have traditionally been classified as albumins, globulins, gliadins and glutenins on the basis of their solubility (Osborne, 1907).

The kinetics of accumulation of the protein fractions were not significantly affected by post-anthesis temperature or drought, whereas N nutrition significantly increased the rate and duration of accumulation of storage proteins. Storage protein first appears in wheat endosperm at about 10 days after anthesis and is located within membrane-bound spherical bodies. These bodies typically 0.5 – 1.5 μm in diameter, derived from the golgi apparatus and appear closely associated with rough endospermic reticulum. During the final stages of grain filling, many protein bodies fuse, forming a continuous, highly compressed protein matrix in which the starch granules are embedded.

Proteins account for 25 percent of fresh weight of mature embryo. According to Jennings and Morton (1963); Panozzo *et al.*,(2003) maximum rate of protein synthesis occurs between 10-16 days after anthesis. Parlov and Kolensink (1974) studied the factors responsible for different levels of protein accumulation in some wheat cultivars and found correlation between nitrogen component and grain wheat. Grain protein has negative association with yield and breeders are likely to concentrate their efforts on yield improvement (Cox *et al.*, 1985). According to Tribol *et al.* (2002) the process of nitrogen partitioning is neither significantly affected by post-anthesis temperature or drought nor by the rate and timing of nitrogen nutrition, at maturity variations in protein fraction composition are mainly because of differences in the total quantity on nitrogen accumulated during grain filling. Calderini and Ortiz-Monasterio (2003) found that lighter grains in wheat (mainly those in distal positions on the spikelet) had lower N concentrations than heavier grains (generally in proximal positions on basal-central spikelet), which suggests that grain position can have considerable impact on nutrient concentration.

## 7. ROLE OF ENZYMES IN REGULATION OF SINK EFFICIENCY

Aimi *et al.* (1958) while working on rice observed that the distribution and rise or fall of sugars, starch and some other enzymes in leaves was directly correlated with the ability of grains to utilize the translocated sugars. They further reported that activity of enzymes in ears somehow guided the physiological processes happening elsewhere in plant. Paleg (1964) demonstrated that exogenously applied gibberellic acid could stimulate α-amylase activity in isolated barley endosperm. Other studies soon showed that it was aleurone layer of the endosperm that was sensitive to gibberellic acid. Macleod and Miller (1962) found that removing aleurone renders the endosperm almost totally insensitive to GA treatment. Subsequent studies showed that GA treatment of isolated aleurone can cause the release of the hydrolytic enzymes required for the digestion of endosperm starch (Paleg, 1964). Brown and Sun (1973) concluded that ABA acts as inhibitor of RNA and protein synthesis in barley aleurone layers. Jacobson and Varner (1967) studied that at least four of the enzymes comprising α-amylase, protease, ribonuclease and B-1, 3-glucanase induced by $GA_3$ treatment arise through de novo synthesis.

Sakai-Wada and Nakata (1987) studied the α-amylase activity and ultrastructure of aleurone cells in seed of oats using seed halves with embryo which had imbibed water with or without $GA_3$, α-amylase activity was detected in the aleurone layers of embryo seeds that had imbibed water with $GA_3$. According to Mitra and Sen (1987) auxin, gibberellin and cytokinin enhanced the activity of α-amylase in the flag leaf cells but protease activity was enhanced only by gibberellins. Gubler *et al.* (1987) studied the release of α-amylase through gibberellin treated barley aleurone cell walls. It indicated that α-amylase is released from aleurone via digested cell wall channels and that except for the inner wall layer unhydrolysed regions are impermeable to the enzyme. Singh and Panwar (1998) studied effect of 2,4-D on α-amylase activity, growth and yield components of wheat cultivars, which showed that 2,4-D enhanced the α-amylase activity, root number, shoot dry weight, leaf area, chlorophyll content, nitrate reductase activity, total biomass production, effective ear number, number of grain per ear and seed size thus resulting in higher grain yield.

There are three modes of formation of α-amylase during grain development which can each led to high activity in harvested grain: first, in the absence of germination and before the grains is mature; second, associated with germination before the onset of dormancy; third, associated with germination after dormancy has broken. In this direction Gold and Duffus (1996) further hypothesized that one of these three factors triggers the onset of gibberellic acid sensitivity and that pre-maturity α-amylase activity only develops if sufficient moisture is present in the grain.

Starch synthetase enzyme plays an important role in the starch synthesis and is one of the factors along with phosphorylase activity which could limit the grain starch precipitation and ultimately resulting in a higher degree of grain shriveling. The poorer grain filling or starch content accumulation in the small grains might be due to either less synthesis or more degradation of starch in the grain causing higher free sugar. From the observations it appears that the phosphorylase, starch synthetase, invertase and α-amylse enzymes present in the grains could be playing a major role in determining sink efficiency or grain carbohydrate filling. Patel and Mohapatra (1996) reported that the activities of the sucrose synthase and invertase were higher and lower respectively, in the endosperm cells of the top spikelet compared with the basal spikelet. They concluded that poor synthesis of the starch leading to partial grain filling in the basal spikelet was due to a lower activity of sucrose synthase. Sucrose synthase is believed to be the dominant enzyme that degrades sucrose for starch synthesis in developing grains (Patel and Mohapatra, 1996). According to (Watanabe *et al.*, 1997) soluble starch synthase (SSS), granule-bound starch synthase (GBSS) and plastidial fructose-1, 6-bisphosphatase (FBPase) were significantly correlated with the starch content. These results suggest that starch branching enzyme is involved in regulation of starch metabolism, possibly in collaboration with other enzymes such as SSS, GBSS and plastidial FBPase in temporary sink organs like the leaf sheath. It was also demonstrated that there is a high correlation between source/sink ratios and amounts of sucrose phosphate synthase (SPS), a key enzyme in the pathway of sucrose synthesis (Seneweera *et al.*, 2002).

Phosphorylase is widely distributed in plant tissues and has been isolated and characterized in barley and triticale. Duffus and Rosie (1973) observed that the phosphorylase activity was mainly associated with endosperm fractions and had peaks of activity between 25-30 days after anthesis and declined later. Baxter and Duffus (1973) detected phosphorylase activity in barley endosperm extracts from 3 days after anthesis. They observed unprimed activity between 2 and 10 days after anthesis which constituted 70-80 per cent of total activity and it declined rapidly as the grain developed. Sastry *et al.* (1979) reported high phosphorylase activity in shrunken and lower activity in bold type of grains in barley. They carried out a comparative study of starch synthetase and phosphorylase enzymes in grains of shriveled Notch-2 and its bold grain parent NP-113. They observed high phosphorylase and low starch synthetase activity in Notsch-2 and the reverse in the parent NP-113. Based on these observations, they suggested some mechanism of regulation of starch metabolism involving both phosphorylase and starch synthetase enzyme system simultaneously may be functioning in barley grains. Pande *et al.* (1981) observed higher phosphorylase activity during early and active phase of starch accumulation suggesting that the enzyme may bring about larger degradation of preformed starch causing grain shriveling in mature kernels of Triticale.

The basic pathway of photosynthetic carbon fixation is the well known Calvin cycle. The enzyme catalyzing its basic reaction, fixation of $CO_2$ to RUBP, is Ribulose 1, 5-bisphosphate carboxylase/oxygenase (Rubisco). Since its discovery by Wildman and Bonner (1947), the attempt to improve the enzyme properties, which in turn will have direct relation with plant productivity (Rintamaiki, 1989), has been followed extensively throughout the world. Rubisco is responsible for annual fixation of about $7x10^{10}$ tons of carbon, equal to $17x10^{10}$ tons of dry matter, indicating, thereby its importance in the overall gamut of plant productivity (Parry *et al.*, 2003). Activation of this enzyme is an ordered and reversible process involving the binding of a $CO_2$ complex. The activity of Rubisco is more or less linearly correlated with the amount of the so called Fraction-I–protein (Kawashima and Mitake, 1969). It is the major soluble protein in green leaves since it constitutes from 23 to 50 percent of the total soluble protein in the leaves of various $C_3$ plants. Wheat grains have very low Rubisco activity. Recent evidence suggests that the carboxylase activity is a property of the large subunits whereas small subunits play a regulatory role (Tabake and Akazawa, 1973). Fraction-I-protein, usually isolated, can probably be considered to be crude Rubisco, since it often contains several other enzymatic activities in addition to that of Rubisco.

During leaf development, the rate of photosynthesis increases rapidly reaching a maximum before full expansion, and subsequently declines. Also there is an increase in the Rubisco activity and a decline upon reaching maturity, and some researchers have suggested that Rubisco may become a limiting factor for photosynthesis (Parry *et al.*, 2003). Rubisco expressed per unit of leaf area has been shown to decrease in the course of senescence in leaves of wheat (Peoples and Dalling, 1978) and rice (Shieh and Liao, 1988). The cause of this decreased Rubisco activity is not entirely clear, most of the studies referred to report a concomitant decrease in the Fraction-I-protein content but in wheat inactivation loss of enzymes appear to be involved, since the specific activity of the enzyme is also reported to decrease. The maximum Rubisco activity in wheat was found mostly prior to or at the end of leaf blade expansion (Brady *et al.*, 1971). Some authors have also found a correlation between the maximum Rubisco activity and photosynthetic rate (Treharne and Eagles, 1979; Medina, 1971; Seneweera *et al.*, 2002).

The loss in Rubisco activity is a significant event in the leaf senescence, since it signals an end to the carbon input into the leaves via photosynthesis. Any further benefit of the leaves to the developing grain can only come from the remobilization of carbon and nitrogen from the leaves (Patterson *et al.*, 1980). Patterson *et al.* (1980) have reported that Rubisco is produced slowly after anthesis and degraded rapidly during senescence. This decline would result from the lack of production of new enzyme as well as the influence of leaf senescence. The Rubisco specific activity was found to remain constant until late senescence by some (Wittenbach, 1979; Shieh and Liao, 1988). According to Vu *et al.* (1998); Seneweera *et al.* (2002) $CO_2$ enrichment in rice plants decreased the activity and content of Rubisco, in these experiment chloroplasts of the high $CO_2$ leaves contained more and larger starch grains than those of the ambient $CO_2$ leaves. Attempts to relate the inactivation of enzyme activity with loss of chlorophyll have yielded controversial results. Parallel losses of photosynthetic activity and chlorophyll content have been reported in wheat (Wittenbach, 1979) and rice (Uchida and Murata, 1982), whereas other investigation found no correlation between the activity and the chlorophyll concentration in leaves of wheat (Patterson and Moss, 1979) and rice (Makino *et al.*, 1983).

All enzymes involved in nitrate assimilation are present in wheat leaves and other parts of the shoot. The rate limiting enzyme nitrate reductase of $NO_3^-$ assimilation that was detected first in leaves (Evans and Nason, 1953) is present in all plant parts including glumes and grains (Kaim *et al.*, 1983) and exhibits a circadian rhythm of activity. Most of the nitrate (75 percent) taken up by wheat plant is reduced in the leaves (Kaim *et al.*, 1983). Nitrate reductase activity can be correlated with protein and grain yield (Abrol *et al.*, 1984). Phytohormones regulate various aspects of plant development including $NO_3^-$ reduction (Knypl, 1979). Induction of nitrate reductase by $NO_3^-$ is stimulated by cytokinins in cereal leaves. Cytokinins in combination with gibberellins replace the requirement of light for the induction of nitrate reductase by $NO_3^-$ in rice (Gandhi and Naik, 1974).

Hageman (1979) strongly argues that nitrate reductase is a sensitive indicator of nitrate flux into the growing plant and seasonal maintenance of this to ensure continuing high levels of nitrate, play an important role in grain and protein production. Biochemical criteria to assess productivity must provide simple reliable and non-destructive selection parameter useful to breeder. Nitrate reductase should be satisfactory yield predictor (Reilly, 1990). Levels of nitrate reductase may sharply and irregularly vary during the life cycle. The problem of redistribution and mobilization of nitrogen (and photosynthate) from vegetation to grain provided the greatest barrier in linking nitrate assimilation by nitrate reductase and yield. Nitrate along with other primary metabolites has a wider regulatory role in growth than the simple provision of substrate for assimilation into nitrogenous plant components (Trewavas, 1983). Islam and Morison (1992) showed that white light stimulated nıtrate reductase m RNA accumulation in an intensity-dependent manner, suggesting that photosynthesis may be involved in nitrate reductase regulation. In barley, a more direct correlation between photosynthesis and light/dark modulation of nitrate reductase at the post-translational level has been observed (Raghuram and Sopory, 1995). They also reported that exogenously added sugar phosphates or experimental conditions that increase sugar phosphate levels *in vivo* protect nitrate reductase from dark-inactivation, presumably by preventing phosphorylation of nitrate reductase by a protein kinase, rather than by activiting a phosphatase.

According to Ricardo (1974); Ricardo and Sovia (1974) vascular invertase participates in the regulation of hexoses levels in mature tissues and in the utilisation of sucrose closed in vacuoles (Leigh *et al.*, 1979). Morris and Arthur (1985); Arai *et al.* (1992) suggested the role of vascular invertase in the sink strength regulation. Eischrich (1980); Morris and Arthur (1985) revealed an important role played by cell wall invertase in the phloem unloading, by creating a steep sucrose concentration gradient, between the source tissue and the sink organs.

According to Pande and Shukla (1996); Mandal and Singh (2000) increased activities of oxidase and peroxidase during late stages of the grain development, enhanced the oxidative conditions in the maturing grains, and oxidized the natural antioxidant present in the grains which are essential for functioning of starch synthetic enzyme. They further revealed that above situation leads to an early cessation of dry matter accumulation in the grain, particularly under higher ambient temperature and late sown conditions. Mandel and Singh (2000) reported that activities of amylase and peroxidase decreased but that of protease increased under salt stress condition in rice. The studies of Chanda and Singh (1997) on three grains of wheat (*Triticum aestivum* L.) differing in the final dry weight and position on spike, showed that soluble IAA oxidase activity increased during elongation stage. While the ionically bound peroxidase activity increased during dry matter accumulation and maturation phases. Whereas, changes in ionically bound peroxidase activity showed inverse correlation with the process of cell elongation, while no relationship between soluble peroxidase activity and grain development was discernible. Foyer *et al.* (1998) studied the activity of mRNA on the coordination of nitrogen and carbon metabolism in maize. They concluded that coordination of nitrogen and carbon metabolism is retained during drought conditions via modulations in the activity of sucrose phosphate synthase.

## 8. ROLE OF VASCULAR SYSTEM OF INFLORESCENCE ON YIELD

In the last century, the grain yield of wheat has increased by about 150 percent as a result of better crop management and the use of cultivars with higher yield potential. The main physiological changes associated with this increase in grain yield has been related with a higher level of saturating light for photosynthesis, a lesser sensitivity to photoperiod, a longer duration of photosynthesis, a decrease in plant height and a higher harvest index, i.e. a better partitioning between grain and vegetative biomass (Austin *et al.*, 1980; Lopez-Garrido *et al.*, 2001). Frequently, the capacity of the vascular system for mobilizing assimilates to the growing grains have been pointed out as one of the causes that could limit the yield in cereals, especially under conditions that promote the development of a high photosynthetic capacity (Evans, 1993). However, little information is available about the modification in the capacity of the vascular system responsible for the translocation of assimilates from the leaves until the grains. The potentiality of a grain to grow and accumulate photosynthetic assimilates has of late been recognized as an important parameter of grain yield in cereals (Yoshida, 1972). Perceival (1921) suggest that the difference of grain size was attributable to either longer period of starch deposition in some grains or competition for assimilates among grains or hormonal interaction amongst them. Investigations during last decades (Rawson and Ruwali, 1972; Venkateswarlu and Visperas, 1987) have shown that sink efficiency is one of the primary determinants for yield in wheat. Dua and Sehgal (1981) have shown that the differences the weight of individual grains growing in the same spike could also be ascribed to the differences in the level of phytohormones. Thus the grains, endowed with better capacity to grow had a higher level of individual as well as total auxin and cytokinins. The growth behaviour and yielding ability of different grains, developing in the same ear vis-à-vis spikelet, showed significant variations in their growth rates depending on their locations.

Each spikelet was conspicuous by two bolder grains and a smaller grain, the former unequivocally showing a significantly higher dry matter precipitation at all stages of grain development. In wheat, anthesis (followed by grain formation and filling) usually begins just above the middle of the ear and proceeds up wards and downwards (Rawson and Evans, 1970; Zadoks, *et al.*, 1975). Moreover spikelets in the middle of the ear are more fertile than those at the base or tip (Hanif and Longer, 1972). It has been shown that only a small proportion of the total number of florets initiated eventually set grain (Beveridge *et al.*, 1965) and that under circumstances, the florets which are fertile are the first to be initiated and reach anthesis. Thus there is a gradation in the number of grains per spikelet from the center of ear towards the apex and base. Also within each spikelet basally situated florets are most likely to carry grain (Walpole and Morgan, 1970). It has been shown, however, that if the basal florets fail to set grain, then the more distal and normally infertile florets are capable of setting grain (Bingham, 1967).

The improvement of yield potential in crops has come largely from increase in the partitioning of assimilates into the harvested organs (Gifford and Evans, 1981). Although this has been achieved by empirical selection by plant breeders, further increase in this direction may be more difficult, and an understanding of the factors which control partitioning could be helpful. As a first step we thought it necessary to study the vascular system of spike and spikelet, on the assumption that any differences in this respect might help to explain variations in grain development according to position in the spike or spikelet.

## 8.1 Vascular System of Spike (Peduncle)

Knowledge of the continuity of vascular bundles and their distribution and size along the rachis is important in understanding the transport system in wheat ear. The rachis, in transverse section, is semi-circular in the lower, and spindle-shaped in the upper internodes. Vascular bundles are embedded in the parenchyma and the larger ones are arranged in a circle or ellipse, while the smaller peripheral ones are located close to chlorophyllous bands (Perceival, 1921). The two largest bundles in the internode, usually at the ends of the long axis of the ellipse, have been termed lateral bundles (Kirby and Rymer, 1974). In their work on barley Kirby and Rymer observed that the two lateral bundles and some of the central bundles branched at every rachis node, one branch entering the spikelet and the other proceeding to the next rachis internode. The bundles on the side opposite the spikelet continued into the next internode without dividing so that the number of central bundles remained the same at successive internodes. Such a configuration is not consistent with the acropetal decrease in the number of vascular bundles in wheat suggested by perceival (1921). According to Lopez-Garrido *et al.* (2001), not all vascular bundles in the peduncle continue into the rachis and that the number of bundles entering each rachis node depends on the position of that node in the rachis. This is contrast to the situation in the stem where the number of bundles is the same at every leaf node (Patrick 1972b; O'Brien *et al.*, 1985). It appears that the development of the vascular system in the rachis is different from that in the stem.

There was a 1:1 relationship between spikelet number per ear and the number, less two, of central vascular bundles at the base of the rachis (Whingwiri *et al.*, 1981). Evans *et al.* (1970) found a linear relationship between the number of spikelets per ear and the number of inner bundles (and Phloem area) in the Peduncle for only one (Late Mexico 120) of the 22 genotypes that they examined. The number of central vascular bundles along the rachis declined at an approximate rate of one bundle per internode. However, there were some exceptions at specific internodes depending on ear size. It was noted that in the basal three internodes, the number declined at a rate of less than one bundle per internode (Whingwiri *et al.*, 1981), and that between internodes 5 and 11 the rate of decline appeared to increase as ear size increased (Whingwiri *et al.*, 1981).

Peripheral bundles are found mostly in the basal internodes. Because the peripheral bundles are immediately adjacent to photosynthesizing chlorophyllus parenchyma cells in the outer extremities of the rachis, they presumably serve a function of transporting assimilates synthesized in these cells into the tissues of the ear (O'Brien *et al.*, 1985).

According to Whingwiri *et al.* (1981) the three central vascular bundles supplying the terminal spikelet two may be surplus, if so, this may be the base number of bundles supplying every spikelet on the ear. These two surplus bundles might well be the lateral bundles portrayed by Kirby and Rymer (1974) in barley, and which they showed as branching at every node. In wheat, branching of bundles as Busby and O'Brien (1979) mentioned is difference of vascular tissues in vegetative nodes. Branching should maintain a constant number of bundles at successive internodes and does not account for the acropetal decrease in the number of central bundles along the rachis. Such a decline requires the entire vascular bundles should be diverted into spikelets without dividing and we have referred to this as "dropping" to contrast it with branching.

Whingwiri *et al.* (1981) demonstrated that branching alone appeared to occur in the basal four internodes and in internodes 13 to14 of an18-spikelet ear. Dropping was observed mainly in internodes 5 and 11, but not exclusively and it appeared that, in this region, there were more bundles dropping in larger than smaller ears. Because of the circumstantial evidence that branching only occurs in the basal spikelets, where there are fewer grains per spikelet, it could be argued that vascular branching is associated with poor grain set. Where, in addition to branching, entire bundles are diverted into spikelets such as between internodes 4 and 11, high grain number may result. For this reason it would appear that the dropping of bundles is advantageous to grain set. The observation that distal spikelets are supplied by smaller bundles suggests that assimilate movement to the distal spikelets will encounter greater resistance than basal spikelets, resistance being determined by bundle size and the length of the vascular channels (Bremner and Rawson, 1978).

Although there has not been a systematic study of the development of the vascular system in the wheat ear, the work on barley by Kirby and Rymer (1974) showed that the vascular connections between the spikelet and the rachis were established at about time of floret initiation. Solansky (1979) has suggested that differences in grain size within the ear are a result of different physiological ages of individual grains. In wheat, the pattern of vascular connections from the rachis to the spikelet might well follow the sequence in- floret initiation.

Lopez-Garrido *et al.* (2001) reported that during the last century, the genetic improvement in grain yield of wheat has not modified the size of the vascular system for the mobilization of assimilates to the grains. This could probably has been due to the fact that in the ancient process of domestication of wheat from wild to cultivated forms, the cross-sectional area of phloem was already increased by over a 16-fold range (Evans *et al.*, 1970). This amount of phloem, therefore, seems to have been sufficient to sustain the maximum rate of grain growth in the actual cultivars. Nevertheless, the existence of a great genotypic variation in the vascular tissues of the peduncle, as revealed in different studies could permit the use of these characteristics to improve the phloem transport capacity in wheat in the future.

There is considerable information on the vascular connections necessary for the transfer of photosynthate within the leaf (Kuo *et al.*, 1972, 1974; Aaltus and Canny, 1982), at the node of insertion of the leaf on the stem (Perceival, 1921; Patrick, 1972b; Busby and O'Brien, 1979), through the stem (Evans *et al.*, 1970), within the inflorescence (Hanif and Langer, 1972; Whingwiri *et al.*, 1981; O'Brien *et al.*, 1985) and grain (Lopez-Garrido *et al.*, 2001). The transfer of photosynthate from the leaf to stem traces is an important component of the link between the leaf and ear. From the work of Lush (1976); Altus and Canuy (1982), it appears that translocation out of a grass leaf may occur in the median and main lateral veins and that the smaller intermediate and cross veins act to collect and transfer photosynthate to the major veins. The smaller intermediate veins entering the stem from the leaf show a degree of cross-linkage with the stem traces at the node before they are diverted to form a ring of peripheral bundles (Patrick, 1972b), where they presumably could also act as collecting agents for photosynthate fixed by the green tissue in the stem. The lateral leaf bundles show some linkage with the stem bundles at the node of entry, but much greater links at the next node below and complete fusion only at the nodal plexus two nodes below the point of entry. The median leaf bundle is unique as it appears to have no link with the stem traces at the node of entry, but passes directly to the next node below, where it shows partial fusion with other stands and, like the lateral bundles fuses completely with the nodal plexus two nodes below the node of insertion (Patrick, 1972b).

Time course studies on movement of $^{14}C$ – labeled photosynthate from the flag leaf to ear of wheat (Wardlow, 1965) indicated a fairly direct transfer between the stem and flag leaf traces at the node of entry. However, if the demand for assimilates was decreased by removing two-thirds of the developing grains, $^{14}C$ reached the node below the point of attachment before appearing in the peduncle. This latter finding suggested that the resistance to lateral transfer was less one node below the node of attachment, where the vascular linkages were more pronounced (Patrick, 1972b). Vascular bundles in the internodal region of the peduncle are not connected by cross veins (unlike those of the leaf) and lateral transfer of $^{14}C$ – labelled photosynthate between them could not be induced by partial phloem blockage which prevented direct vascular connections between the flag node and the ear (Patrick and Wardlaw, 1984). Whingwiri *et al.* (1981) in a comprehensive study have strongly advocated the distribution of vascular bundles in the ear with acropetally diminishing pattern along the rachis. Since the central vascular bundles are of different sizes and any spikelet(s) whenever connected to these would get a preferential treatment over others and this component piloting yield needs to be investigating critically.

### 8.2 Vascular System of Spikelet (Rachilla)

The physiological basis of yield in wheat has been the subject of intensive investigation for many years. Much effort has been devoted to discover what determines the basic components of yield.

It is generally believed that the primary determinant of yield in wheat (sink efficiency) is resolvable into grain size, grain number and number of tillers per plant. Support to the hypothesis that the growing sink is a major determinant in accumulating photosynthates has come from the work of Asana and Bagga (1966); Asana (1968); Bremner and Rawson (1978); Cruz-Aguado *et al.* (1999); Patrick and Offler (2001). Cook and Evans (1978) critically examined the various reports and concluded that though proximity of sink to source conferred a marked advantage but it was the inherent capacity of sink which played a paramount role in securing assimilates from distant sources and in order to execute the same an efficient translocation system may be of immense use (Blum *et al.*, 1988; Pheloung and Siddique, 1991).

Research has been directed towards both ends of the yield system, the source of assimilates derived mainly from the flag leaf, the peduncle and the ear and the sink represented by the sites at which grains develop (Thorne, 1966). Thus Williams (1966) has described the ontogeny of the inflorescence and its parts, while Rawson

(1970) investigated factors controlling spikelet number and its relation to yield per ear.

Rawson and Evans (1970) have demonstrated that early sterilization of either the first or second florets within a central spikelet resulted in the third and forth florets forming grain which would other wise not have occurred, but more distal florets were not induced to contribute to grain yield. Bindraban *et al.* (1998) have drawn attention to the apparent competitive advantage of the lowest floret within the spikelet in attracting supplies of carbohydrate. These and other reports indicate that certain inter-relationships exist among florets of the same spikelet which may determine the likelihood and extent of grain formation and for this reason it seemed important to us to study the morphogenesis of individual florets in different positions on the inflorescence. The vascular system in a wheat floret consists of four tracheids that branch from a basal vascular bundle (Lingle and Chevalier, 1985). One of these tracheids extends through the funiculus to the ovule and the other three extend through the ovary wall, one dorsal and one in each of the flanks. The lateral veins continue to the two stigmas of the flower, while the dorsal one terminates blindly. Shortly after fertilization, these three veins in the ovary wall degenerate (Alexandrov, 1937; Frazier and Appalanaidu, 1963; Lingle and Chevalier, 1985; Ugalde and Jenner, 1990a,b). Zee and O'Brien (1970) found that rich vascularization of the sterile glumes suggests that these organs may be playing an important role in nutrient cycling in the spikelet.

Developing seeds are net importers of organic and inorganic nutrients. Nutrients enter seeds through the maternal vascular system at relatively high concentrations in the phloem. They exit importing sieve elements via interconnecting plasmodesmata and during subsequent symplasmic passage, are sequestered into labile storage pools (Patrick and Offler, 2001). Interconnected sieve tubes form a continuous symplasmic compartment from the sites of loading through which photoassimilates (Offler *et al.*, 1989; Ugalde and Jenner, 1990), amino nitrogen (Ugalde and Jenner, 1990) and mineral elements (Pearson *et al.*, 1995) are imported into developing seeds. Sieve elements in seeds of cereals are symplasmically interconnected with the surrounding ground tissues (Wang and Fisher, 1994; Wang *et al.*, 1994; Patrick *et al.*, 1995; Tegeder *et al.*, 1999; Patrick and Offler, 2001). Circumstantial evidence suggests that these symplasmic routes allow sieve elements to unload water (Murphy, 1989; Wang and Fischer, 1994), sucrose (Offler and Patrick, 1993; Wang and Fisher, 1994), amino acids (Hsu *et al.*, 1984; Fischer and Macnicol, 1986), certain mineral ions (e.g. iron: Marentes and Grusak, 1998) and macromolecules (Fisher *et al.*, 1992).

The very fact, that there are distinct differences in the vascular system of individual florets within the same spikelet. The first three florets have been shown to be supplied by the principal vascular bundles of the rachilla while system of more distal florets consists of sub-vascular bundles, derived from the vascular cylinder formed at the disc of insertion of these florets. Furthermore, florets 1 and 2 appeared to be distinctive by virtue of their proximity to the main vascular system and their broad disc of insertion (Hanif and Langer, 1972; Whingwiri *et al.*, 1981; O'Brien *et al.*, 1985; Lopez-Garrido *et al.*, 2001). Evidence suggests that this pattern seems to be similar in all spikelets within the same inflorescence, with the possible exception of the terminal one and that increasing the supply of nitrogen which usually raises the number of fertile florets also had no effect (Hanif and Langer, 1972).

The apparent constancy of the vascular system supplying florets is in contrast to considerable variability in the number of florets which form a grain. This applies to positional differences on the same inflorescence, in that the basal spikelets often lack fertile florets, unless the nitrogen nutrition of the plant is raised at an early stage of development (Liew, 1968). Nitrogen application at or soon after the double-ridge stage has been shown to increase the number of grain-bearing florets (Beveridge *et al.*, 1965), while nitrogen given upto the emergence of the flag leaf may induce third and fourth florets to form grain thus compensating for low spikelet numbers restricted through previous low levels of supply (Single, 1964).

Availability of assimilates and their partitioning between the various sites within the spikelet have for some time been regarded as having a controlling influence on grain formation (Patrick and Offler, 2001). Ease of translocation presupposes an efficient vascular system, and it is therefore of interest to note that the lower two florets occupied the most preferential positions. The third floret was also found to share the main vascular system of the rachilla, and these three florets have been shown to form grain most reliably in many cases (O'Brien *et al.*, 1985). However, there does not appear to be an invariable relationship between vascular supply and grain-filling capacity. If this were so, one would expect all spikelets to behave uniformly, but in fact those in basal positions contain fewer fertile florets than those near the middle of the spike, even though they all appear to have a similar vascular system. It is also possible to increase the number of grain-bearing florets in the spikelet beyond three, primarily by raising the nitrogen supply, and in this way florets connected only by sub-vascular bundles may be induced to set grain. Other observations with other genotypes and environmental conditions shows that the vascular system of florets bears some relation to their ability to form a grain without, however, setting an upper or a lower limit to the number of fertile florets within the spikelet. It would be difficult to maintain that competition for assimilates is the main factor preventing grain formation, if the florets concerned possess a direct vascular system and if, as has been asserted (Rawson and Evans, 1970), assimilate supply is not limiting. The suggestion made by Zee and O'Brien (1970) that the treachery elements between the pericarp of the caryopsis and the rachilla restrict xylem flow but may assist solute transfer to the sieve tubes could be of some importance in this context.

There is, however, need for more information on the development of florets in different positions at an early stage before assimilate supply for grain formation becomes a critical factor. The contemporary researches in the last decade or so have given a further insight into the differential behaviour of grains perceptibility by looking at the multiple causes rather than a single and thus further expanding the list underlying the causes of a frail sink.

## 8.3 Ultra Structural Studies

Endosperm and embryo of Gramineae have an outer most caryopsis coat comprising of three distinct layers of tissues the inner most nucellus, followed by seed coats (tegmen and testa) and pericarp. The pericarp except in the crease is five or six cell layers thick (Percival, 1921). The epidermis and the underlying one- or two-layered hypodermis make up the outer pericarp. The inner pericarp is made up of thin-walled intermediate cells and underlying cross and tube cell layers.

The nucellus has only one or two cell layers or is merely a hyaline layer. The nucellar projection is a cellular band occurring parallel to the pigment strand. Endosperm is the starchy tissue that is formed during grain development. The endosperm is bounded externally by the aleurone layer, which consists of blocky to somewhat radially elongate cells in contact with the nucellar remnant. The thick aleurone cell wall has two layers distinguishable by differential staining or by transmission electron microscopy (Bacic and stone, 1981). Aleurone cytoplasm lacks starch but is rich in protein granules that are surrounded by lipid droplets. Sub-aleurone, present below the aleurone has been observed to be several cell layers thick, it contains a few starch granules and is rich in protein. Aleurone cells remain alive in mature grains (Bradbury *et al.*, 1956). Starchy endosperm fills most of the grain and its cells are dead at maturity but packed with starch and matrix protein (Campbell *et al.*, 1981). These workers verified that there is a decreasing protein gradient from the aleurone layer to the center of the starchy endosperm.

Plastids lying in close proximity to E.R. are generally acknowledged to synthesize fatty acids. Ultra structural evidence of attachment of E.R. and lipid droplets suggests that the latter accumulate between bilayer of E.R. membrane resulting in the formation of a half unit membrane around the lipid bodies (Warner *et al.*, 1981). Interior to sub-aleurone is the starch endosperm having much larger cells than aleurone and sub-aleurone. Its vacuoles change form and structure during growth and differentiation and accumulate proteins and other macromolecules capable of being degraded when required. Ions and metabolites are in continuous flux between the cytoplasm and vacuoles. Sub-aleurone and starch endosperm in wheat proceed via cell enlargement, development of starch granules and protein deposition in wheat six days after fertilization (Bechtel *et al.*, 1982). Amyloplast synthesize starch grains and they are located peripherally containing single starch granules in wheat (Bechtel *et al.*, 1982). In general amylase activity in cereal grains is high during early grain development and diminishes as caryopsis matures.

## 9. EFFECT OF ALTERNATIVE OXIDATIVE PATHWAY ON SINK EFFICIENCY

The alternative oxidize is a non-proton motive 'alternative' to electron transport through the cytochrome pathway. Despite its wasteful nature in terms of energy conservation, the pathway is likely present throughout the plant kingdom and appears to be expressed in most plant tissues. A small alternative oxidize gene family exists, the members of which are differentially expressed in response to environmental, developmental and other cell signals. The alternative oxidize enzyme possesses tight biochemical regulatory properties that determine its ability to compete with the cytochrome pathway for electrons. Studies show that alternative oxidize can be a prominent component of total respiration in important crop species. All these characteristics suggest this pathway plays an important role in metabolism and/or other aspects of cell physiology (McDonald *et al.*, 2002).

Depending on the species and growth conditions of a plant, 30 to 70 percent of all the carbohydrates fixed in photosynthesis are respired in the same day (McDonald *et al.*, 2002). The terminal part of the respiratory path, which starts with the degradation of carbohydrates via glycolysis, consists of the mitochondrial electron transport pathway, in which, among other components, two terminal oxidizes participate, cytochrome C oxidize and the alternative oxidize. The alternative oxidizes branches from the main electron transport pathway at the ubiquinone pool and beyond the branch point; do not contribute to ATP production. The energy conservation is less than maximal if a part of the respiration proceeds via this non-phosphorylating or uncoupling (alternative) pathway. Because of its energy wasting nature, it is most interesting (scientifically, as well as economically) to investigate under which conditions and to what extent alternative respiration is used, and how its activity is regulated.

The mitochondrial electron transport pathway consists of several dehydrogenases and the cytochrome and alternative oxidizes, which are linked via the ubiquinone pool, and are all located in the inner mitochondrial membrane. The cytochrome pathway consists of a series of three enzymes: Complex III, cytochrome C and Complex IV (cytochrome C oxidase). Cytochrome C oxidaze is sensitive to cyanide; in contrast to the alternative oxidize. The alternative pathway consists of only one protein, the quinol-oxidizing alternative oxidize (AOX). When the electrons from ubiquinon pool are donated to the alternative oxidize, proton pumping of the cytochrome pathway (Complexes III and IV) is bypassed and, therefore, there is no energy conservation via this part of the respiratory pathway. The energy is lost as heat (Millenaar and Lambers, 2003).

Many reports have appeared, in which the activity of the alternative oxidize was assessed with the use of specific inhibitors of the cytochrome (e.g., CN−, azide, antimycin) and alternative (e.g., SHAM, benzhydroxamic acid, propyl gallate) pathways (Lambers, 1997). It was shown that the alternative pathway became active at very high reduction levels of the Q pool (Dry *et al.*, 1989). Studies have shown the activity of the cytochrome pathway will influence AOX activity since the two pathways compete for electrons from the reduced ubiquinone pool (Millar *et al.*, 1995; Ribas-Carbo *et al.*, 1995). The alternative oxidize is more active if the subunits of the dimmer of the enzyme are non-covalently linked (reduced) and it is less active if the sulfur bridges between the dimmer are covalently linked (oxidized) (Umbach and Siedow, 1993). The concentration of alternative oxidizes potentially influences alternative pathway activity. During a variety of stresses the concentration of the alternative oxidizes increases. Infection of *Nicotiana tabacum* leaves with tobacco mosaic virus resulted in an increased concentration of alternative oxidize, however, no change in the activity of the alternative pathway was observed (Lennon *et al.*, 1997).

The concentration of AOX protein often increases after transfer to lower temperatures (Vanlerberghe and McIntosh, 1992; Gonzalez-Meler *et al.*, 1999). In contrast to the alternative pathway capacity (KCN-insensitive, SHAM-sensitive respiration) that increased in lower temperatures (Vanlerberghe and McIntosh, 1992). In mung bean (*Vigna radiata*) growth at 19°C the concentration of the alternative oxidizes increased over two-fold in both hypocotyls and leaves compared with plants grown at 28°C. The plants grown at 19°C maintained a higher activity of the alternative pathway compared with those grown at 28°C. This response, however, was not observed in *Glycine max* cotyledons, despite the increased concentration of alternative oxidize (Gonzalez-Meler *et al.*, 1999). Ribas-Carbo *et al.* (2000) reported that in a chilling-sensitive maize cultivar the activity of the alternative pathway was higher during the recovery period than in a less chilling-sensitive cultivar. According to Millenaar and Lambers (2003) the alternative pathway is inhibited more (higher $k_m$) at low oxygen concentrations compared with the cytochrome pathway. Therefore, the alternative pathway does not have a function at low oxygen concentrations.

The activity of the cytochrome pathway depends on the availability of inorganic phosphate and ADP. If plants are exposed to very low phosphorus supply, the Pi and ADP concentration in the plant cells may become very low. Therefore, it has been postulated that under these conditions the activity of the alternative pathway is increased relative to that of the cytochrome pathway (Millenaar and Lambers, 2003). In leaves of *Phaseolus vulgaris* the AOX concentration increased in P-deficient plants, but it was unchanged in

*Nicotiana tabacum* leaves (Gonzalez-Meler *et al.*, 2001). Some oxygen free radical scavenger enzymes (like catalase and total per oxidize) are more active in P-deficient plants, while others do not change (ascorbate peroxidase and superoxide dismutase) (Juszczuk *et al.*, 2001). Restriction of the cytochrome pathway by phosphours limitation causes an increase in the formation of oxygen free radicals, which can be prevented (partly) by more active alternative pathway. Another proposed function for the alternative oxidize is the prevention of the formation of oxygen free radicals. Oxygen free radicals may lead to severe metabolic disturbances and a wide range of environmentally induced plant disorders, including chilling damage, are mediated by reactive oxygen species (Scandalios, 1993).

While studying the development of grains at basal and distal positions in the middle spikelet of main shoot ear of wheat var. Kalyonsena, Kumari and Ghildiyal (1998) observed that lesser growth of distal grains was associated with higher rate of alternative respiration compared to proximal grains within a spikelet of wheat, which could be one of the reasons of lesser growth of grains at distal position in a spikelet. Gonzalez-Meler *et al.* (1996) reported elevate $CO_2$ concentration inhibited the salicylhydroxamic acid-resistant cytochrome pathway, but had no direct effect on the cyanide-resistant alternative pathway. Elevated $CO_2$ inhibited the activity of cytochrome C oxidize and succinate dehydrogenase (SDH) enzymes and direct inhibition of respiration by $CO_2$ concentration in plants may be due to the inhibition of these enzymes. Rates of dark respiration are directly correlated with leaf nitrogen content. Therefore, when $CO_2$ enrichment leads to a reduction in leaf nitrogen concentration, respiration also declines. Surprisingly, $CO_2$ enrichment increases the average number of mitochondria in each cell, even though leaf respiration rate decreases in response to elevated $CO_2$ across a diverse selection of plant species. This response may be indicative of a shift in plant metabolism and the increased energy demand resulting from higher photosynthetic rates under $CO_2$ enrichment (Woodward, 2002). Dua *et al.* (2003) demonstrated that use of a specific inhibitor (salicylhydroxamic acid) for modulation of alternative oxidative pathway may go along way in achieving a higher economic yield, but it may not be possible to eliminate the disparity between the two types of sinks in the same spike or spikelet of wheat.

The prevalence of AOX respiration in the plants indicates that this pathway plays an important role in metabolism and/or other aspects of cell physiology. The presence of an AOX with sophisticated biochemical regulatory properties provides the cell potential mechanisms to; modulate the rate of ATP production, maintain electron transport under conditions when downstream electron transport in the cytochrome pathway is limited and modulate the reduction state of ETC components, thus controlling the rate of generation of reactive oxygen species (Purvis and Shewfelt, 1993; Simons and Lambers, 1999).

## REFERENCES

Abbate, P. E., Andrade, F.H., Culot, J. P. and Bindraban, P. S. 1997.Grain yield in wheat: Effects of radiation during spike growth period. *Fields Crops Research*, **54:**245-257.

Abrol, Y. P., Kumar, P.A. and Nair, J.V.R. 1984. Nitrate uptake and assimilation and grain nitrogen accumulation. *Adv. Cereal Sci. and Tech.*, **6**: 1-48.

Addicott, F. T., Carns, H. R., Cornforth, J. W., Lyon, J. L., Milborrow, B. V., Ohkuma, K., Rayback, G., Smith, O. E., Thiessen, W, E., and Wareing, P. F. 1968. Abscisic acid: A proposal for the redesignation of abscisin II (dormin). In: *Biochemistry and physiology of plant growth regulators, The Runge Press Ltd., Ottawa, Canada*, Pp: 1527-1529.

Ahmdi, A. and Baker, D. A. 2001. The effect of wheat stress on grain filling processes in wheat. *J. Agri. Sci.*, **136**:256-269.

Ahmdi, A. and Baker, D. A.1999. Effects of abscisic acid (ABA) on grain filling processes in wheat. *Plant Growth Regul.*, 28: 187-197.

Aimi, R., Sawamura, H. and Konna, S. 1958. Physiological studies on the mechanism of crop plants. The effect of the temperature upon behaviour of carbohydrates and some related enzymes during ripening of rice plant. *Proc. Crop Sci. Soc. Japan*, **27**: 405-407.

Alexandrov, V. G. 1937. On the morphology of the grain in cereals. *Comptes Rendus (Doklady) del • Academie Sciences del • URRS,* XVII, 389-391.

Altus, D. P. and Canny, M. J.1982. Loading of assimilates in wheat leaves. I. The specialization of vein types for separate activities. *Aust. J. Plant Physiol.*, **9:**571-581.

Angrish, R., Kumar, B. and Datta, K. S. 2001. Effect of gibberellic acid and kinetin on nitrogen content and nitrate reductase activity in wheat under saline conditions. *Indian J. Plant Physiol.*, **6**. 172-177.

Arai, M., Mori, H. and Imaseki, H. 1992. Expression of gene for sucrose synthase during growth of mungbean seedlings. *Plant Cell Physiol.*, **33**: 503-6.

Archer, M.C., Hansen, D.J. and Floussaert, D. 1982. Corn Yields rate limiting factors and opportunities in plant growth regulation. In: *Chemical manipulation of crop growth and development*. Ed., Mclaren, J.S., *Butterworth London,* Pp: 253-265.

Asana, R.D. 1968. Some thoughts on possible mechanism controlling yield. *Indian J. Genet. Plant Breed.*, **28**: 21-30.

Asana, R.D. and Bagga, A. K.1966. Studies in physiological analysis of yield. VIII. Comparison of development of upper and basal grains of spikelets of two varieties of wheat. *Indian J. plant physiol.*, **9**: 1-21.

Asana, R.D. and Basn, R. N. 1963.Studies in physiological analysis of yield. VI. Analysis of the effect of water stress on grain development in wheat. *Indian J. Plant Physiol.*, **6**: 1-13.

Asana, R.D. and Mani, V.S. 1948. Photosynthesis in the ears of five varieties of wheat. *Nature*, **163**:451-456.

Asana, R.D. and Mani, V.S.1950. Studies in the physiological analysis of yield. I. Varietals differences in photosynthesis in the leaf stem and ear of wheat. *Physiol. Planta*, **3**: 22-39.

Ashton, F.M. 1976. Mobilisation of storage proteins of seeds. *Ann. Rev. Plant Physiol.*, **27**: 95-117.

Atsmon, D., Bush, M. G. and Evans, L. T. 1986.Effects of environmental conditions expression of the 'gigas' characters in wheat. *Aust. J. Plant Pysiol.*, **13**:365-379.

Austin, B. R., and Edrich, J.A. 1975. Effects of ear removal on photosynthesis, carbohydrate accumulation and on the distribution of assimilated $^{14}C$ in wheat. *Ann. Bot.*, **39**:141-152.

Austin, R. B., Bingham, J., Blackwll, R. D., Evans, L. T., Ford, M. A., Morgan, C. L. and Taylor, M. 1980. Genetic improvements in winter wheat yield since 1900 and associated physiological changes. *J. Agric. Sci.*, **94**: 675-689.

Austin, R.B. 1982. A combined genetic and chemical approach to increasing and stabilizing wheat yields. In: *Opportunities for manipulation of cereal productivity,* Hawkins A.F. and Jeffcoat, B. Eds., *British Plant Growth Regulator Group (Monograph No, 7). Wantage, UK*, Pp: 193-203.

Bacic, A. and Stone, B. A. 1981. Isolation and ulrastructure of aleurone cell walls from wheat and barley. *Aust. J. Plant Physiol.,* **8**: 453-474.

Baker, J. T., Allen, L.H. Jr, and Boote K.J. 1990a. Growth and yield responses of rice to carbon dioxide. *Annu. Rev. Plant Physiol. Plant Mol. Biol.,* **44**: 309-332.

Baker, J.T., Allen L.H. Jr, and Boote, K.J. 1990b. Growth and yield response of rice to carbon dioxide concentration. *J. Agri. Sci.*, **115**: 313-320.

Baker, J.T., Allen L.H. Jr, and Boote, K.J. 1992.Temperature effects on rice at elevated $CO_2$ concentration. *J. Exp. Bot.*, **43**: 959-964.

Bangerth, F., Aufhammer, W. and Baum, O. 1985. IAA level and dry matter accumulation at different positions within a wheat ear. *Physiol. Planta.*, **63**:121-125.

Banowetz, G. M. 1992. The effect of endogenous cytokinin content on benzyladenine-enhanced nitrate reductase induction. *Physiol. Planta.,* **86**: 341-348.

Banowetz, G. M. 1997. Cultivars of hexaploid wheat of contrasting stature4 and chlorophyll retention differ in cytikinin content and responsiveness. *Ann. Bot.*, **79**: 185-190.

Banowetz, G. M., Ammar, K. and Chen, D. D. 1999a. Postanthesis temperatures influence cytokinin accumulation and wheat kernel weight. *Plant. Cell. Enviro.,* **22**: 309-316.

Banowetz, G. M., Ammar, K. and Chen, D. D. 1999b.Temperature effect on cytonkinin accumulation and kernel mass in a dwarf wheat. *Ann. Bot.*, **83**: 303-307.

Batten, G. D. and Wardlaw, I. F. 1987. Redistribution of $^{32}P$ and $^{14}C$ from the flag leaf during grain development in wheat. *Aust. J. Plant Physiol.*, **14**: 267-275.

Baxter, E. D. and Duffius, C. M. 1973. Phosphorylase activity in relation to starch synthesis in developing *Hordeum distichum* grain. *Phytochem*, **12**: 2321-2330.

Bechtel, D. B., Gaines, R. L. and Pomeranz, Y. 1982. Early stages in wheat endosperm formation and protein body initiation. *Ann. Bot.,* **50**:507-518.

Bell, A. J. E. and Van Patrick, J. W. 1985. Proton extrusion in seed coats of *Phaseolus vulgaris. Plant Cell Environ.,* **8**: 1-6.

Bell, C.J., and Incoll, L.D. 1990. The redistribution of assimilate in field grown winter wheat. *J. Exp. Bot.*, **229**: 949-960.

Benkova, E., Witters, E., Dongen, W. V., Kolar, J., Motyka, V., Brzobohaty, B., Van Onckelen, H. A. and Machackova, I.1999. Cytokinins in tobacco and wheat chloroplasts. Occurrence and changes due to light/dark treatment. *Plant Physiol.*, **121**:245-251.

Berdahl, J. D., Rasmusson, D. C. and Moss, D. N. 1972. Effect of leaf area, photosynthetic rate, light penetration and grain yield in barley. *Crop Sci.*, **12**: 177-180.

Berry, C.P., Abolina, B.L.D. and Gilles, K.A. 1971. The characterization of Triticale starch and its comparison with starches of rye, durum and *HRS* wheat. *Cereal Chem.*, **48**: 415-427.

Bertin, N. 1995. Competition for assimilates and fruit position affect fruit-size in indeterminate green house tomato. *Ann. Bot.*, **75**: 55-65.

Beveridge, J. L., Jaris, R. H. and Ridgman, W.J., 1965. Studies on the nitrogenous manuring of wheat. *J. Agric. Sci.,* **65**: 379-87.

Bhardwaj, S. N. 1962. Time of sowing an important parameter influencing sink development. *Indian J. Plant Physiol.*, **4**: 28-32.

Bhardwaj, S. N. and Dua I. S. 1975. A study of the competitive inter-relationship between grain setting and growth in aestivum wheats in relation to the production of growth regulating substances. *Indian J. Plant Physiol.*, **18**: 97-103.

Bhardwaj, S. N. and Verma, V. 1985. Hormonal regulation of assimilate translocation during grain growth in wheat. *Ind. J. Exp. Biol.*, **23**: 719-721.

Bhardwaj, S. N. and Verma, V. 1987. Regulation of grain size within a developing ear of bread wheat. *Ind. J. Agri. Sci.*, **57**: 710-714.

Bidinger, F. R., Musgrave, R.B. and Fischer, R.A.1977. Contribution of stored pre-anthesis assimilates to grain yield in wheat and barley. *Nature*, **270**:431-433.

Bindraban, P. S., Sayre, K. D. and Solis-Moya, E.1998. Identifying factors that determine kernel number in wheat. *Field crops Research*, **58**:223-234.

Bingham, J. 1967. Investigations on the physiology of yield in winter wheat by comparison of varieties and by artificial variation in grain number per ear *J. Agri. Sci. Camb.,* **68**: 411-422.

Bingham, J.1972. Physiological objectives in breeding for grain yield in wheat. *Proc. of the 6th Eucaress (Cambrige)*, Pp: 15-29.

Biscoe, P.V., Gallagher, J. N., Littelton, E.J., Monteith, J. L. and Scott, R. K. 1975.Barley and its environment. IV. Source of assimilates for the grain. *J. App. Eco.*, **12:**293-318.

Biswas, A. K. and Choudhuri, M. A. 1977. Regulation of leaf senescence in rice by hormones sprayed at different developmental stages and its effect on yield. *Ind. J. Agri. Sci.*, **47**: 38-40.

Biswas, P. K., Hileman, D. R., Ghosh, P. P., Bhattacharya, N. C. and McCrimmon, J. N. 1996. Growth and yield responses of field grown sweet potato to elevated carbon dioxide. *Crop Sci.,* **36**: 1234-1239.

Black, M., Carbineau, F., Grezesik, M., Giuj, P. and Come, D. 1996. Carbohydrate metabolism in developing and maturing wheat embryo in relation to its desiccation tolerance. *J. Exp. Bot.,* **47**: 161-69

Blade, S. F. and Baker, R. J. 1991. Kernel weight response to source-sink changes in spring wheat. *Crop Sci.*, **31:**1117-1120.

Blum, A.1985. Photosynthesis and transpiration in leaves and ears of wheat and barley varieties. *J. Exp. Bot.*, **39**:432-440.

Blum, A., Mayer, J. and Golan, G. 1988. The effect of grain number per ear (sink size) on source activity and its water relations in wheat. *J. Exp. Bot.*, **39**:106-114.

Blum, A., Sinmena, B., Mayer, J., Golan, G. and Shpiler, L. 1994. Stem reserve mobilization supports wheat-grain filling under heat stress. *Aust. J. plant Physiol.,* **21**:771-781.

Blumenthal, C., Rawson, H. M., McKenzie, E. Gras, P. W., Barlow, E. W. R. and Wrigley, C.W. 1996. Changes in wheat grain quality due to doubling the level of atmospheric $CO_2$. *Cereal Chem.,* **73**: 762-766.

Bonnett, G. D. and Incoll, L. D. 1993a. Effects on the stem of winter barley of manipulating the source and the sink during grain filling. I. Changes in accumulation and loss of mass from internodes. *J. Exp. Bot.*, **44**: 75-82.

Bonnett, G. D. and Incoll, L. D. 1993b. Effects on the stem of winter barley of manipulating the source and the sink during grain filling. II. Changes in the composition of water-soluble carbohydrates of internodes. *J. Exp. Bot.*, **44**:83-91.

Boonstra, A. E. H. R.1937.Der einfluss der verschiedenen assimilierenden teile auf den samentrag von weizen. *Zeitsch. Zucht. Pflanz.*, **21**:115-147.

Boonstra, A.E.H.R. 1929. Differences in vitality of the leaves of four varieties of oats as connected with the yield. *Mede Deal London Whooggach Wageningen. Netherlands*, **33**: 1-25.

Borojeric, S. and Williams, W. A.1982. Genotypes ×environment interaction for leaf area parameters and yield components and their effect on wheat yield crop. *Science*, **22**:1020-1024

Bradbury, D. and MacMasters, M. M. 1956. Structure of the mature wheat kernel. III. Microscopic structure of the endosperm of hard red winter wheat. *Cereal Chem.*, **33**: 361-373.

Brady, C. J., Patterson, B. D., Heng, F. T. and Smillie, R. M. 1971. Protein and RNA synthesis during aging of chloroplasts in wheat leaves. In: *Autonomy and biogenesis of mitochondria and chloroplasts. Amsterdam, London*, Pp: 453-465.

Brain, P. W. and Hemming, H. G. 1955. The effect of gibberellic acid on shoot growth of pea seedlings. *Physiol. Planta.*, **8**: 669-681.

Bremner, P.M. and Rawson, H. M. 1978. The weights of individual grains of the wheat ear in relation to their growth potential, the supply of assimilate and interactien between grains. *Aust. J. Plant Physiol.*, **5**: 61-72.

Bremner, P.M. and Rawson, H.M. 1978. The weights of individual grains of wheat ear in relation to their potential, the supply of assimilate and interaction between grains. *Aust. J. Plant Physiol.*, **5**: 61-72.

Brock-Lehurst, P. A. 1977. Factors controlling grain weight in wheat. *Nature (London)*, **266**:348-349.

Brock-Lehurst, P. A. 1977. Factors controlling grain weight in wheat. *Nature*, **266**: 348-349.

Brooking, I.R. and Kirby, E.J.M.1981. Interrelationships between stem and ear development in winter wheat: The effects of a Norin 10 dwarfing gene, Cai/Rht2. *J. Agric. Sci.*, **97**: 373-381.

Brouwer, R.1962. Distribution of dry matter in the plant. *Neth. J. Agri. Sci.* (Special Issue), Pp: 361-376.

Brown, A.V. and Sun, W. K. 1973. Role of ABA in inhibiting protein synthesis. *Physiol. Planta.*, **23**: 35-42.

Bruckner, P.L. and Frohberg, R.C. 1991.Source-sink manipulation as a postanthesis stress tolerance screening technique in wheat. *Crop Sci.*, **31**: 326-328.

Bruinsma, J.1985. The plant growth regulators in horticulture. *Hort. Sci.*, **36**:1-11.

Busby, C. H. and O'Brien, T.P. 1979.Aspects of vascular anatomy and differentiation of vascular tissues and transfer cells in vegetative nodes of wheat. *Aust. J. Bot.*, **27**: 703-711.

Bush, 1990. Electrogenicity, PH dependence and stiochiometry of proton sucrose symport. *Plant Physiol.*, **93**: 1590-96.

Calderini, D. F. and Ortiz-Monasterio, I. 2003. Grain position affects grain macronutrients and micronutrient concentrations in wheat. *Crop Sci.*, **43**: 141-151.

Campbell, W. P., Lee. J. W., O'Brien, T. P. and Smart, M. G. 1981. Endosperm morphology and protein body formation in developing wheat grain. *Aust. J. Plant Physiol.*, **8**:5-20.

Caputo, C. and Barneix, A. J. 1999. The relationship between sugar and amino acid export to the phloem in young wheat plants. *Ann. Bot.*, **84**: 33-38.

Care, D. J. and Wardlaw, I. F. 1965. The supply of photosynthetic assimilates to the grain from the flag leaf and ear of wheat. *Aust. J. Biol. Sci.*, **18**:711-719.

Chanda, S. V. and Singh, Y. D. 1997. Changes in peroxidase and IAA oxidase activities during wheat grain development. *Plant Physiol. and Biochem.*, **35**: 245-250.

Chatterjee, A., Mandal, R. K. and Sircar, S. M. 1976. Changes in the level of growth substances during grain filling in rice. *Ind. J. Plant Physiol.*, **19**: 254-258.

Chau, N. M. and Bhargawa, S. C. 1998. Different grades of grains occurring during grain filling in short and medium-duration rice. *International Rice Research Notes*, **18**:11-12.

Cheikh, N. and Jones, R. J. 1994. Disruption of maize kernel growth and development by heat stress. *Plant Physiol.*, **106**: 45-51.

Chen, C. M. 1997. Cytokinin biosynthesis and interconversion. *Physiol. Planta.*, **101**: 665-673.

Chikov, V. I., Chemikosova, S. B., Bakirova, G. G. and Gazizova, N. I. 1984. Influence of partial removal of the ear of leaves on assimilates transport and photosynthetic productivity in spring wheat. *Fiziol. Rast.*, **31**: 475-481.

Chinnusamy, V. and Khanna-Chopra. R. 2003. Effect of heat stress on grain starch content in diploid, tetraploid and hexaploid wheat species. *J. Agr. Crop Sci.*, **189**:242-249.

Cholodny, L.1936. IAA increases grain yield in wheat and oats. *J. Appl. Biol.*,**2**: 16-18.

Clark, W. C., Cook, K. H., Marland, G., Weinberg, A. M., Rotty, R. M., Bell, P.R., Allison, L. J. and Cooper, C.L. 1982. In: *The carbon dioxide question: Perspectives for 1982*. Ed., Clark, W.C. *Oxford University Press, New York*. Pp: 3-44.

Cloux, H., Andre, M., Grebaud, A. and Daguenet, A. 1989. Wheat response to CO2 enrichment: Effect on photosynthetic and photorespiratory characteristics. *Photosynthetica*, **23**:145-153.

Cock, J. H. and Yoshida, S. 1972. Accumulation of $^{14}$C-labelledcarbohydrate before flowering and its subsequent redistribution and respiration in the rice plant. *Proc. Crop. Sci. Soc. Jpn.*, **41**: 226-234.

Cohen, J. D. and Bandurski, R. S. 1978. The bound auxins: Protection of indole-3-acetic acid from peroxidase- catalysed oxidation. *Planta*, **139** : 203-208.

Cook H. and Oparka, K. J. 1983. Movement of fluorescence into isolated caryopses of wheat and barley. *Plant cell Environ.*, **6**: 239-242.

Cook, M.G. and Evans, L.T. 1978. Effect of relative size and distance of competing sink on the distribution of photosynthetic assimilates in wheat. *Aust. J. Plant Physiol.*, **5**: 459-509.

Cox, M.C., Qualset, C.O. and Rains, D.W. 1985. Genetic variation for nitrogen assimilation and translocation in wheat. I. Dry matter and nitrogen accumulation. *Crop Sci.*, **25**: 430-35.

Cregan, P. B. and Berkum, V. P.1984.Genetics of nitrogen metabolism and physiological bio chemical selection for increased grain crop productivity. *Theor. Appl. Genet.*, **67**:97-111.

Cruz-Aguado, J. A., Reyes, F., Rodes, R., Perez, I. and Dorado, M. 1999. Effect of source-to-sink ratio on partitioning of dry matter and $^{14}$C-photoassimilates in wheat during grain filling. *Ann. Bot.*, **83**:655-665.

Darroch, B. A. and Baker, R. J. 1995. Two measures of grain filling in spring wheat. *Crop Sci.*, **35**:164-168.

Darussalam, Gole M. A. and Patrick, J. W. 1998. Auxin control of photoassimilate transport to and within developing grains of wheat. *Aust. J. Plant Physiol.*, **25**: 69-77.

Das, H. K. 1963. Effect of nepthalene acetic acid on grain yield in rice. *Plant Physiol.*, **39**: 212-216.

Davey, J. E., and Van Standen, J. 1977. A cytokinin complex in the development fruits of *Lupinus albus*. *Physiol. Planta.*,**39**:221-224.

Davidson, D. J. and Chevalier, T. M. 1992. Storage and remobilization of water soluble carbohydrates in stems of spring wheat. *Crop Sci.*, **32**: 186-190.

Demotes-Mainard, S. and Jeuffroy, M.H.2004. Effects of nitrogen and radiation on dry matter and nitrogen accumulation in the spike of winter wheat. *Field Crops Research*, **87**:221-233.

De-Vos, N.M. 1979.Cultivar differences in plant and crop photosynthesis. In: *Crop physiology and cereal breeding*. Spiertz, J.H.J. and Kramer, T.H. Eds., *Pudoc. Wageningen*, Pp: 71-74.

Dewdeny, S. T. and McMwha, J. A. 1979. The metabolism and transport of abscisic acid during grain filling in wheat. *J. Exp. Bot.*, **29**: 1299-1308.

Dharmasiri, N., Dharmasiri, S., Jones, A. M. and Estelle, M. 2003. Auxin action in a cell-free system. *Current Biology*, **13**: 1418-1422.

Dhiman, S. D., Sharma, H. C. and Singh, R. P. 1980. Association between flag leaf area and grain yield in wheat. *Indian J. Plant Physiol.*, **23**: 282-87.

Dietrich, J. T. Kaminek, M., Blevins, D. G., Reinbott, T. M. and Morris, R.O.1995.Changes in cytokinins and cytokinin oxidase activity in developing maize kernels and the effects of exogenous cytokinin on kernel development. *Plant Physiol. Biochem.*, **33**: 327-336.

Doodson, J. K., Manners, J. G. and Meyers, A. 1964. The distribution pattern of 14carbon assimilated by the third leaf of wheat. *J. Exp. Bot.*, **15**:96-103.

Drake, B. G., Gonzalez-Meler, M. A. and Log, S. P. 1997. More efficient plants: a consequence of rising atmospheric $CO_2$? *Annual Review of Plant Physiology and Molecular Biology*, **48**: 607-637.

Dry, I. B., Moore, A. L., Day, D.A. and Wiskich, J. T. 1989. Regulation of alternate pathway activity in plant mitochondria: Nonlinear relationship between electron flux and the redox poise of the quinone pool. *Archives of Biochemistry and Biophysics*, **273**: 148-157.

Dua, I. S. and Eradatmand Asli, D. 2005. Protomic and non protomic signals in operation during cytokinin action. *Bioinformatics India* (In Press)

Dua, I. S. and Kalsi, P. 1988. Yielding ability of wheat grains of the same ear under nutritional stresses. In: *Abstract of papers. Soc. Plant Physiol. and Biochem.*, New Delhi. Pp: 20.

Dua, I. S. and Sehgal, O. P. 1981. Differential levels of growth promoters in the middle and peripheral grains of the same ear in wheat (*Triticum aestivum* L. emend. Thell). *Proc. Indian Nat. Sci. Acad.*, **B-47**: 543-450.

Dua, I. S., Eradatmand, D. and Sankhyan, P. 2003. Proticity : Its creation and dissipation by a nano pump : A molecular approach to decode the ways and means to improve the biological harvest. *Proc. INAE Conference on Nanotechnology, Chandigarh, India*, Pp: 127-134.

Dua, I.S., Devi, V. and Garg, N. 1990. An appraisal of the hormonal basis of grain growth of buckwheat (*Fagopyrum esculentum* Moench). *Fagopyrum,* **10**: 73-80.

Dua, I.S., Dhir, K.K., Gupta, C.S., Sharma, S.K. and Khatra, G.J. 1982. Dynamics of plant growth regulating substances during grain setting and development in wheat (*Triticum aestivum* L. Amend Thell). *P.U. Research Bulletin,* **33**: 87-100.

Dubey, Y. P. 1990. Effect of soil temperature and moisture on the growth of wheat (*Triticum aestivum* L.). *Indian J. Plant Physiol.*, **28**: 355-58.

Duffus, C. and Rosie, R. 1973. Starch hydrolyzing enzymes in the developing barley grains. *Planta*, **109**: 153-160.

Egli, D. B., Pendleton, J. W. and peters, D.B. 1970. Photosynthetic rate of three soybean communities as related to carbon dioxide levels and solar radiation. *Agro. J.*, **68**:411-414.

Egli, D. B.1998.Seed biology and the yield of grain crops. *CAB Intl. Wallingford.UK.*

Ellen, J. 1990. Effects of nitrogen and plant density on growth, yield and chemical composition of two winter wheats (*Triticum aestivum* L.) Cultivars. *J. Agron. Crop Sci.,* **164**: 174-83.

Ellis, R. H. and Pieta Filho, C. 1992. The development of seed quality in spring and winter cultivars of barley and wheat seed. *Sicence Research,* **2**: 9-15.

Eschrich, W. 1984. Unter suchugen Zur regulation des assimilate transport. *Ber. Dtsch. Bot Ges.,* **93**: 368-78.

Estiarte, M., Penuelas, J., Kimball, B.A., Hendrix, D.L., Pinter Jr, P.J., Wall, G.W., LaMorte, R.L. and Hunsaker, D.J. 1999. Free-air CO2 enrichment of wheat: Leaf flavonoid concentration throughout the growth cycle. *Physiol. Planta.*, **105**:423-433.

Evans, H. J. and Nason, A. 1953. Pyridine nucleotide nitrate reductase from the extracts of higher plants. *Plant Physiol.*, **28**: 233-254.

Evans, L. T., Bingham, J. and Roskams, M. A. 1972. The pattern of wheat. *Aust. J. Biol. Sci.,* **25**: 1-8.

Evans, L.T. 1972. Storage capacity as a limitation on grain yield in rice breeding. *Publ. Int. Rice Res. Sci.,* **16**:627-631.

Evans, L.T. 1993. Crop evolution, adaptation and yield. Cambridge University Press, Cambridge.

Evans, L.T. 1996.Crop evolution, adaptation and yield. *Cambridge Univ. Press. Cambridge.UK.*

Evans, L.T. and Rawson, H. M. 1970. Photosynthesis and respiration by the flag leaf and components of the ear during grain development in wheat. *Aust. J. Biol. Sci.*, **23**:245-254.

Evans, L.T. and Wardlaw, I. F. 1976. Aspects of comparative physiology of grain yield in cereals. *Advance on Agronomy,* **28**: 301-359.

Evans, L.T. and Wardlaw, I. F. 1996. Wheat. In: *Photoassimilate distribution in plants and crops, source-sink relationships*. Eds., Zamski, E. and Schaffer, A.A., *New York, MarcelDekker Inc.*, Pp: 501-518.

Evans, L.T., Bingham, J. and Roskams, M. A. 1972. The pattern of grain set within ears of wheat. *Aust . J .Boil. Sci.*, **25**:1-8.

Evans, L.T., Blundell, C. and King, R. W. 1995. Developmental responses by tall and dwarf isogenic lines of spring wheat to applied gibberellins. *Aust. J. Plant Physiol.,* **22**:365-371.

Evans, L.T., Dunstone, R. L., Rawson, H. M. and Whllms, R.F.1970. The phloem of the wheat stem in relation to requirements for assimilates by the ear. *Aust. J. Biol. Sci.*, **23**:743-752.

Evans, L.T., Wardlaw, I.F. and Fisher, R.A. 1975. Wheat. In: *Crop physiology*, Evans, L.T. Ed. *Cambridge University Press, London*, Pp: 101-150.

Evans, L.T.1975.The physiological basis of crop yield. In: *Crop physiology*. Evans, L.T. Ed., *Cambridge Univ. Press*, Pp: 327-355.

Farrar, J. F. and Williams, M. L. 1991. The effect of increased atmospheric carbon dioxide and temperature on carbon partitioning, source-sink relations and respiration. *Plant Cell Environment*, **14**: 819-830.

Feierabend, J. 1970. Characterisation of cytokinin action on enzyme formation during the development of the photosynthetic apparatus in rye seedlings. Enzymes of the reductive and oxidative pentose phosphate cycles. *Planta*, **94**: 1-15.

Feierabend, J. and Bore, J. D. 1978. Comparative analysis of the action of cytokinin and light on the formation of ribulose bisphosphate carboxylase and plastids biogenesis. *Planta*, **142**: 75-82.

Felker, F. C., Peterson, D. M. and Nelson, O. E. 1984. 14C-sucrose uptake and labelling of starch in developing grains of normal and segi barley. *Plant Phsiol.*, **74**: 43-46.

Fellows, R. J. and Geiger, D. R. 1974. Structural and physiological changes in sugar beet leaves during sink to source conversion. *Plant Physiol.*, **54**: 877-885.

Fischer, R. A. 1975. Yield potential in a dwarf spring wheat and the effect of shading. *Crop. Sci.,* **15**: 607-613.

Fischer, R. A. and Aguilar, M. I.1978. Yield potential in a dwarf spring wheat and the effect of carbon dioxide fertilization. *Agro. J.*, **68**:749-752.

Fischer, R. A. and HilleRisLambers, D.1978.Effect of environment and cultivar on source limitation to grain weight in wheat. *Aust. J. Agric. Res.*, **29**: 443-458.

Fischer, R. A. and Laing, D. R. 1976. Yield potential in a dwarf spring wheat and responses to crop thinning. *J. Agric. Sci.*, **87**:113-122.

Fischer, R. A.1985. Number of kernels in wheat crops and the influence of solar radiation and temperature. *J. Agri. Sci.*, **105**: 447-461.

Fischer, R.A. 1983.Wheat. In: *Potential productivity of field crops under different environments, Los Banos, IRRI*, Pp: 129-154.

Fischer, R.A. and Stockman, Y.M.1980.Kernel number per spike in wheat (*Triticum aestivum* L.): Responses to preanthesis shading. *Aust. J. plant physiol.*,**7**: 169-180.

Fisher, D. B. and Macnicol, P. K. 1986. Amino acid composition along the transport pathway during grain filling in wheat. *Plant Physiol.*, **82**: 1019-1023.

Fisher, D. B. Wu, Y. and Ku, M. S. B. 1992. Turnover of soluble proteins in the wheat sieve tube. *Plant Physiol.*, **100**: 1433-1441.

Fosket, D. E., Volk, M. J. and Goldsmith, M. R. 1977. Polyribosome formation in relation to cytokinin-induced cell division in suspension culture of *Glycine max* L. *Plant Physiol.*, **60**:554-562.

Foyer, C. H., Valadier, M. H., Migge, A. and Becker, T. W. 1998. Drought induced effects on nitrate reductase activity and mRNA and on the coordination of N and C metabolism in maize leaves. *Plant Physiol.*, **117**: 283-292.

Frazier, J. C. and Appalanaidu, B. 1963. The wheat grain during development with reference to nature, location, and role of its translocatory tissue. *American J. Bot.*, **52**: 193-198.

Fujiwara, A. and Ishida, H. 1963. Nutritional physiology in rice plants damaged by cool weather: II. Effects of low temperature and shading treatment at the maximum tillering stage on the nutritional physiology. *J. Sci. Soil Manuree Jpn.*, **34**: 101-106.

Gale, M. D. 1979. Genetic variation for hormonal activity and yield. In: *Crop physiology and cereal Breeding*. Spiertz, J. H. J. and Kramer, T.H. Eds., *Pudoce. Wageningen*, Pp: 29-34.

Gallagher, J. N., Biscoe, P.V. and Hunter, B. 1976. Effects of drought on grain growth. *Nature,* **264**: 541-42.

Gamalei, Y. V. 1998. Photosynthesis and photosynthate export-development of the transport system and source-sink relationships. *Fiziologiya Rastenni* (Moscow), **45**: 614-631.

Gandhi, A. P. and Naik, M. S. 1947. Role of roots, hormones and light in the synthesis of nitrate reductase and nitrite reductase in rice seedlings. *FEBS Lett.*, **40**: 343-345.

Garcia, R. L., Long, S. P., Wall, G. W., Osborne, C. P., Kimball, B. A., Nie, G. Y., Pinter, P. J., Jr., LaMote, R. L. and Wechsung, F.1998. Photosynthesis and conductance of spring wheat leaves: field response to continuous free-air atmospheric CO2 enrichment. *Plant Cell Environ.* **21**: 659-669.

Gebbing, T., Sehynder, H. and Kuchbaud, W. 1999. The utilization of pre- anthesis reserves in grain filling of wheat, *Plant Cell and Environment*, **22**: 851-858.

Geiger, D. R. 1975. Phloem loading. In: *Transport in plants*. I: *Phloem transport*. Zimmerman, M.H. and Milburn, J.A. Ed., *Springer-Verlag, New York*, Pp: 395-431.

Geiger, D. R., Shieh, W. J. and Saluke, R. M. 1989. Carbon partitioning among leaves, fruits and seeds during development of *Phaseolus vulgaris*. *Plant Physiol.*, **91**: 291-297.

Geiger, D.R., Giaquinta, R.T., Sovonick, S.A. and Fellows, R.J. 1973. Solute distribution in sugar beet leaves in relation to phloem loading and translocation. *Plant Physiol.*, **52**: 585-589.

Ghildiyal, M .C. and Sharma-Natu, P. 2000. Photosynthetic accumulation to rising atmospheric carbon dioxide concentration. *Indian J. Exp. Biol.,* **38**: 961-966.

Gibson, S. I. 2004. Sugar and phytohormone response pathways: Navigating a signaling network. *J. Exp. Bot.*, **55**: 253-264.

Gifford, R. M. and Evans, L. T. 1981. Photosynthesis, carbon partitioning and yield. Annu. Rev. *Plant Physiol.*, **32**: 485-509.

Gifford, R. M., Throne, G. H., Hitz, W. D. and Giaquinta, R. P. 1984. Crop productivity and photoassimilate partitioning. *Science*, **225**: 801-808.

Gladun, I. V., Karpov, E. A. and Belozerova, O. L. 1998. Distribution of $^{14}C$-assimilates from the panicle and rice leaves during grain filling. *Fiziologiya Biokhimiya Kul' turnykh Rastenii*, **30**: 342-348.

Gleadow, R. M., Dalling, M. J. and Halloran, G. M. 1982. Variations in endosperm characteristics and nitrogen content in six wheat lines. *Aust. J. Plant Physiol.*, **9**: 539-551.

Gogol, N. and Baruah, K. K. 2000. Effect of cold hardening and GA3 on growth and yield of 'boro' rice. *Indian J. Plant Physiol.*, **5**. 339-343.

Gold, C. M. and Duffus, C. M. 1996. Pre-maturity alpha-amylase production in cereal grains. In: *Pre-harvest sprouting in cereals*. Eds, Noda, K. and Mares, D. J., *Osaka: Center for Academic Societies, Japan*, Pp: 365-369.

Goldbach, E., Goldbach, H., Wagner, H. and Micheal, G. 1975. Influence of N-deficiency on the abscisic acid content of sunflower plants. *Physiol. Planta.*, **34**: 138-140.

Goldbach, H. and Goldbach, E. 1977. Abscisic acid translocation and influence of water stress on grain abscisic acid content. *J. Exp. Bot.,* **28**: 1342-1350.

Goldbach, H. and Micheal, G. 1976. Abscisic acid content of barley grains during ripening as affected by temperature and variety. *Crop Sci.* **16**: 797- 799.

Gonalez-Meler, M. A., Giles, L., Thomas, R. B. and Siedow, J. N. 2001. Metabolic regulation of leaf respiration and alternative pathway activity in response to phosphate supply. *Plant Cell and Environment,* **24**: 205-215.

Gonzalea-Meler, M. A., Ribas-Carbo, M., Giles, L. and Siedow, N.1999. The effect of growth and measurement temperature on the activity of the alternative respiratory pathway. *Plant Physiol.*, **120**:765-772.

Goudriaan, I. and Laar, V. 1994. Modelling potential growth processes. *Dordrecht Kluwer Acad. Pub.*, Pp: 267.

Grabau, L. J., Van Sanford, D. A. and Meng, Q. W. 1990. Reproductive characteristics of winter wheat cultivars subjected to postanthesis shading. *Published in Crop Sci.*, **30**: 771-774.

Gray, J. E., Holroyd, G. H., Van Der Lee, F., Bahrami, A. R. Sijmons, P. C., Woodward, F. I., Schuch, W. and Hetherington, A. M. 2000. The *HIC* signaling pathway links CO2 perception to stomatal development. *Nature*, **408**: 713-716.

Grotenhuis, T. P. and Bugbee, B.1996. Supper-optimal CO2 (>1200$\mu$ mol mol-1) reduces seed yield but not vegetative growth in wheat. *Crop Sci.*, **36**:321-326.

Grundbacher, F. J. 1963. The physiological function of the cereal awn. *Bot. Rev.*, **29**:366-381.

Gubler, F., Ash-Ford, A. E. and Jacobsen, J. V. 1987. The release of •-amylase through gibberellin treated barley aleuron cell walls. An immunochemical study with lowicryl k4m. *Planta (Berlin)*,**172**: 155-161.

Guoping, Z. 1997. Gibberellic acid3 modifies some growth and physiologic effects of paclobutrazol (pp333) on wheat. *J. Plant Growth Regul.*, **16**:21-25.

Gwathmey, C. O., Hall, A. E. and Madore, M. A. 1992. Pod removal effects on cowpea genotypes contrasting in monocarpic senescence traits. *Crop Sci.*, **32**:1003-1009.

Hageman, R. H. 1979. Integration of nitrogen assimilation in relation to yield. In: *Nitrogen assimilation in plants*. Hewitt, E. J. and Cutting, C. V., Eds. *Academic Press, London*. Pp: 591-612.

Hamid, A., Haque, M. M., Khanam, M., Hossain, M. A., Karim, M. A., Khaliq, Q. A., Biswas, D. K., Gomost, A. R., Chowdhury, A. M. and Uprety, D. C.2003.Photosynthesis, growth and productivity of rice under elevated co2. *Indinan J. Plant Physiol.*, **8**:253-258.

Hanif, M. and Langer, R.H.M. 1972. The vascular system of the spikelet in wheat (*Triticum aestivum*). *Ann. Bot. (London)* **36:**721-727.

Hasan, M. A. and Kamal, A. M. A. 1998. Effect of fertilizers on grain yield and grain protein content of wheat. *J. Nat. Sci. Coun. Sri Lanka,* **26**: 1-8.

Havelka, U. D., Wittebach, V. A. and Boyle, M. G. 1984. CO2-enrichment effects on wheat yield and physiology. *Crop Sci.*, **24**: 1163-1168.

Hawker, J.S. and Hatch, M.D. 1985. Mechanism of sugar storage by mature stem of sugarcane. *Plant Physiol.,* **18:** 444-453.

Hayashi, S. P. 1961. Effect of plant growth substances on photosynthesis and photorespiration of souce-sink organs. *Aust. J. Plant Physiol.* **1**:20-23.

Herzog, H. 1982. Relation of source and sink during grain filling period in wheat and some aspects of its regulation. *Physiol. Planta.*, **56**: 155-160.

Heyl, A. and Schmulling, T. 2003. Cytokinin signal perception and transduction. *Current Opinion in Plant Biology*, **6**:480-488.

Ho, L.C. 1988. Metabolism and compartmentation of import sugars in sink organs in relation to sink strength. *Annual Review of Plant Physiology*, **39**: 355-378.

Ho, L.C., and Baker, D. A.1982. Regulation of loading and unloading in long distance transport systems. *Physiol. Planta.*, **59**:225-230.

Horie, T., Ohnish, M.., Angus, J.F., Lwein, L.G., Tsukaguchi, T. and Matano, T. 1977. Physiology characteristics of high yielding rice inferred from cross-location experiments. *Field Crops Res.*, **52**:55-67.

Hsu, F. C., Bennet, A. B., Spanswick, R. M. 1984. Concentrations of sucrose and nitrogenous compounds in the apoplast of developing soybean seed coats and embryos. *Plant Physiol.*, **75**: 181-186.

Hume, D. J. and Criswell, J. G. 1973.Distribution and utilization of 14C-labelled assimilates in soybeans. *Crop Sci.*, **13**: 519-524.

Hunt, L. A., Vander Poorten, G, and Pararajasingham, S. 1991. Post anthesis temperature effects on duration and rate of grain filling in some winter and spring wheats. *Candian J. Plant Sci.*, **71**: 609-617.

Hutley-Bull, P. D. and Schwade, W. W. 1982.Morphogenesis in the wheat apex as influenced by environment and plant growth regulators. In: *Opportunities for manipulation of cereal productivity*, Hawkings, A. F. and Jeffcoat, B. Eds., *British Plant Growth Regulator Group (Monograph, No, 7), Wantage, UK.* Pp: 150-166.

Islam, M. S. and Morison, J. I. L.1992. Influence of solar radiation and temperature on irrigated rice grain yield in Bangladesh. *Field Crops Research*, **30**:13-28.

Itoh, H., Matsuoka, M. and Steber, C. M. 2003. A role for the ubiquitin-26s-proteasome pathway in gibberellin signaling. *Trends in plant Sci.*, **8**: 492-497.

Jacobson, S. and Varner, J. T. 1967. Mechanism of GA mediated enzyme regulation. *Physiol. Planta.*, **18**: 27-35.

James, N. I. and Lund, S. 1960. Meristem development of winter barley as affected by vernalization and potassium gibberellin. *Agron. J.*, **52**: 508-510.

Jedel, P. E. and Hunt, L. A. 1990. Shading and thinning effect on multi- and standard- floret winter wheat. *Crop Sci.*, **30**:128-133.

Jenner, C. F. 1980a. Effects of shading or removing spikelets in wheat: Testing assumptions. *Aust. J. Plant Physiol.*, **7**:113-121.

Jenner, C.F. 1980b. The conversion of sucrose to starch in developing fruits. *Bar. Dtsch. Bot. Res.*, **93**: 249-351.

Jenner, C. F. and Hawker, J. S. 1993. Sink strength, soluble starch synthase as a measure of sink strength in wheat endosperm. *Plant Cell Environ.*, **16**: 1023-1024.

Jenner, C. F., Ugalde, T. D. and Aspinall, D. 1991. The physiology of starch and protein deposition in the endosperm of wheat. *Aust. J. Plant Physiol.*, **18**:211-226.

Jenner, C.F. and Rathjen, A. J. 1972. Limitations to the accumulation of starch in developing wheat grains. *Ann. Bot.*, **36**: 743-754.

Jennings, A. C. and Morton, R. K. 1963. Changes in carbohydrates, protein and non-protein nitrogenous compounds of developing wheat grain. *Aust. J. Biol. Sci.* **16**: 332-41.

Jianchang, Y., Shaobing, P., Romeo M. V., Arnel, L. S., Qingsen, Z. and Shiliang, G. 2000. Grain filling pattern and cytokinin content in the grains and roots of rice plants. *Plant Growth Reghlation*, **30**:261-270.

Jin, S. 1996. Wheat in China. *China Agriculture Press, Beijing.*

Jitla, D. S., Rogers, G. S., Seneweera, S. P., Basra, A. S. Oldfield, R. J. and Conroy, J. P. 1997. Accelerated early growth of rice at elevated CO2: is it related to developmental changes in the shoot apex? *Plant Physiol.*, **115**: 15-22.

Joy, K. W. 1964. Translocation in sugar beet. I. Assimilation of $^{14}CO_2$ and distribution of materials from leaves. *J. Exp. Bot.*, **15**:485-494.

Juszczuk, I., Malusa, E. and Rychter, A. M. 2001. Oxidative stress during phosphate deficiency in roots of bean plants (*Phaseolus vulgaris* L.). *J. Plant Physiology,* **158**: 1299-1305.

Kaim, M. S., Nair, T. V. R. and Abrol, Y. P. 1983. Nitrogen economy of main shoot of wheat (*Triticum aestivum* L.) plants grown at two soil nitrogen levels. III. *In vivo* nitrate reductase activity. *Ind. J. Plant Nutr.*, **2**: 39-50.

Kaminek, M., Dobrea, P., Gaudinova, A., Motyka, V., Malbeck, J., Travnickova, A. and Trckova, M. 2000. Potential physiological function of cytokinin binding proteins in seeds of cereals. *Plant Physiol. Biochem.*, **38**: 79-85.

Kawashima, N. and Mitake, T. 1969. Studies on protein metabolism in higher plants. Part IV. Changes in ribulose diphosphate carboxylase activity and fraction 1-proetin content in tobacco leaves with age. *Agri. Biol. Chem.*, **33**: 539-543.

Keeling, P., Wood, J. r., Tyson, R. H. and Bridge, I. G. 1987. Starch biosynthesis in developing wheat grains. *Plant Physiol.*, **87**: 311-319.

Kemp, D.R., and Whingwiri, E.E.1980.Effect of tiller removal and shading on spikelet development and yield components of the main shoot of wheat and on the sugar concentration on the ear and flag leaf. *Aust. J. Plant Physiol.*, **7**:501-510.

Kende, H. 1971. The cytokinins. *Int. Rev. Cytol.*, **31**: 301-338.

Kimball, B. A. 1983. Carbon dioxide and agricultural yield Assemblage and analysis of 430 prior observations. *Agron. J.,* **75**: 779-788.

Kimball, B. A., Pinter, P. J., Jr. Garcia, R. L., Lamorte, R. L., Wall, G. W., Hunsaker, D. J., Wechsung, F. and Kartschall, T. H. 1995. Productivity and water use of wheat under free-air CO2 enrichment. *Glob Chang Biol.*, **1**: 429-442.

Kinet, J. M., 1977. Effect of light condition on the development of the inflorescence in tomato. *Hort. Sci.*, **6**:27-35.

King, R. W. 1982. Abscisic acid in seed development. In: The physiology and biochemistry of seed development dormancy and germination. Khan, A.A. Ed., Elsevier Biomedical Press, Pp: 157-181.

King, R. W., Morgan, J. M. and Zeng, Z. R. 1967. ABA in developing wheat grains and its relationship to grain growth and maturation. *Planta.* **132**: 43- 51.

Kiniry, J. R. 1993. Nonstructural carbohydrate utilization by wheat shaded during grain growth. *Agron. J.*, **85**:844-849.

Kirby, E. J. M. and Rymer, J. I. 1974. Development of the vascular system in the ear of barley. *Ann. Bot.*, **38**:565-573.

Knypl, J. S. 1979. Hormonal control of nitrate assimilation: do phytohormones and phytochrome control the activity of nitrate reductase? In: *Nitrogen assimilation of plants*. Hewitt, E. J. and Cuttings, C. V. Eds. *Academic press, London*, Pp: 541-556.

Kobata, T. and Moriwaki, N. 1990. Grain growth rate as a function of dry matter production rate: An experiment with two rice cultivars under different radiation environments. *Jpn. J. Crop Sci.*, **59**: 1-7.

Kobata, T., Sugawara, M. and Takatu, S. 2000. Shading during the early grain filling period does not affect potential grain dry matter increase in rice. *Agron. J.*, 92:411-417.

Koch, J. L. and Peterson, D. M. 1991. Carbohydrate deposition during oat (*Avena Sativa*) seed development. *Supplement to Plant physiol.*, **96**: 74-79.

Koch, K. E. 1996. Carbohydrate-modulated gene expression in plants. *Annual Review of Plant Physiology and Plant Molecular Biology*, **47**: 509-540.

Koshkin, E.I. and Tararina, V. V.1989. Yield and source/sink relations of spring wheat cultivars. *Field Crops Resarch*, **22**: 297-306.

Krenzer, E. G. and Moss, D. N. 1975. Carbon dioxide enrichment effects upon

Kridemann, P., 1966. The photosynthetic activity of the wheat ear. *Ann. Bot.,* **30**: 349-363.

Kruk, B. C., Calderini, D. F. and Slafer, G. A. 1997. Grain weight cultivars released from 1920 to 1990 as affected by post-anthesis defoliation. *J. Agri. Sci.*, **128**: 273-281.

Kuiper, D. 1993. Sink strength, established and regulated by plant growth regulators. *Plant Cell Envir.*, **16**: 1025-26.

Kumar, P., Dube, S. D. and Chauhan, V. S. 1998. Relationship among yield and some physiological traits in wheat. *Indian J. Plant Physiol.*, **3**: 220-230.

Kumar, R. and Singh, R. 1980. Enzymes of starch metabolism and their relation of grain size/starch content in developing wheat grains. *J. Sci. Food Agri.*, **32**: 229-234.

Kumari, S. and Ghildiyal, M. C. 1998. Alteration respiration in relation to grain growth within the ear of wheat. *Indian J. Plant Physiol.*, **3**: 287-291.

Kuo, J., O 'Brien, T.P. and Canny, M. J. 1974. Pit field distribution, plasmodesmatal frequency and assimilate flux in the mestome sheath of wheat leaves. *Planta*, **121**:97-118.

Kuo, J., O 'Brien, T.P. and Zee, S. 1972. The transverse veins of the wheat leaf. *Aust. J. Biol. Sci.*, **25**:721-737.

Labrana, X and Araus, J. L. 1991. Effect of foliar application f silver nitrate and ear removal on carbon dioxide assimilation in wheat flag leaves during grain filling. *Field Crops Res.,* **28**:149-162.

Lalonade, S., Boles, E., Hellmann, H., Barker, L., Patrick, J. W., Frommer, W. B. and Ward, J. M. 1999. The dual function of sugar carriers: transport and sugar sensing. *The Plant Cell*, **11**: 707-726.

Lambers, H. 1997. Oxidation of mitochondrial NADH and the synthesis of ATP in Plant Metabolism. Dennis, D.T., Layzell, D.B., Lefebvre, D.D. and Turpin, D.H. Eds., *Singapore, Longman*, Pp: 200-219.

Lang, A. and Reinhard, E. 1961. Gibberellins and flower formation. *Adv. Chem.,* **28**: 71-76.

Langer, R. H. M., Prasad, P. C. and Ashley, D. A. 1973. Effects of kinetin on tiller bud elongation in wheat (*Triticum aestivum* L.). *Ann. Bot.,* **37**: 565-571.

Larsson, R.M. and Hensen, W.K. 1992. Studies on seed quality of Triticale α-amylase and starch. *Agro. Food Industry Hi-Tech.*, **3**: 26-28.

Latimore, M., Gidden, J. and Ashley, D. A. 1977. Effect of ammonium and nitrate nitrogen upon photosynthate supply and nitrogen fixation by soybeans. *Crop Sci.*, **17**:399-404.

Laver, J. G., Munyon, R. L., Ries, S. K. and Wert, V.F. 1983. Growth enhancement of plants by femtomole doses of colloidally dispersed tria contanol. *Science*, **219**:1219-1225.

Lawn, R. J. and Brun, W. A.1974. Symbiotic nitrogen fixation in soybeans. Effect of photosynthetic source-sink manipulations. *Crop Sci.*, **14:** 11-16.

Ledent, J.F. and Stoy, V.1985. Responses to reduction in kernel number or to defoliation in collections of winter wheats. *Agronomie*, **5**: 499-504.

Leigh, R. A. Ap-Rees, T, Fuller, W. A. and Banfield, J. 1979. Location of acid invertase activity and sucrose in the vacuoles of storage roots of beet root (*Beta vulgaris*). *Biochem. Journal*, **178**: 539-47.

Lennon, A.M., Neuenschwander, U.H., Ribas-Carbo, M., Giles L., Ryals, J.A. and Siedow, J.N. 1997. The effects of Salicylic Acid and tobacco mosaic virus infection on the alternative oxidase of tobacoo. *Plant Physiol.*, **115**: 783-791.

Leoppold, A. C. 1949. The control of tillering in grasses by auxin. *Amer. J. Bot.,* **36**: 437-440.

Letham, D. S. 1963. Zeatin, a factor inducing cells division from *Zea mays. Life Sci.,* **8**: 569-571.

Li, A.G., Hou, Y.S., Wall, G. W., Trent, A., Kimball, B. A. and Pinter Jr., P. J. 2000. Free-air CO2 enrichment and drought stress effects on grain filling rate and duration in spring wheat. *Crop Sci.*, **40**:1263-1270.

Liang, J., Zjang, J. and Cao,X. 2001. Grain sink strength may be related to the poor grain filling of *indica-japonica* rice (*Oryza sativa*) hybrids. *Physiol. Planta.*, **112**:470-477.

Lieffering, M., kim, H. Y., Kobayashi, K. and Okada, M. 2004. The impact of elevated CO2 on the elemental concentration of field-grown rice grains. *Field Crops Research*, **88**:279-286.

Liew, F. K. Y. 1968. Nitrogen supply at various stages of development in wheat. M. Agri. Sc. Thesis, University of Canterbury (Lodged in Linoln College Library), Pp: 120.

Lingle, S. E. and Chevalier, P. 1985. Development of the vascular tissue of the wheat and barley caryopsis as related to the rate and duration of grain filling. *Crop Sci.*, **25**: 123-128.

Liu, W. C. and Carns, H. R. 1961. Isolation of abscisin, and abscission accelerating substance. *Science,* **134**: 385-362.

Lopez-Garrido, E., Molina-Quilros, S., Puerta-Lopez, P. G. D. L., Patrick, J.W., Vidal-Bernabe, M. D. R. and Garcia-Moral, L. F. 2001. Quantification of vascular tissues in the peduncle of durum wheat cultivars improved during the twentieth century. *Int. J. Dev. Biol.,* **45**:47-48.

Lupton, F. G. H. and Pinthus, M. J. 1969. Carbohydrate translocation from small tillers to spike producing shoots in wheat. *Nature*, **221**:483-484.

Lupton, F.G.H. 1966. Translocation of photosynthetic assimilates in wheat. *Ann. Appl. Biol.*, **57**: 355-64.

Lush, W. M.1976.Leaf structure and translocation of dry matter in C3 and a C4 grass. *Planta,* **130**:23-244.

Ma, Y.Z., MacKnown, C.T. and VanSanford, D. A. 1990.Sink manipulation in wheat: Compensatory changes in kernel size. *Crop Sci.*, **30**:1099-1105.

MacKown, C. T., Van Sanford, D. A. and Ma, Y.Z. 1989. Main stem sink manipulation in wheat: Effects on nitrogen allocation to tillers. *Plant Physiol.,* **89**:597-601.

Macleod, A. M. and Miller, A. S. 1962. Effect of gibberellic acid on barley endosperm. *J. Inst. Brewing,* **68**: 322-328.

Makino, A., Mae, T. and Ohira, J. 1983. Photosynthesis and ribulose 1, 5 bisphosphate carboxylase in rice leaves. *Plant Phsyiol.,* **73**: 1002-1007.

Makino, A., Nakano, H. and Mae, T. 1994. Response of ribulose-1,5-bisphosphate carboxylase, cytochrome F1 and sucrose synthesis enzymes in rice leaves to leaf nitrogen and their relationships to photosynthesis. *Plant Physiol.*, **105**: 173-179.

MaMillan, J. and Suter, P. T. 1958. The occurrence of gibberellin A1 in higher plants: isolation from the seed of summer bean (*Phaseolous mutliflorus*). *Naturwissenschaften*, **45**: 46-51.

Mandal, M. P. and Singh, R. A.2000. Effect of salt stress on amylase, peroxidase and protease activity in rice (*Oryza sativa L.*) seedlings. *Indian. J. Plant Physiol.,* **5**. 183-185.

Mare, J. and Gifford, R. M. 1984. Floral initiation in wheat, sunflower, and sorghum under carbon dioxide encrichment. *Canadian Journal of Botany,* **62**: 9-14.

Marcelis, L. F. M. and De-Koning, A.N.M.1995.Biomass partitioning in plants. In: Bakker, J.C., Bot, G.P.A., Challa, H. and Van-De- Braak, N.J., Eds., *Greenhouse climate control: An integrated approach. Wageningen, Wageningen, Perss.* Pp: 84-119.

Marentes, E. and Grusak, M .A. 1998. Iron transport and storage within the seed coat and embryo of developing seeds of pea (*Pisum sativum* L.). *Seed Science Research,* **8**: 367-375.

Marshall, B. and Biscoe, P. V. 1982. Environmental and physiological factors affecting assimilate supply during grain growth. In: *Opportunities for manipulation of cereal productivity.* Hawkins, A. F. and Jeffcoat, B. Eds., *British Plant Growth Regulator Group (Monograph No, 7), Wantage, UK.* Pp: 179-192.

Masle, J.2000. The effects of elevated CO2 concentrations on cell division rates, growth patterns, and blade anatomy in young wheat plants are modulated by factors related to leaf position, vernalization, and genotype. *Plant Physiol.*, **122**:1399-1415.

Mason, T. G. and Maskell, E. J. 1928. Studies on the transport of carbohydrate in the cotton plant II. The factors determining the rate of direction of movement of sugars. *Ann. Bot.*, **42**: 571-636.

Mathews, S. and Thomson, W. J. 1984. Growth regulation: Control of growth and development. In: *Cereal production*. Gallagher, E. J. Ed., *Butter worth, London*, Pp: 259-266.

Mathews, S., Koranteng, G. O. and Thomson, W. Y. 1982. Tillering and ear production: Opportunities for chemical regulation. In: *Opportunities for manipulation of cereal productivity*, Hawkins, A. F. and Jeffcoat, B. Eds., *British Plant Growth Regulator Group (Monograph, No, 7) wantage, UK.* Pp: 88-96.

Matile, P. 1987. Biosynthesis of functions of vacuoles. *Ann. Rev. Plant Physiol.*, **29**: 193-213.

Matsushima, S. and Wada, G. 1958.Analysis of developmental factors determining yield and its application to yield prediction and culture improvement of lowland rice: 48.Studies on the mechanism of ripening (9). Relation of the percentage of ripened grains and the yield of grains to the amount of carbohydrates stored by the time of heading that of carbohydrates accumulated after heading and the nitrogen content at heading time. *Proc. Crop Sci. Soc. Jap.,* **27**:201-203

McDonald, A.E., Sieger, S. M. and Vanlerberghe, G.C.2002.Methods and approaches to study plant mitochondrial alternative oxidase. *Physiol. Planta.*, **116**:135-143.

McMaster, G.S., Morgan, J.A., and Willis, W.O.1987. Effects of shading on winter wheat yield, spike characteristics and carbohydrate allocation. *Crop Sci.*, **27**:967-973.

Medina, E. 1971. Relationship between nitrogen level, photosynthetic capacity and carboxydismutase activity in *Atriplex patula* leaves. *Carnegie Inst. Year Book,* **69**: 605-622.

Micheal, G. and Beringer, H. 1980. The role of hormones in yield formation. In: *Physiological Aspects of Productivity. Proc. 15th Collaq. Int. Potash. Inst. Bern.* Pp: 85-116.

Micheal, G. and Seller-Keilbitsch, H. 1972. Cytokinin content and kernel size of barley grains as affected by environmental and genetic factors. *Crop Sci.,* **12**: 162-165.

Micheal, G., Auinger, P. and Wilberg, E. 1970. Eingie aspekte zur hormonelle regulation der korngrosse ber getrude. *Z. Pflanzeernalr Bodnk,* **125**: 24-35.

Mikkonen, A. 1986. Activities of some peptidases and proteases in kidney bean. *Physiol. Planta.*, **67**: 282-86.

Millar, A. H., Atkin, O. K., Lambers, H. Wiskich, J. T. and Day, D. A. 1995. A critique of the use of inhibitors to estimate partitioning of electrons between mitochondrial respiratory pathways in plants. *Physiologia Plantarum,* **95**: 523-532.

Millenaar, F.F. and Lambers, H. 2003.The alternative oxidase: *In vivo* regulation and function. *Plant Biol.*, **5**: 2-15.

Minchin, P. E. H. and Thorpe, M. R. 1984. Apoplastic phloem unloading in stem of bean. *J. Exp. Bot.*, **35**: 538-550.

Minchin, P. E. H. and Thorpe, M. R. 1996. What determines carbon partitioning between competing sinks? *J. Exp. Bott. (Special Issue)*, **47**: 1293-1296.

Mishra, S. D. 1990. Hormonal regulation of plant growth development and productivity. Past Present and Future. *Ind. J. Plant Physiol.*, **42**: 27-32.

Mitcham, E. J. Gross, K. C. and Nq, T. C. 1989. Tomato fruit cell wall synthesis during transpiration induced by (RS) absisic acid. *Nature (London)* **221**:113-116.

Mitra, N. and Sen, S. P. 1987. Hormonal regulation of source-sink relationship: Effect of hormones on excessed source and sink organs of cereals. *Plant Cell Physiol.*, **28**: 1005-1012.

Miyoshi, K. and Sato, T. 1997. The effect of kinetin and gibberellin on the germination of dehusked seeds of indica and japonica rice (*Oryza sativa* L.) under anaerobic and aerobic conditions. *Ann. Bot.*, **80**: 479-483.

Morgan, P.W. 1979. Agricultural uses of plant growth substances. An analysis of present status and future potential. *Proc. 6th Ann. Meeting Plant Growth Regulator Work Group*. Aug. 20-24. Las Vegas, Nevada, Pp: 1-13.

Morris, D. A. and Arthur, E. D. 1985. Effect of gibberellic acid on patterns of carbohydrate distribution and acid invertase activity in *Phaseolus vulgaris. Physiol. Planta.*, **65**: 257-62.

Mothes, K., Engelbrecht, L. and Schutte, H. R. 1961. Uber die akkumulation-aminoisobutter saure in blattgewebe unter dem einfluss von kintein. *Physiol. Planta.*, **14**: 72-75.

Murchie, E. H., Yang, J., Hubbart, S., Horton, P. and Peng, S.2002. Are there associations between grain filling rate and photosynthesis in the flag leaves of field-grown rice? *J. Exp. Bot.*, **53**:2217-2224.

Murphy, R. 1989. Water flow across the sieve tube boundary: estimating turgor and some implications for phloem loading and unloading IV. Root tips and seed coats. *Ann. Bot.*, **63**: 571-579.

Nagato, K. and Chaudhry, F. M.1970. Influence of panicle clipping, flag leaf cutting and shading on ripening of Japonica and Indica rice. *Proc. Crop Sci. Soc. Jpn.*, **39**:204-211.

Nakamura, T., Nagamatu, M. and Hoshikawa, K. 1992.Relatiship between early grain growth, percentage of grain filling and number of endosperm in rice. *Jpn. J. Crop Sci.*, **61** (extra issue): 129-130.

Naryanan, K. R., Mudge, K. W. and Poovaiah, B. W. 1981. Demonstration of auxin binding to strawberry fruit membranes. *Plant Physiol.*, **68**: 1289-1293.

Nautiyal, N., Chatterjee, C. and Sharma, C. P. 1999. Copper stress affects grain filling in rice. *Communications in Soil Sciences and Plant Analysis*, **30**: 11-12.

Nerson, H.1980. Effects of population density and number of ears on wheat yield and its components. *Field Crop Res.*, **3**:225-234.

Nickell, L. G. 1979. Controlling biological behaviour of plants with synthetic plant growth regulating chemicals. In: *Plant growth substances*, Mandava, N. B. Ed., A.C.S. *Symposium Series III, Americana Chemical Society*, Pp: 263-279.

Nickell, L. G. 1985. New plant growth regulator increases grape size. Presented at Plant Growth Gegulator Soc. Amer. Ann. Mtg., Boulder, Colorado.

Nickell, L. G. 1986. Plant growth regulators in retrospect and in prospect. *Food and Fertilizer Technology Center, Chicargo, USA*, **235**: 1-7.

Normanly, J. 1997. Auxin metabolism encompasses transport conjugation- deconjugation conversion and metabolism. *Physiol. Planta.*, **100**: 431-442.

O'Brien, T. P., Sammut, M. E., Lee, J. W. and Smart, M. G.1985. The vascular system of the wheat spikelet. *Aust. J. Plant physiol.*, **12**.487-511.

Offler, C. E. and Partrick, J. W. 1993. Pathway of photosynthate transfer in the developing seed of *vicia faba* L. and structural assessment of the role of transfer cells in unloading from the seed coat. *J. Exp. Bot.*, **44**: 711-724.

Offler, C. E., Nerlich, S. M. and Partrick, J. W. 1989. Pathway of photosynthate transfer in the developing seed of *vicia faba* L. transfer in relation to seed anatomy. *J. Exp. Bot.*, **40**: 769-780.

Okawa, S., Makino, A. and Mae, T. 2003. Effect of irradiance on the partitioning of assimilated carbon during the early phase of grain filling in rice. *Ann. Bot.*, **92**:357-364.

Olszyk, D. M. and Wise, C. 1997. Interactive effects of elevated CO2 and O3 on rice and tomato. *Agric. Ecosys. Envrion.*, **66**: 1-10.

Osborne, T. B. 1907. The proteins of wheat kernel. *Washington, Carnigie Institute Publication*, **84**.1-12.

Oshima, M. 1996. Translocation and redistribution of assimilated $^{14}C$ in rice plant. *Journal of Science of Soil Manure Japan,* **37**: 589-593.

Paleg, L. G. 1964. Cellular localization of the gibberellin induced response of barley endosperm. In: *Regulateurs Naturels*, Nitsch, J. P. Ed., *Paris, CNRS*, Pp: 224-265.

Pande, P. C. and Shukla, D. S. 1996. IAA oxidase and peroxidase activity developing grains of critical. *Ind. J. Plant Physiol.*, **24**: 109-112.

Pande, P. C. Pathak, P. C. Varade, P. B. and Sastry, L. V. S. 1981. Biochemical investigation to improve grain shriveling in triticale and other cereals. *Cereal Res. Commuc.*, **9**: 173-183.

Pandey, I. B., Mishra, S. S. and Singh S. J. 1997. Effect of seed furrow mulching, nitrogen rates, used management practices on growth and yield of late sown wheat (*Triticum aestivum* L.) *Indian J. Agr.*, **42**: 463-467.

Panozzo, J. F., Eagles, H. A. and Wootton, M. 2003. Changes in protein composition during grain development in wheat. *Aust. J. Agri. Res.*, **52**: 485-493.

Pararajasingham, S. and Hunt, L. A. 1991. Postanthesis biomass accumulation and respiration in shaded and unshaded wheat. *(Triticum aestivum L.). Can. J. Plant Sci.*, **71**: 1011-1020.

Parlov, A.V. and Kolensink, T. T. 1974. Factors responsible for different levels of protein accumulation in grains of wheat cultivars with high and low protein contents. *Fiziol. Rast.*, **21**: 329-35.

Parry, M. A. J., Andralojc, P. J., Mitchell, R. A. C., Madgwick, P. J. and Keys, A. J. 2003. Manipulation of rubisco: the amount activity, function and regulation. *J. Exp. Bot.*, **54**: 1321-1333.

Pate, J.S. and Gunning, B. E. S. 1972. Transfer cells. *Ann. Rev. Plant Physiol.*, **23**:173-196.

Pate, J.S., Sharkey, P.J. and Lewis, O.A.M. 1974. Phloem bleeding from legume fruits – A technique for study of fruit nutrition. *Planta*, **120**: 229-243.

Patel, R. and Mohapatra, P. K. 1996. Assimilate partitioning within floret components of contrasting rice spikelets producing qualitatively different types of grains. *Aust. J. Plant Physiol.*, **23**: 85-92.

Patrick, J. W. 1997. Phloem unloading: sieve element unloading and post-sieve element transport. *Annual Review of Plant Physiology and Plant Molecular Biology*, **48**: 191-222.

Patrick, J. W., and Offler, C. E. 1995. Post-sieve element transport of sucrose in developing seed. *Aust. J. Plant Physiol.*, **22**: 681-702.

Patrick, J. W., Offler, C. E. and Wang S. D. 1995. Cellular pathway of photosynthate transport in coats of developing seed of *Vicia faba* L. and *phaseolus vulgaris* L. extent of transport through the coat symplast. *J. Exp. Bot.*, **46**: 35-47.

Patrick, J. W.1972a. Distribution of assimilate during stem elongation in wheat. *Aust.J. Bio. Sci.*, **25**:455-467.

Patrick, J. W.1972b. Vascular system of the stem of the wheat plant. I. Mature state. *Aust. J. Bot.*, **20**:49-63.

Patrick, J.W. and Offler, C. E.2001. Compartmentation of transport and transfer events in developing seeds. *J. Exp. Bot.*, **52**:551-564.

Patrick, J.W. and Wardlaw, I. F. 1984. Vascular control of photosynthate transfer from the flag leaf to the ear of wheat. *Aust. J. Plant Physiol.*, **11**:235-241.

Patrick, J.W.1988. Assimilate partitioning in relation to crop productivity. *Hort. Sci.*, **23**:33-40.

Patterson, T .G. and Brun, W. A. 1980. Influence of sink removal in the senescence patter of wheat. *Crop Sci.*, **25**: 35-47.

Patterson, T. G. and Moss, D. N. 1979. Senescence in field grown wheat. *Crop Sci.*, **19**: 635-640.

Patterson, T. G., Moss, D. N. and Brun, W. A. 1980. Enzymatic changes during the senescence of field grown wheat. *Crop Sci.*, **20**: 15-18.

Paul, A. K., Mukherji, S. and Sircar, S. M. 1971. Metabolic changes in developing rice seeds. *Physiol. Planta.*, **24**: 342-346.

Paul, M. J. and Foyer, C. H. 2001. Sink regulation of photosynthesis. *J. Exp. Bot.*, **360**: 1383-1400.

Pearson, J. N., Rengel, Z, Jenner, C. F. and Graham, R. D. 1995. Transport of zinc and manganese movement in developing what grains. *Aust. J. Plant Physiol.*, **25**: 139-144.

Peltonen-Sainio, P., Rajala, A., Simmons, S., Caspers, R. and Stuthman, D. D.2003.Plant growth regulator and daylenght effects on preanthesis main shoot and tiller growth in conventional and dwarf oat. *Crop Sci.*, **43**: 227-233.

Pena, R. J. 1982. Grain shriveling in secondary hexaploid Tritical, α-amylase activity and carbohydrate control of mature and developing grain. *Cereal Chem.*, **59**: 454-458.

Pendleton, J.W. and Weibal, R.O.1964.Shading studies in winter wheat. *Agron. J.*, **57**:292-293.

Peoples, M. B. and Dalling, M. J. 1978. Degradation of ribulose- 1,5-bisphosphate carboxylase by proteolytic enzymes from crude extracts of wheat leaves. *Planta*, **138**: 153-160.

Percival, J. 1921. The wheat plant. *Duckworth, London*, Pp: 463.

Perez, P., Martinez-Carrasco, R., Molino, M. M. D., Rojo, B. and Ulloa, M.1989.Nitrogen uptake and accumulation in grains of three winter wheat varieties with altered source-sink ratios. *J. Exp. Bot.*, **40**:707-710.

Pernollet, J. C. 1978. Protein bodies of seeds. Ultrastructure, Biology, Biosynthesis and Degradation. *Phytochemistry*, **17**: 1473-1480.

Peterson, D. M. 1983. Effect of spikelet removal and post-heading thinning on distribution of dry matter and N in oats. *Field Crops Res.*, **7**:41-50.

Pheloung, P. C. and Siddique, K. H. M. 1991.Contribution of stem dry matter to grain yield in wheat cultivars. *Aust. J. Plant Phyysiol.*, **18**: 53-64.

Pieta-Filho, C. and Ellis, R. H. 1991. The of seed quality in spring barley in four environments. II. Germination and longevity. *Seed Sci. Res.*, **1**: 163-177.

Pinter, P. J., Jr., Kimball, B. A., Garcia, R. L., Wall, G. W., Hunsaker, D. J. and Lamorte, R. L. 1996. Free-air CO2 enrichment: responses of cotton and wheat crops. In: *Carbon dioxide and the terrestrial ecosystem*. Eds., Koch, G. W. and Mooney, H. A. *Academic Press, San Diego*, Pp: 215-249.

Popova, L. P., Dimitrova, O. D. and Vaklinova, S. G. 1982. Effect of $CO_2$ on the chlorophyll content on the intensity of photosynthetic $CO_2$ fixation and the activity of carboxylating enzymes in $C_3$ and $C_4$ plants. *Biol. Biochem.*, **35**: 797-800.

Porter, H. M. and May, L. H. 1955. Metabolism of radioactive sugars of tobacco leaf discs. *J. Exp. Bot.*, **6**: 43-63.

Prochazka, S. 1978. Effect of indole-3-acetic acid on the translocation of assimilates in wheat in the period of kernel formation. *Acta. Univ. Brne.*, **24**: 99-104.

Purvis, A. C. and Shwwfelt, R. L. 1993. Does the alternative pathway ameliorate chilling injury in sensitive plant Tissues? *Physiol. Planta.*, **88**: 712-718.

Quinlan, J. D. and Sagar, G. R. 1962. An autoradiography study of the movement of 14C-labelled assimilates on the developing wheat plant. *Weed Res.*, **2**:264-273.

Rademacher, W. 1978. Gas chromatographic analyse der veranderungen in hormone gehalt des wachsenden weizenkornes. *Diss, Gottingen.*

Radi, A. F., Shaddad, M. A. K., Fl-Enamy, A. E. and Omran, F. M. 2002. Interactive effects of plant hormones ($GA_3$ or ABA) and salinity on growth and some metabolites of wheat seedlings.*Plant Nutrition*, **92**: 436-437.

Radley, M. 1976. The development of wheat grains in relation to endogenous growth substances. *J. Exp. Bot.*, **27**: 1009-1021.

Radley, M. E. 1982. Some factors affecting grain set in wheat. In: *Opportunities for manipulation of cereal productivity*, Monograph 7. Eds. Hawkins, A. F. and Jeffcoat, B., *British Plant Growth Regulator Group, Wantage*, Pp: 140- 149.

Raghuram, N. and Sopory, S. K. 1995. Light regulation of nitrate reductase gene expression mechanism and signal-response coupling. *Physiol. Mol. Biol. Plants.*, **1**:103-114.

Rahman, M. S., Flood, R. S. and Halloran, G. M. 1981. Grain yield responses to applied gibberellin in normalheight and semi-dwarf wheat. *Aust. J. Agri. Res.*, **32**: 243-248.

Rajewski, J. F. and Francis, C.A. 1991. Defoliation effects on grain fill, stalk rot and lodging of grain sorghum. *Crop Sci.,* **31**:353:359.

Rao, L. V. M., Datta, N., Mahadevan, M., Mukherjee, G. S. and Sopory, S. K. 1984. Influence of cytokinins and phytohormones on nitrate reductase activity in etiolated leaves of maize. *Phyto. Chem.*, **23**: 1875-1879.

Ravi, I., Khan, F. A., Sharma-Natu, P. and Ghildiyal, M. C. 2001. Yield response of tetrapliod and hexaploid wheat varieties to CO2 enrichment. *Indian J. Agr. Sci.,* **71**: 441-449.

Rawson, H. M. 1970. Spikelet number, its control and relation to yield per ear in wheat. *Aust. J. Biol. Sci.,* **23**: 1-15.

Rawson, H. M. and Evans, L. T..1970. The pattern of grain growth within the ear of wheat. *Aust. J. Biol. Sci.,* **23**: 753-764

Rawson, H. M. and Evans, L. T. 1971. The contribution of stem reserves to grain development in a range of wheat cultivars of different height. *Aust. J. Agric. Res.*, **22**:851-863.

Rawson, H. M. and Hofstra, G.1969. Translocation and remobilization of 14C assimilated at different stages by each leaf of the wheat plant. *Aust. J. Bio. Sci.,* **22**:321-331.

Rawson, H. M. and Ruwali, K.N. 1972.Ear branching as a means of increasing grain uniformity in wheat. *Aust. J. Agric. Sci.*, **23**:551-559.

Rawson, H. M., Gifford, R. M. and Bremmer, P. M. 1976. Carbon dioxide exchange in relation to sink demand in wheat. *Planta,* **132**: 19-23.

Rawson, H. M., Hindmarsh, J. H., Fisher, R.A. and Stockman, Y.M. 1983. Changes in leaf photosynthesis with plant ontogeny and relationships with yield per ear in wheat cultivars and 120 progeny. *Aust. J. Plant Physiol.*, **10**: 503-514.

Ray, S. and Choudhuri, M. A. 1980. Effects of plant growth regulators on grain filling and yield of rice. *Indian J. Plant Physiol.*, **46**: 755-758.

Ray, S., Mondal, W. A. and Choudhuri, M. A. 1983. Regulation of leaf senescence, grain filling and yield of rice by kinetin and abscisic acid. *Physiol. Planta.,* **59**: 343-346.

Reilly, M. L. 1984. Functional aspects of cereal structure. In: *Cereal production*, Gallagher, E. J. Ed., *Butterworth, London*, Pp: 137-160.

Reilly, M. L. 1990. Nitrate assimilation and grain yield. In: *Nitrogen in Higher Plants*. Eds. Abrol, Y.P. *Research Studies Press, England*, Pp: 335-336.

Reinecke, D. M. and Bandurski, R. S. 1981. Metabolic conversion of $^{14}C$-indole-3-acetic acid to $^{14}C$-oxindole-3-acetic acid. *Biochem. Res. Commun.*, **103**: 429-433.

Ribas-Carbo, J. A., Yakir, D., Giles, L., Robinson, S. A. Lenon, A M. and Siedow, J. N. 1995. Electron partitioning between the cytochrome and alternative pathways in plant mitochondria. *Plant Physiology,* **109**: 829-837.

Ribas-Carbo, M., Aroca, R., Gonzalez-Meler, M. A., Irigoyen, J. J. and Sanchez-Diaz, M. 2000. The electron partitioning between the cytochrome and alternative respiratory pathways during chilling recovery in two cultivars of maize differing in chilling sensitivity. *Plant Physiology,* **122**: 199-204.

Ricardo, C. P. P. 1974. Alkaline β-fructosidase of tuberous roots; possible physiological function. *Planta*, **118**: 333-343.

Ricardo, C. P. P. and Sovia, D. 1974. Development of tuberous roots and sugars according as related to invertase activity and mineral nutrition. *Planta*, **114**: 43-55.

Richards, R. A. 2000. Selectable traits to increase crop photosynthesis and yield of grain crops. *J. Exp. Bot.*, **51**: 447-458.

Richards, R.A. 1996. Increasing the yield potential in wheat: Manipulating sources and sinks. In: Reynolds, M.P., Rajaram, S. and Mcnab, A. Eds. *Increasing yields potential in wheat: Breaking the barriers*. Mexico, D.F. CIMMYT, Pp: 134-149.

Roeckl, B. 1949. Nechweis eines konzentration shubs zeischen palisaden zellen and siebrohren. *Planta,* **36:** 530-50.

Rogers, G.S., Milham, P.J., Gillings, M. and Conroy, J.P. 1996. Sink strength may be the key to growth and nitrogen responses in N-deficient wheat at elevated $CO_2$. *Aust. J. Plant Physiol.* **23**: 253-264.

Sacher, R. C. and Berry M. 1984. Gene regulation by plant hormones. *Bicohem. Rev*. Pp.15-34.

Saha, S., Nagar, P. K. and Sircar, P. K. 1986. Cytokinin concentration gradient in the developing grains and upper leaves of rice (*Oryza sativa*) during grain filling. *J. Bot.*, **64**: 2068-2072.

Saini, A., Sud, A. and Nanda, R. 1975. Growth regulators in regulation to tillering in wheat. *Ind. J. Plant Physiol.,* **18**: 140-146.

Saini, H. S. and Aspinall, D. A. 1982. Sterility in wheat (*Triticum aestivum* L.) induced by water deficit or high temperature: Possible mediation by abscisic acid. *Aust. J. Plant Physio.*, **76**: 301-306.

Sakai-Wada, A. and Nakata, M. 1987. Effect of GA on the ultrastructure and α- amylase activity of aleuron cells of *Avena sativa* L. Plant Cell Physiol., **28**: 1465-1476.

Sankhla, N. and Huber, W. 1974. Activity of photosynthetic enzymes and $^{14}CO_2$ fixation products in leaves of *Pennisetum typhoides* seedlings. *Physiol. Planta.,* **30**: 291-294.

Sastry, L.V.S., Sundersan, S. and Pande, P. 1979. Variations in phosphorylase and ADP glucose starch, O, 4-glucose transferase activity in relation to grain shriveling in barley activities. *Cereal Res. Comm. Univ.,* **7**: 233-240.

Sayed, H. I. and Gadallah, A. M. 1983. Variation in dry matter and grain filling characteristic in wheat (*Tritucum aestivum)* cultivars. *Field Crop Rec.,* **7**: 61-62.

Scandalios, J. G. 1993. Oxygen stress and superoxide dismutase. *Plant Physiology,* **101**: 7-12.

Schliemann, W. and Schneider, G. 1993. Gibberellins in Gramineae. *Plant Growth Regulation,* **12**: 91-98.

Schmulling, T., Schafer, S. and Romanov, G. 1997. Cytokinins as regulators of gene expression. *Physiol. Planta.,* **100**: 505-519.

Schnyder, H. 1993. The role of carbohydrate storage and redistribution in the source-sink relations of wheat and barley during grain filling-A review. *New Phytologist*, **123:**233-245.

Schussler, J. R., Bremner, M. I. and Brun, W. A. 1984. ABA and its relationship to seed filling in soybeans. *Plant Physiol.*, **76**:301-306.

Seneweera, S., Conroy, J. P., Ishimaru, K., Ghannoum, O., Okada, M., Lieffering, M., Kim, H.Y. and Kobayashi, K. 2002. Changes in source-sink relations during development influence photosynthetic acclimation of rice to free air Co2 enrichment (FACE). *Funct. Plant Biol.*, **29**:945-953.

Sharma-Natu, P., Pandurangam, V. and Ghildiyal, M. C. 2004. Photosynthetic acclimation and producitivity of mungbean cultivars under elevated CO2 concentration. *J. Agron. Crop Sci.*, **190**: 123-128.

Sheldrake, A. R. 1973. Do coleoptile tips produce auxin? *New Phytol.*, **72**: 433-447.

Shieh, Y. J. and Lioa, W. Y. 1988. Changes in leaf photosynthesis and ribulose-1,5 bisphosphate carboxylase from anthesis through maturation in *Oryza sativa* L. *Bot Bull. Academic Sinica.,* **29**: 123-133.

Shiroya, T., Lister, G.R., Nelson, C. D. and Krotkov. 1961. Translocation of 14C in tobacco, at different stages of development following assimilation of 14CO2 by a single leaf. *Can. J. Bot.,* **39**: 855-864.

Shukla, D. S., Kapashi, S. B., Pande, S. C. and Panwar, J. D. S. 1995. Structural component of wheat grain and the post anthesis high temperature tolerance. *Ind. J. Plant Physiol.*, **28**: 151-154.

Siddique, K. H. M. and Whan, B. R. 1993. Ear:stem ratio in breeding populations of wheat: Significance of yield improvement. *Euphytica*, **73**: 241-254.

Siddique, K. H. M., Kirby, E. J. M. and Perry, M. W. 1989. Ear: stem ratio in old and modern wheat varieties. Relationship with improvement in number of grain per ear and yield. *Filed Crops Research,* **21**:59-78.

Sild, E., Younis, S., Pleijel, H. and Sellden, G.1999. Effect of CO2 enrichment on non-structural carbohydrates in leaves, stems and ears of spring wheat. *Physiol. Planta.*, **107**: 60-67.

Simmons, E.W. 1967. Types of life senescence. *Symp. Soc. Exp. Biol.,* **21**:215-230.

Simmons, S. R. and Crookston, R. K. 1979.Rate and duration of growth of kernels formed at specific florets in spikelets of spring wheat. *Crop Sci.*, **19**:690-693.

Simmons, S. R. Crookston, R. K. and Akurle, J. E. 1982. Growth of spring wheat kernels as influenced by reduced kernel number per spike and defoliation. *Crop Sci.*, **22**:983-988.

Simmons, S. R.1987. Growth development and physiology. In Heyne, E.G. Ed., *Wheat and wheat improvement. 2nd Ed. Agronomy,* **13**:77-113.

Simons, B. H. and Lambers, H. 1999. The alternative oxidase: is it a respiratory pathway allowing a plant to cope with stress? In: Lerner, H. R. Ed. *Plant responses to environmental stresses from phytohormones to genome reoranization*. Marcel Dekker Inc., New York, NY, Pp: 265-286.

Singh, B.K. and Jenner, C. F.1984. Factors controlling endosperm cell number and grain dry weight in wheat. Effect of shading on intact plants and of variation in nutritional supply to detached, cultured ears. *Aust. J. Plant Physiol.*, **11**:151-163.

Singh, O. and Panwar, J. D. S. 1998. Effect of 2,4-D on alpha amylase activity, growth and yield components on wheat cultivars. *Indian J. Plant Physiol.*, **3**: 323-325.

Singh, R. and Juliano, B. O. 1977. Free sugars in relation to starch accumulation in developing rice grains. *Plant Physiol.,* **59**: 417-421.

Singh, S. 1963. Hormonal control of grain filling in rice. M.Sc. Thesis. PAU, Ludhiana, Pp: 125.

Singh, S. S. and Singh, G. 1982. Effect of growth regulators on some biochemical parameters in developing grains of rice. *Plant Physiol. Biochem.*, **9**: 68-73.

Single, W. V. 1964. The influence of nitrogen supply on the fertility of wheat ear. *Aust. J. Exp. Agric. Anim. Husb.*, **4**: 165-8.

Sionit, N., Hellmers, H. and Strain, B. R. 1980. Growth and yield of wheat under CO2 enrichment and water stress. *Crop Sci.*, **20**: 687-692.

Slafer, G. A. and Miralles, D. J. 1992. Green area duration during the grain filling period of an Argentine wheat cultivar as influenced by sowing date, temperature and sink strength. *Journal of Agronomy and Crop Science,* **168**: 191-200.

Slafer, G. A. and Rawson, H. M. 1997.CO2 effect on phasic development, leaf number and rate of leaf appearance in wheat. *Ann. Bot.*, **79**:75-81.

Slafer, G.A., Calderini, D. F., Miralles, D.J. and Dreccer, M.F. 1994.Preanthesis shading effects on the number of grains of three bread wheat cultivars of different potential number of grains. *Field Crops Research*, **36**:31-39.

Smart, D. R., Chatterton, N. J. and Bugbee, B. 1994. The influence of elevated $CO_2$ on non-structural carbohydrate distribution and fructan accumulation in wheat canopies. *Plant Cell Environ.,* **17**: 435-442.

Smith, H.R.1933.The physiological relation between tillers and yield in wheat. *J. Counc. Sci. Ind. Res. Aust.,* **6**:32-42.

Sofield, I., Evans, L. T., Cook, M. G. and Wardlaw, I. F. 1997. Factors influencing the rate and duration of grain filling in wheat. *Aust. J. Plant Physiol.*, **4**: 785-797.

Solansky, S. 1979. Distribution system of the stem of the wheat plant. I. *Crop physiology and cereal breeding*, Eds, Spiertz, J. H. J. and Kramer, T.H., Wageningen, Pp: 50-54.

Spiertz, J. H. J. 1979. Weather and nitrogen effect on rate and duration of grain growth and grain yield of wheat cultivars. In: *Crop physiology and cereal breeding*, spiertz, J. H. J. and Kramer, T.H. Eds., *prod. Wageningen,* Pp: 60-64.

Spiertz, J. H. J., Devos, N. M. and Ten Holte, L. 1984. The role of nitrogen in yield formation in cereals. In: *Cereal production*. Gallagher, E. J. Ed., *Butter worth, London*, Pp: 249-258.

Srivastava, G. C. and Goswami, B. K. 1988. Influence of benzyl adenine on leaf senescence and photosynthesis in sunflower (*Helianthus annus* L.). *A. Agro. Crop. Sci.,* **161**: 23-29.

Stahli, D., Perrissin-Fabert, D., Blouet, A. and Guckert, A. 1995. Contribution of the wheat (*Ttiticum aestivum* L.) flag leaf to grain yield in response to plant growth regulatots. *Plant Growth Regulation,* 16: 293-297.

Stapper, M. and Fischer, R. A. 1990. Genotype sowing date and plant spacing influence on high-yielding irrigated wheat in southern. New South walcs. III. Potential yield and optimum flowering dates. *Australian Journal of Agricultural Research*, **41**: 1043-1056.

Sterans, D. J. 1970. Free sugars of wheat aleurone cells. *J. Sci. Agric.*, **21**: 31-34.

Stitt, M. and Krapp, A. 1999.The interaction between elevated carbon dioxide and nitrogen nutrition: the physiological and molecular background. *Plant Cell and Environment*, **22**: 583-621.

Stockman, Y.M., Fischer, R.A. and Brittain, E.G.1983.Assimilate supply and floret development within the spike of wheat (*Triticum aestivum L*). *Aust. J. plant Physiol.*, **10**:585-594.

Stoy, V. 1963. The translocation of 14C-labelled photosynthetic products from the leaf to the ear in wheat. *Physiol. Planta.,* **16**.851-856.

Stoy, V. 1965. Photosynthesis, respiration and carbohydrate accumulation in spring wheat in relation to yield. *Physiol. Planta.*, **4**: 1-25.

Stoy, V. 1979. The storage and remobilization of carbohydrates in cereals. In : *Crop phyisological and cereal breeding*, Spiertz, J.H.J. and Kramer, T.H. Eds., *Pudoc. Wageningen*, Pp: 55-59.

Sun, T. P. 2000. Cibberellin signal transduction. *Current Opinion in Plant Biology,* **3**: 374-380.

Sung, S. S., Shieh, W. J., Geiger, D. R. and Black, C. C. 1994. Growth, sucrose synthase and invertase activities of developing *Phaseoulus vulgaris* L. fruits. *Plant Cell Envir.*, **17**: 419-426.

Sweet, G. B. and Wareing, P. F. 1966. Role of plant growth in regulating photosynthesis. *Nature (London)* **210**: 77-79.

Symons, G.M. and Reid, J.B. 2003.Interactions between light and plant hormones during de-etiolation. *J. Plant Growth Regul.*, **22**:3-14.

Tabake, T. and Akazawa, T. 1973. Catalytic role of subunit A in ribulose-1,5-bisphosphate carboxylase from *Hromatium* strain D[1]. *Arch. Biochem. Biophys.,* **157**: 303-308.

Taiz, L. and Zeiger, E. 2002. Plant physiology, third edition. *Sinauer Association, Inc. U.S.A.* Pp: 690.

Takahashi, T. Tsuchihashi, N., Takaku, T. and Nakaseko, K. 1994, Grain filling mechanisms in spring wheat III. Effect of shading on dry matter accumulation in grain. *Jap. J. Crop. Sci.,* **63**: 313-319.

Takashi, T. and Kanazawa, T. 1996. Grain filling mechanisms in spring wheat: IV. Effects of shadings on number and size of spikes, grains, endosperm cells and starch granules in wheat. *Jap. J. Crop Sci.*, **65**: 277-281.

Talbert, L. E., Lanning, S. P., Murphy, R. L. and Martin, J. M.2001. Grain fills duration in twelve hard red spring wheat crosses. *Crop Sci.,* **41**: 1390-1395.

Tanaka, A. 1961. Studies on the nutrio-physiology of leaves of rice plant. *J. Fac. Agric., Hokkaido Univ.,* **51**: 449-550.

Tanaka, A. and Fujita, K. 1974. Nutrio-Physiological studies on the tomato plant. IV. Source-sink relationship and structure of the source-sink unit. *Soil Sci. Plant Nutr.,* **20**: 305-315.

Tanaka, T. and Matsushima, S. 1971. Analysis of yield-determining process and its application to yield-prediction and culture improvement of lowland rice. CIV. Effects of light intensity and different shading methods during the ripening period on the percentage of ripened grains. *Proc. Crop Sci. Soc. Japan,* **40**: 376-380.

Tanaka, T. and Matsushima, S.1963.Analysis of yield-determining process and its application to yield predication and culture improvement. 64.Studies on the mechanism of ripening.11.Occurrence of abortive kernels which are determined at the early ripening stage and a method for predicting them. *Proc. Crop Sci. Soc. Jpn.*, **32**:35-38.

Tanner, W. 1980. On the possible role of ABA in phloem unloading. *Ber. Dtsch. Bot. Ges.*, **93**:249-351.

Taylor, J. S., Bhalla, M. K., Robertson, J. M. and Piening, L. J. 1989. Cytokinins and abscisc acid in hardening winter wheat. *Can. J. Bot.*, **68**: 1597-1601.

Teare, I. D., Sy, J. W., Waldren, R. P. and Golz, S. M.1972.Comparative data on the rate of photosynthesis, respiration and transpiration of different organs in awned and awnless isogenic lines of wheat. *Can. J. Plant Sci.,* **52**:294-303.

Tegeder, M., Wang, X. D, Frommer W. B., Offler, C. E. and Patrick, J. W. 1999. Sucrose transport into developing seeds of *Pisum sativum* L. *The Plant Journal*, **18**: 151-161.

Tegeder, M., Wang, X. D., Frommer, W. B., Offler, C. E. and Patrick, J. W. 1999. Sucrose transport into developing seeds of *Pisum sativum* L. *The Plant Journal*, **18**: 151-161.

Thorne, G. N. 1966. Physiologyical aspects of grain yield in cereals. *In*: *The growth of cereals and crasses*, Eds., Milthorpe, F. L and Ivins, J. D., *Butterworths,* London, Pp: 88-105.

Thorne, G. N. 1982. Distribution between parts of the main shoot and the tillers of the photosynthates produced before and after anthesis on the top three leaves of main shoots of hobbit and maris huntsman winter wheat. *Ann. Appl. Biol.,* **101**:553-559.

Thorne, G.N. 1963.Distribution of dry matter between ear and shoot of plumage archer and procter barley during growth in the field. *Ann. Bot.*, **27**: 243-252.

Thorne, G.N. 1965. Photosynthesis of ears and flag leaves of wheat and barley. *Ann. Bot.,* **29**: 317-329.

Thorne, G.N., Ford, M.A. and Watson, D.J. 1968. Growth development and yield of spring wheat in artificial climates. *Ann. Bot.,* **32**: 425-466.

Treharne, E. J. and Eagles, C. F. 1979. Effect of temperature on photosynthetic in red clover (*Trifolium pratense* L.). *Nature,* **220**: 457-458.

Treharne, E. J. Stoddart, J. L., Pughe, J., Parnjothy, K. and Wareing, P. F. 1970. Effects of gibberellin and cytokinins on the activity of photosynthetic enzyme and plastid ribosomal RNA synthesis in *Phaseolous vulgaris* L. *Nature,* **228**: 129-131.

Trewavas, A. J. 1983. Nitrate as a plant hormone. In: Interactions between nitrogen and growth regulators in plant development. Jackson, M.B. Ed., British Plant Growth Regulator Group (Monograph No. 9), Wantage, UK Pp: 1-4.

Tribol, E., Martre, P. and Tribol-Blondel, A. M. 2002. Environmentally-induced changes in protein composition in developing grains of wheat are related to changes in total protein content. *J. Exp. Bot.,* **54**: 1731-1742.

Tuba, Z., Szente, K. and Koch J. 1994. Response of photosynthesis, stomatal conductance, water use efficiency and production to long-term elevated $CO_2$ in winter wheat. *J. Plant Physiol.,* **144**: 661-668.

Uchida, N. W. and Murata, Y. 1982. Studies on the photosynthetic activity of a crop leaf during its development and senescence. II. Effect of nitrogen deficiency on the changes in senescing leaf of rice. *Jap. J. Crop Sci.,* **51**: 577-593.

Ugalde, T. D., Jenner, C. F. 1990a. Route of substrate movement into wheat endosperm. I. Carbohydrates. *Aust. J. Plant Physiol.*, **17**: 693-794.

Ugalde, T. D., Jenner, C. F. 1990b. Route of substrate movement into wheat endosperm II, Amino acids. *Aust. J. Plant Physiol.*, **17**: 705-714.

Ulman, p., Catsky, J. and Pospisilova, J. 2000. Photosynthetic traits in wheat grown under decreased and increased CO2 concentration, and after transfer to natural CO2 concentration. *Biol. Planta.*, **43**:227-237.

Umbach, A. L. and Siedow, J. N. 1993. Covalent and noncovalent dimers of cyanide-resistant alternative oxidase protein in higher plant mitochondria and their relationship to enzyme activity. *Plant Physiology,* **103**: 845-854.

Vanderhoff, L. N. and Kosuge, T. T. 1984. The molecular biology of plant hormone-action. *Workshop Summaries-II. Publ. Amer. Soc. Plant Physiolgy, Rockville, Maryland,* Pp: 5-39.

Vanlerberghe, G. C. and McIntosh, L. 1992. Lower growth temperature increases alternative pathway capacity and alternative oxidase protein in tobacco. *Plant Physiology,* **100**: 115-119.

Venkateswarlu, B. 1976. Source- sink interrelationships in lowland rice. *Plant and Soil,* **44**:575-586.

Venkateswarlu, B. and Visperas, R.M. 1987a. Solar radiation and rice productivity. *IRRI Research Paper Series,* **129**:1-29.

Venkateswarlu, B. and Visperas, R.M. 1987b.Source- sink relationships in crop plants. *IRRI Research Paper Series,* **125**:1-19.

Venkateswarlu, B., Parao, F. T. and Vergara, B. S. 1986. Occurrence of good and qualityu grains in panicles of rice (*Oryza sativa* L.). *Crop Sci.*, **28**:123-129.

Voltas, J., Romagosa, I. and Araus, J. L.1998. Growth and final weight of central and lateral barley grains under Mediterranean conditions as influenced by sink strength. *Crop Sci.*, **38**:84-89.

Vreugdenhil, D. 1983. ABA inhibits phloem loading of sucrose. *Plant Physiol.,* **57**: 463-467.

Vu, J. C. V., Baker, J.T., Pennanen, A. H. Allen, L. H. J., Bowes, G. and Boote, K. J.1998.Elevated CO2 and water deficit effects on photosynthesis, ribulose bisphosphate carboxylase-oxygenase, and carbohydrate metabolism in rice. *Physiol. Planta.*, **103**:327-339.

Wall, W. G., Brooks, T. J., Kimball, B. A., LaMorte, R. L., Adamsen, F. J., Pinter, Jr. P. J. 1999. Gas exchange rates of spring wheat leave grown under free- air CO2 enrichment and variable soil nitrogen regimes. *Agronomy Abstracts*, Pp: 19-21.

Walpole, P. R. and Morgan, D. G. 1970. A quantitative study of grain filling in *Triticum aestivum* L. *Ann. Bot.*, **34**: 309-18.

Wang, H. L., Offler, C .E. and Patrick, J. W. and Ugalde, T. D. 1994. The cellular pathway of photosynthates transfer in the developing wheat grain. I. Delineation of a potential transfer pathway using fluorescent dyes. *Plant Cell and Environment,* **17**:257-266.

Wang, N. and Fisher, D.B.1994. Monitoring phloem unloading and post-phloem transport by microperfusion of attached wheat grains. *Plant Physiol.*, **104**:7-16.

Wang, Z., Fu, J., He, M., Tian,Q. and Cao,H.1997. Effects of source/ sink manipulation on net photosynthetic rate and photosynthate partitioning during grain filling in winter wheat. *Biologia Plantarum*, **39**: 379-385.

Wang, Z., Yin, H. M. and Cao, H. 1998. Source-sink manipulation effects of post-anthesis photosynthesis and grain setting on spike in winter wheat. *Photosynthetica* (Prague), **35**: 453-459.

Wang, Z., Yin, Y., He, M., Zhang, Y., Lu, S., Li, Q. and Shi, S. 2003. Allocation of photosynthates and grain growth of two wheat cultivars with different potential grain growth in response to pre- and post- anthesis shading. *J. Agronomy and Crop Sci.,* **186**. 280-285.

Wardlaw, I. F. 1968. The control and pattern of movement of carbohydrates in plants. *Bot. Rev.,* **34**: 79-105.

Wardlaw, I. F. 1990. The control of carbon partitioning in plants. *New Phyto.*, **116**: 341-81.

Wardlaw, I. F. 1994. The effect of high temperature on kernel development in wheat: Variability related to pre-heading and post-heading anthesis conditions. *Aust. J. Plant Physiol.*, **21**: 731-739.

Wardlaw, I. F. and Moncur, L. 1995. The response of wheat to high temperature following anthesies .I. The rate and duration of kernel filling. *Aust. J. Plant Physiol.*, **22**:391-397.

Wardlaw, I. F. and Porter, H. K. 1967. The redistribution of stem sugars in wheat during grain development. *Aust. J. Biol. Sci.*, **20**: 309-318.

Wardlaw, I. F. and Willenbrink, J.1994. Carbohydrate storage and mobilization by the culm of wheat between heading and grain maturity: The relation to sucrose synthase and sucrose-phosphate synthase. *Aust. J. Plant Physiol.*, **21**:255-271.

Wardlaw, I. F.1965. The velocity and pattern of assimilate translocation in wheat plants during grain development. *Aust. J. Biol. Sci.*, **18**:269-281.

Wardlaw, I. F.1970. The early stages of grain development in wheat: Response to light and temperature in a single variety. *Aust. J. Biol. Sci.,* **23**: 765-774.

Wardlaw, I. F.1974. Temperature control of translocation. *R. Soc. N. Z. Bull.*, **12**:533-538.

Wareing, P. F., Khalia, M. M. and Treharne, K. J. 1968. The effects of leaf removal on protein and anzymes activity in wheat. *Nature* (London), **220**: 453-460.

Wareing, P. F., Khalifa, M. M. and Trehearne, K. J. 1968. Rate limiting processes in photosynthesis at saturating light intensities. *Nature,* **220**: 453-457.

Warner, G., Formanek, H. and Theimer, R. R. 1981. The ontogeny of lipid bodies (spherosomes) in plant cell ultrastructural evidence. *Planta*, **151**: 109-123.

Watanabe, Y. Nakamura, Y. and Ishii, R. 1997. Relationship between starch accumulation and activities of the related enzymes in the leaf sheath as a temporary sink organ in rice *(Oryza sativa). Aust. J. Plant Phsiol.*, **24**:563-569.

Waters, S. P., Martin, P. and Lee, B. T. 1984. The influence of sucrose and ABA on the determination of grain number in wheat. *J. Exp. Bot.*, **35**: 829-840.

Watson, D. J. 1946. Leaf growth in relation to yield. In: *The growth of leaves,* Ed, Milthorpe, F. L., *Butterworths, London,* Pp: 178-191.

Watson, D.J., Thorne, G.N. and French, S.A.W.1958. Physiological causes of differences in grain yield between varieties of barley. *Ann. Bot.*, **22**:321-351.

Watson, M.A. and Casper, B.B. 1984. Morphogenetic constraints on pattern of carbon distribution in plants. *Ann. Rev. Ecol. Syst.*, **15**: 233-58.

Watson, R. T., Rodhe, H., Oeschger, H. and Siegenthaler, U. 1990. Green houses gases and aerosols. In: *Climate change,* Eds., Houghton, J. T., Jenkins, G. J. and Ephraums, J. J., *The IPCC Scientific Assessment, Camb. Univ. Press*, Pp: 1-40.

Weigel, H. J., Manderacheid, R., Jager, H. J. and Mejer, G. H. 1994. Effects of season-long CO2 enrichment on cereals. I. Growth performance and yield. *Agric. Ecosys. Environ.,* **48**: 231-240.

Welbank, P.J., French, S. A. W. and Wittis, K. J. 1968. Dependence of yield of wheat varieties on their leaf area duration. *Ann. Bot.,* **30**: 219-299.

Weng, J., Takeda, T., Agata, W. and Hakoyama, S.1982.Studies on dry matter and grain production of rice plants: I. Influence of the reserved carbohydrate until heading stage and the assimilation products during the ripening period on grain production. *Jpn. J. Crop Sci.*, **51**:500-509.

Wheeler, A.W. 1972. Changes in growth substance content during growth of wheat grains. *Ann. Appl. Biol.,* **72**: 327-34.

Wheeler, T., Hong, T., Ellis, R., Batts, G., Morison, J. and Hadley, P. 1996. The duration and rate of grain growth, and harvest index, of wheat (*Triticum aestivum* L.) in response to temperature and CO2. *J. Exp. Bot.*, **47**:623-630.

Whingwiri, E. E., Kuo. J. and Stern, W. R. 1981. The vascular system of the rachis in a wheat ear. *Ann. Bot., (London)* **48**:189-201.

Wicke, H. J., Stock, H.G. and Singh, D.P. 1972. Studies on the effect of sprinkler irrigation on the yield formation in cereal crops. *Arch. Acker, U.P. Flanzeanenbau U. Bodekund,* **16**: 847-55.

Wildman, S. G. and Bonner, J. 1947. The protein of green leaves. I. Isolation, enzymatic properties and auxin content of spinach cytoplasmic protein. *Arch. Biochem.,* **14**: 381-413.

Willey, R.W. and Holliday, R. 1971. Plant population: Shading and thinning studies in wheat. *J. Agric. Sci.*, **77**:453-461.

Williams, L. E., Nelson, S. J. and Hall, J. L. 1992. Characterization of solute transport in plasma membrane vesicle isolated from cotyledons of *Ricinus communis* L. *Planta*, **182**: 540-45.

Williams, R. F. 1966. Development of the inflorescence in Gramineae. In: *The growth of cereals and grasses*, Eds. Milthorpe, F.L. and Ivins, J. D. *Butterworths, London*. Pp: 74-87.

Wittenbach, V. A. 1979. Ribulose 1, 5-biphosphate carboxylase and proteolytic activity in wheat leaves from anthesis through senescence. *Plant Physiol.*, **64**: 884-887.

Wojcieska, U., Wolska, E. and Ruszkowska, M. 1984. Nitrate reductase activity as an indicator for assessing the nitrogen requirement of cereal plants. *Proc. 6th International Colloquium for the optimization of Plant nutrition*, AIONP, Montpellier, France, **4**:1429-1433.

Woodward, F. I. 2002. Potential impacts of global elevated CO2 concentrations on plants. *Current Opinion in Plant Biology*, **5**:207-211.

Yamuda, S. N., Uamane, H. and Takahashi, M. 1963. Effect of auxins on growth of rice plants. *Chem. Regul. Plants,* **1**: 129-138.

Yamuda, S. N., Yamane, H. and Takahashi, M. 1963. Effect of auxins on growth of rice plans. *Chem. Regul. Plants*, **1**: 129-138.

Yin, Y., Wang, Z., He, M., Fu, J. and Lu, S. 1998. Post anthesis allocation of source/sink change. *Biologica Plantarum (Prague)*, **41**: 203-209.

Yoshida, S., and Ahn, S. B. 1968. The accumulation process of carbohydrate in rice varieties to their response to nitrogen in the tropics, *Soil Sci. Plant Nutr.,* **14**: 153-161.

Yoshida, S.1972. Physiological aspects of grain yield. *Ann. Rew. Pl. Physiol.*, **23**:437-441.

Yoshida, S.1981. Fundamentals of rice crop science. *IRRI. los Banos. Philippines.*

Zadoks, J. C., Chang, T. T. and Konzak, C. F. 1975. A decimal code for the growth stages of cereals. *Annual Wheat Newsletter*, **21**: 9-16.

Zamaski, E. and Grunberger, Y. 1995.Short-and long–eard high-yielding hexapolid wheat cultivars: Which has unexpressed potential for higher yield? *Ann. Bot.*, **75**:501-506.

Zee, S. Y. and O'Brien, T. P. 1970. A special type of tracheary element associated with 'xylem discontinuity' in the floral axis of wheat. *Aust. J. Biol. Sci.,* **23**: 783-91.

Zeevart, J. A. D. 1977. Sties of abscisic acid synthesis and metabolism in *Ricinus communis* L. *Plant Physiol.*, **59**: 788-791.

Zhao, F. J., Withers, P. J. A., Evans, E. J. Monaghan, J., Salmon, S. E., Shrewy, P. R. and McGrath, S. P. 1997. Sulphur nutrition: An important factor for the equality of wheat. *Soil Science and Plant Nutrition*, **43**: 1137-1142.

Ziska, L. H. and Teramura, A. H. 1992. Intra specific variation in the response of rice (*Oryza sativa*) to increased $CO_2$. I. Photosynthetic biomass and reproductive characteristics. *Physiol. Plant.,* **84**: 269-274.

Ziska, L. H., Weerakoon, W., Namuco, O. S. and Pamplona, R. 1996. The influence of nitrogen on the elevated $CO_2$ response in field grown rice. *Aust. J. Plant Physiol.,* **23**: 45-52..

## *SECTION — IX*

# *POST-HARVEST PHYSIOLOGY*

*Advances in Plant Physiology*, Vol. 9
Ed. A. Hemantaranjan
Scientific Publishers (India), Jodhpur, 2006 pp. 451-472
E-mail: **info@scientificpub.com** www.scientificpub.com

# 19

# RECALCITRANT SEED STORAGE BEHAVIOUR

S.R. Ambika

Department of Botany, Bangalore University, Bangalore-560056, INDIA

## INTRODUCTION

Forest wealth plays an important role in a country's economic and ecological stability. Forests of tropical countries are declining at an alarming speed due to the interference of many factors like urbanization, industrialization etc. Tropical forest, a storehouse of Biodiversity accounts for 25% of the total forest area of the world. The major causes of loss of Biodiversity are destruction of the habitat i.e. deforestation and genetic erosion or the reduction of diversity within a species. The obvious answers to these problems are to control depletion of forests by sustainable forest management, have more forest reserves and establish genebanks, botanical gardens and arboretum.

Seed is the major source of regeneration and seed storage is an integral part of today's forest regeneration and plantation establishment programs because of the irregular flowering and seed production by trees and the need for continuous supply of high quality seeds for annual nursery production. Successful storage of tree seeds not only requires the knowledge of where, when, and how to collect, handle, process, test and store seeds/fruits but also the knowledge of seed longevity and the optimal storage requirements of various species to preserve their maximum genetic and physiological quality. The life span of seeds varies from a few days to few hundred years.

The developmental stage of seeds at harvest plays an important role in post harvest storage longevity. This longevity depends on plant species and external factors such as moisture content, temperature and composition of the gaseous atmosphere during storage. Seeds eventually deteriorate even under optimal storage conditions. Three main categories of seed storage behaviour are currently recognised (Roberts, 1973; Ellis *et al.,* 1990).

**Orthodox** : Seeds that survive long term dry storage

**Intermediate** : Seeds that can withstand dehydration to a certain extent but have reduced longevity (Ellis *et al.*, 1990).

**Recalcitrant** : Seeds that cannot withstand dehydration.

Most orthodox seeds (Legumes and Cereals) are shed in a relatively dry condition and can be further dried to low moisture contents (< 5%) without losing viability. They can be stored successfully for many years at ambient temperatures. Examples are *zea mays, Hibiscus esculentus, Vigna sesquipedalis, Melia azedarach* etc.

Intermediate seeds survive drying to moderately low moisture contents (8-10% on a fresh weight basis) but are often injured by low temperatures (Ellis *et al.*, 1990; Ellis, 1991). Due to their sensitivity to desiccation and/or low temperatures, intermediate seeds are reputed to have relatively short life span. The essential feature of intermediate seed storage behaviour is that the negative relation between seed longevity in air-dry storage and moisture content is reversed at values below those in equilibrium (at $20^0$C) with about 40-50% RH. This type of seed storage behaviour is found in neem (Gamene *et al.*, 1996 and Sacande *et al.*, 1997). In this species, storage survival also depended on the maturity of the seeds. Yellow fruits (about 13 weeks old) provided better quality seeds than did mature-green or brown fruits. Therefore, one reason for the controversy in the literature concerning the storage behaviour of neem seeds may stem from the fact that seeds of different developmental stages had been harvested.

The intermediate seeds of tropical origin such as arabica coffee and papaya can be stored at moisture contents in equilibrium with 50% RH, 9-10% moisture content at $10^0$ to $20^0$C for upto 5 and 6 years, respectively without loss in viability. The viability of intermediate seeds of temperate origin is also maintained well at moisture contents in equilibrium with about 50% RH but at cooler temperatures of 5-$20^0$C. Wild rice seeds can be maintained in hermetic storage at $-2^0$ or $3^0$C with 9-11.5% moisture content for 9-12 months without loss in viability. Examples of Species producing Intermediate type of seeds are: *Azadirachta indica, Swietenia macrophylla* and *Khaya senegalensis.*

Recalcitrant seeds are characterized by the absence of maturation drying and are shed at moisture contents > 50% (on fresh weight basis). They are readily damaged upon drying and are sensitive to low temperatures (Roberts, 1973; Roberts *et al.*, 1984). Recalcitrant seeds are particularly frequent in the moist humid tropics. Number of economically important tropical and subtropical crops, eg, tea (Berjak *et al.*, 2002), mango, litchi and commercial rubber (Chin and Roberts, 1980) as well as forest and horticultural species are characterized by producing recalcitrant seeds. These are hydrated and metabolically active continuously when shed. As such most are akin to seedlings, and bear little resemblance to the quiescent, desiccated orthodox seeds (Berjak *et al.*, 1990, 2002). Recalcitrant seeds are highly sensitive to desiccation and some are also chilling sensitive, necessitating storage under conditions of high relative humidity and ambient temperatures. This approach, termed wet or hydrated storage is, however, conducive to fungal proliferation that severely curtails the post-harvest lifespan of seeds (Calistru *et al.*, 2000). Examples are: *Avicennia marina, Quercus robus, Aesculus hippocastanum, Araucaria angustifolia, Shorea leprosula, Shorea robusta, S. talura, Artocarpus heterophyllus, Acer pseudoplatanus, Sandoricum koetjape, Carapa guianensis, C. procera, Mangifera indica, Landolphia kirkii, Catanospermum australe, Coffea liberica, Hevea brasiliensis, etc.*

**1.1 To determine Seed storage behaviour**

2. MORPHOLOGY OF RECALCITRANT SEEDS

Among tropical tree species 45 dicotyledonous families are characterized to have recalcitrant seeds. There is wide range of morphological and structural variations among recalcitrant seeds. Most of them are spherical or oval in shape and some are endoespermic. The seeds or the nuts are often surrounded by thick endocarps in which case the structures are botanically true fruits. Many of these fruits are covered with a fleshy or juicy arilloid structure. The fleshy aril of rambutan and Jackfruit are covered by thin papery membrane. In case of Jackfruit this membrane becomes impermeable to water on drying. Recalcitrant seeds show certain peculiarities. For eg., tiny embryos of rambutan seeds are attached at different positions between the two unevenly sized cotyledons (Chin, 1975). The seeds of a number of species like Mango and Mangosteen are polyembryonic.

**2.1 Development in Orthodox and Recalcitrant Seeds**

During the maturation drying processes in orthodox seeds (Pammenter and Berjak, 1999):-

(1) Insoluble reserves accumulate, the volume of water-filled vacuoles reduce
(2) Along with these modifications the de-differentiation of highly
(3) Structured organelles, particularly mitochondria take place
(4) A general switching-off' of metabolism occurs.
(5) The protective mechanisms - vitrification and the deployment of Late Embryogenic Abundant (LEA) Proteins become operative.
(6) Throughout the drying process, anti-oxidant mechanisms are operative.

Finally they acquire desiccation tolerance and undergo maturation drying. But in recalcitrant seeds during development there is some decline in water content and some increase in desiccation tolerance. It has been suggested that recalcitrant seeds show an intermediate developmental pattern in which development is truncated and seeds are shed before fully acquiring desiccation tolerance (Finch-Savage, 1992). Hence, there is an increased metabolism during storage in the seeds and the germination is initiated.

3. PHYSIOLOGY OF RECALCITRANT SEED STORAGE BEHAVIOUR

**3.1. Importance and Role of Seed Water**

Water in seeds is found as free and bound water. The free water is necessary for movement of molecules from one center of metabolism to another. Free water is easily removed from seeds on drying while bound water, which is strongly held, is not. The bound or subcellular water is closely associated with macromolecular surfaces and has a structure imposed upon it. The structured water may possibly ensure the stability of macromolecules and subcellular surfaces and may stabilize membranes in desiccated orthodox seeds (Berjak *et al.*, 1984). Structured water may be involved in ensuring the precise functioning of multienzyme systems (Clegg, 1979). Loss of structured water results in disruption of metabolism; in orthodox seeds this presumably does not occur because of their tolerance to desiccation. In recalcitrant seeds this appear to be the case (*Berjak et al.*, 1984).

Farrant *et al.*, 1988 hypothesized that the recalcitrant seeds of Avicennia *marina* behaved like imbibed orthodox seeds when they are first shed, and could withstand the loss of atleast 18% of their initial content of water and still remain viable. The seeds normally start to germinate on shedding. They become more sensitive to desiccation with the onset of cell division, vacuolation and the early stages of germination (Bewley, 1979). Dehydration of seeds stored for more than 4 days caused a marked decline in germination.

Recalcitrant seeds cannot be dried without damage and so they cannot confirm to the viability equation which describes relation between longevity and air-dry seed storage environments (Roberts, 1973). When fresh recalcitrant seeds begin to dry, viability is first slightly reduced as moisture is lost, but then begins to be reduced considerably at a certain moisture content termed "the critical moisture content". If drying continues further, viability is eventually reduced to zero. Hence, the relationship between the ability to germinate when tested following desiccation and moisture content is typically **S-shaped.**

The water status of recalcitrant seeds has been extensively studied in *Avicennia marina* (Berjak *et al.*, 1990), *Quercus robur* and many other species. Critical moisture content levels vary greatly among species, and even among cultivars and seed lots. They may also vary with the method of drying. The values of the 'lowest safe moisture content' vary between extremes of 23% for *Cocoa sp.* and 61.5% for *Avicennia marina.* Despite great variation in the 'lowest safe moisture content' value among species, these moisture content levels are equivalent to relative humidities of 96-98% (or a seed water potentials of about –1.5 to 5 MPa) although Vertucci and Farrant (1995) have suggested a drier value of –11MPa.

It is also possible that a reduction in moisture content causes loss of membrane integrity and nuclear disintegration as has been shown with rubber seeds on drying (Chin *et al.*, 1981). The seeds of many tropical plants contain high concentration of phenolic compounds and phenolic oxidases. These compounds are normally compartmentalized within the cells; on desiccation the cell membranes are damaged and the phenolic compounds are released. They are then oxidized and protein/phenol complexes are formed with a consequent loss of enzyme activity (Loomis and Battaile, 1966).

### *3.1.2. Properties of water in seeds*

The recalcitrant seeds are characteristic in showing variability among species and within species. There is a wide spectrum of behaviours within the recalcitrant group, from minimally recalcitrant seeds with a relatively long life span and quite tolerant of desiccation to maximally recalcitrant, with short life spans and very sensitive to desiccation (Farrant *et al.*, 1988). They vary in the water content at shedding, dehydration to tolerance, drying rate, storage life span in the hydrated state (from a week or two, to two or three years) and their response to low temperatures. Hence, we should consider the properties of water in seed tissue to know the response of recalcitrant seeds to loss of water (Pammenter and Berjak, 2000).

Vertucci (1990) has identified 5 types of water or levels of hydration in seed tissues. At different hydration levels, the tissue water has different physical properties:

(1) At high water content and high water potential the water has the properties of water in dilute solution

(2) As the water content decreases the water takes on the properties of water in concentrated solution where the interaction between water and solutes become stronger, and the system deviates from ' ideal' behaviour

(3) On removal of more water, the solution becomes so concentrated that it becomes viscous and has the properties of a glass.

(4) Finally, at very low water contents, those characteristics of orthodox seeds, all the remaining water is tightly associated with macromolecular surfaces; its mobility is reduced and it constitutes the so-called "bound water".

At different hydration levels, because the thermodynamic properties of water change, different metabolic processes take place (Pammenter and Berjak, 1999., Vertucci,1989; Vertucci and Farrant, 1995).

(1) At high water contents, full normal metabolism occurs and seeds germinate.

(2) At low water levels though protein and nucleic acid synthesis takes place along with respiration, there is inadequate water for cell growth and germination.

(3) At very low water contents protein and nucleic acid synthesis is also not possible but some respiration can be detected.

(4) At even lower water contents only low level catabolic events occur slowly.

## 4. REASONS FOR DESICCATION SENSITIVITY

Desiccation tolerance in recalcitrant seeds increases during seed development on the mother plant; however, unlike orthodox seeds, maturation drying to low moisture content does not occur, and fresh recalcitrant seeds have high levels of moisture content at maturity/shedding, between for eg, 36% for rubber (Chin *et al.*, 1981) and 90% for Choyote (Ellis, 1991).

Considerable differences in moisture content can be detected among tissues within a particular recalcitrant seed. With the exception of Durian (*Durio zibethinus*) and Jackfruit (*Artocarpus heterophyllus*) the storage tissues of recalcitrant seeds are always at lower moisture content than the embryo axis. Desiccation of excised embryos or embryonic axes has considerable practical potential for the *invivo* conservation of recalcitrant embryos, since embryos are able to survive desiccation to lower water contents than whole seeds (Chin, 1988). For example *Hevea brasiliensis* tolerated desiccation to 20% moisture content but no seeds survived further desiccation to 15% moisture content (Chin *et al.*, 1981). However, 50-80% of their excised embryos (with 55% moisture content) survived desiccation to 14% moisture content when cultured *invitro* (Normah *et al.*, 1986). Reports of survival of excised embryos or embryonic axes to a lower moisture content than their intact seeds are numerous (Finch-Savage, 1992; Chandel *et al.*, 1995).

### 4.1 Sub Ambient Temperature Susceptibility

Sub ambient temperature susceptibility of recalcitrant seeds may be due to the declining fluidity of membrane lipids, which occur during chilling, and this can result in changes in membrane thickness and permeability effecting the membrane bound enzymes. In cocoa the fall in seed viability with declining temperature was abrupt. The possible reasons were:

(1) The presence of some temperature-dependent, rate limiting reaction, the cessation of which caused metabolic disruption.

(2) The absence of protective substances which are not susceptible to chilling

(3) Liberation of toxic material owing to cold induced changes in membrane permeability

(4) In cocoa, Hor (1984) noticed a three-fold increase in leachate conductivity and major ultrastructural changes in the cell membrane systems.

The sensitivity of recalcitrant seeds is also due to:

(1) An insufficient accumulation of dehydrins /LEA Proteins

(2) Lack of protective sugars

(3) An inability to repair desiccation-induced damage upon subsequent rehydration and

(4) An inappropriate proportion or distribution of freezable and nonfreezable (bound) water within the seed (Berjak *et al.*, 1992; Bewley and Oliver, 1992; Vertucci and Ferrant, 1995).

(5) Rapid decrease in the activities of free radical-scavenging enzymes (Ascorbate peroxidase, Peroxidase and Superoxidase dismutase during desiccation (Li and Sun, 1999).

(6) Increases in lipid peroxidation and cellular leakage (Li and Sun, 1999).

(7) Decrease of enzymic protection against desiccation-induced oxidative stresses (Li and Sun, 1999; Greggains *et al.*, 2001).

### 4.2 Cell cycle stage and Seed desiccation Tolerance

Deltour (1985) suggested that the factor related to desiccation tolerance and storage longevity is the stage of cell cycle activity at seed desiccation. Arrest of cell cycle activity at the stage where DNA content per nucleus is the lowest has been thought to render embryos more resistant to water stress conditions. Contradicting this flow cytometry studies conducted on tree species contrasting in seed storage behaviour has indicated that desiccation tolerance is not correlated with cell cycle arrest at any particular DNA level (Connor *et al.*, 1998)

According to Connor *et al.*, 1998, the storage problems of recalcitrant seeds are associated with intact seed moisture content and with lipid composition, metabolism and their distribution in the cells.

### 4.3 Late Embryogenesis Abundant Proteins (LEA Proteins)

Although all recalcitrant seeds are considered to be desiccation sensitive, the degree of water loss tolerated varies with the species. Seeds, spores and pollen are unique in the degree of water loss tolerated; as much as 90 – 95% of the original water is removed during their development leading to a state of metabolic quiescence. Even the whole plant encounters water stress at some time during their life. In order to cope up with the stress the cells undergo physiological changes. One such metabolite is the LEA Proteins of the seeds and stressed plant vegetative tissues. LEA proteins are widely distributed among monocot and dicot species and many different forms have been isolated, cloned and sequenced. The homology among different LEA proteins, the presence of highly conserved domains, their ubiquity and their developmental specificity of their expression strongly imply a fundamental role in desiccation tolerance. They localize in the nucleus and are associated with the cytoskeleton. They have a mode of action similar to the cold shock chaperones, via a role in DNA-binding. Structure analysis indicates that these proteins are rich in hydrophilic aminoacids, possess domains with amphiphilic • -helix structure and are boiling stable. In general they are non-globular and low complexity proteins. Consequently, these proteins may play a structural role as desiccation protectants (Lane, 1991). Considerable evidence suggests that LEA proteins are involved in desiccation resistance, a variety of mechanisms for achieving this end have been proposed including protecting cellular structures from the effects of water loss by retention of water, sequestration of ions, and direct protection of other proteins or membranes (Lane, 1991).

Ingram and Bartels, 1996 have published a summary of the large number of genes with a potential role in drought tolerance. Many of these proteins, or closely related proteins, accumulate in plants in response to any environmental stimulus that has a dehydrative component or is temporarily associated with dehydration, such as drought, low temperature, salinity and seed maturation. Among the induced proteins, dehydrins (LEA, D-11 family) have been the most commonly observed. Dehydrins have detergent and chaperone-like properties that stabilize membrane and protein components of the cell (Close, 1996).

Investigation of dehydrin-gene expression in a wide range of recalcitrant seed types indicated that dehydrin-like proteins are consistently found in recalcitrant seeds of temperate species, but are often absent from recalcitrant seeds of tropical wetland species (Farrant *et al.*, 1996). Finch-savage *et al.*, 1994 detected dehydrins in mature seeds of 5 desiccation- sensitive temperate tree species. *Zizania palustris*, a cold temperate aquatic grass with intermediate seed storage character (Bradford and Chander, 1992; Vertucci *et al.*, 1994), could accumulate dehydrins and ABA during limited dehydration. Hence, the intolerance of *Z. palustris* seeds to dehydration at low temperature does not seem to be due to the absence of dehydrins or an inability to accumulate ABA (Bradford and Chander, 1992). This study indicated that the presence of dehydrins alone is not sufficient to prevent desiccation injury (Blackman *et al.*, 1991; Bradford and Chander, 1992; Finch-savage *et al.*, 1994 and Vertucci and Farrant, 1995). However, these findings are not sufficient to rule out the converse, that the absence of dehydrin-like proteins in certain recalcitrant seeds contribute to their desiccation sensitivity (Farrant *et al.*, 1996).

Dehydrins are present in many recalcitrant seeds of temperate species, though some do not contain in amounts detectable by western blotting (Farrant *et al.*, 1996; Han *et al.*, 1997). In seeds of *Castanospermum australe*, a species of tropical origin, but grown in temperate climate, dehydrin – related proteins are present in the cotyledons and axes. Dehydrins are also produced in response to drying of the axes of *Barringtonia racemosa*, but not of *Avicennia marina* (Farrant *et al.*, 1996).

Since desiccation tolerance is a quantitative feature (Vertucci and Farrant, 1995), the amount of dehydrins/LEAs or the rate at which the proteins accumulate, may determine the level of tolerance. Accumulation of other protectants is likely to be required in addition to Lea proteins (Blackman *et al.*, 1992). Since combinatorial interactions may be necessary to stabilize macromolecules under conditions of water deficit (Close, 1996).

### 4.4 Dehydrin Regulation

Multiple genes encode Dehydrins and their expression is likely to be regulated by a complex mechanism-

(1) In some cases resulting in developmental regulation, tissue/organ specificity and/or stress-specific expression

(2) ABA and water-deficit related stresses might affect plant molecular processes through different pathways and induce different dehydrin genes.

Based on studies conducted in *Arabidopsis thaliana*, the existence of at least two independent signal transduction pathways (ABA responsive and ABA independent), between the environmental stress and the expression of the two genes were proposed (Yamaguchi-Shinozaki and Shinozaki, 1994).

#### *4.4.1 Lea gene expression by ABA*

Recalcitrant seeds have a characteristic feature of absence of maturation drying and an intolerance to desiccation (Ellis, 1991). Apart from the regulation by moisture content, the hormones Abscisic acid and Gibberellic acid also regulate seed germination. During seed maturation, ABA induces dormancy in orthodox seeds and represses germination in recalcitrant seeds (Goldbach, 1979). ABA is a potent inhibitor of cell division (Evans, 1984) and may therefore play a role in the arrest of cell cycle activity in seeds (Durand *et al.*, 1989).

ABA regulates and maintains embryo dormancy. ABA mutants of Arabidopsis germinate precociously and are desiccation intolerant (Karssen *et al.*, 1983); they are also defective in the synthesis of storage proteins and LEA proteins; treatment of ABA avoids precocious germination. Whether dormancy is controlled by the embryo itself or by the surrounding tissues is not clear, though there are evidences for the latter possibility (Krishnamurthy, 1999).

In seeds, ABA rises to a maximum level at mid-development, but may peak more than once during development and maturation, depending on the species. In several species, changes in ABA levels have been associated with seed desiccation (Black, 1991). Although ABA is essential to the expression of desiccation tolerance in somatic embryos, its timing is critical. Somatic embryos of *Medicago sativa* respond to ABA only after the late torpedo stage of development and prior to precocious germination. Earlier or later applications do not induce the expression of desiccation tolerance. The role of ABA seems to be common in many plants including *Medicago, Brassica* (Senaratna *et al.*, 1991*), Pelargonium* (Marsolais *et al.*, 1991), and *Picea* (Attree *et al.*, 1991). One can speculate that ABA may induce the expression of a specific set of genes whose products facilitate survival after subsequent desiccation (Black, 1991).

The expression of Lea/Dehydrin genes appears to be intimately connected with ABA. The extent to which expression is regulated (directly or indirectly) by ABA alone or whether other factors are important requires further investigation.

ABA is not the only regulator involved in the control of all changes in gene expression that occur in response to water stress. In soybean, tomato and pea seedlings, many water-deficit-induced changes in gene expression could not be mimicked by ABA application (Black, 1991).

### 4.5 Oxidative Stress and Antioxidants

Several lines of evidence have suggested the production of highly reactive activated oxygen in desiccation-stressed tissues. Direct measurement of activated oxygen levels using electron paramagnetic resonance techniques have been inconclusive because of the quenching effect of water, but a general increase in their levels during seed aging has been correlated with low viability (Buchvarov and Gantcheff, 1984; Priesley *et al.*, 1985). Stable free radicals accumulate in the embryonic axes from acorns of *Quercus robra* coincident with the loss of moisture and viability in this recalcitrant seed and similar radicals probably quinones have been detected in desiccated mosses and elongating maize radicle (Hendry, 1993). Concentrations of Hydrogen peroxide and Superoxide increase during soybean seed germination in association with increased respiration and coincident with radicle elongation and loss of desiccation tolerance (Simontacchi and Puntarulo, 1992). The development of respiration and the initiation of electron transport reactions may contribute to the loss of desiccation tolerance by providing a continuous supply of activated oxygen (Leprince *et al.*, 1993). Seeds contain large quantities of tocopherols, ascorbate and glutathione that may function as chain breaking antioxidants. The role of enzymic scavengers such as Super oxide dismutase, Catalase, and Ascorbate peroxidase, as protectants in a dry seed is probably negligible because of limitations in diffusion at low water contents. Protection of the dry seed against activated oxygen probably is restricted to the lipid-soluble, chemical antioxidants that are integral components of the membrane lipid matrix (Senaratna *et al.*, 1985a and b). The identity of these antioxidants is not known although tocopherol and phenolics may be candidates (Lai and Mckersie, 1993; Hendry, 1993; Leprince *et al.*, 1993).

Connor and Sowa (2003) concluded from their studies in *Quercus alba* acorns that membrane lipid structure initially exhibited reversible shifts between gel and liquid crystalline phases in response to drying and rehydration; however, reversibility declined as viability was lost. The most sensitive indicator of desiccation damage was the irreversible change in protein secondary structure in embryonic axes and cotyledon tissue and increase in sucrose concentration in the embryonic axes. Hence these authors felt that sucrose does not prevent loss of viability but act as a glycoprotectant against cell collapse and cell wall membrane damage as water stress increases.

### 4.6 Germinating Orthodox and Recalcitrant Seeds similarity

The changes that occur on dehydration of recalcitrant seeds of the mangrove, *Avicennia marina*, are very similar to changes brought about by desiccation of orthodox seeds during the intolerant stage following germination (Farrant *et al.*, 1986). Generally recalcitrant seeds initiate germination-related metabolism after shedding (Vertucci and Farrant, 1995) and in *Avicennia marina*, 10 to 15 days before shedding (Farrant *et al.*, 1993). As germination events progress, the seeds become increasingly sensitive to drying and attempting to store these seeds is akin to air-dry storage of germinated, orthodox seeds (Farrant *et al.*, 1986; 1988).

There is no clear cut event delineating the end of seed development and the start of germination, during both phases, recalcitrant seeds appear to remain metabolically active, although the axes may undergo a very brief period of relative quiescence. If the newly shed recalcitrant seed responds in a manner similar to germinated orthodox seeds, it is important to distinguish between dehydrin production in storage tissues (cotyledons or endosperm) and that occurring in the axis. For eg, the axis of orthodox seeds such as soybean rapidly lose their tolerance to desiccation (brought about by air drying to 10% water content) during the course of germination, while the cotyledons remain tolerant for a considerably longer period (Senaratna and McKersie, 1983a and b).

Table 1. Components of Desiccation tolerance in seeds and their Protective action (Kermode (1995)

| Site/ Process affected | Protective component | Protective action | Possible mode of action |
|---|---|---|---|
| Membranes | Carbohydrates, sucrose, Raffinose, Stachyose | Prevent changes in selective permeability due to lateral phase separation of phospholipids in the bilayer and phase transition from liquid crystalline to gel. | Hydroxyl groups of sucrose replace water on hydrophilic (polar) end groups of membrane. Phospholipids and Oligosaccharides (raffinose /stachyose) inhibit sucrose crystallization during drying, preventing loss of its protective potential. Its scavenging activity increases resistance to free-radical mediated desiccation injury. |
| Structure /Metabolism | Lipid soluble antioxidants (eg. Tocopherol) | Prevent deesterification of membrane phospholipid and free fatty acid accumulation. Prevent whole scale mechanical disruption of cellular components. Prevent loss of tightly bound (vital) water necessary for structural & functional integrity of Biomolecules. | Critical level of reserves in vacuoles/storage bodies confers mechanical strength to whole cell. Water binding capacity of cells enhanced with increased number of sorption sites. |
| | Reserves, Carbohydrates, Lipids, Proteins | | |
| | Hydrophyllic denaturation resistant proteins (Eg. LEA's | As above | Native conformation of protective molecules maintained throughout drying. Bind ions and thereby |

| | | |
|---|---|---|
| and other desiccation-inducible polypeptides) | | counteract damaging effects by increasing ionic strength of cytosol during drying. |
| Repair proteins, Proteases, ubiquitin and extension protein, HSP/ Molecular chaperones, some LEAs. | Rapid reestablishment of structural and metabolic integrity following imbibition | Efficient repair of membranes and other components restores normal functioning, aid proteins in recovering their native conformation and degradation of damaged or denatured proteins. |

### 4.6.1 Influence of fast drying

There are several reports that fast drying allows intact recalcitrant seeds to survive desiccation to lower moisture contents than slow drying (Farrant *et al.*, 1985; Pritchard, 1991). Finch-Savage, 1992 showed that drying rate does not affect the desiccation sensitivity of whole seeds of *Quercus robur*. However, fast drying allowed excised enbryoes of *Araucaria hausteinii*, *Hevea brasiliensis*, *Landolphia kirkii*, *Quercus robur* and *Quercus rubra* to survive desiccation to lower moisture contents than similar embryos dried more slowly within intact seeds (Normah *et al.*, 1986; Pritchard and Prendergast, 1986; Pritchard, 1991; Pammenter *et al.*, 1991; Finch-Savage, 1992). Li and Sun (1999) reported that mature and immature axes of *Theobroma cacao* (cocoa) seeds tolerated desiccation under a rapid drying regime to critical water contents of 1.0 and 1.7 $gg^{-1}dw$, respectively. These critical water contents corresponded to water contents below which activities of free radical-scavenging enzymes (Ascorbate peroxidase, Peroxidase and Superoxide dismutase) decreased rapidly during desiccation. The decline in axis viability below the critical water content was correlated with sharp increases in lipid peroxidation and cellular leakage. Cotyledon tissues were more desiccation-tolerant than axes, with low critical water content of 0.24 $gg^{-1}$ dw. Desiccation sensitivity in cotyledon tissues was also correlated with the decrease in Superoxide dismutase activity and increased lipid peroxidation products. Hence, desiccation sensitivity of recalcitrant cocoa axes seemed to be due to decrease of enzymic protection against desiccation-induced oxidative stresses and not due to sugar-related protective mechanisms as the embryonic axes contained large amounts of sucrose, raffinose, stachyose and traces of reducing monosaccharides. In *Avicennia marina* Greggains *et al.*(2001) reported that the seeds were shed with 65% moisture content (fresh mass basis) and the viability declined as the propagules were dried below 60% moisture content. At 47% moisture content the seeds died. Lipid peroxidation increased with advance of viability loss suggesting that the propagules were experiencing oxidative stress.

### 4.6.2 Longevity of recalcitrant seeds in moist storage

There is no satisfactory method for maintaining the viability of intact recalcitrant seeds over long term. This is because they cannot be dried; neither can they be stored at subzero temperatures because they would then be killed by freezing injury resulting from ice formation. In addition, chilling injury at temperatures of 10-15$^{0}$C and below also damages some tropical recalcitrant seeds. The longevity of recalcitrant seeds is generally short, particularly for species adapted to tropical environments, typically from a few weeks to a few months (King and Roberts, 1979; 1980). However, the longevity of seeds of species adapted to temperate environments can be maintained much longer periods – more than 3 years for oak (*Quercus* spp.) seeds stored moist at –3$^{0}$C.

Table 2. List of plants showing recalcitrant seed storage behaviour

| Name of the tree species | Family | Viability duration at room tempe-rature | Time taken for Germination at room temperature |
|---|---|---|---|
| *Artocarpus heterophyllus* | Moraceae | 10 days | 21 days |
| *Aegle marmelos* | Rutaceae | 21 days | 15-21 days |
| *Syzigium cumini* | Myrtaceae | 10 days | 10 days |
| *Pongamia pinnata* | Papilionaceae | 180 days | 21 days |
| *Michelia champaka* | Magnoliaceae | 30 days | 15 days |
| *Terminalia chebula* | Combretaceae | 240 days | 60 days |
| *Mangifera indica* | Anacadiaceae | _ | _ |
| *Terminalia belarica* | Combretaceae | 150 days | 60 days |
| *Avicennia marina* | Verbenaceae | 4-6 days | 10 days |
| *Hevea brasilensis* | Euphorbiaceae | 120 days | 180 days |
| *Shorea trapezifolia* | Dipterocarpaceae | 7-14 days | 4-10 days |
| *Shorea robusta* | Dipterocarpaceae | 4-8 days | 10 days |
| *Quercus alba* | Fagaceae | _ | 28 days |
| *Quercus muehlenbergii* | Fagaceae | _ | 28 days |
| *Quercus virginiana* | Fagaceae | _ | 28 days |
| *Castanea sativa* | Fagaceae | _ | 21 days |
| *Eugenia brasiliensis* | Myrtaceae | _ | 14 days |
| *Theobroma cacao* | Sterculiaceae | _ | 21 days |
| *Syzygium cuminii* | Myrtaceae | 10 days | 2 days |
| *Cocos nucifera* | Palmaceae | _ | 1 year |
| *Litchi chinensis* | Sapindaceae | _ | 16 days |
| *Euphoria longan* | Sapindaceae | _ | 30 days |
| *Nephelium lappaceum* | Sapindaceae | _ | 14 days |
| *Nephelium malaiense* | Sapindaceae | _ | 16 days |
| *Camellia sinensis* | Theaceae | 2 to 4 weeks | 4 to 5 months |
| *Myristica fragrans* | Myristicaceae | Loose viability imme-diately | _ |

In practical terms, species with recalcitrant seeds can therefore be subdivided into those of **tropical origin**, and those adapted to **temperate climates** (temperate latitudes or high altitudes in the tropics), the latter can be stored at cooler temperatures and for longer Period.

In conclusion, the evidence to date tends to support that desiccation-sensitivity of recalcitrant seeds (particulary those of tropical wet land plant species) is due in part to an inability to accumulate sufficient dehydrins and/or due to the absence of specific LEA. However, much work remains to be done in this area and future studies will determine whether there is a direct relationship between loss of viability during storage and a decline in the ability of the axes of recalcitrant species to synthesize dehydrins. Hence, the underlying basis of desiccation tolerance is diverse and elucidating the role of individual components is indeed a challenging task.

## 5. RECALCITRANT SEED STORAGE BEHAVIOUR- IS IT EVOLUTIONARY OR ENVIRONMENTAL ADAPTATION?

According to the review by Von-Tekhman-1 and Van-Wyk-A-E (1994), 45 dicotyledonous families are associated with bitegmic and crassinucelate ovule with nuclear endosperm development. All these are considered to be ancestral character states of the ovule. Most of these species are also known to have greater seed size, woody habit and tropical habitat; regarded as ancestral character states in dicots. In many species with recalcitrant seeds the predominant storage reserve is carbohydrate. Recalcitrance is significantly associated with exalbuminous type of reserve storage. It is proposed that in large recalcitrant seeds the transfer of main storage function from endosperm to embryo was probably an early development. In many species with recalcitrance, the ovules/seeds are characterized by extensive vascularization of the integument/seed coat or by a patchy chalaza. Patchy chalaza is proposed to be significant functional adaptation for more efficient transfer of nutrients to the embryo/seed. Recalcitrance and some of the other characters proposed to be ancestral in dicots are also present in some Gymnosperms. In relatively advanced dicot families with orthodox seed, recalcitrance persisted only in isolated and relict members. Available evidence supports the reviewer's view that the seed recalcitrance can be regarded as a relatively ancestral character state in Dicots.

There is a substantial literature on the basic physiology and response to desiccation of recalcitrant seeds, but little is known about their ecology. Pammenter and Berjak (2000) have given the following information in their review. The response of early land plants to dehydration is difficult to assess, but it is likely that desiccation tolerance in vegetative tissue and land invasion arose concomitantly. Similarly, it is neither possible to assess the desiccation response of early seeds from the fossil record, nor easy to trace the phylogenic relationships among recalcitrant seed producing species. Available evidence suggests that the first seeds were desiccation-sensitive, but tolerance evolved early and probably a number of times, independently.

The desiccation sensitivity and short life span (generally shorter than the interval between flowering) of recalcitrant seeds have implications in terms of regeneration ecology. Recalcitrant seeds are common in the humid tropics. These conditions generally favor germination and seedling establishment, and there has been no pressure during the evolution of desiccation tolerance or the character has been secondarily lost.

Roberts and King (1980) suggested that there is an association between plant ecology and seed storage behaviour. According to this hypothesis, orthodox species originate from environments subjected to occasional or seasonal drought in which desiccation tolerance of the seeds is essential for seed survival and the continued regeneration of the species. On the other hand, recalcitrant species tend to originate from moist ecosystems in which seeds are subjected to high humidity and are not exposed to desiccation during any part of their life cycle.

Roberts *et al.* (1984) observed that species that produce recalcitrant seeds are of 2 main types:

1. Species from aquatic habitats
2. Most of them are large seeded trees

Many important tropical plantation crops (Like- rubber, cocoa, coconut), tropical fruit crops (mango, jackfruit), tropical timber trees belonging to Dipterocarpaceae and Araucariaceae produce recalcitrant seeds.

Those species of the genus Dipterocarpus that are native to dry habitats show some desiccation tolerance. But those species inhabiting moist evergreen areas tend to be intolerant of desiccation (Tompsett, 1992). Similarly Arabica coffee shows intermediate seed storage behaviour and is native to the dry and cool regions of Ethiopia while Liberica Coffee (*C. liberica*) shows recalcitrant seed storage behaviour and is native to hotter and more humid regions of Liberia (Hong and Ellis, 1998). Species within Palmae that show orthodox seed storage behaviour are native to dry habitats whereas those species showing recalcitrant seed storage behaviour are native to relatively moist habitats. Seeds of *Quercus* sp. (oak) show recalcitrant seed storage behaviour, but seeds of *Q. emoryi*, which is native to Savanna are not recalcitrant (Davies and Pritchard, 1998).

From the information on seed storage behaviour that have now been collected from over 7000 species from more than 250 families of flowering plants, it is evident that species which show recalcitrant seed storage behaviour do not occur naturally in arid habitats, desert and savanna. In these environments, the majority of plant species show orthodox and a few show intermediate seed storage behaviour.

Seed ecologists feel that seeds of climax species of these ecosystems are not found in the soil seed bank, as their life spans are low, while seeds of gap specialists are well represented in the soil seed bank.

The seeds of climax species generally show no dormancy, germinate rapidly (unless they have a hard covering), and are absent from the soil seed bank but persist as a seedling bank. They are generally large and not wind dispersed. Although they are common in mesic tropical forests, they do occur in seasonal habitats evolving specialized regeneration strategies. Many temperate recalcitrant seeds have reduced risk of death by desiccation as the low temperature and general dampness will slow water loss. Hence, recalcitrant species of temperate regions are generally more desiccation tolerant and have longer life span than those of tropical origin.

## 5.1 Association between Taxonomic Classification and Seed Storage Behaviour

In general species of chenopodiaceae, Combretaceae, Compositae, Labiatae, Solanaceae and Pinaceae show orthodox seed storage behaviour, while species in Rhizophoraceae (in which vivipary predominates) are recalcitrant. Species within Leguminosae, Gramineae, Cucurbitaceae, cruciferae and Rosaceae show orthodox seed storage behaviour, but with several notable exceptions. On the other hand, it appears that no member of the Dipterocarpaceae shows orthodox seed storage behaviour, most show recalcitrant and a few intermediate seed storage behaviour. Similarly, although all members of the Sapotaceae were thought to be recalcitrant, there are some species showing intermediate type. And all the three categories of seed storage behaviour can be found among the members of Meliaceae.

Seed storage behaviour can also differ among species within a genus. Desiccation tolerance among *Araucaria* sp. was reported to be geographical and taxonomic in origin (Tompsett, 1983). *Araucaria angustifolia* and *A. araucana* from South America and *A. bidwillii* from Australia mostly show recalcitrant seed storage behaviour, whereas species such as *A. heterophylla*, *A. cunninghamii*, *A. columnaris* are tolerant to desiccation.

Dipterocarpaceae, which make up the major component of the south East Asian tropical forests, is an example of a family where recalcitrant seeds is a major constraint in cultivation and also circumvent *ex situ* storage of seeds. Seeds of *Hopea*, *Cotylelobium* and *Vatica* tend to have greater desiccation tolerance than *Shorea* seeds while most *Dipterocarpus* seeds are relatively intolerant.

Associations between desiccation tolerance and seed size, habitat, seed desiccation rate, and longevity have been observed among the species of the Dipterocarpaceae. Thus seed of three *Shorea* species from different habitats have different desiccation tolerance. The low-rainfall species *S. roxburghii* has seed which can be dried safely to 35% whereas the monsoon or rainforest species *S. almon and S. robusta* cannot be dried below 40% moisture content. Smaller seed dry faster and tend to be more desiccation tolerant. Seeds of *S. roxburghii* has the greatest desiccation tolerance and longevity (Tompsett, 1992).

Farnsworth (2000) reviewed the physiology, morphology, and ecology of desiccation-intolerant, non-dormant lineages and felt that differences in the production and function of plant hormones are implicated in the occurrence of recalcitrance and vivipary in plant families. But there are no research publications to support this.

Before undertaking any 'high tech' research on methods to improve storage of recalcitrant and intermediate species, it is suggested to examine the development pattern of seeds and perform preliminary experiments to determine their desiccation sensitivity as well as to define storage conditions. Even if such experiments do not allow defining long-term storage conditions, increased storage periods may make the whole difference for the use of a species if it allows storage of seeds until the next planting season or transport from the seed source to a prepared planting site.

## 6. COPING WITH DESICCATION

Our present knowledge does not allow us to dry plant cells for germplasm storage or other agronomic purposes. We cannot even treat plants with sucrose, antioxidants or any other Protestants to elicit or impose desiccation tolerance. But methods have been developed in plant cell culture to induce somatic embryo formation from vegetative cells and then to develop and mature those embryos into a dry quiescent state and these may be used as an artificial or synthetic seed (Mckersie and Bowley, 1993). Drying the somatic embryo whether encapsulated or not, allows long-term storage of a clonal propagule. The artificial seed then becomes a true analog of conventional seed to be used for germplasm conservation in seed storage banks (Gray and Purohit, 1991).

The artificial seed is developmentally and morphologically as close as possible to the seed of the species from which it has been derived. Zygotic embryo formation begins with the double fertilization of the egg and polar nuclei, whereas somatic embryos are formed by the differentiation of somatic cells (Gray and Purohit, 1991). Despite this difference in initiation, both somatic and zygotic embryos originate from an organized cell division that establishes polarity. Both zygotic and somatic embryos pass through morphological stages of development, including globular, heart, torpedo and cotyledonary stages (Gray and Purohit, 1991). In late maturation, somatic embryos may acquire desiccation tolerance. To induce desiccation tolerance in these somatic embryos numerous methods have been used, including exogenous ABA and treatment with sublethal stress (Senaratna *et al.*, 1989). In microspore-derived embryos of *Brassica napus* tolerance to dehydration was induced through thermal stress and cold treatment (Anandarajan *et al.*, 1991). In Alfalfa, stresses such as nutrient deprivation, cold stress, thermal treatment and water stress induced using PEG-4000, induced tolerance to desiccation (Senaratna *et al.*, 1989; Attree *et al.*, 1991). Addition of sucrose (6%) in the maturation medium induced desiccation tolerance in *Medicago* somatic embryos. Sucrose and sugar alcohol played a role in desiccation tolerance by acting as osmotic agents (Anandarajan and McKersie, 1990). In numerous confider species, ABA promoted development of somatic embryos and prevented precocious germination. In hybrid larch the maintenance of the somatic embryos beyond 3 weeks on the maturation medium (with 60nM ABA) resulted in a significant decrease in both germination and plantlet frequencies. Simultaneously the ABA content of the somatic embryos increased during the maturation period to reach the highest value at week 5. Hence, further advances in the drying and storage of these clonal propagules will lead to their wider utilization mainly in horticulture and forestry.

## 6.1 Development of Desiccation Tolerance in Seeds

Development of orthodox seeds culminates in the loss of water coinciding with acquisition of desiccation tolerance by the embryo. Though the occurrence and extent of this drying varies, more than 70% of cellular water is lost, and as a direct result the seed becomes quiescent and metabolism stops. Under favorable conditions of moisture, temperature and in some cases light, metabolism is reactivated and germination begins. At this stage, the seed can be dried back to its original moisture level without adversely affecting viability (Simon, 1974). However, at a critical stage, which in many species is coincident with emergence of the radicle from the seed coat, the embryo loses its tolerance of desiccation and during the rest of its life cycle the plant is relatively sensitive to water loss.

Dry seeds tolerate a number of diverse environmental conditions, including extremes in temperature and this has two important ramifications:

1. Apparently many plants have the genetic information that enable them to withstand extreme environments. Their sensitivity to these stressful conditions is therefore a consequence of their stage of development and not the absence of required "stress tolerance" genes.
2. Secondly, because seeds survive these extreme stresses, an understanding of the physiological mechanisms of their desiccation tolerance may enlighten us about the mechanism of tolerance to other stress conditions. Of course, we must take into account the obvious distinction that stress tolerance in seeds occurs in a quiescent tissue that has limited metabolism, whereas stress tolerance in vegetative plants is coincident with active metabolism and growth.

### *6.1.1 Membranes as the site of desiccation injury*

Cell membranes have been regarded for a long time as the site of desiccation injury, mainly because the earliest symptom of injury is enhanced leakage of cytoplasmic solutes during rehydration (Simon, 1974). Ultrastructural studies of dry tissue confirm that membrane disorganization is a common phenomenon, but only if the desiccation is lethal; cells that are dried at a tolerant stage remain organized (Crèvecoeur *et al.*, 1976; Dasgupta *et al.*,1982). The disorganization of membranes and their apparently enhanced permeability may be simply a consequence of physical rupture of the membrane due to tearing during cellular collapse, or alternatively, it may involve more subtle changes in the physical organisation of lipid or protein components in an otherwise intact membrane. Experiments on the kinetics of electrolyte leakage from soybean axes indicate that both events occur (Senaratna and McKersie, 1983a and b).

The phospholipid bilayer of biological membranes is in a liquid-crystalline phase at physiological temperatures and hydration. This state enables rotational and lateral movement of phospholipids and integral proteins within the bilayer. Water loss changes the organisation of the membrane lipids.

There is considerable evidence to indicate that they form a gel phase (Hoekstra *et al.*, 1991). In the gel phase, the rotational and lateral mobility of the lipids is severely restricted, the packing of the acyl chains is tighter, the association between protein and lipid is altered, and permeability is changed. This is significant because the formation of gel or hexagonal phases cause discontinuities in the membrane bilayer that allows leakage of cytoplasmic solutes and cause the disruption of membrane enzyme complexes. Therefore it is highly unlikely that any cell would survive this degree of re-organisation of its membranes.

### *6.1.2. Sugars as stabilizing agents*

Sugar acts as a replacement for water and maintains the hydrophobic-hydrophilic orientation of the membrane phospholipids in the absence of water by interacting with membrane lipids forming hydrogen bonds between the hydroxyl groups of the sugar in the head group of the phospholipid.

Sugars not only preserve membrane integrity in dry organisms, but also afford protection to proteins (Darbyshire, 1974). In the freeze-dried protein, trehalose forms hydrogen bonds with specific polar groups on the protein, thereby preventing formation of hydrogen bonds within the protein that would irreversibly change its three dimensional structure. Extreme desiccation may also promote crystallization of proteins and solutes in the cytoplasm that would severely injure the cell (Leopold, 1990). Vitrification or glass formation within the cytoplasm is a potential mechanism to avoid this. Differential scanning calorimetry and electron paramagnetic resonance have detected glass phases in the cytoplasm of maize embryos at water contents less than 0.2 g $H_2O$/g dry weight (Williams and Leopold, 1989; Bruni and Leopold, 1991). A glass is a liquid solution with the viscosity of a solid and its formation from a liquid involves no chemical or physical change in the solution (Burke, 1986; Williams and Leopold, 1989). A glass phase does not have a thermal transition to a solid (ice) phase at low temperatures. Therefore, even at extremely low temperatures a glass phase will not freeze. The glass phase reversibly melts into a liquid phase without cellular injury simply by the addition of water (Bruni and Leopold, 1991). The glass phase has several additional advantages (Leopold, 1990; Koster, 1991). Because of the high viscosity of the glass, diffusion is impeded; chemical reactions slow to negligible rates and degradative processes are prevented. The glass phase fills space preventing cell collapse in the absence of water, and traps solutes that would otherwise become concentrated, avoiding osmotic and pH shifts. Hydrogen bonding between water in the glass phase and hydrophilic binding sites on macromolecules maintain macromolecular structure. However, in a recent review, Leopold concluded that while the glassy state may contribute to a seed's survival of desiccation, it does not appear to account for desiccation tolerance *per se*. Its major function in dry seeds seems to be its contribution to the stability of the seed components during storage (Leopold *et al.*, 1994). That however does not diminish its importance.

Sugars are suspected to be the solutes that promote vitrification at ambient temperatures as water is removed from seeds (Koster, 1991). Sugar solutions form glass phases at concentrations equivalent to those measured in seeds and the vitrification phase diagrams of maize embryos and sugars are very similar (Williams and Leopold, 1989; Koster, 1991). In addition, the acquisition of desiccation tolerance during maturation and loss during germination is correlated with changes in sugar composition in many species (Leprince *et al.*, 1993). Typically, desiccation tolerant seeds accumulate high amounts of sucrose and lower amounts of the monosaccharide reducing sugars. Although sucrose is a likely protectant, it has the tendency to crystallize in dry conditions. Oligosaccharides such as raffinose inhibit this tendency and their modest accumulation may have mechanistic relevance (Caffrey *et al.*, 1988).

The accumulation of sucrose may have an adaptive advantage because glucose (a reducing sugar) is detrimental to desiccation tolerance from several points of view. Glucose participates in cross-linking reactions with protein by a complex glycosylation reaction between amino and carbonyl groups known as the Maillard reaction (browning) in foods, that may also occur in seeds (Koster and Leopold, 1988). As respiratory substrates, monosaccharides promote respiration and mitochondrial electron transport (Leprince *et al.*, 1992) which would oppose the imposition of quiescence and favour metabolism, energy production and the formation of oxygen radicals. Therefore, sucrose, which has been positively correlated with the degree of stress tolerance in plants for several decades, may function in a variety of ways to aid in the acquisition of desiccation tolerance in seeds. Sucrose can act in water replacement to maintain membrane phospholipids in the liquid-crystalline phase and to prevent structural changes in soluble proteins. Because of its solute properties, sucrose will also promote glass phase formation in the cytoplasm. Sucrose synthesis reduces the quantities of reducing sugars, a prerequisite for quiescence. Nonetheless, sucrose alone does not confer desiccation tolerance to plant cells but is only one of many required components (Farrant *et al.*, 1993; Leprince *et al.*, 1993).

## Conclusions

Extending the storage life of recalcitrant seeds, demand urgent attention to help conserve the genetic resources of many important tropical species. The role of microflora and, particularly, of fungi is emerging as an important factor in determining seed life span, especially as the seeds must be subjected to hydrated storage.

Several studies and observations clearly showed there is no fungal/bacterial infection of recalcitrant seeds if they are subjected to hydrated clean-storage (Calistru *et al.*, 2000). In *Avicennia marina* during hydrated storage the seeds showed a decreased susceptibility to fungal infection; probably the result of inherent, and/or inducible, defense mechanisms developed by those metabolically active seeds, in which germination events have been shown to be well under way. (Farrant *et al.*, 1986; Berjak *et al.*, 1992). Further, in these seeds high level of •-1,3-glucanase and chitinase activity were reported during short term wet storage. This was interpreted as a strategy to create an antimicrobial environment prior to microbial infection, which would also be an effective strategy during germination in the natural environment (Anguelova - Merhar *et al.*, 2003). Similar antifungal role for these enzymes has been suggested for germinating pea and tomato seeds. Inspite of recent advances, it has not proved possible to conserve recalcitrant seeds successfully. There have been some reports on the lengthening of the storage life of rubber and cocoa by a few months based on partial drying and treatment with fungicides.

In our laboratory we could prolong seed viability in Jamun (*Syzygium cuminii*) and Jackfruit (*Artocarpus heterophyllus*) upto 14 months when clean imbibed seeds were stored in Ziplock plastic containers with a mixture of fine sand and sawdust in the ratio of 1:1 at temperature of 22 to 25$^0$C. This medium prevented anoxia and allowed the exchange of water between the seeds and the medium. Hence, the seeds though remained alive did not germinate, as the water content in the medium was insufficient. The seeds also remained fresh and were not infected (Ambika and Pratibha-communicated). Based on our results we could summarize that recalcitrant seeds should be kept at moisture contents above their lowest-safe values and ventilation is needed to remove toxic gases and to prevent anoxia. However, further research is needed for prolonging the life of these seeds.

## REFERENCES

Anandarajah, K. and McKersie (1990). Manipulating the desiccation tolerance and vigor with dry somatic embryos of *Medicago sativa* L. with sucrose, heat shock and abscisic acid. *Plant Cell Rep.* **9**: 451-455.

Anandarajah, K., Kott, L., Beversdorf, W.D. and McKersie, B.D. (1991). Induction of desiccation tolerance in microspore-derived embryos of *Brassica napus* L. by thermal stress. *Plant Sci.* **77**: 119-123.

Anguelora-Merhar, V.S., Calistru, C. and Berjak, P. (2003). A study of some Biochemical and Histopathological responses of wet-stored recalcitrant seeds of *Avicennia marina* infected by *Fusarium moniliforme*. *Annals of Botany*. **92**: 401-408.

Attree, S.M., Moore, D., Sawhney, V.K. and Fowke, L.C. (1991). Enhanced maturation and desiccation tolerance of white spruce [*Picea-Glauca* (Moench) Voss] somatic embryos - Effects of a non-plasmolysing water stress and abscisic acid. *Ann. of Bot.* **68**: 519-525.

Berjak, P. (1989). The basis of the seed storage behaviour of seeds of *Hevea brasiliensis*. *Journal of Natural Rubber Research*. **4**: 195-203.

Berjak, P., Dini, M. and Pammenter, N.W. (1984). Possible mechanism underlying the differing dehydration responses in recalcitrant and orthodox seeds: desiccation- associated subcellular changes in propagules of *Avicennia marina*. *Seed Sci. Technol.,* **12**: 365-384.

Berjak, P., Farrant, J.M. and Pammenter, N.W. (1990). The basis of recalcitrant seed behaviour. Cell biology of the homoiohydrous seed condition. pp 89-108. In: Ed. R.B. Taylorson Recent advances in the development and germination of seeds. Plenum Press, New York.

Berjak, P., Farrant, J.M. and Pammenter, N.W. (2002). Minireview on seed recalcitrance-a current perspectives. *South African Journal of Botany*. **69**(2): 79-89.

Berjak, P., Pammenter, N.W. and Vertucci, C. (1992). Homoiohydrous (recalcitrant) seeds developmental status, desiccation sensitivity and the state of water in axes of *Landolphia kirkii* Dyer. *Planta*. **186**: 249-261.

Bewley, J.D. (1979). Physiological aspects of desiccation tolerance. *Ann. Rev. of Plant Physiol.* **30**:195-238.

Bewley, J.D. and Oliver, M.J. (1992). Desiccation tolerance in vegetative plant tissues and seeds; protein synthesis in relation to desiccation and a potential role for protection and repair mechanisms. pp. 160. In: Somero, G.N. Osmond, C.B. and Bolis, C.L. (Eds). Water and life comparative analysis of water relationships at the organismic, cellular and molecular levels. Berlin, Springer-Verlag.

Black, M. (1991). Involvement of ABA in the physiology of developing and mature seeds. pp. 99-124. In: Eds. W.J. Davies, H.J. Jones. *Abscisic Acid Physiology and Biochemistry*, Bios Scientific Publishers, Oxford.

Blackman, S.A., Wettlanfer, S.H., Obendorf, R.L. and Leopold, A.C. (1991). Maturation proteins associated with desiccation tolerance in soybean. *Plant Physiology*. **96**: 868-874.

Blackman, S.A., Obendorf, R.L. and Leopold, A.C. (1992). Maturation proteins and sugars in desiccation tolerance of developing soybean seeds. Plant Physiology. **100**: 225-230.

Bradford, K.J. and Chandler, P.M. (1992). Expression of 'dehydrin-like' proteins in embryos and seedlings of *Zizania palustris* and *Oryza sativa* during dehydration. *Plant Physiology*. **99**: 488-494.

Bruni, F. and Leopold, A.C. (1991). Glass transition in soybean seed. Relevance to anhydrous biology. *Plant Physiol.* **96**:660-663.

Buchvarov,P. and Gantcheff, T.S. (1984). Influence of accelerated and natural aging on free radical levels in soybean seeds. *Physiol. Plant*. **60**: 53-56.

Burke, M.J. (1986). The glassy state and survival of anhydrous biological systems. pp. 358-363. In: Ed. A.C. Leopold. *Membrane, metabolism and dry organisms.* Cornell University Press, Ithaca, NY.

Caffrey, M., Fonseca, V., and Leopold, A.C. (1988). Lipid-sugar interactions. Relevance to anhydrous biology. *Plant Physiol.* **86**:754-758.

Calistru, C., McLean, M., Pammenter, N.W. and Berjak, P. (2000). The effects of mycofloral infection, the viability and ultrastructure of wet-stored recalcitrant seeds of *Avicennia marina* (Forssk.) Vierh. *Seed Science Research*. **10**: 341-353.

Chandel, K.P.S., Chaudhury, R., Radhamani, J. and Malik, S.K. (1995). Desiccation and freezing sensitivity in recalcitrant seeds of tea, cocoa and jackfruit. *Annals of Botany*. **76**: 443-450.

Chin, H.F. (1975). Germination and storage of rambutan (*Nephelium lappaceum*)seeds. Malay. *Agric. Res.*, **4**: 173-180.

Chin, H.F. (1988). Recalcitrant seeds: A status report. International Plant Genetic Resources Institute, Rome.

Chin, H.F., Aziz, M., Ang, B.B. and Hamzah, S. (1981). The effect of moisture and temperature on the ultrastructur and viability of seeds of Hevea brasiliensis. *Seed Sci. Technology*. **9** : 411-422.

Chin, H.F. and Roberts, E.H. (eds). (1980). Recalcitrant crop seeds. Tropical Press, Kuala Lumpur, Malaysia.

Clegg, J.S. (1979). Metabolism and the intracellular environment. The vicinal water network model. In, Drost-Hansen, W. and Clegg, J.S. (eds). Cell- Associated water; 363-413. Academic Press, New York.

Close, T.J. (1996). Dehydrins: Emergence of a biochemical role of a family of plant dehydration proteins. *Physiologia Plantarum*. **97**: 795-803.

Connor, K.F., Kossmann-Ferraz,I.D., Bonner,T., Vozzo, J.A. (1998). The effects of desiccation on the recalcitrant seeds of *Carapa guianensis and Carapa procera. Seed technology*. **20**(1): 71-82.

Connor, Kristina, F. and Sharon Sowa (2003). Effects of desiccation on the Physiology and Biochemistry of *Quercus alba* acorns. In: *Tree Physiology*. **23**: 1147-1152. Heron Publishing-Victoria, Canada.

Crèvecoeur, M., Deltour, R. and Bronchart, R. (1976). Cytological study on water stress during germination of *Zea mays*. *Planta*. **132**: 31-41.

Dasgupta, J., Bewley, J.D., and Yeung, E.C. (1982). Desiccation-tolerant and desiccation-intolerant stages during the development and germination of *Phaseolus vulgaris* seeds. *J. Exptl. Bot.* **33**:1045-1057.

Darbyshire, B. (1974). The function of the carbohydrate units of three fungal enzymes in their resistance to dehydration. *Plant Physiol.* **54**: 717-721.

Davies,R.I. and H.W. Pritchard. (1998). Seed conservation of dryland palms of Africa and Madagascar: needs and prospects. *Forest Genetic Resources*. **26**: 37- 44.

Deltour, R. (1985). Nuclear activation during early germination of the higher plant embryo. *Journal of cell Science*. **73**: 43-83.

Durand, B.M., Real, M. and Come, D. (1989). Changes in nuclear activity upon secondary dormancy induction by Abscisic acid in apple embryo. *Plant Physiology and Biochemistry*. **27**: 511-518.

Ellis, R.E., Hong and Roberts, E.H. (1990). An intermediate category of seed storage behaviour? 1. Coffee. *Journal of Experimental Botany*. **41**: 1167-74.

Ellis, R. H. (1991). The longevity of seeds. *Hort Science*. **26**: 1119-1125.

Evans, M. L. (1984). Functions of hormones at the cellular level of organization. In: Scott, T.K. (ED). Hormonal regulation of development II, Berlin: Springer-Verlag. 23-79.

Farnsworth, E. (2000). The ecology and physiology of viviparous and recalcitrant seeds. *Ann.Rev.of Ecology and Systematics*. **31**: 107-138.

Farrant, J.M., Berjak, P. and Pammenter, N.W. (1985).The effect of drying on viability retention of recalcitrant propagules of *Avicennia marina*. *S. Afr. J. Bot*. **51**: 432-438.

Farrant, J. M., Pammenter, N.W., Berjak, P., Farnsworth, E.J. and Vertucci, C.W. (1996). Presence of dehydrin-like proteins and levels of abscisic acid in recalcitrant seeds may be related to habitat. *Seed Science Research*. **6**: 175-182.

Farrant, J. M., Pammenter, N. W. and Berjak, P. (1986). The increasing sensitivity of recalcitrant *Avicennia marina* seeds with storage time. *Physiologia plantarum*. **67**: 291-298.

Farrant, J. M., Berjak, P, Cutting, J. G. M., and Pammenter, N.W. (1993). The role of plant growth regulators in the development and germination of the desiccation- a sensitive (recalcitrant) seeds of *Avicennia marina*. *Seed Science Research*. **3**: 55-63.

Farrant, J.M., Pammenter, N. and Berjak, P. (1988). Recalcitrance - a current assessment. *Seed Science and Technology*.**16**: 155-166.

Finch-Savage, W.E. (1992). Embryo water status and survival in the recalcitrant species *Quercus robur* L: Evidence for a critical moisture content. *Journal of Experimental Botany*. **43**: 663-669.

Finch-Savage, W.E., Pramanik, S. K. and Bewley, J. D. (1994). The expression of dehydrin proteins in desiccation-sensitive (recalcitrant) seeds of temperate trees. *Planta*. **193**: 476-485.

Gameni, C.S., Kraak, H.L., Van Pijlen, J.G. and De Vos, C.H. (1996). Storage behaviour of neem (*Azadirachta indica*) seeds from Burkina faso. *Seed Science and Technology*. **24**: 441-448.

Goldbach, H. (1979). Imbibed storage of *Melicocus bijugatus* and *Evgenia brasiliensis* using Abscisic acid as a germination inhibitor. *Seed Science and Technology*. **7**: 403-406.

Gray, D.J., and A. Purohit. (1991). Somatic embryogenesis and development of synthetic seed technology. *Crit. Rev. in Plant Sci.* **10**: 33-61.

Greggains, V., Finch-Savage, W.E., Atherton, N.M. and Berjak, P. (2001). Viability loss and free radical processes during desiccation of recalcitrant *Avicennia marina* seeds. *Seed Sci. Research*. **11** : 235-242.

Hor, Y.L. (1984). Storage of cocoa (*Theobroma cacao*) seeds and changes associated with their deterioration. Ph.D. Thesis. Universiti Pertanian Malaysa, Malaysia.

Hendry, G.A.W. (1993). Oxygen, free radical processes and seed longevity. *Seed Sci. Res.* **3**:141-153.

Han, B., Berjak, P., Pammenter, N., Farrant, J. and Kermode, A.R. (1997). The recalcitrant plant species; *Castanospermum australe* and *Trichilia dregeana*, differ in their ability to produce dehydrin- related polypeptides during seed maturation and in response to ABA or water-deficit-related stresses. Journal of Experimental Botany.

Hoekstra, F.A., Crowe, J.H., and Crowe, L.M. (1991). Effect of sucrose on phase behaviour of membranes in intact pollen of *Typha latifolia* L. as measured with Fourier transform infrared spectroscopy. *Plant Physiol.* **97**:1073-1079.

Hong, T.D. and Ellis, R .H. (1998). Contrasting seed behaviour among different species of Meliaceae. *Seed Science and Technology*. **26**(1): 77-95.

Ingram, J. and Bartels, D. (1996). The molecular basis of dehydration tolerance in plants. *Annual Review of Plant Physiology and Plant Molecular Biology*. **47**: 377-403.

Karssen, C.M., Brinkhorst-Vanderswan, D.L.C., Breckland, A.E. and M.Koorneef. (1983). Induction of dormancy during seed development by endogenous abscisic acid: Studies on abscisic acid deficient genotypes of *Arabidopsis thaliana* (L.) Heynh. *Planta* **157**: 158-165.

Kermode, A.R. (1995). Regulatory mechanisms in the transition from seed development to germination: Interactions between the embryo and the seed environment. pP. 273-332 In: Galili, G and Kigel, J. (Eds). Seed development and germination. New York, Marcel Dekkar, Inc.

King, M.W. and Roberts, E.H. (1979). The Storage of recalcitrant seeds. Acievements and possible approaches . IBGRI, Rome.

King, M.W. and Roberts, E.H. (1980). Maintenance of recalcitrant seeds in storage. In: Chin, H. F. and Roberts, E.H. (Eds). Recalcitrant crop seeds: 53-89. Tropical Press, Kuala Lumpur, Malaysia.

Koster, K.L. (1991). Glass formation and desiccation tolerance in seeds. *Plant Physiol.* **96**: 302-304.

Koster, K.L. and Leopold, A.C. (1988). Sugars and desiccation tolerance in seeds. *Plant Physiol.* **88**: 829-832.

Krishnamurthy, K.V. (!999). Review on Embryos and Embryoids. *Current Science*. **76** (5).647-659

Lai, F.M., and McKersie, B.D. (1993). Effect of nutrition on maturation of alfalfa (*Medicago sativa* L.) somatic embryos. *Plant Sci.* **91**: 87-95.

Lane, B.G. (1991). Cellular desiccation and hydration: developmentally regulated proteins and the maturation and germination of seed embryos. *FASEB Journal*. **5**: 2893-2901.

Leprince, O., McKersie, B.D. and Hendry, G.A. (1993). The mechanisms of Desiccation tolerance in developing seeds. *Seed Sci. Res.* **3**: 231-246

Leprince, O., Van der Werf, A., Deltour, R. and Lambers, H. (1992). Respiratory pathways in germinating maize radicles correlated with desiccation tolerance and soluble sugars. *Physiol. Plant.* **84**: 581-588.

Leopold, A.C. (1990). Coping with desiccation. pp. 57-86. In: Eds. R.G. Alscher and J.R. Cumming. *Stress responses in plants: adaptation and acclimation mechanisms*. Wiley-Liss, Inc., New York.

Leopold A. C., Sun W. Q. and Bernal Lugo, I. (1994). The glassy state in seeds: Analysis and function. *Seed Science Research*. **4**: 267-274.

Li, C.R., Sun, W.Q. (1999). Desiccation sensitivity and activities of free radical scavenging enzymes in recalcitrant *Theobroma cacao* seeds. *Seed Sci. Res.* **9**: 209-217.

Loomis, W.D. and Battaile, J. (1966). Plant phenolic compounds and their isolation of plant enzymes. *Phytochemistry*. **5** : 423-438.

Marsolais, A.A., Wilson, D.P.M., Suijita, M.J.T and Senaratna, T. (1991). Somatic embryogenesis and artificial seed production in zonal (*Pelargonium x domesticum*) geranium. *Can .J. of Bot.* **69**: 1188-1193.

McKersie, B.D. and Bowely, S.R. (1993). Synthetic seeds in alfalfa. pp. 231-255. In: Ed. K. Redenbaugh. Synseeda. Applications of Synthetic seeds to crop improvement. CRC Press. Boca Ralon.

Normah, M.N., Chin, H.F. and Hor, Y.L. (1986). Desiccation and cryopreservation of embryonic axes of *Hevea brasiliensis*. Muell-Arg. *Perlanika*. **9**: 299-303.

Pammenter, N.W., Berjak, P. (1999). A review of recalcitrant seed physiology in relation to desiccation tolerance mechanism. *Seed Sci. Res.* **9**: 13-37.

Pammenter, N.W. and Berjak, P. (2000). Evolutionary and ecological aspects of recalcitrant seed biology. *Seed Science Research.* **10** (3): 301-306.

Priestley, D.A., Wernu, B.G., Leopold, A.C. and Mc Bride, M.B. (1985). Organic free radical levels in seeds and pollen: the effects of hydration and aging. *Physiol. Plant.* **64**: 88-94.

Pritchard, N.W. (1991). Water potential and embryonic axes viability in recalcitrant seeds of *Quercus rubra*. *Ann. Bot.* **67**: 43-49.

Pritchard, H.W. and Prendergast, F.G. (1986). Effects of desiccation on the in vitro viability of embryos of the recalcitrant seed species *Araucaria hunsteinii* K.Shum. *Journal of Experimental Botany*. **37**: 1388-1397.

Roberts, E.H. (1973). Predicting the storage life of seeds. *Seed Science and Technology*.**1**: 499-514.

Roberts, E.H. and King, M.W. (1980). Storage of recalcitrant seeds. In: Withers, L. A. and Williams, J. T. (Eds). Crop Genetic Resources. The Conservation of difficult material: 39-48. International union of Biological Sciences, Serie B42, Paris.

Roberts, E.H., King, M.W. and Ellis, R.H. (1984). Recalcitrant seeds: their recognition and storage. pp.38-52 in Holden, J.H.W. and Williams, J.T. (Eds). Crop genetic resources: Conservation and Evaluation, London, Allen and Unwin.

Sacande', M., Van Pijlen, J.P., De Vos, C.H., Hoekstra, F.A., Bino, R.J. and Groot, S.P.C. (1997). Intermediate storage behaviour of neem tree ( *Azadirachta indica*) seeds from Burkina Faso. Pp. 101-104 in Poulsen, K. Stubsgaard, F. and Onedrogo, A.S. (Eds). Improved methods for the handling and storage of intermediate/recalcitrant tropical forest tree seeds. Rome, IPGRI.

Senaratna, T. and Mckersie, B.D. (1983a). Characterization of solute efflux from dehydration injured soybean (*Glycine max* L. Merr) seeds. *Plant Physiology*. **72**: 911-914.

Senaratna, T. and McKersie, B.D. (1983b). Dehydration injury in germinating soybean (*Glycine max* L.) seeds. *Plant Physiol.* **72**: 620-624.

Senaratna, T., Kott, L., Beversdorf, W.D., and McKersie, B.D. (1991). Desiccation of microspore derived embryos of oilseed rape (*Brassica napus* L.). *Plant Cell Rep.* **10**: 342-344.

Senaratna, T., McKersie, B.D., and Bowley, S.R. (1989). Desiccation tolerance of alfalfa (*Medicago sativa* L.) somatic embryos. Influence of Abscisic acid, stress pretreatments and drying rates. *Plant Sci.* **65**:253-259.

Senaratna, T., McKersie, B.D., and Stinson, R.H. (1985a). Simulation of dehydration injury to membranes from soybean axes by free radicals. *Plant Physiol.* **77**: 472-474.

Senaratna, T., McKersie, B.D., and Stinson, R.H. (1985b). Antioxidant levels in germinating soybean seed axes in relation to free radical and dehydration tolerance. *Plant Physiol.* **78**: 168-171.

Simon, E.W. (1974). Phospholipids and plant membrane permeability. *New Phytol.* **73**: 377-420.

Simontacchi, M. and Puntarulo, S. (1992). Oxygen radical generation by isolated microsome from soybean seedling. *Plant Physiol.* **100**: 1263-1268.

Tompsett, P.B. (1983). The influence of gaseous environment on the storage life of *Araucaria hunstecinii* seed. *Ann. Bot.* **52**: 229-237.

Tompsett, P.B. (1992). A review of the literature on storage of Dipterocarp seeds. *Seed Science and Technology*. **20**(2): 251-267.

Vertucci, C.W. (1989). The effects of low water contents on physiological activities of seeds. *Physiol. Plant.* **77**: 172-176.

Vertucci, C.W. (1990). Calorimetric studies of the state of water in seed tissues. *Biophys. J.* **58**: 1463-1471.

Vertucci, C.W., Crane, J., Porter, R.A. and Oelke, E.A. (1994). Physical properties of water in *zizania* embryos in relation to maturity, water content and temperature. *Seed Sci. Res.* **4**: 211-224.

Vertucci, C. W. and Farrant, J.M. (1995). Acquisition and loss of desiccation tolerance. pp. 701-746 in Galili, G and Kigel, J. (EDs). Seed development and germination. New York, Marcel Dekker, Inc.

Von-Teichman and Van-wyk, A. E. (1994). Structural aspects and trends in the evolution of recalcitrant seeds in dicotyledons. *Seed Science Research*. **4**(2): 225- 239.

Williams, R.J. and Leopold, A. C. (1989). The glassy state in corn embryos. *Plant Physiol.* **89**: 977-981.

Yamaguchi-Shinozaki, K. and Shinozaki, K. (1994). A novel cis- activating element in an *Arabidopsis* gene is involved in responsiveness to drought, low temperature, or high salt stress. *Plant cell.* **6**: 251-264.

*Advances in Plant Physiology*, Vol. 9
Ed. A. Hemantaranjan
Scientific Publishers (India), Jodhpur, 2006 pp. 475-491.
E-mail: info@scientificpub.com www.scientificpub.com

# 20

# CHANGES IN PLANT GROWTH SUBSTANCES DURING FRUIT RIPENING AND POST HARVEST PERIODS

P.K. Nagar
Division of Biotechnology, Institute of Himalayan Bioresource Technology, Palampur, 176061, India

## INTRODUCTION

It has been estimated that at least 25-30% of the harvested horticultural produce is lost due to post harvest deterioration in less developed and sub-tropical countries (Salunkhe and Desai, 1989). Much of these losses are physiological in origin due to ripening, senescence, abscission and eventual regrowth of the tissues. There is an obvious need to delay these ending processes until the harvested crop is consumed. Inhibition of the vital metabolic processes can be achieved in different fruits for period of days, weeks, or even months by chemical treatments and / or storage at low temperature. However, storage problems and treatments are very different depending on the nature of the produce to be stored and this variation in performance has stimulated much interest in post harvest physiology of fruits.

Growth, maturation and senescence are the three important phases in fruit ontogeny. Fruit growth begins with cell division and cell enlargement, which accounts for its final size. Growth and maturation, often referred to as "fruit development", cannot be distinguished very clearly. Wills *et al* (1981) defined 'senescence' as the period when anabolic biochemical processes give way to catabolic processes leading to ageing and final death of the tissue. Ripening, a term reserved for fruit, generally begins during the later stages of maturation and is considered as the beginning of senescence. Biale (1964) referred to fruit ripening as the 'beginning of an end'. Many of the chemical changes that occur to induce alterations in flavor, texture and odour have been well characterized for many fruits. Conversely, the endogenous plant growth substances controlling and regulating such changes remain obscure and very few investigations have been made of their status in fruits especially during ripening and storage. There is little doubt that the whole ageing process, like other growth processes is subject to hormonal control and differential response depends on certain quantitative relationship between the levels of interacting plant growth substances (Davis, 1995).

In working out the hormonal controls of fruit ripening, - several general approaches have been employed. The first and most widely used is external application of plant growth substances to promote or retard ripening. This approach can help to reveal the patterns of hormonal controls only if combined with other data. The second method is to correlate changes in their endogenous levels during ripening and post harvest periods. Unfortunately, majority of the relevant studies on endogenous levels has not aimed at ripening *per se* and therefore, is not closely correlated with ripening, since ripening and post harvest, like many other plant processes appear to involve a shift in the balance of plant growth substances. The present review is mainly aimed to summarize the changes in plant growth substances especially auxins, gibberellins, cytokinins, abscisic acid and ethylene during ripening and post harvest periods in fruits.

## 2. CHANGES IN ETHYLENE DURING FRUIT RIPENING

In terms of ethylene regulation, ripening includes an aging requirement, loss of chlorophyll, increased respiration, ethylene production, softening and abscission. In addition to these processes, there are variations in ripening of different fruits. Some of these reflect the diversity of specific species or cultivars. For example, the colour of ripe apple can be red, yellow or green and different pigments are responsible for these colours.

In fruit physiology, the terms climacteric and non-climacteric are used to describe fruit, which ripen when treated with ethylene, and those that do not although the term climacteric was originally applied to increased fruit respiration, (Biale and Young, 1981) which subsequently included a rise in ethylene production. A climacteric rise in respiration is also observed in other senescing organs like flowers and leaves. The fruits shown in Table 1 have been classified as climacteric or non- climacteric where the variation in rates of ethylene production and respiration can be large. For examples, in certain cultivars of apple, like Cox Orange exhibit a 40 fold rise in ethylene production during ripening white in other cultivars like Golden Delicious the ethylene climacteric is small or absent (Watkins *et al.* 1989). Non-climacteric fruits share a number of attributes, some of these fruits, like pineapple are composed primarily of non-ovarian accessory tissues. Cucumber is harvested and consumed as immature fruits; while others such as citrus are harvested after ripening is complete. It is assumed that the function of increased ethylene production is to insure a maximal rate of ripening, senescence and abscission. Similarly the role of increased respiration is to provide the energy used in catabolic processes. The climacteric has been a useful way of emphasizing the role of ethylene in ripening and senescence and also seems to demarcate fruit development into a period of pre-climacteric growth and subsequent ripening (Fig. 1)

Table 1. List of some fruits characterized as Climacteric and Non-climacteric harvested at commercial maturity.

| Fruit | Reference |
|---|---|
| Climacteric | |
| Apple | Kidd and West, 1925 |
| Banana | McMurchie, *et al.* 1972 |
| Fig | Biale, 1960 |
| Mango | Biale, 1960 |
| Avocado | Rowan *et al* 1958 |
| Peach | Amoros *et al.* 1989 |
| Tomato | Rowan *et al* 1958 |
| Plum | Sekse, 1988 |

| | |
|---|---|
| Non- Climacteric | |
| Bell Pepper | Saltveit, 1977 |
| Sweet-Cherry | Sekse, 1988 |
| Citrus | Aharoni, 1968 |
| Grapes | Biale, 1960 |
| Pineapple | Biale, 1960 |
| Litchi | Joubert, 1986 |
| Strawberry | Biale 1964 |
| Watermelon | Elkashif *et al*, 1964 |

In higher plants, ethylene is produced from L-methionine, which is activated by ATP to form S- adenosyl L methionine (AdoMet) through the catalytic activity of S-adenosylmethionine synthetase (Fig. 2). Starting from SAM, two specific steps result in the formation of ethylene. The first step produces the non-protein amino acid 1-aminocyclopropane 1- carboxylic acid (ACC), which is formed from AdoMet by the action of ACC synthase (ACS) with pyridoxal phosphate acting as a co-factor (Kende 1993). Formation of ACC is the rate-limiting step in ethylene biosynthesis. ACS is encoded by a medium sized multigene family. Various signals, which influence ethylene synthesis, result in increased expression of single members of the ACS gene family. Production of ethylene from ACC is catalyses by ACC oxidase (ACO) and this reaction is oxygen dependent. At anaerobic conditions, ethylene formation is completely suppressed. Two other ethylene regulated genes have been identified that may play a possible role in the methionine cycle, *E4*, a putative methionine sulphoxide reductase protein and *ER69* a putative cobalamine – independent methionine synthase (Zegzonti *et al.* 1999). In this pathway it is well known that biosynthesis is subjected to both positive and negative feedback regulation (Kende, 1993). Positive feedback regulation of ethylene biosynthesis is a characteristic feature of ripening fruits in which exposure to exogenous ethylene or propylene results in a large increase in ethylene production due to the induction of ACS and ACO. Both of these enzymes are encoded by small multigene families and their expression is differentially regulated by various developmental, environmental and hormonal signals (Berry *et al.* 2000; Llop-Tous *et al.* 2000).

At least eight ACS genes have been identified in tomato (Oetikjer *et al.* 1997: Shin *et al.* 1998) and many others have been identified in both climacteric and non-climacteric fruits such as melon, cucumber and citrus (Wong *et al.* 1999). However, the role that ACS plays in ripening has been most widely studied in tomato. This enzyme shows homology to pyridoxal-5'- phosphate (PLP) – dependent amino transferases and also act as dimer (Tarun and Theologis, 1998). Studies of the ACS crystal structure (Capitani *et al.* 1999) and PLP co-factor binding (Huai *et al.* 2001) have confirmed similarity between the ACS catalytic binding site and those of other PLP- dependent transferases. Some studies have also confirmed the presence of *LEACS IA* and *LEACS 6* in tomato fruit before the onset of ripening and that each ACS in fruit has a different expression pattern (Berry *et al.* 2000). As in the case of other plant hormones, ethylene, is thought to bind to a receptor, forming an activated complex, which in turn triggers the primary reaction. This reaction then initiates chain of reactions, including modification of gene expression, thereby leading to a wide variety of physiological reactions (Nagar and Sood, 2004). Thus there are four levels of manipulation one can use to regulate ethylene responses, (i) control the level of ethylene in the tissue by addition or removal of ethylene, (ii) regulate the level on the tissue by stimulating or inhibiting ethylene biosynthesis, (iii) modify the binding characteristics of ethylene to the receptor, or modify the amount of receptor and (iv) manipulating ethylene dependent gene expression.

A small increase in ethylene production has been reported at late ripening stage of strawberry (Perkins- Veezie *et al.,* 1996). As

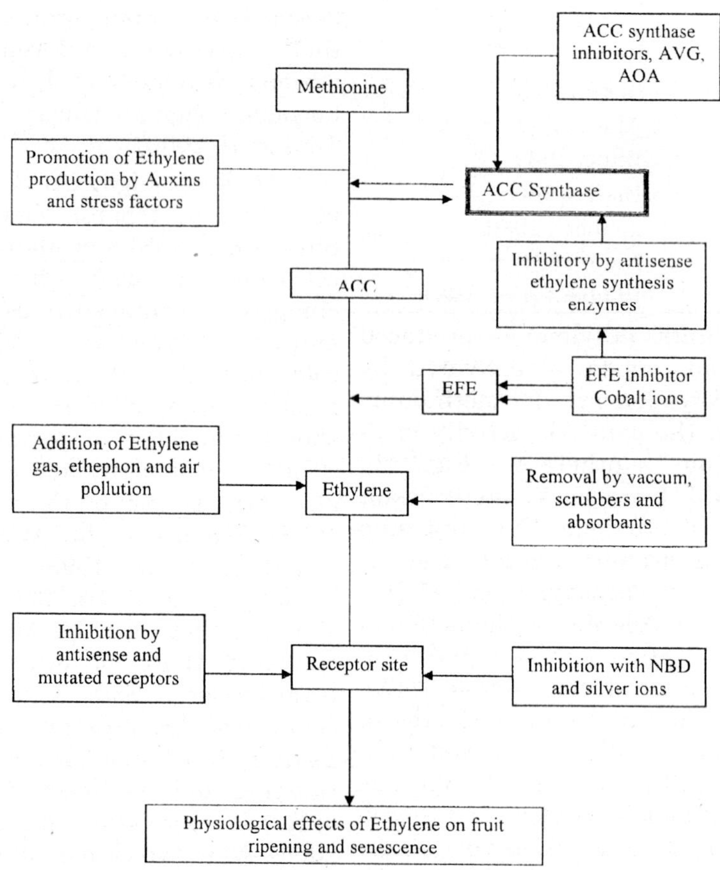


Fig 1. A model for the options available to manipulate behaviour of plants or plant products (fruits, flowers, etc.) by manipulating ethylene physiology

*in vivo,* role for ethylene is supported by the finding that exogenous application of ethylene accelerates strawberry fruit senescence and causes a 50% reduction in fruit firmness (El-Kazzay, *et al,* 1983). Recently, Castillejo *et al* (2004) observed changes in fruit firmness, which could be influenced by changes in FaPE1 activity. The promoter region of the *FaPE1* gene has been shown to contain three putative EREs and this element has previously been found in ethylene- responsive promoters such as the carnation *GST* and the tomato *E4* gene promoter (Itzhaki *et al*, 1994). During strawberry fruit senescence, the inhibition of FaPE1 following ethylene burst would increase the degree of methyl esterification of pectin polysaccharides and thus reduce pectin stabilization mediated by calcium cross links.

Due to the economic importance of post harvest deterioration of fruit crops, enzymes that are implicated in cell wall softening have been examined by transgenic manipulation and *in vitro* assays. Brummell and Harpster (2001) have extensively reviewed this topic hence the present review will rather draw attention to only those enzymes that exhibit ethylene regulation. With progress in ripening process, the pectin rich middle lamella is modified and partially hydrolyzed which in turn affects the final texture of the ripe fruit. In fruits like tomato this process occurs early during ripening (Crookes and Grieson, 1983) while in others it is a late ripening process.

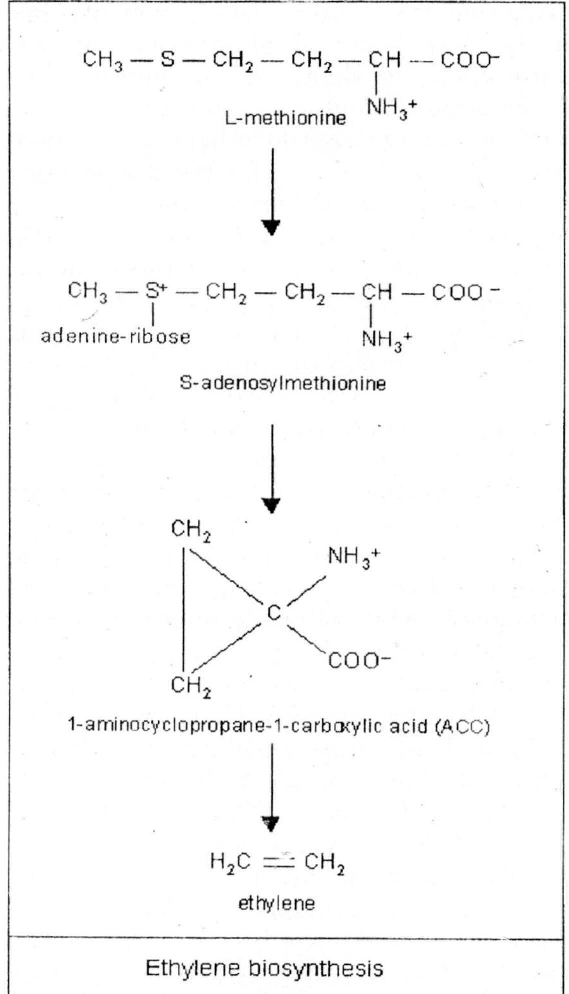


Fig. 2 Ethylene Biosynthesis in higher Plants

In most of the fruits, polyalacturonase (PG) is a major cell wall degrading enzymes. It is transcriptionally activated during ripening (Montgomery *et al.* 1993) and the PG promoter sequence contains ethylene dependent ripening specific control elements (Nicholas *et al.* 1995). Analysis of low ethylene transgenic tomato has shown that induction of *PG* m RNA occurs at a very low ethylene levels (Sitrit and Bennett, 1998). Since during the onset of ripening, PG is synthesized *de novo* (Tucker and Grierson, 1982), this enzyme has been made a good target for antisense suppression (Smith *et al.* 1988). Expression of a truncated PG gene was also found to inhibit the expression of endogenous PG (Smith *et al.* 1990). In homozygous transgenic tomato plant lines PG activity was reduced to 1% of the normal value, the fruit were only slightly firmer which led to the conclusion (Grierson and Schuch, 1993) that PG is not the major determinant of tomato fruit softening. However, low PG fruits are most resistant to splitting, mechanical damage and pathogen infection (Cooper *et al.* 1998). Further, Chun and Huber (2000) showed that transgenic tomato plants suppressed for 38 KDa glycoprotein, known as the PG • - subunit or converter, exhibited increased fruit softening during ripening, higher extractable PG actively and more polyuronide solublization. This indicates that during ripening this protein may be distributed throughout the cell wall to control PG diffusion or action.

During fruit ripening, pectin methylesterase (PME) is responsible for de-esterification of the highly methylesterified polygalacturonase in the cell wall. Esterification drops from 90% in mature green tomato fruit to 35% in red ripe fruit and this makes the polyuronides susceptible to degradation by PG (Carpita and Gibeant, 1993). PME is present as a small gene family in tomato, some members of which are highly homologous. PME protein is found in most plant tissues with three isoforms being specific to fruit, one of which PME1, peaks at breakers stage (Gaffe *et al.* 1997). It was being found by Tiemann and Handa, (1994) that suppression of PME in over ripe fruit caused an almost complete loss of tissue integrity, therefore, PME plays little role in ripening but does act at post harvest periods. Lu *et al.* (2001) has provided evidence to show that ethylene may regulate vesicular transport between different cellular compartments. During early fruit ripening phase, polymeric galactose begins to be broken down into free galactose, with rise in free galactose throughout ripening. The enzyme responsible for this is α- galactosidase, which is encoded by a gene family of at least seven members, all of which display different patterns of expression during fruit ripening (Smith and Gross, 2000). Cell wall localized enzyme, expansins, is thought to cause cell wall loosening by reversible disrupting the hydrogen bonds between cellulose microfibrils and matrix polysaccharides (Cosgrove, 2000). At least six different expansin genes are expressed during tomato fruit development (Brummell *et al*, 1999), one of which *EXP1,* is ethylene regulated and is specific for tomato fruit with RNA transcripts accumulating either just before or at breaker stage (Ross *et al,* 1997). These results indicate that several enzymes are regulated by ethylene during ripening, but their role in fruit ripening remains to be elucidated.

From the above, it can be noticed that fruit ripening and the role that ethylene plays in its regulation is complex. Identification of additional components involved in ethylene signal transduction, the further characterization of ripening mutants and additional studies on the biochemistry of ripening in relation to ethylene are necessary for complete knowledge of fruit ripening. There are some similarities between climacteric and non- climacteric fruit ripening and certain ethylene dependent events in climacteric fruit are observed, apparently in the absence of or with extremely low levels of ethylene, in non-climacteric fruits. Therefore, understanding what controls these processes in non-climacteric ripening may prove pertinent in gaining full understanding of climacteric fruit ripening and post harvest and vice versa.

## 3. CHANGES IN ABA CONTENT

The plant hormone abscisic acid (ABA) modulates a wide spectrum of responses including gene activation and repression, guard cell closure, cell cycle blockage and photosynthesis inhibition under different stress conditions. ABA also plays pivotal role in the developmental programmes of seed maturation, desiccation, dormancy and germination (Chandler and Robertson, 1994). However, in contrast to others, endogenous levels of ABA can rise and fall dramatically in response to different developmental processes. This quantum of changes in ABA levels led to understand ABA action either by applying the compound onto tissues and evaluating the effects or measuring its endogenous levels in specific tissues and correlating it with certain physiological processes like dormancy (Nagar and Kumar, 2000), seed (Bhattacharya *et al.* 2002) and flower development (Sood and Nagar, 2003). The control of fruit maturation and ripening is considered to be due to the interaction and balance between promoting and inhibitory substances. While ethylene is one important promoting factor, ABA appears to be another. Ripening could be hastened in tomato by externally applied ABA (Libermann *et al.* 1977) and advances ripening in grapes if applied during stage 2 of the growth cycle (Coombe and Hale 1973, Cawthan and Morris, 1982). ABA has also been shown to stimulate ethylene production in preclimacteric apples (Libermann *et al.* 1977) Increased levels of ABA have been associated with fruit ripening in citrus (Goldschmidt *et al.* 1973), strawberry (Tsay and Mizuno, 1984) and peach (Looney *et al.* 1974 Noga and Buckovac, 1987).

Gazit and Blumenfeld (1972) reported little changes in ABA levels during fruit development in avocado, and once ripening started, a considerable increase occurred (Adalto *et al.* 1976). These workers also found that the increase in ABA closely followed the ethylene curve, which peaks at the same time. Furthermore, the ratio of free to bound ABA remained the same, indicating that the free ABA was the result of synthesis and not its activation. This is in agreement with the studies of Millborrow and Robinson (1983) who found that ripening avocado could convert labelled mevalonate, which is believed to be the precursor of ABA. Further, Bower *et al.* (1986) found that free ABA content of ripening avocados increased with fruit softening, with the peak at approximately the same degree of softening at which the maximum ethylene peak occurred and declined thereafter. When apple fruit at 92 and 147 days after full bloom were injected with 1mM ABA, it induced an increase both in respiration and ethylene. (Vendrell and Busea, 1989). This effect was specially pronounced in the younger fruits, when not treated, showed no ripening related changes for 3 weeks. It was being suggested that the capacity of fruits to ripen is related to their ABA content and that the effect on ethylene production is indirect. Endogenous ABA content increased markedly in such fruits 2 months before normal ripening and remained high until 4 days after harvest

Immediately after harvest, ABA extracted from fruits of the apple var. 'Golden Delicious' comprised solely the *cis trans* isomers (Bangerth, 1982). During postharvest, however, *trans* ABA accumulated and finally exceeded the level of *cis trans* ABA. However, when the fruits were stored under hypobaric conditions and thus under low O2 partial pressure do not show a noticeable increase in ABA. Thus the question arises why *t* ABA accumulates in ripening apple fruit? In avocado, the *cis* double bond of the side chain is originally formed in the *trans* configuration with subsequent isomerization to the *cis* form. (Robinson and Ryback, 1969). However, Millborrow (1970) was unable to demonstrate a conversion to the *cis* form and concluded that this isomerization must occur at an earlier biosynthetic step before ABA or *t*- ABA is formed. Therefore, it seems unlikely that t-ABA found in ripening apples and some other tissues is an intermediate in ABA biosynthesis as suggested by Jones *et al.* (1976). Millbrrow (1974) suggested that special plant tissues could metabolize ABA to *t*- ABA that is a biologically inactive compound. Concentrations of ABA and phaseic acid (PA), an ABA like biologically active compound, were studied during the ripening and postharvest periods in peach. (Tsuchida *et al.,* 1990). The ABA content immediately after harvest was low increasing to a maximum (20 ug/g) after 5 days at 20C and then decreased again. Both cis and *trans* forms of ABA were detected, but most of the increase in ABA concentration was accounted for by the *trans* isomers. PA content increased as ripening and postharvest periods increased. The authors suggested a possible involvement of PA during ripening and postharvest periods. It was found that the levels of *cis* and *trans* ABA and conjugated *cis* and *trans* ABA showed a general increase towards harvest in the apple seeds (Kardo, *et al.* 1999). Phaseic acid level reached a peak at 41 days after full bloom and decreased until harvest. It was being suggested by the authors that *cis* (-+) ABA treated exogenously to apple fruit might eventually metabolise to *trans* ABA, conjugated *trans* ABA and further metabolites.

## Changes in Auxins and Gibberellins

The nature of the factors that anatagonise the ripening process is still a matter of speculation. There is very little information on the role that auxin and gibberellins (GAs) play in the ripening process and during postharvest periods and very few studies have been conducted on their changes during this period and most of these are related on their exogenous applications. However, indole-3- acetic acid (IAA) which is present in tomato (Catala *et al.* 1992; Riov and Bangerth, 1992) may be involved in the regulation of ripening (Brady, 1987). There are very few studies of changes in IAA levels during the period immediately before and after the onset of ripening in fruits. Auxins in many fruits also peak at various stages of fruit growth but fall off before final fruit maturation and ripening (Nagar and Saha, 1985). IAA acts both as an inhibitor of ripening and also as an inhibitor of ethylene synthesis. The significance of the interaction between IAA and stimulation of increased ethylene synthesis is hard to suggest (Yang, 1985). It may be that it is not significant at the very low physiological levels of IAA (10-mM), but it is potentially an interesting control system with a direct interaction between the levels of inhibitors and promoters. Frenkel *et al.* (1975) envisaged a slow decrease in IAA in pear during the pre-climacteric periods under the influence of an oxidative mechanism with an increase of a ripening promoter 3- methylene oxyindole.

In a given plant tissue, auxin levels are the net result of synthesis, translocation and degradation. An auxin- ethylene feedback system may alter auxin levels by regulating one or more of these processes. The capability of ethylene to reduce auxin levels has been demonstrated in several systems (El-Beltagy *et al.* 1976; Kalinin and Kurchii, 1986). However, other reports have shown that ethylene had no effect (Epstein, 1982), or increased auxin levels in some tissue (Varga *et al.* 1982; Domato and Helwitt, 1983). Increased auxin levels following ethephon or ethylene treatment may be due to the secondary effects of ethylene inhibited auxin transport. In pea seedlings, (Burg *et al.* 1971) the inhibition of growth was accompanied by a 75% reduction in extractable auxin in sub apical tissue and 150% increase in the apical tissue.

The control ripening of non –climacteric species like strawberry is poorly understood. As there is little indication that ethylene is directly involved, auxin has been proposed to be the primary hormone controlling strawberry fruit ripening. Auxin is synthesized in the achenes (true fruit embedded in the receptacle) and positively affects the growth of the receptacle. And a gradual decline in the supply of auxin from achenes in the latter stages of growth has been proposed to be the basis of ripening (Perkins- Veazie, 1995). Most of the ripening associated genes are negatively correlated by auxin, though a few auxins up regulated genes have also been described (Manning, 1998; Ahroni *et al*, 2002). It is known that free and conjugated IAA reaches a peak at the green stage of the fruit development and subsequently declines (Archbold and Dennis, 1984). Recently, Castillejo *et al* (2004) reported cloning of four genes in strawberry, *FaPE1, FaPE2, FaPE3,* and *FaPE4.* Expression of the corresponding cDNAs was examined in vegetative tissues and during fruit development. It was shown that the expression of *FaPE1* was induced by auxin at the onset of fruit ripening and down regulated by ethylene during post harvest periods. It was proposed by the authors that the ethylene mediated decrease of FaPE1 expression could be an important factor determining strawberry fruit post harvest decay. Removal of endogenous auxin source in halved green fruit resulted in a reduced *FaPE1* transcript accumulation, whilst exogenous auxin application caused an enhancement of *FaPE1* expression. It is possible that the repression of *FaPE1* mRNA expression at the late stages of fruit ripening and, particularly during postharvest could be an important factor determining the life of the strawberry fruit.

GA may inhibit ripening by retarding the overall chlorophyll degradation or senescence phase of fruit ripening (Bruinsma *et al.* 1981). They may even delay or promote the reassembly of functional chloroplasts (Goldschmidt, 1980). The relative effectiveness of exogenous GAs and cytokinins in relation to pigmentation of kinnow mandain fruits, particularly at later stages of ripening, may be related to a general decline in their endogenous levels (Nagar, 1993). However, in avocado, Libermann *et al.* (1977) found little or no effect of GAs on fruit ripening at later stages of maturity. Otmani and Coggins (1991) have also reported that GA3 delays ripening in the rind of stored citrus fruits without affecting internal fruit quality. Very few investigations have been carried out to study the effect of ethylene on GA levels and more is known about the interrelationships between ethylene and auxin than between ethylene and GAs. They do not regulate ethylene synthesis; however, it does influence ethylene action. Shechter *et al.* (1989) noted that GA reduced ethylene-induced aging during ripening of tomato and acted as the intermediary hormone for ethylene-induced elongation in rice (Suge, 1985).

## 4. CHANGES IN CYTOKININS

As discussed earlier, fruit development is a complex process that could involve all the known plant growth substances. Thus it is difficult to distinguish these changes, which are only related to fruit maturation. In fruits, chlorophyll loss and the conversion of chloroplast to chmromoplasts are frequently accompanied by a decrease in cytokinin (CK) levels (Monselise *et al.* 1978). The fruits of non-ripening *rin* tomato contains higher CK levels (Davey and Van Staden, 1978) and as the fruit changes colour, the level decreases. In olive fruits (Shulman and Lavey, 1973), CK treatments altered the colour characteristics but not the softening. CKs appear to antagonize many of the physiological processes thought to be medicated, all or in part, by ABA. For example, ABA induced stomatal closure; leaf senescence and leaf and fruit abscission are reversed by exogenous application of Cks while Ck- mediated release of seed dormancy contrasts with ABA inhibition of germination (Salisbury, 1994). This implied antagonism might be the result of metabolic interactions particularly as Cks share, at least in part, a common biosynthetic origin with ABA. CK biosynthesis appears to involve the addition of dimethylallyl pyrophosphate to the N6 position of adenosine monophosphate (AMP) and subsequent conversion of $N^{6\bullet 2}$- isopentenyl AMP to $N^{6\bullet 2}$- isopentenyl adenosine to 2iP, followed by hydroxylation of 2iP to yield 'zeatin like' cytokinins (Binns, 1994). By contrast, ABA is regarded as an apocarotenoid derived from the metabolism of isopentenyl pyrophosphate via xathoxine, 9-cis neoxanthin and xanthoxal, which is then, oxidized to ABA (Cowan and Richardson, 1997; Tan *et al.* 1997), Even so, details of a possible biochemical interaction between cytokinins and ABA during fruit ripening has hitherto remained obscure.

However, research into the metabolic control of avocado fruit ripening has revealed several interesting aspects related to cytokinin -ABA interaction. These include (i) final fruit size is linearly correlated with the endogenous CK: ABA ratio, (ii) mevastanin-induced retardation of avocado fruit growth occurred concomitant with a decline in hydroxy-methylglutaryl coenzyme reducatase (HMGR) activity and increased endogenous ABA concentration responses that were negated in the presence of 2iP; (iii) ABA induced inhibition of cell-to-cell chemical communication was negated by 2iP; and (iv) ABA- induced phenotypic variation was negated in the presence of 2iP (Cowan *et al.* 1997; Moore–Gordon *et al.* 1998). These observations suggest that in addition to supporting the proposed antagonism between cytokinin and ABA, the interactions between the both is exerted at some distance from HMGR in the ABA biosynthetic pathway because 2iP did not fully restore HMGR activation of ABA treated fruit, but reversed ABA- induced retardation of avocado fruit growth (Cowan *et al.* 1997). Cytokinins have also been shown to inhibit the production of ABA by cultures of ABA producing fungi *Botrytis cinera* (Hirai *et al,* 1986) and *Ceorcospore rosicola* (Norman *et al,* 1983) and, in the latter, specifically reduced incorporation of level from [1-14 C] farnesyle pyrophosphate into ABA. The cytokinin analogue and inhibitor of GA biosynthesis, ancymidol, also inhibited ABA biosynthesis in avocado mesocarp and *Ceorcospore rosicola* (Cowan and Railton, 1987). Ancymidol appears to interact directly with cytochrome P-450 dependent mono-oxygenases by binding of electrons from an Sp2 hybridized N- atom in the heterocycle to the protohaem ion thereby displacing the molecular oxygen required for catalysis (Grossman, 1990). Interestingly, the mode of action of cytokinins and ancymidol in avocado mesocarp during ripening has been shown to be similar with respect to the oxidation of ent-kauren, i.e. interaction with cytochrome P-450 (Coolbaugh, 1984). The catabolism of ABA to phaseic acid in higher plants is a cytochrome P-450 mediated reaction, which is catalyzed by ABA 8- hydroxylase (Creelman and Zeewart, 1984). Thus in the presence of inhibitors of cytochrome P-450 activity, an increase in endogenous ABA concentration might be expected.

The endogenous cytokinins present in fruits have been the subject of many studies largely because these structures are very rich sources of these substances (Goodwin, 1978; Nagar and Saha, 1985). In general, cytokinin levels in fruits are high during active cell division and also during rapid fruit growth, which gradually decline to relatively low levels at maturity (Nagar *et al*. 1982). Not all the cytokinin changes in parallel, for, in cherry tomato (Abdel Rahman *et al.* 1975) its riboside and in guava (Nagar and Rao, 1981) its glucoside increases during ripening when total activity decreases. Since fruit or at least seeds are considered to be the active site of auxin, GAs and cytokinin synthesis, it is not clear why they would need to import large quantities of these substances especially the latter. In *Cassia fistula* (Iyer *et al.* 1984), relatively high CK activity was still present in embryo than endosperm when the fruit had reached its maturity. It was being proposed by the authors that the seed may not only be considered as a reservoir of CK activity for developing fruit and a sink for drawing nutrients but also as a reservoir of plant growth substances for germinating seeds. In the competing sink hypothesis it has been proposed (Luckwill, 1977) that these nutrients are needed for building up new tissues. High concentrations of plant growth substances found in fruits seem to be necessary to create a strong sink capable of competing with the rest of the plant and CKs have been implicated in such a hypothesis.

Further, the decline of endogenous cytokinins in matured fruits may be due to metabolism, decrease in biosynthesis or due to diminished transport. The relative importance of each one may vary among tissues and species. Compartmentations of these may explain why some cytokinins especially its glucosides increase during maturity. It may be indicative of some control mechanism whereby these cytokinins considered being the storage forms, will only be hydrolyzing to the freebases and utilizing as and when needed (Letham and Palni, 1983). The question of CK interconversions which involves structural changes and hence a likelihood of a change in biological activity and functions falls beyond the scope of the review. However, it is becoming more and clearer that enough attention has to be given to this aspect if the role of CKs during fruit ripening and post harvest is to be well understood.

## Conclusions

Fruit ripening is a complex, genetically programmed process that culminates in dramatic changes in colour, texture, and aroma of the fruit. Due to economic importance of fruit crop species, these processes have been, and continue to be studied extensively at physiological, biochemical and genetic levels. Further, the economic importance of fruit has served as an incentive to study ripening biochemistry. The transformation of fertilized ovaries and accessories tissues into edible fruit is a multistep process with many parallel and sequential events. Most workers have focused on the economic parameters such as texture, colour and flavour, which are important in fruit quality. Fruit ripening is not only considered as a highly coordinated sequence of independent processes but it also involves a number of relatively independent changes that normally occur at the same time. It is apparent that the performance of fruits after harvest is controlled by the interactions of endogenous growth promoters and inhibitors like auxins, gibberellins, cytokinins, abscisic acid and ethylene and that the balance of these substances can be altered both by preharvest treatment and storage environment. Ethylene besides abscisic acid plays major roles during fruit ripening and post harvest periods. However, in addition to increasing the exploration of gross hormone changes, more concise studies are required to elucidate the way in which these plant growth substances are involved in the basic mechanism controlling post harvest fruit development.

## REFERENCES

Abdel-Rahman, M: Thomas, T.H; doss, G.J, and Howell, I. 1975 Changes in endogenous plant hormones in cherry tomato fruit during ripening, development and maturation. *Plant Physiol.***34**: 39-43.

Adalto, I; Gazit, S; and Blumenfeld, A. 1976. Relationship between changes in abscisic acid and ethylene during ripening of avocado fruit. *Aust. Jour. Plant Physiol.* **3**: 555- 558

G, G; Smit, P; and O'Connell, A.P. 2002. Novel insight into vascular, stress and auxin-dependent and independent gene expression programme in strawberry, a non- climacteric fruit. *Plant Physiol.*, **129**: 1019-1031.

Aharoni, Y. 1968. Respiration of oranges and grape fruits harvested at different stages of development. *Plant Physiol.*, **43**: 99-102.

Amoros, A, Serrano,M., Requelme, F., Romojaro, F. 1989. Levels of ACC and physical and chemical parameters in peach development. *J. Hort. Sci.*, **64**: 673-677.

Archbold, D.D. and Dennis, Jr. F.G. 1984. Quantification of free ABA and free and conjugated IAA in strawberry achene and receptacle tissue during fruit development. *Jour. Amer. Soc. Horti. Sci.*, **109**: 330-335.

Bangerth, F. 1982. Changes in the ratio of *cis-trans* abscisic acid during ripening of apple fruit. *Planta*, **155**: 199-203

Berry, C.S., Llop-Tous, M-I., and Grierson, D. 2000. The regulation of 1- aminocyclopropane-1-carboxylic acid synthase gene expression during the transition from system 1 to system 2 ethylene synthesis in tomato. Plant *Physiol.*, **123**: 979- 986.

Bhattacharya, A., Nagar, P.K. and Ahuja, P.S., 2002. Seed development studies in tea. (*Camellia sinensis* L. (O) Kuntze). *Seed Sci. Res.*, **12**: 39-45.

Biale, J.B. 1960. The post harvest biochemistry of tropical and subtropical fruits. Science. 146: 88-88.

Biale, J.R. 1964. Growth, maturation and senescence in fruits. *Science*, **146**: 880-882

Biale, J.B. and Young, R.E. 1981. Respiration and ripening in fruits- retrospect and prospect, In: *Recent Advances in the Biochemistry of Fruits and Vegetables*. (ED.) Friend, J.and Rhodes, M.J.C. Acad. Press, London

Binns, A.N. 1994. Cytokinin accumulation and action: Biochemical, genetical and molecular approaches. *Annu. Rev. Plant Physiol. and Mol. Biol.*, **45**: 173-196.

Bleecker, A.B. and Kende, H. 2000. Ethylene a gaseous signal molecule in plants. *Annu. Rev. Cell Dev. Biol.*, **16**:1-18.

Bower, J.P. Cutting, G.M; and van Kelyveld, I.J. 1986. Long term irrigation as influencing avocado abscisic acid and fruit quality. *S. Afr. Avocado Grow. Asso. Year Book*, **9**: 43-45

Brady, J. 1987. Fruit ripening. *Annu. Rev. Plant Physiol.*, **38**: 155-171

Bruinsma, J. 1981 Hormonal regulation of senescence, ageing, fading and ripening. *Acta Horti.*, **138**: 141-163

Brummell, D.A., Harpster, M.H., Civello, P.N., Palys, J.M., Bennett, A.B. and Dunsmuir, P. 1999 Modifications of expansin protein abundance in tomato fruit alters softening and cell wall polymer metabolism during ripening. The *Plant Cell*, **11**: 2203-2216.

Brummell, D; and Harpster, M.H. 2001. Cell wall metabolism in fruit softening and quality and its manipulation in transgenic plants. *Plant Mol. Biol.*, **47**: 311-340

Burg, S.P; Apelbaum, A; Eisinger, W. and Kang, B.G. 1971. Physiology and mode of action of ethylene. *Hort Sci.*, **6**: 359-364

Capitani, G., Hohenester, F., Feng, L., Storici, P., Kirsch, J.F. and Jansonius, J.N. 1999. Structure of 1-aminocylopropane-1 carboxylate synthase, a key enzyme in the biosynthesis of the plant hormone ethylene. *J. Mol. Biol.*, **294**: 745-756.

Carpita, N.C. and Gibeant, D.N. 1993. Structural models of primary cell walls in flowering plants: consistency of molecular structure with the physical properties of the walls during growth. *The Plant Jour.*, **3**:1-30.

Castillejo, C; Fuente, J.L; de la; Iannetta, P; Boetella, M. and Valpuesta, V. 2004. Pectin estrase gene family in strawberry fruit. *J. Expt. Bot.*, **55**: 909-918.

Catala, C; Ostin, A; Chamarro, J; Sundberg, G. and Crozier, A. 1992. Metabolism of indole-3-acetic acid by pericarp disc from immature and mature tomato. (*Lycopersicum esculentum* Mill. *Plant Physiol.*, **100**: 1457-1463.

Cawthon, D.L. and Morris, J.R. 1982. Relationship of seed number and maturity to berry development, fruit maturation, hormonal changes and uneven ripening of Concord (*Vitis labrusca* L.) grapes. *J. Amer. Soc. Hort Sci.,* **107**: 1097-1114.

Chandler, P.M. and Robertson, M. 1994. Gene expression regulated by ABA and its relation to stress tolerance. *Annu Rev. Plant Physiol, Plant Mol. Biol.*, **45**: 113-141.

Chun, J.P., and Huber, D.J. 2000. Reduced levels of p-submit protein influence tomato fruit firmness, cell wall ultrastructure and PG-2 mediated pectin hyrolysis in excised pericarp tissue. *J. Plant Physiol.*, **157**:153-160

Coolbough, R, C. 1984. Inhibition of *ent*-kaurene oxidation of cytokinins. *J. Plant Growth Regul.,* **3**: 97-109.

Coombe, B. G. and Hale, C.R. 1973. The hormone content of ripening grape berries and the effects of growth substance treatments. *Plant Physiol.*, **51**: 629-634.

Cooper, W., Bouzayen, M., Hamilton, A., Berry, C., Rosswell, S. and Grierson, D. 1998. Use of transgenic plants to study the role of ethylene and polygalacturonase during infection of tomato fruit by *Colletotrichum glucosporioides*. *Plant Physiol.*, **47**: 308-316.

Cosgrove, D.J. 2000. Loosening of plant cell wall by expansins. *Nature*, **407**: 321-326.

Cowan, A.K. and Railton, I.D. 1987. Cytokinins and ancymidol inhibit abscisic acid. *J. Plant Growth Regul.*, **4**: 211-224

Cowan, A.K; Morre-Gordon, C.S; Burting, I. and Wolstenholme, M.J. 1997. Metabolic control of avocado fruit growth: isopreponyl growth regulators and the reactions catalysed by 3- hydroxy-3-methylglutaryl coenzyme A reductase. *Plant Physiol.*, **114**: 511-516.

Cowan, A.K; and Richardson, G.R. 1997. The biosynthesis of absciscic acid in a cell free system from embryos of *Hordeum vulgare*. *Physiol. Plant.*, **99**: 371-378.

Creelman, R. A. and Zeewart, J.R.D. 1984. Incorporation of oxygen into abscisic acid and phaseic acid from molecular oxygen. *Plant Physiol.*, **75**: 166-169.

Crookes, P.K. and Grierson, D. 1983. Ultrastructure of tomato fruit ripening and the role of polyyalacturonase isozyme in cell wall degradation. *Plant Physiol.*, **72**: 1088-1093.

Davey, J.E. and van Staden, J. 1978. Cytokinin activity in *Lupinus albus*. III. Distribution in fruits. *Physiol. Plant.*, **43**: 87-93.

Davies, P.J. 1995. *Plant Hormone*: Physiology, Biochemistry and Molecular Biology. (Ed). P.J. Davies, Kluwer Academic, The Netherlands

Domato, P.A. and Hewitt, A.A. 1973. Ethephon increases endogenous auxins in seeds of *Prunus salicina. Hort Sci.*, **8**:503-504

El –Kazzay; M.K. Somner, N.F; and Fortlarge, R.J. 1983. Effect of different atmosphere on post harvest decay and quality of fresh strawberry. Phytopath. 73: 282-285.

El. Beltagy, A.S. Hewitt, E.W., and Hall, M.A. 1976. Effect of ethephon on endogenous levels of auxins, inhibitors and cytokinins in relation to senescence and abscission in *Vicia faba. J. Hort. Sci.*, ***51***. 451- 465.

Elkashif, M.E., Huton, D.J. and Brecht, J. K. 1989. Respiration and ethylene production in harvested watermelon fruit: evidence for non-climacteric respiratory behaviour. *J. Amer. Soc. Hort. Sci.*, **114**: 81-85.

Epstein, E.1982. Levels of free and conjugated indole -3- acetic acid in ethylene treated leaves and callus of olive. *Physiol. Plant.*, **56**: 371-373.

Frankel, C; Dyck,C. and Haard, N.E. 1975. Role of auxin in the regulation of fruit ripening. In: Haard, N.E., Salunkhe, D.K. Eds. Postharvest Biology and Handling of fruits and Vegetables. AVI Pub. pp .19-34

Gaffe, J., Tilman, D.M. and Handa, A.K. 1994. Pectin methylesterase isoforms in tomato (*Lycopersicon esculentum*) tissues: effects of expression of a pectin methylesterase gene. *Plant Physiol.*, **105**: 199-203.

Gazit, S; and Blumenfeld, A.1972. Inhibitor and auxin activity in the avocado fruit. *Physiol. Plant.*, **27**: 77-82

Goldschmidt, E.E., Goren R., Even-Chen, Z. and Bittner,S. 1973. Increase in free and bound abscisic acid during natural and ethylene induced senescence of citrus fruit peel. *Plant Physiol.*, **51**: 879-882.

Goldschmidt; E.E. 1980. Pigment changes associated with fruit maturation and their control. In; Thimann, K.V. eds. *Senescence in Plants*. CRC Press, Boca Raton, pp. 207-217.

Goodwin, P.B. 1978. Phytohormones and fruit growth. In: Letham, D.S; Goodwin, P.B. and Higgins eds. *Phytohormons and Related Compounds: A Complete Treatise*. Biomedical Press, Amsterdam, 2: 173-214.

Grierson, D. and Schuch, W. 1993. Control of ripening. *Physiol. Trans. Royal Soc*. London, B **342**: 241-250.

Grossman, K, 1990. Plant growth retardants as a tool for physiological research. *Physiol. Plant.*, **78**: 640-648

Hirai, N; Okamoto, M; and Koshimizu, K. 1986. The 1,4 trans abscisic acid, a possible precursor of abscisic acid in *Botrytis cinera*. *Phytochem.*, **25**: 1865-1868.

Huai, Q; Xia; Chen, Y.Q; Callahan, B; Li, N. and Ke, H.M. 2001. Crystal structure of 1-aminocyclopropane, 1-carboxylate (ACC) synthase in complex with aminoethyoxyvinylglycine and pyridoxal-5- phosphate provide new insight into catalytic mechanisms. *J. Biol. Chem.*, 276: 38210-38216.

Itzhaki, H; Maxson, J.M; Woodson, W.R. 1994. An ethylene responsive enhancer element is involved in the senescence – regulated expression of the carnation glutathione-S- transferase (*GST1*) gene. *Proc. Natl. Acad. Sci.* (USA), **91**: 8925-8929.

Iyer, R.I; Nagar, P.K and Sircar, P.K.1984. Cytokinin changes in embryo and endosperm of *Cassia fistula* L. *J. Plant Physiol.,* **117**: 87-92.

Jones, W.W; Coggins, CH.W. and Embleton, T.W. 1976. Endogenous abscisic acid in relation to bud growth in alternate bearing ' Valencia' orange. *Plant Physiol.,* **58**: 681-682.

Joubert, A.J. 1986. Litchi. In: Monselise, S.P. (ED.). *Handbook of Fruit set and Development.* CRC Press Inc. Boca Raton pp.233-246

Kalinin, F.L.and Kurchii, I.1986 Effects of ethaphon on the content of endogenous phytohormones in winter rye plants. *Fiziol. Biokhim. Kultur. Ras.*, **18**: 182-185.

Kardo, S; Indoc, K. and Tamaki, M. 1999. Changes of abscisic acid and its metabolites during development of apple fruit. *J. Hort. Sci. Biotech.,* **74**: 762-767

Kende, H. 1993. Ethylene biosynthesis. *Annu. Rev. Plant Physiol. Plant Mol. Biol.* **44**: 283-307.

Kidd, F. and West, C. 1925. The course of respiratory activity throughout the life of an apple. *Gt. Brit. Dept. Sci. Ind. Res, Food Invest. Rep.*, 27-33.

Letham, D.S. and Palni, L.M.S. 1983. The biosynthesis and metabolism of cytokinins. *Annu. Rev. Plant Physiol.,* **34**: 214-217.

Liberman, M.J; Backer, E. and Sloga, M. 1977. Influence of plant hormones on ethylene production in apple, tomato and avocado slices during maturation and senescence. *Plant Physiol.*, **60:** 214-217.

Llop-Tous, L.I., Berry, C.S. and Grierson, D. 2000. Regulation of ethylene biosynthesis in response to pollination in tomato flowers. *Plant Physiol.*, **123**: 971-978.

Looney, N.E, MGlasson, W.B., and Coombe, B.G. 1974. Control of fruit ripening in peach, *Prunus persica*: Action of succinic acid 2-2 dimethyl hydrazide (2- chloroethyl) phosphonic acid. *Aust. J. Plant Physiol.*, **1**:77-76.

Lu, C.G., Zainal, Z. Tucker, G.A., and Lycett, G.W. 2001. Developmental abnormalities and reduced fruit softening in tomato plants expressing an antisene *Rubl 1* GTPase gene. *The Plant Cell*, **13**: 1819-1933.

Luckwill, L.C. 1977. Growth regulators in flowering and fruit developments. In: J.R. Plimmer, eds. *Pesticide Chemistry in 20th Century ACS Symposium Series,* **37**: 293-304.

Manning, K. 1998. Isolation of a set of ripening related genes from strawberry: their identification and possible relationship to fruit quality traits. *Planta,* **205**: 622-631.

Mc Murchie, E.L. McGlassn, W.B. and Eaks, E.L. 1972. Treatment of fruits with propylene gives information about the biogenesis of ethylene. Nature 237: 235-236.

Milborrow, B.V. 1970. The metabolism of abscisic acid. *J. Expt. Bot.*, **21**:17-20.

Milborrow, B.V. 1974. The chemistry and physiology of abscisic acid. *Annu. Rev. Plant Physiol.,* **25**: 269-307.

Milborrow, B.V. and Robinson, D.R. 1983. Factors influencing the biosynthesis of abscisic acid. *J. Exp. Bot.*, **24**: 537-548.

Monselise, S.P. Varga, A., Knegl, E. and Bruinsma, J. 1978. Course of zeatin content in tomato fruit and seeds developing on intact or partially defoliated plants. *Z. Pflanzenphysiol.,* **90**: 451-460.

Montgomery, J., Polland, V., Deikman, J., Margossian, L. and Fischer, R.L. 1993. Positive and negative regulatory regions control the spatial distributions of polygalacturonase transcription in tomato fruit pericarp. *The Plant Cell,* **5**: 1049-1062.

Moore –Gordon, C.S; Cowan, A.K; Berting, I; Botha, C E.J. and Cross, R.H.M. 1999. Symplastic solute transport and avocado fruit development: a decline in cytokinin: ABA ratio is related to appearence of the Hass small fruit variant. *Plant and Cell Physiol.,* **24**: 1027-1038.

Nagar, P.K. 1993. Effect of plant growth regulators on natural and ethylene induced pigmentation in kinnow mandarin fruit. *Biol. Plant.,* **35**: 633-636.

Nagar, P.K. and Kumar, A. 2000. Changes in endogenous gibberellins during winter dormancy in tea (*Cammellia senensis* L. (O.) Kuntze. *Acta Physiol. Plant.,* **20**(4): 439- 433.

Nagar, P.K. and Saha, S. 1985. Hormonal Physiology of Fruit Growth. *Indian Rev. Life Sci.,* **5**: 249-276.

Nagar, P.K. and Sood S. 2004. Recent advances in plant hormone receptors. In: *Advances in Plant Physiology,* An International Treatise *Series* (ed. Hemantaranjan, A.), **7**: 383-400.

Nagar, P.K.and Saha, S. 1985. Hormonal Physiology of Fruit Growth. *Indian Rev. Life Sci.,* **5**: 249-276.

Nagar, P.K; and Raja, Rao, T. 1981. Studies in endogenous cytokinins in guava (*Psidium guajava* ) L. *Ann. Bot.*, **48**: 445-452.

Nagar, P.K; Iyer, R.I. and Sircar, P.K. 1982. Cytokinins in developing fruits of *Moringa pterigosperma*. Ca. fruit. *Physiol. Plant.*, **55**: 45-50.

Nicholas, F.J., Smith, C.J.S., Schuch, W., Bird, C.R. and Grierson, D. 1995. High levels of ripening specific reporter gene – expression directed by tomato fruit polygalacturonase gene flanking regions. *Plant Mol. Biol.*, **28**: 423-435.

Noga, G.J., and Bakovac, M.J. 1987. Peach fruit abscission. : Physiological changes induced with foliar application of (2-chloroethyl) methylbis (phynylmethoxy) silane (GA 15281). *Gartenbauwissenchaft.*, **52**: 278-284.

Norman, S.N; Bennett, R.D; Maier, V.P. and Poling, S.M. 1983. Cytokinin inhibits absciscic acid biosynthesis in *Cercospora rosicola. Plant Sci. Letters*, **28**: 255-263

Oetikjer, J.H., Olson, D.C., Shiu, Y., Yang, S.F. 1997. Differential induction of seven 1-aminocyclopropane synthase genes by elicitor in suspension cultures of tomato (*Lycopersicon esculentum*). *Plant Mol. Biol.,* **34**: 275-286.

Perkins- Veazie, P.M. 1995. Growth and strawberry fruit. *Hort. Rev.*, **17**: 267-297.

Perkin-Veazie, P.M; Huber, D.J. and Brecht, J.K. 1996. *In vitro* growth and ripening of strawberry fruit in the presence of ACC, STS or propylene. *Ann. Appl. Biol.,* **128**: 105-116.

Otmani, E.M. and Coggins, C.W. Jr. 1991. Growth regulator effects on retention of quality of stored citrus fruits. *Scientia Hortic.*, **45**: 261-262.

Riov; J., and Bangerth, F. 1992. Metabolism of auxin on tomato fruit tissue. Formation of high molecular weight conjugates of oxindole 3- acetic acid via the oxidation of indole -3- acetylaspartic acid. *Plant Physiol.*, **100**:1396-1416.

Robinson, D.R; and Ryback, G. 1969. Incorporation of tritium from 4(R) 43H mevalonate into ABA. *Biochem J.,* **113**: 895-897

Ross, J.K.C., Lee, H.H., and Bennett, A.B. 1999. Expression of a divergent expansive gene is fruit specific and ripening regulated. *Proc. National Acad. Sci.* (USA), **94**: 5955-5960.

Rowan, K.S., Pratt, H.K. and Robertson, R.N. 1958. Relationship of high-energy phosphate content, protein synthesis and the climacteric rise in the respiration of ripening avocado and tomato fruits. *Aust. J. Biol. Sci.*, **2**: 329- 335.

Sekse, 1988. Respiration of plum (*Prunus domestica* L.) and sweet cherry (*P. sativum*) fruits during growth and ripening. *Acta Agric. Scand.*, **38**: 317-320.

Salisbury, F.B. 1994. The role of plant hormones In: Wilkinson, R.E. ed. *Plant – Environment Interactions.* New York; Marcel, Dekker, 39-81.

Saltveit, Jr. M.E. 1977. Carbon dioxide, ethylene, and colour development in ripening mature green bell peppers. *J. Amer. Soc. Hort.*, **102**: 523-525.

Salunkhe, D.K., and Desai, B.B. 1984. *Postharvest Biotechnology of Fruits*. CRC Press, Boca. Ratio, Florida.

Shechter, S., Goldschmidt, E.E. and Galili, D. 1989. Persistence of (14C} gibberellin A3 and (3H) gibberellins A1 in senescing, ethylene treated citrus and tomato fruit. *Plant Growth Regul.*, **8**: 243-253.

Shin, O.Y., Oeitker, J.H. Yip, W.K., and Yang, S.F. 1998. The promoter of LE-ACS7 and early flooding induced 1- aminocyclopropance –1- carboxylate synthase gene of the tomato is tagged by a SOL3 transpozon. *Proc. Natl. Acad. Sci.* (USA), **95**: 10334-10339.

Shulman, Y. and Lavee, S. 1973. The effect of cytokinins and auxins on anthocyanin accumulation on green Manzamillo olives. *J. Exp. Bot.*, **24**: 655-651.

Sitrit, Y. and Bernett, A.B. 1998. Regulation of tomato fruit polygalacturonase m RNA accumulation by ethylene: a re-examination. *Plant Physiol.*, **116:** 1145-1150.

Smith, C.J.S. Watson, C.F., Ray, J., Brid, C.R., Morris, P.K., Schuch, W. and Grierson, D. 1988. Antisense RNA inhibition of polygalacturonase gene expression in transgenic tomatoes. *Nature,* **334**: 724-726.

Smith, C.J.S., Watson, C.F., Bird, C.R., Raj, J., Schuch, W. and Grieson, D. 1990. *Mol. Gen. Genet.*, **224**: 477-481.

Smith, D.L. and Gross, K.C. 2000. A family of at least seven beta- Galactosidase genes is expressed during tomato fruit development. *Plant Physiol.*, **123**: 1173-1183.

Sood, S. and Nagar, 2003. Changes in abscisic acid and phenols during flower development in two diverse species of rose. *Acta Physiol. Plant.,* **25**(4): 411- 416.

Suge, H. 1985. Ethylene and gibberellins: Regulation of internodal elongation and nodal root development in floating rice. *Plant Cell Physiol.*, **6**: 607-614.

Tan, B.C; Schwartz, S.H; Zeevart, J.A.D. and McMarthy, D.R. 1997. Genetic control of abscisic acid biosynthesis in maize. *Proc. Natl. Acad. Sci.* (USA), **94**: 12235-12240.

Tarun, A.S. and Theologis. A. 1988. Complementation analysis of mutants of 1- aminocyclopropane –1- carboxylate synthase reveals the enzyme is a dimer with shared active sites. *J. Biol. Chem.*, **273**: 12509-12514.

Tielmann, D.M. and Handa, A.K. 1994. Reduction in pectin methylesterase activity modifies tissue integrity and cation levels in ripening tomato *(Lycopersicon esculentam* Mill) fruits. *Plant Physiol.*, **106**: 429-436.

Tsay, L.M. and Mizuno, S. 1984. Changes of respiration, ethylene evolution and abscisic acid content during ripening and senescence of fruits picked at young and mature stage. *J. Jap. Soc. Horti. Sci.*, **52**: 458-463.

Tsuchida, M; Mizuno, S, and Kozukue, N. 1990. Changes in abscisic acid and phaseic acid during ripening and senescence of peach fruits. *J. Jpn. Soc. Hort. Sci.*, **58**: 801-805.

Tucker, G.A., and Grierson, D. 1982. Synthesis of polygalacturonase during tomato fruit ripening. *Planta,* **155**: 64-67.

Varga, M., Rusznak, A., and Nikl. K. 1982. Effect of ethrel on the IAA content and on its distribution in cucumber seedlings. *Biochem. Physiol. Pflanzen.*, **177**: 659- 669.

Vendrell, M, and Busea, A. 1989. Relationship between abscisic acid and ripening of apples. *Acta Horti.*, **283**: 389-396.

Watkins, C.B., Brown, J.H., Walker, V.J. 1989. Assessment of ethylene production by apple cultivars in relation to commercial harvest dates. *New Zealand J. Crop Hort. Sci.,* **17**: 327-331.

Wills, R.H.H; Lee, T.H. Graham, D.; McGlasson, W.B. and Hall, E.G. 1981. *Postharvest: An Introduction to the Physiology and Handling of Fruits and Vegetables.* NSW University Press, Kennington, Australia.

Wong, W.S., Nung, W., Xu, P.L., Kung, S.K., Yang, S.F, and L, N. 1999. Identification of two chilling related 1-aminocyclopropane-1- carboxylate synthase genes from citrus (*Citrus sinensis* O.) fruit. *Plant Mol. Biol.,* **41**: 587-600.

Yang, S.F. 1985. Biosynthesis and action of ethylene. *Hort. Sci.,* **20**: 41-45.

Zegzonti, H., Jones, B., Fresu, P., Marty, C., Maitre, B., Latche, A., Pch, J-C. and Bouzayan, M. 1999. Ethylene- regulated gene expression in tomato fruit: characterization of novel ethylene-responsive and ripening related genes induced by differential display. *The Plant J.,* **18**: 589-600..

# SUBJECT INDEX


1&6$^g$ kestotetraose 358
1,1 nystose 358
1,1,1 kestopentaose 358
1-FEH (Fructan 1-Exo Hydrolase) 353
1-FFT (Fructan: Fructan 1-fructosyl Transferase) 353
1-kestose 358
1-SST (Sucrose: Sucrose 1-fructosyl Transferase) 353
2-(4-carboxyphenyl)-4,4,5,5-tetramethylimidazoline-1-oxyl-3-oxide 24
2,3-enoyl-CoA reductase 235
2-oxoglutarate amino transferase 161
2-peroxiredoxin 27
3-hydroxyacyl-CoA dehydrase 235
3-ketoacyl-CoA reductase 235
3-ketoacyl-CoA synthase 235
4-chloroindole-3-acetic acid 369
5-Cl-dioxindole 370
5-Cl-dioxindole-3-acetic acid 370
6&1-FEH (Fructan Exohydrolase) 353
6,6 nystose 358
6-FEH (Fructan 6-ExoHydrolase) 353
6$^g$, 1 & 6 kestopentaose 358
6$^g$, 6-kestotetraose 358
6G-FFT (Fructan: Fructan 6-fructosyl transferase) 353
6-KEH (6-kestose Exo Hydrolase) 353
6-kestose 358
6-SFT (Sucrose: Fructan 6-fructosyl Transferase) 353
8-bromoadenosins-3,5-cyclic monophosphate 353
α-amylase 173
α-tocopherol 177
β-aminobutyric acid (BABA) 31
β-fructofuranosidase 350
β-fructosidases 350

ABA-induced stomatal closure 8
ABC transporter 241
ABC-type transporter 27
ABI-3 gene 21
Abiotic oxidative stress 49
Abiotic stresses 157
Abscisic acid (ABA) 4, 25, 28, 205, 302, 406, 457
Acanthaceae 141
*Acanthus* 118
*Acanthus ebracteatus* 141
*Acanthus ilicifolius* 120, 151, 153, 154,
*Acanthus volubilis* 141, 152
*Acathus ebracteus* 152
*Acer pseudoplatanus* 452
ACGT-containing G-box motif 21
Acid invertases 25
Acid phosphatase 159, 161, 173
*Acrostichum aureum* 141
Actinorhizal symbiosis signaling 334
Acyl elongation 235
Additive toxicity mode 197
*Aegialitis annulata* 147, 148, 158
*Aegialitis rotundifilia* 141, 151
*Aegiceras* 118
*Aegiceras corniculatum* 119, 141, 146, 147, 148, 149, 151, 153
*Aegilatis* 118
*Aegle marmelos* 461
*Aeluropus lagopoides* 143
*Aeschynomeneae* 337, 339
*Aesculus hippocastanum* 452
*Agrobacterium rhizogenes* 335

*Agrobacterium tumefaciens* 88
Air pollution 49
Aizoaceae 141
Al toxicity 172
Aldehyde dehydrogenase 20
Aldolase 302
Aldolase antisense transgenic rice 305
Alfalfa 144
*Allium cepa* 349
*Allium schoenoprasum* 370
*Allocasuarina verticillata* 329, 333
*Alnus glutinosa* 329, 337
*Alphonsea madraspatana* 141
*Alternanthera paronychoides* 141
*Alternaria brassicicola* 208
Alternative oxidase (AOX) 2, 30
Alternative oxidative pathway 420
Aluminium 172
*Alyssum lesbicum* 180
Amaranthaceae 141
Amarylidaceae 141
Amino acid 159
*Ammora cucullata* 153
*Amoora cuculata* 143
Anacadiaceae 461
Annonaceae 141
Anthraquinone 197
Anthropogenic oxidative stress 52
Antifungal protein 301
Antioxidant biosynthesis 8
Anti-oxidant capacity 157
Antioxidant enzymes 176
Antioxidant response 175, 176
Antioxidants 5, 74, 458
Antioxidative enzymes 113
Apocynaceae 141
Apple 476
Aquaporin(s) 27, 117
Arabidopsis 21, 26, 144, 181, 208, 235, 370
*Arabidopsis thaliama* 30, 183, 260, 351, 355, 457
Araceae 140, 141, 142
*Arachis* 339
*Araucaria angustifolia* 452
*Aristolochia indica* 142
Aristolociaceae 142
*Arthrocnemum indicum* 142
*Artocarpus heterophyllus* 452, 461
Asclepiadaceae 142
Ascorbate (AsA) 27, 75, 177
Ascorbate glutathione cycle 3
Ascorbate peroxidases (APXs) 3, 27, 44
Ascorbic acid (AA) 3
*Asparagillus awamori* 355
Asparagine 24
*Asparagus officinalis* 349, 350
*Asparagus racemosus* 153
*Aspergillus* 347
*Aspergillus sydowi* 348
Asteraceae 142
*Atriplex gmelini* 148
*Atriplex nummularia* 148
AtZEP gene 21
Auxins 401
*Avena sativa* 348, 350
Avicennia 117, 118
*Avicennia alba* 140, 151, 153
*Avicennia germinans* 117, 120, 146, 147, 148
*Avicennia marina* 120, 121, 140, 146, 147, 148, 149, 150, 151, 153, 452, 457, 461
*Avicennia officinalis* 140, 151, 153
Avocado 476
*Azadirachta indica* 452
*Azorhizobium caulinodans* 336

*Bacillus subtilis* 355
Banana 476
*Barringtonia racemosa* 143, 457
Bell pepper 477
*Beta vulgaris* 70, 355
*Betulaceae* 328
Bifurcose 358
Bignoniaceae 142
Binding protein (BiP) 23
Bioactivity 371
Biology of reproduction 107
Biomonitoring 197
Biotic oxidative stress 43
*Borrelia burgdorferi* 269
*Botryococcus brownii* 240
*Botrytis cinerea* 208
*Bradysia* 208
*Brassica* 231
*Brassica juncea* 175
*Brassica oleracea* 232
Brassinosteroids 205
Bronzing 77

*Brownlowia lanceolata* 144
*Brownlowia tersa* 144
*Brucella suis* 269
*Bruguiera* 118
*Bruguiera cylindrica* 140, 151, 153
*Bruguiera gymnorrhiza* 140, 146, 147, 149, 150, 153
*Bruguiera parviflora* 121, 140, 147, 150
*Bruguiera sexangula* 140, 146, 148, 151

C32 acyl moieties 230
$C_4$ cycle 7
$Ca^{2+}$ channels 53
$Ca^{2+}$ homeostasis 42
$Ca^{2+}$ signalling 55
$Ca^{2+}$-dependent protein kinases 41
Cadmium 170
*Caesalpinia bonduc* 142
*Caesalpinia crista* 142, 152, 154
Caesalpiniaceae 142
Calcium 49
Calcium influx 50
Calcium signaling 52
Calcium signature 42, 54
Calvin cycle 7
CaM inhibitors 51
*Camellia sinensis* 461
*Campanula rapunculoides* 353
*Carapa guianensis* 452
*Carapa procera* 452
Carbohydrates 459
Carboxyl-terminal processing peptidase 256
Carboxyl-terminal processing 269
Carboxy-PTIO 24
Carnosic acid 25
Carotenoids 177
Carrier import pathway 272
Caseinolytic protease 258
*Castanea sativa* 461
*Casuarina glauca* 329, 333
*Casuarinaceae* 328
Catalase 3, 30, 159, 171, 176
*Catanospermum australe* 452
Cd toxicity 171
CDT-1 22
Cell cycle 456
Cell membrance integrity 159
Cell permeability 73
*Centaurea nigra* 78
*Cerbera manghas* 141, 152
*Ceriops* 118
*Ceriops decandra* 140, 150, 151, 153
*Ceriops tagal* 140, 146, 148, 150, 151
Cerium perhydroxides 5
Chaperonins 24
Chemical priming 31
Chemically induced ROS 52
Chemiluminescence method 94
Chenopodiaceae 142
*Chenopodium rubrum* 355
Chitinases 301
*Chlamydomonas reinhardtii* 181, 269
Chlorophyll fluorescence 72, 74
Chloroplast(s) 2, 6
Chlorosis 77
Chlrophyll fluorescence 16
Chromium 171
*Cicer arietinum* 78, 79
*Cichorium intybus* 355
*Cistus clusii* 9
Citrus 477
*Citrus limonum* 87
CLC-b chloride channel 27
Cleaved chloroplast 273
*Clerondrom ineme* 144
Cleveland peptide mapping 299
Climacteric 476
Climatic factors 156
$CO_2$ assimilation 16
$CO_2$ compensation point 209
$CO_2$ fixation 15
Coconut palm 155
Coconut seedlings 156
*Cocos nucifera* 461
*Coffea liberica* 452
Coleus 22
*Coleus blumei* 22
Combretaceae 140, 461
*Commelina communis* 52
Compatible solutes 115
Competition 386
Conductometric measure metres 198
Constitutive expressor 32
Convulvulaceae 142
Copper 173
*Coriariaceae* 328

*Crinum defixum* 141, 152
Crop management 156
Crop simulation modeling 315
*Cryptocornye ciliata* 141, 152
Cu toxicity 174
Cu/ZnSOD 48
*Cucumber mosaic virus* 208
Cuticular wax biosynthesis 231
Cuticular wax composition 230
Cuticular wax function 230
Cuticular waxes 229
Cu-Zn-SOD 87, 88, 90
Cyclophilin 303
Cyclosporin 303
*Cynara scolymus* 348
*Cynometra iripa* 142
*Cynometra ramiflora* 142, 152, 154
Cyperaceae 142
*Cyperus cephalotes* 153
Cysteine 7
Cytokinin(s) 205, 353, 404

*Dactylis glomerata* 350, 353
*Dalbergia spinosa* 142, 152, 154
*Datisa glomerata* 328
*Datiscaceae* 328
*Daucus carota* 355
Decarbonylation pathways 239, 240
Defence mechanisms 43
Dehydration 20
Dehydration tolerance 159
Dehydration-specific gene products 19
Dehydrin 30
Dehydrin regulation 457
Dehydrins 117, 457
Dehydroascorbate (DHA) 3
Dehydroascorbate reductase (DHAR) 3
Dehydroascorbate reductase activity 303
*Derris heterophylla* 142, 152, 154
*Derris indica* 142, 152, 154
*Derris scandens* 143
*Derris trifoliate* 143
Desiccation sensitivity 455
Desiccation tolerance 18
Desiccation-tolerant bryopytes 19
Detoxification 5
Digalactosyl diglyceride (DGDG) 74
Dipterocarpaceae 461
*Discaria trinervis* 333
Distribution of mangroves 102
DNA markers in mangroves 109
*Dolichondom spathecia* 142, 152
*Dolichos lablab* 370
*Drosophila melanogaster* 256
Drought 1
Drought index 159
Drought repression 25
Drought stress 4, 157
Drought tolerance 27, 159
Drought tolerance mechanism 13, 22
Dry matter production 159
Duckweeds 193
*Dunaliella salina* 145

*Echinops ritro* 348
Ectomycorrhizas 180
Edaphic factors 156
*Elaegnaceae* 328
Electron microscope 9
Electron spin resonance spectroscopy 94
Electron transport system 71
EMB-rubber model 320
Endogenous reactive 43
Endo-inulinase 349
ENOD genes 334
ENOD40 gene 22
Environmental stress 30
Epicuticular wax 159
*Epitrix hirtipennis* 211
*Erwinia amylovora* 359
*Erysiphe cichoracearum* 208
*Erysiphe orontii* 208
*Escherichia coli* 87, 181, 256
Ethylene 205, 476
Ethylene biosynthesis 479
Ethylene physiology 47, 87
Ethylene signaling 214
*Eugenia brasiliensis* 461
Euphorbiaceae 140, 461
*Euphoria longan* 461
*Excoecaria* 118
*Excoecaria agallocha* 151, 154
*Excoecaria agaocha* 140
Extension protein 460

Fabaceae 142, 327
Fagaceae 461
*Fagus sylvatica* 70, 72
Fatty acid biosynthesis 7
Fatty acids 230
Ferredoxin 6
Fe-SOD 91
Fig 476
Filamentation temperature sensitive protease 262
*Fimbristylis ferrugina* 142
*Finonsonia obovata* 142
Flowering 26
Foliar application 16
Foliar injury 77
*Frankia* 327
Frankia-Actinorhizal plants interaction 331
*Frankliniella occidentalis* 211
Free radical(s) 9, 174
Fructan 347
Fructan exohydrolases 347
Fructan fructan 6G frutosyl transferase 349
Fructan trisachharide 6 kestose 347
Fructosidase enzymes 358
Fruit ripening 26, 475, 476
*Fusarium* 347
*Fusarium oxysporum* 208
Fv/Fm chlorophyll fluorescence 159

Galactinol synthase activity 22
Gas exchange 103
Gene expression 8, 30, 157
Gene regulation 21
Genetic engineering 13
Genetic improvement 27
Genetic transformation 88
Genome 109
*Gentiana asclepiadea* 78
Gibberellic acid 205, 353
Gibberellin(s) 297, 302, 402
*Gluconacetobacter diazotrophicus* 359
Glutathione 3, 7, 27, 75, 175, 177
Glutathione peroxidases 3, 30
Glutathione reductase 44, 171, 174, 176
Glutathione S transferases 301
Glyceraldehyde 3-phosphate dehydrogenase 302
*Glycine max* 70, 370
*Glycine soja* 370
Glycinebetaine 16
Glycolate oxidase 30
Glycolysis 7
Glycophytes 144, 145
Glycoprotein 90
Grain growth 379
Grain yield 381
Grapes 477
GST-expression 54
Guaiacol peroxidase 171, 176
Guard cells 8

$H_2O_2$ 73
$H_2O_2$ sensing 9
$H_2O_2$ sensor 8
$H_2O_2$ signaling 8
Halophytes 102, 145
Harvest index 159
HDZIP genes 22
Heat shock proteins 24
Heat shock transcription factor 21
Heat stress 26, 51
Heavy metal detoxification 177
Heavy metal interaction 181
Heavy metal stress 169
Heavy metal toxicity 193
Heavy metals 174
*Helianthus tuberosus* 349
Herbicides 47
*Heriteria litorallis* 153
*Heriteria macrophylla* 153
*Heritiera fomes* 141, 151, 153
*Heritiera littoralis* 120, 141, 151
*Heritiera macrophylla* 141, 151
*Hevea brasiliensis* 88, 317, 452, 461
*Hibiscus esculentus* 452
*Hibiscus tiliaceus* 143, 146, 147, 152,
*Hibiscus tortuosus* 143
High temperature requirement protease 263
*Hippocratacea* 143
Histidine kinases 8
Histidine rich motifs 240
Homeodomain Leu Zipper family 21
*Horedum vulgare* 72, 145, 209
Hormone signal transduction pathway 205
Hydrogen peroxide 1
Hydrogen Peroxide Scavenging 3

Hydroperoxyl radicals 43
*Hydrophylax maritime* 143
Hydroxycinnamyl alcohols 9
Hydroxyl radicals 73
Hyperaccumulators 176

Indole-3-acetic acid 205
Inefficient antioxidant defense 27
Inflorescence 414
Insoluble invertase 25
*Instia bijuga* 143
Inulin 358
Inulobiose 358
Inulotetraose 358
Invertase 353
Ion channel activity 53
Ion regulation 115
*Ipomea per-caprae* 142
*Ipomea tuba* 142
*Ivr2* expression 25

JA conjugates 208
JA glucosyl 208
Jasmonic acid (JA) signaling 209
Jasmonic acid 205, 207
Jasmonic acid biosynthesis 206
Jasmonic acid mediated induced systemic resistance 209

*Kandelia candel* 140, 146, 147, 148, 150, 151, 153
*Khaya senegalensis* 452

L-4-Cl-tryptophan 370
*Lactobacillus plantarum* 87
*Lactobacillus reuteri* 348
*Lactuca sativa* 70, 73
*Laguncularia* 118
*Laguncularia racemosa* 146
*Landolphia kirkii* 452
*Landoltia punctata* 195, 196
*Lapidium sativum* 370
L-aspartate 161
Late embryogenesis abundant proteins 20, 456
Late embryogenic abundant proteins 453
Lea gene expression 457
LEA genes 20
LEA proteins 20
Lead 173
Leaf anatomy 114
Leaf relative water content (RWC) 14
Leaf senescence 279
Leaf water potential 14, 159
Lecythiadaceae 143
*Lemna gibba* 196
*Lemna minor* 194, 196, 198
*Lemna minuscula* 195, 196
*Lemna paucicostata* 196
*Lemna perpusilla* 195
Lemna test 198
*Lemna trisulca* 196
*Lemna valdiviana* 195, 196
Lemon balm 25
*leontodon hispidus* 53
Levan 358
Levanbiose 358
Light adaptation 109
Light reflection 159
Light sensor 6
Light harvesting complex 105
Lignin polymerization 9
Lignins 9
Linolenic acid 207
*Linum usitatissimum* 370
Lipid metabolism 243
Lipid peroxidation 73, 159, 162, 175
Lipid soluble antioxidants 459
Lipo chitooligosaccharids 330
Lipoxygenase 207
Lipoxygenase activity 47
Litchi 477
*Litchi chinensis* 461
Loganiaceae 143
*Lolium perenne* 348, 350, 352, 353
Lon protease 259
Loranthaceae 143
Lotus effect 231
*Lotus japonicus* 331
Low temperature 50
*Lumnizera littoralia* 140
*Lumnizera racemosa* 140, 158, 151, 154
*Lycopersicum esculentum* 145, 370
Lymphoid cells 303

*Macrosiphum euphorbiae* 211

*Magnaporthe grisea* 213
Magnoliaceae 461
Maize 144
Maize kernels 24
Malic dehydrogenase 161
Malonaldihyde 74
Malvaceae 143
*Mangifera indica* 452, 461
Mango 476
Mangrove environment 103
Mangroves 101, 102, 146
*Marchantia polymorpha* 258
Marciulioniene 197
Mass spectrometry analysis 300
Mechanisms of salt tolerance 114
*Medicago sativa* 243
*Medicago truncatula* 243, 331
Mehler reaction 17
*Melia azedarach* 452
Meliaceae 141, 143
*Melissa officinalis* 25
Melizitose 358
Membrane depolarization 8
Membrane intrinsic protein 27
Membranes 459
*Meropa angulata* 143
*Mesembryanthemun crystallinum* 145
Mesophyll cell walls 5
Mesophyll cells 5
Metabolic engineering 244
Metal toxicity 47
Metallothioneines 178
Metallothionin 300
Methyl viologen (MV) 52
*Michelia champaka* 461
Microsymbionts 336
Mitochondria 2
Mitochondrial processing peptidase 257
Mitogen protein kinase cascade 9
Mixture toxicity index 197
MnSOD 88, 92
Mobilisation transport 384
Molecular chaperones 117, 460
Molecular mechanism of salt tolerance 116
Molecular mechanisms 181
Monodehydroascorbate reductase (MDHAR) 3, 174
Monodehydroascrobate (MDHA) 3
Monogalactosyl diglyceride (MGDG) 74
Moraceae 461
mRNA 71
*Mucuna gigantica* 143
Mycorhiza 180
Myoinositol-6-*O*-methyltransferase activity 23
*Myricaceae* 328
*Myriostachya wightiana* 143
*Myristica fragrans* 461
Myristicaceae 461
Myrsinaceae 141
Myrtaceae 461
*Myzus persicae* 211

$Na^+/Ca^{2+}$-antiporter 27
NADPH oxidase 8
NADP-sedoheptulose-1,7-biphosphate 7
NBS-LRR-type resistant proteins 301
Neokestin 358
Neokestose 358
*Nephelium lappaceum* 461
*Nephelium malaiense* 461
*Neurospora crassa* 87
*Nicotiana tabaccum* 21, 30, 70, 72, 145, 355, 370
Nicotianamine synthase-2 303
Ni-SODs 90, 92
Nitrate reductase activity 173
Nitrate reductase 159
Nitric oxide 23
Nitrogen fixation 24
Nitrogen metabolism 7, 114
NOD factors 329
NOD factors perception 330
NOD factors synthesis 330
Nodule primordium formation 335
Non climacteric 476, 477
Non fructan plants 347, 354
Nuclear DNA content 109
*Nypa fruticana* 140, 154

Octapeptides 272
*Oidium lycopersicum* 208
Oleosin 303
Oligosaccharide metabolism 22
Oligosaccharides 347
*O*-methyl-inositol 23
Optimal leaf temperature 110

Organic acids 180
Organic constituents 107
*Oryza sativa* 27, 79, 145, 298, 355
*Osbornia* 118
Osmoprotectants 117
Osmoregulation 352
Osmotic adjustment 157, 159
Osmotins 117
Oxidative damage 6
Oxidative pentose phosphate cycle 7
Oxidative stress 1, 27, 41, 174, 458
Oxidative stress resistance 159
Oxmotic stress 51
Oxygen species 43
Ozone 49, 67
Ozone pollution 52

Palisade cell diameter 118
Palisade cell length 118
Palmaceae 461
Pandanaceae 143
*Pandanus odoratimus* 143
*Pandanus tectorius* 143
*Panicum miliaceum* 78
Papilionaceae 461
Paraquat 52
*Parasponia* 339
*Parasponia andersonii* 327
Partitioning 389
Pathogenesis related protein 301
*Paxillus involutus* 180
Pb toxicity 173
*Pcea abies* 52
PC-synthase 179
Peach 476
*Penicillium* 347
*Pentatropis capensis* 142
Pentose phosphate pathway enzymes 30
Peroxidase 75, 159, 161, 173
Peroxidation 162
Peroxiredoxins (PrxRs) 3
Peroxisome 2
Pests and diseases 156
Peteridacea 141
*Phaseolus vulgaris* 71, 370
Phenylalamine ammonia lyase 30
Phlein 358
Phloem loading 388
Phloem unloading 389
*Phoenix paludosa* 142
Phosphate metabolism 107
Phosphoenolpyruvate carboxylase 18
Phospholipase 207
Phospholipase D 22
Phosphoribulokinase 7
Photoinhibition 16, 46
Photophosphorylation 71
Photoprotective mechanism 16
Photorespiration 3, 16, 209
Photosynthesis 14, 71
Photosynthesis in mangroves 105
Photosynthetic contribution 383
Photosynthetic efficiency 159, 163
Photosynthetic genes 30
Photosynthetic pathway 116
Photosystem II 15, 105
*Phragmates karka* 143
*Phragmites australis* 148, 175
*Phragmites communis* 148
Phylogenic tree *355*
Physiological basis of yield 379
Phytochelatins 170
Phytochemicals 179
*Phytophthora infestans* 209
Phytoremediation 199
Phytotoxicity 74, 170, 194, 197
Phytotoxins 205
*Pichia pastoris* 351, 356
*Pieris rapae* 211
Pineapple 477
*Pinus sylvestris* 370
*Pisum sativum* 70, 231, 355
Plant dehydroascorbate reductase 303
Plant growth regulators 400
Plant hormones 116
*Plantago major* 69, 73
Plasma membranes 5, 180
Plastid division 7
Plum 476
Plumbagiraceae 141
Poaceae 143
Polyamines 76
Polyphenol oxidase 159, 161
Polyproteins 271
*Pongamia pinnata* 461

*Porteresia coarctata* 143, 153, 154,
*Posidonia oceanica* 181
Post harvest periods 475
Post-harvest physiology 449
Potassium channel ($K^+_{in}$) 53
Potassium transporter 27
*PR-5* 32
Presequence degrading protease (PreP) 257
Protease 27, 460
Protective action 459
Protective component 459
Protein 7
Protein degradation 7
Protein turnover 273
Proteinases 117
Proteins 409
Proteolysis 255, 274
Proteome analysis 299
*Pseudomonas fluorescens* 211
*Pseudomonas syringae* 209
PSII photoinhibition 17
*Pterocarpus officinalis* 146
Pyruvate dehydrogenase complex 18

Quantitative trait loci 27
*Quercus alba* 461
*Quercus muehlenbergii* 461
*Quercus robus* 452
*Quercus virginiana* 461

RAB-18 32
Raffinose 22, 358, 459
RD-29A 32
Reactive hydroxyl radical 3
Reactive oxygen species (ROS) 1, 44, 87, 88, 174
Recalcitrant 452
Recalcitrant seed storage 451
Reddening 77
Regulation of flow of assimilates 389
Rehydration specific proteins 19
Relative humidity 71
Relative water content (RWC) 27
Remote sensing technology 317
Repair proteins 460
Reproductive dry matter 159
Respiration in mangroves 106
Resurrection plant 18
*Rhamnaceae* 328
Rhizobacteria 329
*Rhizobiaceae* 327
Rhizobia-legumes symbiosis 329
*Rhizobium meliloti* 22
Rhizophora 117, 118
*Rhizophora apiculata* 105, 140, 146, 147, 149, 150, 153
*Rhizophora mangle* 117, 146, 147, 148, 149, 150,
*Rhizophora mucronata* 140, 146, 147, 150, 153
*Rhizophora stylosa* 140, 146, 147, 148, 149, 150, 153
Rhizophoraceae 140
*Rhizophoracean mangroves* 147
Rice 27, 144
Rice root proteins 297
Root protein identification 300
*Rosaceae* 328
*Rosmarinus officinalis* 25
Rubber 315
Rubiaceae 141, 143
RUBISCO 71
RuBP carboxylase 209
Rutaceae 143, 451

*Saccharomyces cerevisae* 87, 179
*Saccharomyces pombe* 179
*Saccolobus carinatus* 142
*Saccolobus globosus* 142
*Salacia chinensis* 143
*Salicornia brachiata* 142
*Salicornia fructicosa* 149
Salicyclic acid 205
salicylic acid-dependent *PR-1* 32
Salinity 101
Salinity effect 112, 145
*Salpichroa organifolia* 23
Salt adaptation 112
SalT protein 301, 303
Salt stress 47
Salt tolerance 114
Salt-Overly-sensitive (SOS) signal 51
*Salvadora persica* 144
Salvadoraceae 144
*Salvia officinalis* 5, 5
*Sandoricum koetjape* 452
Sapindaceae 461
*Sarcolobus carinatus* 152

*Saudea maritime* 154
*Saudea monoica* 154
*Scirpus littoralis* 142
*Scyphiphora hydrophyllace* 141
Seed desiccation tolerance 456
Seed germination 26
Seed storage behaviour 453
Seed water 453
Senescence 72
Sesbania 337
*Sesuvium portulacastrum* 141, 152, 154
*Shorea leprosula* 452
*Shorea robusta* 452, 461
*Shorea talura* 452
*Shorea trapezifolia* 461
Signal perception 117
Signal transduction 8
Signal transduction cascades 57
Signaling molecule 7
Singlet oxygen 2
Sink characteristics 386
Sink decreasing 397
Sink efficiency 400, 410, 420
Sink strength 386
SOD assay 92
SOD protein structure 90
Soil moisture 157
Soil moisture conservation 163
Solanaceae 144
*Solanum melanogena* 78
*Solanum trilobatum* 144
*Solanum tuberosum* 355
Soluble protein 161
Somatic chromosome structure 108
*Sonneratia* 118
*Sonneratia alba* 140, 147
*Sonneratia apetala* 140, 153
*Sonneratia caseolaris* 140
*Sonneratia griffithi* 140
Sonneratiaceae 140
Sonneretia 117
Source characteristics 386
Source decreasing 395
Source increasing 392
Source-sink manipulation 390
Source-sink relationship 385
Soybean 24
Spacio temporal shifts 157
*Spaeranthus indicus* 142
*Sphaeranthus indius* 152
Spike 415
Spikelet 417
*Spinifix littoreus* 143
*Spirodela polyrhiza* 194, 196
*Spodoptera exigua* 211
Spongy cell diameter 118
ß-1-4 N-acetylglucosamine 330
Stachyose 22, 358, 459
Starch 408
Starch degradatior 7
Sterculiaceae 141, 461
*Stictocardia tilifolia* 142
Stipple 77
Stomal processing peptidase 256
Stomatal closure 8, 15
Stomatal conductance 14, 16, 69, 70
Stomatal frequency 158
Stomatal regulation 159
Stomatal resistance 159, 161
Strawberry 477
*Streptococcus mutans* 348
Stress adaptation 101, 109
Stress responsive proteins 157
Stress tolerance 25
*Strychnos nux-vomica* 143
*Stylosanthe* 339
*Suaeda maritima* 142, 147, 149
*Suaeda nudijlora* 142
Sub ambient temperature susceptibility 455
Substrate specificities 358
Succinate dehydrogenase 271
Sucrose 358, 459
*Sueada nudiflora* 152
Sugar flux 25
Sugars 408
Sulfinic 7
Sulfiredoxin 7
Sulfonic acids 7
Sulphur metabolism 7
Sunken stomata 157
Superoxide anion 73
Superoxide dismutase 27, 75, 87, 89, 159, 161, 171, 176, 301
Sweet cherry 477
*Swietenia macrophylla* 452
*Synechocystis* 269

*Syzygium cumini* 153, 461

Tamariaceae 144
*Tamarix erecoides* 144
*Tamarix troupii* 144
*Taxithelium* 175
T-DNA 22
Temperate climates 461
Temperature 48
Temperature adaptation 110
*Terminalia belarica* 461
*Terminalia chebula* 461
Tetrazolium salts 94
Thaumatin like proteins 301
Theaceae 461
*Thelephesa terrestris* 180
*Theobroma cacao* 461
Theophylline 353
Thermoregulation 159
*Thermotoga maritima* 354
*Thermus thermophilus* 263
*Thespesia populnea* 143, 152
*Thespesia populnoids* 143
Thiols 7
Thioredoxin 6, 7, 27
Thioredoxin pool 7
Thylakoid processing peptidase (TPP) 256
Tiliaceae 144
Tobacco 8, 30
Tocopherol 459
Tomato 476
Total dry matter 159
Total soluble sugar 159
Transcription factors 117
Transcriptome analysis 26, 28
Transduction 117
Transgenic plants 23
Transgenic tobacco 21
Transketolase 7
Transpiration 70
Transpirational rate 159
*Triticum aestivum* 70, 183, 355
Tropical origin 461
Troposphere 68
Tropospheric $H_2O_2$ 52
Tropospheric ozone 49, 67, 68
Turgidity maintenance 159
Two-dimensional gel electrophoresis 299
*Tylophora tenuis* 142, 152

Ubiquitin 460
*Ulmaceae* 327
Ultra structural studies 419
Ultraviolet radiation 47
Uniconazole 302
Unsaturated fatty acids 73
Ureides 24
*Urochondra setulosa* 143
UV-radiation 50

Vapour pressure deficit (VPD) 15
Vascular system 414
Vegetative dry matter 159
Verbinaceae 144, 461
*Vicia faba* 23, 79, 355
*Vigna catiang* 370
*Vigna radiata* 70, 176
*Vigna sesquipedalis* 452
*Viscum orientate* 143

Water channel proteins 117
Water deficit 8, 14, 24, 157
Water relations 103
Water stress 48
Water use efficiency 159, 163
Watermelon 477
Wheat 144
Wheat fructan 358
*Wolffia arrhiza* 195, 196
*Wolffia globosa* 195, 196
WRKY transcription factor 30

*X. mekongenesis* 151
*Xanthophyllomyces* 347
Xylem vessels (Xv) 5
*Xylocarpous granatum* 153
*Xylocarpus* 117, 118, 120
*Xylocarpus granatum* 141, 147, 151
*Xylocarpus mekongenesis* 141

Yield 385
Yield components 380

*Zea mays* 25, 355, 370

ZEP genes 21
*Zizania palustris* 457
Zn toxicity 175
Zn-binding motif 256
Zn-metalloprotease 256

# CONTENTS OF PREVIOUS VOLUMES
# VOLUMES 1 - 8

## VOLUME - 1


**Section - 1 : BIOLOGICAL NITROGEN FIXATION**

Nitrogen fixation in leguminous crops under saline conditions and the manoeuvrability of their response through plant growth regulators — *Neera Garg and I.S. Dua* 1-55

Biological nitrogen fixation in non-legumes : Cereals — *J.D.S. Panwar and R. Elanchezhian* 57-68

**Section - 2 : PLANT CELL AND TISSUE CULTURE**

Plant tissue culture : Current trends and future prospects — *Minal Mhatre & P.S. Rao* 71-101

Selection of mutants using plant cell and tissue culture — *P. Suprasanna & P.S. Rao* 103-122

**Section - 3 : PLANT METABOLISM**

Leaf Senescence : Physiological and biochemical aspects — *A. Hemantaranjan, O.K. Garg and D.N. Tyagi* 125-156

Signaling molecules in plant metabolism — *S. Naresh Kumar* 157-188

**Section - 4: HERBICIDE PHYSIOLOGY IN RELATION TO NITROGEN FIXATION**

Physiological responses of genetically improved nitrogen- fixing cyanobacteria to agro-chemicalization in relation to paddy culture : Prospect as a source material for engineering herbicide sensitivity and resistance in plants — *A. Vaishampayan.* 191-217

**Section - 5: PLANT GROWTH REGULATORS**

Physiology of grain growth in *aestivum* wheats with special reference to the role played by plant growth regulating substances in modulating the sink efficiency — *I.S. Dua, Bhupinder Singh and K.K. Dhir* 221-271

Salicylic acid : a new PGR in signal transduction — *H.S. Gehlot, Sanjay Purohit, K.K. Bora and S.P. Bohra* 273-289

Triazoles : A new group of promising synthetic plant growth regulators — *R.P. Raghav and Nisha Raghav* 291-320

**Section - 6 : PHYSIOLOGY OF ROOTING**

Physiology of rooting : Effect of some metabolic inhibitors on the rooting response of hypocotyl cuttings of *Phaseolus mungo* and associated biochemical changes — *I.S. Dua, Manjit Singh, Neera Garg and K.K. Dhir* 323-356

**Section - 7 : TREE PHYSIOLOGY**

Role of net carbon balance in flowering and yield of fruit trees — *K.S. Shivankara and C.K. Mathai* 359-371

**Section - 8 : STRESS PHYSIOLOGY**

Relationship between water stress and abundance of Phytophagous insects — *C.P. Srivastava and R.M. Singh* 373-380

Influence of salinity stress on crop plants — *J.P. Srivastava* 381-394

**Section - 9 : GROWTH AND DEVELOPMENT**

Physiology of fruit ripening — *U.S. Prasad* 397-417

Physiology of seed and bud dormancy — *R. Panneerselvam* 419-438

**Section - 10 : TECHNIQUES IN PLANT PHYSIOLOGY**

Analytical improvements in the vibrational spectroscopy for the study of biological systems — *A. Javier Aller* 441-477

Looking into the major achievements in the analytical electrothermal atomic spectrometric techniques — *A. Javier Aller* 479-511

**VOLUME — 2**

**Section - 1 : PLANT METABOLISM**

The Biogenesis of indole Alkaloids — *M. Sottomayor, F. Dicosmo and A. Ros Barcelo* 1-41

**Section - 2 : PHOTOSYNTHESIS AND CROP PRODUCTIVITY**

Photosynthesis in Relation to crop Productivity in Rice — *V.P.Singh & S.R. Voleti* 45-59

**Section - 3 : ENVIRONMENTAL PHYSIOLOGY/STRESS PHYSIOLOGY**

Heat shock Proteins in Plants — *K.K. Bora, P.K. Roy & S.P. Bohra* 63-86

Adaptation to Stressful Environments — Role of Phytohormones Homeostasis — *Sabra Abbas, Parves H. Zaidi and Atar Singh* 87-111

Plant responses and Adaptation to Water Deficits 113-135
— *K.V. Janardhan & R. Bhojaraja*

Plant Response to Elevated Atmospheric Carbon Dioxide 137-151
— *Madhoolika Agrawal & S.S. Deepak*

Variability and Mechanism of Salinity Tolerance in Rice — *J.S. Bohra* 153-158

**Section - 4 : MINERAL NUTRITION**

Mineral Nutrition of ground nut — *A.L. Singh* 161-200

Sulphur Nutrition of oilseed Crops — *A.L. Singh* 201-226

**Section - 5 : IRON NUTRITION AND INTERACTIONS IN PLANTS**

Iron Nutrition I — Soil and Plant Procedures to Control Iron Deficiency 229-250
— *A. Hemantaranjan & A. Wallace*

**Section - 6 : RECENT ADVANCES IN MICRONUTRIENT RESEARCH**

Influence of Succinate on Zinc Toxicity in *Zea mays* — *S. Doncheva et al.* 253-261

**Section - 7 : PHYSIOLOGY OF HOST PARASITE RELATIONSHIP**

Role of Plant Growth Regulators in Disease Resistance 265-277
— *K.K. Bora, R. Ganesh & S.P. Bohra*

**Section - 8 : POST-HARVEST PHYSIOLOGY**

Post-harvest physiology of fruit and vegetable crops — *S.P. Singh & Rajkumar* 281-294

**Section - 9 : METHODOLOGICAL ADVANCES**

Progress in the Biological Significance and Analytical Methodology of Selenium 297-315
— *A.J. Aller*

Methodological Advances in the Determination of Gold — *A.J. Aller* 317-334

**VOLUME — 3**

**Section - 1 : PHOTOCHEMICAL REACTIONS — A MOLECULAR APPROACH**

Photosystem - I — *Jun Sun and Parag R. Chitnis* 1-36

**Section - 2 : PLANT BIOCHEMISTRY AND MOLECULAR PHYSIOLOGY**

Plant Stilbenes : Recent Advances in their Chemistry and Biology 39-70
— *M. Morales, A. Ros Barcelo and M.A. Pedreno*

Metabolic Plasticity of Plant Peroxidases — *A. Ros Barcelo and R. Munoz* 71-92

Prenylquinone Biosynthesis — *Ricardo F. Reategui and Parag R. Chitnis* 93-106

**Section - 3 : RECENT ADVANCES IN PHYTOHORMONE RESEARCH**

Plant Hormones — *G.S.R. Murti and K.K. Upreti* 109-148

An Update on Brassinosteroids — *Sheela Agarwal and H.S. Gehlot* 149-178

Plant Growth Regulators and Nodulation in Legume — Root Nodule Bacterium Symbiosis — *S.S. Dudeja and Kamlesh Kukreja* 179-190

Role of Plant Growth Regulators in Agriculture — A Physiological Approach — *Ajay Arora, V.P. Singh, S.R. Voleti and S.R. Kushwaha* 191-210

**Section - 4 : PHYSIOLOGY OF WEED CONTROL**

Physiology of Herbicide Resistance in Weeds — *Pervez H. Zaidi* 213-238

**Section - 5 : PHYSIOLOGICAL AND MOLECULAR BASIS OF STRESS TOLERANCE**

The Role of Superoxide Dismutase in Stress Tolerant Plants — *Sheela Agarwal and Priya Khimnani* 241-282

Proline as Osmotic Stress - Marker in *in vitro* System — *Gaurab Gangopadhyay and Sangita Basu* 283-304

Isozymes and proteins as osmotic stress-marker in *in vitro* system — *Gaurab Gangopadhyay and Sangita Basu* 305-320

Mechanisms of Aluminum Toxicity and Tolerance in Plants - Perspective and Prospective — *Thiruvambalam Balakumar* 321-333

Physiological Basis of Al Toxicity, Disorders and Dynamics of Tolerance Mechanism in Acid Soils — *Kalyan Singh, Satoshi Mori and Etsuro Yoshimura* 335-351

Mechanism of Tolerance and Crop Production in Acid Soils — *A.L. Singh* 353-394

**Section - 6 : PHYSIOLOGY OF TREE**

Physiology of Mango — *G.S.R. Murti and K.K. Upreti* 397-419

**Section - 7 : FLOWER SENESCENCE**

Physiology of Flower Senescence in Floricultural Crops — *S.R. Voleti, V.P. Singh, Ajay Arora, Narendra Singh and S.R. Kushwaha* 423-439

**Section - 8 : MICRONUTRIENTS AND CROP PRODUCTIVITY**

Silicon Nutrition and its Integrated Management for Sustained Rice Productivity — *Kalyan Singh, Satoshi Mori and U.N. Singh* 443-463

Zinc in Plant Nutrition — *A. Hemantaranjan* 465-495

Physiological Significance of Boron in Plant Nutrition — *A. Hemantaranjan* 497-513

## VOLUME — 4

### Section - 1 : ENVIRONMENTAL STRESS

Photosynthetic responses of $C_3$ plants to drought — *J. Flexas and H. Medrano* 1-56

Influence of high temperature on the photosynthetic apparatus — *Katya Georgieva and Enrico Brugnoli* 57-74

Oxidative stress and damage in chloroplasts from dawn to dusk — *Javier F. Palatnik, Estela M. Valle and Néstor Carrillo* 75-88

Glutathione and oxidative stresses in plants — *Sheela Agarwal* 89-116

Chemical protection of plants against ozone : Role of ethylenediurea (EDU) — *S.B. Agrawal* 117-132

Variable response of legume species and their cultivars to salinity stress during various stages of growth and metabolism — *Neera Garg and Jasleen Chawla* 133-164

Symbiotic nitrogen fixation response to abiotic factors — *V.S.G.R. Naidu, J.D.S. Panwar and S.P. Saikia* 165-175

### Section - 2 : MOLECULAR PHYSIOLOGY / BIOLOGY

$Ca^{2+}$ Signalling systems in plants — The cell as the minimum unity for complexity — *Rui Malhó and Luisa Camacho* 179-203

Molecular markers and tagging of genes in crop plants — *Sujay Rakshit, Pervez, H. Zaidi and S.K. Mishra* 205-223

### Section - III : PHYSIOLOGICAL AND MOLECULAR IMPLICATIONS OF BIOFERTILIZER - THE PHYCOTECHNOLOGICAL ADVANCEMENT

Phycotechnological advances with respect to Bio-N fertilization of rice — *A. Vaishampayan* 227-284

### Section - 4 : IRON NUTRITION AND INTERACTIONS IN PLANTS

Physiological and molecular advances in phytosiderophores — A contrivance for iron nutrition in plants — *A. Hemantaranjan* 287-308

### Section - 5 : PHYSIOLOGICAL PATHOLOGY — A BIOTECHNOLOGICAL APPROACH

Fungal host - pathogen interaction : attack and defense mechanisms — *Balwant Kumar Singh and R.S. Upadhyay* 311-332

### Section - 6 : NITROGEN METABOLISM

Nitrogen metabolism in crop plants : I. Uptake of nitrate and its regulation — *Bandana Bose* 335-346

**Section - 7 : MOLECULAR AND PHYSIOLOGICAL ADVANCES IN BIOLOGICAL NITROGEN FIXATION**

Recent advances in Legume - rhizobium physiology and their inter-relationship with phytohormones — *Neera Garg and Vivekanand Sharma* 349-371

Recent advances in physiology and genetics of nodulation and nitrogen fixation in cereals — *J.D.S. Panwar and V.S. Gopal Rao Naidu* 373-382

**Section - 8 : PHYSIOLOGICAL STRATEGIES FOR SUSTAINABLE AGRICULTURE**

Evaluation of high grain density related agronomic practices for the success of productive, sustainable and nutritious agriculture — *Kalyan Singh, S. Mori, P.C. Sudhakar, Raghavendra Singh and R.S. Chandel* 385-406

**VOLUME — 5**

**Section — 1 : PLANT MINERAL NUTRITION - PHYSIOLOGICAL AND MOLECULAR MECHANISM**

Soil acidity and alkalinity : Two main constraints influencing the behaviour of plants — A review of current knowledge — *H. Zaïd, A. Hemantaranjan, R. Layachi and A. Alem* 1-27

Recent advances in the physiological and molecular mechanism of Al toxicity and tolerance in higher plants — *H. Matsumoto, Y. Yamamoto and B. Ezaki* 29-74

Excess Manganese in plants — *Sheela Agarwal* 75-93

**Section — 2 : PLANT METABOLISM**

The Glyoxylate cycle as an essential step in Carbon reallocation mechanisms — *Pierre Vauclare, Joaquim Cots, Katia Gindro and Francois Widmer* 97-132

**Section — 3 : IMAGING TECHNIQUES IN PLANT PHYSIOLOGY**

Imaging techniques in plant physiology and agronomy : From simple to multispectral approaches — *L. Chaerle, R. Valcke and D. Van Der Straeten* 135-155

**Section —4 : ENVIRONMENTAL STRESSES**

UV-B radiation and the biochemical and ultrastructural responses of plants — *Isabel Santos* 159-180

Leaf photosynthesis in Mediterranean vegetation — *J. Flexas, J. Gulías and H. Medrano* 181-226

Drought as a multidimensional stress affecting photosynthesis in tropical tree crops — *Fábio M. DaMatta* 227-265

Regulation of plant responses to drought : Function of plant hormones and antioxidants — *Leonor Alegre Batlle and Sergi Munné-Bosch* 267-285

Screening plants for environmental fitness: Chlorophyll fluorescence as a "Holy grail" for plant breeders — *Sergey Shabala* 287-340

Morphophysiological and demographic responses of perennial grasses to defoliation under water stress — *C.A. Busso, R.E. Brevedan , A.C. Flemmer and A.I. Bolletta* 341-395

Environmental stresses and their impact on Nitrogen assimilation in higher plants — *Kavita Shah and R.S. Dubey* 397-431

**Section — 5 : CROP PHYSIOLOGY**

Sugarcane yield plateaus and potential for yield increases under tropical and subtropical conditions of India — *R. Snehi Dwivedi* 435-459

**Section — 6 : RECENT ADVANCES IN NITROGEN METABOLISM**

Ammonium and Nitrate as a Nitrogen source for plants — *Surya Kant and Uzi Kafkafi* 463-478

**Section — 7 : PHYSIOLOGICAL PATHOLOGY AND SUSTAINABLE AGRICULTURE**

Immunization of plants against pathogen : Mechanisms and methods — *Neha Verma and R. S. Upadhyay* 481-499

**Section — 8 : PLANT GROWTH REGULATOR**

Plant growth regulating properties of Penicillins — A review — *R.P. Raghava, N. Raghava, S.P. Singh and S.K. Gupta* 503-532

**VOLUME — 6**

**Section — 1 : BIOTECHNOLOGY, GENE EXPRESSION AND GENETICALLY ENGINEERED CROPS**

GE crops : need, potential and contentions — *H.S. Gehlot, N.S. Shekhawat and A. Hemantaranjan* 1-35

Transposable elements and plant gene expression — *W. Siripornadulsil and E. Grotewold* 37-58

**Section — 2 : TRANSGENIC PLANTS — PHYSIOLOGY AND BIOCHEMISTRY**

Approaches to improve the nutritional values of transgenic plants by increasing their methionine content — *Rachel Amir and Gad Galili* 61-77

**Section — 3 : PLANT METABOLISM — BIOCHEMICAL AND MOLECULAR ADVANCES**

Biotin metabolism in plants : biochemical and molecular characterizations — *C. Alban, L. Denis, A. Picciocchi and R. Douce* 81-92

The abundant proteins in sesame seed : storage proteins in protein bodies and oleosins in oil bodies — *Jason T.C. Tzen, Miki M.C. Wang, Sorgan S.K. Tai, Tiger T.T. Lee, Chi-Chung Peng* 93-105

Plant casein kinases — *T. Yupsanis and C. Vergidou* 107-128

**Section — 4 : RECENT ADVANCES IN PLANT SIGNALING, GROWTH AND DEVELOPMENT**

Peptide hormone mediated signaling in plants exhibits mechanistic similarities in animals — *T. Dresselhaus and S. Sprunck* 131-177

Salicylic acid in the plant signaling network — *Thierry Genoud, Marcela Beatriz Trevino Santa Cruz, Jean-Pierre Métraux, Christiane Nawrath* 179-193

Polyamines in plant growth and development — *S.N. Mishra, Kanchan Makkar and Sarita Verma* 195-224

Cavitation fatigue — the weakening of cavitation resistance of xylem and its reversibility — *Uwe G. Hacke, Volker Stiller, John S. Sperry* 225-234

**Section — 5 : PHYSIOLOGY OF STRESSES IN PLANTS**

The physiological mechanism of aluminum tolerance in *Glycine max* L. — *Hideaki Matsumoto, Zhen Ming Yang, Jiang Feng You and Hai Nian* 237-261

Physiological options for mitigating drought and salinity in pulses — *J.D.S. Panwar* 263-275

Root studies under drought: different methodology — *J.D.S. Panwar and A. Bhattacharya* 277-291

**Section — 6 : CROP PHYSIOLOGY AND BIOCHEMISTRY**

Phenology of groundnut — *A.L. Singh* 295-382

**Section — 7 : PHYSIOLOGY, BIOCHEMISTRY AND GENETICS OF NODULATION AND NITROGEN FIXATION**

The synthesis and the roles of glutathione and homoglutathione in legumes — *J. Harrison, A. Puppo and P. Frendo* 385-412

Physiology and genetics of rhizobia — legume nitrogen fixation — *Michael A. Torres and Gary Stacey* 413-441

**Section — 8 : TECHNIQUES IN PLANT PHYSIOLOGY**

Fluorescence imaging : the stethoscope of the plant physiologist — *R. Valcke* 445-462

**Section — 9 : PROTOCOLS OF PLANT CELL VIABILITY**

Plant cell viability assessment — *N. Steward and J.L. Goergen* 465-476

**Section —10 : NITROGEN METABOLISM**

Nitrogen metabolism in plants under $CO_2$ enriched atmosphere — *A.C. Srivastava, A.K. Tiku, Neeru M. Sharma and U.K. Sengupta* 479-493

## VOLUME — 7

**Section – I : PLANT NUTRITION : ADAPTIVE ROOT STRUCTURE**

Cluster roots — a special adaptive root structure — *El Morabet R., H.G. Diem, A. Hemantaranjan, M. Arahou and H. Zaid* 1-33

**Section – II : GERMINATION AND SEEDLING DEVELOPMENT**

Germination and seedling development in maize — a genetic approach — *Giuseppe Gavazzi, Silvana Dolfini, Gabriella Consonni, Anna Giulini and Roberto Pilu* 37-68

**Section – III : PLANT-MICROBE INTERACTIONS AND PATHOGEN ATTACK**

Plant microbe interaction : the *botrytis* grey mould of grapes — biology, biochemistry, epidemiology and control management — *Roger Pezet, Olivier Viret and Katia Gindro* 71-116

Nitric oxide signaling in plant defence responses to pathogen attack — *I. Salgado, E.E. Saviani, L.V. Modolo and M.R. Braga* 117-137

**Section – IV : MOLECULAR ADVANCES IN PLANT METABOLISM**

Inositol metabolism in plant cells — A genomics perspective — *Glenda E. Gillaspy, Mustafa E. Ercetin and Ryan N. Burnette* 141-154

Biosynthesis and signaling of brassinosteroids — *M. Mori* 155-174

Pectin biosynthesis — *Takeshi Ishimizu and Sumihiro Hase* 175-193

Peptide signaling in higher plants — *Heping Yang, Hiroshi Kamada, Kenzo Nakamura and Youji Sakagami* 195-231

Costs of production and physiology of emission of volatile leaf isoprenoids — *Ülo Niinemets* 233-268

**Section – V : ABIOTIC STRESSES : CELLULAR AND MOLECULAR RESPONSES**

Heat shock proteins – biological role and regulation — *Saroj Dua and Gurpreet Kaur* 271-286

Plant cellular and molecular responses to salt stress and their tolerance mechanisms — *Neera Garg and Monika Singhi* 287-308

Plants in relation to excess of heavy metals — *Moinuddin, Shahid Umar and Muhammad Iqbal* 309-336

Aluminium tolerance mechanisms : beyond of $al^{3+}$ – induced organic acids secretion — *Arnoldo Rocha Façanha and Anna L. Okorokova-Façanha* 337-372

**Section – VI : MOLECULAR ADVANCES IN PHYTOHORMONE RESEARCH**

Advances in plant hormone receptors — *P.K. Nagar and Shweta Sood* 375-392

**Section – VII : PLANT BIOMINERALIZATION**


Plant biomineralization — *Paula V. Monje and Enrique J. Baran* 395-410


**Section – VIII : PLANT MOLECULAR BIOLOGY AND BIOTECHNOLOGY**


Quantifying nucleic acids with fluorescent probes — contribution of real-time pcr to applied and fundamental plant research — *Annaick Mingam, Claire Gachon and Bénédicte Charrier* 413-448

Biotechnology in wheat breeding — *Marcelo Helguera and Viviana Echenique* 449-466

Biotechnology in forage crops and pasture plants — *Marina L. Díaz and Viviana Echenique* 467-519


**Section – IX : TECHNIQUES IN PLANT PHYSIOLOGY**


Chemoperception analysis in intact leaves with microsensor-based systems — *Stefan M. Hanstein* 523-536

Precision agriculture — *T. Yellamanda Reddy and P. Umamaheswari* 537-558


**Section – X : PHYSIOLOGICAL BASIS OF YIELD**


Physiological basis of the variations in growth of wheat grains (*Triticum aestivam*) — *I.S. Dua, Neera Garg and Ranju Singla* 561-594


**VOLUME – 8**

**Section – I : PHYSIOLOGICAL AND MOLECULAR ADVANCES IN ENVIRONMENTAL STRESSES**


Physiological and molecular implications of salicylic acid in plants under environmental stresses—*A. Hemantaranjan* 1-21

The physiology of somatic embryo induction — A stressful start — *A. Fehér and K. Ötvös* 23-31

Phenolic compounds in plant cuticles : physiological and ecological aspects — *G. Karabourniotis and G. Liakopoulos* 33-47

Cold acclimation in plants — *Sheela Agarwal* 49-66

Low temperature tolerance in crop plants : strategies for crop improvement — *Arunava Pattanayak and Jiban Mitra* 67-87

*Rhizobium* — Legume symbiosis under salt stress : effects, adaptations and amelioration — *Neera Garg, Anu and Vini Arora* 89-112

Plant peroxidases- A brief note on response to metal pollutants — *Kavita Shah* 113-122


**Section – II : PLANT SIGNALING MOLECULES - PATHWAYS AND MECHANISMS**


Signaling through RHO-type GTPases in plants — *A. Szûcs et al.* 125-132

Cross talk for rapid communication in plants through electrical signalling — *Neeti Sanan Mishra, Sudir K. Sopory and Narendra Tuteja* 133-147

Nitrogen signaling in higher plants — *Ann Ying Chiao and Hon-Ming Lam* 149-168

Auxin as a positional and patterning molecule essential for embryo development in plants — *Christiane Fischer-Iglesias et al.* 169-187

**Section – III : MOLECULAR PHYSIOLOGY AND BIOTECHNOLOGY**

Achene proteins in jelly fig (*Ficus awkeotasang*) and their potential biotechnological application — *Miki M.C. Wang and Jason T.C. Tzen* 191-200

Seed storage proteins : structure, properties and approaches for improvement by genetic engineering — *N. K. Chrungoo et al.* 201-220

Current advances in *Agrobacterium* — Plant interactions and their implications on agricultural biotechnology — *Ajith Anand et al.* 221-242

*Sclerotinia* disease and engineered resistance in oilseed crops — *Xu Hu etal.* 243-262

**Section – IV: PLANT SECONDARY METABOLITE**

Terpenoid metabolism in cotton (*Gossypium* spp.) and qinghao (*Artemisia annua*) — *Shan Lu and Xiao-Ya Chen* 265-292

**Section – V : PLANT DEFENSE MECHANISM AND METABOLISM**

Role of a non-protein tripeptide, glutathione in plant metabolism — *Saroj Dua and Praveen Dobhal* 295-311

Cadmium interaction with thiols and photosynthesis in higher plants — *F. Pietrini, M.A. Iannelli, R. Montanari, D. Bianconi and A. Massacci* 313-326

**Section – VI : PHYSIOLOGICAL BASIS OF YIELD**

Physiological approaches for enhancing yield potential in legumes — *A. Bhattacharya, Vijaylaxmi and J.D.S. Panwar* 329-352

**Section – VII: PHYSIOLOGY OF HORTICULTURAL PLANTS**

Role of calcium in the physiology of horticultural plants — *S.P. Singh et al.* 355-361

**Section – VIII : TECHNIQUES IN PLANT PHYSIOLOGY**

Applications of vibrational spectroscopy to the investigation of plant material — *Enrique J. Baran* 365-392